**SIXTH EDITION**

# STRENGTH OF MATERIALS

## (SI Units)

**Dr. D.S. BEDI**

B.E. (Delhi), M.E. (Roorkee) (Gold) Medalist
Ph.D. (Waterloo Canada)

*Formerly*
Professor Emeritus
Department of Mechanical Engg.
Institute of Engineering and Technology
Bhaddal, Dist. Ropar, Punjab

Professor & Head, Deptt. of Production Engg.
Punjab Engg. College, Chandigarh

Professor & Head, Deptt. of Mech. Engg.
Thapar Institute of Engg. & Technology, Patiala, Punjab

Visiting Professor to Wayen State
University, Detroit, MI, (U.S.A.)

Principal, Baba Banda Singh Bahadur, Engg. College
Fatehgarh Sahib, Punjab

Advisor-cum-Consultant
G.G.S. College of Modern Technology Kharan

Director, Punjab College of Engg. & Technology
Lalru, Punjab

**KHANNA BOOK PUBLISHING CO. (P) LTD.**

Publisher of Engineering and Computer Books

4C/4344, Ansari Road, Darya Ganj, New Delhi-110002
**Phone** : 011-23244447-48 **Mobile:** +91-9910909320
**E-mail** : contact@khannabooks.com
**Website** : www.khannabooks.com

Dear Readers,

To prevent the piracy, this book is secured with HIGH SECURITY HOLOGRAM on the front title cover. In case you don't find the hologram on the front cover title, please write us to at contact@khannabooks.com or whatsapp us at +91-99109 09320 and avail special gift voucher for yourself.

Specimen of Hologram on front Cover title:

Moreover, there is a SPECIAL DISCOUNT COUPON for you with EVERY HOLOGRAM.

How to avail this SPECIAL DISCOUNT:

Step 1: Scratch the hologram
Step 2: Under the scratch area, your "coupon code" is available
Step 3: Logon to www.khannabooks.com
Step 4: Use your "coupon code" in the shopping cart and get your copy at a special discount
Step 5: Enjoy your reading!

**Price : ₹ 575.00**

**ISBN:** 978-93-82609-11-7

**Strength of Materials**
*by* Dr. D.S. Bedi

**First Edition:** 1999
**Sixth Edition:** 2013
**Revised Edition: 2026**



*Published by:*
**Khanna Book Publishing Co. (P) Ltd.**
CIN: U22110DL1998PTC095547

Visit us at: **www.khannabooks.com**
Write us at: **contact@khannabooks.com**

To view complete list of books, Please scan the QR Code:

*Printed in Bharat.*

*This Book is Dedicated To My*

*Wife*

*Smt. Anup Kaur*

## ABOUT THE AUTHOR .....

After graduating from Delhi College of Engineering, Delhi the author started his teaching career as lacturer in Mechnical Engineering at Thapar Engineering College, Patiala in 1958. After availing study leave he completed his Masters Degree in Mechnical Machine Design at the University of Roorkee, Roorkee in the year 1963 under the guidance of Prof. Alvord of University of Michigan, Ann Arbor, USA, who had come to Roorkee as Visiting Professor.

In the year 1966 the author went to University of Waterloo, Waterloo, Canada on Common Wealth scholarship for pursuing Ph.D in the field of Manufacturing. After completing his Ph.D in the year 1969, he worked in the University of Waterloo as Post Doctoral Research Fellow for six months to complete the consultancy project of the sponsoring agency, the Aluminium Co. of Canada. After completing the assingment the author joined back the Thapar Institute as Associate Proffsor in 1970 and shortly there after he was promoted to the post of Prefessior of Head of Deptt. of Mech, Engg. In the year 1971. The author joined Punjab Engg. College, Chandigarh as Professor & Head of Deptt. of Production Engg in 1973. He even started Masters Degree course in production Engg. in the Deptt. In addition, the author had been working as Honorary Consultant to industries like PTL, Mohali, ISGEC, Yamunanagar, Molins of India, Mohali. In the year 1987, looking to his research publications. Wayne State University, Detroit, U.S.A. invited the author to go there as Visiting Professor. After completing the assignment as Visiting Professor for three and a half years the author came back to Punjab Engg. College in January 1991.

In July 1993 the author joined Baba Banda Singh Bahadur Engineering College, Fatehgarh Sahib as the founder principal in which post be continued till his retirement at supperanuation in April 2002. After a short visit to USA he joined G.G.S. College of Modern Technology. Kharar as Adviser cum Consultant in Aug. 2002. After completing the assigned work of setting up various labs he moved to Punjab College of Engg. & Technology as Director cum Consultant in 2005. There after he joined Institute of Engg. & Technology, Bhaddal as Professor Emeritus in which post he continued till sept 1,2008 on which date he had just completed 50 years long spell of teaching.

## PREFACE .....

The sixth edition of the book has been thorougly been modified and enlarged to career the revised syallabii of many universities and other professional Examinations like AMIE and above all to incorporate the suggestions received from the students and faculty alike.

The manner how the equation of static equilibrium can be applied to bodies in motion (linear or rotational) along with the D'Alemberts Principle has been explained and its application has been demonstrated by taking actual day to day live problems.

Similarly, additional problems on two-dimensional complex stress systems have been fully solved by both analytical and Mohr's circle method so that the readers are made aware of the fact that the sign of shear stress on a particular plane has its own important role to play so as to arrive at the correct result which otherwise is normally overlooked or even sometimes neglected. This omission can even give totally absurd results.

The terms "Bending Moment" and "Twisting Moment" have been introduced as vector quantities in order to being out the difference between them. So that the reader can easily decipher each of them and proceed ahead to accomplish the associated objectives. The chapter on Thick Cylinders has been re-written to keep uniformity in sign convention of the stresses throughout the entire text. Further in this chapter the process of authofrettage of a thick cylinder has been introduced along with the "Simplified". Theory of this process has been developed which leads to the creation of two equivalent shrunk cylinders having desired interference fit. This way it eliminates the exuberantly high manufacturing cost of the individual cylinders followed by shrinking process either by heating the other one (which leads to warpage) or cooling the inner one with liquefied nitrogen.

The author has endeavoured to familiarise the readers with the "Yield point phenomenon of Low Carbon Stress", "quantitative definitions of Ductility and Malleability" and "Negative Possions Ratio" which were hitherto not dealt with in most of the texts on the subject.

Two chapters one on "Maxwell's Reciprocal Theorem" and other and "Shear Centre" backed by large number of worked examples have been added in this edition.

On the specific demand of my students almost all the chapaters have been supplemented with objective type questions along with more number of worked examples.

It has been the endeavour of the author to make this edition upto date and free from mistakes but to err is human and the auhtor will feel obliged if any mistake is brought to his notice.

In the end the author extends his sincere thanks to large number of eminent personalities in the field of Tehnical Education for their valuable suggestions, which they have been giving to him from time to time. To mention a few they are Dr. R.S. Sagar and Dr. K.D. Mannan of Baba Bhag Singh Institute of Engg. & Technology, Dr. R.C. Bahl and Dr. B.L. Sethi and Dr. D.K. Aggarwal of Institute of Emerging Technology, Baddi, Dr. N.P. Mehta and Dr. B.L. Sethi of Mulana Engg. College, Mulana, Dr. R.P. Singh of Ryat College of Engg. & Technology Rail Majra, Ropar, Dr. S.S. Wadhwa of Doaba College of Engg. & Technology, Kharar, Prof. S.S. Jolly and S. Grover of Sukhmani Institute of Engg. & Technology Derabassi, Dr. G.D. Bansal, Principal, Chandigarh College Chandigarh, Prof. Balraj Singh Brar of University College of Management & Technology, Malout, Prof. Gurdeep Singh of Institute of Engg. & Technology Bhaddal, Dr. R.S. Prabhakar Principal, Akal College of Engg. and Technology, Baru Sahib, Himachal Pradesh and above all large number of his students.

**D.S. BEDI**

# CONTENTS

## 1. SIMPLE STRESS AND STRAIN ........ 1-42



## 2. COMPLEX STRESSES AND STRAINS ........ 32-105

Chapter–1

# Simple Stress and Strain

## 1.1 FREE BODY DIAGRAM

Free body diagram of a member of a structure, or of an element of a member in static equilibrium, is the diagram of only that member or the element, as if made free from the rest of the structure, with all the external forces acting on it duly marked. These external forces will include those due to the presence of adjacent members in contact with the one in question, the gravitational effect and the inertia force if the member is in motion. Fig. 1.1(*a*) shows the free body diagram of a weight hanging from the end of a crane rope, the gravitational force is balanced by the tension *T* in the rope. The Fig. 1.1(*b*) shows the free body diagram, of the two elements of a tension test specimen under the action of two equal and opposite forces *P*, obtained by taking a section *CC*. The force, $F_2$ on the element $AC_1$ is due to the cohesion of the plane $C_2$, while $F_2$ on the element $C_2B$ is due to the plane $C_1$. Since the test specimen *AB* is in equilibrium any portion of that will also be in equilibrium. For the equilibrium of individual elements $AC_1$ and $C_2B$, $F_1$ and $F_2$ should both be equal to *P*.

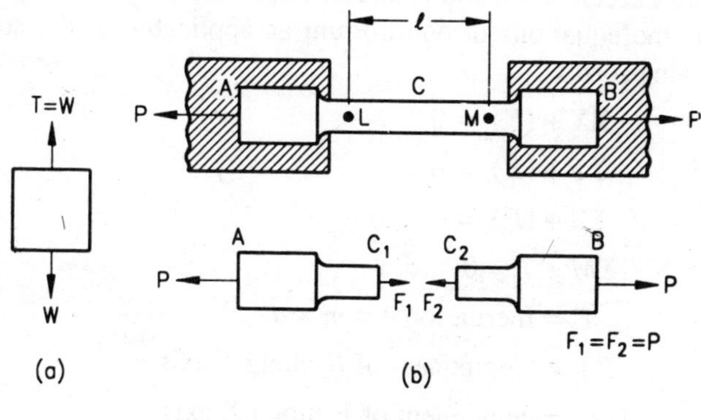


*Fig. 1.1*

### 1.1.1 Equations of Static Equilibrium

Despite the fact that the readers would have been familiar with the equations of static equilibrium, still it is felt necessary to re-write these equations here so that the reader is not at loss to understand the basic concepts involved in this subject properly.

Whenever a solid body under the action of various forces is in static equilibrium or in the state of rest then the algebraic sum of the components of these forces along any chosen three perpendicular axes of reference are separately zero. In addition the algebric sum of the moments of these forces about any point is also zero, if it is not rotating Mathematically these results are expressed in the form of following four equations

$$\Sigma \mathbf{X} = 0 \quad \ldots(A)$$

$$\Sigma Y = 0 \quad \ldots(B)$$

$$\Sigma Z = 0 \quad \ldots(C)$$

$$\Sigma M = 0 \quad \ldots(D)$$

where $\Sigma X$, $\Sigma Y$ and $\Sigma Z$ represent the algebraic sum of the components of forces along $X$, $Y$ and $Z$ direction respectively and $\Sigma M$ the algebraic sum of moments of these individual forces about any point.

## 1.1.2 Equations of Static Equilibrium and Body in Motion

In article 1.1.1 the equations of equilibrium have been introduced and it was pointed out that these equations are derived for the state of rest of the solid body. However D' Alembert pointed out that these equations can be applied to a solid body in motion (linear or rotory) as well, provided the inertia force in the case of linear motion or inertia torque in the case of rotory motion is superimposed on the body in addition to the given forces. The inertia force is equal to the product of mass and acceleration and is directed opposite to the direction of acceleration. Similarly in the case of rotory motion the inertia torque is the product of mass moment of inertia of body and the angular acceleration and is directed opposite to the direction of angular acceleration. Thus the equations of equilibrium as applicable to the solid body in motion take following forms

$$\Sigma X + (F_i)_x = 0 \quad \ldots(E)$$

$$\Sigma Y + (F_i)_y = 0 \quad \ldots(F)$$

$$\Sigma Z + (F_i)_z = 0 \quad \ldots(G)$$

$$\Sigma M + T_i = 0 \quad \ldots(H)$$

where $F_i$ = Inertia force = $m \times a$

$(Fi)_x$ = Component of $F_i$ along $X$ axis

$(Fi)_y$ = component of $F_i$ along $X$-axis

$(Fi)_z$ = component of $F_i$ along $Z$-axis

$T_i$ = Inertia torque = $I_m \alpha$

$M$ = Mass of the solid body

$a$ = Linear acceteration of body

$I_m$ = Mass moment of Inertia

$\alpha$ = Angular acceleration of body

These equations are known as *D'* Alemberts equations of equilibrium for solid bodies in motion and this is known as *D'* Alemberts' Principle.

As a special case if the body is having uniform linear motion or uniform angular motion then both linear acceleration '*a*' and angular acceleration '$\alpha$' will be zero and the equations (*E*), (*F*), (*G*) and (*H*) reduce to (*A*), (*B*), (*C*) and (*D*) respectively. Thus we conclude that when a solid body is at rest or in the state of uniform motion the equations (*A*) through (*H*) apply However when it is accelerating or decelerating then edquations (*E*) through (*H*) will apply.

The utility of the *D'* Alemberts principle as applicable to bodies in motion can better be understood with the help of following example. Let a passanger train consisting of 16 bogies be pulled by an electric locomotive. If it is assumed that the mass of each bogie be $M$ and the air resistance experienced by each bogie be $F_a$ then it is desired to calculate the stress in each coupler which connects two adjacent bogies. The area of cross-section of each coupler is A. For evaluating the stress in the last coupler connecting 16th bogie with 15th bogie we take a section any where along the length of the coupler and draw the free body diagram of the last bogie. Two situations may arise firstly the train be moving with uniform speed and secondly when the train is accelerating.

In first case there will be no inertia force as the acceleration is zero. Thus the free body diagrams of the last two bogies will be as given in figure 1.1 (*c*). Then according to *D'* Alemberts' principle considering the equalibrium of bogie number 16, we have for the horizontal forces

$$F_a = P_{16} \qquad \ldots(F)$$

Similarly for the equalibrium of 15th bogie we will get

$$P_{16} + F_a = P_{15} \qquad \text{or } P_{15} = 2F_a$$

and so on

$$P_{15} + F_a = F_{14} \qquad \text{or } P_{14} = 3F_a$$

$$\vdots \qquad P_2 = 15F_a$$

$$P_2 + F_a = P_1 \qquad \text{or } P_1\ 16F_a$$

Thus the first coupler which connects the first bogie with the locomotive will be stressed maximum and it will be equal to $\dfrac{16F_a}{A}$. While the last coupler will be stressed least $\left(=\dfrac{F_a}{A}\right)$. The max stress $= \dfrac{16F_a}{A}$.

In the second case when the train is accelerating with an acceleration equal to '*a*' then according to *D'* Alemberts principle each bogie will be accelerating and on each the inertia force which is equal to ($M \times a$) will be acting and it will be in the opposite direction to that of the acceleration. The free body diagrams of a few bogies is given in fig (1.1*d*)

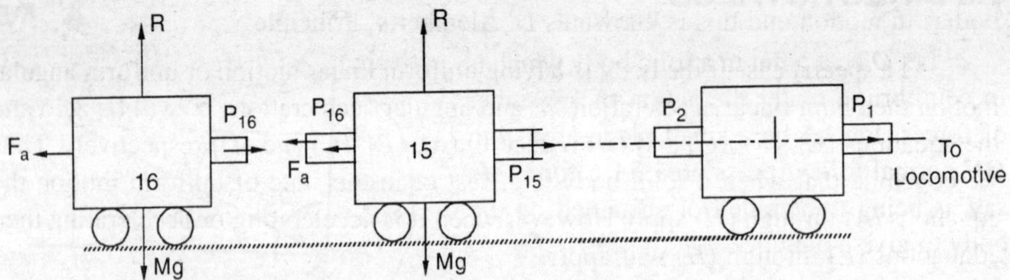


Fig. 1.1c

Then for the equilibrium of bogies number 16

$$F_a + F_i = P_{16}$$

or $$P_{16} = F_a + M.a$$

Similarly for bogie no 15

$$M.a + F_a + P_{16} = P_{15}$$

or $$(M.a + F_a) + (Fa + M.a) = P_{15}$$

or $$2M.a + 2F_a = P_{15}$$

and so on $$F_i + F_a + P_2 = P_1$$

or $$16F_a + 16M.a = P_1$$

The stress in each coupler can then be evaluated by dividing the force transmitted by each coupler with area of cross section *A*.

Thus the max stress will occur in the first coupler which will be equal to $(16F_a + 16M.a)/A$ and likewise the least stressed coupler will be the 16th coupler connecting the 15th and 16th bogies and will be equal to $(F_a + M.a)/A$.

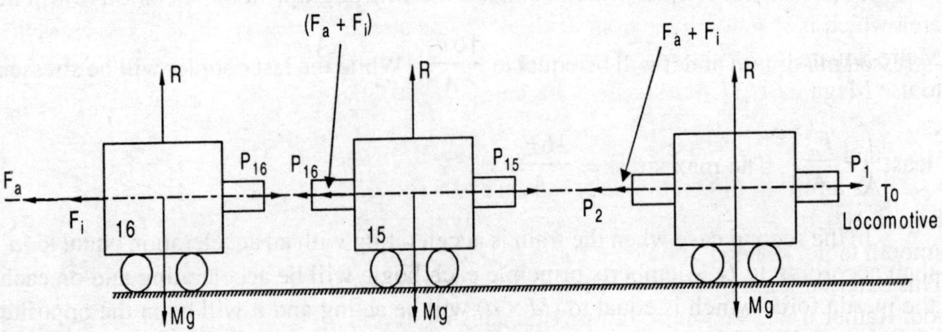


Fig. 1.1 (d)

## 1.2 DIRECT STRESS

Let $O$ be a point in a solid body which is in equilibrium under the action of a number of forces. Let $\delta A$ be a small plane area with $ON$ normal to it. Across area $\delta A$ a force $\delta F$, say, is being transmitted or sustained by the body to give a quantity

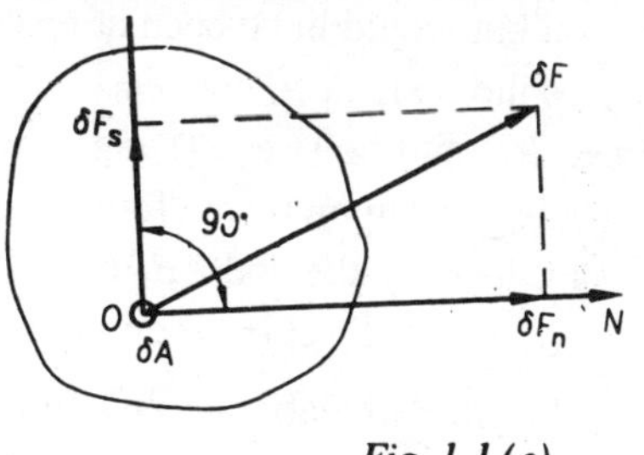


$$\sigma = \underset{\delta A \to 0}{Lt}\left(\frac{\delta F}{\delta A}\right)$$

*Fig. 1.1 (e)*

which is defined as the stress across $\delta A$ at $O$ in the direction $\delta F$. Alternatively other equivalent statements could also be made. Resolve $dF$ normal and parallel to $\delta A$ into, say $\delta F_n$ and $\delta Fs$ as shown in Fig. 1.1(*e*). Then the direct or normal and shear stresses across $dA$ are defined respectively as

$$\sigma_n = \underset{\delta A \to 0}{Lt}\left(\frac{\delta F_n}{\delta A}\right)$$

$$\tau = \underset{\delta A \to 0}{Lt}\left(\frac{\delta F_s}{\delta A}\right)$$

In case $\delta F$ is acting along $ON$, that is normal to $\delta A$, the stress $\sigma$ itself will be the normal or direct stress and the shear stress $\tau$ will be zero. Thus the direct or normal stress at a point of a plane section is defined as the force sustained per unit area of cross-section normal to it.

It is once again emphasized that the solid body must be in equilibrium and it should sustain or transmit force. Evidently to generate stress in a solid body there must be two equal and opposite forces acting on the body and these forces must have their respective points of application separated by some distance. The intervening material of the solid body will then be stressed since that portion of the body will be sustaining the force. As an example in the test specimen of Fig. 1.1(*b*) the direct stress over the section $CC$ will be $F_1/A_0$ (= $P/A_0$) where $A_0$ is the area of cross-section of the specimen since $A_0$ is the area which is sustaining the normal force $P$. The stress is usually measured in Newton/ Metre Square (N/m$^2$) or Pascal (Pa). Since Pascal is a very small unit it is not uncommon to use Mega Pascal (MPa) or Giga Pascal (GPa) as units of stress, where

$$1 \text{ MPa} = 10^6 \text{ Pa}$$

and $$1 \text{ GPa} = 10^9 \text{ Pa}$$

As a further illustration take the case of a rod $AB$ Fig. 1.1(*f*) of area of cross-section, $A_0$, lying on a smooth table, is acted upon by a force $P$ along its axis. This external force $P$ will not generate any stress in the rod. Rather it will set the rod in motion causing acceleration according to Newton's second law of motion. The point to note here is that the force $P$ is not being transmitted or sustained by the rod $AB$.

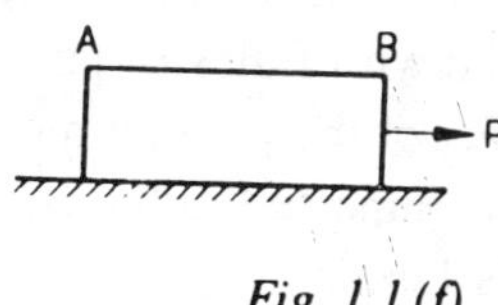


*Fig. 1.1 (f)*

In order to generate stress in it, would require equal force $P$ to be applied at section $A$ in the opposite direction. In that event the rod would have not been set into motion but would have been at rest or in state of equilibrium. It may further he noted that a solid body need not necessarily be in static equilibrium for it to be in the state of stress. It could be stressed while in motion as well. For example if the rod $AB$, Fig. 1.1(*f*), is acted upon by the force $F(<P)$ on the section $A$ towards left, then force $F$ will be sustained by the rod $AB$ to generate normal stress $F/A_0$ in the rod. However the force $(P - F)$ will set the rod into motion according to Newtons second law of motion.

In order to elaborate this point further let the rod $AB$ be acted upon by an other force $P$ acting on the section $B$ but in leftward direction as shown in fig. 1.1 (*g*) and we intend to find the stress at any section $CC$ lying in between $A$ and $B$. Then taking a section $CC$ at the point, the free body diagrams of the two elements $AC_1$ and $C_2B$ are drawn as shown in fig. 1.1(*h*). Evidently if the body $AD$ is in static equilibrium so will be the elements $AC_1$ and $C_2B$. If we consider the static equilibrium of individual elements $AC_1$ and $C_2B$ and resolve the force, along the direction of $P$ we have

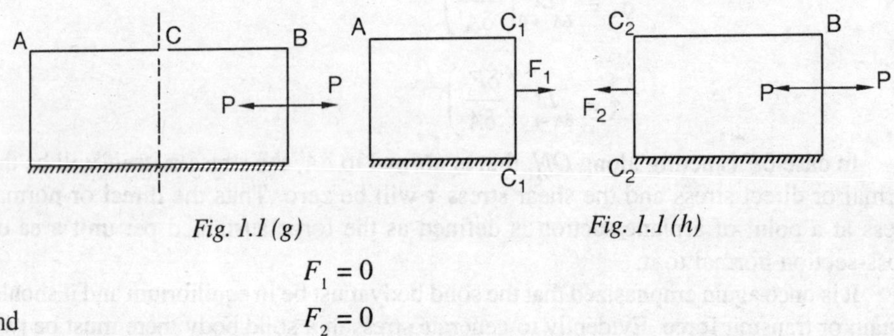


*Fig. 1.1 (g)* *Fig. 1.1 (h)*

$$F_1 = 0$$

and $$F_2 = 0$$

where $F_1$ is the reaction of $C_2$ on $C_1$ and $F_2$ the reaction of $C_1$ on $C_2$ and $F_1 = F_2 = 0$

Thus the force sustained by any cross-section lying between $A$ and $B$ will be zero. In other words the rod material will not be stressed. This proves the statement made above in the previous paragraph *i.e.,* to generate stress two equal and opposite forces must be applied at points at finite distance apart.

A beginner in the subject is sometimes at loss to distinguish between stress and pressure as both have the same units. It is to be noted that stress is encountered in solids and is associated with the direction of the sustained force with respect to the plane passing through the point under consideration. The magnitude of stresses will be different on different planes passing through the point under consideration. This aspect will be dealt in more detail in Chapter 2. On the contrary pressure is the term associated with fluids (liquids and gases) and the force exerted per unit area due to the impact of the moving molecules of the fluid on the walls of the container or on the body immersed in the fluid and is same at a point in the fluid, in all directions (Pascal's Law).

### 1.2.1 Direct Strain

Once again referring to the Fig 1.1(*b*), let there be two points *L* and *M* on the specimen which are at a distance $l_0$ called the gauge length, before the application of the force *P*. After the force is applied the specimen will stretch and the distance between *L* and *M* will increase to say *l*. The change in length per unit gauge length $(l - l_0) / l_0$, is called the axial or direct strain.

## 1.3 THE TENSION TEST

Consider a tension test specimen of round section as shown in Fig. 1.1(*b*) being stretched in a testing machine. The load recorded by the universal testing machine is the reaction of the test specimen. At room temperature the load extension diagram in general, at strain rate less than $2 \times 10^{-3}$/sec, will tend to follow the curve of Fig. 1.2(*a*), which is typical for an annealed metal like aluminium or copper.

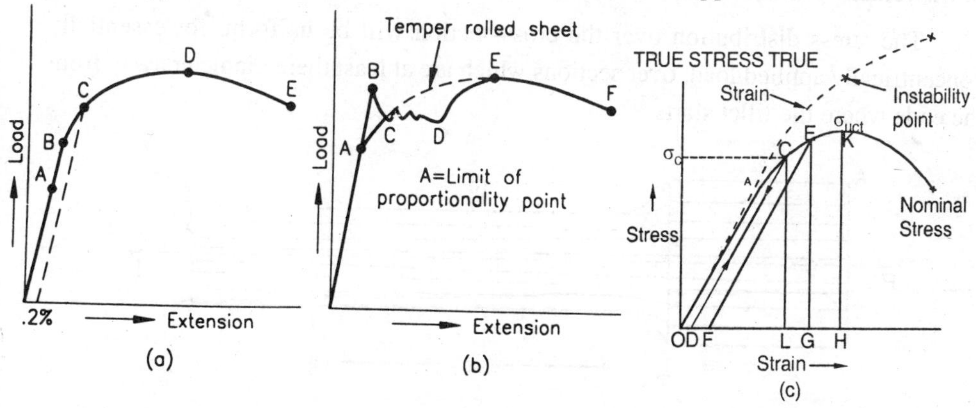


*Fig. 1.2*

Initially the relation between the load and extension is essentially linear—that is, portion *OA* of the curve, where *A* defines the limit of proportionality. On further straining, the relation between the load and extension no longer remains linear but the material is still elastic. That is, if the load is released the specimen will revert back to its original length. The maximum load which can be applied without causing permanent deformation is known as the 'elastic limit' represented by point *B* in Fig. 1.2(*a*). Usually the points *A* and *B* are very near to each other and the real difference can be achieved only by using highly sensitive measuring devices. At the point *B* ends the elastic straining and the 'inelastic' or plastic deformation ensues. Beyond the point B there is a general permanent extension of the test piece until the load attains a maximum value, point *D*, corresponding to which the stress is called the 'ultimate tensile stress' $\sigma_{ult}$ $(= P_{max} / A_0)$. At this load the specimen necks down locally rendering the deformation non-uniform and the load falls off with increasing extension, thereby causing the fracture at a load, point *E*, which is less than that corresponding to the ultimate tensile stress point. In addition another term, called the '*proof stress*' is usually quoted in structural design. This refers to the stress, $\sigma_c$ $(= P_c/A_0)$ required to produce a small strain of 0.2%. This is represented by point C in the Fig. 1.2(*a*).

## 1.3.1 St. Venants' Principle of Uniform Stress

In figure 1.1 *b* is illustrated the equilibrium of the left portion $AC_1$ of the test specimen under the action of the external force *P* at the left end and the tensile stress acting on the cross-section plane $C_1$. For the equilibrium of this portion the stress at this section is calculated as $\sigma = P/A$ and it is assumed to be uniform over the cross-section. But St. Venant observed that this formula is correct only if the cross section under consideration, such as $C_1$ here, is at least three diameters away from the ends of the straight portion where the fillets start. This is due to the fast that the "lines of force flow" are not evenly spaced over different cross-sections taken along the length Fig. (1.2 *d*). It was observed by St. Venant that they become substantially uniformly distributed at points about three diameters away from the end fittings. The statement of St. Venants principle is as follows.

The stress distribution over the cross-section will be uniform, for essentially concentrically applied load, over sections which are at least there diameter away from the ends where the fillet starts.

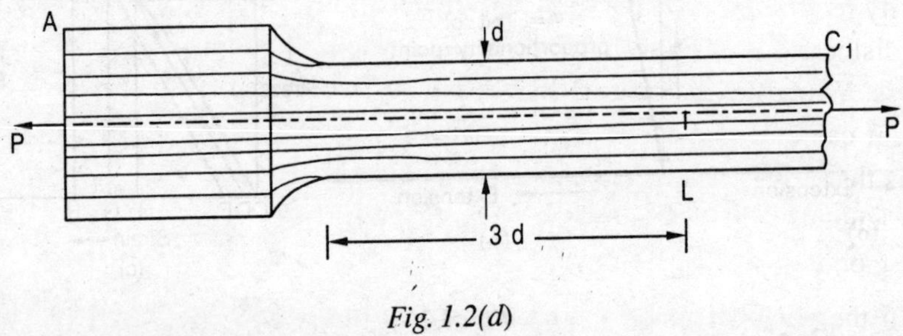


*Fig. 1.2(d)*

Thus while performing the tensile test the gauge length on the tensile test piece is taken in the central portion and its two ends such as *L* and *M* in figure 1.1*b* must be at least three diameters from the respective ends where the fillet starts and on both sides the machine grips should be equal distance from the ends. Then and then only the stress on any cross-section taken any where along the length *LM* will be uniform.

## 1.3.2 The Yield Point In Low Carbon Steel

Typical load extension curve after Murray (1) for annealed low carbon steel (earlier called mild steel) is shown in fig. 1.2 (*b*). The marked difference between this and two diagrams of the materials of the type discussed above is the exhibition of pronounced yield point. Upto the point *B*, the load rises steadily and the material remains elastic. The extension per unit gunge length at this point is of the order $10^{-3}$. At low carbon or nitrogen levels in mild steels, at the point *B* the load suddenly drops and then fluctuates about some constant value of load. The specimen extends considerably at this almost constant load. This sudden extension in mild steels is localized to certain zones in the metal and surface markings, known as "Luder bands" or "Strecteher strains" are produced. Under this situation we have some material which has yielded and some which has not. As the stretching continues upto the point *D* almost the entire specimen gets uniformly deformed. The extension *CD* is of the order of $10^{-2}$ per unit gunge length. Beyond the ponit *D* the load rises again as the specimen work hardens.

The stress at the point *B* at which the load suddenly drops is called the 'upper yield point' and the average constant load in the range CD to which the load has suddenly dropped is called the 'Lower yield point'. This gives rise to a peculiar feature that for low carbon steel (mild steels) that the stress required to maintain plastic flow immediately after yielding has started, is lower than that required to start it. Cottrell (1) (2) has put forth his theory of "Formation of cottrell's Atmosphere" around the dislocation to explain the occurrence of this phenomenon in low carbon steels. According to this theory carbon or nitrogen atoms, in *BCC* ion *i.e.,* in Ferrite phase, diffuse the positions of low energy. The dislocations are pinned or locked with these interstitial solutes (carbon or nitrogen) and additional stress, over that normally required for the movement of dislocations, is needed in order to tear some of the dislocations away from the restraining impurity atoms. This results in increase in stress. When the dislocation line is pulled free from the influence of the solute atoms, it can slip to a lower stress called lower yield point stress. The significant point is that a very small concentration of solute is required to lock or pin along the whole length of dislocation. This phenomenon gives rise to rough surface during drawing or stamping processes.

In the field of metal working, such as in deep drawing of steel blanks, for making steel utensils etc. the presence of Luder Bands is not desirable. The modern method to suppress this phenomenon is to "temper roll" it. The soft annealed sheet of mild steel is subject to light cold rolling (½ to 2 %) reduction in thickness) not more than 24 hours before deep drawing because the yield point suppression is not permanent. Lightly cold rolled sheet, deep drawn after standing for sufficient length of time, does show Luder Bands. Apart from this slow hardening occurs and results in loss of much desired ductility. This process of age-hardening is usually checked by adding aluminium into steel and producing the so called killed ingot.

For temper rolled sheet the stress-strain curve assumes the shape of the one shown in broken lines in fig. 1.2 *b*.

---

1. Murry 6, "The yield point in mild steel, engineering vol. 178, Pp 366, 1954
2. Cotrell; Physical Metallurgy, *Mc Graw* hill.
3. Vijendra Singh, Physical Mettillurgy, Standard Publishers and Distributers, New Delhi.

Modern practice is to add a small percentage of metal like Titanium or columbium into mild steel, which forms carbide with the carbon and the phenomena of lower yield point is eliminated completely and no Luder Bands are produced in the finished product.

Due to wide use of mild steel in industry where permanent change in shape in brought about, the complete stress-strain diagram for the metal is invariably considered. On the other hand its use in structures and machine elements is limited only upto the limit of proportionality point where the emphasis is only on small strains.

## 1.3.3 True Stress Strain Curve

If we now divide the load by the original area of cross-section *A* of the test specimen and the extension by the original length to (which is usually called the gauge length), then the curves shown in fig. (1.2) will be called the nominal or Engg stress-strain curve. These are called nominal or Engg stress and strain because both the parameters *i.e.* stress and strain are obtained by dividing with constant quantities respectively. Actually as the load is slowly or gradually increased the area of cross-section will continuously go on reducing (poissons ratio effect, see chapter 2). Similarly at any instant we take the extension with respect to the original length $\ell_o$. But actually the stretch at any particular instant occurs over the current length. Thus if we consider the current area of cross-sections and divide the current load with it we will get tin "TRUE STRESS". Likewise the change in length over the current length should be taken into account for calculating the TRUE STRAIN.

In the light of above, if *P* is the current load and '*d*' he current diameter of the specimens then

$$\text{True stress } \sigma = \frac{4P}{\pi d^2}$$

Similarly if *dl* is the elemental elongation in current length *l*, at which point the current load is *P*, then the incremental strain $d\varepsilon$ is defined as

$$d\varepsilon = \frac{dl}{l}$$

Integrating this equation we get the total current strain $\varepsilon$

$$\int_0^{\varepsilon} d\varepsilon = \int_{l_0}^{l} \frac{dl}{l}$$

*i.e.*,
$$\varepsilon = \ln\left(\frac{l}{l_0}\right)$$

Thus at any current load *P* the true strain will be

$$\varepsilon = \ln\left(\frac{l}{l_0}\right)$$

In view of the logarithm value of $(l/l_0)$ true strain is also called logarithmie strain or natural strain and associated with this are the names True stress or natural stress of the value $\frac{4P}{\pi d^2}$

The true stress-strain curve will be a monotonically rising curve and it will be higher than the nominal or Engg. Stress strain curve as shown in fig. (1.2C)

Upto the elastic limit point, these curves are very near to each other. However after elastic limit point, the true stress-strain curve is much higher than nominal stress-strain curve.

## 1.3.4 Yield Envelope and Bauschinger Effect

In the fig. (1.2*c*) and portion *AK* of the stress-strain curve is known as the plastic portion of the curve. If the stress on specimen corresponding at any point *C* in this portion of curve $\sigma_C$ is reduced gradually to zero value this reverse stress-strain curve will be a straight line *CD* parallel to the original elastic line *OA*. This phenomenon was observed by Bauschinger and is known as Bauschinger effectAt the point *D* the stress will be zero but residue strain *OD* will be left in the specimen. This strain *OD* is also known as the 'plastic strain' component, $\varepsilon_p$, of the total strain *EC* at *C* which is equal to *OL* the portion *DL* is called the elastic component , $\varepsilon_e$, of total strain

Thus $$\varepsilon_C = OL = OD + DL$$
$$= \varepsilon_p + \varepsilon_e$$

Once again if we restart increasing the stress on the specimen it will follow the path *DC* and then along the curve *EK*. In other words *C* becomes the new yield point and the corresponding stress, $\sigma_c$, the new yield stress. Similarly if this process is repeated when the stress level is $\sigma_E$ then *E* will yet become another new yield point stress, This will continue upto the point *K*, which is the ultimate stress point.

Therefore the portion *AK* is also known as "yield envelope". Another name assigned to the curve *AK* is strain hardening portion of the stress-strain curve.

### 1.3.4.1 Ductility and Malleability

Ductility hitherto, had been defined as the property of the material due to which it could be drawn into wire. It was very well known to the people that gold is more ductile than copper, copper is more ductile than Aluminium and so on. that is, this term was known qualitatively and had not been difined quantitatively. It was in late fifties of the last century that during an international meet, the scientists put their stamp on the proper definition of this term. According to this the strain to which a uniaxially stressed specimen can be subjected prior to instability point or ultimate stress point is called the ductility of the material. That is *OH* (Fig. 1.2 *c*) is the measure of ductility.

## Malleability

Similarly the malleability which is the property due to which the material can be deformed into plate in quantitatively defined as the strain at instability point for that particular material.

## 1.4 UNIFORM STRESS

To generate a uniform stress over a section the force must be applied at its centroid. To prove this statement assume a section shown in Fig.. 1.3, with the centroid at C, under a uniform stress σ. Let the total force acting on the section be P.

If $dA$ be an elemental area at a distance $r$ from $C$.

Then, we have $\int \sigma \, dA = P$ ...(1.1)

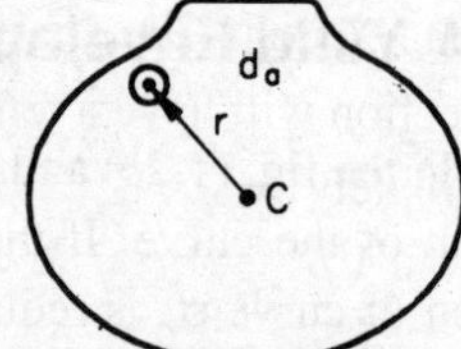

*Fig. 1.3*

Taking moments of elemental forces about the centroid we have if $P$ is at a distance $a$ from $C$

$$P.a = \int r\sigma \, dA = \sigma \int r \, dA \quad \text{(as } \sigma \text{ is constant)}$$

$= \sigma \times$ First moment of area about centroid.

$= 0.$

Now since $P$ is not zero, therefore

$$a = 0.$$

Thus the line of action of $P$ must pass through the centroid of the section.

## 1.5 SHEAR STRESS AND STRAIN

Consider a rectangular block $ABCD$ (Fig. 1.4) of dimensions $a \times b$ and thickness $t$. Let it be subjected to two equal and opposite forces $P$ on the faces $AB$ and $CD$ as shown. Taking a section $XX$ at a height '$y$' from the face $DC$ the free body diagrams of

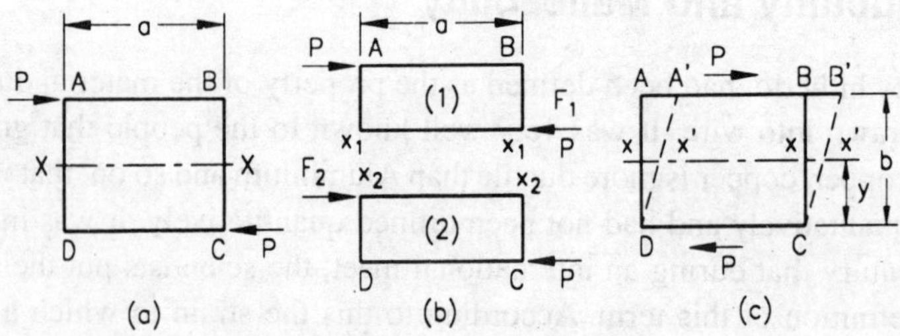

*Fig. 1.4*

the two elements number (1) and (2) are shown in Fig. 1.4(*b*). The face $X_1X_1$ offers a cohesive force $F_2$ on to the section $X_2X_2$ and in its turn the face $X_2X_2$ gives a reaction $F_1$ onto the face $X_1X_1$. For the equilibrium of the two elements separately which follows from that of the original block, we conclude that $F_1 = F_2 = P$. The effect of the force $F_1$ and $F_2$ on the respective faces is to induce a tendency for the face $X_1X_1$ to slide over the face $X_2X_2$ in the direction of the arrowhead $F_2$. The forces $F_1$ and $F_2$ are called the shearing forces and the force per unit area of the cross-section transmitting this force is known as the shear stress. It is represented by $\tau$ and in this case is given by

$$\tau = F_1 / at \; (= P/at) \qquad ...(1.2)$$

It is evident that the shear stress, in this case, is independent of the distance of the section *XX* from *CD*. That is shear stress on any horizontal section will be constant and that section will have a tendency to slide to the right with respect to its adjacent plane immediately below it. The cumulative effect will be to displace the face *AB* to the position *A'B'* prime with respect to *DC*. Likewise then, section *XX* will occupy the position *X'X'*such that at any height *y* from *DC* the displacement at right angles to *y* is *XX'*. This displacement *XX'* over unit height is called the shear strain and is represented by symbol $\phi$ Fig. 1.4(*c*) such that

$$\phi = \frac{XX'}{Y} = \tan ADA' \simeq \text{angle } ADA' \qquad ...(1.3)$$

That is, the shear strain is an angular displacement measured in radian and is constant for any value of *y* between zero and *b*.

## 1.6 HOOKE'S LAW

Hooke's Law states that within elastic limit stress is proportional to strain. Mathematically, for tensile or compressive stress.

$$\sigma/\varepsilon = E \qquad ...(1.4)$$

Where *E* is constant for material and is known as the Young's modulus of elasticity and $\varepsilon$ is called the strain.

For shear stress this ratio is denoted by *G* or *C* and is called the modulus of rigidity.

Thus $$G = \tau/\phi \qquad ...(1.5)$$

**Example 1.1.** *A bar of uniform cross-sectional area 100 mm² carries forces in Newtons as shown in Fig. 1.5. Calculate the relative movement of end A with respect to D.* $E = 200 \times 10^3$ *MN/m²*.

**Solution.** Taking a section anywhere in between *A* and *B*, it is found by drawing the free body diagram that force transmitted through that section is 100 *N*.

Similarly the transmitted force over any section in between $B$ and $C$ is 200 $N$ tensile and that between $C$ and $D$ is 200 $N$ compressive. Accordingly the force distribution diagram is drawn in Fig. 1.5($c$).

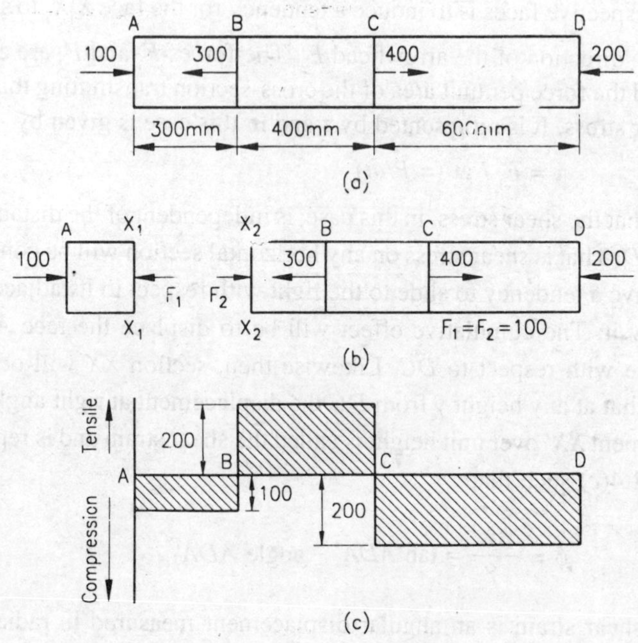


*Fig. 1.5*

Thus the respective deformations are

$$\delta l_{AB} = \frac{1}{E}\sigma_{AB}.l_{AB} = \frac{1}{200 \times 10^3} \times \frac{100}{100} \times 300$$

$$= 0.0015 \text{ mm decrease}$$

$$\delta l_{AB} = \frac{1}{E}\sigma_{BC}.l_{BC} = \frac{1}{200 \times 10^3} \times \frac{200}{100} \times 400$$

$$= 0.004 \text{ mm increase}$$

$$\delta l_{AB} = \frac{1}{E}\sigma_{CD}.l_{CD} = \frac{1}{200 \times 10^3} \times \frac{200}{100} \times 600$$

$$= 0.006 \text{ mm decrease}$$

The movement of '$A$' w.r.t. '$D$'

$$= .004 - (.0015 + .006) = -0.0035 \text{ mm}$$

That is, the points $A$ and $D$ will come closer to each other by 0.0035 mm.

**Example 1.2.** *A turn buckle is used to connect two steel bars each effectively 3 m long. One of the bars is 50 mm dia and has* $1\frac{4}{5}$ *threads per cm over its end, while the other is 75 mm dia with* $1\frac{2}{5}$ *threads per cm on its end.*

*If there is a small tension initially in the rods, find the increase in tension when the turn buckle is tightened through half a turn.*

Take $E = 200\ GPa = 200 \times 10^9\ N/m^2$.

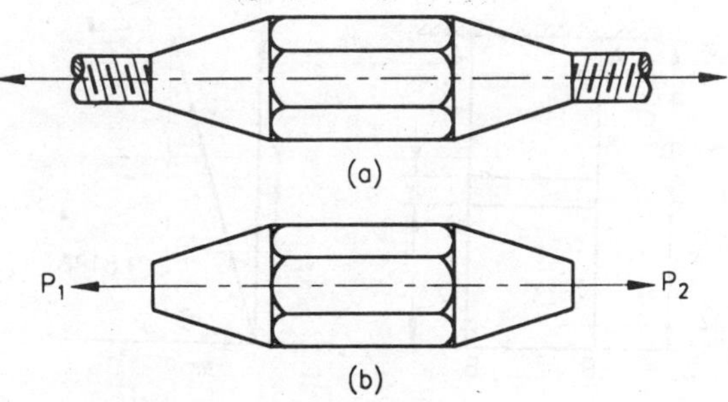


*Fig. 1.6*

**Solution.** In Fig. 1.6(*b*) is drawn the free body diagram of the turn buckle with $P_1$ and $P_2$ respectively as the reactions of the two bars. Evidently for the equilibrium of the turn buckle

$$P_1 = P_2 = P(\text{say}) \quad ...(1.6)$$

Where $P$ is the increase in tension required to be evaluated.

Total stretch in the bars = sum of the movements of the ends due to rotation of turn buckle

$$= \left(\frac{5 \times 10}{9} \times \frac{1}{2} \times \frac{5 \times 10}{7} \times \frac{1}{2}\right) \text{mm}$$

$$= \frac{25 \times 16}{63} \text{mm} \quad ..(1.7)$$

If $\varepsilon_1$ and $\varepsilon_2$ are the respective strains in the two bars, the total stretch is also given by $(\varepsilon_1 l_1 + \varepsilon_2 l_2)$ where $l_1$ and $l_2$ are the respective lengths of the bars. Thus with the help of equations (1.6) and (1.7), we have

$$\frac{25 \times 16}{63} = \varepsilon_1 l_1 + \varepsilon_2 l_2$$

$$= \frac{P_1}{A_1 E} l_1 + \frac{P_2}{A_2 E} l_2$$

$$= \frac{P}{E} l_1 \left(\frac{1}{A_1} + \frac{1}{A_2}\right) \quad \text{as } l_1 = l_2$$

$$= \frac{P \times 3000}{200 \times 10^3}\left[\frac{4}{\pi(50)^2} + \frac{4}{\pi(75)^2}\right]$$

Thus $P = 0.575$ MN .

## 1.7 STRESS IN A BAR DUE TO ITS OWN WEIGHT

Consider a bar of length l metre and uniform cross-section area A m² hanging from the top, Fig. 1.7. The weight of the portion *BC* of the bar will act on the cross section *CC*.

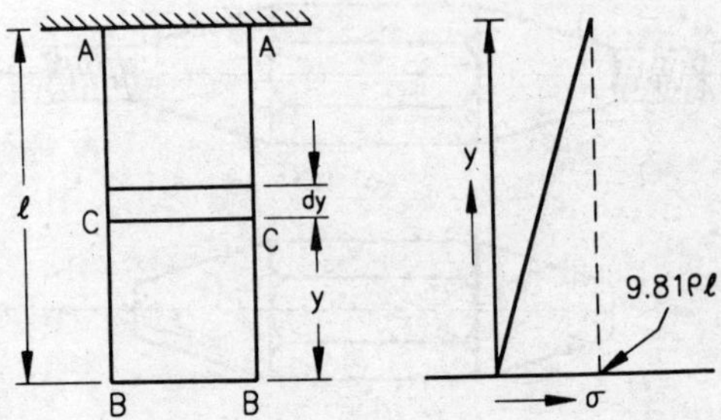


*Fig. 1.7*

If ρ kg/m³ is the density of the material, the stress at a height *y* metre from *B* is given by

$$\sigma = \frac{9.81 \times Ay\rho}{A}$$

$$= 9.81\ \sigma y \text{ N/m}^2 \qquad ...(1.8)$$

Evidently the stress rises linearly with *y* and

$$\sigma_{max} = \sigma_A = 9.81\ \sigma l \text{ N/m}^2 \text{ or Pa} \qquad ...(1.9)$$

The stretch *dl* in the length dy will be given by

$$dl = \frac{\sigma}{E} dy \qquad ...(1.10)$$

Where *E* is the Young's modulus of elasticity.

Total elongation in the bar

$$= \int_0^l (\sigma / E)\, dy = \int_0^l \frac{9.81}{E} \rho y dy$$

$$= \frac{9.81}{2E} \rho l^2 \qquad ...(1.11)$$

## 1.7.1 Tie Bar of Uniform Strength

A tie bar of uniform strength carrying a load of $P$ Newton is shown hanging from the top in the Fig. 1.8. Let the stress everywhere by $\sigma$ N/m$^2$.

The density of the material is $\rho$ kg/m$^3$.

Let $A$ be the area of cross-section $CC$ and $(A + dA)$ that of $DD$. If $A$ varies from $A_1$ to $A_2$ for the sections $BB$ to $AA$ ; we have

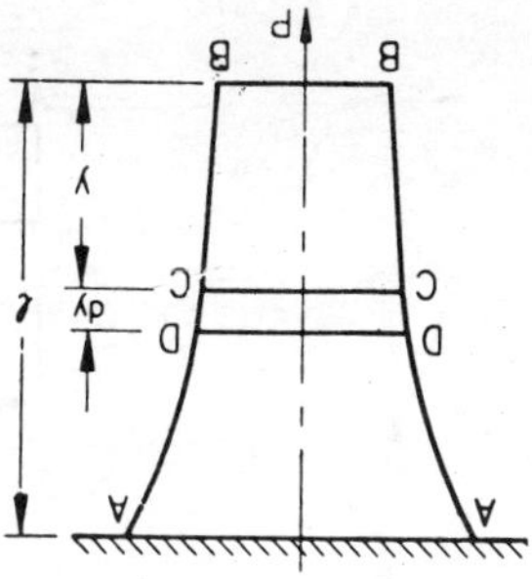

*Fig. 1.8*

For section $BB$,

$$A_1\sigma = P$$

or $$\sigma = P/A_1 \quad ...(1.12)$$

For section $CC$,

$$A\sigma = P + 9.81 \times \text{mass of } BC \text{ in kg} \quad ...(1.13)$$

For section $DD$,

$$(A + dA)\,\sigma = P + 9.81 \text{ mass of portion } BC \text{ in kg} + 9.81(\text{mass of portion } CD \text{ in kg})$$

$$= A\sigma + 9.81\,(\text{mass of portion } CD \text{ in kg})$$

...from equation (1.13)

That is $$A\sigma + dA\sigma = A\sigma + 9.81 \times (Ady)\rho$$

or $$\frac{dA}{A} = \frac{9.81\,\rho}{\sigma}dy \quad ...(1.14)$$

Integrating $$[\ln A]_{A_1}^{A} = \left[\left(\frac{9.81\,\rho}{\sigma}\right)y\right]_0^y$$

or $$\ln(A/A_1) = 9.81\,\rho y/\sigma$$

or $$A = A_1\,e^{9.81\,\rho y/\sigma}$$

At $$y = l\,,\ A = A_2$$

Therefore $$A_2 = A_1 e^{\left[\frac{9.81\,\rho l}{\sigma}\right]} \quad ...(1.15)$$

## 1.8 STRESS IN A BAR DUE TO ROTATION

In the Fig. 1.9 is shown a bar of length $l$ revolving about $Y$-axis at an angular speed of omega rad/sec.

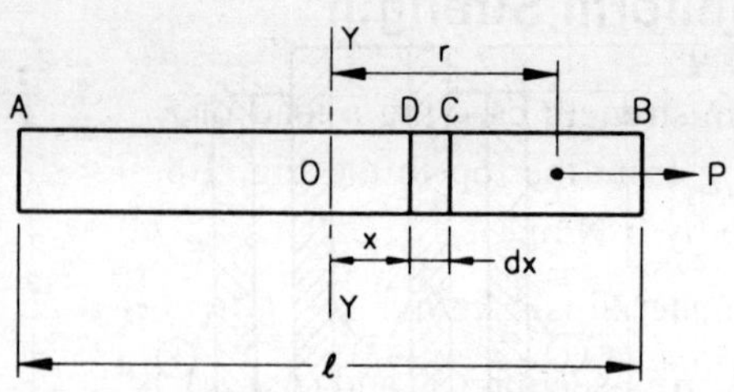


*Fig. 1.9*

Tensile force on the element $DC$ = Centrifugal force $P$ on the part $CB = \omega^2 rM$, where $M$ is the mass of part $CB$.

Thus if $A$ is the area of cross-section in $m^2$ of the bar and $\rho$ the density of the material in $kg/m^3$, then

$$P = \rho A\left(\frac{\ell}{2} - x\right)\omega^2\left[x + \frac{1}{2}\left(\frac{l}{2} - x\right)\right]$$

or
$$P = \frac{1}{2}\rho A\omega^2\,(l^2/4 - x^2)\text{ Newton}$$

*i.e.*
$$\sigma_c = \frac{1}{2}rw^2\,(l^2/4 - x^2)\,\text{N/m}^2 \qquad ...(1.16)$$

Maximum stress will occur at

$$x = 0$$

*i.e.*
$$\sigma_{max} = \frac{1}{8}\rho\omega^2 l^2 \qquad ...(1.17)$$

Extension of the element $DC$

$$= \frac{\sigma_c}{E}dx$$

Thus substituting the value of $\sigma_c$ from equation (1.16) and integrating, the total extension in the bar is given by

$$dl = 2\int_0^{\frac{\ell}{2}}\frac{1}{2E}\rho\omega^2\,(l^2/4 - x^2)\,dx$$

$$= \frac{\rho\omega^2}{E}\left[\frac{l^2}{4}x - \frac{x^3}{3}\right]_0^{\ell/2}$$

$$= \rho\omega^2\, l^3/(12\,E)\text{ metre.}$$

## 1.9 COMPOSITE BARS

A load carrying member which is made up of two different materials is called a composite or compound bar.

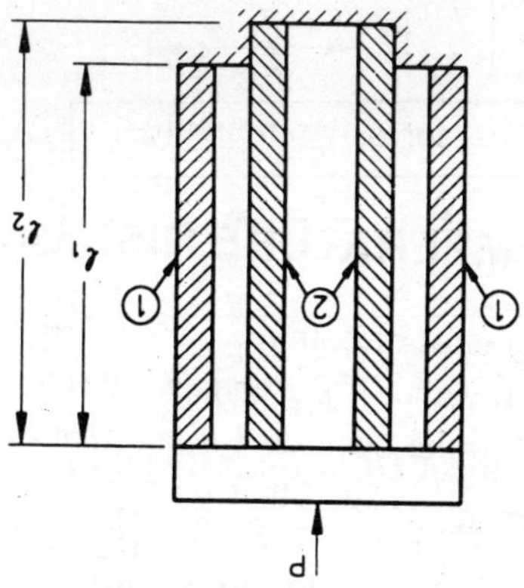


*Fig. 1.10*

In Fig. 1.10 the two materials having the cross-sectional areas $A_1$ and $A_2$, the corresponding modulii of elasticity as $E_1$ and $E_2$ constitute a composite bar.

## 1.9.1 Composite Bar under the Effect of External Load

If the upper ends are rigidly connected to each other, the compression in each will be same

i.e. $$\delta l_1 = \delta l_2$$

or $$\frac{\sigma_1}{E_1} l_1 = \frac{\sigma_2}{E_2} l_2 \qquad \text{...(1.18)}$$

Also the applied load will be equal to the sum of loads sustained by each part

*i.e.* $$P_1 + P_2 = P$$

or $$\sigma_1 A_1 + \sigma_2 A_2 = P \qquad \text{...(1.19)}$$

Then $\sigma_1$ and $\sigma_2$ can be evaluated by solving the simultaneous equations (1.18) and (1.19) to give

$$\sigma_1 = \left(\frac{P}{A_1}\right) \Big/ \left[1 + \frac{E_2 l_1}{E_1 l_2} \frac{A_2}{A_1}\right] \qquad \text{...(1.20}a\text{)}$$

$$\sigma_1 = \left[\frac{P}{A_2}\right] \Big/ \left[1 + \frac{E_1 l_2}{E_2 l_1} \cdot \frac{A_1}{A_2}\right] \qquad \text{...(1.20}b\text{)}$$

If $l_1 = l_2$, equations (1.20) reduce to

$$\sigma_1 = \left(\frac{P}{A_1}\right) \Big/ \left[1 + \frac{E_2}{E_1} \frac{A_2}{A_1}\right] \qquad \text{...(1.21}a\text{)}$$

and $$\sigma_2 = \left(\frac{P}{A_2}\right) \Big/ \left[1 + \frac{E_1}{E_2}\frac{A_1}{A_2}\right] \qquad ...(1.21b)$$

Both $\sigma_1$ and $\sigma_2$ will be compressive in nature.

## 1.10 EFFECT OF TEMPERATURE RISE ON COMPOSITE BAR

To start with, the upper ends of both the bars are rigidly connected to each other so that they are at level $A$. However if the ends were free to expand, the end $A$ of the first bar would have moved to the position $BB$ and that of second one to $CC$, due to temperature rise of $t$°C, (Fig. 1.11) such that

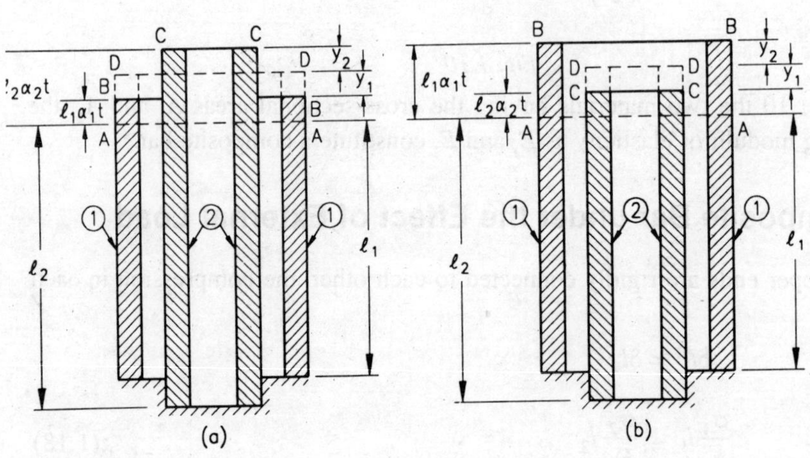


*Fig. 1.11*

$$AB = l_1 \alpha_1 t \qquad ...(1.22)$$

$$AC = l_2 \alpha_2 t \qquad ...(1.23)$$

where $\alpha_1$ and $\alpha_2$ are the respective coefficients of linear thermal expansion.

**Case I.** $l_2 \alpha_2 t > l_1 \alpha_1 t$

Since the two ends are rigidly connected to each other material (1) is forced to extend through a distance $y_1$ to the level $D$ while the material (2) is forced to compress through a distance $y_2$ to the level $D$ Fig. 1.11($a$)

Geometrically, from the Fig. (1.11$a$)

$$y_1 + y_2 = (l_2\alpha_2 - l_1\alpha_1)t$$

or $$\frac{\sigma_1}{E_2} l_1 + \frac{\sigma_2}{E_2} l_2 = (l_2\alpha_2 - l_1\alpha_1) \qquad ...(1.24)$$

Where $\alpha_1$ and $\alpha_2$ are respectively the stresses in the two materials.

Also as no external force is applied at the level $D$ when equilibrium is attained :

Tensile force in material (1) = Compressive force in material (2) *i.e.* $P_1 = P_2$

$$\sigma_1 A_1 = \sigma_2 A_2 \qquad ...(1.25)$$

Solving the simultaneous equations (1.24) and (1.25), we get

$$\sigma_1 = \frac{E_1 t(\alpha_2 - \alpha_1 l_1 / l_2)}{\left[\frac{l_1}{l_2} + \frac{E_1}{E_2} \cdot \frac{A_1}{A_2}\right]} \text{(Tensile)} \qquad ...(1.26a)$$

$$\sigma_2 = \frac{E_2 t\left(\alpha_2 - \alpha_1 \frac{l_1}{l_2}\right)}{\left[1 + \frac{l_1}{l_2} \cdot \frac{E_2}{E_1} \cdot \frac{A_2}{A_1}\right]} \text{(Compressive)} \qquad ...(1.26b)$$

If $l_1 = l_2$, equation (1.26) reduces to

$$\sigma_1 = E_1 t (\alpha_2 - \alpha_1) \Big/ \left[1 + \frac{E_1}{E_2} \cdot \frac{A_1}{A_2}\right] \text{(Tensile)} \qquad ...(1.27a)$$

$$\sigma_1 = E_2 t (\alpha_2 - \alpha_1) \Big/ \left[1 + \frac{E_2}{E_1} \cdot \frac{A_2}{A_1}\right] \text{(Comp.)} \qquad (1.27b)$$

**Case II.** $l_2 \alpha_2 t < l_1 \alpha_1 t$

Fig. 1.11(*b*) refers to this case. For the rigidly connected ends, the material (1) is forced to compress a distance $y_1$ while the material (2) is extended by a distance $y_2$. At the common level *DD* the tensile force experienced by material (2) will be equal to the magnitude of the compressive force in material (1).

Thus $\quad y_1 + y_2 = t\,(l_1\,\sigma_1 - l_2\,\alpha_2\,)$

$$\text{i.e.} \quad \frac{\sigma_1 l_1}{E_1} + \frac{\sigma_2 l_2}{E_2} = t\,(l_1 \alpha_1 - l_2 a_2) \qquad ...(1.28)$$

Equation (1.25) holds good here also. Solving the equations (1.25) and (1.28) for $\sigma_1$ and $\sigma_2$ we get

$$\sigma_1 = \frac{E_1 t\left(\alpha_1 - \frac{l_2}{l_1} \alpha_2\right)}{\left[1 + \frac{E_1 l_2 A_1}{E_2 l_1 A_2}\right]} \text{(Compressive)} \qquad ...(1.29a)$$

$$\sigma_1 = \frac{E_2 t\left[\alpha_1 - \left(\frac{l_2}{l_1} \alpha_2\right)\right]}{\left[1 + \frac{E_1 A_2}{E_2 A_1}\right]} \text{(Tensile)} \qquad ..(1.30a)$$

If $l_1 = l_2$ these reduce to

$$\sigma_1 = \frac{E_1 t(\alpha_1 - \alpha_2)}{\left[1 + \frac{E_1}{E_2} \cdot \frac{A_1}{A_2}\right]} \text{ (Compressive)} \quad ...(1.29b)$$

$$\sigma_1 = \frac{E_2 t(\alpha_1 - \alpha_2)}{\left[1 + \frac{E_1 A_2}{E_2 A_1}\right]} \text{ (Tensile)} \quad ...(1.30b)$$

## 1.11 EFFECT OF BOTH THE TEMPERATURE RISE AND EXTERNAL LOAD

The stresses due to individual effects are evaluated as per the method explained above. The resultant stresses are then calculated by adding the stresses algebraically in each material.

**Example 1.3.** *2A steel bar of uniform diameter of 40 mm is heated to 80°C and then clamped at the ends with the help of two fixtures 4 m apart, and left to cool down to room temperature of 20° C at which temperature the distance between the fixtures was found to be 1.0 mm shorter than that at 80° C. Determine the stress in the bar when it has cooled down to room temperature and the reaction at the fixtures.*

Take $E = 200\ GN/m^2$ and

$\alpha = 1.1 \times 10^{-5}$/deg.C.

**Solution.** $l = 4000$mm

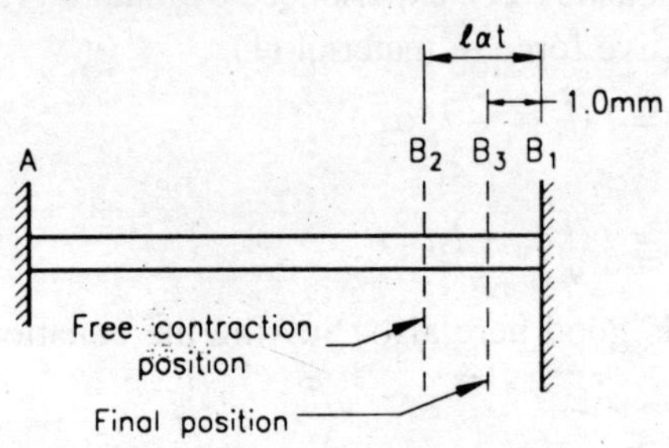


*Fig. 1.12*

Temperature drop

$$t = 80 - 20 = 60°C$$

$$B_1B_2 = \text{Free contraction} = l\,\alpha\,t$$

$$= 4000 \times 1.1 \times 10^{-5} \times 60 = 2.64 \text{ mm}$$

Therefore $\quad B_2B_3$ = Stretch in the bar

$$= B_1B_2 = - B_1B_3$$

$$= 2.64 - 1.0 = 1.64 \text{ mm}$$

Thus the stress in the bar is given by

$$\sigma = \frac{E}{l} \times 1.64$$

$$= \frac{200 \times 10^3}{4000} \times 1.64 = 82 \text{ MN/m}^2 \text{ or N/mm}^2 \text{ (Tensile)}$$

Reaction of Supports

$$= \sigma \times A = 82 \times \frac{\pi}{4}(40)^2$$

$$= 10^3 \times 10^3 \text{ N}$$

$$= 103 \text{ kN.}$$

**Example 1.4.** *A copper sleeve, 21 mm internal and 27 mm external diameter, surrounds a 20 mm steel bolt, one end of the sleeve being in contact with the shoulder of the bolt. After putting a rigid washer on the other end of the tube, a nut is screwed on the bolt until the compressive stress in the sleeve is 80 MN/m². Find the range of the external axial load that can be applied to this assembly if the stress in the bolt is never to be compressive and in the sleeve never to be zero. Take $E_s$ = 200 GN/m² and $E_c$ = 90 GN/m².*

**Solution.** $$A_s = \frac{\pi}{4} \times 20^2 = 10\pi \text{ mm}^2;$$

$$A_c = \frac{\pi}{4}(27^2 - 21^2) = 72\,\pi\ \mu\mu^2$$

When the tube is in compression due to the tightening of the bolt, bolt will be under tension and

$$P_c = P_s$$

or $$\sigma_c A_c = \sigma_s A_s$$

or $$\sigma_s = \frac{\sigma_C A_C}{A_S} = \frac{80 \times 72\pi}{100\,\pi}$$

$$= 57.6 \text{ MN/m}^2 \text{ or (N/mm}^2) \quad ...(1.31)$$

$$P_s = P_c = 57.6 \times 100 \times \pi = 18.1 \text{kN} \quad ...(1.32)$$

$$\varepsilon_s = \frac{\sigma_S}{E_S}$$

$$= \frac{57.6}{200 \times 10^3}$$

$$= 2.88 \times 10^{-4} \text{ (Tensile)} \quad ...(1.33)$$

*Fig. 1.13*

$$\varepsilon_c = \frac{\sigma_C}{E_C} = \frac{80}{90 \times 10^3}$$

$$= 8.89 \times 10^{-4} \text{ (Compressive )} \quad ...(1.34)$$

(*b*) Two cases can arise depending on whether the load, compressive or tensile, is applied on the bolt or on the tube.

(*i*) **Tensile load on the bolt :** In this case the limiting condition will be when the compressive stress already existing in the tube is just brought down to zero. This will happen when the bolt is stretched further to the extent that the additional stretch or strain induced in it is equal to the compression or compressive strain in tube as calcualted in equation 1.34 (as these lengths are equal)

$$\varepsilon_{S_2} = \varepsilon_C = 8.89 \times 10^{-4} \quad ...(1.35)$$

$$P_2 = P_{S_2} = A_S E_S$$

$$= 100\,\pi \times (200 \times 10^{-4})\,(8.89 \times 10^{-4})\text{N}$$

$$= 5.9 \text{ N ( Tensile)}\} \quad ...(1.37)$$

(*ii*) **Compressive load on the bolt :** Here the appleid load will be shared by both bolt and tube and the limiting case will be when the additional induced compressive stress in the bolt becomes equal in magnitude to the existing tensile stress in the bolt as given by equation (1.31).

*i.e.* $$\sigma_{S_3} = \sigma_S \quad ...(1.38)$$

$$P_3 = P_{S_3} + P_{C_3} \quad ...(1.39)$$

and $$\varepsilon_{S_3} = \varepsilon_{C_3}$$

or $$\frac{\sigma_S}{E_S} = \frac{P_{C_3}}{A_C E_C}$$

*i.e.* $$P_{C_3} = \sigma_S A_C E_C / E_S \quad ...(1.40)$$

Thus substituting values in equation (1.39)

$$P_3 = \sigma_S A_S + \sigma_S A_C / E_S$$

$$= \sigma_S [A_S + A_C E_C / E_S]$$

$$= 57.6\left[100\pi + 72\pi \frac{90}{200}\right]$$

i.e $$P_3 = 24 \text{ kN (Compressive)} \quad ...(1.41)$$

(*iii*) **Tensile load on tube :** Here again the load will be shared by both the bolt and the tube and the stress in the tube will be the limiting case and the additional tensile stress induced in the tube will become equal to 80 N/mm$^2$ or 80 MN/m$^2$.

*i.e.* $$\sigma_{C_4} = 80 \text{ N/mm}^2 \quad ...(1.42)$$

$$P_4 = P_{S_4} + P_{C_4} \quad ...(1.43)$$

$$\sigma_{S_4} = \varepsilon_{C_4}$$

or $$\sigma_{S_4} / E_S = \sigma_{C_4} / E_C$$

Thus $$P_4 = \sigma_{S_4} A_S + \sigma_{C_4} A_C$$

$$= \sigma_{C_4} \left[ A_C + \frac{E_S}{E_C} A_S \right]$$

$$= 80 \left[ 72\pi + \frac{200}{90} \times 100\pi \right] \text{N}$$

*i.e.* $$P_4 = 73.9 \text{ kN (Tensile)} \qquad ...(1.44)$$

(*iv*) **Compressive load applied on tube :** In this case the load with be borne only by tube and the limiting condition will be when the stress in bolt reduces to zero. This will happen when the compressive strain induced in tube due to external load is just equal in magnitude to the strain $\varepsilon_S$ in bolt as given in equation (1.33).

Thus $$\varepsilon_{C_5} = \varepsilon_S \qquad ...(1.45)$$

$$P_5 = A_{C_5} A_C$$

$$= 72 \times (2.88 \times 10^{-4}) \times (90 \times 10^3) \text{ N}$$

*i.e.* $$P_5 = 5.86 \text{ kN (Compressive)} \qquad ...(1.46)$$

From equations (1.37), (1.41), (1.44) and (1.46) it is evident that the given conditions to be fulfilled the allowable range of the external load is from 5.86 kN (Compressive) to 55.9 kN (tensile).

**Example 1.5.** *If instead of applying the external load to the bolt and tube assembly of Example 1.4, the temperature is raised through 50°C, find the resultant stresses in the bolt and the tube.*

Take $$\alpha_{steel} = 11 \times 10^{-6}/°C$$

$$\alpha_{copper} = 18 \times 10^{-6}/°C.$$

**Solution.** In this case since the lengths are equal, we have from equation (1.24)0

$$\frac{\sigma_S}{E_S} = \frac{\sigma_S}{E_S} + \frac{\sigma_C}{E_C} = t\,(\alpha_C - \alpha_S) \qquad ...(1.47)$$

Where t is the temperature rise.

Also in equilibrium state the compressive load in copper will be equal to tensile load in bolt, due to the temperature rise effect.

*i.e* $$\sigma_S A_S = \sigma_c A_c$$

or $$\sigma_S = (72/100)\sigma_c \qquad ...(1.48)$$

Substituting equation (1.48) in equation (1.47)

$$\sigma_c \left[\frac{0.72}{E_S} + \frac{1}{E_C}\right] = t(\alpha_c - \alpha_s)$$

or $$\sigma_c \left(\frac{0.72}{200 \times 10^3} + \frac{1}{90 \times 10^3}\right) = 50\ (7 \times 10^{-6})$$

$$= 23.8 \text{ N/mm}^2 \text{ or MPa (Comp.)} \qquad ...(1.49a)$$

and from eqn. (1.48)

$$\sigma_S = .72 \times 23.8$$

$$= 17.2 \text{ MN/m}^2 \text{ or MPa (Tensile)} \qquad ...(14.9b)$$

Thus from equations (1.31) and (1.49)

Resultant stress in bolt

$$= 57.6 + 17.2 = 74.8 \text{ MN/m}^2 \text{ (Tensile)}$$

Resultant stress in tube

$$= 80.0 + 23.8 = 103.8\text{MPa (Comp.)}$$

**Example 1.6.** *Two bars one of aluminum and the other of steel, joined end to end, are placed lengthwise between two rigid supports 600 mm apart. Their axes are horizontal and are along the same straight line. Aluminium bar is 200 mm long and steel bar is 400 mm long. Their cross-sectional areas are 1000 mm² and 200 mm² while Young's modulii are 0.6 × 10⁶kg/cm² (588.6 MN/m²) and 2 × 10⁶ kg/cm² (1962 MN/m²) respectively. The whole system is in un-stressed condition at 10°C. The system is then heated to a temperature of 60°C. Find the stresses developed in the two materials if*

$$a_{steel} = 12 \times 10^{6}/°C$$

*and* $$\alpha_{al} = 24 \times 10^{-6}/°C$$ *(Punjab University)*

**Solution.** If the composite bar were free to expand, the corresponding increase in length of the two portions will be added up to give the total increase in length $\delta l$ as

$$\delta l = \delta l_s + \delta l_a$$

$$= l_s\, \alpha_s (T_2 - T_1) + l_\alpha (T_2 - T_1)$$

$$= (T_2 - T_1)\, [\alpha_s l_s + \alpha_a l_a]$$

$$= (60 - 10)\ (12 \times 10^{-6} \times 400 + 24 \times 10^{-6} \times 200)$$

$$= 50 \times 9600 \times 10^{-6}$$

$$= 0.48 \text{ mm.}$$

Since both the metals have the same axis, they will carry the same compressure axial load, P, if the free expansion is checked by rigid supports. Thus the compression in both the metals due to the load P will add up to *dl*

i.e.,
$$\frac{P}{A_s S_s} l_s + \frac{P}{A_a E_a} l_a = \delta l$$

or
$$P\left[\frac{400 \times 10^2}{(200)(2 \times 10^6)} + \frac{200 \times 10^2}{(1000)(0.6 \times 10^6)}\right] = 0.48$$

or
$$P = 11076.9 \text{ kg}$$

Therefore
$$\sigma_s = \frac{P}{A_S} = \frac{110076.9}{200} \times 10^2$$
$$= 553.85 \text{ kg/ cm}^2 \text{ (comp.)}$$

and
$$\sigma_a = \frac{P}{A_a} = \frac{11076.9}{1000} \times 10^2$$
$$= 1107.70 \text{ kg/cm}^2 \text{ ( comp.)}$$

**Example 1.7.** *A brass rod of 6 mm diameter and 1 m long is joined at one end to a rod of steel 6 mm diameter and 1.3 m long. The compound rod is placed in a vertical position with the steel rod at the top and connected top and bottom to rigid fixings in such a way that it is carrying a tensile load of 3.5 kN.*

*An attachment is fixed at the junction of the two rods and to this a vertical axial load of 1.3 kN is applied downwards. Calculate the stresses in the steel and brass. The temperature is then raised through 30°C. What are the final stresses in steel and brass. Take*

$$E_s = 200 \text{ GN/m}^2$$
$$E_b = 85 \text{ GN/m}^2$$
$$\alpha_s = 12 \times 10^{-6}/°C,$$
$$\alpha_b = 19 \times 10^{-6}/°C.$$

*(Punjab University)*

**Solution.** (*a*) In the first instance, when the composite bar *ABC* is held in fixtures, both steel and brass portions of the rod carry the tensile load of 3500 *N*. Thus the initial tensile stresses in each of the metals will be equal, as their areas of cross sections are equal.

$$A_s = A_b = \frac{\pi}{4}(6)^2$$
$$= 28.274 \text{mm}^2$$

$$\sigma_{s_1} = \sigma_{b_1} = \frac{3500}{28.274}$$
$$= 123.79 \text{N/mm}^2$$
or MN/m² (Tensile)

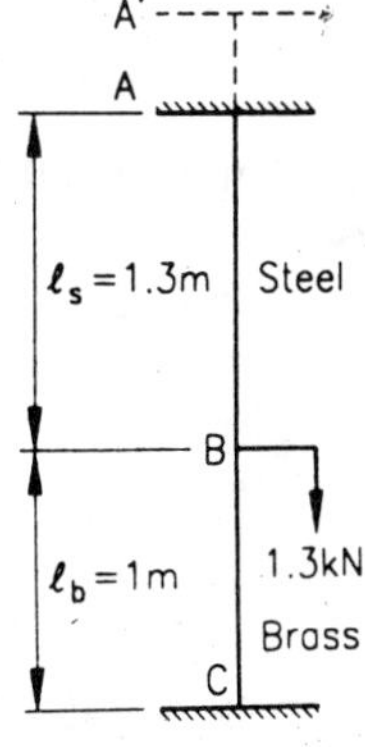


*Fig. 1.14*

(*b*) The load of 1300 *N* applied at the junction point *B* (Fig. 1.14), will induce compressive stress in brass and tensile stress in steel. The magnitudes of the loads $P_s$ and $P_b$ due to the action of 1300 *N* load at *B*, will be such that the stretch in steel will be equal in magnitude to the compression in brass, because the total length *AC* remains constant.

That is $$P_s + P_b = 1300$$

and $$\delta l_s = \delta l_b$$

or $$\frac{P_s l_s}{A_s E_s} = \frac{P_b l_b}{A_b E_b}$$

or $$P_s = \frac{E_s l_b}{E_b l_s} P_b \qquad (\because A_s = A_b)$$

$$= \frac{200}{85} \times \frac{1}{1.3} P_b = 1.81\, P_b$$

Therefore

$$1.81\, P_b + P_b = 1300$$

or $$P_b = 461 \text{ N}$$

Then $$P_s = 1300 - 461$$

$$= 839 \text{ N}$$

Thus the corresponding additional stresses in steel and brass, due to the application of 1300 N load are

$$\sigma_{s2} = \frac{P_s}{A_s} = \frac{839}{28.274}$$

$$= 29.67 \text{ MN/m}^2 \text{ (Tensile)}$$

$$\sigma_{b2} = \frac{P_b}{A_b} = \frac{461}{28.274}$$

$$= 16.3 \text{ MN/ m}^2 \text{ or MPa (Comp.)}$$

Then the resultant stresses are

$$\sigma_b = \sigma_{b1} - \sigma_{b2}$$

$$= 123.79 - 16.3$$

$$= 107.49 \text{MN/m}^2 \text{ or MPa (Tensile)}$$

$$\sigma_s = \sigma_{s1} + \sigma_{s2} = 123.79 + 29.67$$

$$= 153.46 \text{MN/m}^2 \text{ or MPa (Tensile)}$$

(c) In the absence of fixtures, the rod *AC* would have expanded freely due to temperature rise to *CA'*. But the presence of fixtures effectively compresses the rod *AC* by an amount $AA'$ which is equal to $(l_s \alpha_s t + l_b \alpha_b t)$. Thus if σ N/m² is the compressive stress in steel and brass (it will be same as the cross section areas are same) due to temperature rise, then we have

$$\frac{\sigma}{E_s} l_s + \frac{\sigma}{E_b} l_b = l_s \alpha_s t + l_b \alpha_b t$$

or $$\left[\frac{1.3}{200 \times 10^9} + \frac{1}{85 \times 10^9}\right] = 30\,(1.3 \times 12 \times 10^{-6} + 1 \times 9 \times 10^{-6})$$

or $$\sigma = 56.9 \times 10^6 \text{ N/mm}^2$$

$$= 56.9 \text{MN/m}^2 \text{ or MPa}$$

Therefore, the final stresses in steel and brass are

$$\sigma_{s3} = \sigma_s - \sigma = 143.46 - 56.9$$

$$= 96.56 \text{ MN/m}^2 \text{ or MPa (Tensile)}$$

$$\sigma_{b3} = \sigma_b - \sigma = 107.49 - 56.9$$

$$= 0.59 \text{MN/ m}^2 \text{ or MPa (Tensile)}$$

**Example 1.8.** *A bar AD (Fig. 1.15) is pinned at A, and supported by a copper rod 40 mm² in area and 1.5 m long at B, also supported at C by a steel rod of 30 mm² in area and 1 m length. The bar AD is rigid and horizontal before applying a load of 1000 kg (9810 N) at the end D. Determine the stresses and elongations produced in copper and steel rods after the application of above load. Take*

$$E_c = 10 \times 10^3 \text{ kg/mm}^2 \text{ ( } 98.1 \text{GN/m}^2\text{)}$$

$$E_s = 20 \times 10^3 \text{ kg/mm}^2 \text{ (} 196.2 \text{ GN/m}^2\text{).}$$

*(Punjabi University)*

**Solution.** Let the tensile loads sustained by copper and steel rods be $P_c$ and $P_s$ respectively. Then for the equilibrium of the rod **AD**, taking moments of $P$, $P_c$ and $P_s$ about $A$ we have

$$P_c \times 1 + P_s \times 2 = P \times 2.5$$

or $$P_c + 2P_s = 9810 \times 2.5 = 24525 \qquad \text{...(A)}$$

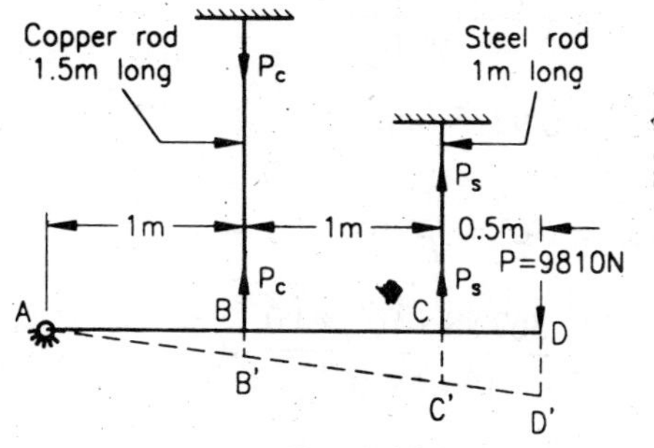


*Fig. 1.15*

Further due to the loads $P_c$ and $P_s$ the copper and steel rods will get stretched by $BB'$ and $CC'$ respectively. However, the extensions will be such that $A$, $B'$, $C'$ and $D'$ will lie in one straight line, as the rod $AD$ is rigid. Therefore from similar triangles $ABB'$ and $ACC'$ we have

$$\frac{CC'}{BB'} = \frac{AC}{AB} = \frac{2}{1}$$

or $$CC' = 2BB'$$

or $$\frac{P_S}{A_S E_S} l_s = 2\frac{P_C}{A_C E_C} l_c$$

or $$P_s = 2\left(\frac{A_s}{A_C}\right)\left(\frac{E_s}{E_C}\right)\left(\frac{l_c}{l_s}\right) P_c$$

$$= 2\left(\frac{30}{40}\right)\left(\frac{196.2}{98.1}\right)\left(\frac{1.5}{1.0}\right) P_c$$

or $$P_s = 4.5\ P_c \qquad \text{...}(B)$$

Substituting ($B$) into ($A$), we get

$$P_c = 2452.5 \text{ N}$$

and $$P_s = 11036.25 \text{ N}$$

Therefore $$\sigma_c = \frac{2452.5}{40}$$

$$= 61.31 \text{ N/mm}^2 \text{ or MPa (Tensile)}$$

and $$\sigma_s = \frac{11036.25}{30}$$

$$= 367.88 \text{N/mm}^2 \text{ or MPa (Tensile)}$$

Also $$\delta l_c = BB' = \frac{\sigma_c}{E_c} \times l_c$$

$$= \frac{61.31 \times 10^6 \times 1.5}{98.1} \text{ m}$$

$$= 0.9375 \text{ mm}$$

Similarly $$\delta l_s = CC' = \frac{\sigma_s}{E_s} l_s$$

$$= \frac{367.88 \times 10^6 \times 1.0}{196.2 \times 10^9} \text{ m}$$

$$= 1.875 \text{ mm.}$$

**Example 1.9.** *Fig. 1.16 shows a round steel rod supported in a recess and surrounded by a coaxial brass tube. The upper end of the rod is 0.1 mm below that of the tube and an axial load is applied to a rigid plate resting on the top of the tube.*

*(a) Determine the magnitude of the maximum permissible load if the compressive stress in the rod is not to exceed 110 MN/m² and that in the tube is not to exceed 80 MN/m².*

*(b) Find the amount by which the tube will be shortened by the load if the compressive stress in the tube is the same as that in the rod,*

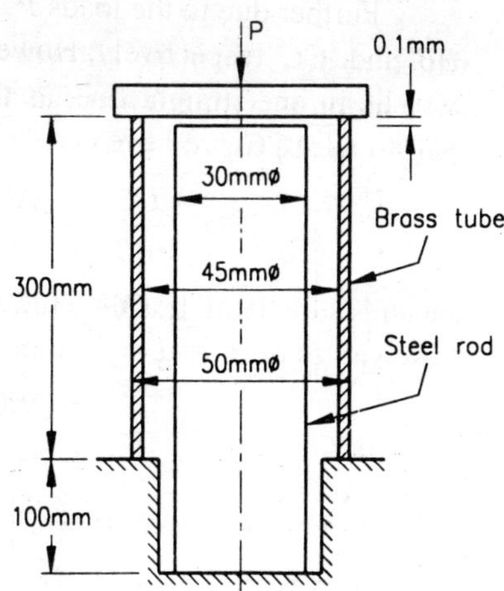


*Fig. 1.16*

*Take* $Es = 200\ GN/m^2 = 200 \times 10^3$ MPa

$E_b = 100$ GN/m² $= 100 \times 10^3$ MPa *(Punjab University)*

**Solution.** (*a*) $A_s = \frac{\pi}{4} \times 706.86\ \text{mm}^2$

$$A_b = \frac{\pi}{4}(50^2 - 45)^2 = 373.07\ \text{mm}^2$$

Let $\sigma_s$ and $\sigma_b$ MN/m² or N/mm² be the stresses induced in the steel rod and brass tube respectively. Since the sum of the loads carried by the steel and brass is equal to the total load

$$P = \sigma_s\ A_s + \sigma_b A_b$$

$$= 706.86\ \sigma_s + 373.07\sigma_b\ \text{N} \quad ...(A)$$

When the top plate makes contact with steel and compresses it, the compression of the brass will exceed that of steel by 0.1 mm.

That is $\delta l_b = \delta l_s + 0.1$

or $$\frac{\sigma_b}{E_b} l_b = \frac{\sigma_s}{E_s} l_s + 0.1$$

or $$\frac{\sigma_b}{100 \times 10^3} \times 300 = \frac{\sigma_s}{200 \times 10^3} \times 400 + 0.1$$

or $$3\sigma_b = 3\sigma_s + 100 \quad ...(B)$$

The maximum compressive stress of 110 MN/m² in rod and 80 MN/m² in tube will not occur simultaneously, rather the magnitudes of induced stresses in the two materials will be related to each other by the equation (*B*). Thus if it is assumed that the stress in steel is the limiting case, then from equation (*B*) for $\sigma_s = 110$ MN/m²,

$$\sigma_b = \frac{2 \times 110 + 100}{3}$$

$$= 106.7 \text{ MN/m}^2$$

which is more than the allowable stress in brass leading to the conclusion that the assumption is wrong. Then again assuming the stress in brass to be the limiting case, then from (*B*) for $\sigma_b = 800 \text{ MN/m}^2$,

$$\sigma_s = \frac{3 \times 80 - 100}{2} = 70 \text{ MN/m}^2$$

which is safe being less than the maximum allowable stress in steel. Thus substituting these values of $\sigma_b$ and $\sigma_g$ in equation (*A*), we get the maximum permissible load *P* as

$$P = (706.86 \times 70 + 373.07 \times 80) \text{ N}$$

$$= 79.325 \text{ kN}$$

(*b*) In this case $\sigma_s = \sigma_b$

Then from equation (*B*)

$$\sigma_b = \sigma_s = 100 \text{ MN/m}^2$$

Therefore $$\delta l_b = \frac{\sigma_b}{E_b} \times l_b$$

$$= \frac{100}{100 \times 10^3} \times 300 \text{ mm}$$

$$= 0.3 \text{ mm.}$$

**Example 1.10.** *A reinforced concrete column is 300 mm in diameter and has 4 steel bars of 12 mm diameter embedded in it. If the allowable stresses in steel and concrete are 65.0 MN/m² and 4.40 MN/m² respectively, calculate the safe axial load which the column can carry. Take $E_s/E_c = 15$.*

**Solution.**

$$\text{Total area of column} = \frac{\pi}{4}(300)^2 = 70686 \text{ mm}^2$$

$$A_s = \text{Area of steel} = 4 \times \frac{\pi}{4}(12)^2 = 452.4 \text{ mm}^2$$

Therefore area of concrete portion

$$= A_c = 70686 - 452.4$$

$$= 70233.6 \text{ mm}^2$$

Since both steel and concrete portions will compress by the same amount and also their original lengths are same, the compressive strains produced by the corresponding loads sustained by each will be same

*i.e.*, $$\varepsilon_c = \varepsilon_s$$

or $$\frac{P_c}{A_c E_c} = \frac{P_s}{A_s E_s} \quad \ldots(A)$$

Also from (*A*)

$$\frac{\sigma_c}{E_c} = \frac{\sigma_s}{E_s}$$

or $$\frac{\sigma_s}{\sigma_s} = \frac{E_s}{E_c} = 15 \qquad \text{...(B)}$$

Thus if it is assumed that the stress in concrete is the limiting case, then for 4 $MN/m^2$ stress to be induced in concrete the stress induced in steel will be from (*B*)

$$\sigma_s = 15 \times \sigma_c = 15 \times 4 = 60 \text{ MN/m}^2$$

(which is less than the allowable stress in steel)

Thus our assumption of limiting stress in concrete is valid. Corresponding loads sustained by concrete and steel are

$$P_c = \sigma_c A_c = 4 \times 70233.6 = 280934 \text{ N}$$

$$= 280.934 \text{ kN}$$

$$P_s = \sigma_s \times A_s = 60 \times 452.4 = 27144 \text{ N}$$

$$= 27.144 \text{ kN}$$

Thus total load which the column can carry

$$= P_c + P_s$$

$$= 280.934 + 27.144$$

$$= 308.078 \text{ kN}.$$

**Example 1.11.** *Two steel plates, each 25 mm thick, are held together with the help of 20 mm diameter steel bolts. If the pitch of the threads is 2.5 mm, find the increase in stress in the shank of the bolt when the nut is turned through $\frac{1}{200}$ of a turn with respect to the head of the bolt, assuming that the plates do not deform. Take $E_s$ = 205 $GN/m^2$.*

**Solution.** Length of the bolt shank = 25 + 25 = 50 mm.

Axial movement of the nut corresponding to rotation of $\frac{1}{200}$ of a turn = $\frac{2.5}{200}$ = 0.0125 mm.

Thus that rotation of the nut will cause 0.0125 mm stretch in 50 mm length of the shank giving rise to strain e given by

$$\varepsilon = \frac{0.0125}{50}$$

Therefore $$\sigma = \varepsilon E = \frac{0.0125}{50} \times (205 \times 10^3)$$

$$= 51.25 \text{ MN/m}^2.$$

**Example 1.12.** *A copper sleeve, 21 mm internal and 27 mm external diameter, surrounds a 20 mm steel bolt, one end of the sleeve being in contact with the shoulder of the bolt. The sleeve is 60 mm long. After putting a rigid washer on the other end of the sleeve, a nut is screwed on the bolt through 10 degrees. If the pitch of the threads is 2.5 mm, find the stresses induced in the copper sleeve and steel bolt. Take $E_s$ = 200 GN/m² and $E_c$ = 90 GN/m².* *(Punjab University)*

**Solution.** Due to 10° rotation of the nut, it will move along the axis of the bolt through a distance δ (say), such that

$$\delta = \frac{10}{360} \times \text{pitch}$$

$$= \frac{10}{360} \times 2.5 = 0.0694 \text{ mm}$$

Further the axial movement of the nut, δ, will be equal to the compression in the copper sleeve plus stretch in the steel bolt.

i.e. $\delta = \varepsilon_c l_c + \varepsilon_s l_s$

or $0.0694 = 60(\varepsilon_c + \varepsilon_s)$ $[\because l_c = l_s = 60 \text{ mm}]$

or $$\frac{0.0694}{60} = \frac{\sigma_c}{E_c} + \frac{\sigma_s}{E_s} \quad ...(A)$$

Also the tube and the bolt, in equilibrium, will carry the same load.

*i.e.*, $P_c = P_s$

$$\sigma_c A_c = \sigma_s A_s$$

or $$\sigma_c = \frac{\sigma_s A_s}{A_c} = \sigma_s \frac{\frac{\pi}{4}(20)}{\frac{\pi}{4}(27^2 - 21^2)} = 1.39\sigma_s \quad ...(B)$$

From (*A*) and (*B*), we get

$$\frac{0.0694}{60} = \left(\frac{1.39}{90 \times 10^3} + \frac{1}{200 \times 10^3}\right)\sigma_s$$

$$= 0.2044 \times 10^{-4}\,\sigma_s$$

or $$\sigma_s = \frac{0.0694}{60} \times \frac{1}{0.2044 \times 10^{-4}} = 56.59 \text{ (Tensile)}$$

From (*B*)

$$\sigma_c = 1.39 \times 56.59$$

$$= 78.66 \text{ MN/m}^2 \text{ (compressive)}$$

## OBJECTIVE TYPE QUESTIONS

1.1 Define the terms, "stress", "pressure".

1.2 Explain the difference between nominal strain and True Strain in a simple Tension Test.

1.3 Explain the difference between Nominal stress and True Stress in a uniaxial stress, test price.

1.4 State whether it is true or false : "If the tensile test piece while in the plastic range of the curve, is unloaded it follows a straight path parallel to the elastic line.

[**Ans.** True]

1.5 (*a*) State whether true or false :

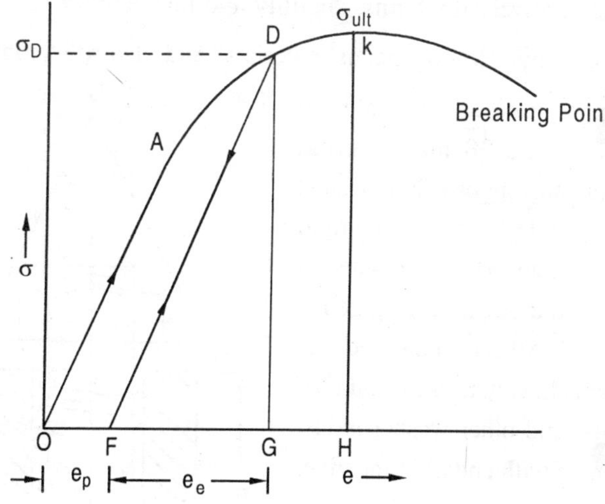


*Fig. 1.17*

When the stress in the simple tension test is at level D *i.e.,* $\sigma_D$ (Fig. 1.17) and the specimen is unloaded to no stress state the left over strain OF is equal to the plastic strain component of the total strain of D and FG is the elastic strain component or elastic recovery.

[**Ans.** True]

(*b*) If the test piece is reloaded or restressed from the point *F* it follows the path *FD*. Then $\sigma_D$ is the new yield stress. Similarly if this process is repeated with reference to point *L* will, then the new yield point corrspond to $\sigma_L$?

[Yes]

1.6 Define Ductility and Malleability.

1.7 Will the strain at Instability point in the case of Aluminium be more then that for steel?

[Yes]

1.8 In a straight bar two equal and opposite forces of 100 kN each are acting at the mid length section. What is stress in the cross section at ¾th length. Take the area of x-section as 0.01 m$^2$. **[Ans.** Zero]

1.9. (*a*) State *D'* Alemberts Principle.

(*b*) Can be body in linear motion be in a state of stress ? State yes of No.

If yes, give example.

1.10 Can the equilibrium equations for a body in static equilibrium be applied to the body in motion? If so explain the phenomenon.

1.11 State Hookes law.

1.12 Define proof stress.

1.13 Define quantitatively the terms, ductility and malleability.

1.14 What is meant by "Homogeneous materials" and "Isotropic materials"?

1.15 (*a*) A copper tube 25 mm internal diameter and 30 mm external diameter surrounds a 20 mm steel bolt Fig. (1.18). After putting rigid washers *A* and *B* on two sides of the tube the nut is just tightened with hand so that the bolt head side washer is in contact with end *C* of the tube and other washer is just in contact with end *D* of the tube. The nut is further tightened so that it rotates ¼th of a rotation with respect to the bolt head. If the pitch of the threads is 1.5 mm find the stresses induced in bolt and copper tube. Take $E_S$ = 200 × 10$^3$ MPa and $EC$ = 90 × 10$^3$ MPa. Take length of tube = 300 mm.

(*b*) The washers *A* and *B* are then pulled apart with the help of wire ropes. Find the value of the pull such that the stress in copper take becomes zero. What is then the stress in steel bolt ?

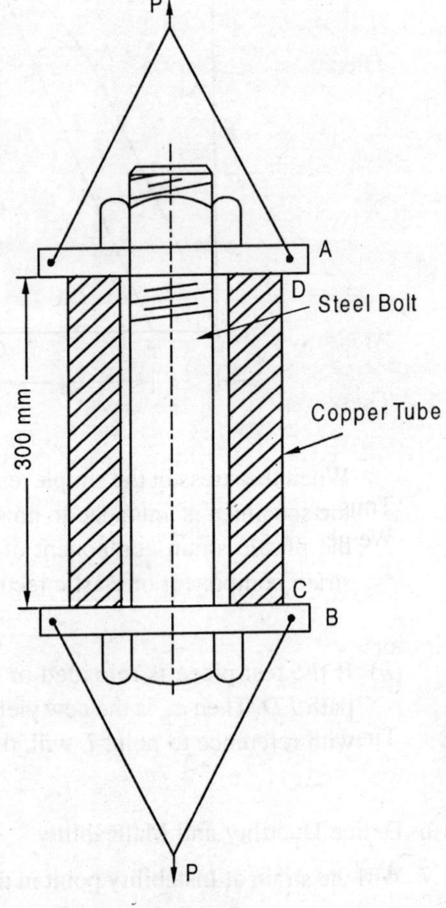


*Fig. 1.18*

**Solution.** Area of cross-section of bolt

$$A_s = \frac{\pi}{4} \times (20)^2 = 100\pi \text{ mm}^2 = 314.159 \text{ mm}^2$$

Area of cross-section of the tube

$$A_C = \frac{\pi}{4}(30^2 - 25^2)$$

$$= \frac{\pi}{4} \times 275 = 215.98 \text{ mm}^2$$

Free movement of nut with respect of head

= Pitch × no of terms

$$= 1.5 \times \frac{1}{4} = 0.375 \text{ mm}$$

= compression in copper × stretch in steed ...(*A*)

If *W* is the axial compressive load in copper

Then compression in Cu $= \varepsilon_C l_C$

$$= \left(\frac{W}{A_C}\right)\frac{l_C}{E_C}$$

$$= \frac{W}{215.98} \times \frac{300}{90 \times 10^3} \text{ mm}$$

$= 0.0154 \times 10^{-3}$ W mm

At equilibrium tensile load in steel = compressive load in Cu = W

Therefore Stretch in steel $= \frac{W}{A_S}\frac{1}{E_S}l_s = \frac{W}{314.119} \times \frac{1}{200 \times 10^3} \times 300 \text{ mm}$

$= 0.00477 \times 10^{-3}$ W mm

Thus substituting the values in equation (*A*)
We get

$$0.375 = 0.154 \times 10^{-3} W + .00477 \times 10^{-3} W$$

or $$W = \frac{0.375 \times 10^3}{0.02017} = 18.59 \times 10^3 \text{N} = 18.59 \text{ kN}$$

Therefore $\sigma_s$ stress in steel $= \frac{W}{A_S} = \frac{18.59 \times 10^3}{314.159} = 0.0592 \times 10^3 \text{ N/mm}^2$

= 59.2 MPa (Tensile)

And stress in copper $= \sigma_c = \frac{W}{A_c} = \frac{18.59 \times 10^3}{215.98} = 0.08614 \times 10^3 \text{ N/mm}^2$

= 86.14 MPa (comp.)

Then Strain in copper $= \dfrac{\sigma_C}{E_C} = \dfrac{86.14 \text{ MPa}}{90 \times 10^3 \text{ MPa}}$

*i.e.,* $\varepsilon_c = \dfrac{86.14}{90 \times 10^3} = 0.957 \times 10^{-3}$

$$\delta l_c = 300 \times 0.957 \times 10^{-3}$$

$$= 0.287 \text{ mm}$$

**Solution.** (*b*) In the part (*b*) the load *P* will act as pull only on the bolt. Its value is to be found such that the resultant stress in copper is zero. In other words if the bolt stretches further by an amount 0.287 mm then resultant compression stress in copper will become zero because when additional stretch in steel equals 0.287 mm at that time the washers will just leave contact with tribe.

Thus additional stress in steel will be due stretch of 0.287 mm

$$\delta'_S = \varepsilon'_S \, l_s$$

*i.e.,* $\varepsilon'_S = \dfrac{\delta'_S}{l_s} = \dfrac{0.287}{300}$

Therefore $\sigma'_S = \varepsilon'_S E_S = \dfrac{.287}{300} \times 200 \times 10^3 \text{ MPa}$

$$= 191.33 \text{ MPa (tensile)}$$

That is additional load $P'_S = \sigma'_S A_s = \left(\dfrac{0.287}{300} \times 200 \times 10^3\right) 314.159$

$$= 60 \times 10^3 \text{ N} = 60 \text{ kN}$$

Thus we conclude that 60 kN pull applied in the bold will bring the stress in copper to zero.

Then the resultant stress in steel bolt

$$= 59.2 + 191.33 = 250.53 \text{ MPa (tansile)}$$

1.16 A brass tube 20 mm internal dia and 20 mm external diameter surrounds a 18 mm shank of a steel bolt. After putting rigid washers on the two sides of the tube the nut is rotated such that it is just tight and no load is exerted on the tube by the washers. The length of the tube is 200 mm. The temperature is then raised to 270°C. If the room temperature is 20°C, find the increase in the stresses in brass and steel.

Take $\alpha_b = 18.7 \times 10^{-6}/°C$

$$\alpha_s = 11.6 \times 10^{-6}/°C$$

$$E_b = 100 \text{ GPa}$$

$$E_S = 200 \text{ GPa}$$

**Solution.** Rise in temperature = $t$ = 220 – 20 = 200°C

Free expansion of brass = $l_b \alpha_b t = \delta_b$

or $$\delta_b = 200 \times 18.7 \times 10^{-6} \times 200 = 0.748 \text{ mm}$$

Free expansion of steel = $l_S \alpha_S t = \delta_S$

or $$\delta_S = 200 \times 11.6 \times 10^{-6} \times 200 = 0.464 \text{ mm}$$

Because of this difference in the free expressions steel will get pulled up while brass will get compressed due to temperature rise.

Thus if $\sigma_S$ and $\sigma_b$ are the stresses induced in two materials then

$$(\delta_b - \delta_s) = \frac{\sigma_b}{E_b} l_b + \frac{\sigma_s}{E_S} l_s$$

But $l_b = l_s$

$$(0.748 - 0.464) = \left(\frac{\sigma_b}{E_b} + \frac{\sigma_S}{E_S}\right) l_S$$

or $$0.284 = \left[\frac{\sigma_b}{E_b} + \frac{\sigma_S}{E_S}\right] 200$$

or $$\frac{\sigma_b}{E_b} + \frac{\sigma_S}{E_S} = \frac{.254}{200} = 0.00142 \qquad \ldots(A)$$

Also at equilibrium $$P_S = P_b$$

$$\sigma_S A_S = \sigma_b A_b$$

or $$\sigma_S = \left(\frac{A_b}{A_S}\right) \sigma_b \qquad \ldots(B)$$

But $$A_b = \frac{\pi}{4}(25^2 - 20^2)$$

$$= \frac{\pi}{4} \times 227 = 176.71 \text{ mm}^2$$

$$A_S = \frac{\pi}{4} \times 18^2 = 254.469 \text{ mm}^2$$

$\therefore$ $$\sigma_S = \frac{176.71}{254.469} \times \sigma_b = 0.6938\, \sigma_b$$

Thus from equations ($A$) and ($B$) we get

$$\frac{\sigma_b}{(100 \times 10^3)} + \frac{0.6938\,\sigma_b}{(200 \times 10^3)} = 0.00142$$

or $$\sigma_b\,[2 + 0.6938] = 0.00142 \times 200 \times 10^3$$

$$\sigma_b = \frac{0.00142 \times 200 \times 10^3}{2.6938} = 0.528 \text{ MPa}$$

Then $$\sigma_s = 0.6938\,\sigma_b$$

$$= 0.6938 \times 0.528 = 0.3663 \text{ MPa}$$

## PROBLEMS FOR PRACTICE

1. A rigid cross bar is supported horizontally by two vertical bars, *A* and *B* of equal length and hanging from their tops. The bars *A* and *B* are 0.6 m apart. The cross bar stays horizontal even after a vertical force of 6000 *N* is applied to it at a point 0.4 m from *B*. If the stress in *A* is 200 MN/m², find the stress in B and the areas of cross section of the two rods.

   Take $$E_A = 200 \text{ GN/m}^2$$

   and $$E_B = 130 \text{ GN/m}^2$$

   **[Ans.** : 130 MN/m², 20.0 mm² ; 15.385 mm²]

2. A rigid bar *AD*, 1 m long, is hinged at *A* and is held horizontal by supporting it at *B* and *C* as shown in the Fig. 1.19. The area of cross section of the steel rod *BE* is 16 mm² and that of copper rod *CF* is 12 mm². The bar *AD* is acted upon by a vertical force of 500 N at *D*. Find the stresses induced in the rods *BE* and *CF*. Take

   $$E_s = 200 \text{ GN/m}^2$$

   and $$E_C = 90 \text{ GN/m}^2.$$

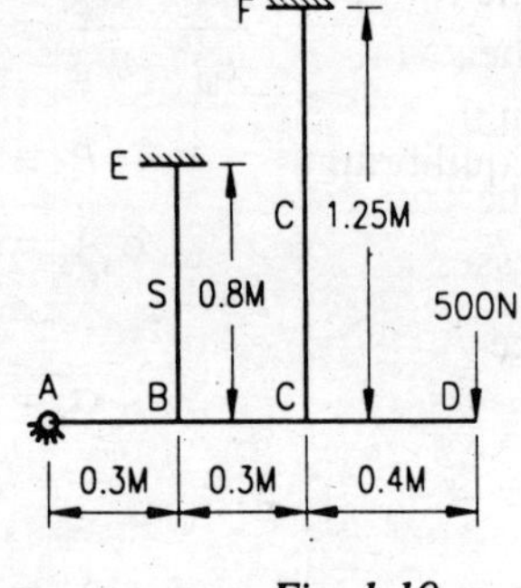


*Fig. 1.19*

   **[Ans.** 56.0 MN/m², 32.2 MN/m²]

3. A steel rod, 20 mm diameter, passes centrally through a steel tube 25 mm internal diameter and 40 mm external diameter. The tube, 750 mm long, is closed by two rigid washers of negligible thickness which are fixed onto the tube by tightening the nuts on the threaded rod. If the pitch of the threads is 2.5 mm, find the increase in the stresses in the tube and the rod when one nut is turned through one quarter of a turn with respect to the nut on the other end. Take

   $$E_{steel} = 205 \text{ GN/m}^2.$$

   **[Ans.** 49.8 MN/m² ; 121.2 MN/m²]

4. A cast iron bracket as shown in Fig. 1.20 is held fast to the foundation with the help of 4 steel bolts. Find the increase in tensile stresses of the four bolts due to the load P of magnitude 1000 N. The bolts are all 12 mm in diameter.

   [**Ans.** For *A* and *B* 0.596 MN/m$^2$ ; For *C* and *D* 3.576 MN/m$^2$].

   [**Hint** : The bracket will have a tendency of turning about the line *RS* and the bolts will have a tendency to stretch proportional to their distances from *RS*].

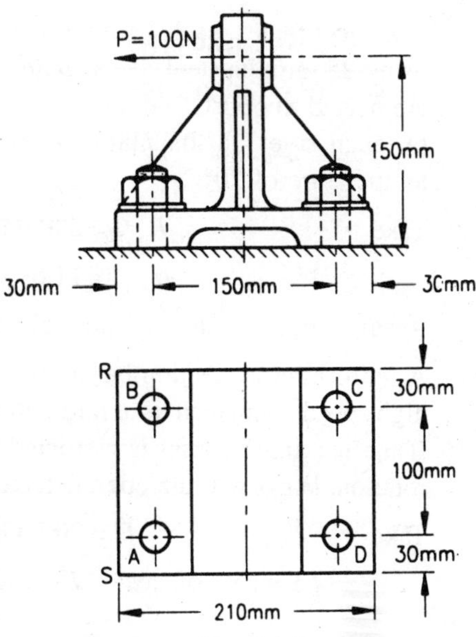


*Fig. 1.20*

5. A 25 mm diameter brass bar is enclosed in a steel tube of 25 mm internal diameter and 50 mm external diameter. Both of them are 1m long at room temperature and fastened rigidly to each other at the ends.

   If the room temperature is 20°C, find to what temperature the assembly should be heated so as to generate a compressive stress of 48.7 MN/m$^2$ in brass. What is then the stress in steel ?

   If the composite bar is then pulled with a force of 50000 N, find the resulting stresses in both the metals.

   Take
   $$E_{steel} = 200 \text{ GN / m}^2$$
   $$E_{brass} = 100 \text{ GN / m}^2$$
   $$\alpha_{steel} = 11.6 \times 10^{-6}\ /°\text{C} ; \alpha_{brass} = 18.7 \times 10^{-6}/°\text{C} .$$

   [**Ans.** 80°C, 16.23 MN/m$^2$ tensile ; 34.17 MN/m$^2$ comp ; 45.3 MN/m$^2$ tensile]

6. A 28 mm diameter steel bar, 0.4 long, is placed centrally within a brass tube having an inside diameter of 30 mm and outside diemeter of 40 mm. The bar is shorter in length than the tube by 0.12 mm. While the bar and the tube are held vertically on a rigid horizontal platform, a compressive force of 60000 N is applied at the top of the tube through a rigid plate. Determine the stresses induced in both the bar and the tube.

   Take
   $$E_{steel} = 200 \times 10^3 \text{ MN/m}^2$$
   $$E_{brass} = 100 \times 10^3 \text{ MN/m}^2.$$

   [**Ans.** 48.85 MN/m$^2$ ; comp. 54.42 MN/m$^2$ comp.]

7. Two steel plates, each 25 mm thick, are riveted together forming a lap joint by using 25 mm dia steel rivets. Before placing the rivets in the drilled holes, they are heated to 600°C and then the heads formed. Find the gripping force exerted by each rivet on the plates when the assembly has cooled down to room temperature of 20°C.

   Take $E_{steel} = 200\ \text{GN/m}^2$

   $\alpha_{steel} = 11.6 \times 10^{-6}/°\text{C}$ .

   Assume that the plates do not deform. **[Ans.** 660.52 kN]

8. A uniform cross section bar of diameter 10 mm is 1 m long and revolving at a high speed about an axis at mid-length and perpendicular to the length of the bar. If the maximum stress is restricted to 65 MN/m$^2$, find the maximum speed of rotation. What will then be the total extension of the rod ?

   Take $E = 200\ \text{GN/m}^2$

   Density of steel = 7800 Kg/m$^3$. **[Ans.** 2465.7 r.p.m. ; 0.2167 mm]

Chapter–2

# Complex Stresses and Strains

## Introduction

In Chapter 1, we considered the direct stresses on the planes where the force being sustained was normal to the sectional plane and passing through the centroid of the section. Further the stress was acting only along the longitudinal direction. In other words the stress system was uni-axial. However in actional practice in general the stresses at a point in a material will be acting in all directions and the forces, due to these stresses can be resolved along three axes of reference which can be arbitrarily chosen. Thus the system of stresses can be reduced to three dimensional stress system. But to start with we shall take up the cases where the stresses are acting in two perpendicular directions and along the third perpendicular direction these will be assumed to be zero. Such a stress system is called. TWO DIMENSIONAL STRESS SYSTEM". In majority of the machine elements which a mechanical engineer comes accross in his day to day work, be it a staft or a beam, or a bracket, or a pressure vessel, or a spanner, or a wrench only two dimingional stress systems are encountered. In this chapter we will assume that stresses along two perpendicular axes are given and along the third axis it is zero. Even if their is no mention of the this in the third direction it will be assumed to be zero.

## 2.1 Stresses on Oblique Planes in a Bi-axial Stress-System

Fig. 2.1(a) shows a rectangular specimen of areas of cross-section $A_x$, $A_y$ and $A_z$ normal to *X*, *Y* and *Z* axes under the action of bi-axial tensile forces $P_x$ and $P_y$ as shown such that the section normal to *x* and *y* axes carry stresses $(P_x / A_x)$ and $(P_y/A_y)$ respectively. Let these stresses be represented by $\sigma_x$ and $\sigma_y$. Around the point *O* consider a rectangular element *ABCDEFGH* of the dimensions *a*, *b* and *t* which are very small but finite. The stress on the faces *BCGF* and *ADHE* or as a matter of fact on any plane section parallel to these faces will be $\sigma_x$ and on the faces parallel to *DCGH*, $\sigma_y$. In actual practice the material may be under the action of a variety of forces such that the stress on the sections normal to some axis (say *x*-axis), will not be constant but may depend on the *y*-coordinate of the point O and likewise the stress on the section normal to the *y*-axis may depend on the *x* co-ordinate of the point *O*. But still, the dimensions of the element *ABCDEFGH* being very small, it can be assumed that the stress on the opposite faces of such an element will be same. The diagram given in Fig. 2.1(*b*) is called the point diagram and it depicts the state of stress around the point $O$. Let us now proceed to find the stresses, direct as well as shearing, on any plane $KLMN$ passing through the point $O$, normal to the $z$-plane ($ABCD$) and inclined at an angle $\theta$, with the $y$-plane ($DCGH$).

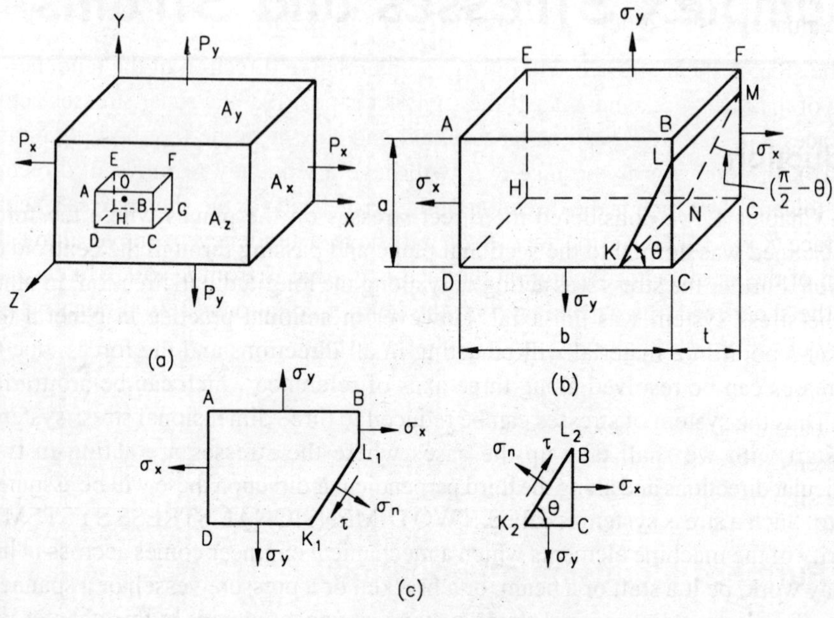


*Fig. 2.1*

In Fig. 2.1($c$) are drawn the free body diagrams of the two elements obtained by taking a section along the plane $KLMN$. $\sigma_n$ and $\tau$ are the direct and shear stresses on the faces $K_1L_1M_1N_1$ and $K_2L_2M_2N_2$ (to be written in short as $K_1L_1$ and $K_2L_2$) induced due to the normal and tangential components of the cohesive force between them. Considering the equilibrium of the element number 2 and resolving the force normal and along the plane $K_2L_2$, we get

$$\sigma_n (K_2L_2 \times t) - \sigma_x (L_2C \times t) \sin\theta - \sigma_y (K_2C \times t) \cos\theta = 0$$

$$\tau (K_2L_2 \times t) + \sigma_x (L_2C \times t) \cos\theta - \sigma_y (K_2C \times t) \sin\theta = 0$$

Dividing the above equations by $(K_2L_2 \times t)$ and writing $\sin\theta$ and $\cos\theta$ for $(L_2C/K_2L_2)$ and $(K_2C/K_2L_2)$ respectively, we get

$$\sigma_n = \sigma_x \sin^2\theta + \sigma_y \cos^2\theta \qquad ...(2.1a)$$

$$= \frac{1}{2}(\sigma_x + \sigma_y) - \frac{1}{2}(\sigma_x - \sigma_y)\cos 2\theta \qquad ...(2.1b)$$

And

$$\tau = -(\sigma_x - \sigma_y)\sin\theta\cos\theta \qquad ...(2.2a)$$

$$= -\frac{1}{2}(\sigma_x - \sigma_y)\sin 2\theta \qquad ...(2.2b)$$

## 2.1.1 Sign Convention

The positive value of $\sigma_n$ as obtained from equation (2.1) will mean that the direct stress is tensile otherwise compressive. Similarly the signs of $\sigma_x$ or $\sigma_y$ are also to be taken as positive if they are tensile, and negative if compressive while $\sigma_n$ and $\tau$ are being evaluated with the help of equations (2.1) and (2.2).

The sign of shear stress $\sigma$ will only give the relative direction of the tendency of slipping of the faces $K_1L_1$ and $K_2L_2$. If the sign of $\tau$ is positive, the shear stresses acting on the faces $K_1L_1$ and $K_2L_2$ will be as assumed and shown in the free body diagrams (Fig. 2.1*c*). In other words the face $K_1L_1$ will have a tendency to move in direction from $K_2$ towards $L_2$ which is the direction of the arrowhead for the shear stress $\tau$ acting on the face $K_2L_2$. Likewise the face $K_2L_2$ will have a tendency to move along the direction of the arrowhead $\tau$ acting on the face $K_1L_1$, that is from $L_1$ to $K_1$. In case the sign of the shear stress as calculated with the help of equations (2.2) in negative the tendency of sliding along the section *KL* will be reversed.

Secondly, it may be seen from the Fig. 2.1(*c*) that when $\tau$ is positive, the moment of the shearing force, due to which $\tau$ is induced on the faces $K_1L_1$ and $K_2L_2$, about any point taken within the material of the respective element will be clockwise. For the negative sign of $\tau$, the moments will become anti-clockwise.

## 2.1.2 Further Discussion on Equations (2.1) and (2.2)

For algebraic maximum or minimum value of shear stress, differentiating tau and equating it to zero, we get cos 2θ = 0.

For which we have two values of θ given by

$$\theta_1 = \pi/4 \text{ and } \theta_2 = 3\pi/4$$

Thus from equation (2.2),

When $\quad \theta = \theta_1 = \pi/4 \quad \tau_1 = -\frac{1}{2}(\sigma_x - \sigma_y) \qquad$ ...(2.3)

When $\quad \theta = \theta_2 = 3\,\pi/4 \quad \tau_2 = \frac{1}{2}(\sigma_x - \sigma_y) \qquad$ ...(2.4)

From equations (2.3) and (2.4) it is evident that $\tau_1$ and $\tau_2$ are both of the same magnitude but of different signs. That is, both of them are in a way maximum, the only difference being the nature of the moment of the shearing force due to shearing stress about a point within the material. The least value of the shear stress will in fact be zero and this occurs on the planes given by

$$\tau = -\frac{1}{2}(\sigma_x - \sigma_y)\sin 2\theta = 0$$

*i.e.* $\quad \sin 2\theta = 0 \qquad$ ...(2.5)

For which, again, we have two planes given by

$$\theta = \theta_3 = 0 \text{ and } \theta = \theta_4 = \pi/2$$

The planes, on which, there is no shear stress acting, are known as *'principal planes'* and the stresses acting on them the *'principal stresses'*. Evidently $\theta_3$ and $\theta_4$ planes correspond to $Y$ and $X$ planes respectively.

Thus from equation (2.1)

When $\theta = \theta_3 = 0$

$$\sigma_n = \frac{1}{2}(\sigma_x + \sigma_y) - \frac{1}{2}(\sigma_x - \sigma_y) = \sigma_y \qquad ...(2.6)$$

When $\theta = \theta_4 = \pi/2$

$$\theta_n = \frac{1}{2}(\sigma_x + \sigma_y) + \frac{1}{2}(\sigma_x - \sigma_y) = \sigma_x \qquad ...(2.7)$$

Equations (2.6) and (2.7), evidently reveal that the $Y$ and $X$ planes are also principal planes and $\sigma_y$ and $\sigma_x$ are the principal stresses.

In addition, from equations (2.3) to 2.6) we see that the magnitude of the maximum shear stress is half of the algebraic difference between the principal stresses. Secondly the two principal planes are inclined at right angles to each other. The two planes which carry the maximum shear stress (here $\theta_1$ and $\theta_2$) are also inclined to each other at 90° and they also bisect the angle between the principal planes and vice-versa.

For evaluating the maximum and minimum values of the direct stress, differentiate the equation (2.1*b*) with respect to θ and equate to zero,

$$\frac{1}{2}(\sigma_x - \sigma_x)\, 2 \sin 2\theta = 0$$

*i.e.*. $\sin 2\,\theta = 0$ ...(2.7)

Equation (2.8) is just the same as equation (2.5). The inference drawn from this is, that the principal planes carry either the maximum or minimum direct stress. Also since the principal stresses $\sigma_x$, $\theta_y$ are not equal, the algebraically greater of the two will be the maximum principal stress and the other the minimum principal stress.

## 2.2 Complementary Shear Stresses

Consider a small rectangular block of length *l*, height h and thickness t under the action of shearing stresses $\tau_0$ and $\tau'$ as shown in the Fig. 2.2, such that corresponding forces *P* sub 0 and *P* given by

$$P_0 = \tau_0\,(l.t) \qquad ...(2.9)$$

$$P = \tau'\,(h.t) \qquad ...(2.10)$$

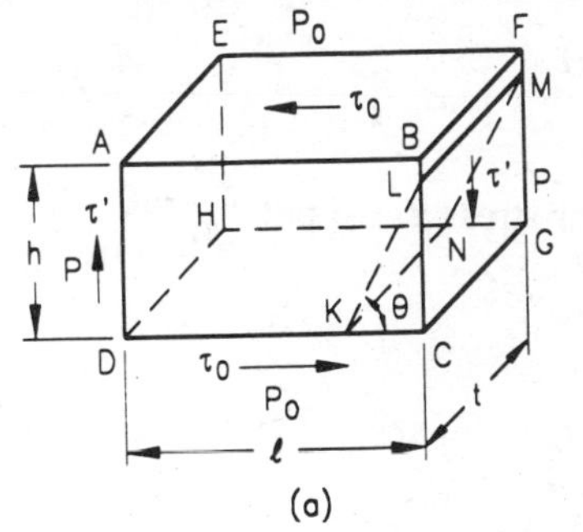


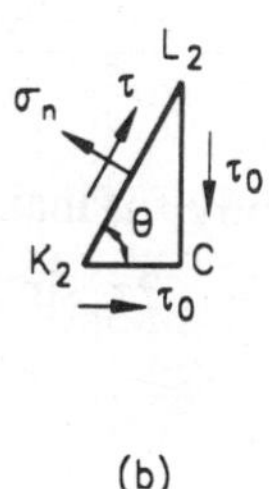


*Fig. 2.2*

The shearing force $P_0$ on the face *ABEF* is equal in magnitude and opposite in direction to $P_0$ on the face *DCGH*. But these are separated by the distance h. Similarly the shearing force *P* on the face *BCGF* is equal in magnitude to that in the face *DHEA* but opposite in direction and separated by the distance *l*. Evidently the forces in the horizontal and vertical directions are balanced. Therefore, if the block has to be in equilibrium, the turning moments due to the forces *P* and $P_0$ must be equal in magnitude and opposite in nature to each other.

*i.e.* $$P \cdot l = P_0 \cdot h$$

Substituting the values of *P* and $P_0$ from equations (2.9) and (2.10)

$$(\tau' \, ht)\, l = (\tau_0 \, lt)\, h$$

or $$\tau' = \tau_0 \qquad ..(2.11)$$

Equation (2.11) shows that a shear stress $\tau_0$ must be accompanied by another shear stress $\tau'$ of equal magnitude but of opposite sense, acting on planes at right angles to the planes of the original stress $\tau_0$. The shear stress $\tau'$ is called the complementary shear stress and this system is known to be under pure shear.

## 2.2.1 Stresses on Oblique Planes Due to Pure Shear

Let a section be taken along the plane *KL* at an angle θ with the plane *DCGH*. The free body diagram of the element is shown in Fig. 2.2(*b*). For the equilibrium of the element, resolving the forces along the normal to the plane $K_2L_2$, if the normal and shear stresses on the plane $K_2L_2$ are respectively $\sigma_n$ and $\tau$

$$\sigma_n (K_2L_2 \cdot t) - \tau_0 (K_2C \cdot t) \sin\theta - \tau_0 (L_2C \cdot t) \cos\theta = 0$$

dividing by $(K_2L_2 \cdot t)$

$$\sigma_n = \tau_0 (K_2C/K_2L_2) \sin\theta + \tau_0 (L_2\, C/K_2L_2) \cos\theta$$

$$= 2\, \tau_0 \sin\theta \cos\theta$$

*i.e.* $$\sigma_n = \tau_0 \sin 2\theta \qquad ...(2.12)$$

and resolving the forces along the plane $K_2L_2$ we get

$$\tau\,(K_2L_2 \,.\, t) + \tau_0\,(K_2C \,.\, t) \cos\theta - \tau_0\,(L_2C \,.\, t) \sin\theta = 0$$

dividing by $(K_2L_2 \,.\, t)$

$$\tau = \tau_0\,(L_2C\,/\,K_2L_2) \sin\theta - \tau_0\,(K_2C\,/\,K_2L_2) \cos\theta$$

$$= \tau_0 (\sin^2 \theta - \cos^2 \theta) \quad \text{...(2.13)}$$

*i.e.* $\tau = -\tau_0 \cos 2\theta$

Evidently for maximum or minimum shear stress $\tau$, we put $\frac{d\tau}{d\theta} = 0$

*i.e.,* $2\tau_0 \sin 2\theta = 0$

$\sin 2\theta = 0$

*i.e.* Either $\theta = 0$ for which $\tau = -\tau_0$ ...(2.14*a*)

or $\theta = \frac{\pi}{2}$ for which $\tau = \tau_0$ ...(2.14*b*)

These values correspond to the given complementary shear stresses.

Principal planes will be located by setting $\tau$ equal to zero in equation (2.13)

*i.e.* $\cos 2\theta = 0$

Therefore either

$\theta = \pi/4$ forwhich form equation (2.12)

$\sigma_n = \tau_0$ ...(2.15*a*)

or $\theta = 3\,\pi/4$ for which $\sigma_n = -\tau_0$ ...(2.15*b*)

Thus the magnitude of the principal stresses is same as that of the applied shear stresses. On the first of the two principal planes which are inclined at 45° and 135° anticlockwise with the plane *DCGH* the principal stress is tensile while on the other it is compressive in nature.

## 2.3 Resultant Stress on a Plane

Let $P_n$ and $P_t$ be the normal and the tangential forces acting on a plane. The resultant force *P* per unit area of the plane is called the resultant stress and represented by $\sigma_r$. Thus if *A* is the area of the plane.

$$(\sigma_r A)^2 = P_n^2 + P_t^2$$

or $$\sigma_r^2 = (P_n / A)^2 + (P_t / A)^2$$

or $$\sigma_r = (\sigma_n^2 + \tau^2)^{1/2} \quad \text{...(2.16)}$$

The inclination of the resultant stress, $\beta$, with the normal to the plane is given by

$$\tan \beta = \frac{P_t}{P_n} = \frac{\tau A}{\sigma_n A} = \frac{\tau}{\sigma_n} \quad \text{...(2.17)}$$

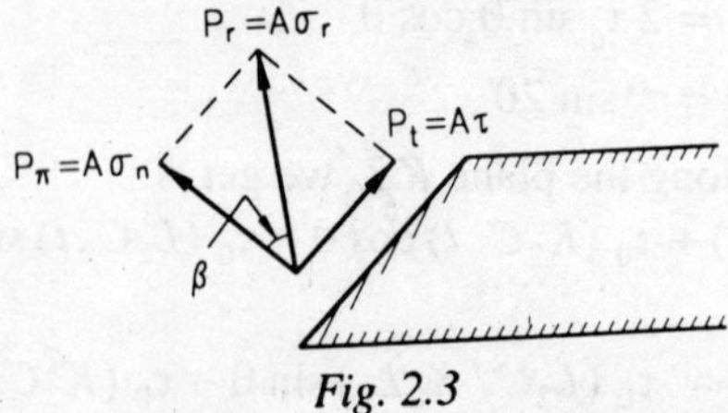


*Fig. 2.3*

$\beta$ is also known as the obliquity of the resultant stress on the plane in question.

**Example 2.1.** *For a two dimensional stress system having $\sigma_x$ and $\sigma_y$ as the principal stresses find the plane across which the resultant stress is most inclined with the normal to that plane.*

**Solution.** For the maximum value of $\beta$, tan $\beta$ will also be maximum.

*i.e.* $$\frac{d}{d\theta}(\tan\beta) = 0$$

or $$\frac{d}{d\theta}(\tau/\sigma_n) = 0$$

*i.e.* $$\sigma_n d\tau/d\theta = \tau\, d\,\sigma_n/d\theta \qquad ...(2.18)$$

Substituting equations (2.1*b*) and (2.2*b*) in equation (2.18)

$$\frac{\sigma_n}{\tau} = \frac{\frac{1}{2}(\sigma_x - \sigma_y)\,2\sin 2\theta}{-\frac{1}{2}\left(\sigma_x - \sigma_y\right)2\cos 2\theta} = -\tan 2\theta$$

or $$\cot\beta = \sigma_n/\tau = -\tan 2\theta$$

or $$\tan 2\theta = -\cot\beta = \tan\left(\frac{\pi}{2} + \beta\right) \qquad ...(2.19)$$

Substituting equation (2.19) in equation (2.17)

$$\frac{\sin\beta}{\cos\beta} = \frac{-\frac{1}{2}(\sigma_x - \sigma_y)\sin\left(\frac{\pi}{2} + \beta\right)}{\frac{1}{2}\left(\sigma_{x'} + \sigma_y\right) - \frac{1}{2}(\sigma_x - \sigma_y)\cos\left(\frac{\pi}{2} + \beta\right)}$$

which on simplification reduces to

$$\sin\beta = \frac{(\sigma_y - \sigma_x)}{(\sigma_y + \sigma_x)} \qquad ...(2.20)$$

**Example 2.2.** *The principal stresses at a point in a material, under the action of two dimensional stress system are 60 MN/m² tensile and 40 MN/m² compressive. Determine the normal and shear stresses on a plane inclined at 30° to the line of action of the tensile stress. Also find the resultant stress on this plane. What is the inclination of the resultant stress with the line of action of the given tensile stress ?*

**Solution.** $\sigma_x$ = 60MN/m² ; $\sigma_y$ = – 40 MN/m² and $\theta$ = 30°

From equation (2.1*b*)

$$\sigma_n = \frac{1}{2}(60 - 40) - \frac{1}{2}(60 + 40)\cos 60 = 10 - 25$$

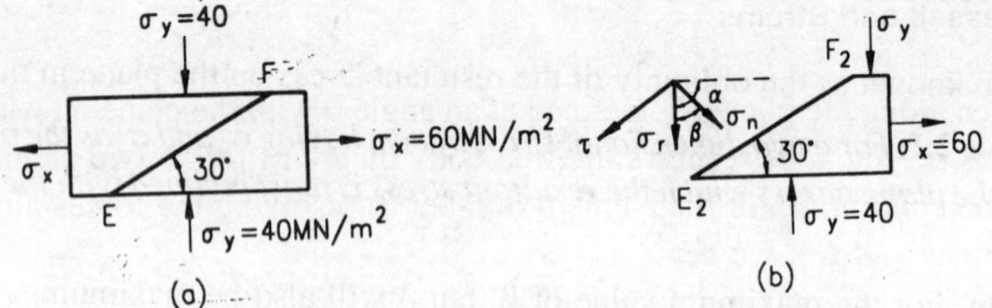


Fig. 2.4

or $\quad \sigma_n = -\ 15\ \text{MN/m}^2$ (*i.e.* comp.)

$$\tau = -\frac{1}{2}(60 + 40)\sin 60 = -\ 43.3\ \text{MN}/\text{m}^2$$

$$\sigma_r = [(-15)^2 + (-\ 43.3)^2]^{1/2} = 45.7\ \text{MN}/\text{m}^2$$

In Fig. 2.4(*b*)

$$\tan\beta = \frac{|\tau|}{|\sigma_n|} = \frac{43.3}{15} = 2.89;\ i.e.,\ \beta = 71°$$

The inclination of the resultant stress sigma sub r, with the direction of 60 MN/m² stress is alpha Fig. 2.4(b) and

$$\alpha = \beta + 60° = 71° + 60° = 131° \text{ clockwise.}$$

## 2.4 General Case of Two Direct Stresses Accompanied by Shear Stresses

Consider a small rectangular element *ABCD* of thickness t normal to the plane of paper under a general two dimensional stress system as shown in Fig. 2.5(*a*). Let a

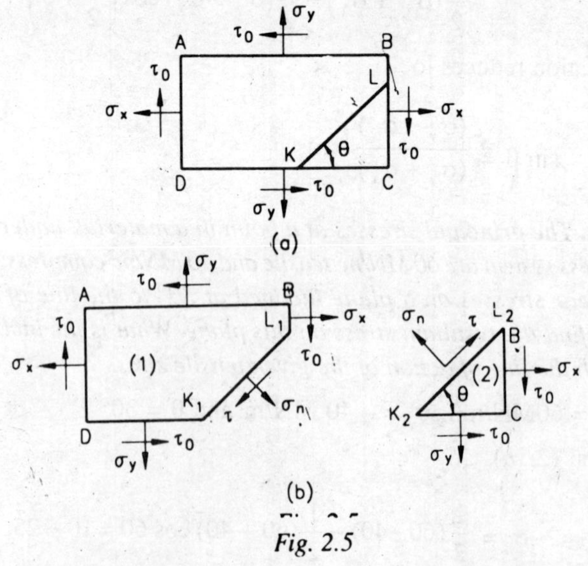


Fig. 2.5

section be taken along the plane $KL$ inclined at an angle $\theta$ with the plane on which the normal stress is $\sigma_y$. Fig. 2.5($b$) shows the free body diagrams of the two elements so formed. For the equilibrium of the element number 2, resolving the forces along the normal to the plane ($K_2L_2$), we get,

$$\sigma_n(K_2L_2 \,.\, t) - \sigma_x\,(L_2C \,.\, t)\sin\theta - \tau_0(L_2C \,.\, t)\cos\theta$$
$$- \tau_0\,(K_2C \,.\, t)\sin\theta - \sigma_y\,(K_2C \,.\, t)\cos\theta = 0$$

Dividing by ($K_2L_2 \,.\, t$), we get

$$\sigma_n = \sigma_x \sin^2\theta + \sigma_y \cos^2\theta + 2\tau_0 \sin\theta\cos\theta$$

$$= \frac{1}{2}\sigma_x\,(1 - \cos 2\theta) + \frac{1}{2}\sigma_y\,(1 + \cos 2\theta) + \tau_0 \sin 2\theta$$

or $$\sigma_n = \frac{1}{2}(\sigma_x - \sigma_y) - \frac{1}{2}(\sigma_x - \sigma_y)\cos 2\theta + \tau_0 \sin 2\theta \qquad ...(2.21)$$

Similarly resolving the forces along the plane $K_2L_2$

$$\tau\,(K_2L_2\,.t) + \sigma_x\,(L_2C\,.t)\cos\theta - \tau_0\,(L_2C\,.t)\sin\theta$$
$$- \tau_0\,(K_2C\,.t)\cos\theta - \sigma_y\,(K_2C\,.t)\sin\theta = 0$$

Dividing by ($K_2\,L_2 \,.\, t$)

$$\tau = -\sigma_x \sin\theta\cos\theta + \sigma_y \sin\theta\cos\theta + \tau_0\,(\sin^2\theta - \cos^2\theta)$$

or $$\tau = -\frac{1}{2}(\sigma_x - \sigma_y)\sin 2\theta - \tau_0 \cos 2\theta \qquad ...(2.22)$$

## 2.4.1 Maximum and Minimum Direct Stresses

For maximum and minimum value of $\sigma_n$, $d\sigma_n/d\theta = 0$

Thus differentiating equation (2.21), with respect to $\theta$, we get

$$-\frac{1}{2}(\sigma_x - \sigma_y)(-2\sin 2\theta) + 2\tau_0 \cos 2\theta = 0$$

or $$\tan 2\theta = -\frac{2\tau_0}{(\sigma_x - \sigma_y)} \qquad ...(2.23)$$

Equation (2.23) will give two values of theta, say $\theta_1$ and $\theta_2$ given by

$$2\theta_1 = \tan^{-1}\left[-\frac{2\tau_0}{(\sigma_x - \sigma_y)}\right] \qquad ...(2.24)$$

and $$2\theta_2 = \pi + 2\theta_1 \; i.e., \theta_2 = \frac{\pi}{2} + \theta_1 \qquad ...(2.25)$$

Then from equations (2.24) and (2.25)

$$\sin 2\theta_1 = \frac{2\tau_0}{\sqrt{(\sigma_x - \sigma_y)^2 + 4\tau_0^2}} \qquad ...(2.26a)$$

$$\cos 2\theta_1 = \frac{-(\sigma_x - \sigma_y)}{\sqrt{(\sigma_x - \sigma_y)^2 + 4\tau_0^2}} \qquad ...(2.26b)$$

and $\sin 2\theta_1 = \sin(\pi + 2\theta_1) = -\sin 2\theta_1$

$$= -2\tau_0 / [(\sigma_x - \sigma_y)^2 + 4\tau_0^2]^{1/2} \qquad ...(2.27a)$$

$$\cos 2\theta_2 = \cos(\pi + 2\theta_1) = -\cos 2\theta_1$$

$$= (\sigma_x - \sigma_y) / [(\sigma_x - \sigma_y)^2 + 4\tau_0^2]^{1/2} \qquad ...(2.27b)$$

If $a_{n1}$ and $\sigma_{n2}$ are the direct stresses acting on the planes inclined at $\theta_1$ and $\theta_2$ respectively with *Y*-plane we have from equations (2.21) and (2.26).

$$\sigma_{n1} = \frac{1}{2}(\sigma_x + \sigma_y) + \frac{\frac{1}{2}(\sigma_x - \sigma_y)^2}{\sqrt{(\sigma_x - \sigma_y)^2 + 4\tau_0^2}} + \frac{2\tau_0^2}{\sqrt{(\sigma_x - \sigma_x)^2 + 4\tau_0^2}}$$

$$= \frac{1}{2}(\sigma_x + \sigma_y) + \frac{1}{2}\sqrt{(\sigma_x - \sigma_y)^2 + 4\tau_0^2}$$

or $$\sigma_{n1} = \frac{1}{2}(\sigma_x + \sigma_y) + \sqrt{\left(\frac{\sigma_x - \sigma_y}{2}\right)^2 + \tau_0^2} \qquad ...(2.28)^*$$

Similarly equations (2.22) and (2.27) give

$$\sigma_{n2} = \frac{1}{2}(\sigma_x + \sigma_y) - \frac{\frac{1}{2}(\sigma_x - \sigma_y)^2}{\sqrt{(\sigma_x - \sigma_y)^2 + 4\tau_0^2}} \frac{2\tau_0^2}{\sqrt{(\sigma_x - \sigma_y)^2 + 4\tau_0^2}}$$

or $$\sigma_{n2} = \frac{1}{2}(\sigma_x + \sigma_y) - \sqrt{\left(\frac{\sigma_x - \sigma_y}{2}\right)^2 + \tau_0^2} \qquad ...(2.29)$$

---

*. **Note :** In equation (2.26*b*) the negativesign had been associated with cos $2\theta_1$ thereby resulting in the positive and neative signs of the last terms in equations (2.28) and (2.29) respectively.Had the negative sign of tan $2\theta$ (equation 2.23) been associated with sin $2\theta_1$ in equation (2.26*a*), it can easily be seen that the signs of the last terms in equations (2.28) and (2.29) would have been negative and positive respectively. It is, therefore, cautioned that $\sigma_{n1}$ or $\sigma_{n2}$ must not directly be linked to $\theta_1$ or $\theta_2$ without checking. It will be advisable to first evaluate $\theta_1$ and $\theta_2$ and then calculate the direct stresses on these planes with the help of equation (2.21) to decide the maximum diret stress rather than directly applying the equations (2.28) and (2.29).

From equations (2.28) and (2.29) it is evident that $\sigma_{n1}$ is greater than $\sigma_{n2}$ algebraically. Thus $\sigma_{n1}$ will correspond to the maximum value of the direct stress and $\sigma_{n2}$ to the minimum value.

## 2.4.2 Principal Planes and Principal Stresses

As defined earlier the principle planes are those planes on which there is no shearing stress. Therefore setting $\tau$ (equation 2.22) equal to zero, we have

$$-\frac{1}{2}(\sigma_x - \sigma_y)\sin 2\theta - \tau_0 \cos 2\theta = 0$$

or
$$\tan 2\theta = -\frac{2\tau_0}{(\sigma_x - \sigma_y)} \quad ...(2.30)$$

Equation (2.30) is just the same as equation (2.23). Thus the solution is same as before *i.e.* $\theta_1$ and $\theta_2$. In other words planes having maximum and minimum direct stresses are also principal planes. Thus if $\sigma_1$ and $\sigma_2$ are the principal stresses they are given by

$$\sigma_1 = \sigma_{n1} = \frac{1}{2}(\sigma_x + \sigma_y) + \sqrt{\left(\frac{\sigma_x - \sigma_y}{2}\right)^2 + \tau_0^2} \quad ...(2.31)$$

$$\sigma_2 = \sigma_{n2} = \frac{1}{2}(\sigma_x + \sigma_y) - \sqrt{\left(\frac{\sigma_x - \sigma_y}{2}\right)^2 + \tau_0^2} \quad ...(2.32)$$

## 2.4.3 Maximum Shear Stress

For maximum shear stress, $\frac{d\tau}{d\theta} = 0$. Therefore differentiating equation (2.22) with respect to $\theta$, we get

$$-\frac{1}{2}(\sigma_y - \sigma_y)(2\cos 2\theta) + \tau_0 (2\sin 2\theta) = 0$$

or
$$\tan 2\theta = \frac{(\sigma_x - \sigma_y)}{2\tau_0} \quad ...(2.33)$$

Equation (2.33) will have two solutions theta sub 3 and theta sub 4, say, given by

$$2\theta_3 = \tan^{-1}\left[\frac{\sigma_x - \sigma_y}{2\tau_0}\right] \quad ...(2.34)$$

and
$$2\theta_4 = \pi + 2\theta_3 \ \textit{i.e.}\ \ \theta_4 = \frac{\pi}{2} + \theta_3 \quad ...(2.35)$$

Then
$$\tan 2\theta_3 = \frac{1}{2}(\sigma_x - \sigma_y) \Big/ \left[\left(\frac{x_y - \sigma_y}{2}\right)^2 + \tau_0^2\right]^{1/2} \quad ...(2.36a)$$

$$\cos 2\theta_3 = \tau_0 \Bigg/ \left[\left(\frac{\sigma_x - \sigma_y}{2}\right)^2 + \tau_0^2\right]^{1/2} \qquad \text{...(2.36}b\text{)}$$

$$\sin 2\theta_4 = \sin(\pi + 2\theta_3) = -\sin 2\theta_3$$

$$= -\frac{1}{2}(\sigma_x - \sigma_y) \Bigg/ \left[\left(\frac{\sigma_x - \sigma_y}{2}\right)^2 + \tau_0^2\right]^{1/2} \qquad \text{...(2.37}a\text{)}$$

$$\cos 2\theta_4 = \sin(\pi + 2\theta_3) = -\cos 2\theta_3$$

$$= -\tau_0 \Bigg/ \left[\left(\frac{\sigma_x - \sigma_y}{2}\right)^2 + \tau_0^2\right]^{1/2} \qquad \text{...(2.37}b\text{)}$$

Thus if $\tau_3$ and $\tau_4$ are the shear stresses on planes $\theta_3$ and $\theta_4$, we have from equations (2.22), (2.36) and (2.37).

$$\tau_3 = \frac{-\frac{1}{4}(\sigma_x - \sigma_y)^2}{\left[\left(\frac{\sigma_x - \sigma_y}{2}\right)^2 + \tau_0^2\right]^{1/2}} - \frac{\tau_0^2}{\left[\left(\frac{\sigma_x - \sigma_y}{2}\right)^2 + \tau_0^2\right]^{1/2}}$$

*i.e.*

$$\tau_3 = -\sqrt{\left(\frac{\sigma_x - \sigma_y}{2}\right)^2 + \tau_0^2} \qquad \text{...(2.38)}$$

and

$$\tau_4 = \frac{\frac{1}{4}(\sigma_x - \sigma_y)^2}{\left[\left(\frac{\sigma_x - \sigma_y}{2}\right)^2 + \tau_0^2\right]^{1/2}} + \frac{\tau_0^2}{\left[\left(\frac{\sigma_x - \sigma_y}{2}\right)^2 + \tau_0^2\right]^{1/2}}$$

*i.e.*

$$\tau_4 = \sqrt{\left(\frac{\sigma_x - \sigma_y}{2}\right)^2 + \tau_0^2} \qquad \text{...(2.39)}$$

Evidently $\tau_3$ and $\tau_4$ are both of the same magnitude but of different signs. If view of the sign convention for the shear stress it is obvious that both of them are maximum in magnitude (rather they are complimentary shear stresses as will be shown below). Even though in the strict algebraic sense $\tau_3$ appears to be minimum, yet the least value of the shear stress will be zero which will occur on the principal planes.

### 2.4.4 Complementary Nature of $\tau_3$ and $\tau_4$

From a well known equation of trigonometry, we have for any $X$.

$$\tan^{-1} X + \tan^{-1}\frac{1}{X} = \frac{\pi}{2} \qquad ...(2.40)$$

Substituting $\left(\frac{\sigma_x - \sigma_y}{2\tau_0}\right)$ for $X$ in equation (2.40) we obtain

$$\tan^{-1}\left(\frac{\sigma_x - \sigma_y}{2\tau_0}\right) + \tan^{-1}\left(\frac{2\tau_0}{\sigma_x - \sigma_y}\right) = \frac{\pi}{2}$$

and with the help of equations (2.24) and (2.34) it reduces to

$$2\theta_3 - 2\theta_1 = \frac{\pi}{2}$$

or
$$\theta_3 - \theta_1 = \frac{\pi}{4} \qquad ...(2.41)$$

Equation (2.41) can also be written as

$$\left(\theta_3 + \frac{\pi}{2}\right) - \left(\theta_1 + \frac{\pi}{2}\right) = \frac{\pi}{4} \qquad ...(2.42)$$

and with the help of equations (2.25) and (2.35) it boils down to

$$\theta_4 - \theta_1 = \frac{\pi}{4} \qquad ...(2.42)$$

From equations (2.25), (2.35), (2.41) and (2.42), it is evident that the maximum shear stress planes bisect the angle between the principal planes and vice-versa. In addition the principal planes are at right angles to each other and so are the maximum shear stress planes to each other.

From equations (2.28), (2,29) and (2.39), we get

$$\sigma_{n1} - \sigma_{n2} = 2\sqrt{\left(\frac{\sigma_x - \sigma_y}{2}\right)^2 + \tau_0^2} = 2\tau_4 = 2\tau_{max}$$

or
$$\sigma_1 - \sigma_2 = 2\tau_{max}$$

*i.e.* Max. shear stress = $\frac{1}{2}$ (difference in the two principal stresses) ...(2.43)

Also the direct stress on planes $\theta_3$ and $\theta_4$ can be calculated form equations (2.21), (2.36) and (2.37) as

$$\sigma'_{n3} = \frac{1}{2}(\sigma_x + \sigma_y) - \frac{1}{2}(\sigma_x - \sigma_y)\frac{\tau_0}{\sqrt{\left(\frac{\sigma_x - \sigma_y}{2}\right)^2 + \tau_0^2}} + \frac{\tau_0 \frac{1}{2}(\sigma_x - \sigma_y)}{\sqrt{\left(\frac{\sigma_x - \sigma_y}{2}\right)^2 + \tau_0^2}}$$

*i.e.* $$\sigma_{n3} = \frac{1}{2}(\sigma_x + \sigma_y) \quad \text{...(2.44a)}$$

Similarly

$$\sigma_{n3} = \frac{1}{2}(\sigma_x + \sigma_y) - \frac{1}{2}(\sigma_x - \sigma_y)\frac{-\tau_0}{\sqrt{\left(\frac{\sigma_x - \sigma_y}{2}\right)^2 + \tau_0^2}} + \frac{\tau_0\left(-\frac{1}{2}\right)(\sigma_x - \sigma_y)}{\sqrt{\left(\frac{\sigma_x + \sigma_y}{2}\right)^2 + \tau_0^2}}$$

*i.e.* $$\sigma_{n4} = \frac{1}{2}(\sigma_x + \sigma_y) \quad \text{...(2.44b)}$$

Also from equation (2.28) and (2.29)

$$\sigma_{n1} + \sigma_{n2} = (\sigma_x + \sigma_y) \quad \text{...(2.45)}$$

Therefore from equations (2.44) and (2.45)

$$\sigma_{n3} = \sigma_{n4} = \frac{1}{2}(\sigma_{n1} + \sigma_{n2}) \quad \text{...(2.46)}$$

That is the direct stress on the maximum shear stress planes is equal to half the algebraic sum of the principal stresses.

**Example 2.3.** *At a point in an elastic material the stresses on three mutually perpendicular planes are as follows :*

*First plane : 50 MN/m² tensile and 40 MN/m² shear*

*Second plane : 30 MN/m² compressive and complementary shear stress of 40 MN/m² or MPa*

*Third plane : No stress.*

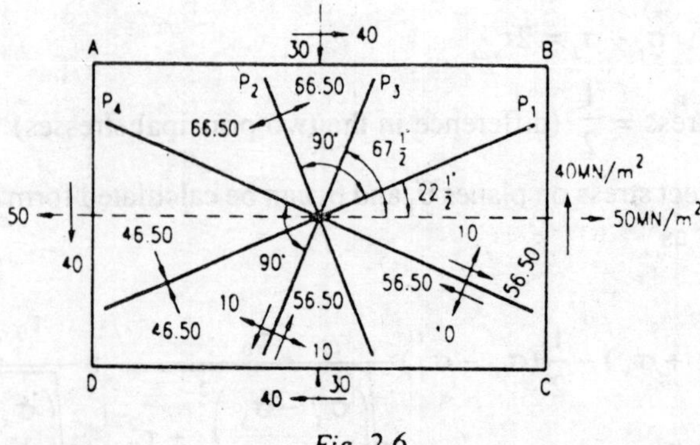


*Fig. 2.6*

Find (*a*) the position of principal planes and magnitude of principal stresses, (*b*) the position of the planes on which maximum shear stress acts and calculate the normal and sheer stresses on them.

**Solution.** Here $\sigma_x = 50$, $\sigma_y = -30$ and $\tau 0 = -40$.

(It must be noted here that had the arrow heads for the shear stresses on various planes been reversed only then the value of $\tau_0$ would have been + 40).

The location of principal planes is given by the equation (2.24)

$$\tan 2\theta = \left(\frac{2\tau_0}{\sigma_y - \sigma_x}\right) = \frac{2(-40)}{-30-50} = 1$$

*i.e.* $\theta_1 = 22½°$ and $\theta_2 = 90 + 22½ = 112½°$

The two principal planes $P_1$ and $P_2$ are inclined with the plane *DC* at 22½° and 112½° respectively as shown in Fig. 2.6. The principal stresses on these planes as discussed earlier, will be calculated with the help of equation (2.22).

For $\theta_1$ $$\theta_{n1} = \frac{1}{2}(\sigma_x - \sigma_y) - \frac{1}{2}(\sigma_x - \sigma_y)\cos 2\theta_1 + \tau_0 \sin 2\theta_1$$

$$= \frac{1}{2}(50-30) - \frac{1}{2}(50+30)\cos 45° - 40 \sin 45°$$

*i.e.* $\sigma_{n1} = -46.5$ MN/m² (*i.e.*comp.)

For $\theta_2$ $$\sigma_{n2} = \frac{1}{2}(\sigma_x + \sigma_y) - \frac{1}{2}(\sigma_x - \sigma_y)\cos 2\theta_2 + \tau_0 \sin 2\theta_2$$

$$= 10 - 40\left(-\frac{1}{\sqrt{2}}\right) + 40\left(\frac{1}{\sqrt{2}}\right)$$

*i.e.* $\sigma_{n2} = 66.5$ MN/m² or MPa (*i.e.* Tensile )

(*b*) Maximum shear stress planes bisect the angle between the principal planes. Therefore if $\theta_3$ and $\theta_4$ are the inclinations of these planes with the plane *DC*, then

$$\theta_3 = \theta_1 + 45° = 67½°$$
$$\theta_4 = \theta_2 + 45° = 157½°$$

Therefore $$\tau_3 = -\frac{1}{2}(\sigma_x - \sigma_y)\sin 2\theta_3 - \tau_0 \cos 2\theta_3$$

$$= -\frac{1}{2}(50+30)\frac{1}{\sqrt{2}} + 40\left(-\frac{1}{\sqrt{2}}\right)$$

$$= -56.5 \text{ MN/m}^2 \text{ or MPa}$$

The complementary shear stress $\tau_4$ will be = 56.5 MN/m$^2$ or MPa

The direct stresses on the maximum shear stress planes is given by equation (2.46)

$$\sigma_{n3} = \sigma_{n4} = \frac{1}{2}(\sigma_{n1} + \sigma_{n2})$$

$$= \frac{1}{2}(-46.5 + 66.5) = 10\text{MN/m}^2 \text{ or MPa (Tensile)}$$

In case the principal stresses would have not been evaluated, $\sigma_{n3}$ and $\sigma_{n4}$ could be found with the help of equation (2.21).

All these stresses on various planes are shown in Fig. 2.6.

## 2.5 Mohr's Stress Circle

Refer to the stress system of Fig. 2.7(*a*) it can been seen that on any plane inclined at an angle θ (anticlockwise) with the plane *DC* the direct and shear stresses are given by equations (2.21) and (2.22). For convenience these are being re-written here as

$$\sigma_n = a - b\cos 2\theta + \tau_0 \sin 2\theta \qquad ...(2.47)$$

$$\tau = -b\sin 2\theta - \tau_0 \cos 2\theta \qquad ...(2.48)$$

where $$a = \frac{1}{2}(\sigma_x + \sigma_y) \qquad ...(2.49a)$$

$$b = \frac{1}{2}(\sigma_x - \sigma_y) \qquad ...(2.49b)$$

If *a* point is plotted on a graph paper with *x* and *y* coordinates as $\sigma_n$ and $\tau$ respectively, the point will represent the state of stress on the plane θ for which direct and shear stresses have been calculated from equations (2.47) and (2.48). Since equations (2.47) and (2.48) are parametric equations with '2θ' as the parameter, the locus of the general point ($\sigma_n$ , τ) will be obtained by eliminating '2θ' from these equations. The coordinates of any point on this locus will give the direct and shear stresses on the plane represented by that point.

Squaring and adding equations (2.47) and (2.48), we get

$$(\sigma_n - a)^2 + \tau^2 = (-b\cos 2\theta + \tau_0 \sin 2\theta)^2 + (-b\sin 2\theta - \tau_0\cos 2\theta)^2$$

$$= b^2(\cos^2 2\theta + \sin^2 2\theta) + \tau_0^2(\sin^2 2\theta + \cos^2 2\theta)$$

$$= b^2 + \tau_0^2$$

*i.e.* $$(\sigma_n - a)^2 + \tau^2 = r^2 \qquad ...(2.50)$$

where $$r = \sqrt{b^2 + \tau_0^2} \qquad ...(2.51a)$$

or $$r = \sqrt{\left(\frac{\sigma_x - \sigma_y}{2}\right)^2 + \tau_0^2} \qquad \text{...(2.51}b\text{)}$$

Equation (2.50) is evidently the equation of *a* circle with centre lying on the *x* aixs at ( *a* ,0) or $\left(\frac{\sigma_x + \sigma_y}{2}, 0\right)$ and radius given by equation (2.51). This circle, which is known as the Mohr's stress circle, is drawn in Fig. 2.7(*b*).

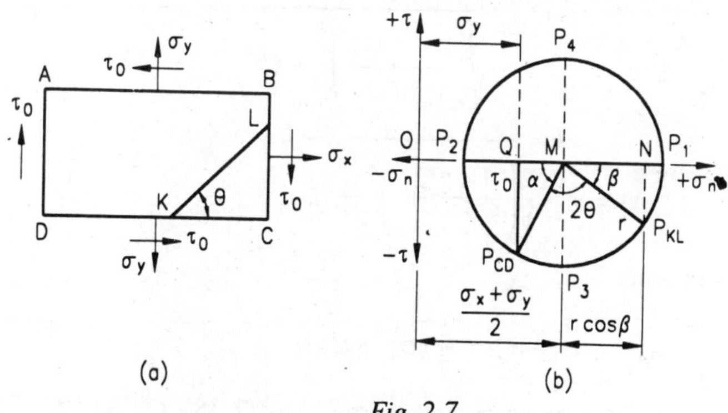


*Fig. 2.7*

The sign convention as given in article 2.1.1 will be adhered to ; that is, tensile direct stresses are taken as positive and compressive as negative and plotted accordingly along the *x*-axis.

Similarly the positive or clockwise-moment forming (about any point within the material) shear stresses are plotted on the positive side of the *y*-axis and negative shear stresses on the negative side. For example, the point representing the plane *DC* or *AB* or any plane parallel to *DC*, will have the coordinates ($\sigma_y$, $-\tau_0$).

Coming back to the Mohr's circle it is evident that at the point of its intersection with the *x*-axis, the *y*-coordinate will be zero. That is on the planes, represented by the points $P_1$ and $P_2$ the shear stress will be zero. In other words these points represent the principal planes.

Therefore from the circle diagram the *x*-coordinate of $P_1$

$$\sigma_{n1} = OP_1 = OM + MP_1 = OM + r$$

$$= \frac{1}{2}(\sigma_x + \sigma_y) + \sqrt{\left(\frac{\sigma_x - \sigma_y}{2}\right)^2 + \tau_0^2}$$

which is same as given by equation (2.28).

Also from the circle diagram, the *x*-coordinate of $P_2$

$$\sigma_{n2} = OP_2 = OM - MP_2 = OM - r$$

$$= \frac{1}{2}(\sigma_x + \sigma_y) - \sqrt{\left(\frac{\sigma_x - \sigma_y}{2}\right)^2 + \tau_0^2}$$

which is same as given by equation (2.29).

Again the points $P_3$ and $P_4$ obviously refer to the planes on which the magnitude of the shear stress is maximum and the *y*-co-ordinate of these points is equal to *r* in magnitude. For the plane represented by $P_3$ the shear stress will be negative and that for $P_4$ positive. Thus, from equation (2.51*b*)

$$\tau_3 = -r = -\sqrt{\left(\frac{\sigma_x - \sigma_y}{2}\right)^2 + \tau_0^2}$$

and $$\tau_4 = r = \sqrt{\left(\frac{\sigma_x - \sigma_y}{2}\right)^2 + \tau_0^2}$$

Also for these points

$$\sigma_{n3} = \sigma_{n4} = a = \frac{1}{2}(\sigma_x + \sigma_y)$$

These are once again same as given by equations (2.38), (2.39) and (2.44).

Further the points $P_1$ and $P_2$ subtend an angle of 180° at the centre *M* of the circle and same is the case with the points $P_3$ and $P_4$. That is, the angle subtended by the points at the centre is double the actual angle between the corresponding planes in the point diagram. This is also true for the points $P_1$ and $P_4$ as in the circle diagram $MP_4$ is inclined at an angle of 90° (anticlockwise) with $MP_1$ against the actual inclination of 45° (anticlockwise) between the planes represented by $P_1$ and $P_4$. Even though from the above it appears that the angle subtended by the points in the circle diagram at the centre is double the actual angle between the corresponding planes represented by these points, it needs further checking for any general point.

For this we first plot the point corresponding to the plane *DC* with coordinates $(\sigma_y, -\tau_0)$. Let it be represented by $P_{CD}$ in the circle diagram. If we set $P_{CD}MP_{KL} = 2\theta$ anticlockwise, the co-ordinates of the point $P_{KL}$ should give the direct and shear stresses on the plane *KL*.

*X*-coordinate of $P_{KL}$ Fig. 2.7(*b*)

$$= ON = OM + r\cos\beta$$

$$= OM + r\cos[\pi - (\alpha + 2\theta)] \qquad [\because \alpha + \beta + 2\theta = \pi]$$

$$= OM - r\cos\alpha\cos 2\theta + r\sin\alpha\sin 2\theta$$

$$= \frac{1}{2}(\sigma_x + \sigma_y) - (QM)\cos 2\theta + (QP_{CD})\sin 2\theta$$

$$= \frac{1}{2}(\sigma_x + \sigma_y) - [OM - OQ]\cos 2\theta + \tau_0 \sin 2\theta$$

$$= \frac{1}{2}(\sigma_x + \sigma_y) - \frac{1}{2}[\sigma_x - \sigma_y)\cos 2\theta + \tau_0 \sin 2\theta$$

$Y$-coordinate of the point $P_{KL}$

$$= -NP_{KL} = -r \sin \beta$$

$$= -r \sin [\pi - (\alpha + 2\theta)]$$

$$= -r \sin \alpha \cos 2\theta - r \cos \alpha \sin 2\theta$$

$$= -\tau_0 \cos 2\theta - QM \sin 2\theta$$

$$= -[OM - OQ] \sin 2\theta - \tau_0 \cos 2\theta$$

$$= -\frac{1}{2}(\sigma_x - \sigma_y)\sin 2\theta - \tau_0 \cos 2\theta$$

These expressions are same as given in equations (2.47) and (2.48). Thus it proves the rule cited above.

Further since $ON$ and $NP_{KL}$ are at right angles and they represent the direct and the shear stresses on the planes $KL$, it is obvious the line $OP_{KL}$ will represent the resultant stress on the plane $KL$. The angle $NOP_{KL}$ will be the inclination of the resultant stress on the plane $KL$ with the normal to this plane. Thus the maximum obliquity will be obtained by drawing the two tangents to the circle from the origin $O$ and measuring the angle which these tangents make with the $x$-axis. The coordinates of the tangential points will give the magnitude of the direct and shear stresses on the planes represented by the points. The actual location of the corresponding planes in the point diagram can be ascertained by measuring the angle which these points subtend individually at the centre of the circle with respect to any of the known points on the circle. Half of this will be the actual inclination of the plane in question with the plane represented by the point on the circle diagram with respect to which measurements have been made.

The above discussion can be summarized as follows :

(*i*) The centre of Mohr's circle lies on $x$-axis.

(*ii*) The radius of the circle is of the magnitude of maximum shear stress.

(*iii*) The circle can be drawn if the position of the centre and the radius are known.

(*iv*) If the direct and shear stresses are given on two planes, the corresponding points can be plotted. After joining these points its right bisector is drawn to meet the x-axis at the centre of the circle. If the given planes are at right angles to each other the line joining the corresponding points, itself will intersect the x-axis at the centre of the circle. Moreover the line will divide the distance between the feet of the perpendiculars drawn from these points on the $x$-axis, into two equal parts. The circle can then be drawn with the radius equal to the distance between the centre and one of the plotted points.

(v) If the stresses, direct as well as shear, on one plane are known and the maximum shear stress is given, the centre of the circle can be located by drawing an arc of radius equal to maximum shear stress with centre at the point representing the given plane. The intersection of the arc with the x-axis is the required centre. There will in general be two possible solutions.

(vi) After having drawn the Mohr's stress circle any required information can be had from it.

(vii) If the given principal stresses are equal and are of same sign the Mohr's circle will shrink to a point and there will be no shear stress (equation 2.2).

**Example 2.4.** *(a) Solve the Example 2.3 with the help of Mohr's circle method.*

*(b) Also find the planes on which the resultant stress is most inclined with the normal on the planes.*

**Solution.** With the origin at $O$, to some convenient scale plot the points $P_{BC}$ and $P_{DC}$ with co-ordinates (50, – 40) and (– 30, 40) respectively. These are corresponding to the direct and shear stresses on the planes $BC$ and $DC$. Join the points $P_{BC}$ and $P_{DC}$. The right bisector of the line $P_{BC}\ P_{DC}$ will intersect the $x$-axis, along which $\sigma_n$ is plotted, at $M$. With $M$ as centre and $MP_{BC}$ as radius draw the Mohr's stress circle intersecting the $x$-axis at $P_1$ and $P_2$. The $x$- coordinates of the points $P_1$ and $P_2$ represent the principal stresses on the corresponding planes. These are

$OP_1$ = 46.5MN/m, (Comp.); $OP_2$ = 66.5MN/m² ( Tensile)

For finding the location of the planes $P_1$ and $P_2$ with respect to any given plane, say $DC$ here, the point which represents the given plane, $P_{DC}$ here, is joined to the centre of the circle. The bearing of the radial line $P_1\ M$ is then measured from this line ( $MP_{DC}$). In this case it comes out to be 45° anticlockwise. The inclination of the plane $P_1$ with the plane $DC$ in the point diagram is then $\frac{1}{2} \times 45 = 22\frac{1}{2}°$. Similarly since the radial line $MP_2$ is at angle of 180° anticlockwise with radial line $MP_1$, the principal plane $P_2$ will be at an angle of $\frac{1}{2} \times 180 = 90°$ anticlockwise from the plane $P_1$.

The points representing the planes of maximum shear stress are $P_3$ and $P_4$. The magnitude of maximum shear stress is given by $MP_4$ = + 56.5 MN/m, on the plane $P_4$ and $MP_3$ = – 56.5 MN/m² on plane $P_3$. The direct stress on these planes is = $OM$ = 10 MN/m² (tensile). The location of the plane $P_3$ with respect to plane $DC$ in the point diagram is = $\frac{1}{2} \angle P_{DC}MP_3 = \frac{1}{2} \times 135 = 67\frac{1}{2}°$ anticlockwise. The bearing of plane $P_4$ with respect to plane $DC$ is = $\frac{1}{2} \angle P_{DC}\ M\ P_4 = \frac{1}{2} \times 45° = 22\frac{1}{2}°$ clockwise.

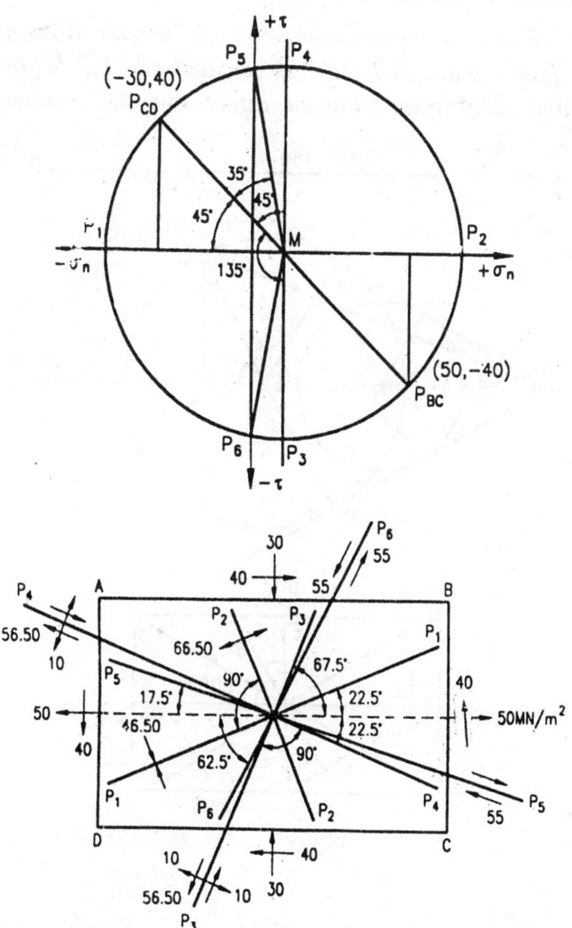


*Fig. 2.8*

(*b*) Normally for finding the planes where the resultant is most inclined to the normal, tangents to circle are to be drawn from the origin. But in this case the origin lies within the circle. Under these circumstances the maximum obliquity will become 90° so that the normal stress on these planes will be zero and the resultant stress will be equal to the shear stress. Such planes will correspond to the points of intersection of the circle with the *y*-axis. These are $P_5$ and $P_6$. The inclination of plane $P_5$ = $\frac{1}{2}\angle P_{BC}\ M\ P_5 = \frac{1}{2}(35°) = 17½°$ clockwise with the plane *DC*. Likewise the inclination of the plane $P_6$ is = $\frac{1}{2}\angle P_{DC}\ M\ P_6 = \frac{1}{2}(123°) = 61½°$ anticlockwise with plane *DC*. Also $\tau_6 = OP_6 = -55$ MN/m² and $\tau_5 = OP_5 = 55$ MN/m². It may be noted that the shear stresses on planes $P_5$ and $P_6$ are equal and opposite but they are not complementary as the planes are not at right angles to each other.

**Example 2.5.** *In a two dimensional stress system, the direct and shear stresses at a point on the plane A are 110 MN/m/l² tensile and 30.7 MN/m² respectively. The obliquity is limited to 30°. If the allowable direct stress in the material is limited to 150*

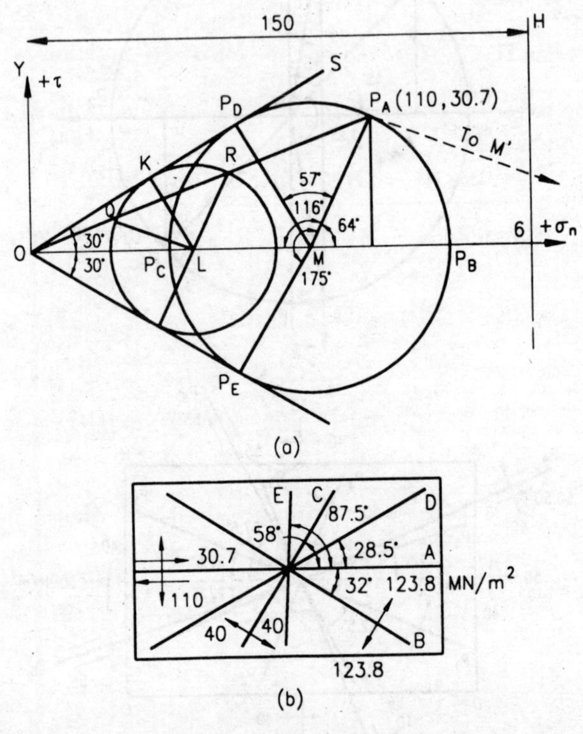


Fig. 2.9

*MN/m² draw the Mohr's stress circle and find the inclination of the principal planes with the plane A and the magnitude of the stresses on them. Find the resultant stress on the planes on which the resultant stress is most inclined with the normal.*

**Solution.** Draw a line $OS$ inclined at an angle of 30° with the $x$-axis. Plot the point $P_A$ (110, 30.7) to some scale, representing state of stress on the plane $A$. Join $P_A$ to $O$. Take any point $L$ on the $x$-axis and draw a circle with $L$ as centre and $LK$ as radius such that it touches the line $OS$ at point $K$. This circle intersects the line $OP_A$ in $Q$ and $R$. Join both $Q$ and $R$ to $L$. Draw the two lines $P_A M$ and $P_A M'$ parallel to $RL$ and $QL$ respectively meeting the $x$-axis at $M$ and $M'$. With centres $M$ and $M'$ and radii $MP_A$ and $M'P_A$ respectively two possible Mohr's stress circle can be drawn. But since the allowable stress is limited to 150 MN/m², the Mohr's circle for the given stress system must lie in the space bounded within the lines $OY$ and $GH$. The circle with centre $M'$, being outside this region, will be ignored. Required information can then be had from the circle with centre $M$.

(*i*) Stress on the principal plane *B*

$$= OP_B = 123.8 \text{ MN/m}^2 \text{ (Tensile)}\}$$

Plane *B* is inclined at $\frac{1}{2} \angle P_A \, MB_B = \frac{1}{2} \times 64° = 32°$ clockwise with Plane *A*.

(*ii*) Stress on the principal plane $C = OP_C = 40$ MN/m² (Tensile)

Plane *C* is inclined at $\frac{1}{2} \angle P_A \, MP_C = \frac{1}{2} \times 116° = 58°$ anticlockwise with Plane *A*

(*iii*) Resultant stress on planes *D* and *E*, on which the resultant is most inclined with the normal

$$= OP_D = OP_E = 71 \text{ MN/m}^2 \text{ or MPa}$$

The plane *D* is inclined at $\frac{1}{2} < P_A \, M \, P_D = \frac{1}{2} \times 57° = 28½°$ anticlockwise from Plane *A*.

The plane *E* is inclined at $\frac{1}{2} < P_A \, MP_E = \frac{1}{2} \times 175° = 76½°$ anticlockwise from Plane *A*.

**Example 2.6.** *At a point in a material the stresses are as shown in Fig. 2.10(a). One of the principal planes, KL, is found to be at 60° anticlockwise with the plane DC. Find the value of $\sigma_x$ and the principal stress on the plane KL.*

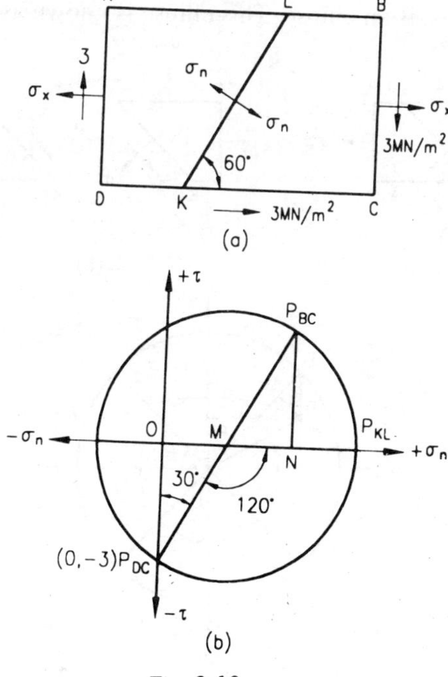


*Fig. 2.10*

**Solution.** Plot the point, $P_{DC}$ (0,– 3), corresponding to the plane $DC$. Since the principal plane $KL$ is inclined at 60° anticlockwise to the place $DC$, the angle $P_{DC}MP_{KL}$ in the circle diagram will have to be 120°, if $M$ is the centre of the circle. To achieve this, evidently, set a line $P_{DC}M$ making an angle of 30° with $OP_{DC}$ to meet the $x$-axis in $M$. With $MP_{DC}$ as radius and centre at $M$ draw the Mohr's circle. Produce the line $P_{DC}M$ to intersect the circle in $P_{BC}$. Then

$$\sigma_x = ON = 3.46 \text{ MN/m}^2$$

$$(\sigma_n)_{KL} = OP_{KL} = 5.196 \text{ MN/m}^2.$$

## 2.6 Ellipse of Stresses

The principal stresses in a two-dimensional stress system are shown in Fig. 2.11($a$).

Let a section be taken along the plane $KL$ which is at right angles to the plane of the paper and on which the stresses have to be found.

Draw two concentric circles with radii proportional to $\sigma_x$ and $\sigma_y$. Draw the line $K_2L_2$ at angle $\theta$ with the horizontal line or parallel to the given plane $KL$.

Draw $OGE$ perpendicular to $K_2L_2$ to meet the $\sigma_x$ and $\sigma_y$ circles in $E$ and $G$ respectively. From the points $E$ and $G$ draw perpendicular lines to the respective axes *i.e.* from $E$ to $X$-axis and from $G$ to $Y$-axis to meet in $H$. Then $OH$ represents the resultant stress in direction and magnitude on the plane $K_2L_2$. The normal and shearing components of stress are $OM$ and $MH$. It will be shown below that the locus of the point $H$, as $\theta$ changes, is an ellipse. This ellipse is known as ellipse of stresses.

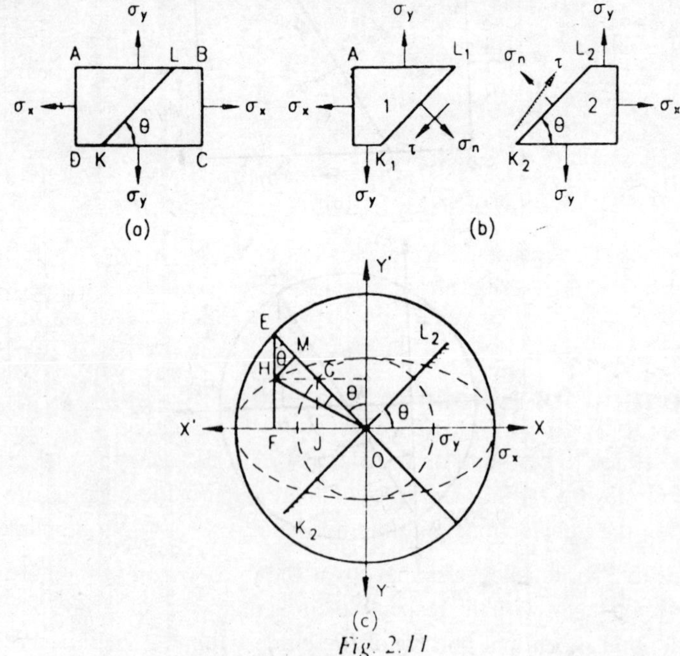

Fig. 2.11

**Proof :** Geometrically

$$\angle MHG = \theta$$

Normal stress $= \sigma_n = OM$

$= OG + GM$

$= \sigma_y + GH \sin (\angle MHG)$

$= \sigma_y + (EG \sin \theta) \sin \theta$

$= \sigma_y + (\sigma_x - \sigma_y) \sin^2 \theta$

$= \sigma_x \sin^2 \theta + \sigma_y \cos^2 \theta$

This result is same as obtained in equation (2.1).

Shear Stress $= \tau = - MH$

$= - EH \sin \theta$

$= - (EG \cos \theta) \sin \theta$

$= - (\sigma_x - \sigma_y) \sin \theta \cos \theta$

This also is same as obtained in equation (2.2).

Thus vector $\vec{OH}$ represents the resultant stress on the plane $K_2L_2$ in direction and magnitude. The normal and shear stress components are represented by $\vec{OM}$ and $\vec{MH}$ respectively.

Now the component of $\vec{OH}$ along $X$-axis

$= OF = OE \sin \theta$

$= \sigma_x \sin \theta = x$ (say ) ...(2.52*a*)

and the component of $\vec{OH}$ along $Y$-axis

$= FH = JG = \sigma_y \cos \theta = y$ (say ) ...(2.52*b*)

Equations (2.52) are parametric equations of an ellipse with θ as the parameter i.e. as θ changes $H$ will describe an ellipse. Hence the method is named after this property.

## Sign Convention for Ellipse of Stresses

If the stresses are to be found on the plane $K_2L_2$ of the element number 2, those half portions of circles (of radii $\sigma_x$ and $\sigma_y$) which are outside the material are taken as positive for taking the intersections with the normal drawn at $O$ on the plane $K_2L_2$.

For the positive (or tensile) principal stress the intersection of the normal to the line $K_2L_2$ at $O$ is to be taken with the respective circle on the positive side otherwise on the opposite side, that is, on that portion of the circle which lies within the material.

The perpendiculars are then drawn from these points to the respective axes to get the point $H$.

**Example 2.7.** *Solve the Example 2.2 with ellipse of stresses method.*

**Solution.** Draw the circles with radii $\sigma_x$ and $\sigma_y$ to some scale. Set $E_2\,OF_2$ parallel to $EF$. Draw the normal $AOG$ to $E_2\,OF_2$. Since $\sigma_x$ is tensile, the intersection of the normal with the circle is taken on its positive side *i.e.* at point $A$ but as $\sigma_y$ is negative the intersection is taken on the negative half of $\sigma_y$ circle *i.e.* at the point $G$ and not $M$. From the points $A$ and $G$ draw perpendiculars to $x$ and $y$ axes respectively to give the point of their intersection as $H$.

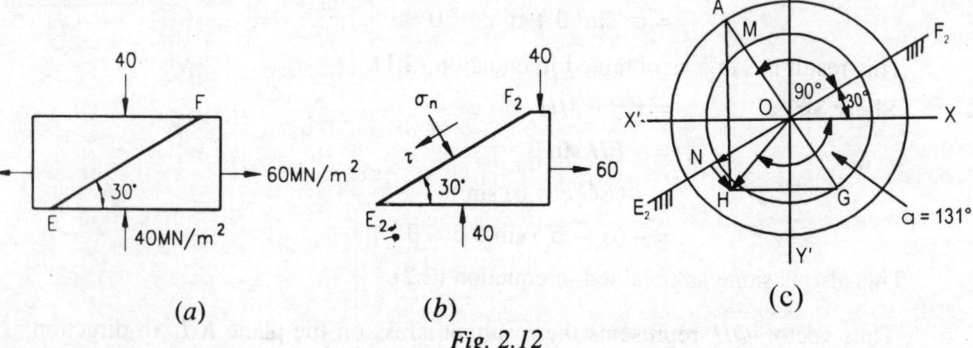


*Fig. 2.12*

$\vec{OH}$ is the resultant stress vector ; its magnitude being 45.7 MN/m², it is inclined at $\alpha = 131°$ with the direction of $\sigma_x$. Also $\sigma_n = -NH = -$ 15 MN/m² (*i.e.* comp.) and $\tau = -ON = -$ 43.3 MN/m².

## 2.7 Poisson's Ratio and Principal Strains

When a body is acted upon by a direct stress, tensile or compressive, the dimensions will increase or decrease along the line of action of the stress applied. The change in length per unit length along this direction is known as the longitudinal or primary strain. Accompanied with the strain in the longitudinal direction there is the strain in the lateral directions known as lateral or secondary strain but these strains are of the opposite signs. That is, if there is positive strain in the primary direction there will be negative strains in the lateral directions and vice versa. For a material stressed within elastic limit the ratio of the magnitudes of lateral to primary strain is constant, known as Poisson's ratio. It is usually represented by ν. Consider a rectangular element under the action of three tensile principal stresses $\sigma_x$, $\sigma_y$ and $\sigma_z$ as shown in Fig. 2.13. Let the modulus of elasticity and Poisson's ratio of the material be $E$ and ν. Due to the stress $\sigma_x$ there will be strain in the $x$-direction =

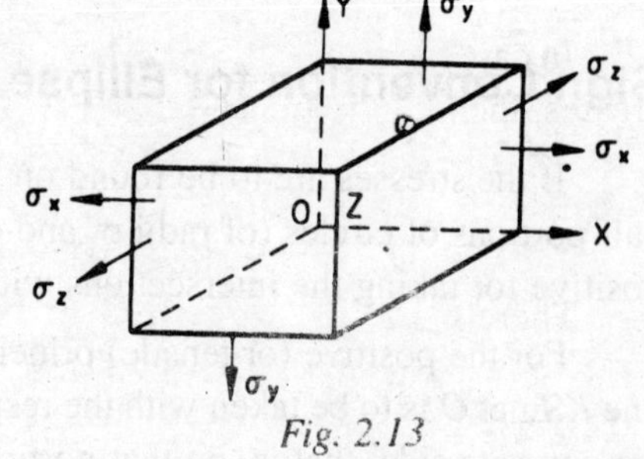


*Fig. 2.13*

$\sigma_x/E$. In addition due to $\sigma_x$ there will be contraction along $Y$ and $Z$ axes and the induced strain in both these directions will be $(-\nu\sigma_x/E)$. Similarly the effect of $\sigma_y$ will be to induce a strain of $\sigma_y/E$ along $y$-axis and $(-\nu\sigma_y/E$ along both $X$ and $Z$-axes. Likewise the effect of $\sigma_z$ will be to induce a strain of $\sigma_z/E$ in the $Z$ direction and $(-\nu\sigma_z/E)$ along the $X$ and $Y$ directions.

Thus the resultant strains in the three directions will be given by

$$\varepsilon_x = \frac{\sigma_x}{E} - \frac{\nu}{E}(\sigma_y + \sigma_z) \quad ...(2.53)$$

$$\varepsilon_y = \frac{\sigma_y}{E} - \frac{\nu}{E}(\sigma_z + \sigma_x) \quad ...(2.54)$$

$$\varepsilon_z = \frac{\sigma_z}{E} - \frac{\nu}{E}(\sigma_x + \sigma_y) \quad ...(2.55)$$

In case $\sigma_x$, $\sigma_y$ and $\sigma_z$ are all equal to $\sigma$, equations (2.53) to (2.55) reduce to

$$\varepsilon_x = \varepsilon_y = \varepsilon_z = \frac{\sigma}{E} - \frac{2\nu\sigma}{E}$$

$$= \frac{\sigma}{E}(1 - 2\nu) \quad ...(2.56)$$

## 2.7.1 Principal Strains for a Given Two-dimensional Stress System

If in equations (2.53) through (2.55) $\sigma_z$ is equated to zero, the resultant strains in $x$ and $y$- directions will be

$$\varepsilon_x = \frac{\sigma_x}{E} - \frac{\nu\sigma_y}{E} \quad ...(2.57)$$

$$\varepsilon_x = \frac{\sigma_y}{E} - \frac{\nu\sigma_x}{E} \quad ...(2.58)$$

$$\varepsilon_x = -\frac{\nu}{E}(\sigma_x + \sigma_y)$$

In case the strains $\varepsilon_x$ and $\varepsilon_y$ are known and it is required to find the corresponding stresses, then multiplying equation (2.58) by $\nu$ and adding to equation (2.57)

$$\varepsilon_x + \nu\varepsilon_y = \frac{\sigma_x}{E}(1 - \nu^2)$$

Therefore, $$\sigma_x = \frac{E}{(1 - \nu^2)}(\varepsilon_x + \nu\varepsilon_y) \quad ...(2.60)$$

Similarly, multiplying equation (2.57) by $\nu$ and adding to equation (2.58)

$$\sigma_y = \frac{E}{(1 - \nu^2)}(\varepsilon_y + \nu\varepsilon_x) \quad ...(2.61)$$

## 2.8 Relation between *E*, G and *K*

Consider a cube *ABCD* under the action of shear stress $\tau_0$ as shown in Fig. 2.14(*a*). The movement of the corner *A* is very small and then the shear strain $\phi$ is given as

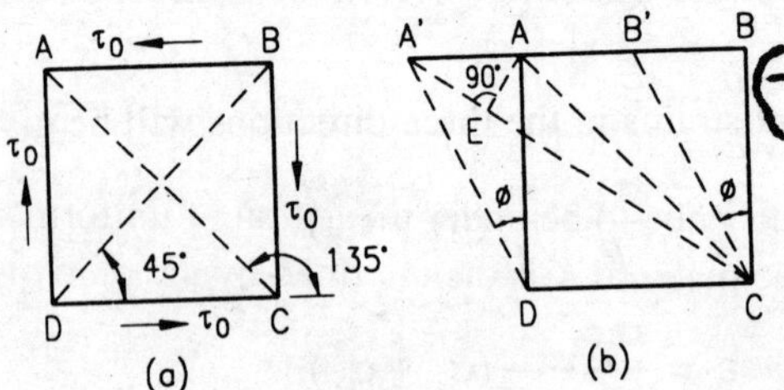


*Fig. 2.14*

$$\phi = \frac{AA'}{DA} = \frac{\tau_0}{G} \qquad \text{...(2.62)}$$

where *G* is the modulus of rigidity.

From *A* drop *AE* perpendicular to *CA'* then *CA* is approximately equal to *CE*.

Then the linear strain along *CA* is given by $\varepsilon_{CA}$ as

$$\varepsilon_{CA} = \frac{CA' - CA}{CA}$$

$$= \frac{EA'}{CA} \simeq \frac{AA' \cos 45°}{DA\sqrt{2}}$$

$$= \frac{1}{2}\frac{AA'}{DA} \qquad \text{...(2.63)}$$

From equations (2.62) and (2.63)

$$\varepsilon_{CA} = \frac{\tau_0}{2G} \qquad \text{...(2.64)}$$

Also with the help of article (2.2), it is seen that the planes *DB* and *CA* are principal planes and the principal stresses are given by equations (2.15).

For $\theta = \frac{p}{4}, \sigma_1 = \tau_0$ (Tensile stress acting along the direction *CA* and on plane *DB*)

For $\theta = \frac{3\pi}{4}, \sigma_2 = -\tau_0$ (*i.e.*, Comp. stress acting along the direction *DB* on plane *CA*)

Thus the strain along the line *CA* can also be found from stress considerations as

$$\varepsilon_{CA} = \left(\frac{\sigma_1}{E} - \nu\frac{\sigma_2}{E}\right) = \frac{\tau_0}{E} + \nu\frac{\tau_0}{E}$$

*i.e.*
$$\varepsilon_{CA} = \frac{\tau_0}{E}(1+v) \quad \text{...(2.65)}$$

Thus from equations (2.64) and (2.65), we get the relation between $E$ and $G$.

$$\frac{\tau_0}{2G} = \frac{\tau_0}{E}(1+v)$$

or
$$E = 2G(1+\nu) \quad \text{...(2.66)}$$

Again consider this cube to be under the action of uniform stress $\sigma$ on all sides. The bulk modulus of the material $K$ is then defined by the ratio of $\sigma$ to the volumetric strain caused by $\sigma$.

$$K = \frac{\sigma}{\text{volumetric strain}} \quad \text{...(2.67)}$$

For the stress $\sigma$ along $x$, $y$ and $z$ directions of the cube, the strains in these three directions are given by

$$\varepsilon_x = \varepsilon_y = \varepsilon_z = \frac{\sigma}{E} - \nu\frac{\sigma}{E} - \nu\frac{\sigma}{E}$$

$$= \frac{\sigma}{E}(1-2\nu) \quad \text{...(2.68)}$$

If the original dimensions of the cuboid $x$, $y$ and $z$ change by $dx$, $dy$ and $dz$ respectively under the action of uniform stress $\sigma$, then

Original volumie $V = x \, . \, y \, . \, z$

Change in volume $dV = (x+dx)(y+dy)(z+dz) - xyz$

$= xydz + yzdx + zxdy$

neglecting the product of two small quantities.

Therefore volumetric strain

$$= \frac{dV}{V} = \left(\frac{xydz + yzdx + zxdy}{xyz}\right)$$

$$= \frac{dx}{x} + \frac{dy}{y} + \frac{dz}{z}$$

$$= \varepsilon_x + \varepsilon_y + \varepsilon_z \quad \text{...(2.69)}$$

Thus from equations (2.68) and (2.69)

$$\frac{dV}{V} = 3\frac{\sigma}{E}(1-2v) \quad \text{...(2.70)}$$

Therefore substituting equation (2.70) in equation (2.67)

$$K = \frac{\sigma}{3\sigma(1-2v)/E}$$

or
$$E = 3K(1-2\nu) \quad \text{...(2.71)}$$

Thus from equations (2.66) and (2.71)

$$E = 2G(1 + \nu) = 3K(1 - 2\nu) \quad ...(2.72)$$

## 2.8.1 Limiting Values of Poisson's Ratio

For an isotropic (a material which does not have a preferred orientation) perfectly plastic material the stress-strain curve will be horizontal. Thus the value of the Youngs' modules, E, which is equal to the slope of the stress-strain curve will be zero.

Therefore, from equation 2.72 we get from the first part of this equation.

$$\nu = -1$$

From the second part of this equation

$$\nu = 0.5$$

Thus the limiting values of $\nu$ are – 1 to 0.5. In other words 0.5 is the maximum value of the Poisson's ratio.

Since ordinary materials contract laterally when stretched and expand laterally when compressed axially, Poisson's ration for such materials is positive. It is 0.45 for lead, 0.33 for aluminium and 0.27 to 0.33 for common steels. For cellular solids such as typical foams its value lies within the range 0.1 to 0.4.

For cork the value of Poisson's ratio is nearly zero. It is due to this property that cork stopper is used while bottling champagne. The cork must be easily inserted and removed, yet it also must withstand the pressure from within the bottle. Rubber, with a poisson's ratio of nearly 0.5 could not be used for this purpose because it would expand when compressed in to the neck of the bottle and would jam. Cork, by contrast with Poisson's ratio of nearly zero, is ideal in this application.

Biological tissues and various rubbery materials have values of Poisson's ratio approaching 0.5. Therefore from equation (2.71), it is seen that

$$G = \frac{3K(1 - 2\nu)}{[2(1 + \nu)]}$$

And when $\nu \rightrightarrows 0.5$ the value of $G$ approaches zero.

Thus for rubbery materials the shear modulus $G$, is much less than the bulk modulus, $K$. In physical sense these materials readily undergo shear deformation but resist volumetric (or bulk) deformation.

On the other extreme a solid with Poisson's ratio close to – 1 would have the value of K from equation (2.71) approaching to zero because

$$K = \frac{2G(1 + \nu)}{[3(1 - 2\nu)]}$$

When $\nu \rightrightarrows -1, K \rightrightarrows 0$

In physical sense their behaviour will be opposite of rubber. That is they are very difficult to shear but easy to deform volumetrically. Mathematically for such solids the value of shear modulus, $G$ will be much greater than the bulk modulus $K$. Therefore these materials may be called ANTI-RUBBER materials.

## 2.8.2 Negative Poisson's Ratio Materials

Till 1968 it was believed by many that materials with negative values of Poisson's ratio were unknown (Young 1968), (Love 1944) had reported a single example of "Single Crystal" pyrite in having a Poisson's ratio of –0.14 in his book published in 1944. He further suggested that this effect might have been result from a twinned crystal. Analysis of the tensorial elastic constants (*R*. Hill) of anisotropic single crystal cadmium suggests Poisson's ratio may attain negative values in some directions (Li 1976). Anislotropic macroscopic, two-dimensional flexible models of certain honeycomb structures (not materials) have exhibited negative. Poisson's ratio in some directions (Gibson et. El 1982) These known examples of negative Poisson's ratio all depend on the presence of high degree of anisotropy, the effect only occurs in some directions and may be dominated by coupling between stretching force and shear deformation.

Polymer foams of honey-comb structures, having negative Poisson's ratio have been developed. It is anticipated that these foams may be used in such applications as sponges, shock absorbing materials, air filters, fasteners, car and air craft seats. For having additional information the reader is advised to refer to the article entitled "Advances in negative Poisson's ratio materials" published in "Advanced Materials, 5, pp 229–296 (1993).

## 2.9 Strains on Oblique Planes

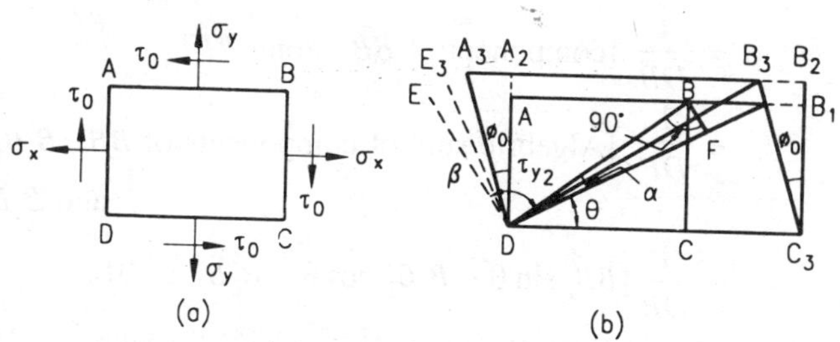


*Fig. 2.15*

In the Fig. 2.15(*a*) it is shown a rectangular block *ABCD* under the action of general stress system, Fig. 2.15(*b*) shows the way this block will get deformed.

Let $\varepsilon_x$ and $\varepsilon_y$ be the linear strains along $x$ and $y$-axes such that

$$\varepsilon_x = \frac{\sigma_x}{E} - \nu\frac{\sigma_y}{E} \qquad ...(2.73)$$

$$\varepsilon_y = \frac{\sigma_y}{E} - \nu\frac{\sigma_x}{E} \qquad ...(2.74)$$

Due to linear strain $\varepsilon_x$ the edge $AB$ will elongate to $AB_1$ and then under the action of strain $\varepsilon_y$ the point $B_1$ will move to $B_2$ and $A$ to $A_2$. Ultimately due to the shear strain $\phi_0$ on the faces $AB$ and $BC$, the point $B_2$ will move to $B_3$ and $A_2$ to $A_3$ so that the final shape of the block will be $DC_3B_3A_3$.

It is required to find the linear and shear strains along the direction B inclined at an angle $\theta$ with the $x$-axis.

The movement of $B$ along $DB$

$= $ Projection of $BB_3$ on $DB$

$= $ (Projection of $BB_1$ along $DB$) + (Projection of $B_1B_2$ along $DB$) + (Projection of $B_2B_3$ along $DB$)

$$\approx BB_1 \cos\theta + B_1B_2 \sin\theta - B_2B_3 \cos\theta \qquad ...(2.75)$$

(neglecting the effect of very small angle between $DB$ and $DB_3$)

Now $\quad BB_1 = DC\varepsilon_x = DB\cos\theta\,\varepsilon_x \qquad ...(2.76)$

$\quad B_1B_2 = BC\varepsilon_y = DB\sin\theta\,\varepsilon_y \qquad ...(2.77)$

and $\quad B_2B_3 = B_2C_3\,\phi_0 = BC\,(1+\varepsilon_y)\,\phi_0 \qquad ...(2.78)$

$\quad \approx DB\,(\sin\theta)\,\phi_0 \qquad ...(2.78a)$

(neglecting the product $\varepsilon_y\,\phi_0$)

Thus, substituting equations (2.76) to (2.78) in equation (2.75) and dividing by $DB$, we get the linear strain along $DB$ as

$$\varepsilon_\theta = \varepsilon_x \cos^2\theta + \varepsilon_y \sin^2\theta - \phi_0 \sin\theta\cos\theta$$

or

$$\varepsilon_\theta = \frac{1}{2}(\varepsilon_x + \varepsilon_y) + \frac{1}{2}(\varepsilon_x - \varepsilon_y)\cos 2\theta - \frac{1}{2}\phi_0 \sin 2\theta \qquad ...(2.79)$$

Now if the clockwise rotation of $DB$ is $\alpha$ $(= \angle BDB_3)$, then

$$\alpha = \frac{1}{DB}\,[\text{component of } BB_2 \text{ along } BF]$$

$$= \frac{1}{DF}\,[\text{Algebric sum of components of } BB_1,\ B_1B_2 \text{ and } B_2B_3 \text{ along } BP]$$

$$= \frac{1}{DB}\,[BB_1 \sin\theta - B_1B_2\cos\theta - B_2B_3\sin\theta]$$

Therefore substituting values from equations (2.76) through (2.78).

$$\alpha = (\varepsilon_x - \varepsilon_y)\sin\theta\cos\theta - \phi_0\sin^2\theta \qquad ...(2.80)$$

Since the shearing strain is defined by the change in angle between two mutually perpendicular planes, $\theta$ will be replaced by $\left(\frac{\pi}{2} + \theta\right)$ in equation (2.80) to evaluate the clockwise rotation $\beta$ of the line $DE$ and then finding the change of the angle between $DB$ and $DE$.

That is, $\beta = (\varepsilon_x - \varepsilon_y) \sin\left(\frac{\pi}{2} + \theta\right) \cos\left(\frac{\pi}{2} + \theta\right) - \phi_0 \sin^2\left(\frac{\pi}{2} + \theta\right)$

or $\beta = -(\varepsilon_x - \varepsilon_y) \sin\theta \cos\theta - \phi_0 \cos^2\theta$ ...(2.81)

Thus the shear strain along the plane *DB* is

$$\phi_\theta = \alpha - \beta$$

$$= 2(\varepsilon_x - \varepsilon_y) \sin\theta \cos\theta - \phi_0 (\cos^2\theta - \sin^2\theta)$$

*i.e.*, $\frac{1}{2}\phi_\theta = \frac{1}{2}(\varepsilon_x - \varepsilon_y) \sin 2\theta - \frac{1}{2}\phi_0 \cos 2\theta$ ...(2.82)

The maximum and minimum values of $\varepsilon_\theta$ are obtained by setting $\frac{d}{d\theta}(\varepsilon_\theta) = 0$ to give

$$\tan 2\theta = \frac{-\phi_0}{(\varepsilon_x - \varepsilon_y)} \quad ...(2.83)$$

There are two solutions of equation (2.83) given by

$$2\theta_1 = \tan^{-1}[-\phi_0 / (\varepsilon_x - \varepsilon_y)] \quad ...(2.84)$$

and $2\theta_2 = \pi + 2\theta_1$

or $\theta_2 = \left(\frac{\pi}{2} + \theta_1\right)$ ...(2.85)

Such that

$$\sin 2\theta_1 = \phi_0 / [(\varepsilon_x - \varepsilon_y)^2 + \phi_0^2]^{1/2} \quad ...(2.86)$$

$$\cos 2\theta_1 = (\varepsilon_x - \varepsilon_y) / [(\varepsilon_x - \varepsilon_y)^2 + \phi_0^2]^{1/2} \quad ...(2.87)$$

$$\sin 2\theta_2 = \sin(\pi + 2\theta_1) = -\sin 2\theta_1 \quad ...(2.88)$$

$$\cos 2\theta_2 = \cos(\pi + 2\theta_1) = -\cos 2\theta_1 \quad ...(2.89)$$

Substituting values from equations (2.86) through (2.89) in equation (2.79), we get

Along $\theta_1$ $\varepsilon_{max} = \frac{1}{2}(\varepsilon_x + \varepsilon_y) + \sqrt{\left(\frac{\varepsilon_x - \varepsilon_y}{2}\right)^2 + \left(\frac{\phi_0}{2}\right)^2}$ ...(2.90)

Along $\theta_2$ $\varepsilon_{min} = \frac{1}{2}(\varepsilon_x + \varepsilon_y) - \sqrt{\left(\frac{\varepsilon_x - \varepsilon_y}{2}\right)^2 + \left(\frac{\phi_0}{2}\right)^2}$ ...(2.91)

Again from equation (2.83)

$$\tan 2\theta = -\left(\frac{\tau_0}{G}\right) \Bigg/ \left[\left(\frac{\sigma_x}{E} - \frac{\nu\sigma_y}{E}\right) - \left(\frac{\sigma_y}{E} - \nu\frac{\sigma_x}{E}\right)\right]$$

$$= \frac{E}{G(1+\nu)}\left[\frac{-\tau_0}{(\sigma_x - \sigma_y)}\right]$$

Now from equation (2.72)

$$E = 2G\,(1 + \nu)$$

Therefore $$\tan 2\theta = \frac{-2\tau_0}{(\sigma_x - \sigma_y)} \qquad ...(2.92)$$

Equation (2.92) is just the same as equation (2.30).

Thus the principal strains and principal stresses act along the same directions.

## 2.9.1 Mohr's Strain Circle

The equations (2.79) and (2.82) are just similar to the equations (2.21) and (2.22) for stresses. Thus if the "*Linear strain*" and "*half of the shear strain*" in any direction are plotted along the x and y axis respectively the locus of this point for different values of θ will be a circle with centre lying on the x-axis. It is known as Mohr's strain circle.

**Example 2.8.** *In a two dimensional stress system, the direct strains along x and y directions are + 8 × $10^{-6}$ and – 4 × $10^{-6}$ respectively. These are also accompanied by shear strain of 4 × $10^{-6}$. Find graphically the principal strains and the maximum shear strain.*

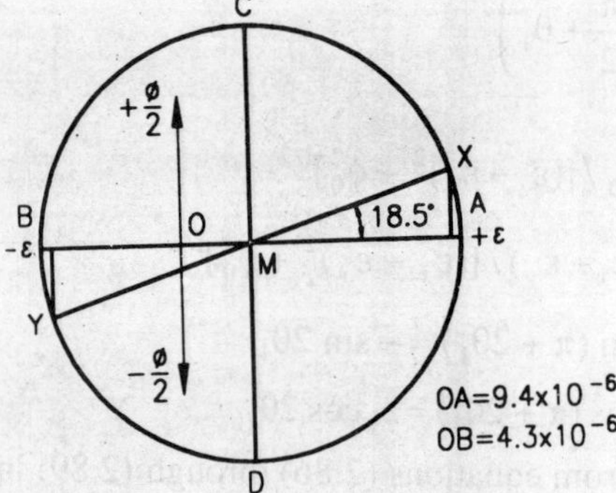


*Fig. 2.16*

**Solution.** Plot the points *x* and *y*-corresponding to *x* and *y* planes with coordinates (8, 2) × $10^{-6}$ and (– 4, – 2) × $10^{-6}$ respectively. Join *X* and *Y* to intersect the *x*-axis at *M*. With *MX* as radius and *M* as centre draw the Mohr's circle which meets the *x*-axis at *A* and *B*.

From the diagram Max. principal strain

$$= OA = 8.32 \times 10^{-6}$$

Min. principal strain

$$= OB = -\,4.32 \times 10^{-6}$$

Max. Shear Strain

$$= 2MC = 2MD = 12.64 \times 10^{6}$$

Inclination of plane *A* with plane *X*

$$= \frac{1}{2}(18° \ 30') = 9° \ 15' \text{ clockwise}$$

Inclination of plane $C$ with plane $X$

$$= \frac{1}{2}(71° \ 30') = 35° \ 45' \text{ anticlockwise}$$

**Example 2.9.** *Strains were measured with the help of strain gauges in two perpendicular directions at 30° to the directions of principal stresses as 420 × 10*$^{-6}$ *and 80 × 10*$^{-6}$ *respectively, both tensile. Find the normal stresses in the directions of measured strains. Also determine the principal stresses Take E = 200 GN/m², v = 0.3.*

**Solution.** If $\varepsilon_x$ and $\varepsilon_y$ are the principal strains and $\sigma_x$ and $\sigma_y$ principal stresses, then

$$\varepsilon_{30°} = 420 \times 10^{-6} = \frac{1}{2}(\varepsilon_x + \varepsilon_y) + \frac{1}{2}(\varepsilon_x - \varepsilon_y)\cos 60°$$

$$\varepsilon_{120°} = 80 \times 10^{-6} = \frac{1}{2}(\varepsilon_x + \varepsilon_y) + \frac{1}{2}(\varepsilon_x - \varepsilon_y)\cos 240°$$

Let $\frac{1}{2}(\varepsilon_x + \varepsilon_y) = a$ and $\frac{1}{2}(\varepsilon_x - \varepsilon_y) = b$

Then $a + b.\frac{1}{2} = 420 \times 10^{-6}$

and $a - b.\frac{1}{2} = 80 \times 10^{-6}$

Solving these simultaneous equations

$$a = 250 \times 10^{-6} \text{ and } b = 340 \times 10^{-6}$$

That is, $\varepsilon_x + \varepsilon_y = 420 \times 10^{-6}$

and $\varepsilon_x - \varepsilon_y = 680 \times 10^{-6}$

Which give $\varepsilon_x = 590 \times 10^{-6}$ and $\varepsilon_y = -90 \times 10^{-6}$

Therefore
$$\sigma_x = \left(\frac{E}{1 - \nu^2}\right)(\varepsilon_x + \nu\varepsilon_y)$$

$$= \frac{200 \times 10^9}{1 - 0.09}(590 \times 10^{-6} - 0.3 \times 90 \times 10^{-6})$$

$$= 123.5 \text{ MN/m}^2 \text{ or MPa (Tensile)}$$

and
$$\sigma_y = \left(\frac{E}{1 - \nu^2}\right)(\varepsilon_y + \nu\varepsilon_x)$$

$$= \frac{200 \times 10^3}{1 - 0.09}[-90 \times 10^{-6} + 0.3 \times 590 \times 10^{-6}]$$

$$= 19.1 \text{ MN/m}^2 \text{ or MPa (Tensile)}$$

If $\sigma_{30}$ and $\sigma_{120}$ are the normal stresses along these planes respectively, then

$$\frac{\sigma_{30}}{E} - \nu\frac{\sigma_{120}}{E} = 420 \times 10^{-6}$$

and $$\frac{\sigma_{120}}{E} - \nu\frac{\sigma_{30}}{E} = 80 \times 10^{-6}$$

Solving these equations, we get

$$\sigma_{30} = 97.8 \text{ MN/m}^2 \text{ or MPa}$$

and $$\sigma_{120} = 45.7 \text{ MN/m}^2 \text{ or MPa.}$$

## 2.10 Modified Modulus of Elasticity

In article 1.6, Young's Modulus of elasticity $E$ has been defined as the ratio of stress to strain in the case of an uniaxially loaded specimen. In other words the stress $\sigma_y$ is applied only along the axis of the specimen (say $y$-axis). There are no stresses applied in the other two directions. In other words along $X$ and $Z$ axes the specimen is free to contract or expand depending on whether the stress along $y$-axis is tensile or compressive in nature.

However if the expansion or contraction along either $X$ or $Z$ directions or along both $X$ and $Z$ directions is not free but restricted completely or partially due to the presence of some casing or fixture then on the $X$ and $Z$ directions will be acting stresses $\sigma_x$ and $\sigma_z$. Thus it will then be case of three dimensional stress system (in general). Depending upon the extent of allowed expansion or contractions (or strains) in these two drictious the stresses $\sigma_x$ and $\sigma_z$ can be evaluated. Then, knowing the values of threee stresses, the resultant strain $\varepsilon_y$ along the $y$-direction can be calculated due to the combined effect of $\sigma_x$, $\sigma_y$ and $\sigma_z$. The ratio of $\sigma_y$ to $\varepsilon_y$ is called the Modified Young's Modules of Elasticity.

**Example 2.10.** *Determine the ratio between the Young's modulus for compression and the modified modulus when lateral expansion in one direction is entirely prevented. Take Poisson's ratio as ν.*

**Solution.** Let the applied compressive stress be $\sigma_y$ in both the cases. In the first case when there is free expansion in lateral directions.

$$\varepsilon_z = \varepsilon_x = \frac{\nu\sigma_y}{E} \qquad \text{...(1)}$$

and $$\varepsilon_y = \frac{-\sigma_y}{E}$$

or $$E = -\frac{\sigma_y}{\varepsilon_y} \qquad \text{...(2)}$$

In the second case let the expansion in $z$-direction be prevented then in this direction compressive stress $\sigma_z$ will be acting.

Then $\varepsilon'_z = 0 = -\dfrac{\sigma_z}{E} + \dfrac{\nu\sigma_y}{E}$ ...(3)

*i.e.,* $\sigma_z = \nu\sigma_y$

Also in this case $\varepsilon'_y = -\dfrac{\sigma_y}{E} + \dfrac{\nu\sigma_z}{E}$

$$= -\frac{\sigma_y}{E} + \nu^2\frac{\sigma_y}{E}$$

$$= -\frac{\sigma_y}{E}(1-\nu^2)$$

or $-\dfrac{\sigma_y}{\varepsilon'_y} = \dfrac{E}{(1-\nu^2)}$ ...(4)

Thus if $E'$ is the modified modulus in the second case

$$E' = -\sigma_y/\varepsilon'_y = \frac{E}{1-\nu^2}$$

Therefore $\dfrac{E}{E'} = (1-\nu^2)$

**Example 2.11.** *(a) In a two dimensional stress system the principal stresses at a point are 10 MPa (Tensile) and 2 MPa (Tensile). Find analylically the direct and sheaat r stress on a plane inclined at 30° clockwise with the first principal plane then find the magnitude and the inclination of the resultant stress on this plane with the direction of the 10 MPa principle stress.*

*(b) Also find the location of the planes on which the obliquity is maximum, with respect to the plane on which the major principal stress of 10 MPa acts.*

*(c) What is the value of* $\tau_{max}$ *?*

**Solution.** Let the $X$ plane (plane $BC$) be the principal plane which carries the principal stress of 10 MPa and other principal plane $AB$ carries the stress 2 MPa.

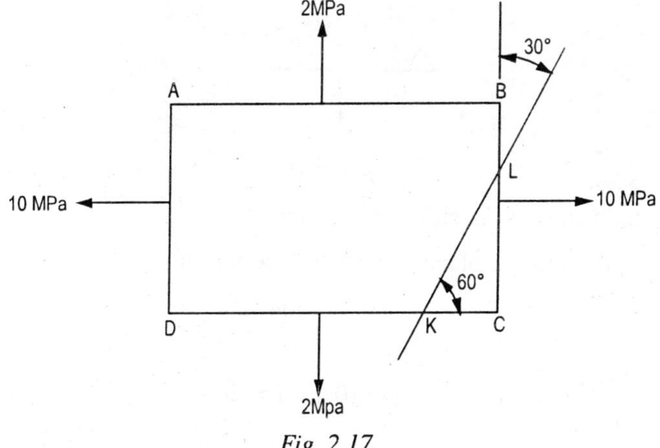


*Fig. 2.17*

(*a*) The plane *KL* is inclined at 30° clockwise with the plane *BC* as shown. Its inclination with the *y*-plane *DC* is 60°

Therefore from equations (2.1) and (2.2), where

$$\sigma_x = 10 \text{ MPa}$$

$$\sigma_y = 2 \text{ Mpa}$$

$$\theta = 60°$$

Then the normal stress on the plane *KL* $\sigma_n$ is given by

$$\sigma_n = \left(\frac{\sigma_x + \sigma_y}{2}\right) - \left(\frac{\sigma_x - \sigma_y)}{2}\right) \cos 2\theta$$

$$= \frac{10+2}{2} - \left(\frac{10-2}{2}\right) \cos (120°)$$

$$= 6 - 4 \cos 120° = 6 - 4\,(-0.5)$$

$$= 6 + 2 = 8 \text{ MPa}$$

And shear stress on plane *KL* is given by $\tau$

$$\tau = -\left(\frac{\sigma_x - \sigma_y}{2}\right) \sin 2\theta$$

$$= -\left(\frac{10-2}{2}\right) \sin 120° \; = -\,3.464 \text{ MPa.}$$

These stress are shown in figure (2.18).

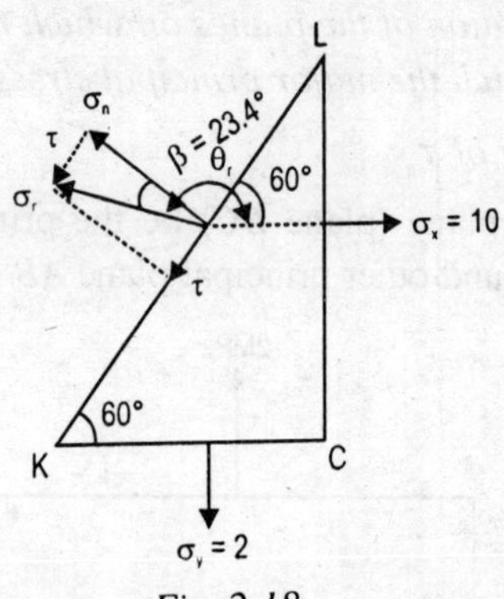


*Fig. 2.18*

The resultant stress $\beta_r$ is inclined at an angle $\beta$

with the normal to the plane or with the diction of $\sigma_n$ where

$$\tan \beta = \frac{|\tau|}{|\sigma n)} = \frac{3.464}{8} = 0.433$$

or $$\beta = \tan^{-1}(0.433) = 23.4°$$

From figure (2.18) therefore inclination of $\sigma_r$ with the direction of $\sigma_r$ (= 10 MPa) is

$$\theta_r = 60° + 90° = 23.4° = 173.4° \text{ anticlockwise}$$

Also $$\sigma_r^2 = \sigma_n^2 + \tau^2 = (8)^2 + (3.464)^2 = 75.999$$

or $$\sigma_r = 8.7177 \text{ MPa.}$$

(*b*) Let $\beta_{max}$ be the maximum in clination of the resultant stress with the normal to that plane. Then from eqn. 2.20

$$\sin \beta_{max} = \frac{-\sigma_x + \sigma_y}{\sigma_n + \sigma_y} = \frac{-10+2}{10+2} = \frac{-8}{12} = \frac{-2}{3}$$

or $$\beta_{max} = 221.81° \text{ or } 318.79°$$

If θ is the corresponding inclination of these planes having maximum obliquity with the plane *DC* on which minor principal stress acts, then from eqn. 2.19

$$2\theta = \frac{\pi}{2} + \beta_{max}$$

when $$\beta_{max} = 221.81°$$

$$\theta_1 = \frac{\pi}{4} + \frac{1}{2}\beta_{max} = 45° + \frac{221.81}{2} = 155.90° \approx (180 - 24.95°)$$

anticlockwise with plane *DC*

or It is equivalent to 24.9° clockwise with *DC*

when $$\beta_{max} = 318.19°$$

$$2\theta_2 = \frac{\pi}{2} + \beta_{max}$$

or $$\theta_2 = 45° + \frac{318.19}{2} = 204.09° = 180 + 24.095$$

*i.e.,* 204.09° anticlockwise with *DC*

or it is equivalent to 24.095° anticlockwise with *DC*

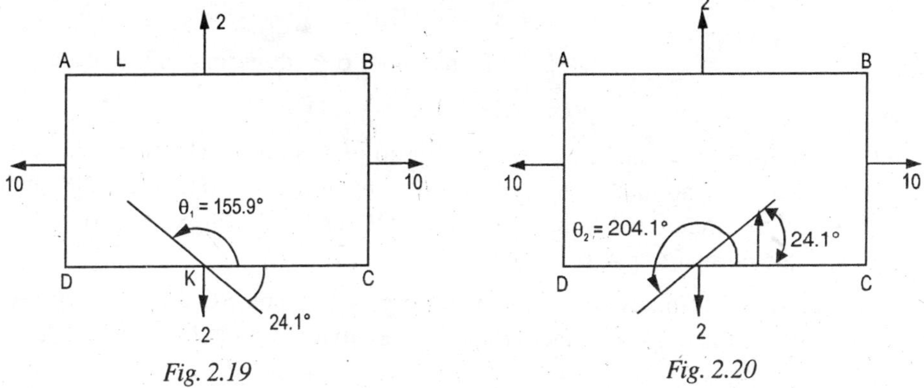


*Fig. 2.19* *Fig. 2.20*

(c) $$\tau_{max} = \frac{\sigma_x - \sigma_y}{2} = \frac{10-2}{2} = 4 \text{ MPa}.$$

**Example 2.12.** *Solve the problem of example 2.13 with the help of Mohr's circle Method.*

**Solution.** Corresponding to the point diagram fig 2.17, plot the point $P_{BC}$ and $P_{CD}$ representing the planes *AB* and *CD* with co-ordinates (10,0) and (2, 0) an a suitable scale. With $P_{BC}$ $P_{CD}$ as diameter draw the Mohr's circle with M as the centre where M is the mid point of diameter $P_{BC}$ $P_{CD}$.

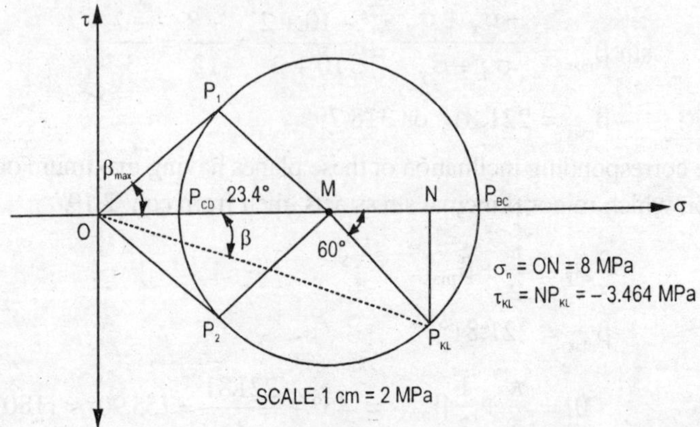


*Fig. 2.21*

The point $P_{KL}$ representing the plane *KL* which is inclined at 30° clockwise with plane *BC* is found on the Mohr's circle by drawing a radial line $MP_{KL}$ at 60° clockwise with the radial line $MP_{BC}$. Draw a perpendicular $P_{KL}$ N on X-axis.

Then *ON* represents direct or normal stress $\sigma_n$ and $NP_{KL}$ the negative shear stress $\tau$ on the plane *KL* on the same scale on which line diagram has ben drawn.

From Mohr's circle $\sigma_n = ON = 8$ MPa

and $\tau_{KL} = NP_{KL} = -3.464$ MPa

The resultant stress $(\sigma_r)_{KL}$ is $= OP_{KL}$

$= 8.7177$ MPa

and $\beta$ = Inclination of $\sigma_r$ with normal to the plane

point = 23.4°

Location of point $P_1$ and $P_2$ corresponding to planes on which the resultant is most inclined with the normal to the plane, is found by drawing two tangents $OP_1$ and $OP_2$ to the Mohr's circle. The inclination of lines $OP_1$ and $OP_2$ with x-axis gives $\beta_{max}$ = 41.8° with the normal to plane.

The indication of planes $P_1$ and $P_2$ with respect to plane *BC* can be found by measuring the angles, $<P_1M\ P_{CD}$ and $<P_2M\ P_{CD}$ and dividing it by 2.

Thus inclination of plane $P_1$ is $\frac{1}{2}(< P_{CD}MP_1) = \frac{1}{2} \times 48.2° = 24.1°$ anticlockwise and that of plane $P_2$ is 24.1° clockwise with the plane *BC*.

(*b*) Similarly the plane 2 in the point diagram, which is represented by point $P_2$ on Mohr's circle and is inclined at $\frac{1}{2}(< P_{CD}MP_2) = 24.1°$ clockwise with plane *CD*.

(*c*)

$$\tau_{max} = \text{Radius of Mohr circle}$$

$$= \frac{1}{2} (\text{diameter } P_{BC}P_{CD})$$

$$= \frac{1}{2} \times 8 = 4 \text{ MPa.}$$

**Example 2.13.** *In a two dimensional stress system the principal stresses are 2 MPa comp and 10 MPa comp. Find direct stress, shear stress and resultant stress on two planes which are inclined at 30 clockwise and 30° anticlockwise respectively with the plane on which stress of 2 MPa acts. What is the inclination of resultant stress with the plane on which normal stress is 10 MPa. Also find* location of planes on which $\beta_{max}$ occurs.

$$\sigma_n = -\ 2 \text{ MPa}$$

$$\sigma_y = -\ 10 \text{ MPa}$$

**Solution.** (*a*) From the plane *KL* inclined at 30° clockwise with plane *BC* angle θ, inclination of *KL* with *y*-plane is.

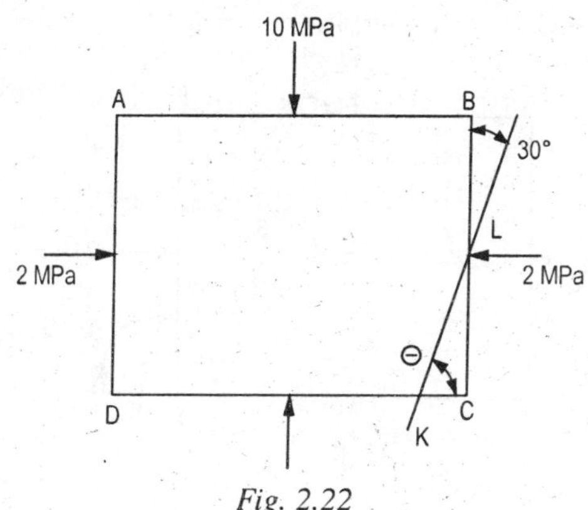


*Fig. 2.22*

$$\theta = 90 - 30 = 60°$$

Therefore $\tau = -\left(\frac{\sigma_x - \sigma_y}{2}\right) \sin 120°$

$$\sigma_n = \left(\frac{-2-10}{2}\right) - \left(\frac{-2+10}{2}\right) \cos 120°$$

$$= -6 - 4(-0.5) = -6 + 2 = -4$$

*i.e.,* $\sigma_n = 4$ MPa (comp.)

and $\tau = -\left(\dfrac{\sigma_x - \sigma_y}{2}\right) \sin 120°$

$$= -4 \sin 120° = -3.464 \text{ MPa.}$$

$$\sigma_r^2 = \sigma_r^2 + \tau^2$$

$$= 16 + (-3.464)^2 = 27.999$$

$$\sigma_r = 5.29 \text{ MPa}$$

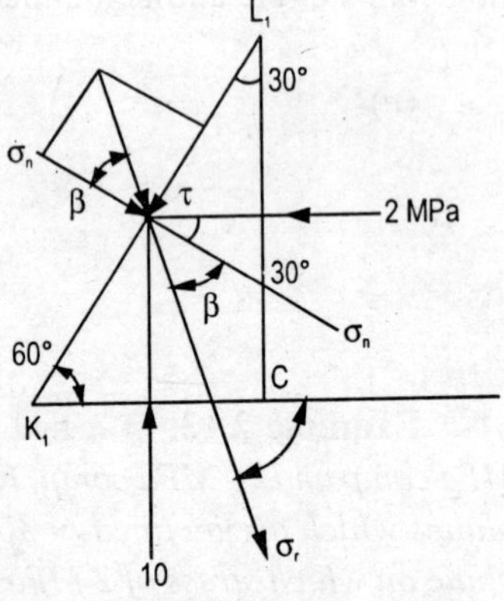

Fig. 2.23

If β is inclination of $\sigma_r$ with normal as shown in fig. 2.23.

$$\tan \beta = \frac{|\tau|}{|\sigma_n|}$$

$$= 40.89°$$

Inclination of $\sigma_r$ vector with plane *DC*

$$= 30 + \beta$$

$$= 30 + 40.89$$

$$= 70.90° \text{ (clockwise)}$$

(*b*) When the plane $K_2L_2$ is inclined at 30° anticlockwise with the plane *BC*, the inclination θ of plane $K_2L_2$ with the *y* plane is

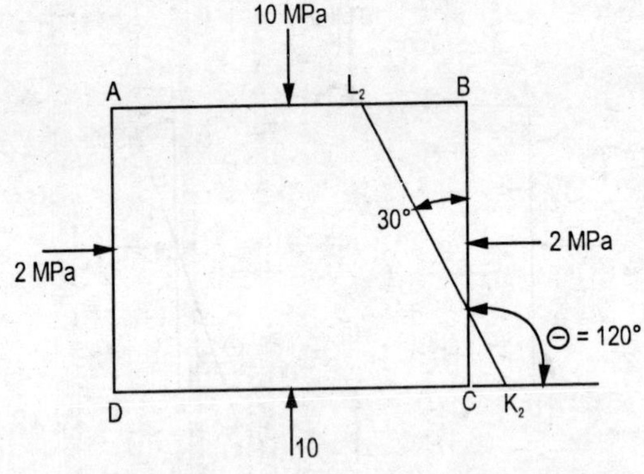

*Fig. 2.24*

$$\theta = 120°$$

Then $\sigma_n = \dfrac{\sigma_x + \sigma_y}{2} - \dfrac{\sigma_x - \sigma_y}{2} \cos 2\theta$

$$= \left(\frac{-2-10}{2}\right) - \left(\frac{-2+10}{2}\right) \cos 240°$$

$$= -6 - 4(-0.5) = -4 \text{ i.e., } 4 \text{ MPa comp.}$$

$$\tau = \left(\frac{\sigma_x - \sigma_y}{2}\right) \sin 2\theta$$

$$= -\left(\frac{-2+10}{2}\right) \sin 240$$

$$= -4(-.866) = 3.464 \text{ MPa.}$$

$$\sigma_r^2 = \sigma_n^2 + \tau^2$$

$$= (-4)^2 + (3.464)^2$$

$$= 27.999$$

$$\sigma_r = 5.29 \text{ MPa}$$

In figure 2.25 $\tan\beta = \dfrac{|\tau|}{|\sigma_n|}$

$$= \frac{3.464}{4} = .866$$

or $\beta = 40.89°$

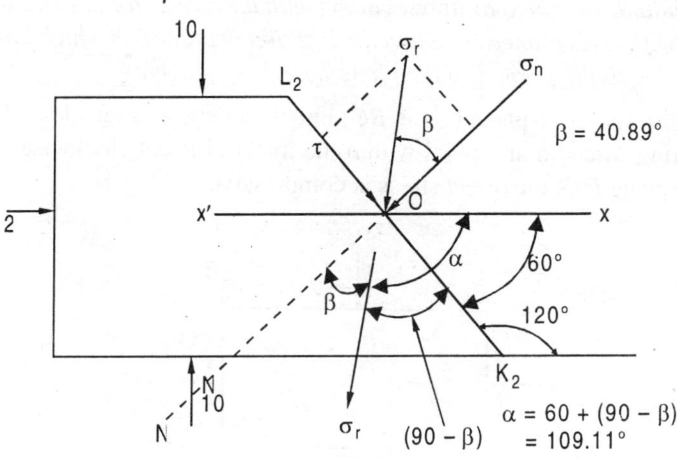


*Fig. 2.25*

Inclination of $\sigma_r$ with plane $DC = \alpha$

$$= 90 + 60 - \beta$$

$$= 150 - 40.89°$$

$$= 109.11° \text{ (clockwise)}$$

(*c*) If $\beta_{max}$ is max value of $\beta$, then from equation 2.2 we have

$$\sin\beta_{max} = \frac{-\sigma_x + \sigma_y}{\sigma_x + \sigma_y}$$

$$= \frac{-2+10}{-2-10} = -\frac{8}{12} = -\frac{2}{3}$$

$$\text{Sin } \beta_{max} = -0.667$$

$$\beta_{max} = 318.2° \text{ or } 221.8°$$

When $\beta_{max} = 318.2°$

Then $2\theta_1 = \frac{\pi}{2} + \beta_{max}$

$$\theta_1 = \frac{\pi}{4} + \frac{318.2}{2} = 45° + 159.1°$$

$$= 204.1° \text{ with } y\text{-plane}$$

And when $\beta_{max} = 221.8°$

$$\theta_2 = \frac{\pi}{4} + \left(\frac{221.8}{2}\right)$$

$$= 45° + 110.9° = 155.9° \text{ with y-plane.}$$

**Example 2.13.** *Figure 2.26 is the point diagram of a two dimensional stress system. Find by analytical method the location of principal, Max shear stress planes and the magnitude of principal stresses along with the direct stresses and shear stresses on Max shear stresses planes. Also find the location of planes on which line resultant is most inclined with the normal to the plane i.e. where $\beta$ is max.*

**Solution.** On the *x*-plane (plane *BC*) line direct stress is tensile and the moment due to shearing force on any point within the material is anticlockwise. Similarly on the *x*-plane (plane *DC*) the direct stress is compressive

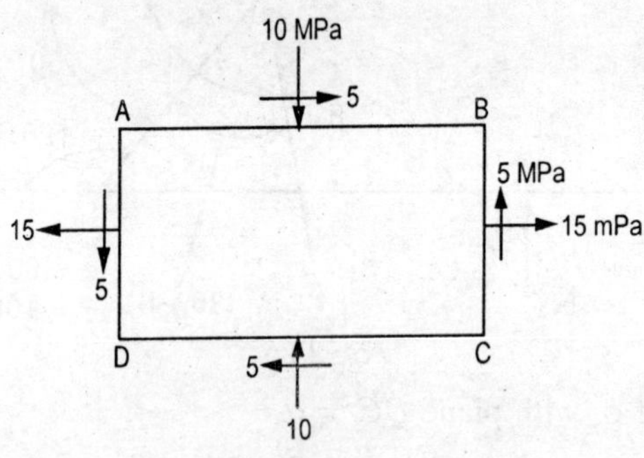


*Fig. 2.26*

Therefore is equation (2.21) through (2.29)

$$\sigma_x = 15 \text{ MPa}$$

$$\sigma_y = -10 \text{ MPa}$$

$$\tau_o = -5 \text{ MPa}$$

The inclination of principal planes with the plane *DC* is given by $\theta_1$ and $\theta_2$ which are solution of equation

$$\tan 2\theta_1 = \frac{-2\tau_0}{(\sigma_x - \sigma_y)}$$

$$= \frac{-2(-5)}{(15+10)} = \frac{10}{25}$$

$$2\theta_1 = 21.8°$$

or $$\theta_1 = 10.9°$$

$$\theta_2 = 90 + 10.9° = 100.9°$$

$$(\sigma_n)_{\theta_1} = \sigma_1 = \left(\frac{\sigma_x + \sigma_y}{2}\right) - \left(\frac{\sigma_x - \sigma_y}{2}\right)\cos(21.8°) - 5\sin(21.8°)$$

$$= 2.5 - 12.5 \times .9285 - 5 \times .3714$$

$$= 2.5 - 11.606 - 1.857$$

$$= -10.963 \text{ MPa (comp)}$$

*i.e.,* $\sigma_1 = 10.963$ MPa (Comp) on principal plane which is inclined to plane *DC* at 10.9° anticlockwise.

For the second principal plane

$$\theta_2 = 100.9°$$

Corresponding direct stress on the plane

$$(\sigma_n)_{\theta_2} = \sigma_2 = \left(\frac{\sigma_x + \sigma_y}{2}\right) - \left(\frac{\sigma_x - \sigma_y}{2}\right)\cos 2\theta_2 + \tau_0 \sin 2\theta_2$$

$$= \left(\frac{15-10}{2}\right) - \left(\frac{15+10}{2}\right)\cos(201.8°) - 5\sin(201.8°)$$

$$= 2.5 - (12.5)(-0.9281) - (0.3714)$$

$$= 2.5 + 11.606 + 1.857$$

$$= 15.963 \text{ MPa (Tensile)}$$

If $\theta_3$ and $\theta_4$ are the inclinations of the max shear stress planes with plane *DC*

$$\theta_3 = \theta_1 + 45° = 10.9745 = 55.9°$$

$$\theta_4 = \theta_2 + 45° = 100.9 + 45 = 145.9°$$

$$(\sigma_n)_{\theta_3} = (\sigma_n)_{\theta_4} = \frac{\sigma_x + \sigma_y}{2} = \frac{\sigma_1 + \sigma_2}{2}$$

$$= \frac{15-10}{2} = 2.5 \text{ MPa}$$

$$\tau_{max} = \pm\sqrt{\left(\frac{\sigma_x - \sigma_y}{2}\right)^2 + \tau^2}$$

$$= \pm\sqrt{\left(\frac{15+10}{2}\right)^2 + (-5)^2}$$

$$= \pm\sqrt{(12.5)^2 + 25}$$

$$= 13.463 \text{ MPa}$$

Alternatively $\tau_{max} = \left|\left(\frac{\sigma_1 - \sigma_2}{2}\right)\right|$

$$= \left|\frac{-10.963 - 5.963}{2}\right|$$

$$= -|13.463| \text{ MPa}$$

However in order to decide on which of the two planes $\theta_3$ and $\theta_4$, the shear stress is positive or negative. The values of $\theta_3$ and $\theta_4$ are to be substituted in equation 2.22

Here $2\theta_3 = 2 \times 55.0° = 111.8$

*i.e.* $(\tau)_{\theta 3} = -\left(\frac{\sigma_x - \sigma_y}{2}\right) \sin 2\theta_3 - \tau_0 \cos 2\theta_3$

$$= -\left(\frac{15+10}{2}\right) \sin (111.8°) + 5 \cos (111.8°)$$

$$= -11.606 - 1.857$$

$$= -13.463 \text{ MPa}$$

$$2\theta_4 = 2 \times 145.9° = 291.8°$$

$$(\tau)_{\theta 4} = -\left(\frac{15+10}{2}\right) \sin 291.8° + 5 \cos (291.8°)$$

$$= 11.606 + 1.857$$

$$= 13.463 \text{ MPa}$$

Thus it is seen that shear stress is positive on plane $\theta_4$ and negative on plane $\theta_3$

The final result is shown in figure 2.27.

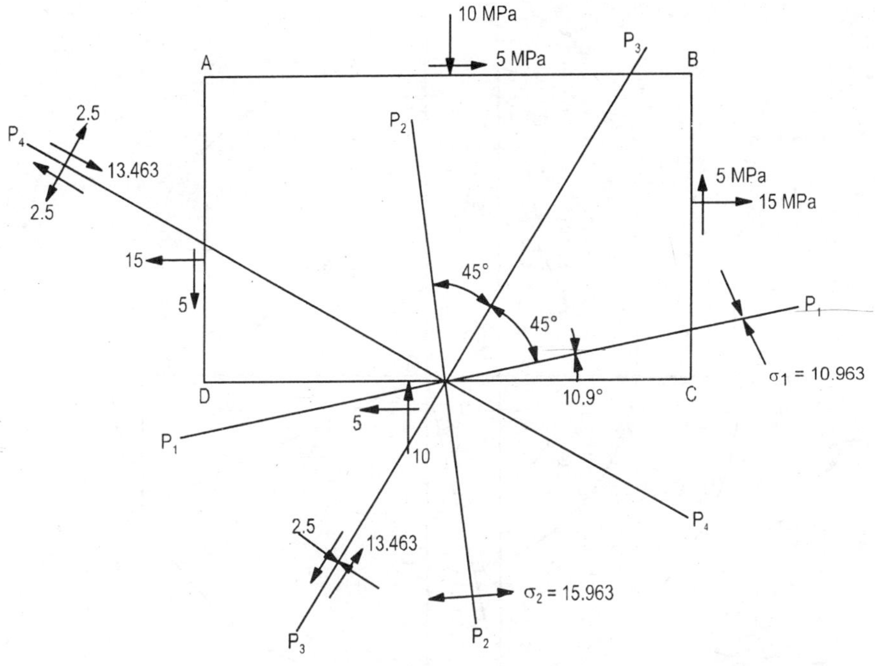


*Fig. 2.27*

**For finding $\beta_{max}$**

$$\sin \beta_{max} = \frac{\sigma_1 - \sigma_2}{\sigma_1 + \sigma_2} = \frac{15 + 10}{15 - 10} > 1$$

Thus the solution fails and the analytical method can not be applied when the principal stresses are of opposite nature.

**Example 2.14.** *Solve the example 2.13 by Mohr's circle method.*

**Solution.** Mark the $\sigma$ and $\tau$ axes with origin as $O$ and select suitable scale. The points presenting the planes $DC$ and $BC$ is $P_{CD}$ and $P_{BC}$ are plotted with co-ordinates (– 10,5) and (15, – 5) respectively. The line joining the points $P_{BC}$ and $P_{CD}$ will intersect the $\sigma$ or $X$ axis at $M$, which will be the centre of Mohr's circle with M as centre and $MP_{DC}$ as radius draw the circle. The points of intersection of the X-axis with Mohr's circle $P_1$ and $P_2$ will correspond to principal planes. The point $P_1$ is located with respect to radial line $M\ P_{CD}$ at 21.8° anticlockwise. Therefore the inclination of the principal plane $P_1$ is $\frac{1}{2} \times 21.8 = 10.9°$ anticlockwise from plane $DC$ and the principal stress $\sigma_1$ is equal to $OP_1$ (comp) which when converted to stress units by multiplying with the scale of the diagram has the value 10.963 MPa. The other principal plane $P_2$ is at right angles to $P_1$ and carries the tensile stress of magnitude $OP_2$ which when converted into stress units by taking into account the scale of the diagram.

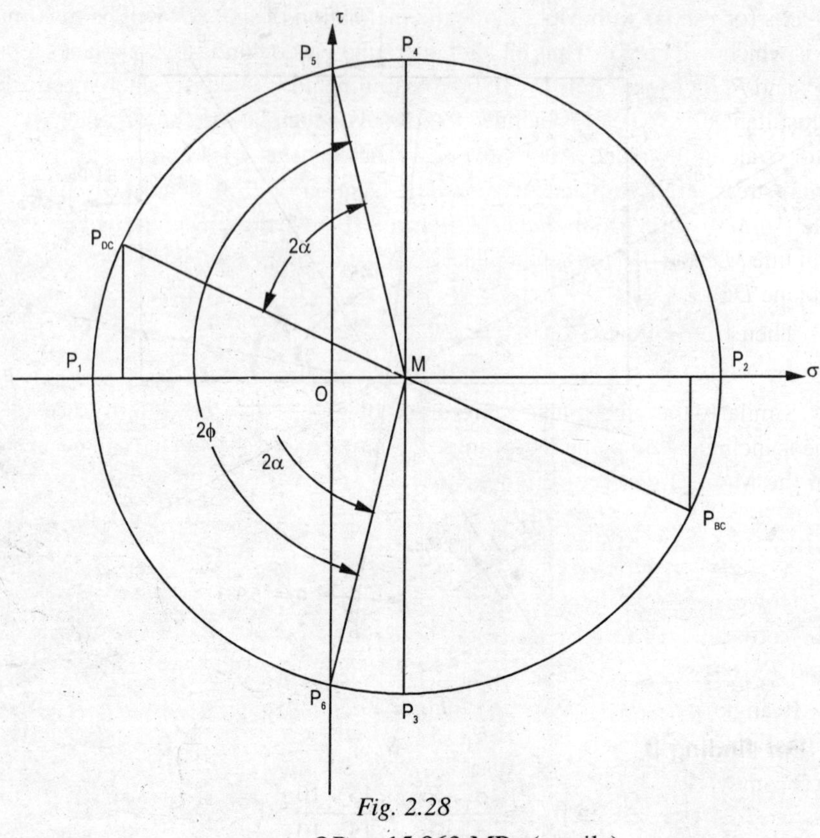


*Fig. 2.28*

we get $\sigma_2 = OP_2 = 15.963$ MPa (tensile)

Then draw a vertical diameter $P_3\ P_4$ of the Mohr's circle passing through $M$. The points $P_3$ and $P_4$ represent the locations on Mohr's circle of the planes which carry maximum shear stress. $MP_3$ line subtends angle of 90° with the radial line $P_1M$, anticlockwise. The max shear stress plane represented by $P_3$ carries τmax (*–ve*) of magnitude $OP_3$ equal to radius of the circle which when multiplied with scale will give the value of $\tau_{max} = -13.462$ MPa

However in the plane represented by $P_4$

$$(\tau)_{max} = 13.463 \text{ MPa}.$$

(*b*) For finding the location of planes having maximum value of β, normally we draw tangents from the origin $O$ to the Mohr circle. The point of tangency then represent the location of such planes which will have $\beta_{max}$. However in this case the origin lies within the circle and apparently the method fails. But on the basis of the procedure to find the value of β for any plane we join the corresponding point $P$ on the Mohr circle with the origin then the inclination of this line (say $OP$) with the σ-axis (or x-axis) ives β. The length $OP$ represents the magnitude of $\sigma_r$ on the plane represented by $P$. Evidently when $P$ moves along the Mohr circle the inclination of line $OP$ with $x$-axis will also change. When $P$ will occupy the positions $P_5$ and $P_6$ which are the points of intersection of $y$-axis (or $\tau$-axis) with Mohr circle the inclination of line $OP$ will be maximum with $x$-axis which will be 90°. Thus on logical ground we conclude that on planes represented by $P_5$ and $P_6$ on Mohr circle $\beta$ will be maximum and $\beta_{max}$ = 90° which means that $\sigma_r$ or the resultant stress vector is inclined to normal on the plane at 90°. In other words the $\sigma_r$ vector is along the plane. Alternatively we can say that $\sigma_r$ is purely shear stress or the normal stress on these planes is zero. The location of $P_5$ is decided by measuring the angle $P_{CD}\ MP_5$. If on Mohr circle $P_5$ is located at angle $2\alpha$ clockwise with respect to the radial line $MP_{CD}$. Then the actual plane $P_5$ in the material is inclined at $\alpha$-clockwise with the plane $DC$.

Then from figure 2.28 $2\alpha = 57°$

or $\alpha = 28.5°$ (clockwise)

Similarly the other plane represented by $P_6$ has also maximum value of $\beta$. This plane is inclined with plane $DC$ at angle 2v anticlockwise. Its value can be as measured from the Mohr circle. From figure 2.28

$$2\gamma = 99°$$

or

$$\gamma = 49.5°$$

However it may be said that the radial line $OP_1$ bisects the angle $P_5\ M\ P_6$. In other words the plane $P_1$ in the point diagram must bisect the angle between the planes $P_5$ and $P_6$.

If angle $P_5$ M $P_6$ = $2\phi$, then these planes are at angle of $\phi$ on either side of principal plane $P_1$

From Mohr circle $< P_5\ MP_6 = 2\phi = .56°$

Then $$\phi = \frac{156}{2} = 78°$$

The plane $P_5$ is at an angle $\phi$ clockwise from plane $P_1$ and $P_6$ is at angle $\phi$ anticlockwise from plane $P_l$ as shown in figure 2.28.

## 2.11 Hydrostatic Stress

Consider a two dimensional stress system where two principal stresses are both equal in magnitude and nature.

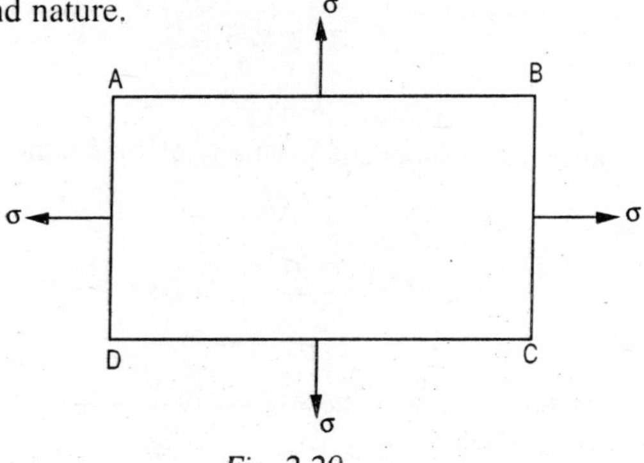


*Fig. 2.29*

The eqn. (1.2) gives the value of shear stress on any plane inclined at an angle θ with plane *DC* given by

$$\tau = -\frac{1}{2}(\sigma_x - \sigma_y)\sin 2\theta$$

Since $\sigma_x = \sigma_y = \sigma$

$$\tau = 0$$

That is there is no shear stress what-so-ever in the material and all the planes will carry the same direct stress as principal stress. In other words there are infinite principal planes and the direct stress in all direction is same. If this problem is extended to three dimensional stress system (chapter 18), where in the *z*-direction also $\sigma_z = \sigma$. Then there will be no shear stress in the material and the direct stress at any point within the material will be same in all directions and equal to σ. This situation is synonymous with hydrostatic pressure at a point in a static fluid. Therefore the principal stress σ is called HYDROSTATIC STRESS.

**Example 2.15.** *Figure 2.30 shows a two dimensional stress system. Find the max shear stress in the material.*

**Solution.** In the Figure 2.30 the principal stresses $\sigma_1$ and $\sigma_2$ are given as

$$\sigma_x = -20 \text{ MPa}$$
$$\sigma_y = -20 \text{ MPa}$$

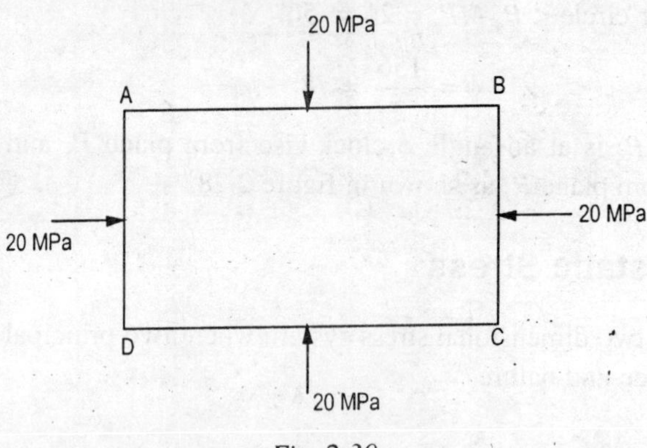


*Fig. 2.30*

Then from equation (2.2) shear stress on any plane *KL* inclined at an angle θ with *y*-plane is

$$\tau = -\left(\frac{\sigma_x - \sigma_y}{2}\right)\sin 2\theta = 0$$

That is there is no shear stress what so ever in the material and all the planes will be principal planes.

**Example 2.16.** *Solve the problem of example 2.15 by Mohr's circle method.*

**Solution.** The coordinates of the points $P_{CD}$ are $P_{BC}$ on $\sigma - \tau$ space will be (– 20, 0) each. When plotted these will overlap each other. Further the centre of Mohr circle is obtained by drawing the right bisector of the line $P_{BC}$ $P_{CD}$ to meet the $x$-axis at the centre of the circle.

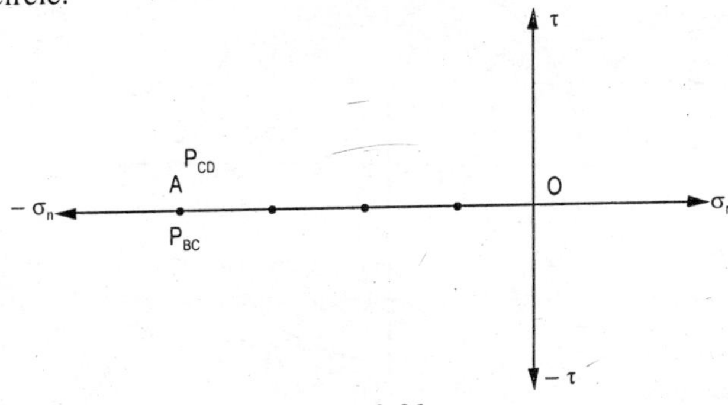


*Fig. 2.31*

However here since the points $P_{BC}$ and $P_{CD}$ overlap each other say at $A$. The Mohr's circle will shrink into a point and that too lying on $x$-axis and the radius of Mohr circle will be zero. Evidently since the $y$-coordinate will be zero for all the points lying on the circle, there would be no "Shear Stress" what so ever in the material. As a matter of fact the material will have no shear stress.

**Example 2.17.** *Figure 2.32 represents a two dimensional stress system. Find the magnitude of the principal stresses and inclination of principal planes with respect to the plane BC. Use Mohr circle method.*

**Solution.** Make the σ and τ axes of reference with $O$ as origin and choose some suitable scale.

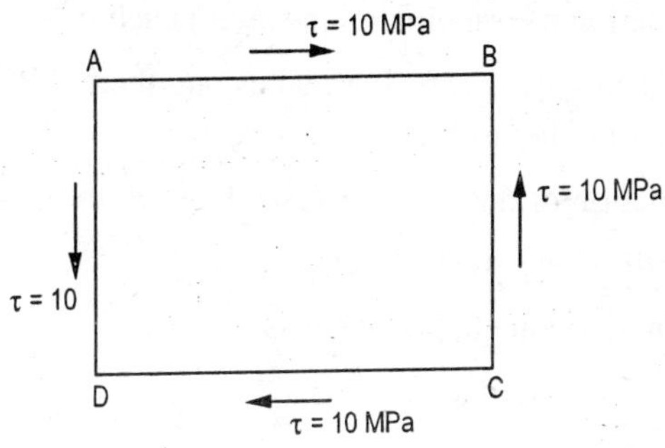


*Fig. 2.32*

The co-ordinates of the points $P_{AB}$ and $P_{BC}$ representing the planes $AB$ and $BC$ are (0, 10) and (0, – 10) respectively with $P_{AB}$ $P_{BC}$ as diameter and O as centre draw the circle. Evidently origin $O$ will be the centre of circle. The principal planes are represented on Mohr circle by $P_1$ and $P_2$, The points of intersection of the circle with X-axis. In the circle diagram the point $P_1$ subtends an angle of 90° anticlockwise with

$OP_{BC}$. Thus in actual material or in point diagram first principal plane $P_1$ will be at 90/ 2 = 45° anticlockwise from plane $BC$.

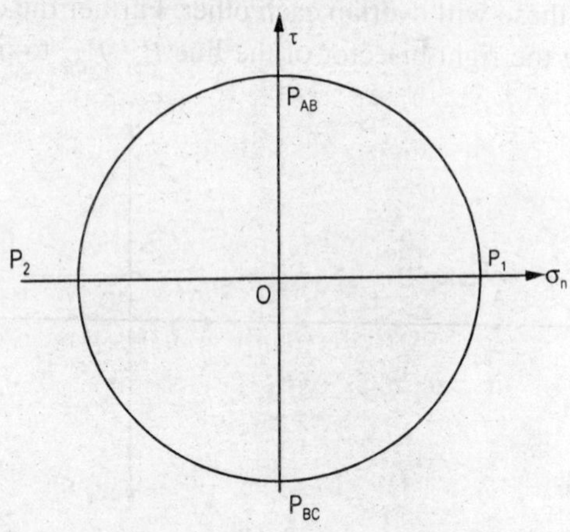


*Fig. 2.33*

Similarly the second principal plane $P_2$ will be at 45° clockwise from the $BC$.

The principal stress $\sigma_1$ on plane $P_1 = OP_1 = 10$ MPa (tensile)

and on Plane $P_2 = OP_2 = 10$ MPa *i.e.*, (Comp).

## REVIEW QUESTIONS

2.1 Define principal planes. How many such planes are there in 2-$D$ stress system?

2.2 Do the principal planes carry maximum and minimum direct stress ?

2.3 What is the inclination of principal planes with each other ?

2.4 How many max shear-stress carrying planes one there ? What is the inclination of these planes among themselves ?

2.5 What is the inclination of max shear stress planes with principal planes ?

2.6 On which planes the shear stress is zero?

2.7 What is the minimum value of shear stress ?

2.8 Define obliquity of stress on a plane.

2.9 Define complimentary shear stresses.

2.10 State whether following statement true or false. The direct stress on two planes having complimentary shear stresses is average of the principal stresses."

2.11 Define Mohrs stress circle.

2.12 What do the co-ordinates of a point lying an Mohrs circle represent ?

2.13 State whether the following are true or false.

(*i*) The centre of Mohrs circle lies on the x-axis.

(*ii*) Diameter of circle is equal to the value of maximum shear stress, as per the scale of the circle diagram. (False)

(*iii*) The angle subtended by two points on the circumference of Mohrs circle at its centre, is twice the actual angle between the planes, represented by these points, in the point diagram. (True)

(*iv*) The locations of the points representing the principal planes are the points of intersection of Mohrs circle with X-axis. (True)

(*v*) If the principal stress are of same nature and different values, the location of the planes having maximum obliquity is determined by drawing two tangents to the circle from origin. (True)

2.14 In case when the two principal stresses are of same magnitude and of the same nature, what will be the radius of Mohr circle ? Then how many principal planes are there ? (Zero, Infinite)

2.15 In the case when the principal stresses are of opposite nature, then explain how to locate the points on the Mohr circle which correspond to plane, having maximum obliquity. (Intersection of circle with Y-axis)

2.16 The average of the direct stresses on two plane, represented by the points of interaction of any diameter of the Mohr circle, is constant and is equal to the average of principal stresses, state whether false of true. (True)

2.17 Distance of any point on Mohrs circle from the origin is equal to the megnitude of the resultant stress on the plane represented by the point state whether true or false. (True)

2.18. In a two dimensional stress system on two planes *AB* and *BC* direct stress 100 MPa (tensile) and 50 MPa tensile are acting as shown in fig. 2.34*a* and these are accomparied by 25 MPa shear stress. Find analytically as well as by Mohrs method the direct and shear stresses on a plane *KL* which is unclined at 40° anticlockwise with plane *CD*.

**Solution.** In the question it is not mentioned whether the shear stress on the plane *BC* is positive or negative. Therefore depending upon the point diagram the solution will depend we will take both the possibilities.

**Case I : When $\tau_0$ shear stress on plane *BC* is + ve**

Here $\sigma_x = 50$ MPa

$\sigma_y = 100$ MPa

$\tau_0 = 25$ MPa

$\theta = 40°$

then from equation (2.21) and (2.22) we have

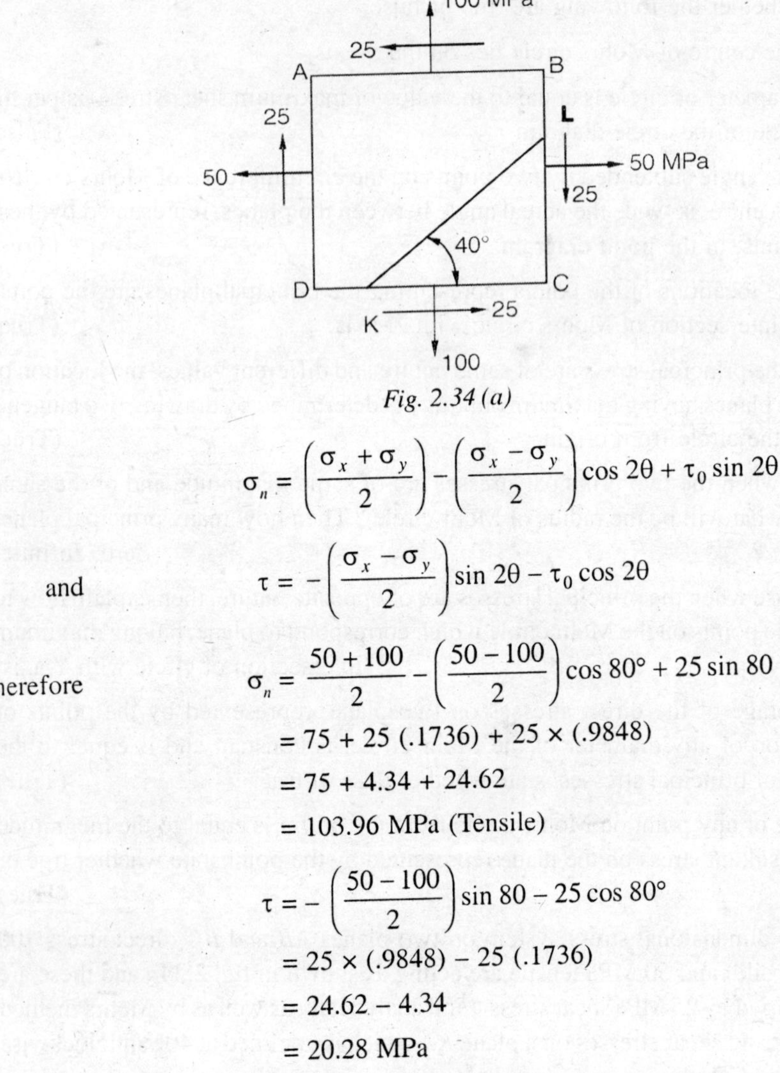


*Fig. 2.34 (a)*

$$\sigma_n = \left(\frac{\sigma_x + \sigma_y}{2}\right) - \left(\frac{\sigma_x - \sigma_y}{2}\right)\cos 2\theta + \tau_0 \sin 2\theta$$

and
$$\tau = -\left(\frac{\sigma_x - \sigma_y}{2}\right)\sin 2\theta - \tau_0 \cos 2\theta$$

therefore
$$\sigma_n = \frac{50 + 100}{2} - \left(\frac{50 - 100}{2}\right)\cos 80° + 25 \sin 80$$

$$= 75 + 25\ (.1736) + 25 \times (.9848)$$

$$= 75 + 4.34 + 24.62$$

$$= 103.96 \text{ MPa (Tensile)}$$

$$\tau = -\left(\frac{50 - 100}{2}\right)\sin 80 - 25 \cos 80°$$

$$= 25 \times (.9848) - 25\ (.1736)$$

$$= 24.62 - 4.34$$

$$= 20.28 \text{ MPa}$$

Altei native solution by Mohr circle method

The direct stress on plane *BC* being tensile is + ve and the shear stress in the plane forms clockwise moment on any point within the material. Therefore it is also positive. Thus the co-ordinates of the point $P_{BC}$ corresponding to plane *BC* will be (50, 25). Similarly the co-ordinate of the point $P_{CD}$ corresponding to plane *DC* will be (100, 25).

After marking the axes of reference is $\sigma_n$ along *X*-axes and $\tau$ along *Y*-axis, the points $P_{BC}$ and $P_{CD}$ are plotted by taking some scale join the points $P_{BC}$, and $P_{CD}$ and draw the right bisector of the line $P_{BC}\,P_{CD}$ to meet the *X*-axis in point *M*. Draw exile with radius $MP_{CD}$. and centre M.

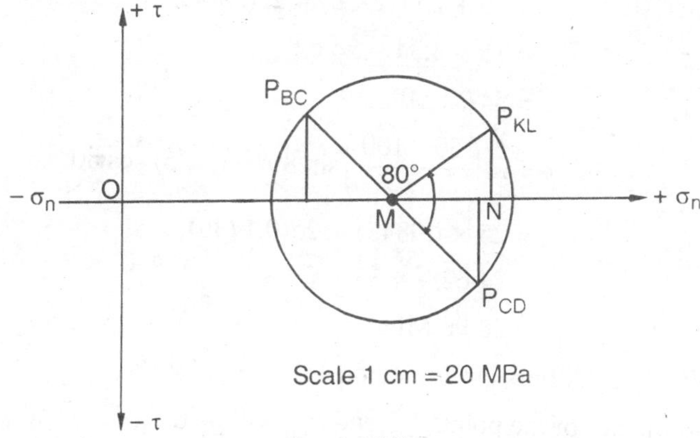


*Fig. 2.34 (b)*

In order to locate the point $P_{KL}$ draw line $MP_{VC}$ at (2 × 40 = 80°) anticlockwise with $MP_{CD}$.

Draw a normal $P_{KL}$ $N$ on $X$-axis. Measure $ON$ and $NP_{KL}$ to read the values of direct and shear stresses on plane $KL$. These values are

$$\sigma_n = NP_{KL} = 103.96 \text{ PKa}$$

$$\tau = +\,20.28 \text{ KPa}$$

**Case II : In case** $\tau_0$ is taken negative on plane $BC$, then the point diagram will be as shown in figure 2.34$C$

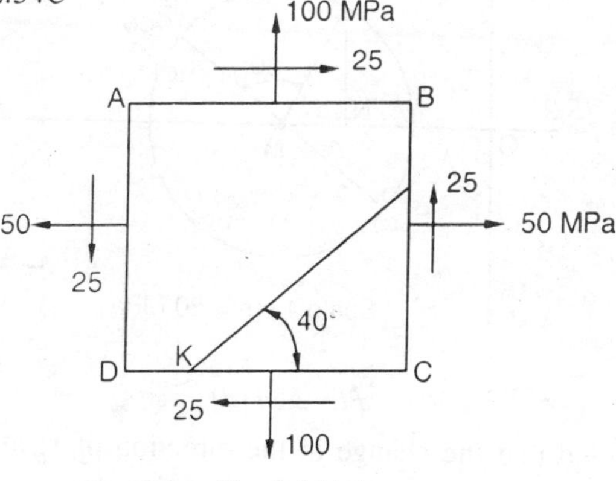


*Fig. 2.34 (c)*

In this case

$$\sigma_x = 50 \text{ MPa}$$

$$\sigma_y = 100 \text{ MPa}$$

$$\tau_u = -\,25 \text{ MPa}$$

$$\theta = 40°$$

then
$$\sigma_n = \left(\frac{50+100}{2}\right) - \left(\frac{50-100}{2}\right)\cos 80° + (-\,25)\sin 80°$$

$$= 75 + 25\,(.1736) - 25(.9848)$$

$$= 75 + 4.34 - 24.62$$

$$= 54.72 \text{ MPa}$$

and
$$\tau = -\left(\frac{50-100}{2}\right)\sin 80° - (-25)\cos 80°$$

$$= 25 \times (.9848) + 25(0.1736)$$

$$= 24.62 + 4.34$$

$$= 28.96 \text{ MPa}$$

Alternatively by Mohrs circlr method

The co-ordinates of the points $P_{BC}$ and $P_{CD}$ will be (50, – 25) and (100, 25) mark the $\sigma_n$ and $\tau$ axes to some scale.

Plot the points $P_{BC}$ and $P_{CD}$. Join them and draw its right bisector to meet the X-axis at $M$. With $M$ as centre draw circle of radius equal to $MP_{CD}$. Then draw the line $MP_{KL}$ at 80° anticlockwise with $P_{CD}M$, the co-rodinates of $P_{KL}$ give the direct and shear stresses on plane $KL$ as

$$\sigma_n = ON = 54.72 \text{ MPa}$$

$$\tau = P_{KL}N = 28.96 \text{ MPa}$$

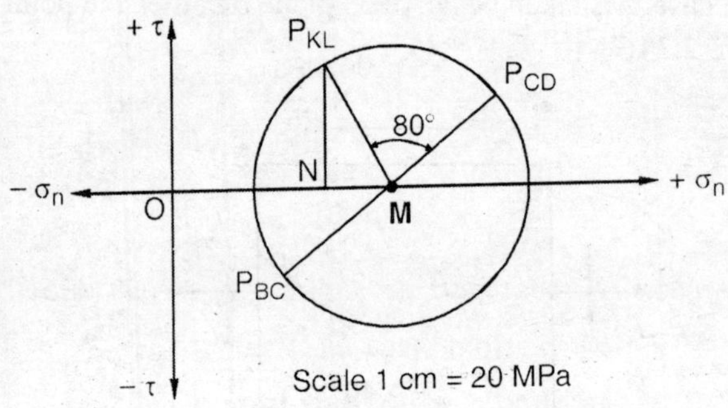


*Fig. 2.34 (d)*

It may be noted that the change in the direction of $\tau_0$ alters the value of the stresses considerably. It is once again emphasized that the answer depends upon the point diagram.

**2.19.** In a stressed material the stresses are as shown in point diagram (Fig. 2.35 a) find the maximum and minimum direct stresses and max shear stress alongwith the location of these planes with the plane $CD$.

**Solution.** Here on plane $BC$, $\tau_0$ is – ve

$$\sigma_x = 200 \text{ MPa}$$

$$\sigma_y = 100 \text{ MPa}$$

$$\tau_0 = -50 \text{ MPa}$$

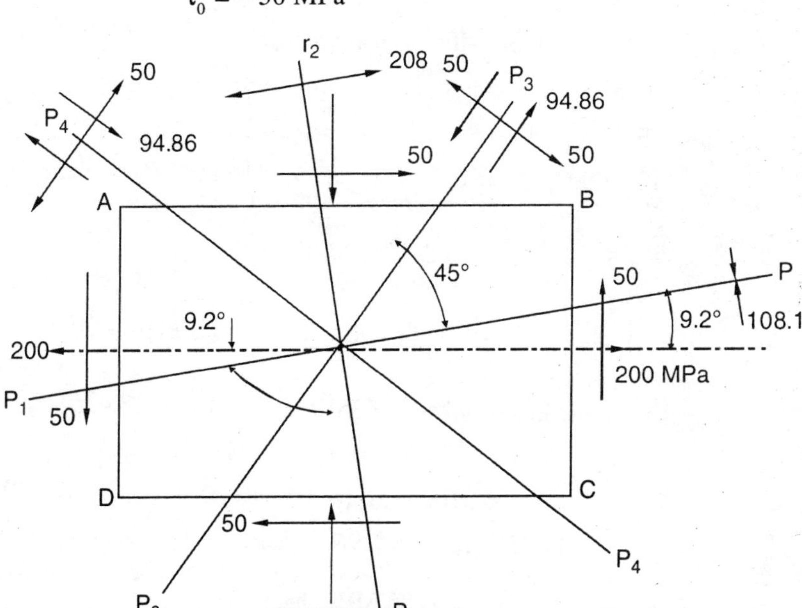


*Fig. 2.35 (a)*

The location of principal planes is given by equation (2.23)

$$\tan(2\theta) = -\frac{2\tau_0}{(\sigma_x - \sigma_y)} = -\frac{2(-50)}{(200+100)}$$

$$= \frac{100}{300} = \frac{1}{3} = \tan(18.435)$$

or $\quad 2\theta_1 = 18.435$

or $\quad \theta_1 = 9.2175°$ from plane $DC$

and $\quad \theta_2 = 90 + 9.2175 = 99.2175°$ from plane $DC$

then
$$(\sigma_n)_{\theta_1} = \sigma_1 = \frac{\sigma_x + \sigma_y}{2} - \left(\frac{\sigma_x - \sigma_y}{2}\right)\cos 2\theta_1 + \tau_0 \sin 2\theta$$

$$= \frac{200-100}{2} - \frac{(200+100)}{2}\cos 18.435° - 50 \sin 18.435°$$

$$= 50 - 150\,(0.94868) - 50\,(0.3162)$$

$$= 50 - 142.302 - 15.8114$$

$$= -108.1134 \text{ or } 108.1134 \text{ MPa (comp)}$$

$$(\sigma_n)_{\theta_2} = \sigma_2 = \frac{\sigma_x + \sigma_y}{2} - \left(\frac{\sigma_x - \sigma_y}{2}\right)\cos(2 \times 99.2155°)$$

$$+ \tau_0 \sin(2 \times 99.2155°)$$

$$= 50 + 142.302 + 15.8114$$

$$= 208.1134 \text{ MPa (Tensile)}$$

If $\theta_3$ and $\theta_4$ are max shear stress plane inclinations

then $\theta_3 = \theta_1 + 45° = 9.175 + 45 = 54.2155°$

and $\theta_4 = 90 + \theta_3 = 90 + 54.21555 = 144.2155°$ from plane $DC$

and
$$(\tau_{max})_{\theta_3} = -\left(\frac{\sigma_x - \sigma_y}{2}\right)\sin 2\theta_3 - \tau_0 \cos 2\theta_3$$

$$= -\left(\frac{200+100}{2}\right)\sin(90+2\theta_1) + 50\cos(90+2\theta_1)$$

$$= -150 \sin(90 + 18.435) + 50 \cos(90 + 18.435)$$

$$= -150 \sin 108.435° + 50 \cos 708.435°$$

$$= -150 \times 0.3162 + 50(-0.9468)$$

$$= -47.43 - 47.43 = -94.86 \text{ MPa}$$

$$(\tau_{max})_{\theta_4} = -(\tau_{max})_{\theta_3} = 94.86 \text{ MPa}$$

$$(\sigma_n)_{\theta_3} = (\sigma_n)_{\theta_3} = \frac{1}{2}(\sigma_1 + \sigma_2)$$

$$= \frac{1}{2}(-108.1134 + 208.1134)$$

$$= \frac{1}{2}(100) = 50 \text{ MPa}$$

Also
$$(\sigma_n)_{\theta_3} = (\sigma_n)_{\theta_4} = \frac{1}{2}(\sigma_1 + \sigma_2) = \frac{1}{2}(\sigma_x + \sigma_y)$$

$$= \frac{1}{2}(\sigma_x + \sigma_y)$$

$$= \frac{1}{2}(200 - 100)$$

$$= 50 \text{ MPa}$$

The planes $\theta_1$, $\theta_2$, $\theta_3$ and $\theta_4$ along with the stresses on them are shown in the point diagram.

Alternatively the problem can be solved by Mohr's method as follows :

Mark the axes and plot the point $P_{CD}$ and $P_{BC}$ on some scale.

The co-ordinates of the points $P_{BC}$ and $P_{CD}$ are (200 – 50) and (– 100 + 50) respectively

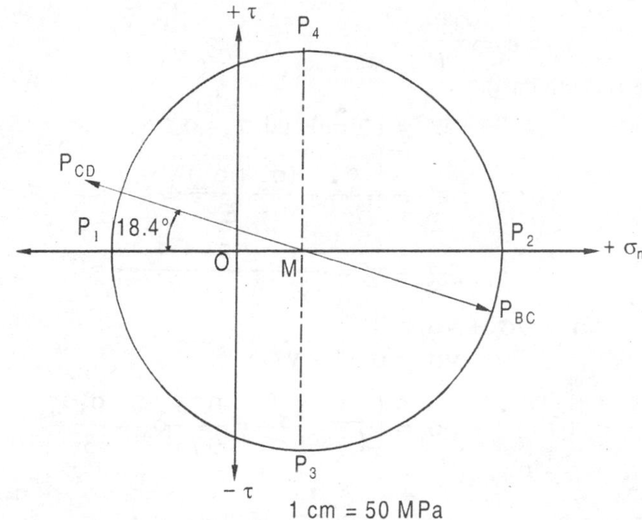


*Fig. 2.35 (b)*

Draw the right bisector of the line $P_{BC}\,P_{CD}$ to intersect the x-axis at $M$ which is the centre of the circle. With M as centre and $MP_{CD}$ as radius draw the circle, which intersects x-axis at $P_1$ & $P_2$

then $(\sigma_n)_1$ = principal stress on plane $P_1 = OP_1 = 108.1134\ MPa$ (comp)

$(\sigma_n)_2$ = Pcincipal stress on plane $P_2$

$= OP_2 = 208.1134\ MPa$ (tensile)

$$2\theta_1 = 18.435 \quad \therefore\ \theta_1 = 9.2135$$

$$\theta_2 = 90 + \theta_1 = 99.2135$$

on $P_3$ $\quad \tau_3 = -MP_3 = -94.86\ MPa$

and $\quad \theta_3 = \theta_1 + 45° = 9.2175 + 45$

$$= 54.2175°$$

$$\tau_4 = MP_4 = 94.86\ MPa$$

$$(\sigma_n)_3 = (\sigma_n)_4 = OM = 50\ MPa$$

**2.20.** A cylindrical test piece is compressed in the axial direction while the lateral expansion is completely checked by a rigid casing all along the length 8 cm of test piece. If Poisson's ratio is 0.25 find the ratio of the axial strain to that in the cylinder with unrestricted lateral expansion.

**Solution.** In the first case : When the lateral exparsion is free

Let axial compression stress be $\sigma_y$

then $\quad \varepsilon_y = \dfrac{-\sigma_y}{E}$

and $\qquad \varepsilon_x = \varepsilon_z = \dfrac{\nu\sigma_y}{E}$

In case II, with casing

Given $\qquad \varepsilon'_x = \varepsilon'_z = 0$ and $\sigma_x = \sigma_z$

But $\qquad \varepsilon'_y = \dfrac{-\sigma_y}{E} + \dfrac{(\sigma_z + \sigma_x)\,\nu}{E}$

and $\qquad \varepsilon'_x = \varepsilon'_z = \dfrac{-\sigma_x}{E} + \dfrac{(\sigma_y + \sigma_y)\nu}{E} = 0$

or $\qquad -\sigma_x + \nu\sigma_y + \nu\sigma_x = 0$

$$\nu\sigma_y = \sigma_x (1 - \nu)$$

*i.e.,* $\qquad \sigma_x = \left(\dfrac{\nu}{1-\nu}\right)\sigma_y = \dfrac{0.25}{0.75}\sigma_y = \dfrac{\sigma_y}{3}$

Therefore when $\qquad \sigma_x = \sigma_z = \dfrac{\sigma_y}{3}$, then

$$\varepsilon'_y = -\frac{\sigma_y}{E} + \frac{\nu}{E}(\sigma_z + \sigma_x)$$

$$= -\frac{\sigma_y}{E} + \frac{\nu}{E}(2\sigma_x)$$

$$= -\frac{6y}{E} + \frac{2\nu}{\in}\left(\frac{\sigma_y}{3}\right)$$

$$= -\frac{\sigma_y}{E}\left(1 - \frac{2}{3}\right) = -\frac{1}{3}\frac{\sigma_y}{E}$$

$\therefore \qquad \dfrac{\varepsilon'_y}{\varepsilon'_y} = \dfrac{-\dfrac{1}{3}\left(\dfrac{\sigma_y}{E}\right)}{-(\sigma_y / E)} = \dfrac{1}{3}$

## PROBLEMS FOR PRACTICE

1. The principal stresses at a point in a material are 75 MPa (tensile) and 45 MPa (tensile). Find for a plane inclined at 40° to that of former stress the magnitude of the normal and tangential stress components. What is the obliquity of the resultant stress on this plane ? Show its direction by means of a sketch.

[**Ans.** 62.7 MPa, 14.8 MPa, 13.5°]

2. (*a*) At a point in an elastic material direct stresses of 60 MN/m$^2$ compression and 80 MPa tension are applied on planes at right angles to each other. If the greater principal stress in the material is limited to 100 MPa tension, calculate

the maximum shear stress that could be applied to the material on the given planes. What will then be the maximum shearing stress induced in the material at the point ?

(*b*) Solve the problem with the help of Mohr's diagram method also.

[**Ans.** 56.6 MPa, 90 MPa]

[**Hint :** Refer to paragraph (iv) of summary in § 2.5]

3. The minor principal stress at a point in the cross-section of a beam is 20 MN/m$^2$ compressive and the maximum shear stress is 80 MN/m$^2$. Draw the Mohr's circle for this point. From the diagram find :

(*i*) The direct stress on the planes having maximum shear stress.

(*ii*) The direct and shear stresses on the plane making an angle of 60° clockwise with the plane of minor principal stress.

(*iii*) The magnitude of the major principal stress.

(*iv*) The location of planes where the resultant stress is most inclined to the normal. [**Ans.** 60 MN/m$^2$ ; 100 MN/m$^2$ ; 69.3 MN/m$^2$ ; 140 MN/m$^2$ comp. ; 20½° on either side of the plane of minor principal stress]

4. Passing through a point in a material there are two planes *A* and *B*. *B* is inclined to *A* at 45° clockwise. The direct and shear stresses on the plane *A* are 60 MPa tensile and 40 MPa respectively. On the plane *B* there is a tensile stress of 120 MPa and an undetermined shearing stress. Find :

(*a*) The values of the principal stresses at the point.

(*b*) The value of the maximum shearing stress.

(*c*) The positions of principal planes relative to *A*. Also find the principal strains.

Take $E$ = 200 GPa and $\nu$ = 0.3

Show all the planes with the help of a diagram.

[**Ans.** 267.6 MPa tensile, 52.4 MPa tensile, 107.9 MPa, 79° – 6′ anticlockwise, 10° – 54′ clockwise from *A*]

5. Solve the problem No. 4 if the plane *B* were inclined to *A* at 45° anticlockwise, analycally as well as graphically.

[**Ans.** 125.7 MN/m$^2$ tensile, 34.3 MN/m$^2$ tensile, 45.8 MN/m$^2$, 58½° anticlockwise and 31½ clockwise from *A*]

[**Hint :** For graphical solution plot the plane *A* (60, – 40) join this point with *C* (60, 40). Draw a line *CB* making $\angle ACB$ = 45° to intersect line $\sigma$ = 120 in *B*.]

6. A circle of 100 mm diameter is drawn on a steel plate. It is then stressed along two perpendicular directions by stresses $\sigma_x$ and $\sigma_y$. A shear stress + $\tau_0$ also acts on

the plane on which $\sigma_x$ acts. If the magnitudes of $\sigma_x$, $\sigma_y$ and $\tau_0$ are 600 MN/m$^2$ tensile, 120 MN/m$^2$ tensile and 180 MN/m$^2$ respectively and the circle deforms into an ellipse, calculate the major and minor axes of the ellipse and also find their directions. Take $\nu = 0.3$ and $E = 200$ GN/m$^2$.

[**Ans.** Major axis = 100.2265 mm and minor axis = 99.8365 mm. along 71°54′ anticlockwise and 18° 6′ clockwise with the direction of $\sigma_x$ respectively].

7. A cylindrical test piece is compressed in its axial direction while the lateral expansion is checked by a casing extending all over its length to one quarter of the amount it would otherwise be. Find the ratio of the axial strain to that in a cylinder with unrestricted lateral expansion. Take Poisson's ratio = $\nu$.

$$\left[\textbf{Ans. } 1 - \frac{3\nu}{2(1-\nu)}\right]$$

8. A rectangular element is under the action of a compressive stress of 100 MPa along the $x$-axis. Find the stress to be applied along the $y$ and $z$ axes to prevent any change in the dimensions of the cross-section normal to $x$-axis. Also determine the ratio in which the strain along $x$-axis has been reduced. Assume $\sigma_y = \sigma_z$ and $\nu = \frac{1}{3}$.

$$\left[\textbf{Ans. } 50 \text{ MPa Comp.}; \frac{2}{3}\right]$$

9. A steel bar is subjected, in its longitudinal direction to a direct stress. With the help of lateral stresses its strains in the two directions at right angles to its length are reduced to on-half and one-third respectively of those of the free expansion. If $E = 200$ GN/m$^2$ and $\nu = 0.25$ find the value of the modified modulus of elasticity. [**Ans.** 221.3 GN/m$^2$]

10. In a two dimensional stress system, the direct strains along $x$ and $y$ directions are $7 \times 10^{-6}$ tensile and $6 \times 10^{-6}$ compressive. The faces normal to these axes also carry shear strains of $3 \times 10^{-6}$. Find graphically or otherwise the principal strains.

If E = 200 GN/m$^2$ and v = 0.25, calculate the principal stresses.

[**Ans.** $7.09 \times 10^{-6}$ tensile, $6.09 \times 10^{-6}$ compressive, 1.19 MN/m$^2$ tensile, 0.494 MN/m$^2$ compressive]

11. In a structural member the direction of the principal stresses are known. With an idea to determine these stresses two strain gauges are fixed at 30° and 120° to the known direction of one of the principal stresses. The strains observed in these directions are $+ 455 \times 10^{-6}$ and $- 32 \times 10^{-6}$ respectively.

Find the magnitude of the principal stresses. If $E = 200$ GN/m$^2$ and $\nu = 0.3$.

[**Ans.** 135 MN/m$^2$ tensile, 14.45 MN/m$^2$ comp.]

12. At a point in a stressed material the strains along three directions $OA$, $OB$ and $OC$ were measured with the help of a strain rosette as $\varepsilon_A = + 400 \times 10^{-6}$, $\varepsilon_B = + 150$

$\times 10^{-6}$ and $\varepsilon_C = -40 \times 10^{-6}$, $\angle BOA = 30°$ ; $\angle COA = 75°$ and lne $OB$ lies in between $OA$ and $OC$.

Find the principal strains in magnitude and directions. What will be the strains along the line $OD$ inclined at 45° to $OA$ ?

[**Ans.** $470 \times 10^{-6}$ at 44° clockwise with $OA$, $-47 \times 10^{-6}$ at 46° anticlockwise with $OA$, $32 \times 10^{-6}$]

13. The readings on the three strain gauges of a rosette are : $\varepsilon_0° = +592 \times 10^{-6}$ ; $\varepsilon_{45}° = 308 \times 10^{-6}$ and $\varepsilon_{90}° = 432 \times 10^{-6}$, angles being measured anticlockwise from the axis of the first gauge. Find the principal strains and their directions relative to the axis of the first gauge. Assuming $E = 200$ GN/m² and $\nu = \frac{1}{2}$, determine the principal stresses. [**Ans.** $+640 \times 10^{-6}$, $-480 \times 10^{-6}$, 168°, 78°, + 109 MN/m², – 63.3 MN/m²]

## REFERENCES


1. Y.C. Young 1968, Foundation of solid mechanics, Prentice Hall p. 353.
2. A.F.H. Love (1944), A Treatise on the Mathematical Theory of Elasticity, Dover, 4th edition.
3. R. Hill Theory of Plasticity.
4. Y. Li (1976), Phys. Solids, 38 pp 171.
5. J. Gibson, M.F. Ashby, G.s. Schajer and C.J. Robertson (1982), Proc. Royal Society, London, A 382, 25.

Chapter–3

# Shearing Force and Bending Moment in Beams

## 3.1 Beams and their Classification

Any member of a machine or structure under the action of lateral forces in the axial plane is called a beam. The effect of these forces will be to cause bending of the member. Depending upon the end conditions there are different types of beams. A beam with both the ends rigidly fixed is known as a fixed or encastre beam. If one of the ends is rigidly fixed and the other is absolutely free it is known as a cantilever beam. If both the supports at the ends are such that the beam is capable of taking any angular position at the end and not the lateral movement, the beam is called to be a freely or simply supported beam ; in actual practice this requirement is met by providing a well lubricated hinge at one end and a support on rollers at the other.

If such a beam has ends extending beyond the supports, it is called an overhanging beam. A beam which is supported at more than two supports is called a continuous beam.

As beams are invariably horizontal, and the external forces are weights, the beams will be taken horizontal and the external forces as vertical, although the same conclusions will hold good in other cases. Members of structures and machines transmit axial loads in addition to lateral forces acting upon them. However in the present chapter only transverse forces will be considered.

A concentrated or point load is one which is applied to the beam through a knife edge. A uniformly distributed load is that which is spread evenly over a certain length of the beam so that the load carried by a portion of the beam is proportional to the length of that portion. A distributed load will be called a varying load when the rate of loading varies from one point to another.

### 3.1.1 Moment of a Force, Plane of Moment and Axis of Moment

Moment of a force about a point is the product of the moment and the perpendicular distance of the point from the force vector. Plane containing the force vector and the point about which moment is taken is called in PLANE OF MOMENT. And the axis passing through the point and normal the plane of moment is called the axis of the moment. Since force is a vector quantity the moment will also be a vector quantity as the distance is a scalar quantity. This force vector is represented by an arrowhead the length of which to some chosen scale represents the magnitude of the force and the arrow head indicates the direction of the force. Similarly to represent a moment vector with an arrow head right handed screw rule is followed. The length of the arrowhead will represent the magnitude of the moment to some scale. However before we describe the method to decide the placement of the arrow and the direction to which it is to point we have to see whether the moment about the given point is clockwise moment a anticlockwise moment depending upon the location of the point with respect to the direction of the arrow head representing the force vector. If the point lies on the left side of force vector arrow head as seen along the arrow head then the moment will be reckoned as anticlockwise. Similarly if the point in question lies on the right hand side of the force vector then it will be called the clockwise moment because this moment will tend to rotate the minutes arm of the clock, placed at the point where the moment is taken, in the clockwise direction or vice versa. After having decided whether the moment is clockwise or anticlockwise, we proceed to apply the right hand screw rule. The right handed screw rule states that with respect to a fixed nut placed in the plane of the moment, the direction of axial movement of the right handed screw when acted upon by the given moment, will be the direction of the arrow head representing the moment vector. Evidently the axial movement of the screw with respect to fixed nut which is placed in the plane of the moment, will be normal to in plane of the moment. Thus if the plane of the moment is vertical and is visible in elevation, the normal will be visible in the plan view and will be normal to plane of the moment. It may be noted that the arrow head will be normal to the plane of the moment and it will be parallel to the axis of the moment.

## 3.1.2 Classification of Moment : Bending Moment and Twisting Moment

The next question arises as to whether the moment acting on any section of a beam or shaft is bending moment or twisting moment. When the plane of the moment is normal to the cross sectional plane of the beam or shaft on which the moment is acting, the moment is called Bending moment as it will tend to bend the beam or shaft. Alternately moment will be a Bending Moment if the axis of the moment is parallel to or lies in the cross sectional plane of the beam.

Similarly if the plane of the moment is parallel to or contains the cross sectional plane of the beam or shaft it will be called twisting moment as it tends to twist the shaft. Alternately the moment will be a twisting moment if the axis of the moment is normal to the cross-sectional plane of the beam or shaft or if the axis of the moment coincides with the longitudinal axis if the shaft.

## 3.2 Shearing Force and Bending Moment

Consider a simply supported beam $AB$ under the action of vertical forces $P_1$, $P_2$, $P_3$ and $P_4$ with reactions $R_A$ and $R_B$ shown in the Fig. 3.1.

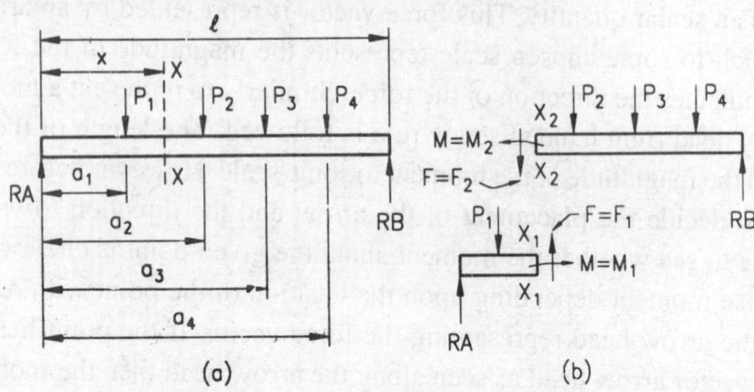


*Fig. 3.1*

Since the beam is in equilibrium the algebraic sum of all the vertical forces acting on the beam will be zero

*i.e.* $$R_A + R_B = P_1 + P_2 + P_3 + P_4 \quad ...(3.1)$$

Again taking a section along $XX$ at a distance $x$ from $A$ and considering the equilibrium of the two portions separately, if $F_1$ and $M_1$ are the force and resisting moment along the section $X_1 X_1$ offered by element No.2 and $F_2$ and $M_2$ are the force and moment offered by element No. 1 on to element No.2 along the section $X_2X_2$, then $F_1 = F_2$ (= $F$ say) and also

$$F_1 = F = - R_A + P_1 \quad ...(3.2)$$

$$F_2 = F = P_2 + P_3 + P_4 - R_B \quad ...(3.3)$$

If $R_A < P_1$ then from equation (3.1) $R_B < (P_2 + P_3 + P_4)$ and the values of F from equations (3.2) and (3.3) will be positive. In other words the forces $F_1$ (= $F$) and $F_2$ (=$F$) on the cross–section $XX$ will be as shown in the free body diagrams of the two elements (Fig. 3.1b), that is, the section $X_1X_1$ of element No.1 will have a tendency to move downwards (in the direction of $F_2$) and section $X_2$ $X_2$ of the element No. 2 will tend to slide upwards with respect to $X_1$ $X_1$. Again if $R_A > P_1$ the sliding tendency will be reversed. The force $F$ (= $F_1 = F_2$ ) is known as the shearing force over the section $XX$ and is defined with the help of equations (3.2) and (3.3) as the algebraic sum of all the external lateral forces on one side of the section. If we start from the left hand side and the algebraic sum is positive (*i.e.* $R_A < P_1$) the direction of shearing force $F_2$ will be downward and that of $F_1$ upward, it is said to be a positive shearing force otherwise negative Fig. 3.2.

Once again for the equilibrium of the two elements separately

$$M = - R_A\, x + P_1(x - a_1) \quad ...(3.4)$$

$$M = - R_B\, (1 - x) + P4\, (a_4 - x) + P_3(a_3 - x) + P_2\, (a_2 - x) \quad ...(3.5)$$

This moment $M$ is called the bending moment over the section $XX$ as its axis is perpendicular to the longitudinal axis of beam.

From equations (3.4) and (3.5) it is evident that the bending moment over a section is the algebraic sum of the moments about that section of the forces on one side of the section. If the moment is positive as evaluated from equations (3.4) and (3.5), the directions will be as shown in Fig. 3.1(*b*) and is called a sagging moment (after the shape of a sagging wire held over two vertical poles), as shown in Fig. 3.2.

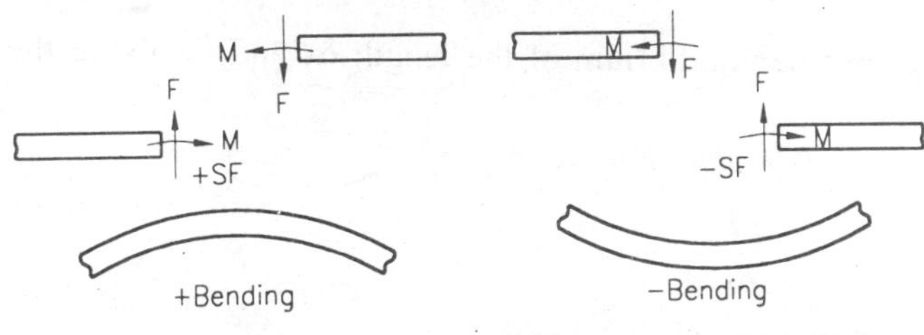


*Fig. 3.2*

## 3.3 Shearing Force and Bending Moment Diagrams

The values of the shearing force and bending moment will go on varying from point to point along the length of the beam. The curves showing the variation of shearing force and bending moment along the length of the beam are known as the shearing force and bending moment diagrams respectively.

The point on the beam where the bending moment changes sign is known as the point of contraflexion. Evidently at this point the B.M. curve will intersect the base line ; that is, the value of bending moment will be zero.

For finding the sign of the Bending Moment due to any force (Load or reaction) over the section *XX* following procedure may be conveniently adopted :

Consider the beam clamped at the section *XX* and forget about all other forces except the one in question due to which the moment is to be found (say the reaction ($R_A$). The tendency of $R_A$ will be to move the point *A* upwards with respect to *X* and the shape of the portion *XA* of the beam will be similar to that of a sagging wire on two electric poles. Thus the moment ($R_A\,x$) will be called a sagging moment and rendered negative. Similarly the moment $P_1\,(x - a_1)$ is positive.

## 3.4 Relation between Bending Moment and Shearing Force

Consider a small length $\delta x$ of a beam situated at a distance $x$ from some chosen origin. Let the beam by carrying some distributed load which may or may not be uniformly distributed load. But $\delta x$ being sufficiently small the rate of loading w MN/m run may be taken as constant over the length $\delta x$.

The shearing force F and $(F + \delta F)$ along with the bending moments $M$ and $(M + \delta M)$ on either end of this element are assumed to be as shown in Fig. 3.3 conforming to the positive signs as per the convention adopted in article 3.2.

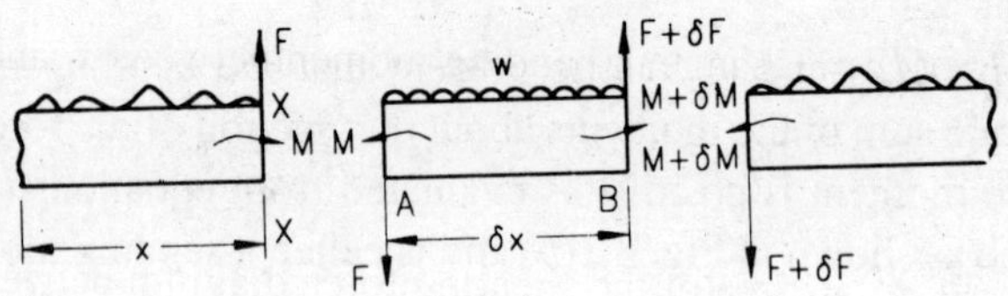


*Fig. 3.3*

Considering the equilibrium of the length $\delta x$ and resolving the forces in the vertical direction

$$F + \delta F = F + w\delta x$$

i.e.
$$\frac{\delta F}{\delta x} = w$$

In the limiting case when $\delta x \to 0$

$$\frac{dF}{dx} = w \qquad ...(3.6)$$

That is, the rate of change of shearing force or the slope of the shearing force curve at $XX$ is numerically equal to the rate of loading at that point.

Again for the equilibrium of the length $\delta x$, equating moments of all the forces about any point in the section $A$.

$$(M + dM) + (w\delta x).\frac{\delta x}{2} = M + (F + \delta F).\delta x$$

Neglecting the product of two small quantities, it gives

$$\frac{\delta M}{\delta x} = F$$

and in the limiting case when delta $x \to 0$, we have

$$\frac{dM}{dx} = F \qquad ...(3.7)$$

That is, the rate of change of bending moment or the slope of the bending moment curve at $XX$ is equal to the shearing force at that section.

Evidently from equations (3.6) and (3.7), we have

$$w = \frac{dF}{dx} = \frac{d^2M}{dx^2} \qquad ...(3.8)$$

Again integrating equations (3.6) and (3.7) between sections $x_0$ to $x$

$(F - F_0)$ = Change in the shearing force between $x_0$ and $x$

$$= \int_{x_0}^{x} w dx \qquad ...(3.9a)$$

= Total load between $x_0$ and $x$ ...(3.9$b$)

and $(M - M_0)$ = Change in bending momentbetween $x_0$ and $x$

$$= \int_{x_0}^{x} F dx \qquad ...(3.10)$$

= Area of the shearing force diagram between $x_0$ and $x$.

## 3.5 Shearing Force and Bending Moment Diagrams

### 3.5.1(*a*) Beam Simply Supported at its Ends and Carrying a Load at The Mid–Span :

Consider a beam $AB$ supported at its ends and a load $W$ at $C$. The reactions $R_A$ and $R_B$ will both the equal to $\frac{W}{2}$ as the beam is loaded symmetrically.

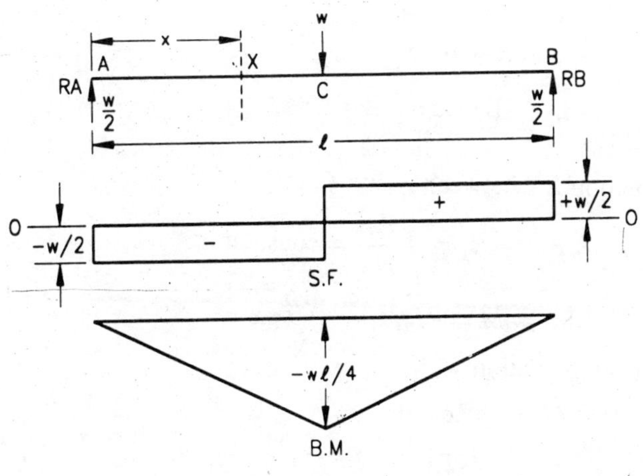


*Fig. 3.4*

For the section $XX$ lying in between $A$ and $C$ at a distance $x$ from chosen origin at $A$ and according to the adopted sign convention of Fig. 3.2

$$\text{S.F.} = -R_A = -\frac{W}{2}$$

For the section $XX$ lying in between $C$ and $B$

$$\text{S.F.} = -R_A + W$$

$$= -\frac{W}{2} + W = \frac{W}{2}$$

B.M. at the section $XX$ between $A$ and $C$

$$= -\frac{W}{2}x,$$

which is straight line giving the value of B.M. as zero at $A$ and $-\dfrac{Wl}{4}$ at $C$.

B.M. at any section between $C$ and $B$

$$= -\frac{W}{2}x + W\left(x - \frac{1}{2}\right)$$

$$= \left(\frac{Wx}{2} - W\frac{l}{2}\right)$$

which is again a straight line ; variation being $-\dfrac{Wl}{4}$ and $C$ and zero at $B$.

The two diagrams are shown in Fig. 3.4.

## (*b*) Point Load Not at the Mid Span

Consider the load $W$ to be acting at the point $C$ which is at distance '$a$' from $A$ and '$b$' from $B$.

The reactions at $A$ and $B$ will be

$$R_A = \frac{Wb}{l};\ R_B = \frac{Wa}{l}$$

Taking the section $XX$ between $A$ and $C$

$$\text{S.F.} = -R_A = -\frac{Wb}{l} = \text{constant}$$

and
$$\text{B.M.} = M = -R_A x + -\frac{Wb}{l}.x$$

which is a straight line variation with

$$M = 0 \text{ at } x = 0$$

and
$$M = -\frac{Wba}{l} \text{ at } x = a$$

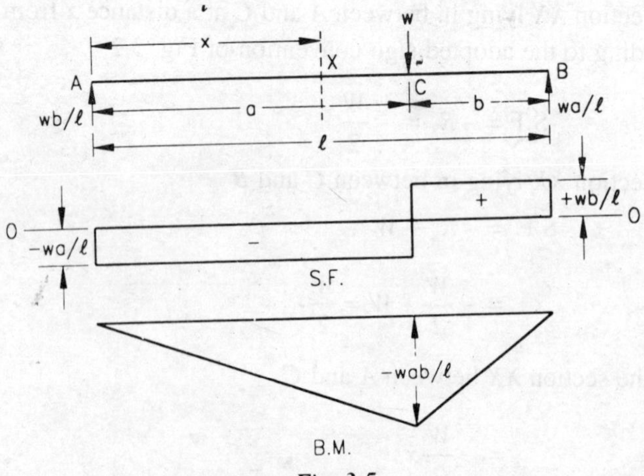


*Fig. 3.5*

For the section $XX$ between $C$ and $B$

$$\text{S.F.} = -R_A + W = -\frac{Wb}{l} + W = \frac{Wa}{l}$$

$$= \text{constant}$$

$$M = -R_A x + W(x-a)$$

$$= -\frac{Wba}{l}x + W(x-a)$$

which is again a straight line such that

$$M = -\frac{Wba}{l}. \quad \text{at } x = a$$

and $$M = 0 \quad \text{at } x = l.$$

The shearing force and bending moment diagrams are shown in Fig. 3.5.

## 3.5.2 Simply Supported Beam with Uniformly distributed Load

Let the beam $AB$ of span $l$ be acted upon by a uniformly distributed load of $w$ Newton per metre length as shown in the Fig. 3.6.

The beam is loaded symmetrically. Therefore

$$R_A = R_B = \frac{wl}{2}$$

For any section $XX$ at a distance x from $A$

$$\text{S.F.} = -R_A + wx$$

$$R = -\frac{wl}{2} + wx$$

which is a straight line in variation with $x$ such that

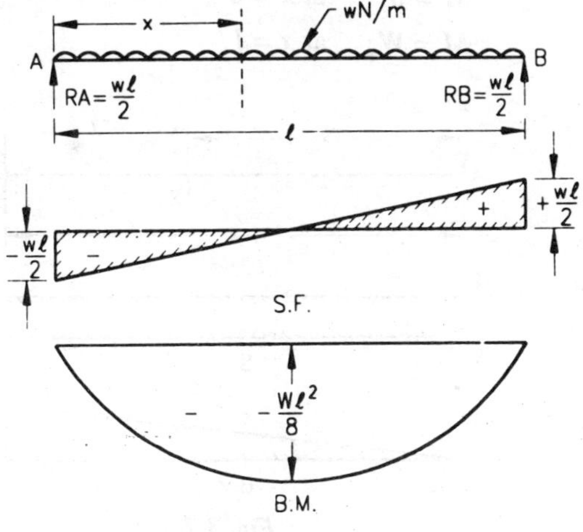


*Fig. 3.6*

$$\text{S.F.} = -\frac{wl}{2} + wx \quad \text{at } x = 0$$

and $$\text{S.F.} = +\frac{wl}{2} \quad \text{at } x =_{l}$$

Also $$M_x = -R_A x + wx\left(\frac{x}{2}\right)$$

$$= -\frac{wl}{2}x + \frac{wx^2}{2}$$

which is parabolic variation with $x$

$$M_x = 0$$

at $x = 0$ as well as at $x = l$.

The magnitude of $M_x$ is maximum at $x = \frac{1}{2}$ where it is equal to $\left(-\frac{wl^2}{2}\right)$

The magnitude of B.M. can be determined for a few more values of x and the curve is then plotted as shown in Fig. 3.6.

## 3.6 Shearing Force and Bending Moment Diagrams for Cantilevers

### 3.6.1 With Point Load at Free End

Taking a section $XX$ at a distance $x$ from $A$, the SF at $x$ is given by

$$\text{S.F.} = W = \text{constant.}$$

And $$\text{B.M.} = M_X = W_x,$$

which is a straight line variation with $x$ such that

$$M = 0 \quad \text{at } x = 0$$

$$M = W_X \quad \text{at } x = l$$

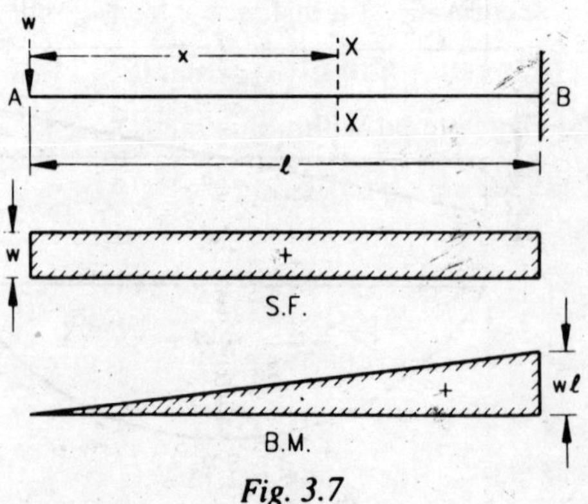


*Fig. 3.7*

The shearing force and bending moment diagrams are shown in Fig. 3.7.

### 3.6.2 Load Distributed Over Whole Span

Taking a section *xx* at a distance *x* form *A*, Fig. 3.8 the SF at *x* is given by

$$SF = wx$$

white is a straight line variation with SF = 0 at $x = 0$ and SF = $wl$ at $x = l$.

And bending moment *M* is given by

$$M = wx \cdot \frac{x}{2} = \frac{wx^2}{2}$$

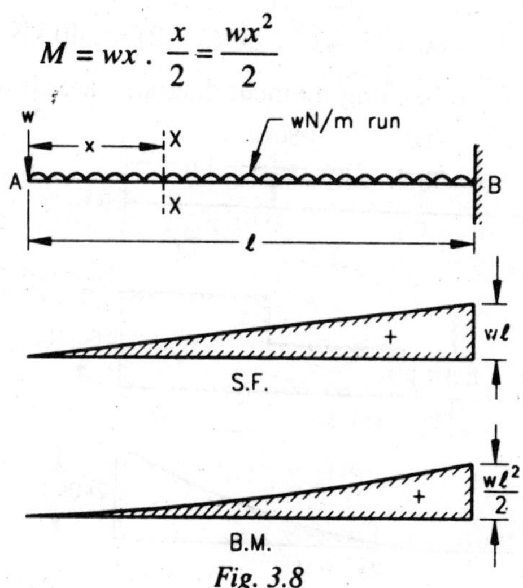


*Fig. 3.8*

which is parabolic in variation with $M = 0$ at $x = 0$ and $M = \dfrac{wl^2}{2}$ at $x = l$.

The shearing force and bending moment diagrams are shown in Fig. 3.8.

**Example 3.1.** *A cantilever 3 m long carries two point loads each 80 kN, one placed at distance of* 1m *and the other at 2 m from the fixed end. Draw to scale the shearing force and bending moment diagrams for the cantilever.*

**Solution.** Take the section *XX* at a distance *x* from the assumed origin at *A*.

**(*i*) When *XX* lies between *A* and *C* *i.e.* when 0 < *x* < 1**

As there is no load encountered within this range

$$SF = 0$$

$$M = 0$$

(*ii*) When $1 < x < 2$

$$SF = 80 \text{ kN} = \text{constant}$$

$$M = 80\,(x - 1) \qquad \text{(Straight line variation.)}$$

Therefore $\quad M = 0 \quad$ at $x = 1$ m

and $\quad M = 80$ kN-m $\quad$ at $x = 2$ m

(*iii*) When $2 < x < 3$

$$SF = 80 + 80 = 160 \text{ kN} = \text{constant}$$

$$M = 80\,(x - 1) + 80\,(x - 2) \qquad \text{Straight line}$$

Therefore at $x = 2$ m

$$M = 80 \text{ kN} - \text{m}$$

and at $x = 3$ m

$$M = 80\,x\,(3 - 1) + 80\,(3 - 2) = 240 \text{ kN} - \text{m}$$

The shearing force and bending moment diagrams are shown in Fig. 3.9.

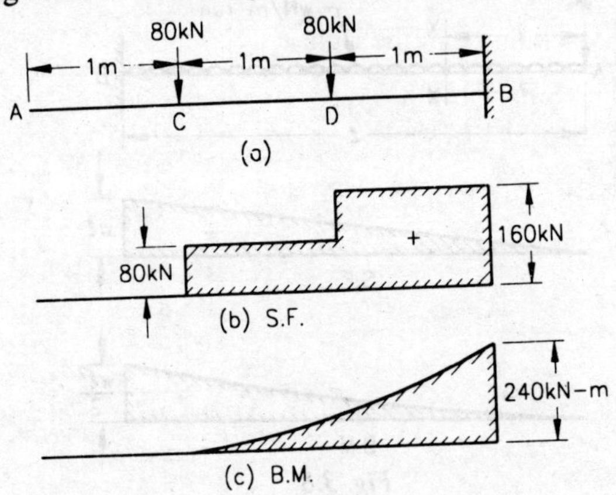


*Fig. 3.9*

**Example 3.2.** *Draw the S.F. and B.M. diagrams for the cantilever shown in Fig. 3.10.*

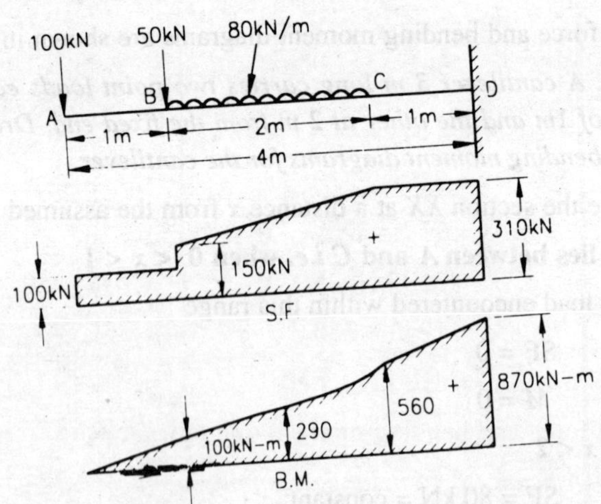


*Fig. 3.10*

Solution. Taking a section *XX* at a distance *x* from the origin at *A*,

(*i*) When $0 < x < 1$

$$SF = 100 \text{ kN} = \text{constant}$$

$$M = 100\,x$$

This is a straight line such that at

$$x = 0,\ M = 0$$

and at $x = 1$ m, $M = 100$ kN – m

(*ii*) When $1 < x < 3$

$$SF = 100 + 50 + 80\,(x - 1)$$

$$= 70 + 80x \qquad \text{Straight line}$$

at $x = 1$ m (approaching from the right),

$$\text{S.F.} = 150 \text{ kN}$$

at $x = 3$ m, S.F. $= 70 + 240$

$$= 310 \text{ kN}$$

$$M = 100x + 50\,(x - 1) + \frac{80}{2}\,(x - 1)^2$$

$$= -\,50 + 150x + 40\,(x - 1)^2 \text{ Parabola}$$

at $x = 1$, $M = 100$ kN – m

at $x = 2$, $M = 290$ kN – m

at $x = 3$, $M = 560$ kN – m

(*iii*) When $3 < x < 4$

$$SF = 100 + 50 + 80 \times 2$$

$$= 310 \text{ kN} = \text{constant}$$

$$M = 100x + 50\,(x - 1) + (80 \times 2)\,(x - 2)$$

$$= 310x - 370 \qquad \text{Straight line}$$

at $x = 3$ $M = 930 - 370$

$$= 560 \text{ kN – m}$$

at $x = 4$ $M = 1240 - 370$

$$= 870 \text{ kN – m}$$

The shearing force and bending moment diagrams are shown in Fig. 3.10.

**Example 3.3.** *A beam AB, 15 m long, is simply supported at points C and D, 10 m apart, the overhangs AC and DB being 2m and 3 m respectively. The beam carries a distributed load of 20 kN/m over the whole length along with a load of 50 kN at A and 80 kN at B. Draw to scale the S.F. and B.M. diagrams giving maximum values and stating where these occur. Also find the points of contraflexion.*

**Solution.** For the equilibrium of the beam, taking moments about *D*.

$$10\,R_C = 50 \times 12 + (20 \times 15)\,(12 - 7.5) - 80 \times 3$$

*i.e.* $R_C = 171$ kN

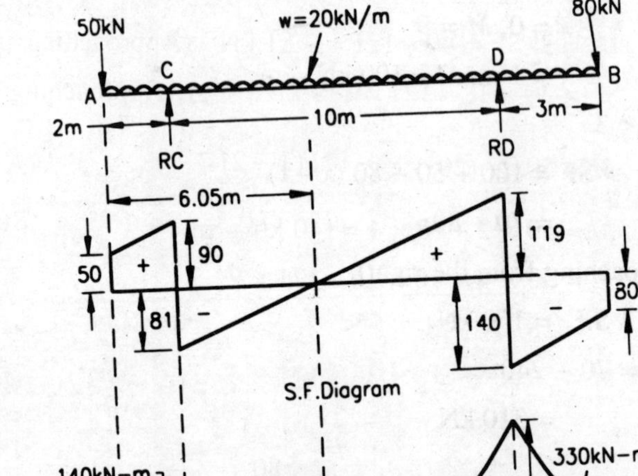


*Fig. 3.11*

Therefore $R_D = 50 + (20 \times 15) + 80 - 171$

$= 259$ kN.

Shearing force and B.M. are calculated by taking a section *XX* at a distance *x* from A as follows :

(*i*) When $0 < x < 2$

$$SF = 50 + 20x \quad \text{(A straight line)}$$

at $x = 0$, SF = 50 kN

at $x = 2$, SF = 90 kN (Approaching from left)

$$M = 50x + (20x)\frac{x}{2} \quad \text{(A parabola)}$$

$$= 50x + 10x^2$$

at $x = 0$, $M = 0$

at $x = 1$, $M = 50 + 10 = 60$ kN – m

at $x = 2$, $M = 50 \times 2 + 10 \times 4 = 140$ kN – m

(*ii*) When $2 < x < 12$

$$SF = 50 + 20x - 171$$

$$= 20x - 121 \qquad \text{(A straight line)}$$

at $x = 2$, SF $= 40 - 121 = -81$ kN (Approaching from right)

at $x = 12$, SF $= 20 \times 12 - 121 = 119$ kN (Approaching from left)

And $$M = 50x + (20x)\frac{x}{2} - 171(x-2) \qquad \text{(A parabola)}$$

at $x = 2$, $M = 50 \times 2 + 10 \times 4 = 140$ kN – m

at $x = 4$, $M = 50 \times 4 + 10 \times 16 - 171 \times 2$

$= 18$ kN – m

at $x = 6$, $M = 50 \times 6 + 10 \times 36 - 171 \times 4$

$= -24$ kN – m

at $x = 8$, $M = 50 \times 8 + 10 \times 64 - 171 \times 6$

$= 14$ kN – m

at $x = 10$, $M = 50 \times 10 + 10 \times 100 - 171 \times 8$

$= 132$ kN – m

at $x = 12$, $M = 50 \times 12 + 10 \times 144 - 171 \times 10$

$= 330$ kN – m

Evidently as $x$, increases from 2 to 12, *BM* after showing a decrease, starts increasing. In other words there must be maximum value of $M$, corresponding to which $x$ can be evaluated by equating

$$\frac{dM}{dx} = 0 \text{ or } SF = 0$$

Thus equating

$$SF = 20x - 121 = 0$$

$$x = \frac{121}{20} = 6.05 \text{ m}$$

For which $M = 50 \times 6.05 + 10\,(6.05)^2 - 171\,(6.05 - 2)$

$= -24.02$ kN – m

Also since the value of $M$ is changing sign, the points of contraflexion with be within this range and the location can be found by equating $M$ to zero and solving for $x$

That is $50x + 10x^2 - 171\,(x - 2) = 0$

or $10x^2 - 121x + 342 = 0$

Therefore $x = \frac{1}{20}\left[121 \pm (121 \times 121 - 4 \times 342 \times 10^{\frac{1}{2}}\right]$

*i.e.* $x = 4.5$ m or 7.6 m

Thus the points of contraflexion are at 4.5 m and 7.6 m from *A*.

(*iii*) When $12 < x < 15$

$$SF = 50 + 20x - 171 - 259$$
$$= 20x - 380 \quad \text{(A straight line)}$$

at $x = 12$, $SF = 20 \times 12 - 380 = -140$ kN (Approaching from right)

at $x = 15$, $SF = 20 \times 5 - 380 = -80$ kN

And $M = 50x + (20x)\frac{x}{2} - 171\,(x-2) - 259\,(x-12)$

$$= 50x + 10x^2 - 171\,(x-2) - 259\,(x-12) \quad \text{(A parabola)}$$

at $x = 12$, $M = 50 \times 12 + 10 \times 144 - 171 \times 10$
$= 330$ kN – m

at $x = 13$, $M = 50 \times 13 + 10 \times 169 - 171 \times 11 - 259 \times 1$
$= 200$ kN – m

at $x = 14$, $M = 50 \times 14 + 10 \times 196 - 171 \times 12 - 259 \times 2$
$= 90$ kN – m

at $x = 15$, $M = 0$

The complete *SF* and *BM* diagrams are shown in Fig. 3.11, from which the maximum *BM* = 330 kN–m occurring at point *D* and maximum shearing force is 119 kN occurring just on the left of point *D*.

**Example 3.4.** *A beam of length l, carrying a uniformly distributed load over its whole length, is symmetrically supported over two simple supports. Find the position of the supports to keep the maximum bending moment on the beam as low as possible.*

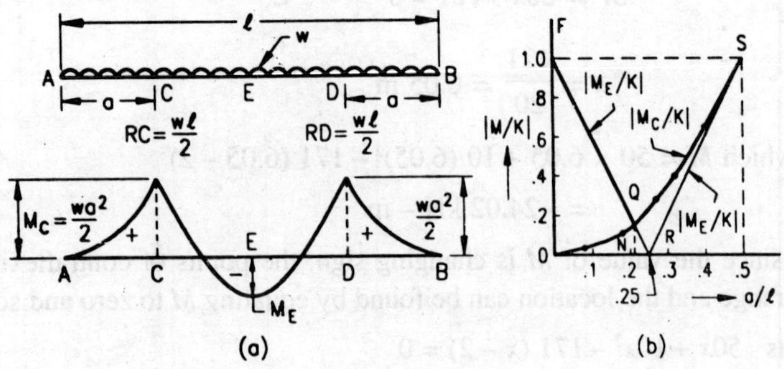


*Fig. 3.12*

**Solution.** Fig. 3.12(*a*) shows the bending moment diagram for some value of '*a*'. For any value of '*a*', the bending moments at the supports and the mid–point *E* are to be evaluated and the greater of the two will correspond to the maximum bending moment on the beam.

Thus for the load *w* kN/m run and the configuration of Fig. 3.12.

$$M_E = \left(\frac{wl}{2}\right)\frac{l}{4} - \frac{wl}{2}\left(\frac{l}{2} - a\right)$$

$$= -\frac{wl^2}{8}\left(1 - 4\frac{a}{l}\right)$$

or $$\frac{M_E}{K} = -1 + 4\frac{a}{l} \qquad \text{...}(A)$$

Where $$K = \frac{wl^2}{8} = \text{constant}$$

and $$M_C = \frac{wa^2}{2}$$

or $$\frac{M_C}{K} = 4\frac{a^2}{l^2} \qquad \text{...}(B)$$

Depending upon the magnitude of $\left|\frac{M_C}{K}\right|$ or $\left|\frac{M_E}{K}\right|$ we have to pick up the greater of the two, but in turn these values depend upon the ratio (*a*/*l*) which has the extreme values of zero and 0.5 corresponding to the supports being at the ends or merging at the midpoint *E* respectively. For this range of the values of (a/*l*), $\left|\frac{M_C}{K}\right|$ and $\left|\frac{M_E}{K}\right|$ have been plotted in Fig. 3.12(*b*) as *OQS* and *PRS* respectively.

**TABLE 3.1**

| $\frac{a}{l}$ | $\frac{M_C}{K}$ | $\frac{M_E}{K}$ | $\left\lvert\frac{M_E}{K}\right\rvert$ |
|---|---|---|---|
| 0.0 | 0.00 | – 1.0 | 1.0 |
| 0.1 | 0.04 | – 0.6 | 0.6 |
| 0.2 | 0.16 | – 0.2 | 0.2 |
| 0.25 | 0.25 | 0.0 | 0.0 |
| 0.3 | 0.36 | 0.2 | 0.2 |
| 0.4 | 0.64 | 0.6 | 0.6 |
| 0.5 | 1.00 | 1.0 | 1.0 |

Evidently for the values of $(a/l)$ between $O$ and $N$ the maximum value of the B.M. is to be given by the curve $PQ$ (equation $B$) and for any value of $(a/l)$ between $N$ and 0.5 the maximum $BM$ on the beam will be picked up from the curve $QS$ (equation $A$). Thus the curve $PQS$ is the curve which gives the magnitudes of the maximum $B.M.$ for any value of $(a/l)$. Clearly the point $Q$ which is the intersection of the curves of equations ($A$) and ($B$) gives the least value of the maximum $B.M.$ on the beam and the corresponding value of $(a/l)$ can be found by equating the magnitudes of moments as obtained from equations ($A$) and ($B$) at the same time keeping in view that $M_E$ is actually negative for the portion of the curve $PR$ on which lies the point $Q$.

Thus for the greatest B.M. to be as small as possible, we have

B.M. at supports = Maximum $BM$ between supports

$$|M_C| = |M_E|$$

$$K.4\frac{a^2}{l^2} = -K\left(-1+4\frac{a}{l}\right)$$

or $$4\frac{a^2}{l^2}+4\frac{a}{l}-1 = 0$$

Therefore $$\frac{a}{l} = \frac{1}{8}[-4 \pm (16+16)^{1/2}]$$

*i.e.* $$\frac{a}{l} = 0.207 \text{ or } -1.207$$

The only admissible value is

$$\frac{a}{l} = 0.207$$

or $\quad a = 0.207l.$

**Example 3.5.** ***A beam AB, 10 m long, is carrying a uniformly distributed load w N/m. The beam is to be supported freely at two piers 6 m apart in such a way that the bending moment on the beam is as small as possible. The piers may not necessarily be placed symmetrical. Find the position of the piers with respect to the ends of the beam.***

***(Jaipur University)***

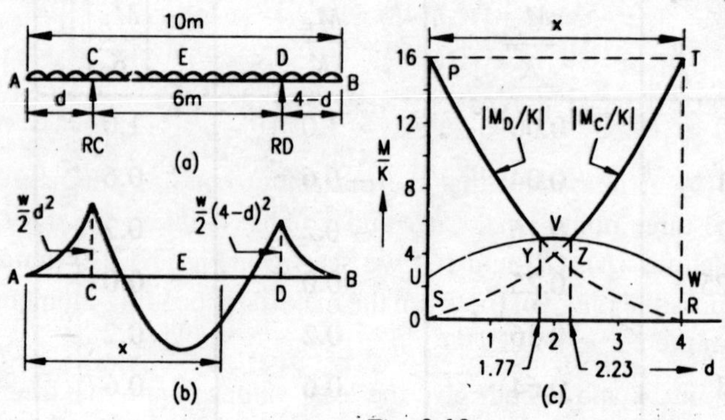


*Fig. 3.13*

**Solution.** Let the left hand support $C$ be situated at a distance '$d$' from $A$. The distance of the other support $D$ will then be $(4 - d)$ from $B$ as shown in Fig. 3.13. Then for this configuration taking moments about $D$.

$$6\,R_C = 10w\,[(6 + d) - 5]$$

i.e.
$$R_C = \frac{5}{3}w\,(1 + d)$$

Also
$$M_C = \frac{w}{2}d^2$$

and
$$M_D = \frac{w}{2}(d - 4)^2$$

Thus if
$$K = \frac{w}{2},\ \frac{M_C}{K} = d^2 \qquad ...(A)$$

and
$$\frac{M_D}{K} = (d - 4\,)^2 \qquad ...(B)$$

Also B.M. at section $XX$ which is at a distance $x$ from $A$ and lying between $C$ and $D$.

$$M_x = \frac{wx^2}{2} - \frac{5}{3}w(l + d)\,(x - d)$$

Now the magnitude of $M_x$ will be maximum when

$$dM_x/dx = 0$$

That is when $wx - \dfrac{5}{3}w\,(1 + d) = 0$

or
$$x = \frac{5}{3}(1 + d)$$

Thus the maximum bending moment between the supports is given by $M_E$ as

$$M_E = \frac{w}{2}\left[\frac{5}{3}(1 + d)\right]^2 - \frac{5}{3}w(1 + d)\left[\frac{5}{3}(1 + d) - d\right]$$

$$= \frac{5}{18}w\,(1 + d)\,(d - 5)$$

Therefore
$$\frac{M_E}{K} = \frac{5}{9}(1 + d)\,(d - 5) \qquad ...(C)$$

From Fig. 3.13($b$), it is evident that the maximum bending moment will be obtained at one of the three points, viz. $C$, $D$ and $E$. Thus plotting the three curves corresponding to equations ($A$), ($B$) and ($C$) we see from Fig. 3.13($c$) that for any value of d within the possible range of1m to 4m the maximum bending moment on the beam is obtained from the curve $PYVZT$.

Evidently the points $Y$ and $Z$ both give the least values of the maximum B.M. acting on the beam. Points $Y$ and $Z$ are intersections of the equation ($C$) with equations

($B$) and ($A$) respectively. It may also be pointed out that the actual values of the moments given by equation ($C$) are negative.

Thus equating $|M_C|$ to $M_E$ and $|M_D|$ to $M_E$ separately we get the location of points $Y$ and $Z$ respectively as under :

$$Kd^2 = -K\left[\frac{5}{3}(1+d)(d-5)\right]$$

*i.e.* $14d^2 - 20d - 25 = 0$

Then the admissible solution is

$$d = 2.23 \text{ m} \qquad ...(D)$$

And from equations ($B$) and ($C$)

$$K(d-4)^2 = -K\left[\frac{5}{9}(1+d)(d-5)\right]$$

*i.e.,* $14d^2 - 92d + 119 = 0$

Therefore the admissible solution is

$$d = 1.77 \text{ m} \qquad ...(E)$$

The result of equation ($E$) could, also, be obtained from equation ($D$) because the points $Y$ and $Z$ are symmetrically placed about $d = 2.0$ which is evident from the Fig. 3.13($c$).

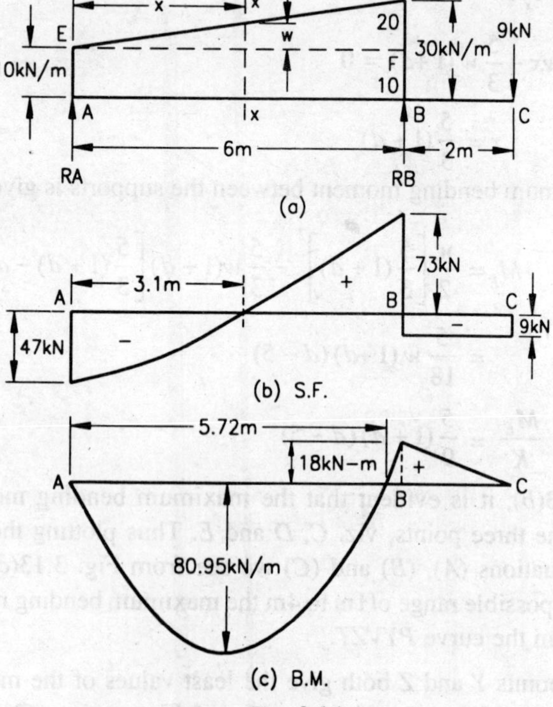


*Fig. 3.14*

Thus the smallest $M_{max}$ is obtained from equation (*E*) by substituting 2.23 or1.77 for *d*.

*i.e.* Least $$|M_{max}| = -\frac{5}{9}\frac{w^2}{2}(1+1.77)(1.77-5)$$

$$= 2.485w \text{ Nm}.$$

**Example 3.6.** *An overhanging beam ABC, carrying a varying load and point load as shown in the Fig. 3.14, is freely supported at A and B.*

Draw to scale the shearing force and bending moment diagrams.

**Solution.** The varying load can be split into a triangular load and a rectangular (uniformly distributed) load as shown in the Figure.

Taking moments about *B*

$$6R_A = (10\times 6)\ \frac{6}{2} + \left(\frac{1}{2}\times 6\times 20\right)\frac{6}{3} - 9\times 2$$

*i.e.* $$R_A = 47 \text{ kN}$$

and $$R_B = \frac{1}{2}(10+30)\,6 - 47 + 9 = 82 \text{ kN}.$$

For the section *XX* at a distance *x* from *A* (since we have from similar triangles $w = \frac{20}{6}x$)

$$SF = R_A + 10x + \frac{1}{2}xw$$

$$= -47 + 10x + \frac{5}{3}x^2 \qquad \text{(a parabola)}$$

at $x = 0$, $SF = -47$

at $x = 2$, $SF = -20.33$

at $x = 4$, $SF = +19.67$

at $x = 6$, $SF = +73$

Also $SF = 0$ when $\frac{5}{3}.x^2 + 10x - 47 = 0$

or when $x = 3.1$ m

This will also correspond to Max B.M. position.

Shearing force for any section between *B* and *C* will be

$$= -47 + \frac{1}{2}(10+30)\times 6 - 82$$

$$= -9 \text{ kN} \qquad \text{(constant)}$$

For the section *XX* between *A* and *B* at a distance *x* from *A*

$$M = -R_A x + (10.x) \times \frac{x}{2} + \left(\frac{1}{2} x . w\right) \frac{x}{3}$$

$$M = -47x + 5x^2 + \frac{5}{9}x^3$$

at $x = 0$, $M = 0.0$

at $x = 1$, $M = 36.45$

at $x = 2$, $M = -69.56$ kN – m

at $x = 3.1$, $M_{max} = -80.95$ kN – m

at $x = 4$, $M = -72.45$ kN – m

at $x = 6$, $M = +18$ kN – m

For the point of contraflexion

$$M = 0$$

or $$-47x + 5x^2 + \frac{5}{9}x^3 = 0$$

Since $x \neq 0$ for the point of contraflexion it reduces to

$$\frac{5}{9}x^2 + 5x - 47 = 0$$

The admissible solution of this equation is

$$x = 5.72 \text{ m.}$$

For the section $XX$ between $B$ and $C$ at a distance $x$ from $A$.

$$M = -R_A x + (10 \times 6)(x-3) + \left(\frac{1}{2} \times 6 \times 20\right)(c-4) - R_B (x-6)$$

*i.e.* $$M = (72 - 9x) \qquad \text{(A straight line)}$$

at $x = 6$, $M = 18$ kN – m

at $x = 8$, $M = 0.0$ kN – m

The shearing force and bending moment diagrams are drawn in Fig. 3.14.

**Example 3.7.** *4A beam AB, 8 m long, is simply supported at two points A and B 6 m apart. It is loaded with two distributed loads and a point load at C as shown in Fig. 3.15. Draw the shearing force and bending moment diagrams.*

**Solution.** Taking moments about $B$

$$6R_A - 104 - (20 \times 2) \times 1 + (60 \times 2) \times 1 = 0$$

*i.e.* $$R_A = 4 \text{ kN}$$

Therefore $R_B = 60 \times 2 + 20 + 20 \times 2 - R_A$

$$= 176 \text{ kN}$$

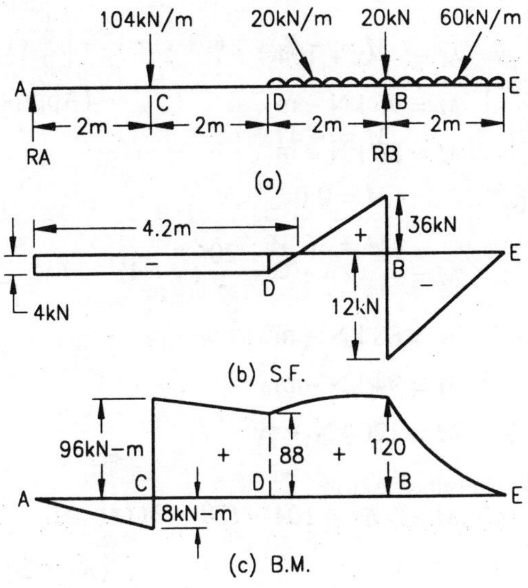


*Fig. 3.15*

(*a*) **Shearing Force at any section** bold *XX* **at a distance** bold *x* **from** bold *A*

(*i*) $0 < x < 4$

$$SF = -R_A = -4 \text{ kN} = \text{constant}$$

(*ii*) $4 < x < 6$

$$SF = -4 + 20(x - 4) \qquad \text{(A straight line)}$$

at $x = 4$, $SF = -4$ kN

at $x = 4.2$, $SF = 0$ (Maximum B.M. occurs here)

at $x = 6$, $SF = +36$kN (Approaching from left)

(*iii*) $6 < x < 8$

$$SF = -4 + 20 \times 2 + 20 - 176 + 60(x - 6)$$

$$= -120 + 60(x - 6 \qquad \text{(A straight line)}$$

at $x = 6$, $SF = -120$ (Approaching from right)

at $x = 8$, $SF = 0$

(*b*) **B.M. at any section** *XX* **distant** *x* **from** A

(*i*) $0 < x < 2$

$$M = -R_A x = -4x \qquad \text{(A straight line)}$$

at $x = 0$, $M = 0$

at $x = 2$, $M = -8$ kN – m (Approaching from the left)

(*ii*) $2 < x < 4$

$$M = -4x + 104 \qquad \text{(A straight line)}$$

at $x = 2$, $M = 96$ kN – m (Approaching from the right)

at $x = 4$, $M = 88$ kN – m

(*iii*) $4 < x < 6$

$$M = -4x + 104 + \frac{20}{2}(x+4)^2 \qquad \text{(Parabola)}$$

at $x = 4$, $M = 88$ kN – m

at $x = 5$, $M = 94$ kN – m

at $x = 6$, $M = 120$ kN – m

(*iv*) $6 < x < 8$

$$M = -4x + 104 + (20 \times 2)(x - 5) - (R_B - 20)(x-6) + \frac{60}{2}(x-6)^2$$

$$= 30(8 - x)^2 \qquad \text{(Parabola)}$$

at $x = 6$, $M = 120$ kN – m

at $x = 7$, $M = 30$ kN – m

at $x = 8$, $M = 0$.

The S.F. and B.M. diagrams are drawn in Fig. 3.15.

**Example 3.8.** *The shearing force diagram for a simply supported beam is given in Fig. 3.16. The beam is supported at A and B. Find the distance between B and D. Draw the B.M. diagram for the beam.*

**Solution.** Since the supports are at *A* and *B*, the beam is an overhanging one. The S.F. remains constant and positive for the portion *CA*. Hence there is a load of 50 kN at the point *C*. At the point *A* there is a vertical drop of *SF* from + 50 to – 40 kN. Thus the reaction $R_A = 50 - (-40) = 90$ kN.

From the points *A* to *B* the S.F. changes gradually from – 40 kN to 100 kN. Thus there will be a uniformly distributed load. If its intensity is *w* kN/m then

$$w = \frac{dF}{dx} \text{ Slope of the shearing force line}$$

$$= \frac{100 - (-40)}{4} = \frac{140}{4} = 35 \text{ kN/m}$$

Again shearing force drops suddenly from + 100 kN to – 80 kN at the point *B*. That is, there is a reaction $R_B = 100 - (-80) = 180$ kN at the support *B*. Further the S.F. remains constant at 80 kN for the portion *BD* which signifies that there will be a point load of 80 kN at the point *D*.

The load diagram is shown in Fig. 3.16(*b*).

The value of '*a*' is found by taking moments about *B*. That is

$$-50 \times 5 + 90 \times 4 - (35 \times 4) \times 2 + 170 \times a = 0$$

*i.e.* $a = 1$ m.

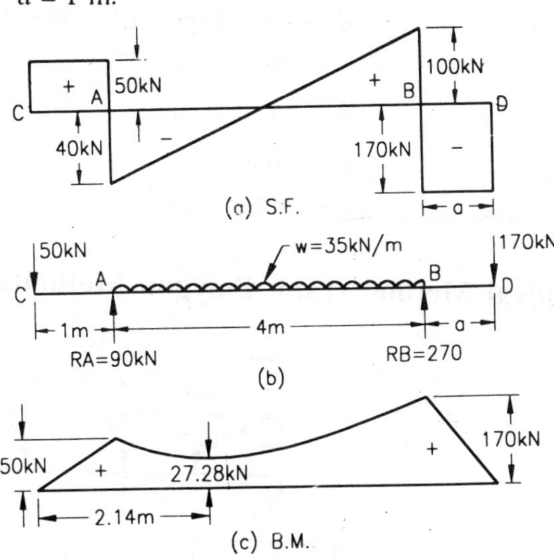


*Fig. 3.16*

For the B.M. diagram take a section *XX* at a distance x from the point *C*.

(*i*) When $0 < x < 1$

$$M = 50x \qquad \text{(A straight line)}$$

at $x = 0$, $M = 0$

at $x = 1$, $M = 50$ kN – m

(*ii*) When $1 < x < 5$

$$M = 50x - 90(x - 1) + \frac{35}{2}(x - 1)^2$$

at $x = 1$, $M = 50.0$ kN – m

at $x = 2$, $M = 27.5$ kN – m

at $x = 3$, $M = 40.0$ kN – m

at $x = 4$, $M = 87.5$ kN – m

at $x = 5$, $M = 770.0$ kN – m

Evidently the value of B.M. decreases within the range $x = 1$ to 2 m and then starts rising. To find the location of least B.M. within this range set

$$\frac{dM}{dx} = 0$$

*i.e.* $50 - 90 + 35\,(x - 1) = 0$

or $x = 2.143$ m

and at $x = 2.143$ $M = 27.28$ kN – m

(*iii*) When $5 < x < 6$

$$M = 50x - 90\,(x - 1) + (35 \times 4)\,(x - 3) - 270\,(x - 5)$$

$$= 170\,(6 - x) \qquad \text{(A straight line)}$$

at $x = 5$, $M = 170$ kN – m

at $x = 6$, $M = 0.0$

The B.M. diagram is shown in Fig. 3.16(*c*).

## 3.7 Graphical Method (Link Polygon Method)

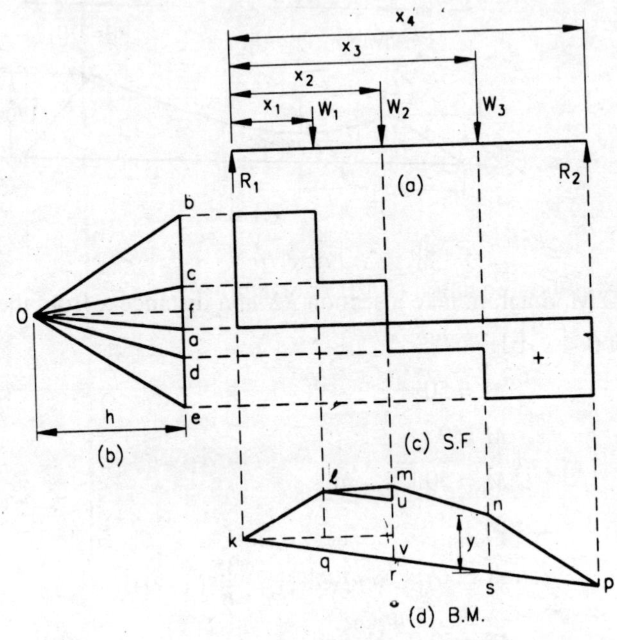


*Fig. 3.17*

Fig.3.17(*a*) shows the beam carrying the point loads for which it is desired to draw the S.F. and B.M. diagrams. For this to start with, the spaces are named using the Bow's notation e.g. *A*, *B*, *C*.... *E* etc as shown in the Fig.. Choose a pole *O* after drawing the load line bcde (Fig. 3.17*b*) such that to some scale $bc = W_1$, $cd = W_2$ and $de = W_3$. Join *O* to *b*, *c*, *d* and *e*. This is called the polar diagram.

Choose any point *k* on the line of action of $R_1$, draw *kt* parallel to *Ob* in the space '*B*', *lm* parallel to *Oc* in the space '*C*', mn parallel to *Od* in the space '*D*' and similarly *np* parallel to *Oe* in the space '*E*'. Join *kp*. Draw *Oa* parallel to *kp*. Then *ab* and *ea* represent the reactions $R_1$ and $R_2$ and $k\ l\ m\ n\ p$ the bending moment diagram on base *kp*, the bending moment *M* being proportional to vertical ordinates like *ql*, *rm* etc.

## Proof

In Fig. 3.17(*b*), $W_1$ is represented by *bc* and it acts through the point l (Fig. 3.17*d*). Since *kl* is parallel to *Ob* and *lm* is parallel to *Oc*, $W_1$ can be replaced by forces *bO* along *lk* and *Oc* along *lm* (law of triangle of force). Similarly $W_2$ at m is equivalent to two forces represented by *cO* along *ml* and *Od* along *mn*. *In* the same way $W_3$ at *n* is replaced by *dO* along *nm* and *Oe* along *np*. Evidently all these forces nullify each other except *bO* along *lk* and *Oe* along *np*. The two forces must be balanced by $R_1$ and $R_2$ as the beam is in static equilibrium. This is only possible if $R_1$ at *k* is replaced by *Ob* along *kl* and *aO* along *pk*, because $R_2$ at *p* is equivalent to *eO* along *pn* and *Oa* along *kp*. Thus $R_1$ is given by *ab* and $R_2$ by *ea*.

– Now coming to the bending moments since the triangles *Oba* and *klq* are similar

$$lq = ba.kq/Oa \qquad ...(3.11)$$

Drawing *Of* and *kt* perpendicular to be and *lq* respectively we have from the similar triangles *Oaf* and *kqt*

$$kqlOa = x_1/h \qquad ...(3.12)$$

Thus substituting equation (3.12) in equation (3.11)

$$lq = ba \,.\, x_1/h \qquad ...(3.13)$$

But as proved above, *ba* represents $R_1$. Therefore the product $ba.\ x_1$ $(= R_1 \,.\, x_1)$ is proportional to the bending moment at the point where $W_1$ acts. Hence from equation (3.13), for a given position of the pole *O, lq* will represent to a certain scale the bending moment at a distance $x_1$ from the left hand support.

Again draw *lu* parallel to *kp* to meet *mr* in *u*, and the perpendicular *qv* will meet *mr* in v. Evidently triangles *cOa* and *mlu* are similar and

$$mu = ca \,.\, lu/Oa \qquad ...(3.14)$$

Also since triangles qvr and *Ofa* are similar

$$lu = qr = Oa \,.\, qv/Of$$

or $$lu/Oa = qv/Of \qquad ...(3.15)$$

Thus from equations (3.14) and (3.15)

$$mu = ca \,.(x_2 - x_1)/h \qquad ...(3.16)$$

Therefore $$mr = mu + ur$$

$$= mu + lq$$

$$= ca \,.\, (x_2 - x_1)/h + ba \,.\, x_1/h \qquad \text{From (3.13)}$$

$$= (ba - bc)(x_2 - x_1)/h + ba \cdot x_1/h$$

$$= \frac{1}{h}[(R_1 - W_1)(x_2 - x_1) + R_1 x_1]$$

*i.e.* $$mr = \frac{1}{h}[R_1 x_1 - W_1(x_2 - x_1)] \qquad ...(3.17)$$

Thus for a given position of the pole $O$, for which h is fixed, equations (3.13) and (3.17) give that the ordinates $ql$ and $rm$ of the diagram 3.17$d$ represent to some scale the bending moments at these points. Similarly it can be shown that at any other point the ordinate of the diagram in Fig. 3.17($d$) will give the bending moment at that point according to the same scale, which is determined as follows.

Let the scale of load in polar diagram be 1 mm = $S_F$ Newton, that of the length in Fig. 3.17($a$) as 1 mm = $S_L$ metre and of the bending moment in Fig. 3.17($d$) as 1 mm = $S_M$ Nm. Then from equation 3.13 the length.

$$ql = \left(\frac{R_1}{S_F} \cdot \frac{x_1}{S_L}\right) \Big/ h \qquad ...(3.18)$$

Also length $$ql = M_1/S_M \qquad ...(3.19)$$

Thus since $M_1 = R_1 x_l$, we have from equations (3.18) and (3.19)

$$S_M = h \cdot S_F S_L \qquad ...(3.20a)$$

Thus for any point if the ordinate in Fig. 3.17($d$) is $y$ mm then B.M.= $yS_M$ = $y \cdot h \cdot SF \cdot S_L$ Nm, where h is measured in mm.

For drawing the S.F. diagram, horizontal lines are projected from the points $b$, $c$, $d$,$e$ and a across from the polar diagram. The S.F. diagram is obtained with the horizontal line from '$a$' as the base line.

This method can also be used for distributed load by dividing it into a large number of very small point loads concentrated at the centre of gravity of each small strip.

## REVIEW QUESTIONS

3.1 Explain the terms "Moment of a force about point A" and "Plane of the Moment".

3.2 State the right handed screw rule for drawing the "moment vector" How is the vector represented ?

3.3 Differentiate between "Banding Moment" and "Twisting Moment".

3.4 Write down the equation showing the relationship between Bending Moment shearing force and rate of loading at a particular point.

3.5 Define point of contraflexion.

3.6 Draw the *SF* and *BM* diagram of the cantilever shown in adjoining figure 3.18 (*a*) choosing the free end of the cantilever *D*, as origin and measuring *x* positive towards left.

For the range $0 < x < \frac{l}{2}$

$F_x$ = weight of the portion *CX*

$$= wx = \frac{2W}{l}x \qquad \text{(st. line)}$$

Therefore $(F_x)_{x=0} = 0$

At $x = \frac{l}{2}$,

$$F = \frac{2W}{l}\frac{l}{2} = W$$

For the range $\frac{l}{2} < x < l$

$$F_n = w\frac{l}{2} + W = 2W$$

$$= \frac{2W}{l}\cdot\frac{1}{2} + W = W + W$$

(constant)

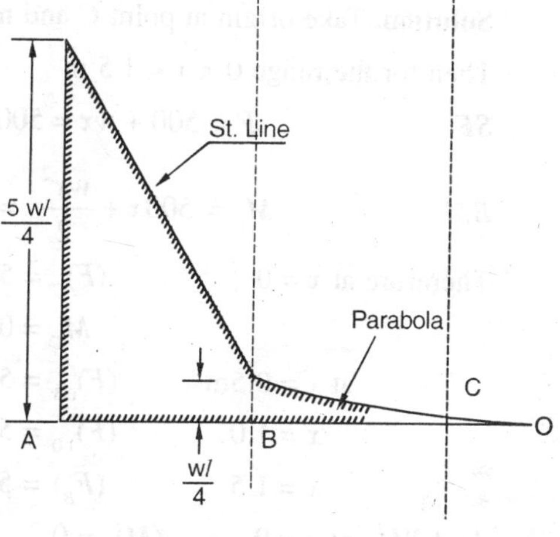


*Fig. 3.18*

**For *BM* diagram**

For the range $0 < x < \frac{l}{2}$

$$M_x = \frac{wx^2}{2} = \frac{2W}{l}\frac{x^2}{2} = \frac{Wx^2}{l} \qquad (A)$$

At $x = 0$ $\qquad M_x = 0$

$x = \frac{l}{2}$ $\qquad M_B = \frac{w}{2}\left(\frac{l}{4}\right)^2 = \frac{w}{l}\frac{l^2}{16} = \frac{Wl}{16}$

For range $\frac{l}{2} < x < l$

$$M_x = \left(\frac{wl}{2}\right)\left(x - \frac{l}{4}\right) + W\left(x - \frac{l}{2}\right)$$

$$= \left(\frac{2W}{l}\frac{l}{2}\right)\left(x - \frac{l}{4}\right) + W\left(x - \frac{l}{2}\right)$$

$$= W\left(x - \frac{l}{4}\right) + w\left(x - \frac{l}{2}\right)$$

$$= 2Wx - \frac{3l}{4}W \qquad \text{(st. line)}$$

At $x = \frac{l}{2}$ $\qquad M_B = 2W\frac{l}{2} - \frac{3l}{4}W = \frac{Wl}{4}$

$x = l$ $\qquad M_A = 2Wl - \frac{3l}{4}W = \frac{5W}{4}l$

3.7 Draw the *SF* and *BM* diagrams of the cantilever shown in the figure 3.19 (*a*).

**Solution.** Take origin at point *C* and measure distance *x* toward left as positive.

Then for the range $0 < x < 1.5$ m

*SF* $\qquad F = 500 + wx = 500 + 100x$ (st. line)

*BM* $\qquad M_x = 500x + \frac{wx^2}{2} = 500x + 50x^2$ (parabola)

Therefore at $x = 0$ $\qquad (F)_C = 500$

$M_C = 0$

at $x = 0.5$m $\qquad (F)_{0.5} = 500 + 100 \times .5 = 550$ kN

$x = 1.0$ $\qquad (F)_{1.0} = 500 + 100 \times 1 = 600$ kN

$x = 1.5$ $\qquad (F_B) = 500 + 100 \times 1.5 = 650$ kN

And *BM* at $x = 0$ $\qquad (M)_0 = 0$

*at* $x = .5$ $\qquad M_{0.5} = 500 \times .5 + 50 \times (.5)^2 = 250 + 12.5 = 282.5$ kN-m

$x = 1.0$ $\qquad (M)_{1.0} = 500 \times 1.0 + 50\ (1)^2 = 550$ kN-m

$x = 1.5$ $\qquad (M)_{1.5} = M_B = 500 + 1.5 + 50 \times (1.5)2 = 750 + 112.5$

$= 862.5$ kN-m

**For the range** $1.5 < x < 2$

*SF* $\qquad F = 500 + 100 + 1.5 = 650$ kN const.

*BM* $\qquad M = 500x + w\ (1.5)\ (x - 0.75)$

$= 500x + 150\ (x - 0.75)$ (st. line)

Therefore $\qquad M_{1.5} = 500 + 1.5 + 150\ (1.5 - .75)$

or $\qquad M_B = 750 + 150 \times 0.75$

$= 750 + 112.5 = 862.5$ *K*N-m

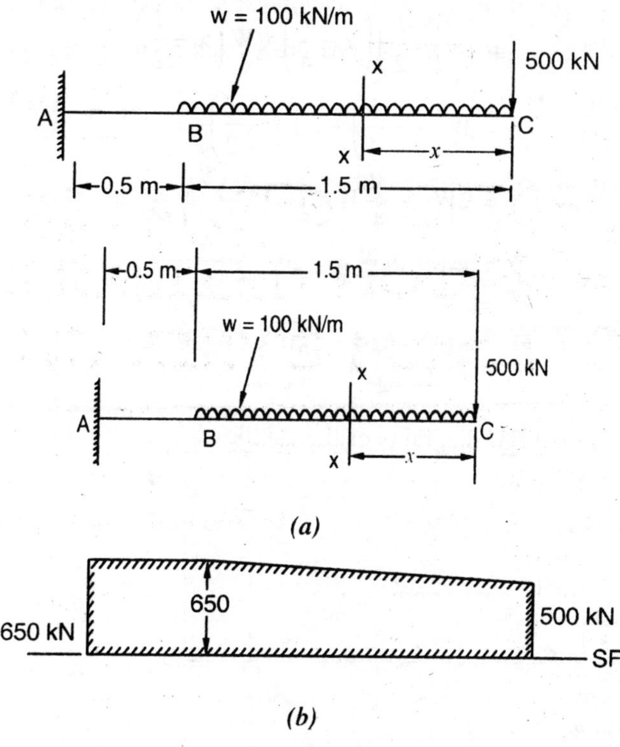


*(a)*



*(b)*

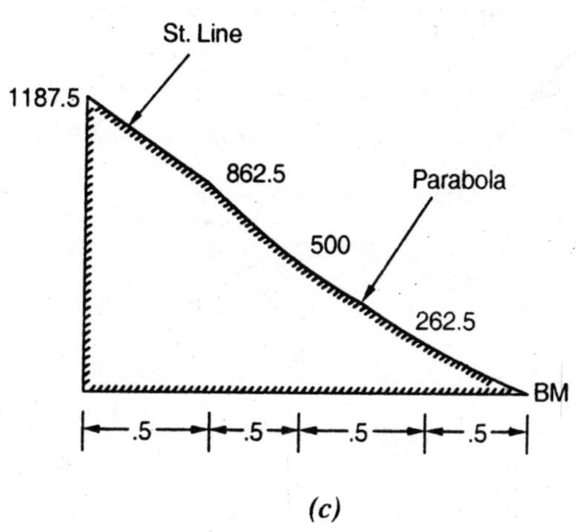


*(c)*

*Fig. 3.19*

Similarly at $x = 2.0$

$$M_A = 500 \times 2 + 150\,(2 - .75)$$

$$= 1000 + 150 \times 1.25 = 1187.5 \text{ kN-m}$$

The SF and *BM* diagrams are as shown in figure 3.19

3.8 Draw $SF$ and $BM$ diagram for the cantilever shown in figure (3.20)

**Solution.** (Take origin at $D$ and measure $x$ from $D$, left words positive)

For Range $0 < x < 0.5$)

$F = 0, M = 0$

because there is no load sustained by portion $CD$

For the range $0.5 < x < 1.5$ m

$$F_x = w(x - 0.5)$$

At $x = .5$ $\quad F_C = 0$

At $x = 1.5$ $\quad F_B = w(1.5 - .5) = 500 \times 1 = 500$ kN-m

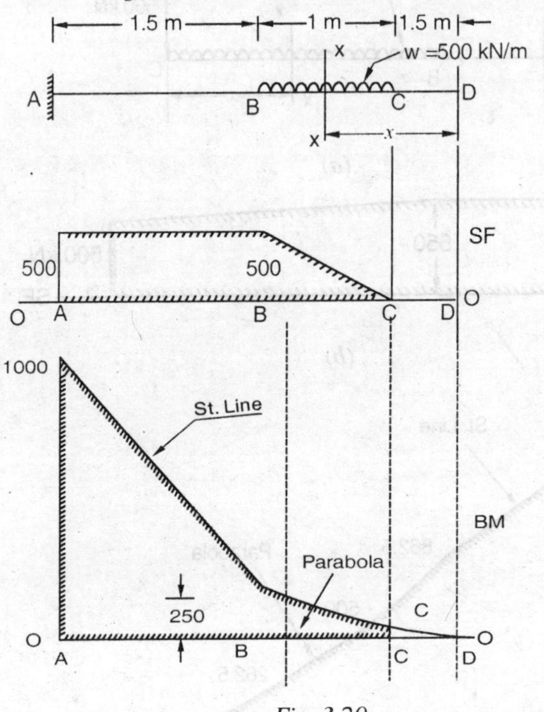


*Fig. 3.20*

$$M_x = \frac{w}{2}(x - 0.5)^2 \qquad \text{(parabola)}$$

At $x = .5$m $\quad M_C = 0$

At $x = 0.6$ $\quad (M)_{.6} = \dfrac{500}{2}(0.6 - 0.5)^2 = \dfrac{500 \times .01}{2}$

$= 2.5$ kN m

At $x = 0.7$ $\quad (M)_{.6} = 250 \times (0.7 - .5)^2 = 250 \times 0.04$

$= 10.0$ kN-m

At $x = 0.8$ $\quad (M)_{.8} = 250 \times (.8 - .5)^3 = 22.5$ kN-m

At $x = 1.0$ $\quad (M)_{1.0} = 250 \times (1.2 - .5)^2 = 250 \times 0.49 = 122.5$ kN m

at $x = 1.5$ $\quad (M)_B = 250\ (1.5 - .5) = 250.0$ kN-m

For Range $1.5 < x < 3.0$

$$F = w \times 1 = 500 \text{ kN constant}$$

$M = (w \times 1)\ (x - 1)$ (because the *CG* of *u/d* load is at mid pint of *BC* is it is at 1.0 m from *D*)]

$$= 500\ (x - 1) \text{ (st. line)}$$

Therefore at $\quad$ at $x = 1.5$

at $x = 3$

$$M_B = 500 \times (1.5 - 1) = 250 \text{ kN-m}$$

$$M_A = 500\ (3 - 1) = 100 \text{ kN-m}$$

Accordingly *SF* and *BM* diagrams are drawn.

3.9 For the beam carrying varying load, shown is figure draw *SF* and *MB* diagrams. Rate of loading increases from 0 at *A* to 100 kN-m at *C* and gradually falls to zero at *B*.

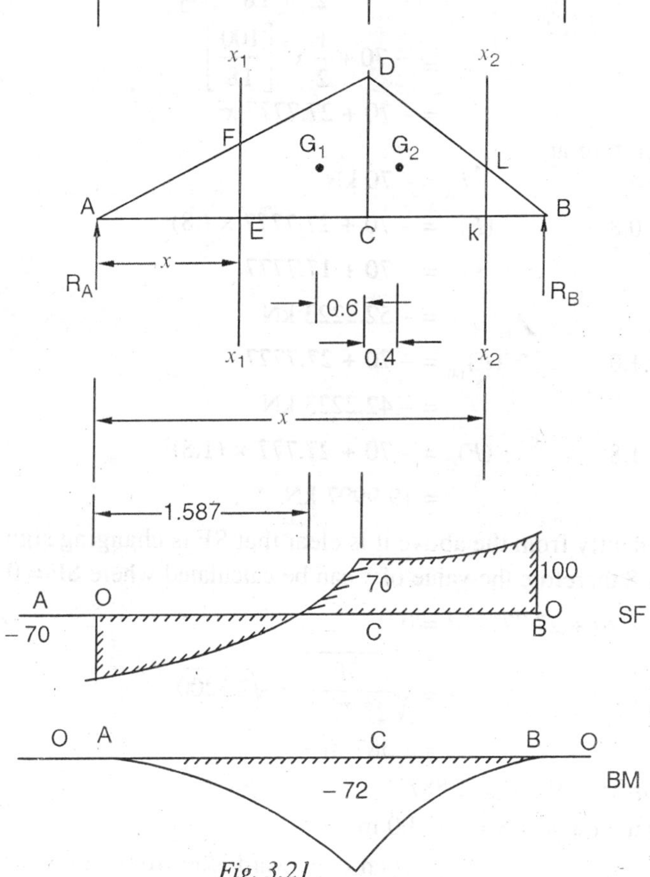


*Fig. 3.21*

Distance $AC = 1.8$ m

and $BC = 1.2$ m

**Solution.** Splitting the load into two triangular loads $ACD$ and $DCB$. Taking moments about $B$ for the static equilibrium of the beam, we get

$$R_A\ (1.8 + 1.2) - \frac{1}{2}(1.8)100\ (0.6 + 1.2)\ = -\frac{1}{2}(.12)\ 100\ (0.8) = 0$$

$$3RA - 90 \times 1.8 - 48 = 0$$

$$R_A = \frac{210}{3} = 70 \text{ kN}$$

For the range $0 < x < 1.8$ m with origin at $A$

$$F_x = -R_A + \text{triangular weight } AEF$$

$$= -70 + \frac{1}{2}x(EF)$$

$$= -70 + \frac{1}{2}x\left[\frac{CD}{1.8} \times x\right]$$

$$= -70 + \frac{1}{2}x^2\left[\frac{100}{1.8}\right]$$

$$= -70 + 27.7777\, x^2 \qquad \text{(parabola)}$$

Therefore at

$x = 0$ $\qquad F_A = -70$ kN

$x = 0.8$ $\qquad (F)_{.8} = -70 + 27.7777 \times (.8)^2$

$= -70 + 17.7777$

$= -52.2223$ kN

$x = 1.0$ $\qquad (F)_{1.0} = -70 + 27.7777$

$= -42.2223$ kN

$x = 1.8$ $\qquad (F)_C = -70 + 27.777 \times (1.8)^2$

$= 19.9997$ kN

Evidently from the above it is clear that SF is changing sign within the range $0 < x > 1.8$ therefore the value of $x$ can be calculated where SF = 0 as under

$$-70 + 27.7777\, x^2 = 0$$

or $$x = \sqrt{\frac{70}{27.7777}} = \sqrt{2.5200}$$

or $$x = 1.587 \text{ m}$$

Thus $SF = 0$ at $x = 1.587$

For the range $1.8 < x < 3.0$ m

$$F_x = \text{Total downward load from a to } k$$

*i.e.,* $F_x = -R_A$ + Trapezium load *KLDC*

and triangular load *ADC*

Now $\dfrac{KL}{KB} = \dfrac{CD}{1.2}$ (From similar $\Delta$s *BKL* and *BCD*]

or $KL = (KB)\dfrac{CD}{1.2} = (3 - x)\dfrac{100}{1.2}$

Also $CK = AK - AC = (x - 1.8)$

Therefore $F_x = -70 + \dfrac{1}{2}(1.8) \times 100 \ + \dfrac{1}{2}\left[CD + (3 - x)\dfrac{100}{1.2}\right](x - 1.8)$

$= -70 + 90 + \dfrac{1}{2}[100 + 50\,(x - 1.8)\ \ 41.66\,(3 - x)\,(x - 1.8)$

$\therefore$ At $x = 1.8$ $\quad F_C = 20 + \dfrac{1}{2}(100 + 0 + 0] = 70$ kN

At $x = 2$m $\quad (F)_{2.0} = 20 + \dfrac{1}{2}[100 + 50 \times .2 + 41.66\,(3.2)\,(0.2)$

$= 20 + \dfrac{1}{2}[110 + 8.332] = 79.166$ kN

At $x = 2.5$ $\quad (F)_{2.5} = 20 + \dfrac{1}{2}[100 + 50\,(2.5 - 1.8) + 41.66\,(0.5)\,(0.7)]$

$= 20 + \dfrac{1}{2}[135 + 14.581] = 94.79$ kN

At $x = 3.0$ $\quad F_B = 20 + \dfrac{1}{2}[100 + 50 \times 1.2 + 41.66 \times 0 \times 1.2]$

$= 20 + \dfrac{1}{2}[160] = 100$ kN

For BM diagram

for range $0 < x < 1.8$ m

$$M_x = -R_A x + \frac{1}{2}(AE)\,(EF)\left(\frac{1}{3}x\right)$$

$$= -70x + \frac{1}{2}x\left(\frac{CD}{1.8} \times x\right)\frac{x}{3}$$

$$= -70x + \frac{50}{1.8}x^2\left(\frac{x}{3}\right)$$

$$= -70x + 9.259x^3$$

At $x = 0$ $\quad M_A = 0$

At $x = 0.8$ $\quad (M)_{0.8} = -70 \times .8 + 9.259 \times (.8)^3$

$= -56 + 4.746 = -51.2594$ kN-m

At $x = 1$ $\quad (M)1.0 = -70 \times 1 + 9.259$

$= -68.74$ kN-m

At $x = 1.8$ $\quad (M)_C = -70 \times 1.8 + 9.259 \times (1.8)3$

$= -126 + 53.9985$

$= -72$ kN-m

At $n = 1.187$ $\quad (M)_{max} = -70 \times 1.587 + 9.259\,(1.587)^3$

$= 111.09 + 37.008$

$= -74.82$ kN-m

For range 1.8 < x < 3

$$M_X = R_A x + \frac{1}{2} AC \times CD \times (x - 1.2)$$

$$+ \left(\frac{1}{2} NL \times ND\right)\left(\frac{2}{3} CK\right) + \frac{1}{2}(LK \times CK)\left(\frac{1}{2} CK\right)$$

$$= -70x + \frac{1}{2}(1.8)\,(100)\,(x - 1.2)$$

$$+ \frac{1}{2}(x - 1.8)\left[100 - \frac{100}{1.2}(3 - x)\right] \times \left(\frac{2}{3}\right)(x - 1.8) + \frac{1}{2}\left[(3 - x)\frac{100}{1.2}\right][x - 1.8]^2$$

$$= -70x + 90(x - 1.2) + \frac{1}{2}(x - 1.8)\,[100]\left[1 - \frac{1}{1.2}(3 - x)\right]$$

$$\times \frac{2}{3}(x - 1.8) + \frac{1}{2} \times \left[\frac{100}{1.2}(3 - x)\right](x - 1.8)^2$$

$\therefore$ At $x = 1.8$ $\quad M_C = -70 \times 1.8 + 90 \times 0.6 + 0 + 0 + 0$

$= -126 + 54 = -72$ kN-m

At $x = 2$

$$(M)_{x=2} = -70 \times 2 + 90(2 - 1.2) + \frac{1}{2}(2 - 1.8)\,[100]\left[1 - \frac{1}{1.2}\right]$$

$$\times \frac{2}{3}(2 - 1.8) + \frac{50}{1.2}(3 - 2)\,(2 - 1.8)^2$$

$$= -140 + 72 + (0.2 \times 50)\left(\frac{1.2 - 1}{1.2}\right) \times \frac{2}{3} \times 0.2 + \frac{50}{1.2}(0.2)^2$$

$= -68 + .2222 + 1.6666$

$= -66.1112$ kN-m

At $x = 2.5$

$$(M)_{2.5} = -70 \times 2.5 + 90\,(2.5 - 1.2) + \frac{1}{2}(2.5 - 1.8)\,100$$

$$\times \left[1 - \frac{(3 - 2.5)}{1.2}\right] \times \frac{2}{3}(2.5 - 1.8) + \frac{50}{1.2}(3 - 2.5)\,(2.5 - 1.8)^2$$

$$= -\ 175 + 117 + 9.52778 + 10.2083$$
$$= -\ 38.2639 \text{ kN-m}$$

At $x = 3$ $\quad (M)_B = -\ 70 \times 3 + 90(3 - 1.2) + \frac{1}{2}(3 - 1.5)$

$$\times\ 100\,(1.0)\,\frac{2}{3}\,(3 - 1.8) + \frac{50}{1.2}\,(0)\,(3 - 1.8)^2$$

$$= -\ 210 + 162 + 48 = 0$$

Thus the *SF* and *BM* are obtained by plotting these points

3.10 A beam 5m long is simply supported at its ends. It caries a u/d load of 25 kN-m over *AC* and a point load at *D* 1m from end *B* draw *SF* and *BM* diagrams for this beam.

**Solution.** For finding the reactions take moments about point B for the equilibrium of beam.

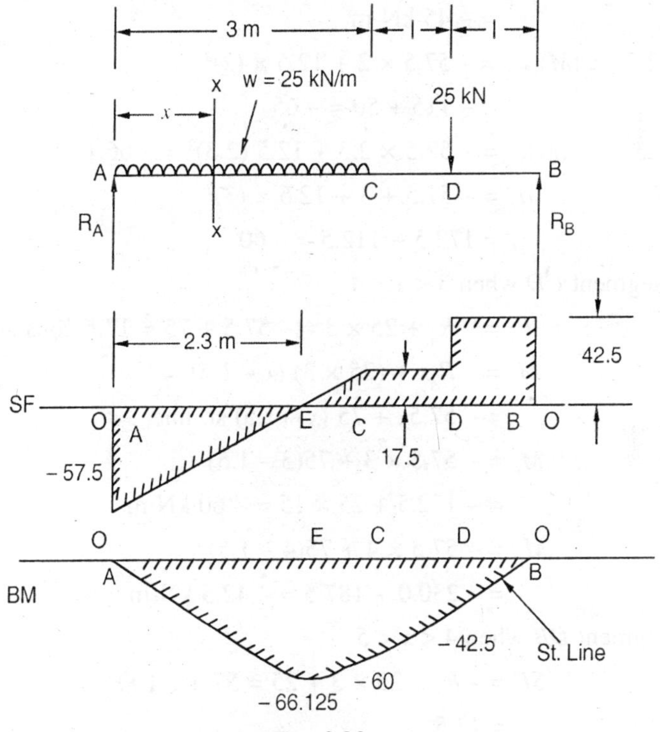


*Fig. 3.22*

$$5R_A - (25 \times 3)\,(3.5) - 25 \times 1 = 0$$

or
$$R_A = \frac{1}{5}\,[262.5 + 25] = 57.5 \text{ kN}$$

$$R_B = \text{Total load} - R_A$$
$$= 25 \times 3 + 25 - 57.5 = 42.5 \text{ kN}$$

Take origin at $A$ and measure $x$ from $A$.

For the segment $AC$ where $0 < x < 3$

$$F_x = -R_A + wx = -57.5 + 25 \times x \text{ st line}$$

$$M_x = -R_A x + 25\frac{x^2}{2} \qquad \text{(parabola)}$$

$$(F)_0 = -57.5 \text{ kN}$$

$$(F)_{x=3} = F_C = -57.5 + 25 \times 3 = 17.5 \text{ kN}$$

It changes sign and $F_C = 0$ when $x = \dfrac{57.5}{25} = 2.3$ m

$$(M)_A = 0 \text{ at } x = 0$$

At $x = 1$ $\quad (M)_{x=1} = -57.5 \times 1 + 12.5 \times (1)^2$

$$= -45 \text{ kN-m}$$

At $x = 2$ $\quad (M)_{x=2} = -57.5 \times 2 + 12.5 \times (2)^2$

$$= -115 + 50 = -65$$

At $x > 2.3$ $\quad (M)_{2.3} = -57.5 \times 2.3 + 12.5\,(2.3)^2 = -66.125$

At $x = 3$ $\quad M_C = -57.5 + 3 + 12.5 \times (3)^2$

$$= 172.5 + 112.5 = -60$$

For the segment $CD$ when $3 < x < 4$

$$F_x = -R_A + 25 \times 3 = -57.5 + 75 = 17.5 \text{ constant}$$

$$M_x = -R_A x + (25 \times 3)\,(x - 1.5)$$

$$= -57.5x + 75\,(x - 1.5) \text{ st. line}$$

At $x = 3$ $\quad M_C = -57.5 \times 3 + 75(3 - 1.5)$

$$= -172.5 + 25 \times 15 = -60 \text{ kN m}$$

At $x = 4$ $\quad M_D = -57.5 \times 4 + 75(4 - 1.5)$

$$= -230.0 + 187.5 = -42.5 \text{ kN-m}$$

For the segment $DB$ when $4 < x < 5$

$$SF = -R_A + 25 \times 3 + 25 = 57.5 + 100$$

$$= 42.5$$

$$M_x = -R_A x + (25 \times 3)\,(x - 1.5) + 25\,(x - 4) \text{ st line}$$

$$(M_x)_{x>4} = M_D = -57.5 \times 4 + 75\,(4 - 1.5) + 0$$

$$= -42.5 \text{ kN-m}$$

At $x = 5$ $\quad (M_B) = -57.5 \times 5 \times 75\,(5 - 1.5) + 25\,(5.4)$

$$= 0 - 287.5 + 262.5 + 25 = 0$$

With the above information the $SF$ and $BM$ diagrams are drawn as shown.

## PROBLEMS FOR PRACTICE

1. A beam, 10 m long, is simply supported over a span of 5 m, the overhangs on the left and right hand supports being 3 m and 2 m respectively. The left and right hand overhangs carry uniformly distributed loads of 60 kN/m and 105 kN/m respectively. The latter load also extends upto left hand support. Draw to scale the shearing force and bending moment diagrams for the beam. Calculate the value of the least B.M. between the supports and find the position where it occurs. [**Ans.** 41.85 kN–m at 3.9 m from the left hand support]

2. An overhanging beam *ABC*, 9 m long, is simply supported at points *A* and *B*, 7 m apart. The beam carries a load of 225 kN uniformly distributed over its whole length together with three point loads of 120 kN, 70 kN and 150 kN at 1 m, 4 m and 6 m respectively from the point *C*.

   Draw to scale the S.F. and B.M. diagrams giving the maximum value of B.M.

   [**Ans.** 394.3 kN–m at 3 m from *A*]

3. A wooden beam 5 m long and 0.4 m × 0.4 m in section, floating in horizontal position is loaded with a point load at the centre just sufficient to immerse it completely. If the mass of 1 cubic metre of wood is 800 kg, draw and S.F. and B.M. diagrams for the beam. What are the maximum values of S.F. and B.M. and where do they occur ? [**Ans.** 1570 N and 981 Nm both at the centre]

4. If the beam in question 3 were to be loaded with two weights placed symmetrically at distance '*a*' from the ends so that the maximum bending moment to which the beam is subjected is as small as possible, find the value of each weight and its position. What is then the magnitude of the maximum bending moment ?

   [**Ans.** 785 N ; 1.035 m from each end ; 169 Nm]

5. A concrete pile, of length L and uniform cross section, is to be lifted with the help of a crane by putting two slings symmetrically around it so that it always stays horizontal. Find the most suitable positions for putting the slings if it is assumed that the pile is most adversely affected by the bending moments caused by self–weight and not by shearing forces. [**Ans.** 0.207 *L* from each end]

6. A horizontal beam *AB*, 5 m long is supported freely at the ends. It carriers a vertical load of 15 kN at a distance of 4.0 m from *B*. In addition there acts, at a section 3 m from *B*, clockwise couple of magnitude 5 kN–m on the beam. The plane of the couple being vertical and passing through the axis of the beam, draw the B.M. and S.F. diagrams. Where does the maximum B.M. occur ? Find its value. [**Ans.** 3 m from *B* ; 12 kN–m]

7. A beam *AB*, 7 m long, is simply supported at its ends. It carries a load of 30 kN, applied through a *L*–shaped bracket *CDE* which is fixed to the beam at C, distant 2 m from *A*, such that *CDE* lies in the vertical plane passing through the axis of the beam and the portion *DE* is towards the side of point *B*. If *DE* = 0.5 m. Draw

the S.F. and B.M. diagrams for the beam *AB* giving the values of maximum S.F. and B.M. $\left[\textbf{Ans. } \frac{135}{7} \text{ kN} ; \frac{375}{7} \text{ kN} - \text{m just to the right of C}\right]$

**8.** A cantilever, 2 m long, carries a load varying from zero at the free and to 50 kN/m run at the fixed end. Draw the S.F. and B.M. diagrams. What are the values of maximum S.F. and bending moment ?

[**Ans.** 50 kN ; 33.33 kN–m both at the fixed end]

**9.** An overhanging beam *ABCD*, 18 m long, is simply supported at *A* and *C* 12 m apart. It carries a load of 15 kN concentrated at *B* which is at a distance of 4 m from *A*. In addition there acts a distributed load with intensity varying from zero at *A* and *D* to 15 kN/m run at *C*.

Draw the S.F. and B.M. diagrams and calculate the position and magnitude of maximum B.M. Find from diagram the position of the point of inflexion.

[**Ans.** 5.3 m from A ; 121.6 kN–m ; 10.55 m]

**10.** The shearing force diagram, for a beam, simply supported at *A* and *B* is given in Fig. 3.23. Draw the B.M. diagram for the same.

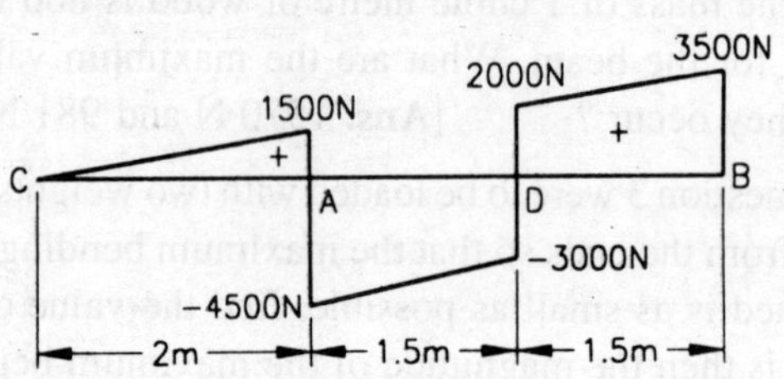


*Fig. 3.23*

**11.** For an overhanging beam *ABCD*, simply supported at *B* and *C*, the S.F diagram is given in Fig. 3.24. Draw the B.M. diagram.

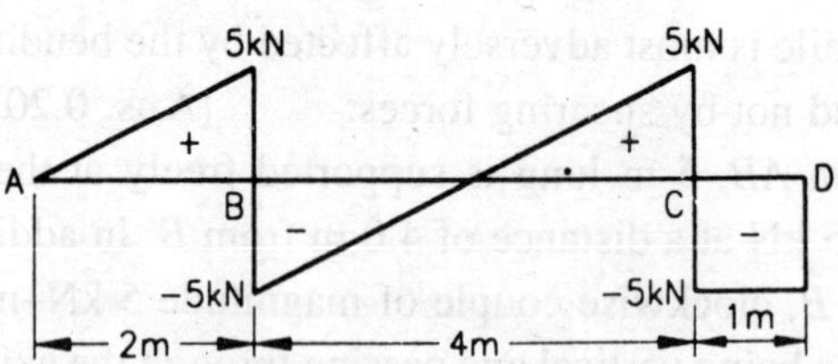


*Fig. 3.24*

Chapter–4

# Bending Stresses

## 4.1 Simple or Pure Bending

It was seen in article 3.4 that at any point of a laterally loaded beam, the rate of change of bending moment is numerically equal to the shearing force on the beam at that point. If the bending moment is constant over a length of a beam, evidently the shearing force over that length will be zero. The bending moment over portion of the beam where it is not accompanied by shearing force, is called pure bending moment and the bending of the beam as pure or simple bending. However in actual practice such a situation rarely arises and invariably the bending moments are accompanied by shearing forces, but the distortion caused by the shearing action is usually very small. The aim of the present chapter will be to study the effect of bending moments only over a section of a beam in as much as to how the material is stressed at various points of a section of the beam.

## 4.2 Theory of Simple Bending

If an initially straight beam is acted upon by a pure bending moment M the beam will get bent as shown in Fig. 4.1. Evidently the fibres in the upper portion will be extended and those in the lower part will get compressed. Alternatively the fibres which are nearer to the centre of the curvature of the bent beam will be under compression and those away in tension. In between there will be some surface NS the fibres along which will neither extend nor shorten. This surface is known as the neutral surface and the intersection of this surface with any cross sectional plane is called the neutral axis of the section. It is represented by NA in Fig. 4.1. In evaluating the tensile or compressive stresses and the curvature of the beam following assumptions will be made :

1. The beam is initially straight and the radius of curvature of the bent beam is very large as compared to the dimensions of the cross–section.

2. The material of the beam is homogeneous and isotropic, that is, its properties are alike in all directions and at all points.

3. The material is not stressed beyond the elastic limit and obeys Hooke's law.

4. The values of a Young's modulii of elasticity in tension and compression are equal.

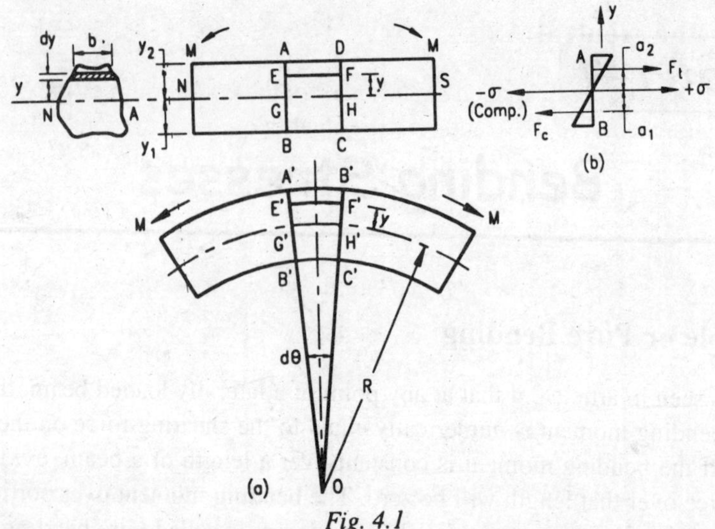


*Fig. 4.1*

5. Transverse plane sections remain plane even after bending.

6. Each layer is free to expand or contract independently without being constrained by the adjoining fibres.

Let the radius of curvature of the neutral surface of the beam at the section *AB*, be *R* such that $EF = GH = G'\ H' = R\ (d\ \theta)$. Considering the change in length of the fibre *EF* at a distance '*y*' from the Neutral Surface the tensile strain along *EF* is

$$\varepsilon = \frac{E'F' - EF}{EF} = \frac{[(R+y)\,d\theta - Rd\theta)]}{Rd\theta}$$

$$= \frac{y}{R} \qquad \text{...(4.1)}$$

If σ is the direct stress along the fibre *EF* and *E* the Young's modulus of the material, then equation (4.1) gives

$$\frac{\sigma}{E} = \frac{y}{R}$$

or

$$\frac{\sigma}{y} = \frac{E}{R} \qquad \text{...(4.2)}$$

From equation (4.2) it is evident that the direct stress at a point in the cross–section increases with y linearly. Its value will be zero at any point lying on the neutral axis. Similarly if *y* is measured from the neutral axis to any point in the region nearer to the centre of curvature the direct stress there will be compressive and increase linearly with *y*. Since the value of the Young's modulus *E* has been assumed to be same for tension and compression, the stress distribution curve in the two regions will have the same slope, rather extension of the same line in the two regions Fig. 4.1(*b*).

## Position of the Neutral Axis

If there is no externally applied axial force, for the equilibrium of the portion of the beam lying on the left of the section $AB$ the algebraic sum of the forces in the axial direction will be zero. That is, if the total force in the tension region is $F_t$ and that in the compression region $F_c$ then

$$F_t + F_c = 0 \quad ...(4.3)$$

Now if an elemental area $dA$ is taken at a distance $y$ from the neutral axis

$$F_t = \int_0^{y_2} \sigma dA$$

and

$$F_c = \int_{-y_1}^{0} adA$$

Where $y_1$ and $y_2$ are the maximum values of $y$ in compression and tension regions respectively. Substituting the value of $\sigma$ from equation (4.2).

$$F_t = \frac{E}{R}\int_0^{y_2} \sigma dA$$

$$= \frac{E}{R} \text{(First moment about } NA \text{ of sectional area above } NA) \quad ...(4.4)$$

and

$$F_c = \frac{E}{R}\int_{-y_1}^{0} ydA$$

$$= \frac{E}{R} \text{(First moment about } NA \text{ of sectional area below } NA) \quad ...(4.5)$$

From equation (4.3) through (4.5) we obtain

$$= \frac{E}{R} \text{(First moment about } NA \text{ of the cross section)} = 0 \quad ...(4.6)$$

But $E/R$ cannot be zero, hence from equation (4.6) the first moment about $NA$ of the cross section must be zero. This will only be true if the neutral axis $NA$ passes through the centroid of the section.

## Evaluation of Radius of Curvature

From equation (4.3), it is evident that the two forces $F_t$ and $F_c$ will be equal and opposite. The couple produced by these forces acting at a distance $(a_1 + a_2)$ apart, Fig. 4.1($b$), will balance the moment $M$ for the equilibrium of the portion of the beam to the left of the section $AB$. The moment $F_t(a_1 + a_2)$ is called the resisting moment of the section.

Thus $\qquad M = F_t (a_2 + a_1)$

$$= F_t a_2 + F_c a_1 \qquad \text{(as } F_t = F_c\text{)}$$

= Moment of the tensile forces about NA

+ Moment of compressive force about *NA*

$$= \int_0^{y_2} (\sigma dA)\, y + \int_{-y_1}^{0} (\sigma dA)\, y$$

$$= \int_0^{y_2} \left(\frac{yE}{R} dA\right) y + \int_{-y_1}^{0} \left(\frac{E}{R} ydA\right) y \qquad \text{(From Eq. 4.2)}$$

$$= \frac{E}{R}\left[\int_0^{y_2} (dAy)^2 + \int_{-y_1}^{0} (dAy^2)\right]$$

$$= \frac{E}{R}\int_{-y_1}^{y_2} (y^2 dA)$$

$= \dfrac{E}{R}$ × (Second moment of the area about NA which passes through the centroid)

$$= \frac{E}{R} . I$$

or $\qquad M/I = E/R \qquad$ ...(4.7)

Where $I$ is the second moment of the area of cross–section about a centroidal axis *NA*, which is also parallel to the axis of the moment $M$.

The equations (4.2) and (4.7) can also be written in the usual form as

$$\frac{\sigma}{y} = \frac{M}{I} = \frac{E}{R} \qquad \text{...(4.8)}$$

Knowing the values of $M$ and $I$ at any section of a beam the radius of curvature $R$ of the neutral surface at that section can be found with the help of equation (4.8).

Again from equation (4.8)

$$\sigma_{max} = \frac{M}{I/y_{max}} = \frac{M}{Z}$$

Where $Z = I/y_{max}$. is known as section modulus.

### 4.2.1 Second Moment of Area, Theorem of Perpendicular axes

It is defined as the summation of the product of elemental areas and the square of their respective distances from a given axes, for the whole area.

Let $dA$ be an elemental area at a distance $y$ from $x$-axis. The integral of $\oint y^2\, dA$ for the entire area is called the second moment of area and represented by $I_{xx}$.

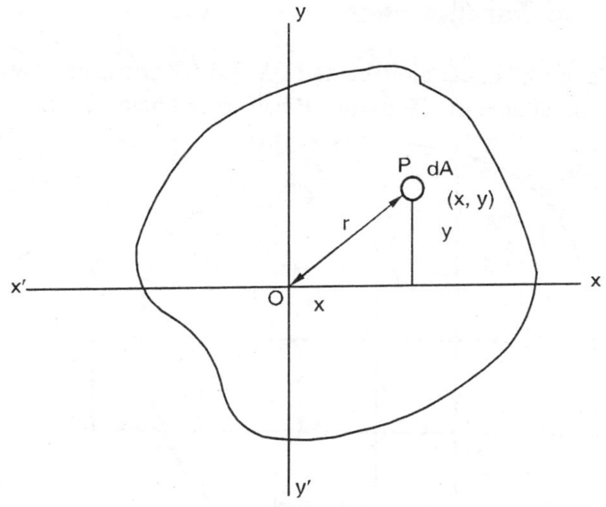


*Fig. (4.1a)*

Thus, $$I_{xx} = \oint y^2 dA \quad ...(A)$$

Similarly if the distance of $dA$ is $x$ from $y$-axis, then second moment of area about $y$-axis will be

$$I_{yy} = \oint x^2 dA \quad ...(B)$$

However if we take $Z$-axis which is normal to the plane $X - Y$ and passing through 0 then $I_{zz}$, which is called the polar second moment of area is defined as

$$I_{zz} = \oint r^2 dA \quad ..(C)$$

When $r$ is the distance of $dA$ from $zz$ axis

Expression ($C$) can also be $w$-written as

$$I_{zz} = \oint (x^2 + y^2)\, dA$$

$$= \oint y^2 dA + \oint x^2 dA$$

$$= I_{xx} + I_{yy} \quad ...(D)$$

Equation ($D$) is known as theorem of perpendicular axes. That is the polar second moment of area of any section about an axis $OZ$, which is normal to the a in equal to the sum of second moment of areas about two perpendicular axes $XX$ and $YY$ lying in the cross-sectional plane.

Till the middle of twentieth century the term I had been called as moment of Inertia. The Scientists from all over the world collectively decided to call this as second moment of area and the MassMoment of Inertia for the second moment of the elemental mass. But still some of the authors continue with and misnomer the call it moment of inertia rather than second moment of area.

## 4.2.2 Theorem of Parallel Axes

Let 0 be the centroid of certain area and *OX* and *OY* two axes of reference figure [4.1 (*b*)]. The elemental area of *dA* around *P* has co-ordinates (*x*, *y*).

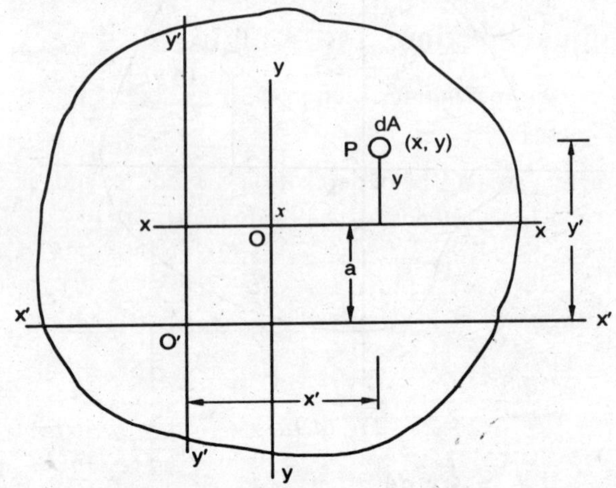


*Fig. (4.1b)*

The second moment of entire area Å about *XX* is given by

$$I_{xx} = \oint y^2 dA \qquad \text{...(D)}$$

Let $X'O'Y'$ be another set of axes such that the co-ordinates of point *P* with respect to $X'O'Y'$ axes are $(x', y')$

Then $$I_{x'x'} = \oint (y')^2 dA \qquad \text{...(E)}$$

$$= \oint [y + (y' - y)]^2 dA$$

Let $$y' - y = a$$

Then $$I_{x'x'} = \oint (y + a)^2 dA$$

$$= \oint (y^2 + a^2 + 2ay)\, dA$$

$$= \oint y^2 dA + a^2 A + 2a \oint y dA$$

But $\oint ydA$ is first moment of the area about centroid 0 of the section. It must be zero as *X*-axis passes through centroid of the section

Thus $$I_{x'x'} = I_{xx} + Aa^2 \qquad \text{...(F)}$$

Equation (*F*) is known as theorem of parallel axes. It states that second moment of area of any section about *X'X'* which is parallel to *XX* passing through centroid, is equal to $I_{xx}$ about the centroidal axis plus area times square of the distance between these axes.

### 4.2.3 Second Moment of Areas of Standard Sections

In the following paragraphs, expressions for the second moment of area of various for the second moment of area of various commonly used sections, have been derived.

### 4.2.4 Rectangular Section

Let *ABCD* be the rectangular section with its centroid at 0. [Fig. 4.1(*c*)] with width as '*b*' and depth '*d*'.

Let the elemental area ***dA*** be of depth *dy* and width '*b*' and taken at a distance '*y*' from the axis *XX* and $I_{XX}$ is intended to be evaluated.

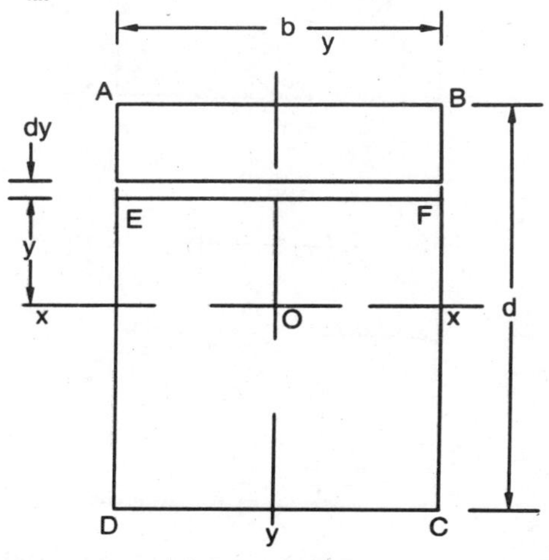


*Fig. (4.1c)*

Then
$$I_{XX} = \int_{\frac{-d}{2}}^{d/2} y^2 dA = \int_{\frac{-d}{2}}^{d/2} y^2 b dy = \left[\frac{y^3}{3}\right]^{d/2} = \frac{bd^3}{12}$$

Similarly by taking the elemental area strip parallel to *y* axes at a distance *x*, Fig. (4.1*d*) we have

$$I_{YY} = \int_{\frac{-b}{2}}^{b/2} x^2 dA$$

$$= \int_{\frac{-b}{2}}^{b/2} x^2 d \,.\, dx = \frac{db^3}{12}$$

Also
$$I_{DC} = I_{XX} + (bd)\left(\frac{d}{2}\right)$$

$$= \frac{bd^3}{12} + \frac{bd^3}{4} = \frac{bd^3}{3}$$

*Fig. (4.1d)*

It is important to note that is the expressions ($g$) and ($h$) the side of the rectangle which is parallel to the axis under consideration appears with exponent as unity and the perpendicular side with exponent 3.

### 4.2.4.1 Second Moment of Area of Square Section

If the two sides of the rectangular station are both equal to '$a$', '*say*' fig (4.1$e$)

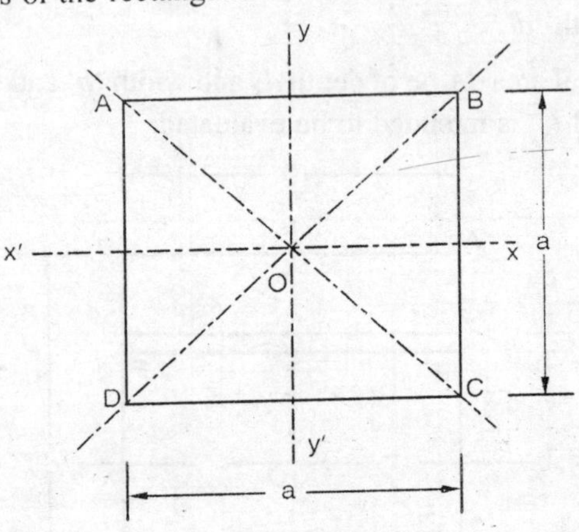


*Fig. (4.1e)*

Then $$I_{XX} = I_{YY} = \frac{a \times a^3}{12} = \frac{a^4}{12} \qquad ...(E)$$

Further since the two diagonals of square section are at right angles to each other, applying. Theorem of perpendicular axes, we get

$$I_{ZZ} = I_{XX} + I_{YY} = I_{AC} + I_{DB}$$
$$= 2I_{XX} = 2I_{DC}$$

Thus $$I_{XX} = I_{YY} = I_{AC} = I_{BD}$$

or $$I_{BD} = I_{AC} = I_{XX} = \frac{a^4}{12} \qquad ...(I)$$

Further by applying theorem of parallel axes we get

$$I_{DC} = I_{XX} + (\text{Area } ABCD) \cdot \left(\frac{1}{2}BC\right)^2$$

$$= \frac{a^4}{12} + a^2\left(\frac{a}{2}\right)^2$$

$$= \frac{a^4}{12} + \frac{a^4}{4} = a^4\left(\frac{1}{12} + \frac{1}{4}\right) = \frac{a^2}{3} \qquad ...(J)$$

## 4.2.5 Hollow Circular Section

Let $R_1$ and $R_2$ be the external and inner radii of a hollow circular section Fig (4.1 *f*)

Let an annular ring of radius $r$ and radial thickness $dr$ be taken as elemental area. This area is at a distance $r$ from the *z*-axis $Oz$ normal to the area of cross-section.

Then the polar second moment of area of this ring will be

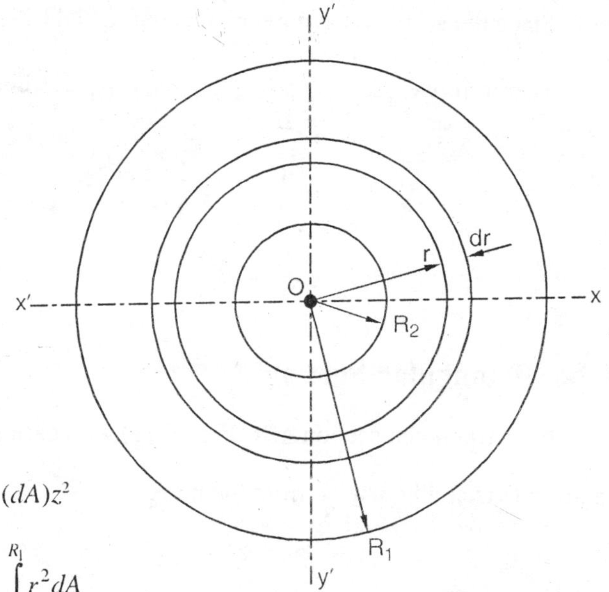

$$dI_{zz} = (dA)z^2$$

or
$$I_{zz} = \int_{R_2}^{R_1} r^2 dA$$

*Fig. (4.1f)*

But
$$dA = 2\pi r\, dr$$

$\therefore$
$$I_{zz} = \int_{R_2}^{R_1} r^2 (r\pi r dr)$$

$$= 2\pi \int_{R_2}^{R_1} r^3 dr$$

$$= 2\pi \left[\frac{r^4}{4}\right]_{R_2}^{R_1} = 2\pi \left(\frac{R_1^4}{4} - \frac{R_2^2}{4}\right)$$

or
$$I_{zz} = \frac{\pi}{2}\left(R_1^4 - R_2^2\right) \quad ...(K-1)$$

$$= \frac{\pi}{32}\left(D_1^4 - D_2^4\right) \quad ...(K-2)$$

For the solid shaft the diameter $D$ the value of $I_{zz}$ is obtained by putting $D_2 = 0$ and replacing $D_1$ with $D$

$$I_{zz} = \frac{\pi}{32} D^4 \quad ...(L)$$

But we know, with the help of theorem of perpendicular axes that

$$I_{ZZ} = I_{XX} + I_{YY}$$

Since the section is symmetrical about $XX$ and $YY$ axes

$$I_{XX} = I_{YY}$$

Therefore we get

$$I_{ZZ} = 2I_{XX}$$

or $$I_{XX} = \frac{1}{2} I_{ZZ} = \frac{1}{2}\left(\frac{\pi D^4}{32}\right)$$

*i.e.,* $$I_{XX} = I_{YY} = \frac{\pi D^4}{64} \quad \text{...}(M)$$

### 4.2.6 Triangular Section

For a triangular section $ABC$ Fig. (4.1$g$) with base $BC = b$ and height $AD = h$, the centroid $G$ is at a height $\frac{h}{3}$ from the base.

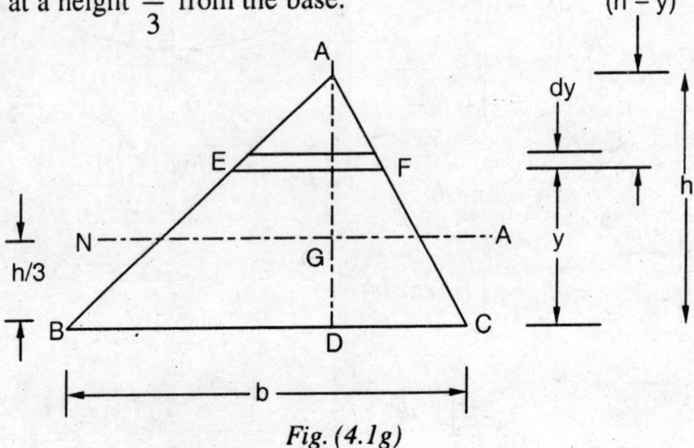


*Fig. (4.1g)*

Let an elemental area $dA$ be at a height '$y$' from the base and it thickness be $dy$

Then $$I_{BC} = \int_O^h (dAy^2) = \int_O^h y^2 (EF)\, dy$$

But $$EF = \frac{BC}{h}.(h - y) = \frac{b}{h}(h - y)$$

Therefore $$I_{BC} = \int_0^h y^2 \frac{b}{h}(h - y)\, dy$$

or $$I_{BC} = b\int_0^h \left(y^2 - \frac{y^3}{h}\right) dy$$

$$= b\left[\frac{y^3}{3} - \frac{y^4}{4h}\right]_0^h$$

$$= b\left[\frac{h^3}{3} - \frac{h^3}{4}\right] = \frac{bh^2}{12} \qquad ...(N)$$

Neutral axis passes through $G$, the centroid of the section

Invoking theorem of parallel axes we have

$$I_{BC} = I_{NA} + (\text{Area } ABC)\left(\frac{h}{3}\right)^2$$

$$\therefore \qquad I_{NA} = I_{BC} - \left(\frac{1}{2}bh\right)\left(\frac{h}{3}\right)^2$$

$$= \frac{bh^3}{12} - \frac{bh^3}{18} = \frac{bh^3}{36} \qquad ...(P)$$

## 4.2.7 Trapeziodal Section

Let $ABCD$ be the trapeziodal section of dimensions as shown in figure (4.1$h$)

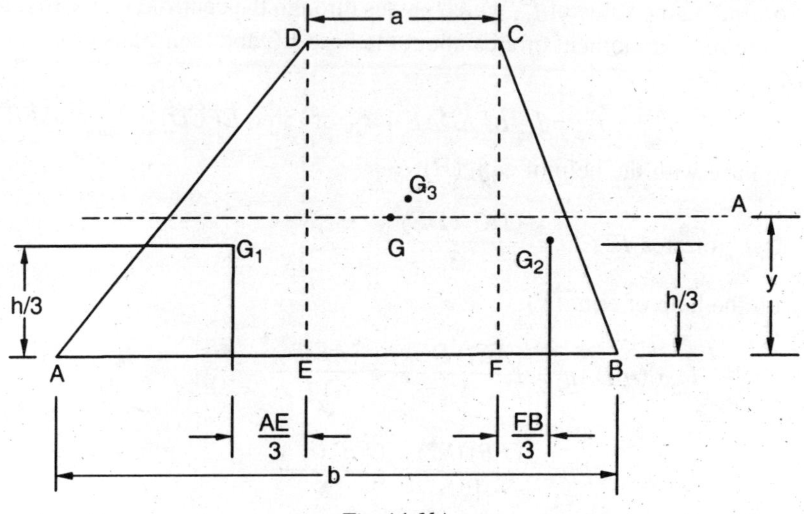


*Fig. (4.1h)*

*i.e.,* $\quad AB = b$

$CD = a$

and $\quad$ height $= DE = h$

For finding the location or specifically the height $y$ of centroid of the trapezium from the base $AB$, we take the moments of areas $ADE$, $DEFC$ and $CFB$ about the base

$AB$ (since the height of $G_1$ and $G_2$, the centroids of $\Delta$s $DEA$ and $BFC$ respectively, from base $AB$ is $h/3$ and centroid $G_3$ of rectangle $DCFE$, is at height $h/2$ from $AB$)

$$\text{Then(Area } ABCD)\,\bar{y} = \text{Area } (AED)\,\frac{h}{3} + (\text{Area } DEFC)\,\frac{h}{2} + (\text{area } CFB)\,\frac{h}{3}$$

or
$$\left[\frac{1}{2}(a+b)h\right]\bar{y} = \left(\frac{1}{2}hAE\right)\frac{h}{3} + (ah)\frac{h}{2} + \frac{1}{2}h\,(FB)\frac{h}{3}$$

$$= \frac{h^2}{6}(AE + FB) + \frac{ah^2}{2}$$

$$= \frac{h^2}{6}(b-a) + \frac{ah^2}{2}$$

$$= \frac{h^2}{6}(b - a + 3a) = \frac{h^2}{6}(b + 2a)$$

Therefore
$$\bar{y} = \frac{h^2}{6}(b+2a)\frac{2}{(a+b)\,h}$$

or
$$\bar{y} = \frac{h}{3}\left(\frac{b+2a}{b+a}\right) \qquad \text{...}(Q)$$

For finding the value of $I_{NA}$ which passes through the centroid $G$, we first find the value of the second moment of area about the base $BC$ and then transfer to centroidal axis $NA$

$$I_{AB} = I_{AB} \text{ of } \Delta DAE + I_{AB} \text{ of (area } EFCD) + I_{AB} \text{ of } \Delta FBE$$

We have with the help of eqn. ($H$)

$$I_{AB} \text{ of area } EFCD = \frac{(DC)\,(DE)^3}{3}$$

with the help of eqn. ($N$)

$$I_{AB} \text{ of } \Delta DAE = \frac{(AE)\,(DE)^3}{12} = \frac{(AE)h^3}{12}$$

and
$$I_{AB} \text{ of } \Delta CFB = \frac{(FB)\,(h^3)}{12} = \frac{(FB)\,(h^3)}{12}$$

Therefore adding these three

$$I_{AB} \text{ of trapezium} = \frac{ah^3}{3} + \frac{(AE)h^3}{12} + \frac{(FB)h^3}{12}$$

$$= \frac{ah^3}{3} + \frac{h^3}{12}(AE + FB)$$

$$= \frac{ah^3}{3} + \frac{h^3}{12}(b - a) \qquad (\because\ b = AE + FB + EF)$$

or
$$I_{AB} = \frac{h^3}{12}(4a + b - a)$$

or
$$I_{AB} = \frac{h^3}{12}(3a + b) \qquad ...(q)$$

Now from theorem of parallel axes

$$I_{AB} = I_{NA} + \text{Area } ABCD \times (\bar{y})^2$$

where $\bar{y}$ distance of centroid $G$ from $AB$

or
$$I_{NA} = I_{AB} - \text{Area } ABCD \times (\bar{y})^2$$

$$= \frac{h^3}{12}(3a + b) - \frac{1}{2}(a + b)\,h\left[\frac{h}{3}\frac{(b + 2a)}{(b + a)}\right]^2$$

$$= \frac{h^3}{12}(3a + b) - \frac{(a + b)\,h^3}{18}\frac{(b + 2a)^2}{(b + a)^2}$$

$$= \frac{h^3}{12}(3a + b) - \frac{h^3}{18}\frac{(b + 2a)^2}{(b + a)}$$

$$= \frac{h^3}{36}\left[3(3a + b) - \frac{2(b + 2a)^2}{(b + a)}\right]$$

$$= \frac{h^3}{36\,(b + a)}[3(3a + b)\,(b + a) - 2(b + 2a)^2]$$

$$= \frac{h^3}{36\,(b + a)}[3(3ab + 3a^2 + b^2 + ab) - 2(b^2 + 4a^2 + 4ab)]$$

$$= \frac{h^3}{36\,(b + h)}[9ab + 9a^2 + 3b^2 + 3ab - 2b^2 - 8a^2 - 8ab]$$

$$= \frac{h^2}{36\,(b + a)}[4ab + b^2 + a^2]$$

$$= \frac{h^3}{36\,(b + a)}[(b + a)^2 + 2ab]$$

or
$$I_{NA} = \frac{h^3}{36}\left[(a + b) + \frac{2ab}{(a + b)}\right] \qquad ...(R)$$

Equation ($R$) finds its wide application while determining the stresses in crane Hooks of Trapezoidal section. (Chapter 14).

**Example 4.1.** *A beam of I–section shown in Fig. 4.2 is freely supported at its ends. Over what span can a uniform load of 5000 N/m run be carried if the maximum stress is limited to 50 MN/m².*

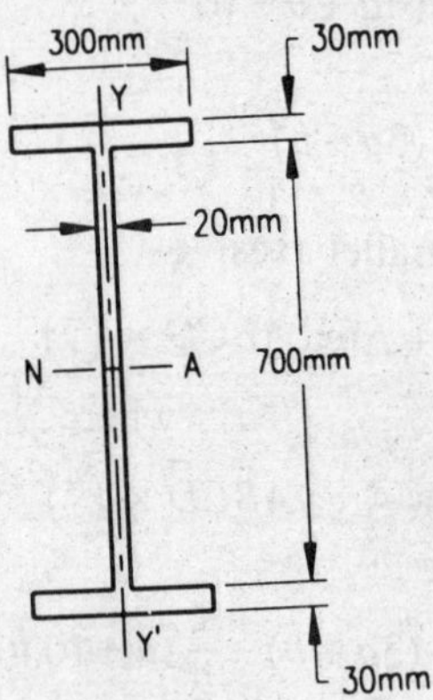


*Fig. 4.2*

**Solution.** $$I_{NA} = \frac{300 \times 760^3}{12} - \frac{280 \times 700^3}{12}$$

$$= 0.297 \times (10)^{10}\ (\text{mm})^4$$

Now $$\mathbf{M}_{max} = \frac{\sigma_{max}}{y_{max}} \cdot I_{NA} = \frac{50 \times 2}{760} \times 0.297 \times 10^{10}$$

$$= 391 \times 10^8\ \text{N– mm}$$

$$= 3.91 \times 10^5\ \text{Nm}$$

For a simply supported beam with span $l$ and a distributed load $w$, we have at the mid span

$$M_{max} = \frac{1}{8} w l^2 = \frac{5000 \times l^2}{8} = 3.91 \times 10^5$$

*i.e.* $$l^2 = \frac{3.91 \times 10^5 \times 8}{5000} = 625.6$$

or $$l = 25.01\text{m}.$$

**Example 4.2.** *A strongest beam of rectangular section is to be cut out of a round wooden log of diameter D. Find the dimensions of the rectangular beam.*

**Solution.** Let the dimensions of such a beam be $b \times d$ (Fig. 4.3)

So that $$b^2 + d^2 = D^2 \qquad \text{...(4.10)}$$

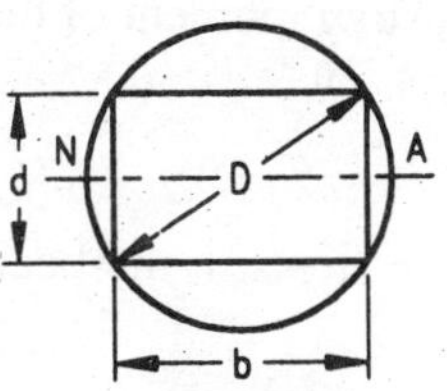


*Fig. 4.3*

**TABLE 4.1**

| *S.No.* | *Section* | *Area* | $I_{NA}$ |
|---|---|---|---|
| 1 | | $bd$ | $\frac{bd^3}{12}$ |
| 2. | | $a^2$ | $\frac{a^4}{12}$ |
| 3. | | $\pi d^2/4$ | $\frac{\pi d^4}{64}$ |
| 4. | | $\frac{\pi}{4}(D^2 - d^2)$ | $\frac{\pi}{64}(D^4 - d^4)$ |
| 5. | | $\pi ab$ | $\frac{\pi ab^3}{4}$ |
| 6. | | $(BD - bd)$ | $\frac{(BD^3 - bd^3)}{12}$ |
| 7. | | $\frac{1}{2}bh$ | $\frac{bh^3}{36}$ |

For a strongest beam the resisting moment of the section must be greatest. From Equation (4.9).

$$M = \sigma Z$$

That is, for a given allowable stress in the beam, $M$ will be greatest when $Z$ is maximum

Now $$Z = \frac{I}{y_{max}} = \frac{bd^3 \times 2}{12d} = \frac{bd^2}{6}$$

Substituting equation (4.10) we obtain

$$Z = \frac{b}{6}(D^2 - b^2)$$

For Z to be maximum

$$\frac{dZ}{db} = 0$$

i.e. $$\frac{1}{6}(D^2 - 3b^2) = 0$$

or $$b = D / \sqrt{3}$$

and from equation (4.10)

$$d = \sqrt{\frac{2}{3}}D.$$

**Example 4.3.** *A beam of built up section shown in Fig. 4.4 is simply supported at its ends with a span of 5 m. It carries a concentrated load of 10 kN at a distance of 2 m from the left hand support. Calculate the maximum bending stress induced in the beam.*

**Solution.** $I_{NA} = 2\,[I_{NA}$ of top plate + $I_{NA}$ of one channel]

Applying principle of paralel axes $I_{NA}$ of top plate.

$$I_{NA} = 2\left[\frac{200 \times 5^3}{12} + (200 \times 5)(52.5)^2\right]$$

$$\cdot\ 2\left[\frac{40 \times 100^3}{12} - \frac{35 \times 90^3}{12}\right]$$

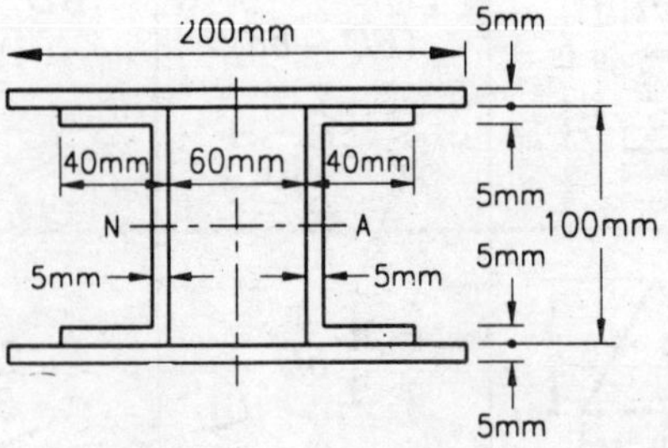

*Fig. 4.4*

$$= 7.904 \times 10^6 \text{ mm}^4$$

From § 3.5.1 we have

$$M_{max} = \frac{10 \times 2\,(5-2)}{5}$$

$$= 12 \text{ kN-m} = 12 \times 10^6 \text{ N-mm}$$

$$= 12 \text{ kN–m} = 12 \times 10^6 \text{ N–mm}$$

$$\sigma_{max} = y_{max}\frac{M_{max}}{I_{NA}} = \frac{55 \times 12 \times 10^6}{7.904 \times 10^6} = 83.7 \text{ N / mm}^2$$

$$= 83.7 \text{MN/m}^2 .$$

**Example 4.4.** *An electric pole, standing 10 m above the ground level, is square in section throughout. The dimensions being 200 mm × 200 mm at the base and tapering uniformly to 100 mm × 100 mm at the top. The resultant force due to the tension in the transmission lines is 50 N in the direction along the diagonal of the section at the top. Determine the maximum stress induced in the poles.*

**Solution.** Consider a section *XX* at a distance *x* metre from the top. The side '*a*' of this section will be given by

$$a = \left(100 + \frac{x}{10} \times 100\right) = 100 + 10x \text{ mm}$$

If '*d*' is diagonal, then

$$d = \sqrt{2}a = \sqrt{2}\,(100 + 10x) \text{ mm}$$

B.M. at *XX* $\quad M_x = 50x$ N–m

$$= 50 \times 10^3 x \text{ N–mm}$$

$I_{XX}$ = Second moment of area about diagonal of the section $= a^4/12$

$$= \frac{(100 + 10x)}{12}$$

$$y_{max} = \frac{d}{2} = (100 + 10x)/\sqrt{2}$$

Thus $\quad \sigma_x = (M_x)(y_{max})/I_{XX}$

Thus $\quad \sigma = \dfrac{100 + 10x}{\sqrt{2}} \cdot \dfrac{50 \times 10^3 \times 12}{(100 + 10x)^4} = \dfrac{60 \times 10^4 x}{\sqrt{2}\,(100 + 10x)^3}$

The location of the section having maximum bending stress will be found by setting

$$\frac{d\sigma_x}{dx} = 0$$

$$\frac{d\sigma_x}{dx} = \frac{60 \times 10^4}{\sqrt{2}}\left[\frac{(100 + 10x)^3 - x \times 3\,(100 + 10x)^2 \times 10}{(100 + 10x)^6}\right] = 0$$

*i.e.* $(100 + 10x)^3 - 30x\,(100 + 10x)^2 = 0$

or $x = 5$ m

Therefore $\sigma_{max} = \dfrac{60 \times 10^4 \times 5}{\sqrt{2}\,(100 + 50)^3} = 141.42\text{N}/\text{mm}^2(\text{MN}/\text{m}^2)$

## 4.3 Composite or Flitched Beams

A composite or flitched beam is one whose section is built up of two different materials rigidly fixed to each other throughout its length, such as a timber beam reinforced by steel plates called the flitches. Since the two materials are connected rigidly the strains at the common surfaces are equal. In addition both the materials will bend to the same radius.

Following are the two cases which arise due to the placement of the flitches with respect to the neutral axis of the main beam.

### Case I

The flitches are placed symmetrically about the neutral axis of the beam such that the flitches alone will have the same neutral axis. Examples of such cases are shown in Fig. 4.5.

If $M_1$ and $M_2$ are the resisting moments of the two materials respectively, the resisting moment of the composite section will be given by

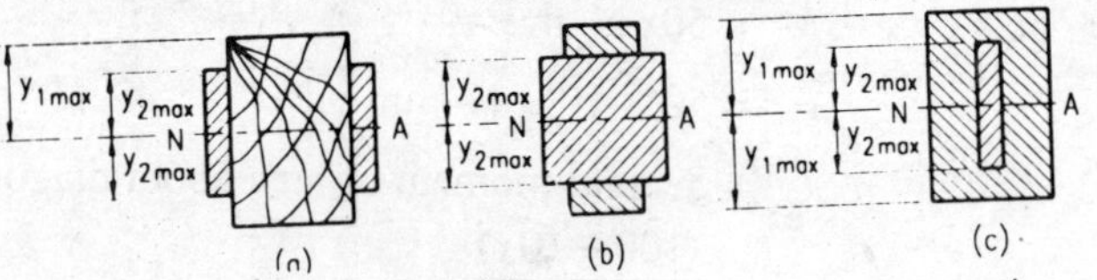

*Fig. 4.5*

$$M = M_1 + M_2 \qquad ...(4.11)$$

Also the radii of curvature of the two parts are same

*i.e.* $R_1 = R_2$

or $\dfrac{E_1 I_1}{M_1} = \dfrac{E_2 I_2}{M_2}$

or $\dfrac{M_1}{M_2} = \dfrac{E_1 I_1}{E_2 I_2}$

or $$M_2 = k\frac{I_2}{I_1}M_1 \qquad ...(4.12)$$

Where $k$ is called the Modular ratio and given by

$$k = E_2/E_1 \qquad ...(4.13)$$

Also if $\sigma_{1max}$ and $\sigma_{2max}$ are the maximum induced stresses in the respective materials

$$\sigma_{1\,max} = \frac{M_1}{I_1} y_{1\,max} \qquad ...(4.14)$$

and

$$\sigma_{2\,max} = \frac{M_2}{I_2} y_{2\,max} \qquad ...(4.15)$$

Dividing equation (4.14) with (4.15)

$$\frac{\sigma_{1\,max}}{\sigma_{2\,max}} = \frac{M_1}{M_2} \cdot \frac{I_2\, y_{1\,max}}{I_1 y_{2\,max}}$$

$$= \frac{1}{k}\left(\frac{y_{1\,max}}{y_{2\,max}}\right) \qquad ...(4.16)$$

Thus if the total moment acting on the composite section is known, the moments taken up, by the respective materials, can be evaluated with the help of equations (4.11) and (4.12). Further the induced stresses can then be evaluated with the help of equations (4.14) and (4.15).

Alternately if the permissible stresses for the two materials are given as $\sigma_1$ and $\sigma_2$ respectively and the resisting moment of the composite section is to be evaluated then first of all with the help of equation (4.16) the limiting stress out of $\sigma_1$ and $\sigma_2$ is ascertained such that the induced stresses in the two materials do not cross the respective permissible values. After having evaluated the maximum induced stresses the moments $M_1$ and $M_2$ are calculated with the help of equations (4.14) and (4.15) to get $M$.

## Case II

The plate is placed in such a way that if alone its neutral axis will not coincide with that of the timber beam taken alone. Fig. 4.6(*a*) illustrates such a case.

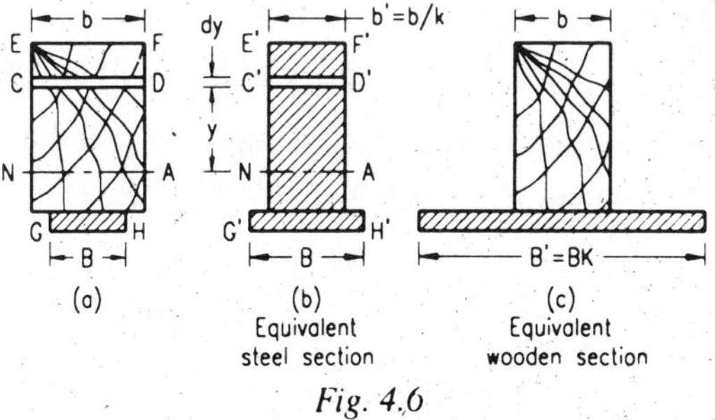


*Fig. 4.6*

The neutral axis of the composite beam will evidently not pass through the centroid of the wooden portion. This type of problem is tackled by replacing the given section with an equivalent section of one material (either of steel or wood) as shown in Fig.

4.6(*b*) and 4.6(*c*). The equivalent section is such that it will bend to the same radius and extend to the same height as the original beam. Under these conditions the strains in any fibre *CD* and the corresponding fibre *C'D'* in the equivalent steel section will be same.

Thus if the height of the fibres *CD* or *C'D'* is *y* from the neutral axis, location of which is yet to be found, then assuming plane sections remaining plane after bending.

$\varepsilon_{CD}$ = Longitudinal Strain at a distance y from N.A., in timber portion.

$$= \sigma_{CD}/E_t \quad ...(4.17)$$

$\varepsilon_{C'D'}$ = Longitudinal strain at a distance y from N.A., in the equivalent steel portion.

$$= \sigma_{C'D'}/E_S \quad ...(4.18)$$

Where $E_t$ and $E_s$ re the Young's modulii for timber and steel respectively. As these strains are equal

$$\sigma_{C'D'}/E_s = \sigma_{CD}/E_t$$

$$\sigma_{C'D'} = \left(\frac{E_s}{E_t}\right)\sigma_{CD}$$

$$= k\sigma_{CD} \quad ...(4.19)$$

where $k$ is the Modular ratio $E_s/E_t$.

Thus the stress at any point in the equivalent steel section is $k$ times the stress at the corresponding point in the original section.

Also since the resisting moment of the equivalent steel portion should be same as that of the original timber portion the normal force on the elements *CD* and *C'D'* must be same

That is $\quad \sigma_{CD}\,(bdy) = \sigma_{C'D'}(b'dy)$

$= (K\,\sigma_{CD})\,(b'dy)$ (From equation 4.15)

*i.e.*, $\quad b = kb' \quad ...(4.20)$

Thus the width of the equivalent steel portion will be $1/k \times$ the width of the original timber portion.

Since $k(= E_s/E_t)$ is greater than unity, the width $b'$ will be smaller than $b$.

After having found the equivalent steel section the position of the neutral axis and the value of second moment of area about neutral axis can be found. This can then be used to evaluate the stress $\sigma_{C'D'}$ with the help of equation $\sigma_{C'D'}/y = \dfrac{My}{I}$ if the total resisting moment $M$ of the section is known. Alternatively if the total resisting moment is to be evaluated with allowable stresses being given for timber and steel, the stress $\sigma_{E'F'}$ is calculated with the help of equation (4.19) and is used to evaluate $M$ as indicated above. The value of $M$ is also checked for the allowable stress in steel at $G'H'(\sigma_{GH} = \sigma_{G'H'})$. It must be noted that for the corresponding points in the steel portion of the two sections the stress is equal or in other words in equation (4.19) modular ratio $k$ is reduced to unity. The smaller of the two values of $M$ is then the resisting moment of the section.

As an alternative an equivalent timber section could be found as shown in Fig. 4.6($c$) and the required quantities evaluated.

**Example 4.5.** *Fig. 4.7(a) shows the cross–section of a compound bar which has been formed by brazing a brass strip 25 mm × 12 mm to a steel strip 25 mm × 6 mm. The compound bar is bent to a circular shape with neutral axis parallel to BC. Find the position of the neutral axis and the ratio of the maximum stress in steel to that in the brass. Also find the radius of curvature of the neutral surface and the resisting moment of the section when the maximum stress in steel becomes 66 MN/m².*

Take $E_{steel} = 200 \times 10^3$ MN/m² and $E_{brass} = 80 \times 10^3$ MN/m²

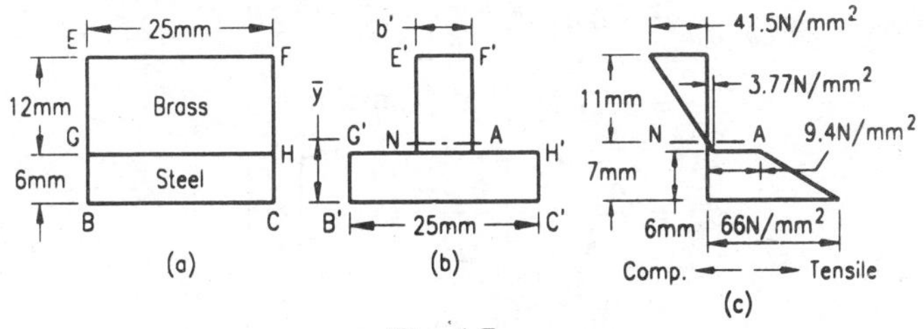

Fig. 4.7

**Solution.**
$$k = E_S / E_b = \frac{200 \times 10^3}{80 \times 10^3} = 2.5$$

$$b' = b/k = 25/2.5 = 10 \text{ mm}$$

Taking moment about $B'C'$ Fig. 4.7($b$)

$$(25 \times 6 + 12 \times 10)\bar{y} = (25 \times 6) \times 3 + (12 \times 10)(6 + 6)$$

*i.e.*
$$\bar{y} = 7.0\text{mm}$$

Now from bending equation $\sigma/y = E/R$, we have

$$\sigma_{B'C'}/\sigma_{E'F'} = y_{B'C'}/y_{E'F'}$$
$$= 7/(18 - 7) = 7/11$$

But
$$\sigma_{BC} = \sigma_{B'C'} \text{ and } \sigma_{EF} = \sigma_{E'F'}/k$$

Therefore

$$\sigma_{BC}/\sigma_{EF} = \frac{7}{11} \times k = \frac{7}{11} \times 2.5 = 1.59$$

It must be noted here that if $\sigma_{BC}$ were tensile in nature then $\sigma_{EF}$ would be compressive in nature.

(*b*) For evaluating of the radius of curvature, we have

$$R = Ey/\sigma$$

and since the maximum stress in steel occurs at $BC$ or $B'C'$

$$R = E_s y_{B'C'}/\sigma_{B'C'} = \frac{200 \times 10^3 \times 7}{66} \text{ mm}$$

The stress distribution on the section can also be found

In the steel $$\sigma_{GH} = \sigma_{G'H'} = \frac{\sigma_{B'C'}}{y_{B'C'}} . y_{G'H'} = \frac{66 \times 1}{7}$$

$$= 9.428 \text{MN/m}^2 \text{ (Tensile)}$$

In Brass $$\sigma_{GH}\sigma_{G'H'} = \sigma_{G'H'} / k = \frac{1}{k}\left[\frac{\sigma_{B'C'}}{y_{B'C'}} . y_{G'H'}\right] = \frac{9.428}{2.5}$$

$$= 3.77 \text{ MN/m}^2 \text{ (Tensile)}$$

$$\sigma_{EF} = \frac{\sigma_{E'F'}}{k} = \left[\frac{\sigma_{B'C'}}{y_{B'C'}} . y_{E'F'}\right] \times \frac{1}{k}$$

$$= \frac{1}{2.5} \times \frac{66}{7} \times 11$$

$$= \frac{66}{1.59} = 41.5 \text{ MN / m}^2 \text{ (Comp.)}$$

The stress distribution diagram is shown in Fig. 4.7(*c*). The resisting moment of the section is given by

$$M/I_{NA} = E/R$$

*i.e.* $$M = EI_{NA}/R$$

Applying theorem of parallel axes

Now $$I_{NA} = \left[\frac{25 \times 6^3}{12} + (25 \times 10)\, 4^2\right] + \left[\frac{10 \times 12^3}{12} + (10 \times 12)\, 5^2\right]$$

$$= 8890 \text{ (mm)}^4$$

Therefore $$M = \frac{200 \times 10^3 \times 8890}{2100} = 84700 \text{ N / mm}$$

$$= 84.7 \text{ Nm} .$$

**Example 4.6.** *If the brass and steel strips of Example 4.5 were not brazed together and just placed over each other as shown in Fig. 4.7(a) so that they could bend independently and the bending stresses in brass and steel were limited to 66 MN/m² and 100 MN/m² respectively find the resisting moment of the section.*

**Solution.** In this case each strip will bend independently about its own neutral axis. Also it can be assumed for all practical purposes, that the radii of curvatures of the two neutral surfaces are same. Then for the individual strips.

$$\sigma_s/y_s = E_s/R_s = M_s/I_s$$

and $$\sigma_b/y_b = E_b/R_b = M_b/I_b$$

Now $$I_s = \frac{25 \times 6^2}{12} = 450 \text{ mm}^4$$

$$I_b = \frac{25 \times 12^2}{12} = 3600 \text{ mm}^4$$

From the bending equations, the maximum induced stresses in the two materials are related as

$$\frac{1}{R_S} = \frac{1}{R_b} = \frac{\sigma_s}{y_s E_s} = \frac{\sigma_b}{y_b E_b}$$

*i.e.,* $$\sigma_s = \sigma_b \cdot \frac{y_s}{y_b} \cdot \frac{E_S}{E_b}$$

$$= \sigma_b \left(\frac{3}{6}\right)(2.5)$$

or $$\sigma_s = 1.25\, \sigma_b$$

Thus if the stress in steel were the limiting case then the induced stress in brass would have been

$$\sigma_b = 100/1.25 = 80 \text{ N/mm}^2 \text{ or } 80 \text{ MN/m}^2.$$

This is not admissible as the stress in brass is limited to 66 $MN/m^2$.

Therefore the stress in brass will be the limiting stress corresponding to which the stress induced in steel will be

$$= 66 \times 1.25 = 82.5 \text{ N/mm}^2 \text{ or } 82.5 \text{ MN/m}^2$$

For these stresses in steel and brass the respective resisting moments are

$$M_s = \frac{\sigma_s I_s}{y_s} = \frac{82.5 \times 450}{3} = 12375 \text{ M} - \text{mm}$$

and $$M_b = \frac{\sigma_b I_b}{y_b} = \frac{66 \times 3600}{6} = 3960 \text{ N / mm}$$

Thus total resisting moment

$$= M_s + M_b = 51975 \text{ N–m}$$

$$= 51.975 \text{ N–m}.$$

**Example 4.7.** *A composite beam is made of two timber joists each 100 mm wide by 300 mm deep, with a steel plate 25 mm thick and 300 mm deep firmly fixed between them. Determine the resisting moment of the composite section if the maximum stress in timber is limited to 7.5 MN/m². Also find the maximum centrally applied load which the beam can carry on a simple supported span of 4 m. Take* $E_{steel} = 200 \times 10^3 MN/m^2$ *and* $E_{timber} = 10 \times 10^3 MN/m^2$.

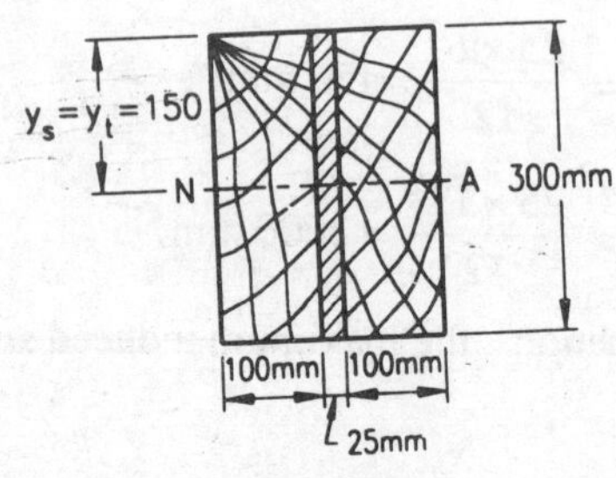


*Fig. 4.8*

**Solution.** Since both timber and steel are going to bend to the same radius of curvature, we have

$$\frac{\sigma_s}{y_s} = \frac{M_s}{I_s} = \frac{E_s}{R}$$

and
$$\frac{\sigma_t}{y_t} = \frac{M_t}{I_t} = \frac{E_t}{R}$$

where suffix *t* stands for timber and s for steel

That is
$$\frac{\sigma_s}{y_s E_s} = \frac{\sigma_t}{y_t E_t}$$

or
$$(\sigma_s)_{max} = (\sigma_s)_{max} \frac{(y_s)_{max}}{(y_t)_{max}} \cdot \frac{E_s}{E_t}$$

$$= 7.5 \times \frac{150}{150} \left[ \frac{200 \times 10^3}{10 \times 10^3} \right]$$

$$= 150 \text{ N/mm}^2 \text{ or MN/m}^2$$

Now
$$I_s = \frac{25 \times (300)^3}{12} = 56.25 \times 10^6 \text{ mm}^4$$

$$I_t = \frac{2 \times 100 \times (300)^3}{12} = 450 \times 10^6 \text{ N / mm}^4$$

Therefore
$$M = \frac{I_s \sigma_s}{y_s} = \frac{56.25 \times 10^6 \times 150}{150}$$

$$= 56.25 \times 10^6 \text{ N–mm}$$

and $$M_t = \frac{I_t \sigma_t}{y_t} = \frac{450 \times 10^6 \times 7.5}{150}$$

$$= 22.50 \times 10^6 \text{ N–mm}$$

∴ Resisting moment of section

$$= 56.25 \times 10^6 + 22.50 \times 10^6$$

$$= 78.75 \times 10^6 \text{ N–mm} = 78.75 \times 10^3 \text{ Nm}$$

(*b*) For a centrally applied load *W* on a simply supported beam of span 4 m, the maximum bending moment will be $(W \times 4)/4 = W$ Nm. (See Fig. 3.4)

This must be equal to the resisting moment of the section

That is $W = 78.75 \times 10^3$ N.

**Example 4.8.** *A wooden beam 100 mm wide by 300 mm deep is reinforced by two similar steel plates each 12 mm thick, one being secured to the upper face of the beam and the other to the lower face. Determine the necessary width of plates so that the composite section thus formed should be able to resist a bending moment of 8300 N–m in the vertical plane without the stress in wood exceeding 5.0 MN/m². Take* $E_{steel} = 200 \times 10^3$ *MN/m² and* $E_{wood} = 20 \times 10^3$ *MN/m².*

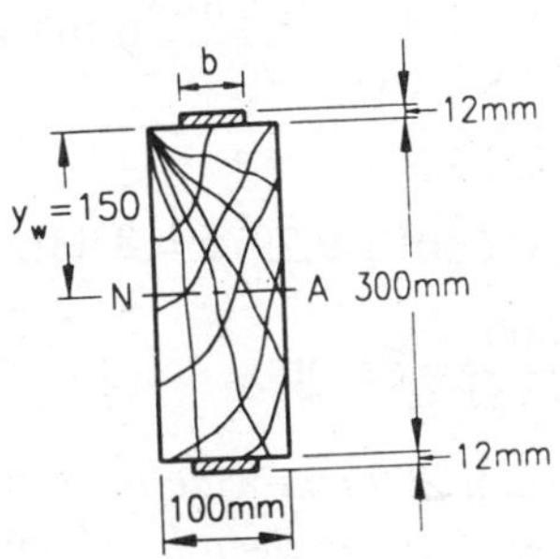


*Fig. 4.9*

**Solution.** Since both wood and steel will have the same radius of curvature, we have

$$\frac{\sigma_s}{y_s} = \frac{M_s}{I_s} = \frac{E_s}{R}$$

and $$\frac{\sigma_w}{y_w} = \frac{M_w}{I_w} = \frac{E_w}{R}$$

where the suffix *s* stands for steel and *w* for wood.

Therefore

$$\frac{\sigma_s}{y_s E_s} = \frac{\sigma_w}{y_w E_w}$$

or $$(\sigma_s)_{max} = \sigma_w \frac{(y_s)_{max}}{(y_w)_{max}} \frac{E_s}{E_w}$$

$$= \frac{5.0 \times 162}{150} \times \frac{200 \times 10^3}{20 \times 10^3}$$

$$= 54 \text{ N/mm}^2 \text{ or } 54 \text{ MN/m}^2$$

Now if $b$ mm is the width of the steel plates then by invoking principle of parallel axes

$$I_s = 2\left[\frac{b \times 12^3}{12} + (b \times 12)(156)^2\right] \text{mm}^4$$

$$I_w = \frac{100 \times 300^3}{12} = 225 \times 10^6 \text{ mm}^4$$

Again from bending equation

$$M_w = \frac{\sigma_w I_w}{y_w} = \frac{5 \times 225 \times 10^6}{150} = 7.5 \times 10^6 \text{ N-mm}$$

$$= 7.5 \times 10^3 \text{ N-m}$$

$$M_s = \frac{54 \times 27.89 \times 10^3 \, b}{162} = 9.297 \times 10^3 \, b \text{ N-mm}$$

$$= 9.297b \text{ N-m}$$

Thus $$M_w + M_s = 7.5 \times 10^3 + 9.297b = 8300$$

*i.e.* $$b = \frac{800}{9.297} = 86 \text{ mm}.$$

**Example 4.9.** *A timber beam 200 mm wide by 300 mm deep is reinforced with two rectangular steel plates each 250 mm deep by 10 mm thick. The steel plates are fixed rigidly one on each side of the beam as shown in the Fig. 4.10(a). Find the position of the neutral axis and the resisting moment of the composite section when the stresses in steel and timber are limited to 85 MN/m² and 5.5 MN/m² respectively. Take modular ratio as 20.*

Solution. Since the modular ratio is 20, the equivalent steel width for 200 mm wide timber section will be 200/20 = 10 mm. Accordingly the equivalent steel section is shown in Fig. 4.10($b$). Taking moments of the area about the line $G'H'$.

$$\bar{y}\,(10 \times 300 + 2 \times 10 \times 350) = 10 \times 300 \times 150 + 2 \times 10 \times 250 \times 125$$

*i.e.* $$\bar{y} = 134.4 \text{ mm}$$

Now $$\frac{\sigma}{y} = \frac{M}{I} = \frac{E}{R}$$

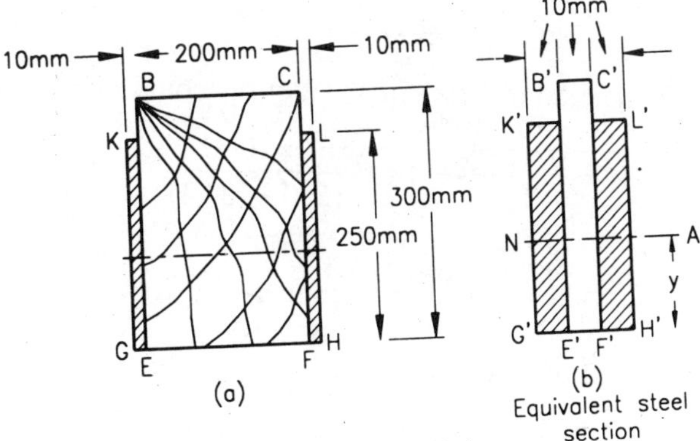


*Fig. 4.10*

Therefore $\dfrac{\sigma_{B'C'}}{y_{B'C'}} = \dfrac{\sigma_{K'L'}}{y_{K'L'}}$

*i.e.* $\sigma_{B'C'} = \sigma_{K'L'} \dfrac{y_{B'C'}}{y_{K'L'}} = \dfrac{(300-134.4)}{(250-134.4)} \sigma_{K'L'}$

or $\sigma_{B'C'} = \dfrac{165.6}{115.6} \sigma_{K'L'}$

Now $\sigma_{KL} = \sigma_{K'L'}$

and $\sigma_{B'C'} = k\sigma_{BC} = 20\sigma_{BC}$

Therefore $\sigma_{BC} = \dfrac{1}{20}\left(\dfrac{165.6}{115.6}\right)\sigma_{KL} = 0.071\,\sigma_{KL}$

Assuming the stress in steel to be the limiting stress

i.e. sigma sub KL = 85 MN/m sup 2

Then corresponding induced stress in timber will be

$$= 0.071 \times 85 = 6.04 \text{ MN/m}^2$$

This is more than the allowable stress in timber. Therefore the stress in timber will be the limiting case *i.e.* the stress

$$\sigma_{BC} = 5.5 \text{ MN/m}^2$$

and the corresponding induced stress in steel will be

$$\sigma_{KL} = \sigma_{K'L'} = 5.5/0.071 = 77.5 \text{ MN/m}^2$$

From Fig. 4.12(*b*), using theorem of parallel axes we get

$$I_{NA} = 2\left[\frac{10 \times 250^3}{12} + 10 \times 250\,(134.4-125)^2\right] + \left[\frac{10 \times 300^2}{12} + 10 \times 300\,(150-134.4)^2\right]$$

$$= 26.44 \times 10^6 + 23.23 \times 10^6$$
$$= 49.67 \times 10^6 \text{ mm}^4$$

Therefore if $M$ is the total resisting moment of the composite section, then

$$M = I_{Na}\left(\frac{\sigma_{K'L'}}{y_{K'L'}}\right) = \frac{49.67 \times 10^6 \times 77.5}{(250 - 134.4)}$$

$$= \frac{49.67 \times 10^6 \times 77.5}{115.6}$$

$$= 33.5 \times 10^6 \text{ N– mm}$$
$$= 35.5 \text{ kN–m.}$$

**Example 4.10.** *A metallic tube of internal diameter 25 mm and 5 mm wall thickness is simply supported on a span of 1 m and the load at the mid span just sufficient to bring the stress to elastic limit level, is found to be 875 N.*

*Four such tubes are firmly clamped to each other to form a single beam, the centres of the tubes forming a square of 35 mm side with two sides of this square horizontal. Calculate the maximum central load which this composite beam can carry if the stress is not to exceed the elastic limit.* *(Punjab University)*

**Solution.** For the single tube with load at mid span

$$M_{max} = \frac{W \times 1}{4} \text{ Nm}$$

$$\sigma_{max} = \frac{M_{max}}{I_{XX}} \cdot \frac{D}{2}$$

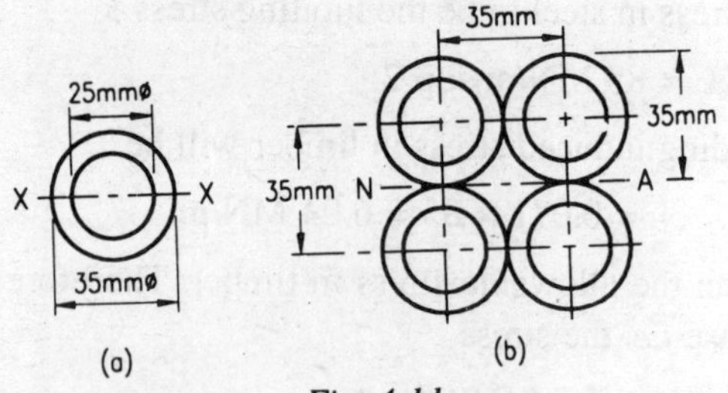

*Fig. 4.11*

Now $$I_{xx} = \frac{\pi}{64}(35^4 - 25^4)$$

and $$Z = I_{xx} \cdot \frac{2}{D} = \frac{\pi}{64}(35^4 - 25^4) \times \frac{2}{35}$$

$$= 3120 \text{ mm}^3$$

Therefore $$\sigma_{max} = \frac{875 \times 1000}{4 \times 3120} \text{ N/mm}^2 \text{ or MN/m}^2$$

For the composite section

$$I_{NA} = 4\left[\frac{\pi}{64}(35^4 - 25^4) + \frac{\pi}{4}(35^2 - 25^2)\left(\frac{35}{2}\right)^2\right]$$

$$\therefore \qquad Z' = \frac{I_{NA}}{35} = 2.245 \times 104 \text{ mm}^3$$

Thus if $W'$ is the load which can be carried by the composite tube

$$\frac{(W' \times 1000)}{4} = Z' \sigma_{max}$$

$$= 2.245 \times 10^4 \times \left(\frac{875 \times 1000}{4 \times 3120}\right)$$

i.e. $$W' = \frac{2.45 \times 10^4}{3120} \times 875 = 6.3 \times 10^3 \text{ N}$$

$$= 6.3 \text{ kN.}$$

**Example 4.11.** *A 25 mm diameter bronze rod is enclosed tightly in a steel tube of outside diameter of 50 mm. Find the resisting moment of the composite section assuming that the allowable stresses is steel and bronze do not exceed 150 MN/m² and 120 MN/m² respectively.* $E_{steel}$ *= 200 × 103 MN/m² and* $E_{bronze}$ *= 100 × 10³ MN/m².*

**Solution.** The two metals bend to the same radius of curvature. Therefore from the bending equation

$$\frac{1}{R} = \frac{\sigma_s}{E_s y_s} = \frac{\sigma_b}{E_b y_b}$$

i.e. $$\sigma_s = \sigma_b \left(\frac{E_s}{E_b}\right)\left(\frac{y_s}{y_b}\right) = \sigma_b\,(2)\,(2) = 4\sigma_b$$

Thus if the stress in steel is the limiting case

i.e. $$\sigma_s = 150 \text{ MN/m}^2$$

then maximum induced stress in bronze will be

$$\sigma_b = 150/4 = 37.5 \text{ MN/m}^2$$

which being less than the allowable stress in bronze of 120 MN/m² is admissible.

Thus $$M_s = I_s \sigma_s / y_s$$

$$= \frac{\pi}{64}(50^4 - 25)^4 \frac{150 \times 2}{50}$$

$$= 1.715 \times 10^6 \text{ N–mm}$$

and $$M_b = I_b \sigma_b / y_b = 0.0575 \times 10^6 \text{ N–mm}$$

Therefore total resisting moment of the section

$$= M_s + M_b = (1.715 + .0575)\ 10^6 \text{ N–mm} = 1.7725 \text{ kN–m}.$$

## 4.4 Combined Direct and Bending Stresses

When a member of a structure is under the action of both a bending moment and axial force, the bending and direct stresses induced due to their action may be calculated separately and resultant stress at any point found by adding them algebraically. Thus if $\sigma$ is the resultant stress at any point of the section, then

$$\sigma = \pm \sigma_0 \pm \sigma_b \qquad \text{...(4.21)}$$

Where $\sigma_0$ is the axial force divided by the area of cross section and $\sigma_b$ is the bending stress as calculated from the bending equation (4.8). The signs of $\sigma_0$ and $\sigma_b$ being taken positive or negative depending upon whether they are tensile or compressive.

## 4.5 Eccentric Longitudinal Loads

Let the line of action of a force $P$ be parallel to the axis of a bar and intersect the axis of symmetry $XX$ of the cross section $ABCD$ at a distance $x_1$ from the centroid $O$ Fig. 4.12.

The equilibrium of the column will not be disturbed by introducing two equal and opposite forces $P_1$ and $P_2$ (both equal to $P$) at the centroid $O$ as in the Fig. 4.12. The forces $P$ and $P_2$ will constitute a bending moment $Px_1$ with the axis of bending as $YY$. Thus the original force $P$ applied at the point $Q$ is equivalent to a bending moment $Px_1$ about $YY$ and a direct load $P_1$ (= $P$) applied at the centroid of the section. The direct load $P_1$ will induce a direct compressive stress which will be uniform over the cross section.

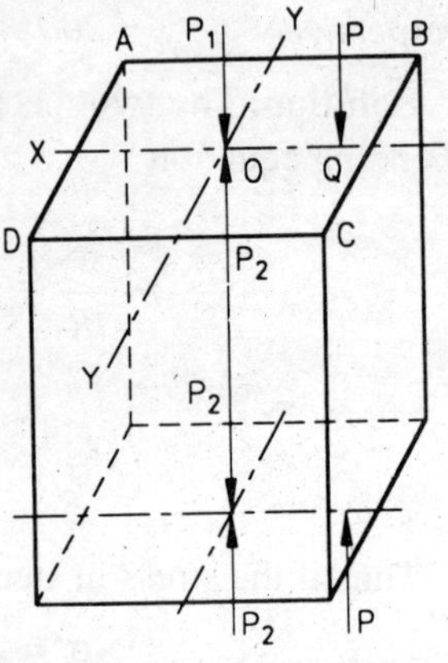

*Fig. 4.12*

$$\sigma_0 = -\frac{P}{bd} \qquad \text{...(4.22)}$$

where $b$ and $d$ are the breadth and the depth of the cross section.

The effect of the bending moment $Px_1$ will be to bend the column in the plane passing through the point $O$ and containing the force $P$, such that the centre of curvature will lie on the right side of the centroidal bending axis $YY$. Thus as discussed in article 4.2, the bending stress at any point of the section lying to the right of $YY$ will be compressive and all the points of the section to the left of $YY$ will experience tensile bending stresses.

The magnitude of the bending stress at any point at a distance $x$ from the $YY$ line will be

$$\sigma_b = x \times [(Px_1)/I_{YY}] = \frac{12\, Pxx_1}{db^3} \qquad ...(4.23)$$

Thus $\sigma_b$ in magnitude increases linearly with $x$. The maximum value of the bending stress will occur when $x$ is greatest that is along the lines $AD$ or $CB$ where $x = b/2$

*i.e.* $$(\sigma_b)_{BC} = \frac{12Px_1}{db^3} \cdot \frac{b}{2} = \frac{6Px_1}{db^2} \qquad \text{(Comp.)}$$

also $$(\sigma_b)_{AD} = \frac{6Px_1}{db^2} \qquad \text{(tensile)}$$

Therefore the magnitude of the resultant maximum and minimum compressive stresses over the section are given by

$$\sigma = \frac{-P}{db} \pm \frac{6Px_1}{db^2} \qquad ...(4.24)$$

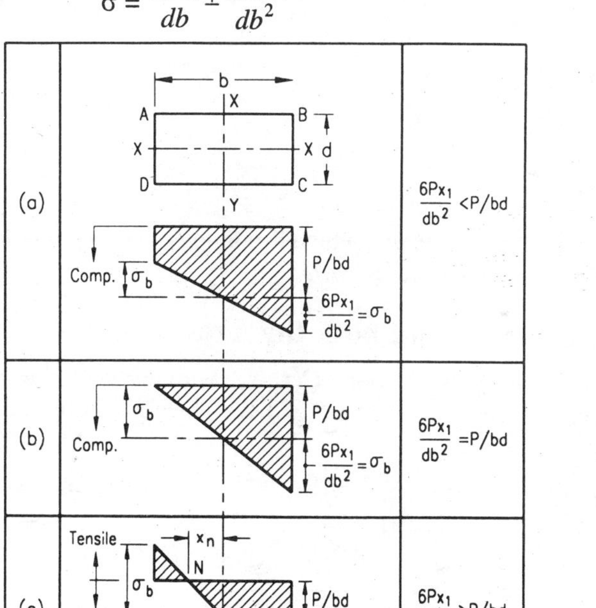


*Fig. 4.13*

Fig. 4.13(*a*), (*b*) and (*c*) show the possible forms of the resultant stress distribution curves. It is evident that in the case (*c*) some portion of the section will experience a tensile stress even though the nature of the applied load at $Q$ is compressive. This type of situation is called "*the reversal of stress*". In the cases (*a*) and (*b*) the entire section is having the stress of type of the applied load. As a matter of fact for the non–

reversibility of the stress (as in the case of cast iron or a masonary structure), the maximum eccentricity of the load is given by the condition.

$$\frac{6Px_1}{db^2} = \frac{P}{bd}$$

or $$x_1 = b/6$$

Since this eccentricity can be on either side of $YY$, for the non–reversibility of stress the load point $Q$ must lie within the middle third of the width of the column.

Similarly if the point of application of the load were lying on the other axis of symmetry *i.e.* on $YY$ at a distance $y_1$ from $O$, the bending will take places about $XX$. Here also to avoid reversal of stress the load point must be within the middle third of the depth $d$.

The position of the neutral axis of the section in the case of combined direct and bending stresses can be found graphically by projecting a line parallel to $YY$ on the section from the point $N$ (Fig. 4.13*c*) because the stress at all the points of the section on this line will be zero. Alternatively if the neutral axis is lying at a distance $x_N$ on the left of $YY$ (side opposite to that of load) then from the definition of the neutral axis

$$\sigma_{NA} = 0$$

or $$(\sigma_b)_{NA} + (\sigma_0)_{NA} = 0$$

*i.e.,* $$\frac{12Px_1\, x_N}{db^3} = \frac{P}{bd}$$

or $$x_N = b^2/(12\, x_1) \quad \text{...(4.25)}$$

When $x = b/6$, $x_N$ becomes $b/2$, which is nothing but the case (*b*) discussed above, where the neutral axis coincides with $AD$.

## 4.6 Load Eccentric about both the Axes

Let the load $P$ be acting at a point $Q(x_1, y_1)$, with respect to the axes of symmetry of the rectangular section as shown in Fig. 4.14. The equilibrium of the column will

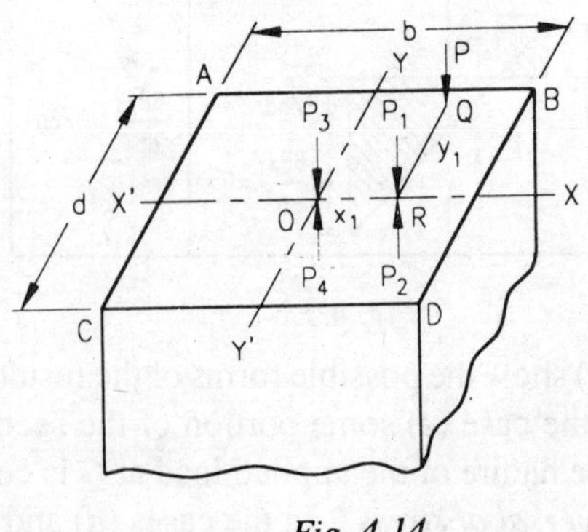


*Fig. 4.14*

not be affected by the introduction of two sets of equal and opposite forces at $R$ and $O$ as shown in Fig. 4.14 such that $P_1 = P_2 = P_3 = P_4 = P$.

From the figure it is evident that

$$P \text{ at } Q \equiv \text{Moment } Py_1 \text{ about } XX + P_1 \text{ at } R$$

$$\equiv \text{Moment } Py_1 \text{ about } XX + \text{Moment } Px_1 \text{ about } YY + P_3 \text{ at } O$$

Thus $P$ at $Q \equiv P$ at the centroid O + Moment $Py_1$ about $XX$ + Moment $Px_1$ about $YY$ ...(4.26)

Let $(h, k)$ be the co–ordinates of any point $S$ at which the stress is to be found. The effect of $P$ at the centroid will be to induce a compressive stress of uniform intensity $\sigma_0 = (P/bd)$ all over the cross section including the point $S$ (article 1.4).

Effect of bending moment $Py_1$ about $XX$ will be to give rise to a compressive bending stress at the point $S$, lying in first quadrant.

$$\sigma_{S_1} = -k\frac{Py_1}{I_{xx}} = -\frac{12Py_1k}{d^3b}$$

Similarly the effect of the other moment $Px_1$ about $YY$ will be to induce a compressive bending stress at $S$.

$$\sigma_{S_2} = -h\frac{Px_1}{I_{yy}} = -12\frac{Px_1h}{b^3d}$$

Thus the resultant stress at $S$ is given by

$$\sigma_S = \sigma_0 + \sigma_{S_1} + \sigma_{S_2}$$

$$= -\frac{P}{bd} - \frac{12\,Py_1k}{d^3b} - \frac{12\,Px_1h}{b^3d} \qquad \text{...(4.27)}$$

For the positive values of $h$, $k$, $x_1$ and $y_1$ the value of $\sigma_S$ will be negative signifying that the resultant stress at $S$ will be compressive. If the point $S$ is in the third quadrant both $h$ and $k$ will be negative which will render the last two terms in equation (4.27) positive. Therefore depending upon the values of other quantities appearing in equation (4.27) the sign of $\sigma_S$ may turn out to be positive. Under these circumstances there will be the reversal of stress over the section. Since for the point $D$ of the section the respective magnitudes of $h$ and $k$ are greatest, though negative, the reversal of stress will occur first at this point. Thus the limiting condition for the non–reversibility of the stress will be to equate the resultant stress at the point $D$ to zero.

That is

$$\sigma_D = \frac{P}{db} - \frac{12Py_1}{d^3b}\left(\frac{-d}{2}\right) + \left[\frac{12Px_1}{b^3d}\right]\left(\frac{-b}{2}\right) = 0$$

or

$$\frac{x_1}{(b/6)} + \frac{y_1}{(d/6)} = 1 \qquad \text{...(4.28)}$$

Thus generalising relation (4.28), the locus of the point $(x_1, y_1)$, lying in the first quadrant only, for the non–reversibility of stress will be the straight line of equation.

$$\frac{x}{(b/6)} + \frac{y}{(d/6)} = 1 \qquad \text{...(4.29)}$$

This straight line is shown in Fig. 4.15 by *EF*.

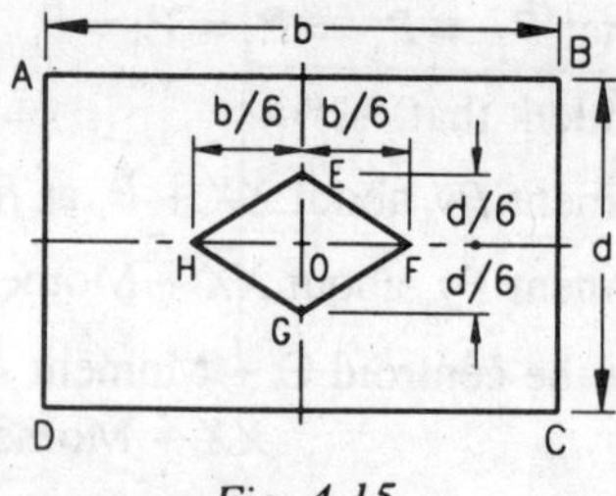


*Fig. 4.15*

Thus if the load application point lies anywhere within the area *EOF* there will be no reversal of stress on the section. Similarly it can be shown that if the point *Q* is not necessarily within the first quadrant but goes in the other three quadrants as well, the condition for the non–reversibility of the stress will be that the point *Q* lies within the area *EFGH*. This area *EFGH* is known as the Core or Kernel of the rectangular section *ABCD*.

## Stress Distribution Curve

Let the point of application of the load, *Q*, be outisde the core of the section and has the coordinates $(x_1, y_1)$. Then the stress at any point *S* with coordinates (*h*, *k*) is given by equation (4.27). The neutral axis of the section can be found by finding the locus of *S* such that the stress $\sigma_s = 0$.

Therefore, from equation (4.27) by setting $\sigma_s$ equal to zero we get the relation

$$\frac{-12y_1k}{d^2} - \frac{12x_1h}{b^2} = 1 \qquad \text{...(4.30)}$$

Thus, generalising the relation (4.30), the equation of the neutral axis of the given rectangular section is

$$\frac{x}{(-b^2/12x_1)} + \frac{y}{(-d^2/12y_1)} = 1 \qquad \text{...(4.31)}$$

This again is an equation of a straight line having intercepts on *x* and *y* axes as $(-b^2/12x_1)$ and $(-d^2/12y_1)$ respectively. This is shown by *NN′* in Fig. 4.16. Evidently the stress on any point on *NN′* will be zero. The resultant stress on any point on the portion *ANN′ CB* will be compressive and on the portion *NN′ D* tensile. Moreover the iso–stress lines will be parallel to *NN′*. In order to draw the stress distribution curve for the section produce *NN′* to a convenient point $M_N$ and draw $M_DM_NM_B$ the base line of the stress distribution curve, at right angles to *NN′* $M_N$, $OM_O$, $DM_D$ and $BM_B$ are perpendiculars on the line $M_DM_B$. From equation (4.27), it is evident that at the centroid *O*, of the section, the resultant stress will only be the direct stress (P/*bd*). To some convenent scale measure $M_0T_0$ to represent $\sigma_0 (= P/db)$. Join $M_NT_0$ to meet the line $BM_B$ and $DM_D$ at $T_B$ and $T_D$ respectively. $T_DM_NT_B$ is the required resultant stress distribution curve as shown in Fig. 4.16.

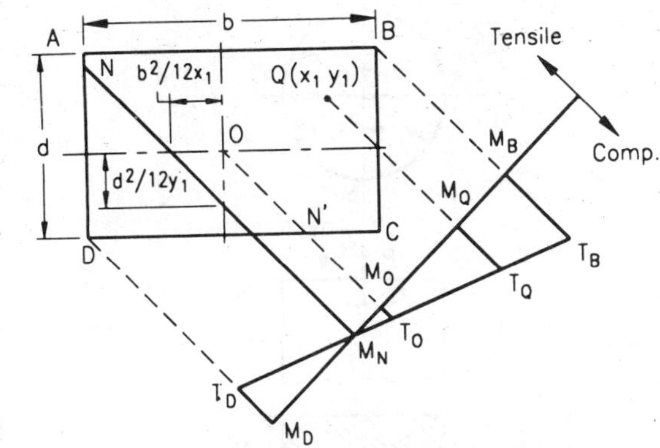


*Fig. 4.16*

The stress at any point $Q$ is then represented by $M_Q T_Q$ which can be measured.

## 4.7 Eccentric Loading of Solid Circular Short Column

Let the load $P$ parallel to the axis of the bar be acting at any point $Q$. Marking the $x$–axis along the line $OQ$ so that the eccentricity of loading is $x_1$ about the $YY$ axis.

As discussed in article (4.5) the force $P$ at $Q$ is equivalent to a bending moment $Px_1$ about $YY$ and a direct load $P_1$ (= $P$) applied at the centroid of the section. The direct load $P_1$ will induce a direct compressive stress, uniform over the cross–section

*i.e.* $$\sigma_0 = -\frac{P}{A} = -\frac{4P}{\pi D^2} \qquad ...(4.32)$$

where A is the area of cross–section of the circular section of diameter $D$.

The effect of the bending moment $Px_1$ will be to bend the column in the plane containing the force $P$ and passing through the point $O$, such that the centre of curvature will lie on the right side of the centroidal bending axis $YY$. Thus as discussed in article (4.1), the bending stress at any point of the section lying to the right of $YY$ will be compressive and all the points of the section lying to the left of $YY$ will experiences tensile bending stresses.

As per the discussion of the article (4.5), evidently the condition for non–reversibility of stress will be obtained by setting the resultant stress at A equal to zero.

*i.e.* $$\sigma_A = \sigma_0 + \frac{D}{2}\left(\frac{Px_1}{I_{yy}}\right) = 0$$

or $$\frac{-4P}{\pi D^2} + \frac{D}{2}\,\frac{P.x_1 \times 64}{\pi D_4} = 0$$

or $$x_1 = \frac{D}{8} \qquad ...(4.33)$$

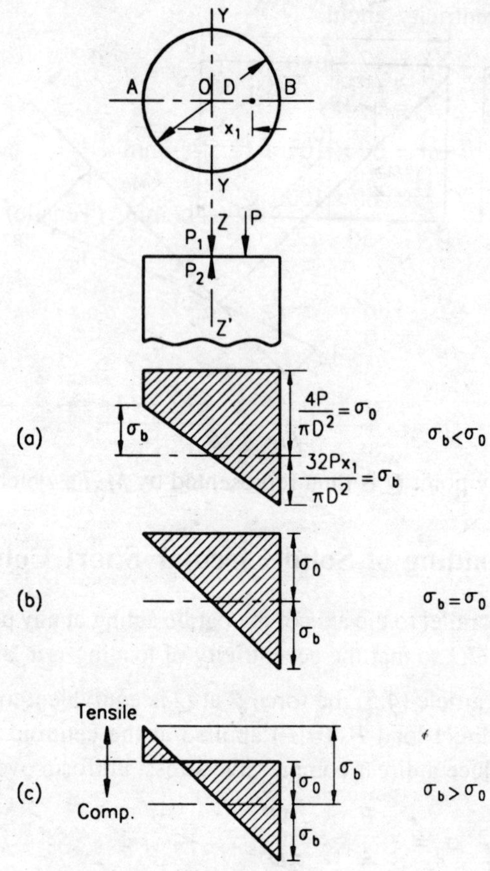


*Fig. 4.17*

Thus for the non–reversibility of the stress the load point must lie within the circle of radius $D/8$, *i.e.* within the middle quarter of the given column section.

For any given load position the stress distribution curves can be drawn on the lines of article 4.5. These diagrams are shown in Fig. 4.17.

**Example 4.12.** *A 50 mm × 12.5 mm flat steel bar was placed in a testing machine and subjected to a 60 kN load acting as shown in Fig. 4.18. An extensometer placed in line with the load recorded an extension of 0.16 mm on a gauge length of 200 mm. Calculate the maximum and minimum stresses set up, and the value of Young's modulus.*

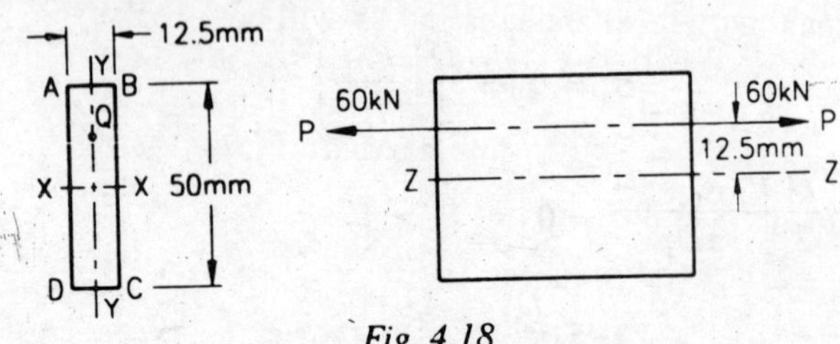


*Fig. 4.18*

**Solution.** Eccentricity about

$$XX = 12.5 \text{ mm}$$

$$P = 60 \text{ kN}$$

$$M = 60 \times 10^3 \times 12.5 \text{ N–mm}$$

Direct stress $\sigma_0 = \dfrac{60 \times 1000}{50 \times 12.5} = 96 \text{ N / mm}^2$ (Tensile)

Bending stress at $Q$

$$= (\sigma_b)_Q = \frac{M}{I_{xx}} \cdot y_Q$$

$$= \frac{(60000 \times 12.5) \times 12.5 \times 12}{12.5 \times (50)^3}$$

$= 72$ N/mm² orMN/m² ( Tensile)}

Therefore resultant stress along the load line

$$= 96 + 72$$

$= 168$ N/mm² or MN/m²

Therefore strain along load line

$$= \frac{168}{E}$$

Equating this to the given value, we have

$$\frac{168}{E} = \frac{0.16}{200}$$

or $$E = \frac{200 \times 168}{200} = 210000 \text{ N/mm}^2$$

or $$= 210 \times 10^3 \text{ MN/m}^2$$

The bending stresses at $AB$ and $DC$ will be equal in magnitude and are given by

$$(\sigma b)_{AB} = (\sigma b)_{CD} = \frac{(60000 \times 12.5)}{12.5 \times (50)^3} \times 12.25$$

$= 144$ N/mm²

Thus the maximum resultant stress at $AB$ will be

$$= 96 + 144$$

$= 240$ N/mm² or MN/m² (tensile)

and the minimum stress at $DC$ will be

$$= 144 - 96$$

$= 48$ N/mm² or MN/m² (Comp.)

**Example 4.13.** *Fig. 4.19 shows the boom of a cantilever type hydraulic crane. A load of 200 kN is supported by a single rope which passes over two pulleys, mounted on the boom as shown.*

*After passing through the hollow rectangular pillar of the crane, it is then connected to the hydraulic apparatus. The axis of the pulley D is vertically above the centroid of the section AB. The outside dimensions of the hollow rectangular section AB are 760 mm × 380 mm and the material is 30 mm thick all round, and the longer dimension is in the direction AB. Find the maximum tensile and compressive stresses induced in the section AB. Also find the location of the neutral axis.*

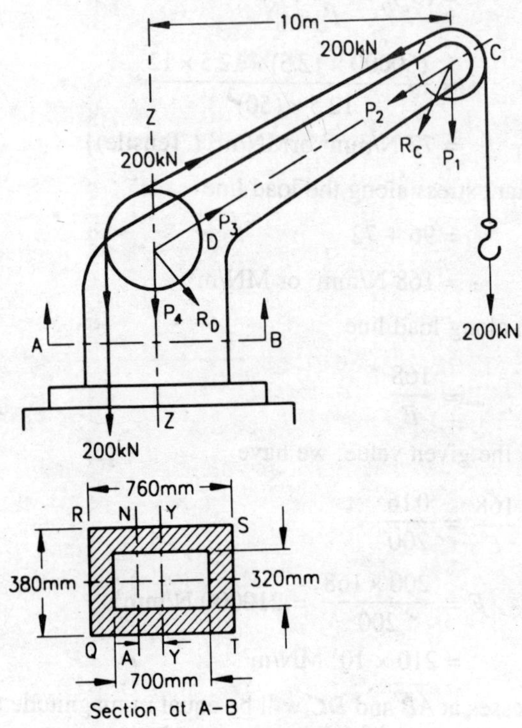


*Fig. 4.19*

**Solution.** Neglecting friction, the tension in the rope will be 200 kN everywhere. By drawing free body diagrams of the two pulleys it can be seen that due to the tension in the rope, the reactions on the bearings of the pulleys at $C$ and $D$ will be $R_C$ and $R_D$ respectively. These reactions can be split up into loads, $P_1$ and $P_2$ at $C$, and $P_3$ and $P_4$ at $D$ as shown in Fig. 4.19 such that $P_1 = P_2 = P_3 = P_4 = 200$ kN. Out of these four effective loads on the boom, $P_2$ and $P_3$ being equal and opposite will cancel out each other's effect on the section $AB$. Thus the effect of the load supported by the hook, on the section $AB$ is equivalent to the combined effect of the loads $P_1$ and $P_4$ acting at the points $C$ and $D$ of the boom respectively.

**Effect of $P_4$.** Its effect will be to induce a uniform direct compressive stress on the section *AB* of magnitude

$$(\sigma_0)_{P_4} = -\frac{200}{A} \qquad ...(1)$$

**Effect of** $P_1$. Its effect will be to induce direct as well as bending stresses.

Direct compressive stress

$$(\sigma_0)_{P_1} = -\frac{200}{A} \qquad ...(2)$$

Bending moment about *YY*

$$= M = 200 \times 10 \text{ kN–m}$$

$$I_{yy} = \frac{0.380 \times (0.760)^3}{12} - \frac{0.320\,(0.700)^3}{12}$$

$$= 0.00475 \text{ m}^4$$

$$A = (0.760 \times 0.38) - (0.7 \times 0.32) = 0.0648 \text{ m}^2$$

Max. bending stress

$$= (\sigma_b)_{max} = \frac{My}{I_{yy}} = \frac{2000 \times 0.38}{0.00475}$$

$$= 1.6 \times 10^5 \text{ kN/m}^2$$

This bending stress is compressive at any point on *ST* and tensile on *QR*.

Now from (1) and (2)

$$(\sigma_0)_{P_4} = (\sigma_0)_{P_1} = -\frac{200}{0.0648} = -3080 \text{ kN/m}^2$$

Therefore the resultant stress along *ST*

$$= (\sigma_0)_{P4} + (\sigma_0)_{P_1} - (\sigma_b)_{max}$$

$$= -3080 - 3080 - 1.6 \times 10^5$$

$$= -166160 \text{ kN/m}^2$$

Or the maximum compressive stress = 166.16 MN/m$^2$

The resultant stress along

$$\text{RQ} = (\sigma_b)_{max} + (\sigma_0)_{P_4} + (\sigma_0)_{P_1}$$

$$= 1.6 \times 10^5 - 3080 - 3080$$

$$= 153840 \text{ kN/m}^2$$

Or the max. tensile stress along

$$RQ = 153.84 \text{ MN/m}^2.$$

**Location of Neutral Axis.** Let it be at a distance x to the left of *YY*. Then as the resultant stress along neutral axis is to be zero the magnitudes of the direct and bending stresses will be equal.

*i.e.*
$$\frac{Mx}{I_{yy}} = (\sigma_0)_{P_1} + (\sigma_0)_{P_4}$$

or
$$\frac{2000 \times x}{0.00475} = 3080 + 3080$$

*i.e.*
$$x = \frac{6160 \times 0.00475}{2000} = .01465 \text{ m}$$

Thus the neutral axis is 14.65 mm to the left of *YY* and parallel to it.

**Example 4.14.** *Fig. 4.20 shows a forged crank with a section along the line AB. The thrust P of the connecting rod acts on the crank at a angle θ as shown. Find the maximum allowable value of the thrust P if the stress at the section AB were not to exceed 45 MN/m². Assume the axis ZZ to pass through the centroid of the section AB.*

**Solution.** Area of the section *AB*

$$= 150 \times 75 - 30 \times 100$$

$$= 8250 \text{ mm}^2$$

$$I_{NA} = \frac{75 \times (150)^2}{12} - \frac{30 \times (100)^3}{12}$$

$$= 1.86 \times 10^7 \text{ (mm)}^4$$

The thrust *P* can be resolved into two components, *viz.* $P_z = P \cos\theta$ giving direct compressive stress on the section and $P_y = P \sin\theta$ giving rise to a bending moment of magnitude ($P_y \times 100$ ) over the section *AB*, due to which bending stresses will be induced.

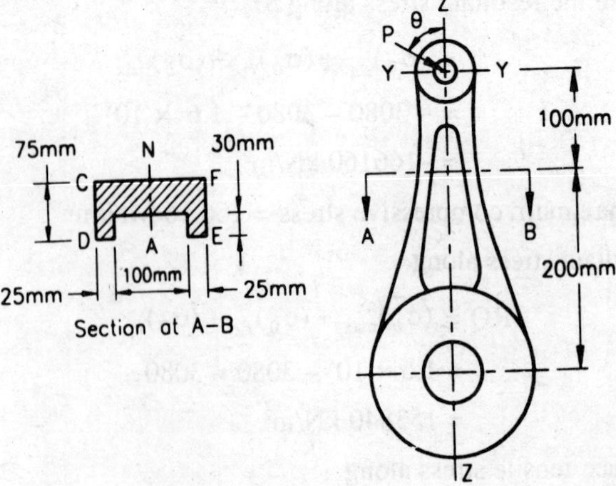


*Fig. 4.20*

For any value of θ the resultant compressive stresses along the line *FE* will be greatest and its magnitude is given by

$$|\sigma_{EF}| = \frac{P_Z}{A} + \frac{100\,P_y}{I_{NA}} \cdot \left(\frac{150}{2}\right)$$

$$= \frac{P}{A}\cos\theta + \frac{7500}{I_{NA}}\,P\cos\theta = 0$$

For getting the absolute maximum value of $\sigma_{EF}$, $(d\sigma_{EF}/d\theta)$ should be zero

*i.e.,* $$-\frac{P}{A}\sin\theta + \frac{7500}{I_{NA}}\,P\cos\theta = 0$$

or $$\tan\theta = \frac{7500\,A}{I_{NA}}$$

$$= \frac{7500 \times 8250}{1.86 \times 10^7}$$

$$= 3.33$$

or $$\theta = 73°\,18'$$

Substituting this value of θ, we have

$$\sigma_{max} = \frac{P}{A}\cos 73°\,18' + \frac{7500}{I_{NA}}\,P\sin 73°\,18'$$

$$= \frac{P}{8250} \times 0.2874 + \frac{7500}{1.86 \times 10^7} \times P \times 0.9578$$

or $$45 = 4.21 \times 10^{-4} P$$

*i.e.* $$P = 106500 \text{ N or } 106.5 \text{ kN.}$$

**Example 4.15.** *A, 100 mm diameter, round steel bar is acted upon by a longitudinal pull whose line of action is parallel to the axis of the bar. At a certain section are marked three points A, B and C around the section equally spaced from each other. The strains along the longitudinal direction at these points are measured with the help of three strain gauges attached on the surface of the bar at these points in the direction parallel to the axis of the bar. The respective strains at these points are italic* $\varepsilon_A = 45 \times 10^{-5}$ *(tensile),* $\varepsilon_B = 15 \times 10^{-5}$ *(tensile) and* $\varepsilon_C = 37.5 \times 10^{-5}$ *(tensile).*

*If the value of Young's modulus of steel is 200 GN/m² determine (a) the magnitude and location of the greatest and least stresses at the section ; (b) the magnitude and eccentricity of the pull.*

**Solution.** The cross–section of the bar is shown in Fig. 4.21. The longitudinal stresses at the points *A*, *B* and *C* are

$$\sigma_A = E\varepsilon_A$$

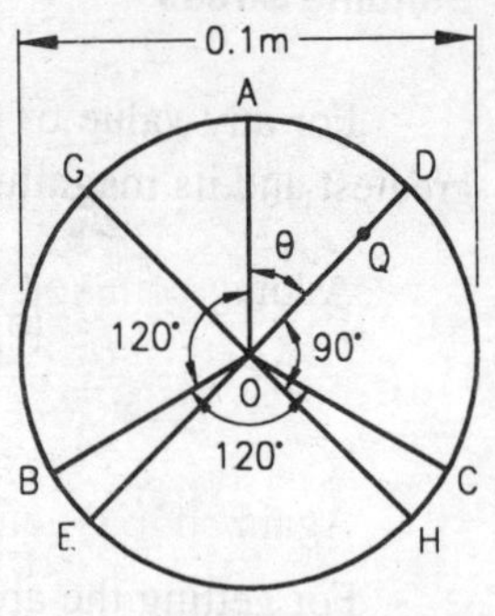


$$= (200 \times 10^3)\ (45 \times 10^{-5})$$
$$= 90 \text{ MN/m}^2$$
$$\sigma_B = E\varepsilon_B$$
$$= (200 \times 10^3)\ (15 \times 10^{-5})$$
$$= 30 \text{ MN/m}^2$$
$$\sigma_C = E\varepsilon_C$$
$$= (200 \times 10^3)\ (37.5 \times 10^{-5})$$
$$= 75 \text{ MN/m}^2$$

*Fig. 4.21*

Since the stresses at $A$ and $C$ are higher than that at $B$, it is expected that the point of application of the load, $Q$, will be on the radial line $OD$ lying between $OA$ and $OC$.

Let the load be $P$ with the eccentricity $OQ$ equal to $x$ such that the axis of bending is $GH$. Assuming the inclination between $OA$ and $OD$ as $\theta$, we have

## For the point A

$$y_A = OA \cos\theta \text{ and } I_{GH} = \frac{\pi d^4}{64}$$

$$\sigma_A = \frac{4P}{\pi d^2} + \frac{P \times x \times 64}{\pi d^2} \cdot \left(\frac{d}{2}\cos\theta\right) = 90$$

or $$\frac{4P}{\pi\,(0.1)^2} + \frac{32\,Px\cos\theta}{\pi\,(0.1)^3} = 90$$

or $$\frac{400\,P}{\pi}(1 + 80x\cos\theta) = 90$$

or $$1 + 80\,x\cos\theta = \frac{9\pi}{40P} \qquad \ldots(A)$$

**Similarly for the point B**

$$\sigma_B = \frac{400P}{\pi}[1 + 80x\cos(120° + \theta)] = 30$$

or $$1 + 80x\left[-\frac{1}{2}\cos\theta - \frac{\sqrt{3}}{2}\sin\theta\right] = \frac{3\pi}{40P}$$

or $$1 + 40x\,[-\cos\theta - \sqrt{3}\sin\theta] = \frac{3\pi}{40P} \qquad \ldots(B)$$

**For the point C**

$$\sigma_C = \frac{400P}{\pi}[1 + 80x\cos(120° - \theta)] = 75$$

or $$1 + 80x\left[-\frac{1}{2}\cos\theta + \frac{\sqrt{3}}{2}\sin\theta\right] = \frac{75\pi}{400P} \qquad ...(C)$$

Adding equations (*B*) and (*C*)

$$2 - 80x\cos\theta = \frac{10.5\pi}{40P} \qquad ...(D)$$

Again adding equations (*A*) and (*D*)

$$3 = \frac{19.5\pi}{120} = 0.51 \text{ MN}$$

or $$P = \frac{19.5\pi}{120} = 0.51 \text{ MN}$$

Also subtracting equation (*B*) from (*C*)

$$80\sqrt{3}\,x\sin\theta = \frac{4.5\pi}{40 \times 0.51} = 0.693 \qquad ...(E)$$

and from equation (*A*)

$$80x\cos\theta = \frac{9\pi}{40 \times 0.51} - 1 = 0.386 \qquad ...(F)$$

Therefore dividing equations (*E*) and (*F*)

$$\tan\theta = \frac{0.693}{80\sqrt{3} \times 0.7193} = 0.0096$$

*i.e.* $$\theta = 46^\circ$$

From equation (*E*)

$$x = \frac{0.693}{80\sqrt{3} \times 0.7193} = 0.00695 \text{ m}$$

$$\sigma_{max} = \sigma_D = \frac{4P}{\pi(0.1)^2} + \frac{64P\,.\,x}{\pi(0.1)^4}\left(\frac{0.1}{2}\right)$$

$$= \frac{400\,P}{\pi}(1 + 80x)$$

$$= \frac{400 \times 0.51}{\pi}(1 + 80 \times 0.00695)$$

$$= 101 \text{MN/m}^2 \text{ (tensile)}$$

$$\sigma_{min} = \sigma_E = \frac{400\,P}{\pi}(1 - 80x)$$

$$= \frac{400 \times 0.51}{\pi}(1 - 80 \times 0.00695)$$

$$= 28.9 \text{ MN/m}^2 \text{ (tensile).}$$

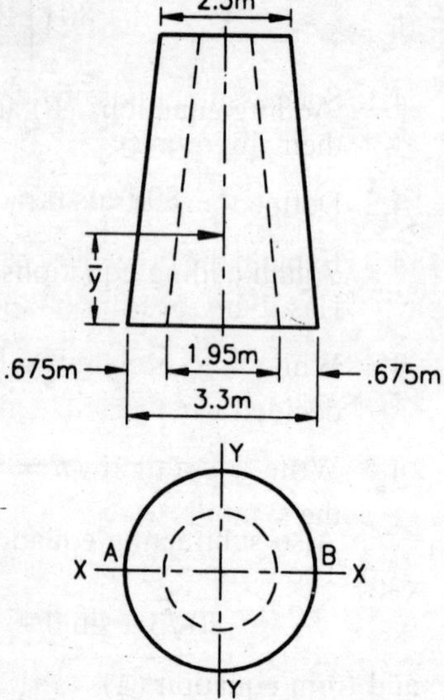

Fig. 4.22

**Example 4.16.** *A 30 m high brick chimney is 2.3 m outside diameter at the top and tapers uniformly to 3.3 m outside diameter at the bottom. The chimney weighs 2200 kN and is 0.675 m thick at the base. If a uniform horizontal wind pressure of 1 kN/m² acts on the projected area of the chimney, determine the maximum and minimum normal stresses on the base.*

**Solution.** Total wind force $F$

$$= 1 \times \text{Projected area.}$$

$$= 1 \times \frac{1}{2}(3.3 + 2.3) \times 30)$$

$$= 84 \text{ kN}$$

The force $F$ acts at a height $\bar{y}$ from the base of the chimney and passes through the centroid of the projected area which is similar to the elevation as drawn in Fig. 4.22. To evaluate $\bar{y}$, take moments of the projected area about the base $AB$

$$\left[\frac{1}{2}(3.3 + 2.3) \times 30\right]\bar{y} = (2.3 \times 30)\,15 + 2\left[\frac{1}{2} \times .5 \times 30 \times \frac{30}{3}\right]$$

*i.e.* $\bar{y} = 14.1$ m

Now $$I_{YY} = \frac{\pi}{64}[(3.3)^4 - (1.95)^4]$$

$$= 5.1 \text{ m}^4$$

$$A = \frac{\pi}{4}(3.3^2 - 1.95^2) = 5.56 \text{ m}^2$$

$$M = F\bar{y} = 84 \times 14.1 \text{ kN} - \text{m}$$

$$(\sigma_b)_B = \frac{(84 \times 14.1)}{5.1} \times \left(\frac{3.3}{2}\right) = 383 \text{ kN/m}^2 \text{ (comp.)}$$

and $(\sigma_b)_A = 383$ kN/m² (tensile)

Direct stress $$\sigma_0 = \frac{2200}{A} = \frac{2200}{5.56}\ 395 \text{ kN/m}^2 \text{ (comp.)}$$

Therefore max. stress at $B$

$$\sigma_B = 383 + 395 = 778 \text{ kN/m}^2 \text{ (comp.)}$$

and minimum stress at

$$\sigma_A = 395 - 383 = 12 \text{ kN/m}^2 \text{ (comp.)}$$

## REVIEW QUESTIONS

4.1 State assumptions made while deriving the equation for bending stress and explain their importance.

4.2 Define the neutral surface and neutral axis.

4.3 Explain the procedure as to how the neutral axis is fixed in position and direction. Has it any bearing on the axis of moment ? If yes, how ?

4.4 While determining the bending stresses at any point on the section, how it is decided whether it is tensile of compressive ?

4.5 Write down the expression for bending stress and radius of evaluature. Explain the symbols used.

4.6 The cross-section of a rectangular beam is 200 mm wide by 250 mm deep Fig (4.22*a*). This beam is simply supported at its ends and carries a central vertical load of 100 kN. If the span is 3m. Find the maximum tensile and max compressive stresses induced in the beam and give the location of the max stress points.

**Solution.** Here $b = 200$ mm

$d = 250$ mm

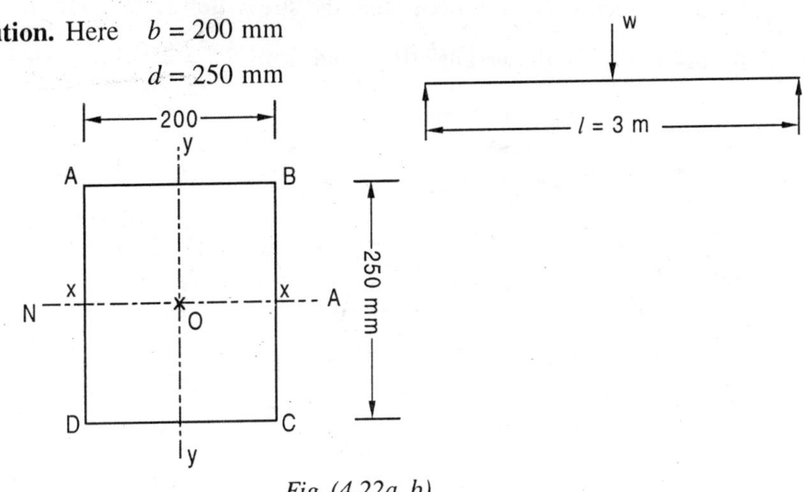


*Fig. (4.22a, b)*

$$I_{NA} = \frac{bd^3}{12} = \frac{200 \times (250)^3}{12} \text{ (mm}^4\text{)}$$

$$= 260.4166 \times 10^6 \text{ (mm}^4\text{)}$$

$\sigma_{max}$ will occur along *AB* where it will be (comp) while along *CD* it will be maximum tensile

$$\sigma_{AB} = \frac{yM}{I}$$

Here $$M = \frac{Wl}{4}$$

$$= \frac{100\,(\text{kN}) \times 3m}{4}$$

$$= 75 \text{ kN-m} = 75 \times 106 \text{ N-mm}$$

Now $\sigma_{max} = \sigma_{AB} = \dfrac{yM}{I}$

Here $y = \dfrac{1}{2}$ depth = 125 mm

$\therefore$ $\sigma_{AB} = 125 \times \dfrac{75\ 10^6}{260.4166 \times x^6}$ N/mm²

$$= 36.00 \text{ MPa}$$

$$= 36 \text{ MPa}$$

It will be compressive on points along the line *AB*. On the contrary it will be tensile on points along *DC* because it is crying sagging moment.

4.7 Find the value of $I_{NA}$ for the section shown in Fig. (4.22*c*) and location of *NA*. If *B.M.* on the section is 1150 kN-m find the stress along *AD*.

**Solution.** Let $\bar{y}$ be the height of centroid from *BC* = *h*

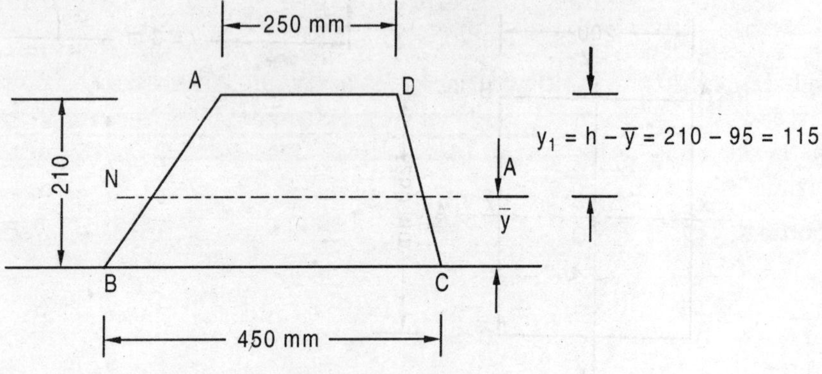


*Fig. (4.22c)*

Here $a = 250$ mm

$b = 450$ mm

$h = 210$ mm

Then from equation (*Q*) of article 4.2.7)

We have $\bar{y} = \dfrac{h}{3}\left(\dfrac{b+2a}{b+a}\right)$

$$= \frac{210}{3}\left[\frac{450+500}{450+250}\right]$$

$$= 70 \times \frac{950}{700} = 95.0 \text{ mm}$$

From eqn. ($R$) of § 4.2.7.

$$I_{NA} = \frac{h^3}{36}\left[(a+b) + \frac{2ab}{(a+b)}\right]$$

$$= \frac{(210)^3}{36}\left[(250+450) + \frac{2 \times 250 \times 450}{700}\right] \text{mm}^4$$

$$= \frac{(210)^3}{36}\left[700 + \frac{500 \times 450}{700}\right]$$

$$= \frac{(210)^3}{36}(700 + 321.4286) = \frac{1021.4286 \times (210)^3}{36}$$

$$= 2627.7625 \times 10^6 \text{ mm}^4$$

($b$) Let the stress along the line $AD$ due to bending moment $M$ be $\sigma_{AB}$ and then

$$\sigma_{AD} = \frac{M}{I_{NA}} \cdot y_{AD}$$

Now $y_{AD}$ = 115 mm and M = 1150 kN-m = 1150 × 106 kN

Then $$\sigma_{AD} = y_{AD}\frac{M}{I_{NA}} = \frac{115 \times (1150 \times 10^6)}{2627.7625 \times 10^6}$$

$$= 50.328 \text{ N/mm}^2 = 50.328 \text{ MPa}$$

4.8. An I-section 120 mm × 240 mm (Fig 4.22$d$) beam, 3m long, is simply supported at its ends Fig. (4.22$b$) The flange thickness is 20 mm and web thickness 10 mm. The beam carries a load of 120 kN at a distance 2m from one end. Find the maximum stress induced in the beam.

**Solution.**

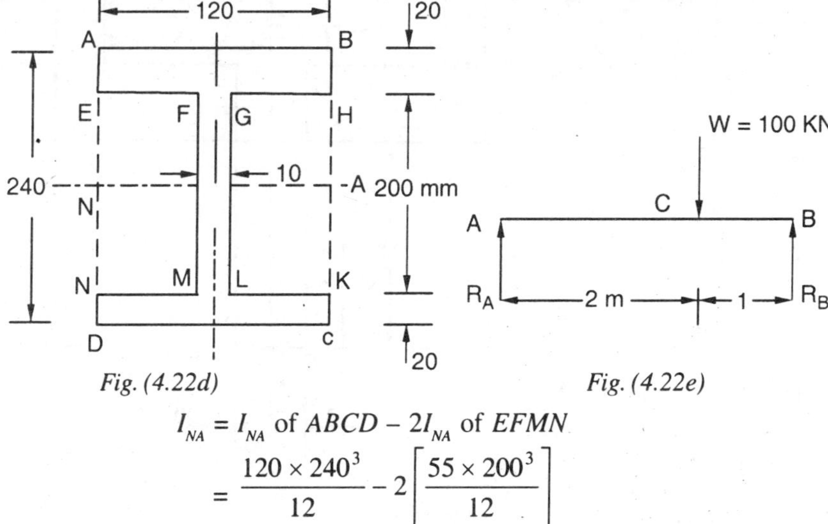


*Fig. (4.22d)* *Fig. (4.22e)*

$$I_{NA} = I_{NA} \text{ of } ABCD - 2I_{NA} \text{ of } EFMN$$

$$= \frac{120 \times 240^3}{12} - 2\left[\frac{55 \times 200^3}{12}\right]$$

$$= 13824 \times 10^4 - 7.333333 \times 10^4$$

$$= 6491 \times 10^4 \text{ mm}^4$$

$$M_{max} = \frac{Wab}{l}$$

when $W = 120$ kN

$a = 2$m

$b = 1$m

$l = 3$m

*i.e.,*

$$M_{max} = \frac{120\, kN \times 2m \times 1m}{6\, m}$$

$$= 40 \text{ kN-m} = 40 \times 10^6 \text{ N-mm}$$

and

$$y_{max} = \frac{240}{2} = 120 \text{ mm}$$

Therefore

$$(\sigma_n)_{max} = y_{max} \frac{M_{max}}{I_{NA}} = \frac{120 \times 40 \times 10^6}{6491 \times 10^4} \text{ N/mm}^2$$

= 73.949 MPa (comp. along *AB* and tansile along *DC*)

4.9. The cross-section of a cantilever beam is shown in figure (4.22*f*). It carries a load of 300 kN at the free end. The length of the cantilever is 1.2 m. Find the maximum tensile and compressive stress induced in the section.

**Solution.** For finding the location of the centroid, take moments of areas about the line *AB*. Let NA be at a distance $\bar{y}$ from *AB*

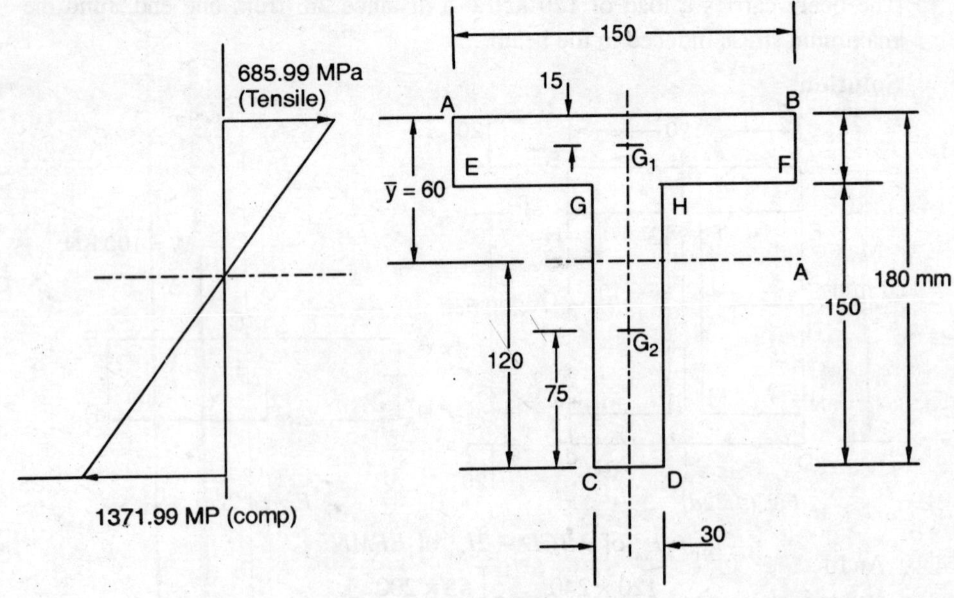


*Fig. (4.22f)*

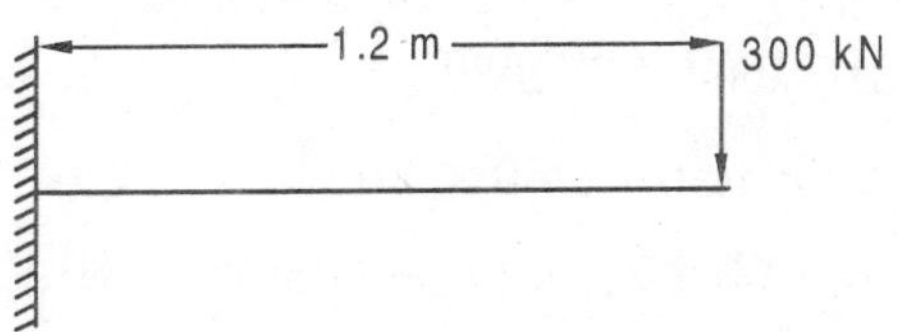


*Fig. (4.22g)*

$$\bar{y}\,[(150 \times 30) + (150 \times 30)] = (150 \times 30 + 15) + 150 \times 30\,(30 + 75)$$

or $$\bar{y} = 60 \text{ mm}$$

Then $$I_{NA} = I_{NA} \text{ of } ABFE + I_{NA} \text{ of } GHCD$$

$$= \left[\frac{150 \times 30^3}{12} + (150 \times 30)\left(\bar{y} - 15)^2\right)\right]$$

$$+ \left(\frac{30 \times 150^3}{12} + (150 \times 30)\,(180 - \bar{y} - 75)^2\right)$$

$$= 31.487 \times 10^6 \text{ mm}^4$$

$$M_{\max} = (300 \times 10^3 \text{ N})\,(1.2 \times 10^3 \text{ mm})$$

$$= 360.0 \times 10^6 \text{ N.mm (Hogging)}$$

The max tensile stress will occur along $AB$ where

$$y_{AB} = \bar{y} = 60 \text{ mm}$$

$\therefore$ $$\sigma_{AB} = y_{AB}\,\frac{M_{\max}}{I_{NA}} = \frac{60 \times 360 \times 10^6}{31.487 \times 10^6}$$

$$= 685.987 \text{ N/mm}^2$$

*i.e.,* $$\sigma_{AB} = 685.997 \text{ MPa (tensile)}$$

Max compressive stress will occur along any point on $CD$ where $y_{CD} = 180 - \bar{y}$ $= 120$ mm

$$\sigma_{CD} = \frac{M_{\max}\,y_{CD}}{I_{NA}}$$

$$= \frac{(360 \times 10^6 \text{ N} - \text{mm}) \times 120 \text{ mm}}{31.487 \times 10^6\,(\text{mm}^4)}$$

$$= 1371.99 \text{ MPa (comp.)}$$

4.10 An I-beam of the section shown in figure (4.22*h*) is 3m. long and simply supported at its ends. It carries a u/d load of 10 kN-m, over its entire span. Calculate the max tensile and max compressive stresses induced in the material.

**Solution.** If $\bar{y}$ is the distance of *NA* from *NM*, then taking moments of Areas *ABCD, EFGH* and *LKMN* about *MN* gives

$$\bar{y}\,[(100 \times 20) + (20 \times 100) + (200 \times 30)$$

$$= (100 \times 20) \times 140 + (100 \times 20)\ 80 + (200 \times 30) \times 15$$

or $$\bar{y} = 53 \text{ mm}$$

$$I_{NA} = I_{NA} \text{ of } ABCD + I_{NA} \text{ of } EFGH + I_{NA} \text{ of } KLNM$$

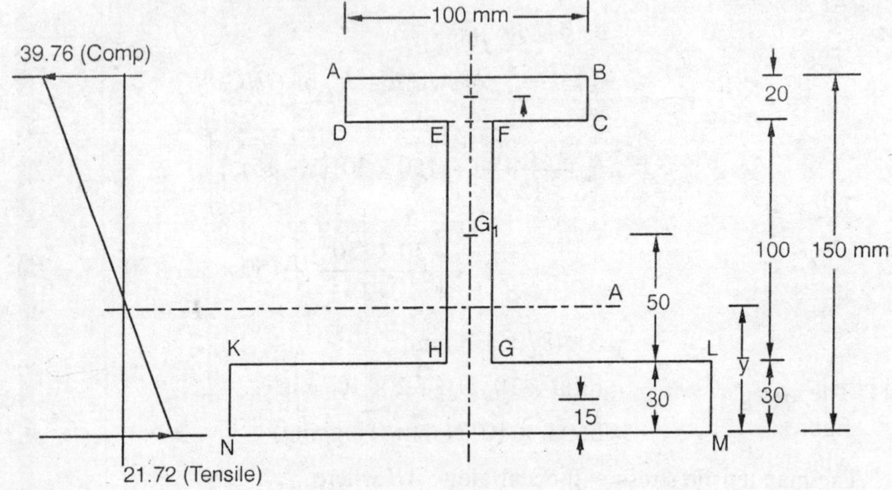


*Fig. (4.22h)*

or $$I_{NA} = \left[\frac{100 + 20^3}{12} + (100 \times 20)(140 - 53)^2\right]$$

$$+\left[\frac{20 \times 100^3}{12} + (20 \times 100)(80 - 53)^2\right]$$

$$+\left[\frac{200 \times 30^3}{12} + (200 \times 30)(53 - 15)^2\right]$$

$$I_{NA} = 6.66 \times 10^2 + 1513.8 \times 10^2 + 165.667 \times 10^2$$

$$+ 145.8 \times 10^2 + 866.4 \times 10^2$$

$$= 2744.33 \times 10^4 \text{ mm}^4$$

$$\text{Max } BM = \frac{wl^2}{8}$$

where $$w = 10 \text{ kN-m}$$

$$l = 3\text{m}$$

w = 10 KN \ m

A B

3 m

*Fig. (4.22 i)*

$$= \frac{10\ (3)^2}{8}\ \text{kN} - \text{m}$$

$$= 11.25\ \text{kN-m} = 11.25 \times 10^6\ \text{N-mm}$$

Now $\sigma = \frac{M}{I_{NA}}$

At $AB$ $\sigma_{AB} = y_{AB}\frac{M}{I_{NA}} = (150 - 53)\frac{11.25 \times 10^6}{2744.33 \times 10^4}$

$$= \frac{97 \times 11.25 \times 10^6}{2744.33 \times 10^4} = 39.76\ \text{N/mm}^2$$

$= 39.76$ MPa (comp.)

$$\sigma_{MN} = y_{MN}\frac{M}{I_{NA}} = \frac{53 \times 11.25 \times 10^6}{2744.33 \times 10^4}$$

$= 21.724\ \text{N/mm}^2$

$= 21.724$ MPa (tensile)

4.11 The cross-section of hollow rectangular pillar of a hydraulic crane is a shown in figure (4.22*j*). While lifting the rated load the axis of the bending moment to which this section is subjected is perpendicular to *X*-axis. If the allowable stresses in tension and compression for the material an 20 MPa and 85 MPa respectively, find the bending moment which this section can withstand without crossing the allowable stresses in tension as wall as in compression. While lifting the load the crane can rotate so that the load can be on the left side of the vertucle *N-A* as well.

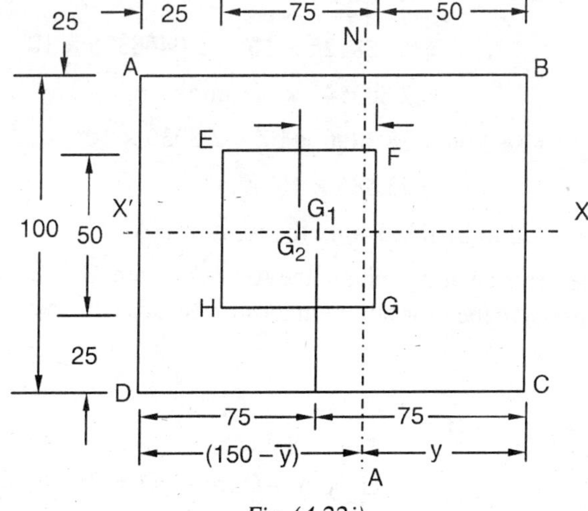


*Fig. (4.22j)*

**Solution.** From the problem it is evident that the load to be lifted is such that the axis of *BM* is to be at right angles to *XX*. That is the centroidal bending axis or *N-A* of the section will be vertical.

In order to locate the centroid *G* of the hollow section take moments of areas about edge *BC*. Let its distance from *BC* be $\bar{y}$. Then taking the moment of areas *ABCD* and *EFHK* and subtracting the later from the former we have

$$(150 \times 100 - 75 \times 50)\bar{y} = \left[(150 \times 100)\frac{150}{2}\right] - (75 \times 50)(50 + 37.5)$$

$$11250\,\bar{y} = 1.125 \times 10^6 - 0.328125 \times 10^6$$

or
$$\bar{y} = 70.8 \text{ mm}$$

$$I_{NA} = I_{NA} \text{ of } ABCD - I_{NA} \text{ of } EFHK$$

Let $G_1$ be the centroid of *ABCD* and $G_2$ centroid of *EFHK*

Then
$$GG_1 = \frac{150}{2} - \bar{y} = 75 - 70.8 = 4.2 \text{ mm}$$

and
$$GG_2 = (50 + 37.5) = 16.7 \text{ mm}$$

Then
$$I_{NA} \text{ of } ABCD = \frac{(150)^3 \times 100}{12} + \text{Area } ABCD \times GG_1^2$$

$$= \frac{150^3 \times 100}{12} + (150 \times 100)(4.2)^2$$

$$= 28.125 \times 10^6 + 0.2646 \times 10^6$$

$$= 28.3896 \times 10^6 \text{ mm}^4$$

$$I_{NA} \text{ of } EFHK = \frac{50 \times 75^3}{12} + (50 \times 75)(16.7)^2$$

$$= 1.758125 \times 10^6 + 1.0458375 \times 10^6$$

$$= 2.803646 \times 10^6 \text{ mm}^4$$

There $I_{NA}$ of given section $= 28.3896 \times 10^6 - 2.8036 \times 10^6$

$$= 25.586 \times 10^6 \text{ mm}^4$$

**Case I :** When load to be lifted is on the right of NA

In this case the stress on any point on line AD will be tensile and that on any point on BC will be compressive then for failure in tension because the material is weaker in tension

$$M = \frac{\sigma_{AD}\, I_{NA}}{y_{AD}}$$

Where $y_{AD}$ = distance of *AD* from *NA* = 79.2 mm

$$\sigma_{AD} = 20 \text{ MPa} = 20 \text{ N/mm}^2$$

Therefore $$M = \frac{20\,(25.586 \times 10^6)}{79.2}$$

$$= 6.46 \times 106 \text{ N-mm} = 6460 \text{ N-m}$$

**Case II :** When the load is on the left of NA, the stress on any point on BC will be tensile and on AD compressive taking failure in tension,

We have $$M = \frac{\sigma_{BC} \times I_{NA}}{y_{BC}}$$ Here $y_{BC} = \bar{y}$, $\sigma_{BC} = 20$ MPa

$$= \frac{20 \times 25.586 \times 10^6}{70.8}$$

$$= 7.2277 \times 106 \text{ N-mm} = 7227.7 \text{ N-m}$$

Thus the safest moment which this section can take up is smaller of the two values *i.e.*, 6460 N-m.

4.12 A short hollow cylindrical column, 250 mm outside dia and 150 mm inside dia, carries an eccentric load of 500 kN whose line of action is parallel to cylinder axis but at a distance of 70 mm from the axis of the cylinder. Find the maximum values of the tensile and compressive stresses induced in the material.

**Solution.** Let the electricity = ***e*** = 70 mm of given compressive load

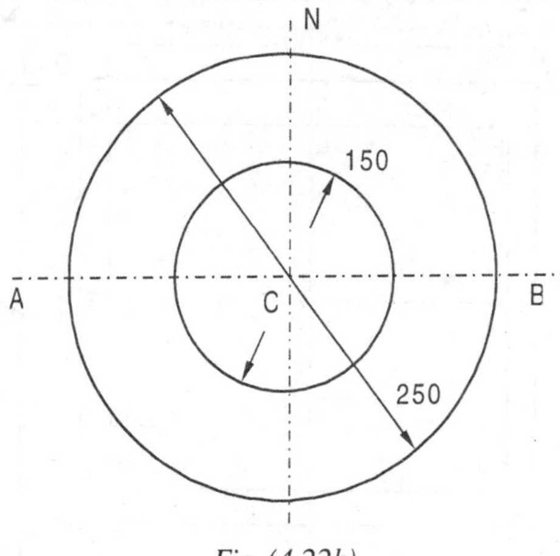


*Fig. (4.22k)*

Given load $P = 500$ kN

$A$ = Area of the cross-section

$$= \frac{\pi}{4}(250^2 - 150^2)$$

$$\frac{\pi}{4}(62500 - 22500) = 40000 \text{ mm}^2$$

$$I_{NA} = \frac{\pi}{64}(D^4 - d^4) = \frac{\pi}{64}(250^4 - 150^4) = 166.897 \times 10^6 \text{ mm}^4$$

$$\sigma_0 = \text{direct stress} = \frac{P}{A} = \frac{500 \text{ kN}}{40000} = 12.5 \text{ MPa (Comp)}$$

$$M = P \times e = (500 \text{ kN}) \times 70$$

$$= 35000 \text{ kN-mm}$$

Resultant stress at $B$

$$(\sigma_r)_B = \sigma_0 + \sigma_b = (12.5) \text{ MPa (comp.)}$$

But
$$\sigma_B = y_B \cdot \frac{M_B}{I_{NA}} = \frac{250}{2} \times \frac{(35 \times 10^6)}{166.897 \times 10^6}$$

$$= 26.2137 \text{ MPa (comp.)}$$

Thus the resultant max comp stress at $B$

$$= 12.5 \text{ MPa} + 26.2137 \text{ MPa}$$

$$= 38.7137 \text{ MPa (comp)}$$

At the point A the bending stress

$$= 26.2137 \text{ MPa Tensile}$$

Thus resultant tensile stress at A

$$= 26.7137 - 12.5 \text{ MPa}$$

$$= 13.7137 \text{ MPa (tensils)}$$

4.13. The cross-section of a hollow pier is as shown in figure (4.22l)

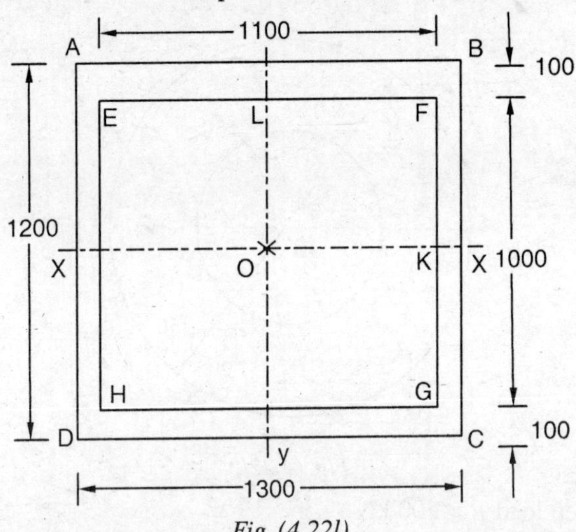


*Fig. (4.22l)*

A Compressive load of 100 kN normal to section acts at the point $F$.

Find the stresses at the four corners $A$, $B$, $C$ and $D$ of the section.

Let the load be = $P$ = 100 kN = $10^5$ N

The load line is eccentric with both the axes

$$M_{XX} = BM \text{ about } XX = P \times FK$$
$$= (100 \times 10^3) \times 500 \text{ N-mm} = 5 \times 10^7$$

$$M_{yy} = \text{BM about yy} = \left(10^5 \times \frac{1100}{2}\right) = 5.5 \times 10^7 \text{ N-mm}$$

$$I_{XX} = \frac{1300 \times 1200^3}{12} - \frac{1100 \times 1000^3}{12}$$
$$= 1872 \times 10^8 - 916.7 \times 10^8$$
$$= 955.3 \times 10^8 \text{ mm}^4$$

and
$$I_{YY} = \frac{1200 \times (1300)^3}{12} - \frac{1000\,(1100)^3}{12}$$
$$= 2197 \times 10^8 - 1109.2 \times 10^8$$
$$= 1087.8 \times 10^8 \text{ mm}^4$$

Direct stress over the entire section = $\dfrac{P}{\text{Area of the section}}$

$$\sigma_0 = \frac{P}{(1300 \times 1200) - (1100 \times 1000)}$$

$$= \frac{10^5 N}{46 \times 10^4 \text{ mm}^2} = 0.2174 \text{ N/mm}^2 \text{ (comp.)}$$

**Stress at ponit *B***

(*i*) $\sigma_B$ =Direct stress $\sigma_0$ + stress due to $M_{XX}$ + Stress Due to $M_{YY}$

(*ii*) Bending stress due $M_{XX}$

$$(\sigma_B)_{YY} = \frac{(\text{distance of B from } XX)}{I_{yy}} M_{YY}$$

$$= \frac{600 \times (5 \times 10^7)}{(955.3 \times 10^8} = 0.314 \text{ N/mm}^2 \text{ (comp.)}$$

$$(\sigma_B)_{XX} = \frac{(\text{distance } of\ B \text{ from } yy)_{M_{yy}}}{I_{yy}}$$

$$= \frac{(500)(5.500 \times 10^7)}{(1087.8 \times 10^8)} = 0.2528 \text{ N/mm}^2 \text{ (comp.)}$$

Therefore
$$\sigma_B = (-2174) - (0.314) - (0.2528)$$
$$= 0.7842 \text{ MPa (comp) N/mm}^2 \text{ or MPa}$$

**(*ii*) Stress at the point *C***

$\sigma_{BC}$ = Resultant of direct stress + Boarding stress at $C$ due $M_{XX}$ + Bending stress at $C$ due to $M_{YY}$

Now $\sigma_0 = 0.2174$ MPa (comp.)

Bending stress at $C$ due $M_{XX}$= (Bending stress at $B$ are to $M_{XX}$ because the distance of $C$ from $XX$ is same as that of $B$.]

*i.e.*, $(\sigma_C)_{XX} = (\sigma_B)_{XX} = 0.314$ MPa (tensile)

Bending stress at *C* due $M_{YY}$

$$= (\sigma_C)_{YY} = (\sigma_B)_{YY} = 0.2528 \text{ MPa (comp.)}$$

Therefore resultant of these

$$\sigma_C = -0.2174 + 0.314 - 0.2528 = -0.1562 \text{ MPa}$$

or $= 0.1562$ MPa (comp)

**(*iii*) Stress at Point D**

*i.e.*, $\sigma_D = (\sigma_0) + (\sigma_D)_{XX} + (\sigma_D)_{YY}$

Now $\sigma_0 = -0.2174$ (comp.)

And $(\sigma_D)_{XX} = (\sigma_C)_{XX} = 0.314$ MPa (tensile)

Also $(\sigma_D)_{YY} = -(\sigma_C)_{YY} = -(-0.2528) = 0.2528$ MPa (tensile)

$\therefore$ $\sigma_D = -0.2174 + 0.314 + 0.2528$

$= 0.3494$ MPa = MPa (tensile)

**(*iv*) Stress at point *A*** $= \sigma_0$ + stress due to $M_{XX}$ + stress due to $M_{YY}$

$= -0.2174 - 0.314 + 0.2528$

$= -0.2786$ MPa

$= 0.2786$ MPa (comp.)

4.14. A wooden beam of hollow section as shown in figure (4.22*m*) is simply supported at its ends with span of 3m. A load of 10kN acts at a distance 2m from left hand support. Calculate in maximum tensile and compressive stress on the mid-section which 1.5 m from the left support

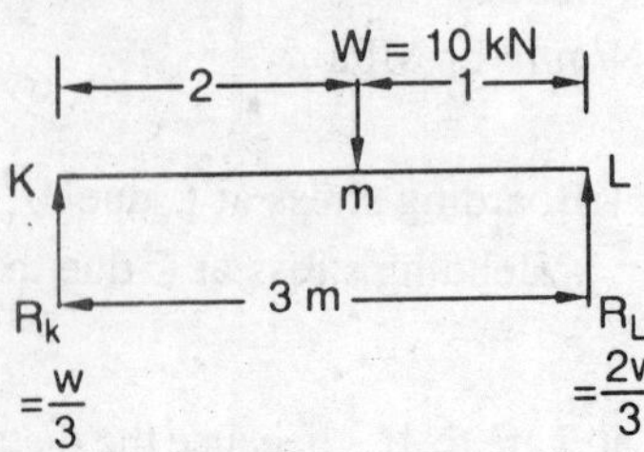


*Fig. (4.22 n)*

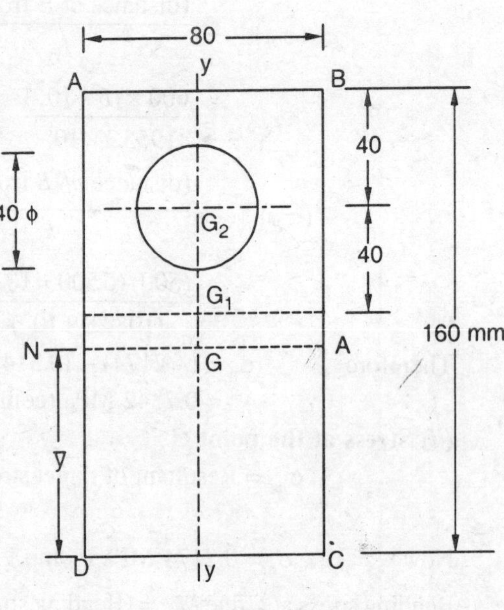


*Fig. (4.22m)*

Here $R_K = \dfrac{W}{3} = \dfrac{10}{3} kN$

and $R_L = \dfrac{2W}{3} = \dfrac{20}{3} kN$

**Solution.** Let $M$ be the BM at mid-section

$$M = \frac{l}{2} R_K = R_k \times \frac{l}{2} = \frac{10}{3} \times \frac{3}{2} \text{ kN} - \text{m} = 5 \text{ kN-m}$$

Let $\bar{y}$ be the distance of the centroid of the given section from $DC$. Then taking moments of the individual areas about $DC$ we get

$\bar{y}$ [Area $ABCD$ – Area of circular hole]

= Moment of area $ABCD$ about $DC$

– moment of circular area about $DC$

*i.e.,* $$\bar{y}\left[(80 \times 160) - \frac{\pi}{4}(40)^2\right] = \left[(80 \times 160) \times 80 - \frac{\pi}{4}(40)^2 (80 + 40)\right]$$

Therefore $\bar{y} = 75.6455$ mm

Then $$I_{NA} = \left[\frac{80 \times 160^3}{12} + (80 \times 160)(GG_1)^2\right] - \left[\frac{\pi \times (40)^4}{64} + \frac{\pi}{4}(40)^2 \times (G_2G)^2\right]$$

$= 24.95.558 \times 10^6$ mm$^4$

For the simply supported beam the bending moment at the middle section $N$ is

$$M_n = R_K \cdot \frac{l}{2} = \frac{W}{3} \cdot \frac{l}{2} = \frac{Wl}{6}$$

$$= \frac{(10^4 N)\ 3m}{6} = \frac{(10^4 . N) \times 3000\,(mm)}{6}$$

$= 5 \times 10^6$ N.mm

Therefore $$\sigma_{AB} = y_{AB} \frac{M}{I_{NA}}$$

where $y_{AB}$ = distance of $AB$ from $G = (160 - \bar{y})$

$= (160 - 75.6455) = 84.355$ mm

or $$\sigma_{AB} = \frac{(84.355) \times 5 \times 10^6}{24.951558 \times 10^6} \text{ N/mm}^2$$

$= 16.904$ N/mm$^2$ or MPa (comp.)

$$\sigma_{CD} = y_{CD} \frac{M}{I_{NA}} \quad \text{Here } y_{CD} = \bar{y}$$

$$= 75.6455 \times \frac{5 \times 10^6}{24.951558 \times 10^6}$$

$$= 15.158 \text{ N/mm}^2 \text{ or MPa (tensile)}$$

4.15. Cast iron channel 500 mm × 320 mm (Fig. 4.22p) carries water. The channel is 10 m long and is simply supported at its ends. If the bending stresses in the channel are not to exceed 15.5 MPa in tension and 45.5 MPa in compression, estimate the depth of water in the channel. Density of *CI* is 7000 kg/m³ and that of water 1000 kg/m³.

**Solution.** Let the NA of channel section be at a distance of $\bar{y}$ from *EF*.

Then taking the moments of areas *ABCD, NCLM* and *DEFK* about *FE*

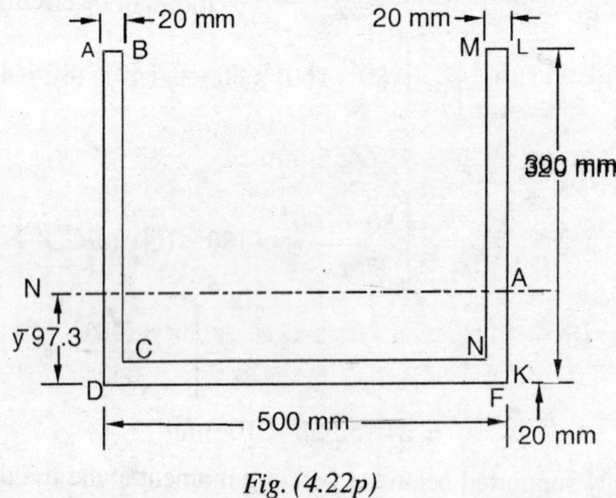


*Fig. (4.22p)*

We get $\bar{y}[(500 \times 20) + 2\,(300 \times 20)]$

$$= (500 \times 20) \times 10 + 2\,(300 \times 20)\,(20 + 150)$$

$$\bar{y} = 97.3 \text{ mm}$$

Then $\qquad I_{NA} = I_{NA}$ of area $DEFK + 2I_{NA}$ of $NKLM$

$$= \left[\frac{500 \times (20)^3}{12} + (500 \times 20)\,(97.3 - 10)^2\right]$$

$$+ 2\left[\frac{20 \times (300)^3}{12} + (20 \times 300)\,(170 - 97.3)^2\right]$$

or $\qquad I_{NA} = 3.3333 \times 10^5 + 762.2129 \times 10^5$

$$+ 2\,[450 \times 10^5 + 317.1174 \times 10^3]$$

$$= 2299.6971 \times 10^5 \text{ mm}^4$$

Resisting moment of channel section in compression is given by $M_C$ as

$$M_C = \frac{\sigma_{AB}\, I_{NA}}{y_{AB}}$$

But $\sigma_{AB} = 45.5$ MPa

and $y_{AB} = 320 - \bar{y}$

$$= (320 - 97.3) = 222.7 \text{ mm}$$

Therefore $$M_C = \frac{45.5 \times 2299.697 \times 10^5}{222.7} \text{ N-mm}$$

$$= 469.853 \times 10^5 \text{ N-mm}$$

$$= 46.9853 \text{ kN-m}$$

Let Resisting moment of section in tension = $M_t$

Given $\sigma_{DF} = 15.5$ MPa

$y_{DF} = \bar{y} = 97.3$

$$M_T = \frac{\sigma_{DF}}{y_{DF}} . I_{NA} = 15.5 \times \frac{2299.697 \times 10^5}{97.3}$$

$$= 366.344 \times 10^5 \text{ N-mm}$$

$$= 36.6344 \text{ kN-m}$$

Smallar of the two will be safer. Thus in actional practice

$\sigma_{DF} = 15.5$ MPa

$\bar{y} = 97.3 \text{ mm} = y_{EF}$

will prevail. Thus if wkN-m is the weight per unit length of the channel (including that of water present at home) for the beam in question, then

$$M_{max} = \frac{wl^2}{8} \quad \text{where } l = 10 \text{ m}$$

$$= \frac{w \times 10^2}{8} = 12.5w \text{ kN-m}$$

Equating this moment to the resisting moment

We get $12.5w = 36.6344$

or $w = 2.9307$ kN-m

Now weight of CI channel per unit length (excluding water)

$$= [2(20 \times 300) + (20 \times 500)] \times 1000 \times \frac{7000}{(10)^9} \times 9.8$$

$$= 1510.74 \text{ N/m} = 1.51074 \text{ kN-m}$$

Therefore weight of 1 m length of water column

$$= w - w_{CI} = 2.9307 - 1.51074$$

*i.e.,* $$w_{\text{water}} = 1.4133 \text{ kN-m}$$

But $$w_{\text{water}} = \left(\frac{h}{1000} \times \frac{460}{1000} \times \frac{1000}{1000}\right) \times (1000 \times 9.81) \text{ K/m}$$

$$= 4.5126h \times 10^{-3} \text{ kN/m}$$

Thus $$(10^{-3} \times 4.5126)h = 1.4133$$

$$h = 0.31318 \times 10^3 \text{ mm}$$

$$= 313.18 \text{ mm}$$

## PROBLEMS FOR PRACTICE

1. A cast iron beam of section shown in Fig. 4.23, is simply supported at its ends and carries a load of 18 kN at mid span. Find the maximum allowable span if the stress due to bending is not to exceed 30 MN/m² tension. Neglect the weight of the beam.

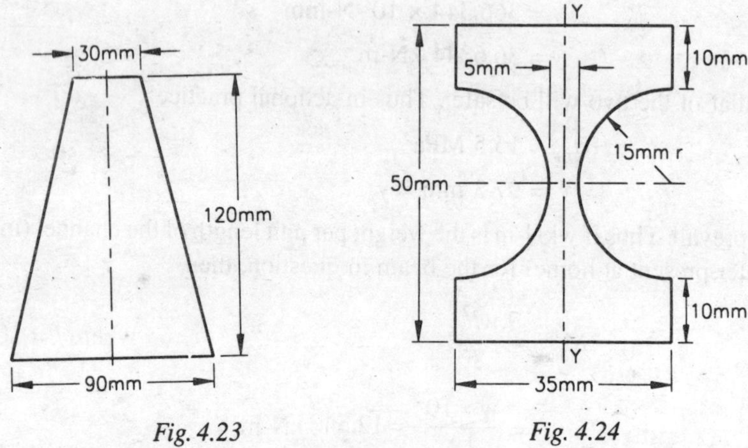


*Fig. 4.23* *Fig. 4.24*

What will be the maximum compressive stress ? **[Ans.** 1.056 m ; 42 MN/m²]

2. Fig. 4.24 shows the section of a beam. What is the ratio of its moment of resistance to bending in the plane *YY* to that for bending in the plane *XX* if the maximum stress due to bending is the same in both cases ?

   For a semi–circle of radius *r* the centroid is at a distance of 4 *r*/(3π) from the centre. **[Ans.** 2.85]

3. Find the ratio of the resisting moments of a square section beam when in one case the two sides are vertical and in the other a diagonal is vertical. **[Ans.** $\sqrt{2}$ :1]

4. Prove that the moment of resistance of a beam of square section with its diagonal in the plane of bending is increased by flattening the top and bottom corners, as shown in Fig. 4.25, and that the moment of resistance is a maximum when

$$y = \frac{8}{9}Y$$

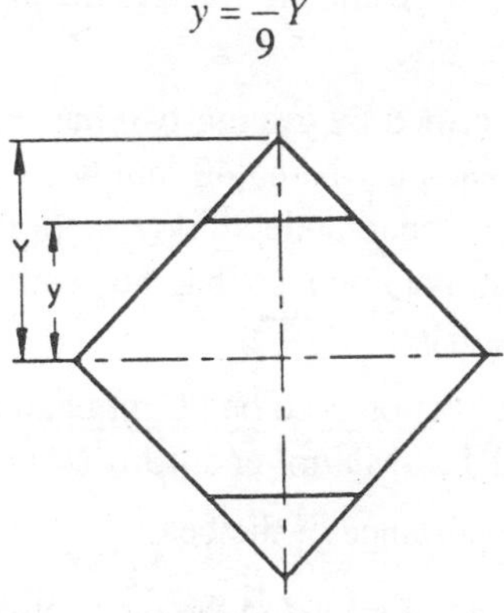


*Fig. 4.25*

5. A cantilever specimen for a fatigue testing machine is of circular cross section throughout its length, but in a length of 75 mm the diameter decreases from 10 mm at the fixed end to 6.25 mm at the free end. Calculate the maximum stress due to bending when a static load of 300 N is applied at the free end in a direction perpendicular to the length of the specimen. [**Ans.** 231.4 MN/m$^2$]
6. An I–section girder, 240 mm deep, has a web 20 mm thick. The top flange is 120 mm × 20 mm and the bottom flange is 160 mm × 20 mm. If the girder is simply supported at its ends, find the maximum span which can be used if the total distributed load per metre run is 6000 N/m, and the maximum stress is limited to 70 MN/m$^2$ . [**Ans.** 8.666 m]
7. A symmetrical I–section steel beam 100 mm × 100 m × 12.5 mm thick is stored by placing it on two simple supports in such a manner that the maximum bending moment to which it is subjected is the least possible. Determine the maximum length of the beam which can be supported in this manner without the stress crossing the allowable limit of 240 MN/m$^2$. The density of steel is 7800 kg/m$^3$.

   [**Ans.** 66.9 m]
8. A timber beam of rectangular section 150 mm wide × 300 mm deep is to be strengthened by two steel flitches 1 mm thick firmly attached to the sides and symmetrically placed about the horizontal centre line of the beam section. The beam is required to carry a central concentrated load of 52.5 kN on a span of 3 m, being simply supported at the ends. If the maximum stress in timber is not to exceed 8.5 MN/m sup 2 find the depth of the steel flitches required. $E_{steel} = 20\ E_{timber}$. [**Ans.** 220 mm]
9. A timber beam 80 mm wide × 160 mm deep is to be strengthened with 5 mm thick steel plates. For the same value of maximum bending stress in the timber,

compare the moments of resistance of the composite sections in the following two cases :

1. The plates are 160 mm deep and fixed to the vertical sides of the timber beam,
2. The plates are 80 mm wide and are fixed to the top and bottom surfaces of the timber beam. **[Ans.** 0.703]

10. A compound beam is formed by joining two bars rigidly together, one of steel and the other of brass, each bar being 40 mm wide. The bars are of thickness $t_1$ and $t_2$ respectively so that the total depth is $(t_1 + t_2)$. If E for the steel is twice that for the brass, find the ratio of $t_1$ to $t_2$ so that the neutral axis of the section is at the dividing line of the two bars.

    If the total depth of the section is to be 20 mm and the stresses in the steel and brass are not to exceed 110 MN/m$^2$ and 40 MN/m$^2$ respectively, determine the maximum moment of resistance of the beam. **[Ans.** 1 : $\sqrt{2}$ ; 125 Nm]

11. Two rectangular bars, one of steel and the other of brass, each 30 mm × 10 mm are placed together to form a beam 30 mm wide × 20 mm deep on two supports 0.75 m apart. The brass bar is on the top of the steel bar.

    Determine the maximum central load if the bars are, (*i*) separate and can bend independently, (ii) firmly secured to each other throughout their length.

    The maximum permissible stresses in brass and steel are 70 MN/m$^2$ and 105 MN/m$^{-2}$ respectively. $E_B = 85 \times 10^3$ MN/m$^2$ and $E_S = 300 \times 10^3$ MN/m$^2$.
    **[Ans.** 373.3 N, 825 N]

12. A bar of rectangular section 75 mm wide by 25 mm thick is subjected to an axial tensile force of 100 kN. By how much can the width of the section be safely reduced by removing material from one edge only, if the maximum stress over the reduced width may not exceed 100 MN/m$^2$. **[Ans.** 12.05 mm]

13. Fig. 4.26 shows a C–clamp which is tightened up to a load of 540 N. Calculate the stresses at the bottom and top edges of the section *XX*. Draw the stress distribution diagram.

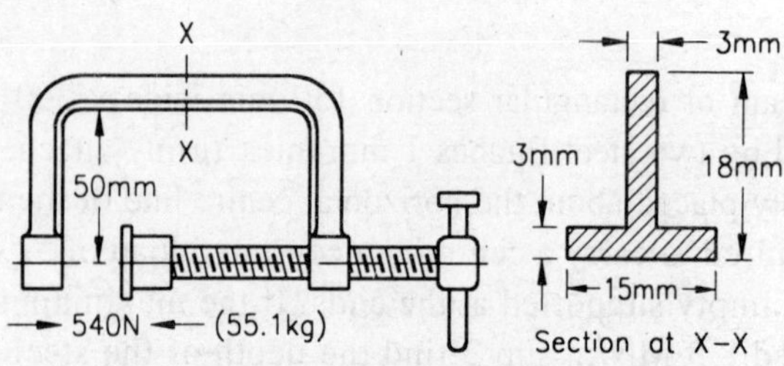


*Fig. 4.26*

**[Ans.** 63.6 MN/m$^2$ tensile, 109.2 MN/m$^2$ comp.]

14. A short, hollow, cylindrical column carries a compressive load of 490.5 kN. Find the maximum permissible eccentricity of this load, if the maximum permitted stresses are 75 MN/m$^2$ compressive and 20 MN/m$^2$ tensile. The diameters are 200 mm external and 124 mm internal. [**Ans.** 63 mm from the axis]

15. A pillar 1.5 m high is of rectangular section 50 mm thick and tapers longitudinally from a width of 150 mm at the base to 50 mm at the top, as shown in Fig. 4.27. A compressive load of 100 kN acts through the centroid at the top end and is parallel to the vertical edge. Determine the magnitude of the maximum compressive stress and the cross-section at which it will occur.

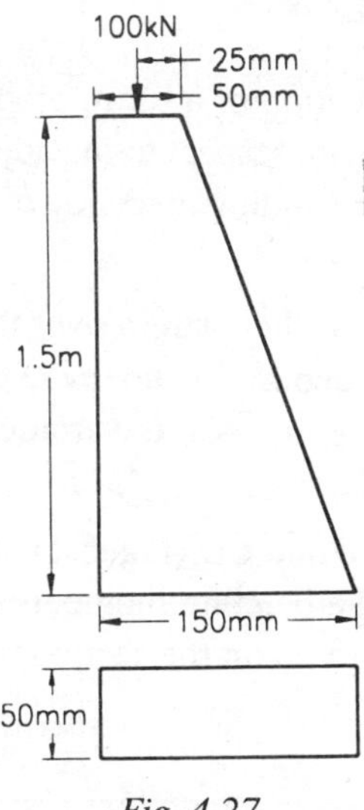


*Fig. 4.27*

[**Ans.** 55.33 MN/m$^2$ ; 0.375 m from top]

16. A short hollow pier, 1.2 m × 1.2 m outside and 0.75 m × 0.75 m inside, supports a vertical point load of 120 kN located on a diagonal and 0.69 m from the vertical axis of the pier. Neglecting the self–weight of the pier, calculate the normal stresses at the four outside corners on a horizontal section of the pier.

[**Ans.** 616.8 kN/m$^2$ comp. ; 136.8 kN/m$^2$ comp. ;
136.8 kN/m$^2$ comp. ; 343.2 kN/m$^2$ Tensile]

Chapter–5

# Shear Stresses in Beams

## 5.1 Shear Stress Distribution

The shearing force at any section of a beam produces a sliding tendency of the transverse elements of the beam with respect to the adjacent fibres. In addition due to the complementary shear stress the sliding tendency of the longitudinal elements also results.

As will be seen herenuder the shear stress over the cross-section of the beam is not uniform, but varies along the depth. It is not even constant along the width of the section. However in the following analysis its variation across the width will not be considered.

Let there be two sections, *AB* and *CD* (Fig. 5.1) of a uniform cross-section beam at a distance *dx* apart such that the bending moments on them are *M* and (*M* + *dM* ) respectively and let the shearing force on the section *AB* be *F*, then from chapter 3.

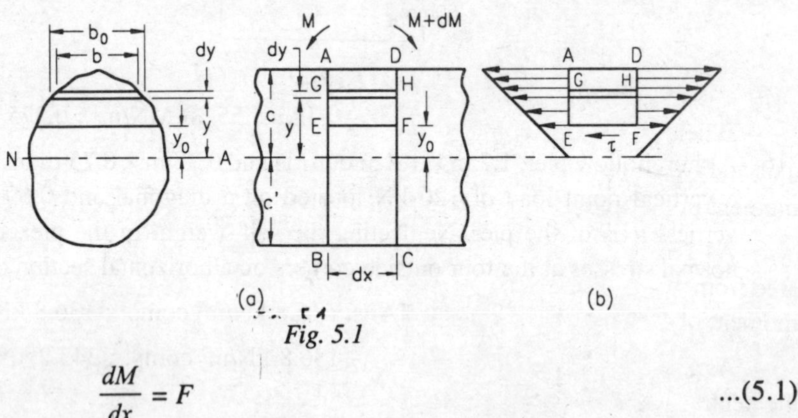


Fig. 5.1

$$\frac{dM}{dx} = F \qquad ...(5.1)$$

As a first step the shear stress on the plane *EF*, at a distance $y_0$ from the neutral surface, will be evaluated.

Bending stress at the end *G* of the element *GH′* (Chapter 4)

$$\sigma_b = \frac{M}{I}y$$

Therefore horizontal force towards the left hand side on the end $G$of the element $GH$

$$\delta F_G = \frac{M}{I} y\,(bdy)$$

Similarly the horizontal force towards the right hand side on the end $H$ of the element $GH$

$$\delta F_H = \frac{(M + dM)}{I} y(bdy)$$

Thus, if the shear stress on the plane $EF$ is $\tau$, considering the equilibrium of the element $EFDA$, we have

$$\int_{y_0}^{c} \frac{M + dM}{I} y\,(bdy) - \int_{y_0}^{c} y\,(bdy) = \tau b_0 dx$$

or
$$\int_{y_0}^{c} \frac{dM}{I} bydy = \tau b_0 dx$$

or
$$\left(\frac{dM}{dx}\right) \frac{1}{Ib_0} \int_{y_0}^{c} bydy = \tau \qquad ...(5.2)$$

Therefore from equations (5.1) and (5.2) the shear stress on the horizontal plane $EF$ at a distance $y_0$ from neutral surface is given by

$$\tau = \frac{F}{Ib_0} \int_{y_0}^{c} bydy \qquad ...(5.3)$$

This expression can also be written in a much more convenient form

$$\tau = \frac{F}{Ib_0} (A\bar{y}) \qquad ...(5.4)$$

Where $A$ is the area of the cross-section taken from height $y_0$ to $c$ from the neutral axis and $\bar{y}$ the distance of the centroid of this area from the neutral axis, $(A\bar{y})$ is the first moment of the area of cross section taken from height $y_0$ to $c$ about $NA$.

As a matter of fact the evaluation of this moment can also be done by taking the area from $y_0$ to $(-c')$ because in the absence of any external longitudinal force the first moment of the entire area of cross-section about the neutral axis must be zero.

As stated above the shear stress $\tau$ acts on the horizontal plane at a distance $y_0$ from the neutral surface. A complementary shear stress of the same magnitude will also be acting on the cross-sectional plane $AB$ at the height $y_0$ from $NA$.

### 5.1.1 Shear Stress Distribution in a Rectangular Section

Let the rectangular section be of width $b$ and depth $d$ as shown in Fig. 5.2. Then the first moment of the area $EFBA$ is given by

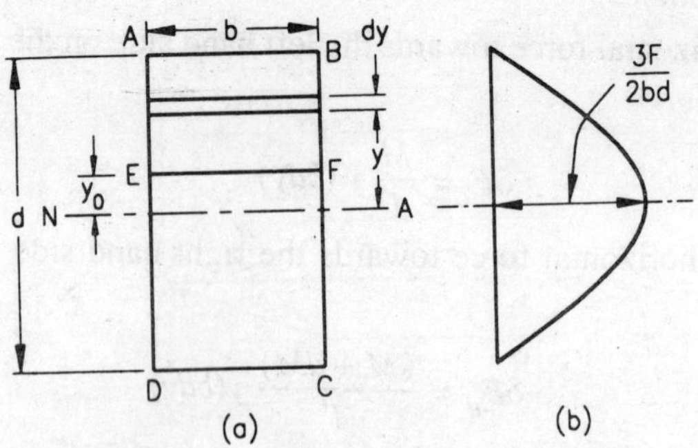


Fig. 5.2

$$A\bar{y} = \int_{y_0}^{d/2} (b.dy)\, y = \frac{b}{2}\left[y^2\right]_{y_0}^{\frac{d}{2}} = \frac{b}{2}\left(\frac{d^2}{4} - y_0^2\right)$$

Thus from equation (5.3), since $b_0 = b$ and $I = bd^3/12$

$$\tau = \frac{12F}{bd^2 b} \cdot \frac{b}{2}\left(\frac{d^2}{4} - y_0^2\right)$$

$$= \frac{6F}{bd^2}\left(\frac{d^2}{4} - y_0^2\right) \qquad ...(5.5)$$

Evidently the variation of shear stress $\tau$ with $y_0$ is parabolic. The magnitude at the top and bottom points, when $y_0 = \pm\, d/2$, will be zero and it will be maximum at $NA$ where $y_0 = 0$

Thus $$\tau_{max} = \frac{6F}{bd^2} \cdot \frac{d^2}{4} = \frac{3}{2}\frac{F}{bd} \qquad ...(5.6)$$

*i.e.*, $$\tau_{max} = \frac{3}{2} \text{ (Average shear stress over the section)}$$

## 5.1.2 Shear Stress Distribution in a Solid Circular Section

For a shearing force $F$ over the transverse solid circular section of diameter $D$, the shear stress at a height $y_0$ is

$$\tau = \frac{F}{Ib_0}\int_{y_0}^{D/2} (b.dy)\, y \qquad ...(5.7)$$

From the Fig. 5.3, let

$$y_0 = \frac{D}{2}\sin\theta_0\,; \quad y = \frac{D}{2}\sin\theta\,; \quad dy = \left(\frac{D}{2}\cos\theta\right)d\theta$$

and $$b_0 = \left(\frac{D}{2}\cos\theta_0\right); \quad b = 2\left(\frac{D}{2}\cos\theta\right)$$

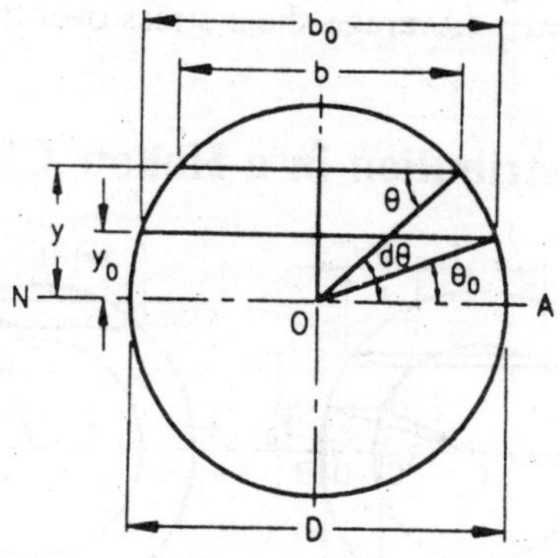


*Fig. 5.3*

Therefore substituting these values in equation (5.7)

$$\tau = \frac{64F}{\pi D^5 \cos\theta_0}\int_{\theta_0}^{\pi/2} (D\cos\theta)\left(\frac{D}{2}\sin\theta\right)\left(\frac{D}{2}\cos\theta\right)d\theta$$

$$= \frac{-16F}{\pi D^2 \cos\theta_0}\int_{\theta_0}^{\pi/2} \cos^2\theta\,(-\sin\theta)\,d\theta$$

$$= \frac{-16F}{\pi D^2 \cos\theta_0}\left[\frac{\cos^3\theta}{3}\right]_{\theta_0}^{\pi/2}$$

$$= \frac{16F}{3\pi D^2}\cos^2\theta_0$$

$$= \frac{16F}{3\pi D^2}(1-\sin^2\theta_0)$$

*i.e.*
$$\tau = \frac{16F}{3\pi D^2}\left(1-\frac{4y_0^2}{D^2}\right) \qquad \text{...(5.8)}$$

Again the variation of shear stress with $y_0$ is parabolic. The value of the shear stress at top and bottom fibres, where

$$y_0 = \pm\frac{D}{2}$$

is zero. The shear stress at the neutral axis where $y_0 = 0$ is evidently maximum such that

$$\tau_{max} = \frac{16F}{3\pi D^2} = \frac{4}{3}\left[\frac{F}{\left(\frac{\pi}{4}D^2\right)}\right] \qquad \text{...(5.9)}$$

*i.e.* $\tau_{max} = \frac{4}{3}$ (Average shear stress over the section)

## 5.1.3 Shear Stress Distribution in a Hollow Circular Section

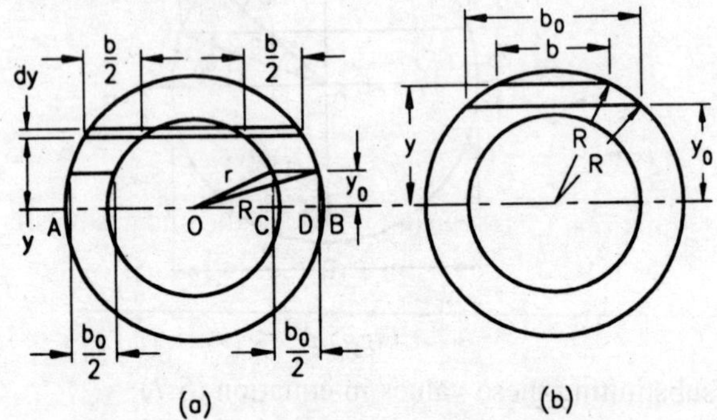


*Fig. 5.4*

Fig. 5.4 shows the hollow circular section. From equation 5.3 the shear stress at the height $y_0$

$$\tau = \frac{F}{Ib_0}\int_{y_0}^{R} (b.dy)y \qquad ...(5.10)$$

Case I. When $0 < y_0 < r$

$$\tau = \frac{F}{Ib_0}\left[\int_{y_0}^{r} bydy + \int_{r}^{R} bydy\right] \qquad ...(5.11)$$

Here $$b_0 = 2(OD - OC) = 2\left[\sqrt{R^2 - y_0^2} - \sqrt{r^2 - y_0^2}\right]$$

Similarly $$b = 2\left[\sqrt{R^2 - y^2} - \sqrt{r^2 - y^2}\right]$$

Now $$\int_{y_0}^{r} bydy = 2\int_{y_0}^{r}\left[\sqrt{R^2 - y^2} - \sqrt{r^2 - y^2}\right] ydy \quad \text{[From Fig. 5.4 (a)]}$$

$$= -\int_{y_0}^{r} (R^2 - y^2)^{1/2}(-2y)\,dy + \int_{y_0}^{r} (r^2 - y^2)^{1/2}(-2y)\,dy$$

$$= \left[\frac{2}{3}(R^2 - y^2)^{3/2}\right]_{y_0}^{r} + \left[\frac{2}{3}(r^2 - y^2)^{3/2}\right]_{y_0}^{r}$$

$$= -\frac{2}{3}\left[(R^2 - r^2)^{3/2} - (R^2 - y_0^2)^{3/2}\right]$$

$$+ \frac{2}{3}\left[(r^2 - r^2)^{3/2} - (r^2 - y_0^2)^{3/2}\right]$$

$$= \frac{2}{3}\left[(R^2 - y_0^2)^{3/2} - (r^2 - y_0^2)^{3/2} - (R^2 - r^2)^{3/2}\right] \quad ...(5.12)$$

and $$\int_r^R by\,dy = -\int_r^R (R^2 - y^2)^{1/2}(-2y)\,dy \quad \text{[Fig. 5.4}b\text{]}$$

$$= -\left[\frac{2}{3}(R^2 - y^2)^{3/2}\right]_r^R$$

$$= \frac{2}{3}(R^2 - r^2)^{3/2} \quad ...(5.13)$$

Thus from equations (5.10) through (5.13), the shear stress is

$$\tau = \left(\frac{F}{3I}\right)\frac{(R^2 - y_0^2)^{3/2} - (r^2 - y_0^2)^{3/2}}{(R^2 - y_0^2)^{1/2} - (r^2 - y_0^2)^{1/2}} \quad ...(5.14)$$

Case II. When $r < y_0 < R$

In this case from Fig. 5.4(*b*),

$$b_0 = 2(R^2 - y_0^2)^{1/2} \quad ...(5.15)$$

$$\int_{y_0}^R by\,dy = \int_{y_0}^R 2(R^2 - y^2)^{1/2}\,y\,dy = -\int_{y_0}^R (R^2 - y^2)^{3/2}(-2y)dy$$

$$= -\left[\frac{2}{3}(R^2 - y^2)^{3/2}\right]_{Y_0}^R$$

$$= \frac{2}{3}(R^2 - y_0^2)^{3/2} \quad ...(5.16)$$

Thus substituting equations (5.15) and (5.16) in equation (5.10)

$$\tau = \frac{F}{I\left\{2\left(R^2 - y_0^2\right)^{\frac{1}{2}}\right\}}\left[\frac{2}{3}\left(R^2 - Y_0^2\right)^{3/2}\right]$$

or $$\tau = \frac{F}{3I}(R^2 - y_0^2) \quad ...(5.17)$$

Evidently from equations (5.14) for the Case I the maximum shear stress will occur when

$$y_0 = 0$$

*i.e.* $$\tau_{max} = \frac{F}{3I}\left[\frac{R^3 - r^3}{R - r}\right] = \frac{4F}{3\pi(R^4 - r^4)} \cdot \frac{R^3 - r^3}{R - r} \quad ...(5.18)$$

Also $$\tau_{mean} = \frac{F}{\pi(R^2 - r^2)}$$

Therefore $$\frac{\tau_{max}}{\tau_{mean}} = \frac{4}{3(R^4 - r^4)} \frac{(R^3 - r^3)}{(R - r)}.(R^2 - r^2)$$

$$= \frac{4}{3} \frac{(R^2 + Rr + r^2)}{(R^2 + r^2)} \qquad ...(5.19)$$

Case III. Very thin tube

Here $r \Rightarrow R$

Therefore from equation 5.19

$$\frac{\tau_{max}}{\tau_{mean}} \simeq \frac{4}{3}\left(\frac{3R^2}{2R^2}\right)$$

$$= 2.$$

**Example 5.1.** *Draw the shear stress distribution curve for an I-section shown in Fig. 5.5 due to a shearing force of 100 kN.*

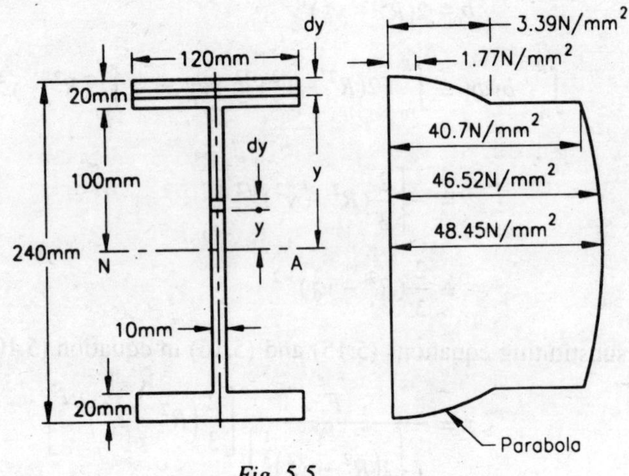


*Fig. 5.5*

**Solution.** Refering to Fig. 5.5

$$I = \frac{1}{12}(0.12) \times (.24)^3 - \frac{1}{12}(.11 \times (.20)^3$$

$$= 64.81 \times 10^{-6} \text{ m}^4$$

The shear stress is given by

$$\tau = \frac{F}{Il_0}(A\bar{y})$$

Case I. When $0.1 < y < 0.12$ Distribution of shear stress in the flange is as follows

Here $b_0 = 0.120$ m

$$A = (0.12 - y) \times 0.12$$

$$\bar{y} = \frac{(.12 - y)}{2} + y = \frac{1}{2}(.12 + y)$$

Therefore $$\tau = \frac{100 \times (.12 - y) \times .12 \times \frac{1}{2}(.12 + y)}{(64.81 \times 10^{-6}) \times 0.12}$$

$$= 771\,(0.0144 - y^2)\ \text{MN/m}^2 \text{ (Parabolic)}$$

At $y = 0.12$, $\tau = 0$

$y = 0.11$, $\tau = 1.77\ \text{MN/m}^2$

$y = 0.10$, $\tau = 3.39\ \text{MN/m}^2$

Case II. When $0 < y < 0.1$ Distribution of shear stress in the web is as follows

Here $b_0 = .01$ m

$$[A\bar{y}] = (0.02 \times 0.12)\,(0.1 + 0.01) + \left[0.01 \times (0.1 - y)\left\{\frac{0.1 - y}{2} + y\right\}\right]$$

$$= 2.64 \times 10^{-4} + 0.005\,(0.01 - y^2)$$

Therefore $$\tau = \frac{100 \times [2.64 \times 10^{-4} + .005\,(0.01 - y^2)]}{(64.81 \times 10^{-6}) \times 0.01}\ \text{kN/m}^2$$

$$= 15.43\,[2.64 + 50\,(0.01 - y^2)]\ \text{MN/m}^2 \qquad \text{(Parabolic)}$$

At $y = 0.01$, $\tau = 40.7\ \text{MN/m}^2$

$y = 0.05$, $\tau = 46.52\ \text{MN/m}^2$

$y = 0$, $\tau = 48.45\ \text{MN/m}^2$.

**Example 5.2.** *MIA shearing force of 300 kN acts over a T-section shown in Fig. 5.6. Draw the shear stress distribution curve. What is the value of the maximum shear stress ?*

*Calculate the percentage shear force carried by the vertical web.*

**Solution.** Taking moments of area about *AB* distance of *NA* from *AB* is given by *h* as

$$h = \frac{(150 \times 30 \times 15) + 150 \times 30\,(30 + 75)}{(150 \times 30) + (150 \times 30)} = 60 \text{ mm}$$

$$I_{NA} = \left(\frac{150 \times 30^3}{12} + 4500 \times 45^2\right) + \left(\frac{30 \times 150^3}{12} + 4500 \times 45^2\right)$$

$$= 31.487 \times 10^6\ \text{mm}^4.$$

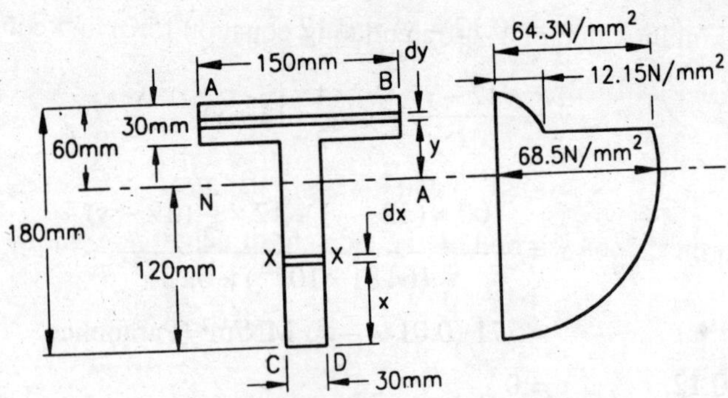


*Fig. 5.6*

Case I. When $30 < y < 60$ Shear stress in the flange is given below

Here $$b_0 = 150 \text{ mm} \quad \text{and} \left(A\bar{y}\right) = 150\,(60 - y).\,\frac{1}{2}\,(60 + y)$$

Therefore $$\tau = \frac{300 \times 1000}{31.487 \times 10^6 \times 150} \times \left[150 \times (60 - y) \times \frac{1}{2}(60 + y)\right]$$

$$= 4.76 \times 10^{-3}\,(3600 - y^2) \text{ N/mm}^2 \text{ or MN/m}^2$$

At $y = 60$ mm, $\tau = 0$

$y = 50$ mm, $\tau = 5.34 \text{ N/mm}^2$

$y = 40$ mm, $\tau = 9.56 \text{ N/mm}^2$

$y = 30$ mm, $\tau = 12.85 \text{ N/mm}^2$.

Case II

For the distribution of shear stress in the web it is much convenient to take any section at a distance $x$ from the bottom of the web as shown in the Fig. 5.6.

Here $b_0 = 30$ mm and $(A\bar{y})$ is taken for the area between $XX$ and $CD$.

*i.e.,* $$(A\bar{y}) = 30x\left(120 - \frac{x}{2}\right)$$

Therefore $$\tau_x = \frac{300 \times 1000}{31.48 \times 10^6 \times 30}\left[30x\left(120 - \frac{x}{2}\right)\right]$$

$$= 4.76 \times 10^{-3}\,x\,(240 - x) \text{ N/mm}^2 \qquad \text{...(5.20)}$$

At $x = 0$, $\tau_x = 0$

$x = 30$ mm, $\tau_x = 30.3 \text{ N/mm}^2$

$x = 60$ mm, $\tau_x = 51.5 \text{ N/mm}^2$

$x = 120$ mm, $\tau_x = 68.5 \text{ N/mm}^2$

$x = 150$ mm, $\tau_x = 64.3 \text{ N/mm}^2$

For maximum value of $\tau_x$ differentiating equation (5.2) and equating to zero

$$240 - 2x = 0$$

*i.e.,* $$x = 120$$

Thus $$\tau_{max} = 68.5 \text{ N/mm}^2 \text{ (along the N.A.)}$$

(*b*) The shear force carried by the element of height $dx$

$$dF_1 = \tau_x \,(30\, dx)$$

$$= 4.76 \times 10^{-3}\, x\,(240 - x)$$

Therefore total shear force carried by web

$$= 4.76 \times 10^{-3} \int_0^{150} x\,(240 - x)\, dx$$

$$= 4.76 \times 10^{-3} \left[ 240\frac{x^2}{2} - \frac{x^3}{3} \right]_0^{150}$$

$$= 7497 \text{ N} = 7.497 \text{ kN}$$

Therefore percentage of shear force carried by web

$$\therefore \quad \frac{F_1}{F} \times 100 = \frac{7.497}{300} \times 100 = 2.5\%.$$

**Example 5.3.** *A beam of square section of side 20 mm is placed so that the plane of bending is parallel to the diagonal. The shearing force at a section is 20 kN. Calculate the values of the maximum shear stress and the mean shear stress. Draw the shear stress distribution diagram for the section.*

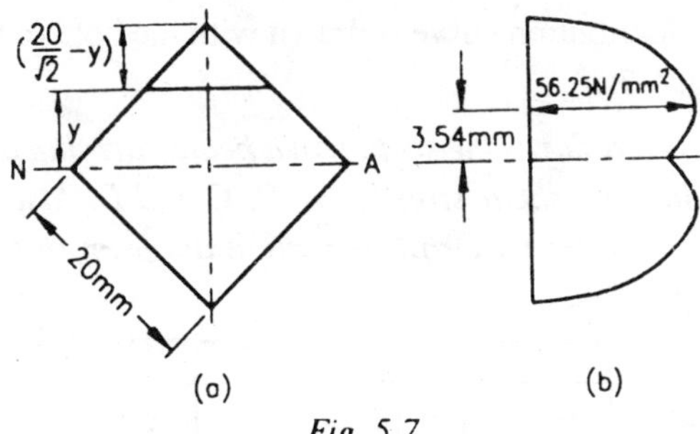


*Fig. 5.7*

**Solution.** The shear stress at a height $y$ from the neutral axis is given by

$$\tau = \frac{F}{Ib_0}(A\bar{y})$$

Here $$A = \frac{1}{2} \times 2\left(\frac{20}{\sqrt{2}} - y\right) \times \left(\frac{20}{\sqrt{2}} - y\right)$$

$$b_0 = 2\left(\frac{20}{\sqrt{2}} - y\right)$$

$$I = \frac{(20)^4}{12}$$

Therefore $$\tau = \frac{20 \times 1000}{\left(\frac{20^4}{12}\right) \times 2 \times \left(\frac{20}{\sqrt{2}} - y\right)}\left(\frac{20}{\sqrt{2}} - y\right)^2 \times \left[y + \frac{1}{3}\left(\frac{20}{\sqrt{2}} - y\right)\right]$$

*i.e.* $$\tau = 0.75\left(10\sqrt{2} - y\right) \times \left(10\sqrt{2} + 2y\right) \qquad ...(5.21)$$

For the maximum shear stress,

$$d\tau/dy = 0$$

That is $$\left(10\sqrt{2} - y\right) \times 2 + \left(10\sqrt{2} + 2y\right)(-1) = 0$$

or $$y = \frac{10\sqrt{2}}{4} = 3.54 \text{ mm}$$

Therefore $$\tau_{max} = 0.75\left(10\sqrt{2} - \frac{10\sqrt{2}}{4}\right)\left(10\sqrt{2} + \frac{10\sqrt{2}}{2}\right)$$

$$= 168.75 \text{ N/mm}^2$$

Also $$\tau_{mean} = \frac{20 \times 1000}{20 \times 20} = 50 \text{ N/mm}^2$$

The shear stress distribution curve is drawn with the help of equation (5.21). The curve is shown in Fig. 5.7(*b*).

**Example 5.4.** *Fig. 5.8 shows the section of a beam carrying a transverse shearing force of 20 kN. Calculate the shear stress at A, B, C and D. Also find the ratio of the maximum to average shear stress. Draw to scale a diagram to show the variation of shear stress across the section.*

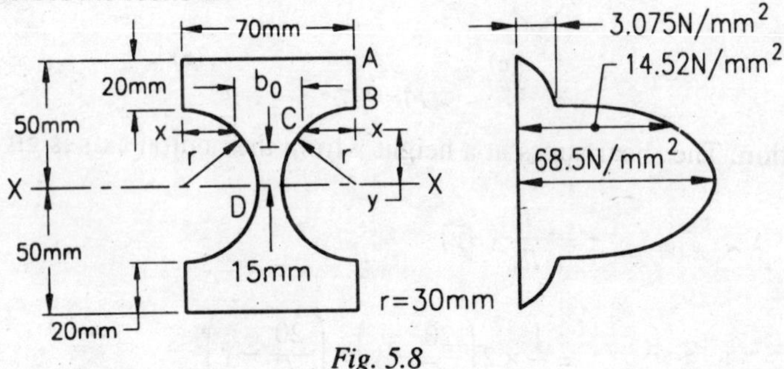


*Fig. 5.8*

**Solution.** $$I = \frac{70 \times 100^3}{12} - \frac{\pi}{64} \times (60)^4$$

$$= 5.197 \times 10^6 \text{ mm}^4$$

For the point $A$

$$(A\bar{y}) = 0, \quad \tau_A = 0$$

For the point $B$

$$(A\bar{y}) = (70 \times 20)(30 + 10) = 56000 \text{ mm}^3$$

Therefore $$\tau_B = \frac{(20 \times 1000) \times 56000}{(5.197 \times 10^6) \times 70} = 3.075 \text{ N/mm}^2$$

For the point $C$

It will be convenient to find the stress at any general point $Y_2$ distant $y$ from the neutral axis

$$b_0 = 70 - 2\sqrt{30^2 - y^2}$$

and $$(b_0)_C = 70 - 2\sqrt{30^2 - 15^2} = 18 \text{ mm}$$

$$(A\bar{y})_Y = 70 \times (50 - y) \times \left\{ y + \left( \frac{50 - y}{2} \right) \right\} - \int_y^{30} (2dy)y$$

$$(A\bar{y})_C = 79625 - \int_{15}^{30} (30^2 - y^2)^{1/2} (2y)\, dy$$

$$= 79625 + \frac{2}{3}\left[ (30^2 - y^2)^{3/2} \right]_{15}^{30}$$

$$= 79625 - 11691 = 67934 \text{ mm}^3$$

Therefore $$\tau_C = \frac{20 \times 1000}{5.197 \times 10^6 \times 18} \times 67934 \text{ mm}^3$$

$$= 14.52 \text{ N/mm}^2 \text{ orMN/m}^2.$$

**For the point $D$**

$$b_0 = 70 - 60 = 10 \text{ mm}$$

$$(A\bar{y}) = (70 \times 50 \times 25) - \int_0^{30} (30^2 - y^2)^{1/2} (2y)\, dy$$

$$= 87500 + \frac{2}{3}[30^2 - y^2)^{3/2}]_0^{30}$$

$$= 69500 \text{ mm}^3$$

Therefore $$\tau_D = \frac{20 \times 1000}{5.197 \times 10^6 \times 10} \times 69500$$

$$= 26.7 \text{ N/mm}^2 \text{ or MN/m}^2$$

The stress at $D$ is also the maximum shear stress

Average shear stress

$$\tau_{av} = \frac{20 \times 1000}{\left(70 \times 100 - \frac{\pi}{4} \times 3600\right)}$$

$$= 4.79\text{N/mm}^2$$

Therefore

$$\tau_{max} / \tau_{av} = \frac{26.7}{4.79} = 5.58.$$

**Example 5.5.** *At some instant the reaction P due to the line shaft on a wall bracket is as shown in Fig. 5.9. Find the principal stresses, at the point Q when θ = 60°.*

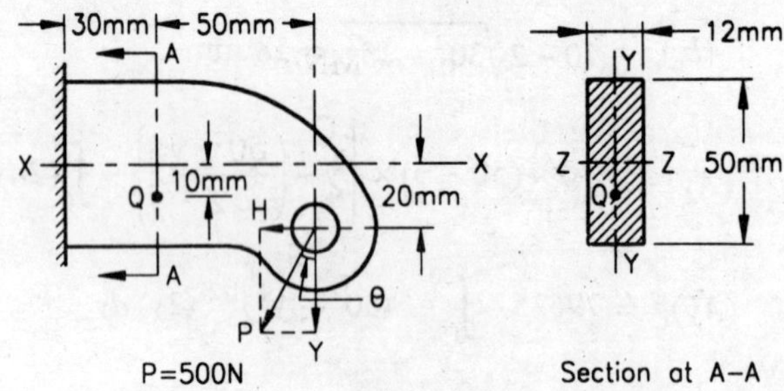


*Fig. 5.9*

**Solution.** Resolving $P$ into horizontal and vertical components

$$H = P \sin 60° = 500\sqrt{3}/2 = 4340 \text{ N}$$

$$V = P \sin 30° = 500 \times \frac{1}{2} = 2500 \text{ N}.$$

(*i*) *Effect of V*. Due to component $V$ the section $AA$ will experience a bending moment of magnitude $V \times 50$ along with a shearing force of $V$ over the section.

Due to the shearing force $V$ the shear stress over the section at $Q$ is given by

$$\tau_Q = \frac{V}{I_{zz} b_0}(A\bar{y})$$

$$= \frac{2500 \times 12}{12 \times 50^3 \times 12}[(15 \times 12)(10 + 7.5)]$$

$$= 5.25 \text{ N/mm}^2 \text{ or MN/m}^2$$

Due to the bending moment 50$V$, there will a compressive bending stresses on the cross-section at the point $Q$ given by

$$(\sigma_b)_Q = \frac{50\,V\,(10)}{I_{zz}} = \frac{50 \times 2500 \times 12}{12 \times 50^3} \times 10$$

$$= 10 \text{ N/mm}^2 \text{ orMN/m}^2 \text{ (comp.)}$$

(*ii*) **Effect of *H*.** For the section $AA$ the component $H$ will be an eccentric load, giving rise to direct compressive stress and bending compressive stress over the section $AA$ at the point $Q$

$$\sigma_0 = \frac{H}{\text{Area}} = \frac{4340}{12 \times 50}$$

$$= 7.233 \text{ N/mm}^2 \text{ orMN/m}^2 \text{ (comp.)}$$

$$(\sigma_b)_Q = \frac{20H}{I_{zz}} \times 10$$

$$= \frac{4340 \times 20 \times 10 \times 12}{12 \times (50)^3}$$

$$= 6.94 \text{ N/mm}^2 \text{ or MN/m}^2 \text{ (comp.)}$$

Thus net direct stress over the section at $Q$

$$= 10 + 7.233 + 6.74$$

$$= 4.173 \text{ N/mm}^2 \text{ orMN/m}^2 \text{ (comp.)}$$

The principal stresses are then given by

$$\sigma_{1,2} = \frac{1}{2}(-24.173) \pm \sqrt{\left(\frac{-24.173}{2}\right)^2 + (5.25)^2}$$

$$= -12.086 \pm 13.1$$

*i.e.* $\sigma_1 = 25.186$ N/mm² (comp.)

and $\sigma_2 = 1.014$ N/mm² orMN/m² (tensile).

## PROBLEMS FOR PRACTICE

1. A R.S.J. is of I-section of overall height 200 mm and flange width 125 mm. The web thickness is 7 mm and the flange thickness is 11 mm. The standard taper on the flanges may be neglected and all corners may be assumed sharp. The beam is subjected to transverse loads acting parallel to the web, and at one section the shear force is 100 kN. Determine the maximum vertical shearing stress in the web at this section. What proportion of the total shear force is carried by the web? [**Ans.** 80.8 MN/m² ; 94.8%]

2. Calculate the ratio of maximum to mean shear stress in an I-beam, 200 mm wide × 350 mm deep, having the flanges 25 mm thick and web 12.5 mm thick. Also find the percentage of the total shearing force carried by the web.

[**Ans.** 3.57, 92.9%]

3. A *T*-section beam, symmetrical about the vertical axis has the flange 120 mm × 10 mm and the web 100 mm × 10 mm. What is the percentage of the shearing force carried by the web ? [**Ans.** 93.5%]

4. Prove that for a hollow circular section the ratio of maximum shear stress to mean stress lies between 4/3 and 2 for all values of *R*/*r*.

5. A rod of circular section is subjected to a shearing force on a plane perpendicular to its axis. Find the maximum shearing stress in terms of the shearing force and the rod diameter.

   If the rod is used as a beam simply supported at its ends and having a central concentrated load, find the ratio of the maximum bending stress to the maximum shear stress in terms of the span and the rod diameter.

$$\left[\textbf{Ans. } \frac{16F}{3\pi d^2}; \frac{3l}{2d}\right]$$

6. A beam of hollow square section with inner dimensions as 75 mm × 75 mm and wall thickness of 37.5 mm is simply supported at its ends on supports 3 m apart. Determine the greatest bending and transverse shear stresses in the beam at 37.5 mm from the neutral axis due a point load of 100 kN applied at a point which is 1 m away from one support. Find the principal stresses and the maximum shear stress at this point. [**Ans.** 65 ; 7, 66, 33 MN/m$^2$]

7. A tube of hollow square section, 50 mm square outside and 6 mm uniform thickness, is subjected to a shearing force of 50 kN (5100 kg) acting in the direction of a diagonal. Find the maximum shearing stress produced.

[**Ans.** 105.9 MN/m$^2$]

8. A beam has a symmetrical triangular section of breadth *B* and depth *D* and is subjected at a certain section to a vertical shearing force *S* acting in the direction of the axis of symmetry. Deduce in terms of *B*, *D* and *S* the shearing stress τ at any depth *d* from the vertex of the triangular section. Plot a graph showing how τ varies over the depth of the section and find the ratio of the average shearing stress over the section to the maximum shearing stress.

$$\left[\textbf{Ans. } \frac{12\,dS\,(D-d)}{BD^3}; \frac{2}{3}\right]$$

Chapter–6

# Deflection of Beams

## 6.1 Introduction

In Chapter 4 a study has been made about the induced stresses due to the lateral loads on beams. In addition to the stresses, the lateral loads cause the beam to bend ; that is, the neutral surface of the beam before the application of the lateral loads moves to a new position after the load is applied. The displacement of the neutral surface at any section is called the deflection of the beam at that section due to the given lateral loads on the beam. For the smooth running of various machine members it is essential that the deflection of various members, where there are lateral loads, is not excessive. Even in the case of structures there is a limit to the deflection so that their original shape is not changed considerably.

There are various methods of calculating the deflection of beams ; a few which are most common, are discussed below.

## 6.2 Macaulay's Method (Integration Method)

**Relation between Curvature, Slope and Deflection.** It was shown in Chapter 4

That $$\frac{E}{R} = \frac{M}{I} \qquad \text{...(6.1)}$$

If $(x, y)$ are the co-ordinates of a point on the neutral surface of the beam, in the deflected position with respect to certain chosen origin, the radius of curvature $R$ is given by the expression

$$\frac{1}{R} = \frac{d^2y/dx^2}{\left[1+\left(\frac{dy}{dx}\right)^2\right]^{3/2}} \qquad \text{...(6.2)}$$

In practice, when the stresses in the beam do not cross the elastic limit, the slope $i(= dy/dx)$ is very small and the square of this quantity can be neglected in comparison to unity. Equation (6.2) will then reduce to

$$\frac{1}{R} \simeq \frac{d^2y}{dx^2} \qquad ...(6.3)$$

Substituting equation (6.3) in (6.1), we obtain

$$EI\frac{d^2y}{dx^2} = M \qquad ...(6.4)$$

Also we know that

$$\frac{dM}{dx} = F \qquad ...(6.5a)$$

and

$$\frac{dF}{dx} = w \qquad ...(6.5b)$$

Thus from equations (6.4) and (6.5)

$$EI\frac{d^3y}{dx^3} = F \qquad ...(6.6a)$$

and

$$EI\frac{d^4y}{dx^4} = w \qquad ...(6.6b)$$

In most of the cases the bending moment $M$, for a given load can be expressed as a function of $x$, the distance of any section of the beam from the chosen origin. Then by successive integration of equation (6.4) the values of $EI\ (dy/dx)$ and $EI\ y$ can be obtained. The constants of integration are determined by using the end conditions. The product $EI$ is called the flexural rigidity. In case the cross-section of the beam is not uniform and it will be available as a function of $x$, equation (6.4) should be expressed in the form

$$E\frac{d^2y}{dx^2} = \frac{M}{I} \qquad ...(6.7)$$

Equation (6.7) can then be integrated and constants of integration found with the help of end conditions.

## 6.3 Standard Cases of Beam Deflections

### 6.3.1 Cantilever with a Concentrated Load at the Free End

Take the origin at $A$, the fixed end of the cantilever $AB$ of length $l$ and uniform section having $I$ as the second moment of area. Thus for the section $xx$ where the bending moment is $W\ (l - x)$ we have from equation (6.4).

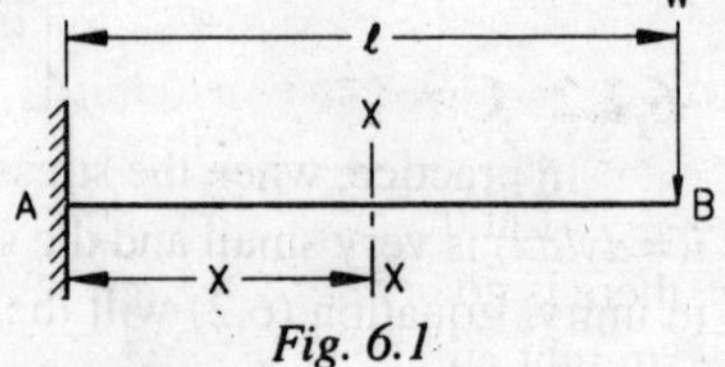

*Fig. 6.1*

$$EI = \frac{d^2y}{dx^2} = W(l - x)$$

Integrating we get

$$EI\frac{dy}{dx} = W\left(lx - \frac{x^2}{2}\right) + C_1 \qquad ...(6.8)$$

We know that at the fixed end $A$ the slope is zero and $x = 0$, so that from equation (6.8)

$$C_1 = 0$$

Therefore from (6.8), the slope equation is

$$EI\frac{dy}{dx} = W\left(lx - \frac{x^2}{2}\right) \qquad ...(6.9)$$

Again integrating (6.9), we have

$$EIy = W\left(\frac{lx^2}{2} - \frac{x^3}{6}\right) + C_2 \qquad ...(6.10)$$

But when $\qquad x = 0, \quad y = 0\,;$

so that $\qquad C_2 = 0$

Thus the deflection equation is

$$EI\,y = W\left(\frac{lx^2}{2} - \frac{x^2}{6}\right) \qquad ...(6.11)$$

Evidently from equations (6.9) and (6.11) the slope and deflection are maximum at the free end $B$ of the cantilever

where $\qquad x = l$

That is $\qquad (dy/dx)_{max} = (dy/dx)_B = \dfrac{Wl^2}{2EI} \qquad ...(6.12a)$

and $\qquad y_{max} = y_B = \dfrac{Wl^3}{3EI} \qquad ...(6.12b)$

Slope and deflection can be calculated at any section of the cantilever with the help of equations (6.9) and (6.11) respectively.

### 6.3.2 Cantilever with a Concentrated Load not at the Free End

Let the concentrated load $W$ be acting at the point $C$ of the cantilever Fig. 6.2. As there is no bending moment over the portion $CB$ of the cantilever, this portion will stay straight and the portion $AC$ only will bend.

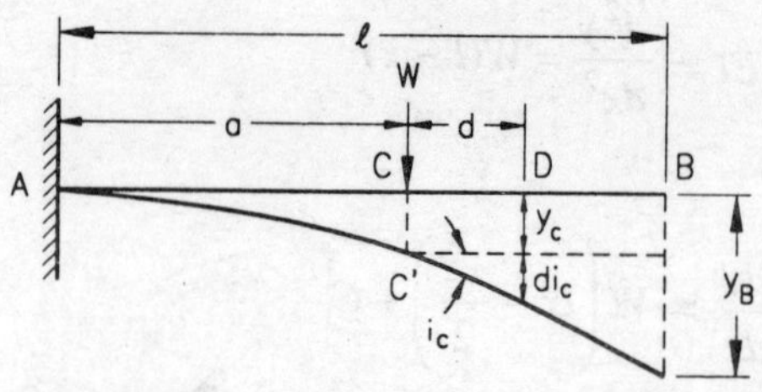


*Fig. 6.2*

For the slope and deflection of the portion $AC$ equations (6.9) and (6.11) will apply, the only change being that '$a$' replaces '$l$' in these equations. That is for the portion $AC$ $(0 < x < a)$

$$EI\frac{dy}{dx} = W\left(ax - \frac{x^2}{2}\right) \qquad \text{...(6.13}a\text{)}$$

and

$$EI\,y = W\left(\frac{ax^2}{2} - \frac{x^3}{6}\right) \qquad \text{...(6.13}b\text{)}$$

Also $$(dy/dx)_C = i_C = \frac{Wa^2}{2EI};$$

and

$$y_C = \frac{Wa^3}{3EI} \qquad \text{...(6.14)}$$

For any section $D$, at a distance '$d$' from $C$ (Fig. 6.2) the deflection is given by

$$y_D = y_C + di_C$$

$$= \frac{Wa^3}{3EI} + \frac{Wa^2 d}{2EI} \qquad \text{...(6.15)}$$

and for the point $B$,

$$d = (l - a)$$

$$y_B = \frac{Wa^3}{3EI} + \frac{Wa^2\,(l - a)}{2EI} \qquad \text{...(6.16)}$$

The slope at any point between $CB$ is constant and is equal to $i_C$

### 6.3.3 Cantilever with a Uniformly Distributed Load

In this case, referring in Fig. 6.3

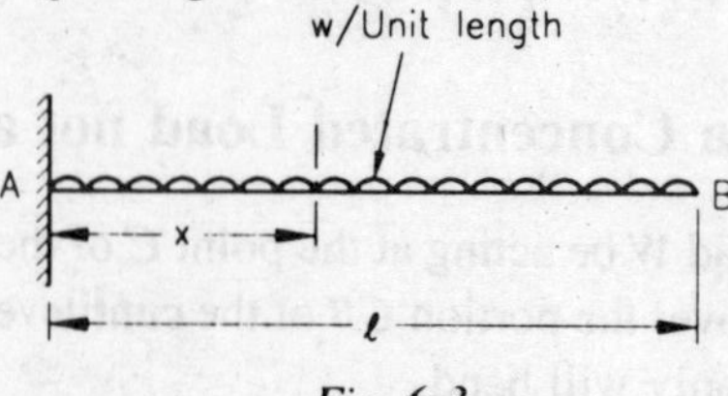


*Fig. 6.3*

$$EI\frac{dy}{dx} = \frac{w}{6}(l-x)^3 + C_1 \qquad ...(6.17)$$

Integrating equation (6.17)

$$EI\frac{dy}{dx} = -\frac{w}{6}(l-x)^3 + C_1 \qquad ...(6.18)$$

when $x = 0, \dfrac{dy}{dx} = 0$

so that from (6.18) $C_1 = \dfrac{wl^3}{6}$ ...(6.19)

Therefore from equations (6.18) and (6.19) we get the slope equation as

$$EI\frac{dy}{dx} = -\frac{w}{6}(l-x)^3 + \frac{wl^3}{6} \qquad ...(6.20)$$

Again integrating equation (6.20)

$$EI\,y = \frac{w}{24}(l-x)^4 + \frac{wl^3}{6}x + C_2 \qquad ...(6.21)$$

When $x = 0, y = 0$

so that from equation (6.21), we get

$$C_2 = -\frac{wl^4}{24} \qquad ...(6.22)$$

The deflection equation (6.21), therefore, reduces to

$$EI\,y = \frac{w}{24}(l-x)^4 + \frac{wl^3}{6}x - \frac{wl^4}{24} \qquad ...(6.23)$$

At $x = l$, equation (6.23) gives

$$y_B = \frac{1}{EI}\left[\frac{wl^4}{6} - \frac{wl^4}{24}\right] = \frac{wl^4}{8EI} \qquad ...(6.24)$$

Also from equation 6.20, at $x = l$ the slope at $B$ is

$$i_B = \frac{wl^3}{6EI} \qquad ...(6.25)$$

### 6.3.4 Uniformly Distributed Load over Part Length of a Cantilever

Again since there is no bending moment over the portion $CB$ of the cantilever, this portion will stay straight and the portion $AC$ only will bend as shown in the Fig. 6.4.

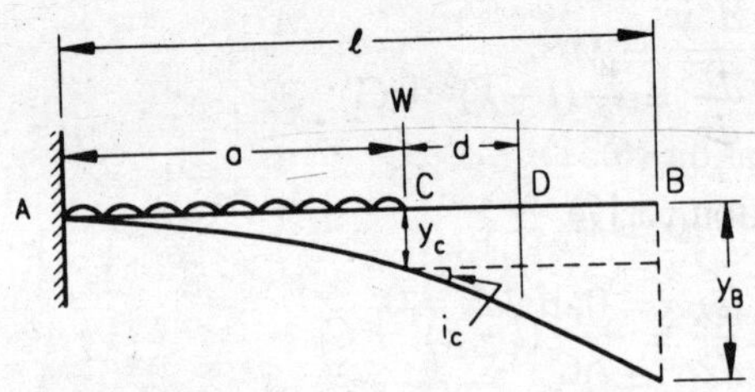


*Fig. 6.4*

From equations (6.20) and (6.23) the slope and deflection equations for the portion $AC$ $(0 < x < a)$ are obtained by replacing '$l$' by '$a$'

$$EI\frac{dy}{dx} = -\frac{W}{6}(a-x) + \frac{wa^3}{6} \qquad \text{...(6.26)}$$

$$EI\,y = \frac{w}{24}(a-a)^4 + \frac{wa^3}{6}x - \frac{wa^4}{24} \qquad \text{...(6.27)}$$

Also $$(dy/dx)_C = i_C = \frac{wa^3}{6EI} \qquad \text{...(6.28}a\text{)}$$

and $$y_C = \frac{wa^4}{8EI} \qquad \text{...(6.28}b\text{)}$$

For any section $D$, at a distance '$d$' from $C$, Fig. 6.4.

$$y_D = y_C + d\,i_C$$

$$= \frac{wa^4}{6EI} + \frac{wa^3 d}{6EI} \qquad \text{...(6.29)}$$

For the point $B$,

where $$d = (l-a)$$

$$y_B = \frac{wa^4}{8EI} + \frac{wa^3 d}{6EI} = \frac{wa^4}{8EI} + \frac{wa^3\,(l-a)}{6EI} \qquad \text{...(6.30)}$$

The slope at any point between $C$ and $B$ is constant and is equal to $i_C$.

### 6.3.5 Cantilever with a Couple $M$ at the Free End

The bending moment all through the length of the cantilever $AB$ will be equal to $M$, a constant. Therefore for any section $XX$

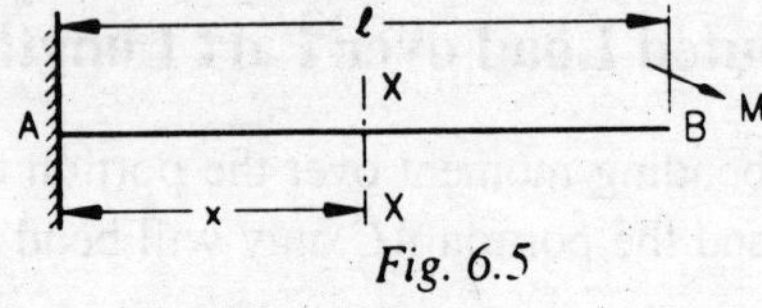


*Fig. 6.5*

$$EI\frac{d^2y}{dx^2} = M \qquad ...(6.31)$$

Integrating equation (6.31)

$$EI\ dy/dx = Mx + C_1$$

When $x = 0,\ dy/dx = 0,$

so that $C_1 = 0$

Thus $EI\ dy/dx = Mx \qquad ...(6.32)$

Integrating again

$$EI\ y = \frac{1}{2}Mx^2 + C_2$$

But at $x = 0,\ y = 0\ ;$

so that $C_2 = 0$

Therefore the deflection equation is

$$EI\ y = \frac{1}{2}Mx^2 \qquad ...(6.33)$$

Thus from equations (6.32) and (6.33)

$$i_B = (dy/dx)_{max} = \frac{Ml}{EI} \qquad ...(6.34a)$$

and

$$y_B = (y)_{max} = \frac{Ml^2}{2EI} \qquad ...(6.34b)$$

### 6.3.6 Central Point Load on Simply Supported Beam

This particular case can be dealt with very conveniently by taking the origin at the point $C$. Equations for slope and deflection for the portion $CB$ will be found and by symmetry they can be found for any point in the portion $AC$ keeping in mind that the slope in portion $AC$ is of opposite sign to that of the portion $CB$. For the Section $XX$

$$EI\frac{d^2y}{dx^2} = M_X = -\frac{W}{2}\left(\frac{l}{2} - x\right)$$

Integrating with respect to '$x$'

$$EI\frac{dy}{dx} = -\frac{W}{2}\left(\frac{l}{2}x - \frac{x^2}{2}\right) + C_1$$

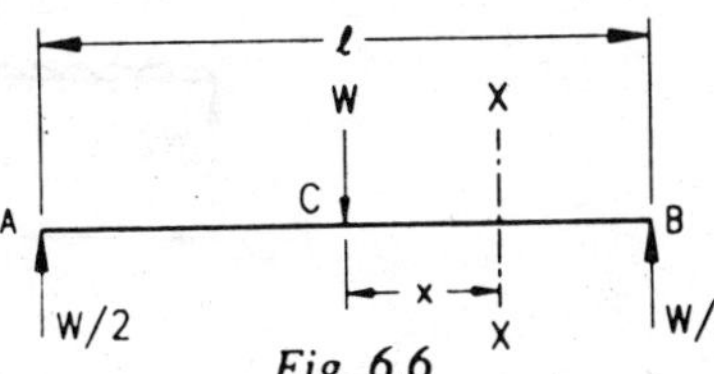

Fig. 6.6

By symmetry when $x = 0,\ \frac{dy}{dx} = 0\ ;$

so that $C_1 = 0$

Therefore for the portion *CB* the slope equation is

$$EI\frac{dy}{dx} = -\frac{W}{2}\left(\frac{1}{2}x - \frac{x^2}{2}\right) \quad ...(6.35)$$

Again by integrating equation (6.35)

$$EI\,y = -\frac{W}{2}\left(\frac{lx^2}{4} - \frac{x^3}{6}\right) + C_2 \quad ...(6.36)$$

When $x = l/2,\ y = 0$

so that from equation (6.36)

$$C_2 = \frac{Wl^3}{48}$$

Therefore, the deflection equation becomes

$$EI\,y = -\frac{W}{2}\left(\frac{lx^2}{4} - \frac{x^3}{6}\right) + \frac{Wl^3}{48} \quad ...(6.37)$$

The maximum deflection will occur at $x = 0$

*i.e.* $$y_{max} = y_C = \frac{Wl^3}{48\,EI} \quad ...(6.38)$$

The maximum slope will occur at the supports *i.e.* at $x = l/2$ and from equation (6.35)

$$i_B = (dy/dx)_{max} = -\frac{W}{2EI}\left(\frac{l}{2}\cdot\frac{l}{2} - \frac{l^2}{8}\right)$$

$$= -\frac{Wl^2}{16EI} \quad ...(6.39)$$

It must be noted that

$$i_A = \frac{Wl^2}{16EI}$$

### 6.3.7 Uniformly Distributed Load over a Simply Supported Beam

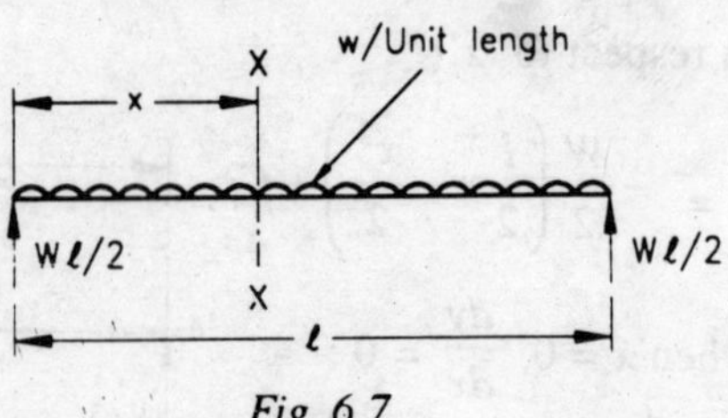


*Fig. 6.7*

For the section $XX$ the bending moment is

$$M_x = -\frac{wl}{2}x + \frac{wx^2}{2}$$

Therefore, we get

$$EI\frac{d^2y}{dx^2} = -\frac{wl}{2}x + \frac{wx^2}{2}$$

By integrating

$$EI\, dy/dx = -\frac{wl}{4}x^2 + \frac{wx^3}{6} + C_1 \qquad ...(6.40)$$

By symmetry

when $x = \frac{l}{2}$,

$$\frac{dy}{dx} = 0,$$

so that from equation (6.40)

$$C_1 = \frac{wl^3}{24}$$

Thus the slope equation reduces to

$$EI\, dy/dx = -\frac{wl}{4}x^2 + \frac{wx^3}{6} + \frac{wl^3}{24} \qquad ...(6.41)$$

Integrating equation (6.41) once again, we get

$$EI\, y = -\frac{wl}{12}x^3 + \frac{wx^4}{24} + \frac{wl^3}{24}x + C_2$$

When $x = 0, y = 0,$

so that $C_2 = 0.$

Thus the deflection equation becomes

$$EI\, y = -\frac{wl}{12}x^3 + \frac{w}{24}x^4 + \frac{w}{24}l^3x \qquad ...(6.42)$$

The maximum deflection will occur at $x = \frac{l}{2}$

*i.e.* $$y_{max} = \frac{1}{EI}\left[-\frac{wl^4}{96} + \frac{wl^4}{384} + \frac{wl^3}{48}\right] = \frac{5}{384}\frac{wl^4}{EI}$$

Magnitude of the maximum slope can be calculated by substituting zero for $x$ in equation (6.41). Thus

$$i_{max} = i_A = \frac{wl^3}{24EI}$$

### 6.3.8 Point Load not at the Centre of a Simply Supported Beam

Taking the origin at $A$

For the range $0 < x < a$

$$M_x = -\frac{Wb}{l}x \qquad \text{...(6.43)}$$

and for the range $a < x < l$

$$M_x = -\frac{Wb}{l}x + W[x-a] \qquad \text{...(6.44)}$$

For the given case two end conditions are known for the support points where the deflection is zero. If either of the equations (6.43) and (6.44) is individually equated to $\left(EI\frac{d^2y}{dx^2}\right)$ and integration is done twice the evaluation of two constants of integration will not be possible because of the limitations for $x$ for the two portions coupled with the fact that the two end conditions are known in the different regions. Under these circumstances, Macaulay suggested that equation (6.44) can be taken as representative equation for the entire range with the condition that the contribution of the last term be ignored in equation (6.44) when $x < a$. The last term is usually written in square brackets.

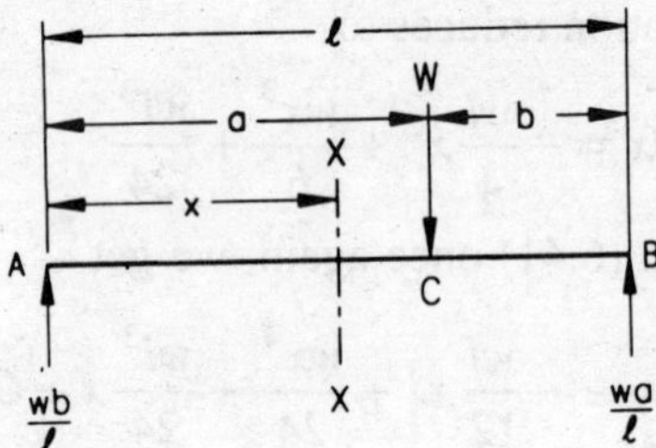


*Fig. 6.8*

Thus with the condition that when $x < a$, the effect of last term in the following equations be ignored, we have

$$EI\frac{d^2y}{dx^2} = -\frac{Wb}{l}x + W[x-a]$$

Integrating equation (6.44) under the conditions stated above

$$EI\frac{dy}{dx} = -\frac{Wb}{2l}x^2 + C_1 + \frac{W}{2}[x-a]^2 \qquad \text{...(6.45)}$$

Integrating once again

$$EI\,y = -\frac{Wb}{6l}x^3 + C_1x + C_2 + \frac{W}{6}[x-a]^3 \qquad \text{...(6.46)}$$

Now when $x = 0$ for which the last term in equation (6.46) is to be ignored,

$$y = 0$$

so that $C_2 = 0$

Also at $x = l,$

$y = 0,$

so that from equation (6.46), we have

$$0 = -\frac{Wb}{6l}.l^3 + C_1 l + \frac{W}{6}(l - a)^3$$

*i.e.* $$C_1 = \frac{Wb}{6l}(l^2 - b^2) = \frac{Wab}{6l}(a + 2b) \qquad ...(6.47)$$

(because $a + b = l$)

Thus the slope and deflection equations (6.45) and (4.46) reduce to

$$EI\frac{dy}{dx} = -\frac{Wl}{2l}x^2 + \frac{Wab}{6l}(a + 2b) + \frac{W}{2}[x - a]^2 \qquad ...(6.48)$$

and $$EI\,y = -\frac{Wb}{6l}x^3 + \frac{Wab}{6l}(a + 2b)\,x + \frac{W}{6}[x - a]^3 \qquad ...(6.49)$$

with the condition that the square bracket terms are to be ignored whenever $x < a$.

Thus for finding the deflection under the load point we have from equation (6.49) for $x = a$

$$EI\,y_C = -\frac{Wb}{6l}a^3 + \frac{Wa^2b}{6l}(a + 2b)$$

*i.e.* $$y_C = \frac{Wa^2b^2}{3EIl} \qquad ...(6.50)$$

## Maximum Deflection

The maximum deflection occurs where the slope is zero and, for $a > b$, this point will evidently lie between $A$ and $C$. Hence omitting the term in square brackets and equating $dy/dx$ to zero we get from equation (6.48)

$$0 = -\frac{Wb}{2l}x^2 + \frac{Wab}{6l}(a + 2b)$$

*i.e.* $$x = \sqrt{\frac{a(a + 2b)}{3}} = \sqrt{\frac{l^2 - b^2}{3}} \qquad ...(6.51)$$

Substituting this value of $x$, the maximum deflection from equation (6.49), again ignoring the term in square brackets, is

$$EI\,y_{max} = -\frac{Wb}{6l}\left[\frac{a}{3}(a + 2b)\right]^{3/2} + \frac{Wab}{6l}(a + 2b) \times \left[\frac{1}{3}(a + 2b)\right]^{1/2}$$

$$= -\frac{Wb}{l}\left[\frac{a}{3}(a+2b)\right]^{3/2}\left(\frac{1}{6}-\frac{1}{2}\right)$$

$$= \frac{Wb}{l}\left[\frac{a}{3}(a+2b)\right]^{3/2}$$

or $$y_{max} = \frac{wb}{9\sqrt{3}\,EIl}[a(a+2b)]^{3/2}$$

$$= \frac{wb}{9\sqrt{3}EIl}(l^2 - b^2)^{3/2} \qquad \text{...(6.52)}$$

The position of the maximum deflection point is given by relation (6.51). Also $b$ can take the values from 0 to $l/2$ (since $a > b$). Therefore the maximum value of $x$ in this case will be found by putting

$b = 0$ in equation (6.51)

That is $$x_{max} = \sqrt{\frac{l^2}{3}} = 0.5773l \qquad \text{...(6.53)}$$

On the other extreme, minimum value of $x$ occurs when

$$b = \frac{l}{2}$$

which gives from equation (6.51)

$$x_{min} = \frac{l}{2} \qquad \text{...(6.54)}$$

Thus the point experiencing the maximum deflection lies within the range 0.5 $l$ to 0.5773 $l$. Alternatively the maximum deflection is always within 0.0773 $l$ ($\simeq l/13$) of the centre of the beam.

### 6.3.9 Distributed Load over a Part Length of a Simply Supported Beam

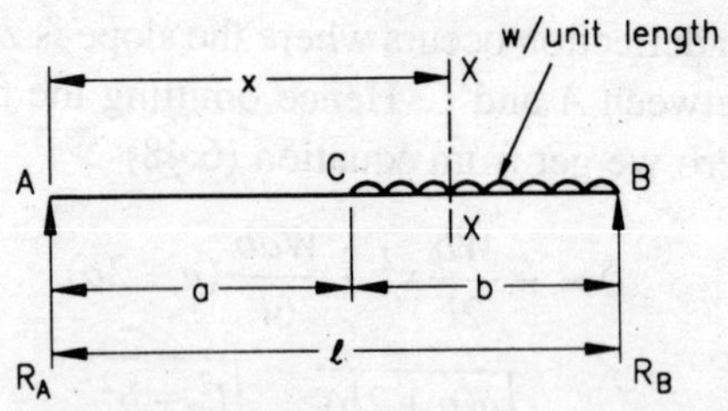


*Fig. 6.9*

Taking moments about the point $B$ (Fig. 6.6) the reaction at the left hand support can be found as

$$R_A = \frac{w(l-a)^2}{2l} \qquad \text{...(6.55)}$$

The representative equation for the bending moment for the entire span of the beam is

$$M_x = -R_A x + \frac{w}{2}[x-a]^2$$

the effect of the term with square brackets to be omitted when $x < a$.

Thus the differential equation of flexure becomes

$$EI\frac{d^2y}{dx^2} = -R_A x + \frac{w}{2}[x-a]^2$$

Integrating this, the slope is given by

$$EI\frac{dy}{dx} = -R_A \cdot \frac{x^2}{2} + C_1 + \frac{w}{6}[x-a]^3 \qquad \text{...(6.56)}$$

Integrating again the deflection is given by

$$EI\,y = -R_A\frac{x^3}{6} + C_1 x + C_2 + \frac{w}{24}[x-a]^4 \qquad \text{...(6.57)}$$

Now, when $x = 0, y = 0$

so that $C_2 = 0$

as the last term in equation (6.57) will be omitted.

Also when $x = l$,

$y = 0$,

so that from (6.57)

$$0 = -\frac{w}{2l}(l-a)^2\left(\frac{l^3}{6}\right) + C_1 l + \frac{w}{24}(l-a)^4$$

*i.e.*

$$C_1 = \frac{wb^2}{24l}(2l^2 - b^2) \qquad \text{...(6.58)}$$

(Since $l - a = b$)

Thus substituting the values of $R_A$ and $C_1$ in equations (6.56) and (6.57) the slope and deflection equations are

$$EI\frac{dy}{dx} = -\frac{wb^2}{4l}x^2 + \frac{wb^2}{24l}(2l^2 - b^2) + \frac{w}{6}[x-a]^3 \qquad \text{...(6.59)}$$

and

$$EI\,y = -\frac{wb^2}{12l}x^3 + \frac{wb^2}{24l}(2l^2 - b^2)x + \frac{w}{24}[x-a]^2 \qquad \text{...(6.60)}$$

The terms with square brackets to be omitted when $x < a$.

### 6.3.10 Couple Applied at a Point

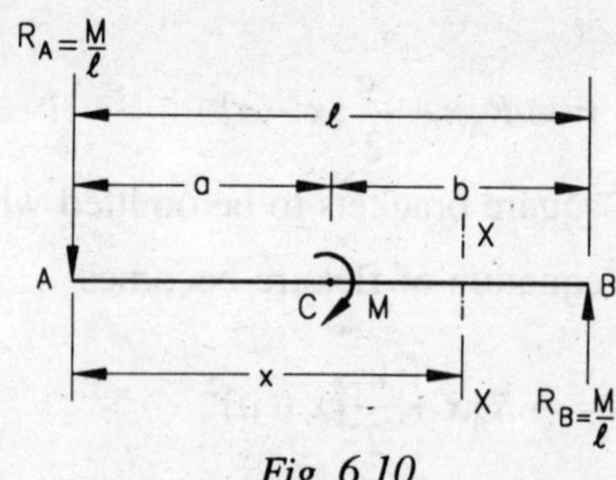


*Fig. 6.10*

For the equilibrium of the beam the end reactions $R_A$ and $R_B$ will both be equal in magnitude but of opposite sign such that

$$R_A = R_B = \frac{M}{l} \quad ...(6.61)$$

Taking origin at $A$, the bending moment $M_X$ at any section $XX$ will have the following values.

For the range $o < x < a$

$$M_X = \frac{M}{l}.x \quad ...(6.62)$$

For the range $a < x < 1$

$$M_X = \frac{M}{l}x - M \quad ...(6.63)$$

To make the equation (6.63) representative for the entire range $o < x < l$ the effect of the last term in equation (6.63) should be removed, when $x \leq a$. Thus could be achieved by writing the equation (6.63) in the form

$$M_X = \frac{M}{l}x - M[x-a]^0 \quad ...(6.64)$$

Thus the differential equation of flexure becomes

$$EI\frac{d^2y}{dx^2} = \frac{M}{l}x - M\,[x-a]^0$$

with the condition that the effect of the term with square bracket is to be neglected when $x < a$. At $x = a$ its contribution should automatically be zero. Therefore integrating this we get the slope equation

$$EI\frac{dy}{dx} = \frac{M}{2l}x^2 + C_1 - M\,[x-a] \quad ...(6.65)$$

Integrating once again

$$EI\;y = \frac{Mx^3}{6l} + C_1x - \frac{M}{2}[x-a]^2 + C_2 \quad ...(6.66)$$

When $x = 0,$

$y = 0,$

so that from equation (6.66), omitting the term with square brackets, we get

$$C_2 = 0$$

Again when $x = l,$

$y = 0\ ;$

so that from equation (6.66), we have

$$0 = \frac{Ml^2}{6} + C_1 l - \frac{M}{2}(l - a)^2$$

or

$$C_1 = \frac{M}{6l}(l^2 - 3b^2) \qquad \text{...(6.67)}$$

(Since $l = a + b$)

Thus the slope and deflection equations reduce to

$$EI\frac{dy}{dx} = \frac{M}{2l}x^2 - \frac{M}{6l}(l^2 - 3b^2) - M[x - a] \qquad \text{...(6.68)}$$

and

$$EI\,y = \frac{M}{6l}x^3 - \frac{M}{6l}(l^2 - 3b^2)\,x - \frac{M}{2}[x - a]^2 \qquad \text{...(6.69)}$$

the necessary condition being that the terms with square brackets are to be neglected when $x < a$.

## 6.4 Bending Moment Area Method

Fig. 6.11(*a*) shows a case of a certain beam *AB* with a known general type of loading corresponding to which is shown, the bending moment diagram in Fig. 6.11(*b*). The types of the end supports are not being specified for this general case.

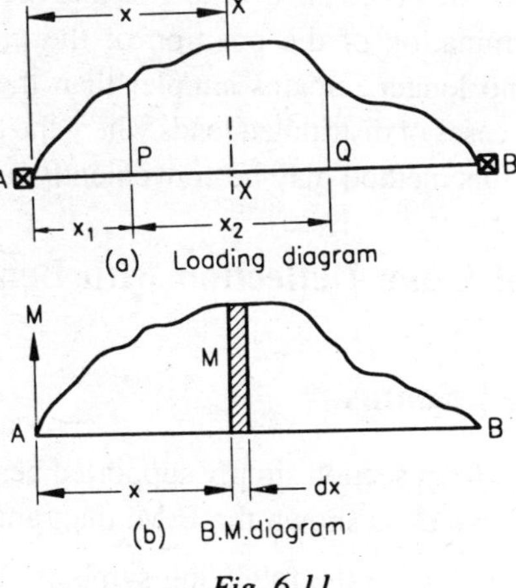


*Fig. 6.11*

Taking the origin at the point $A$, let the bending moment at any section $XX$ be $M$. Then we know

$$EI\frac{d^2y}{dx^2} = M \qquad \text{...(6.70)}$$

Integrating equation (6.70) within two sections $P$ and $Q$

$$\left[\frac{dy}{dx}\right]_P^Q = \int_P^Q \frac{M}{EI}dx$$

or

$$(i_Q - i_P) = \frac{1}{EI}\int_P^Q Mdx$$

i.e.

$$\begin{bmatrix}\text{Change of slope}\\ \text{between points}\\ P \,\&\, Q\end{bmatrix} = \frac{1}{EI}\begin{bmatrix}\text{Area of B.M.}\\ \text{diagram between}\\ \text{point } P \text{ and } Q\end{bmatrix} \qquad \text{...(6.71)}$$

Again multiplying equation (6.70) with $x$ and then integrating by parts within points $P$ and $Q$.

$$\int_P^Q x\frac{d^2y}{dx^2}dx = \int_P^Q \frac{M}{EI}x\,dx$$

$$\left[x\frac{dy}{dx} - y\right]_P^Q = \frac{1}{EI}\int_P^Q Mxdx$$

$$[xi - y]_P^Q = \frac{1}{EI}\times\begin{bmatrix}\text{Moment of B.M. area}\\ \text{about origin}\end{bmatrix}_P^Q \qquad \text{...(6.72)}$$

By appropriate location of the origin $xi$ at both the limits and $y$ at one limit can usually be made zero, thereby leaving the value of $y$ at the other limit.

This method is particularly convenient in case of beams acted upon with point loads in which case the B.M. area consists of triangles and rectangles. In the case of distributed loads the determination of the position of the centroid itself involves integration and as such it no longer remains simpler than the Macaulay's method. However in certain standard cases of distributed loads where the position of the centroid of the B.M. areas is known this method may be conveniently used.

## 6.5 Standard Cases of Beam Deflection with Bending Moment Area Method

### 6.5.1 Simply Supported Beams

Fig. 6.12($a$) shows a uniform section simply supported beam with a general type of loading, corresponding to which is shown the B.M. diagram in Fig. 6.12($b$).

Taking the origin at $A$ and using the following symbols

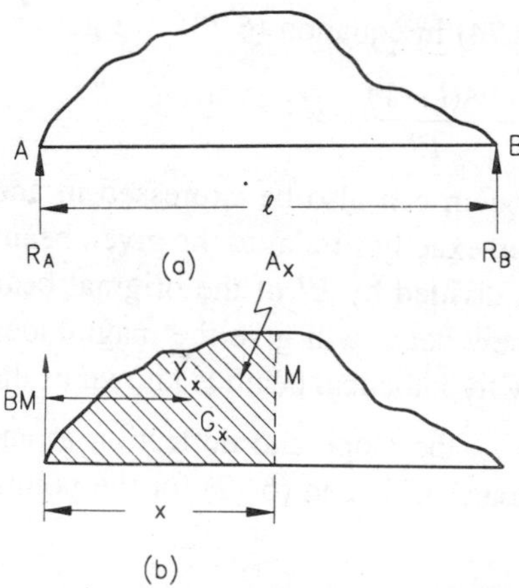


*Fig. 6.12*

A = Area of the B.M. diagram from $A$ to $B$, depending on the sign of the B.M. diagram this may be positive or negative.

$A_x$ = B.M. area from origin to the section$XX$

$\bar{x}$ = Distance of the centroid of the entire bending moment area from $A$ and $B$, from the origin.

$\bar{x}_x$ = Distance of the centroid of the bending moment area from $A$ to $X$ from the origin.

$i$ = Slope at any section

$y$ = Deflection at any section.

We have by applying equation (6.71) from the points $A$ to $B$.

$$[i]_A^B = \frac{1}{EI} \text{ (Bending moment area from } A \text{ to } B)$$

*i.e.,*

$$i_B - i_A = \frac{A}{EI} \qquad ...(6.73)$$

Again using equation (6.72) for the points $A$ to $B$, we get

$$[xi - y]_A^B = \frac{1}{EI}\left[\begin{matrix}\text{Moment of B.M. are} \\ \text{about origin}\end{matrix}\right]_A^B$$

or

$$(x_B\, i_B - y_B) - (x_A\, i_A - y_A) = \frac{1}{EI}(A\bar{x})$$

or

$$li_B - 0 - 0 + 0 = \frac{A\bar{x}}{EI}$$

*i.e.,*

$$i_B = \frac{A\bar{x}}{EIl} \qquad ...(6.74)$$

Substituting equation (6.74) in equation (6.73), we get

$$i_A = -\frac{A(l-\bar{x})}{EI} \quad \text{...(6.75)}$$

The end slopes of the beam can also be expressed in another form. Consider another simply supported beam exactly similar to the given beam *AB*. Let it be loaded by the bending moment area divided by *EI* of the original beam then evidently the reactions $R_A'$ and $R_B'$ of the new beam will give the magnitudes of the slopes at the corresponding ends. This new hypothetical beam is known as the conjugate beam.

After finding the slope $i_A$, the slope and deflection at any section *XX* can be found by applying the equations (6.71) and (6.72) for the points *A* to *X*.

That is $$[i]_A^X = \frac{A_x}{EI}$$

or $$i_x - i_A = \frac{A_x}{EI}$$

*i.e.*, $$i_x = i_A + \frac{A_x}{EI} \quad \text{...(6.76)}$$

And $$[xi - y]_A^X = \frac{1}{EI}\begin{bmatrix}\text{Moment of B.M. area} \\ \text{about origin}\end{bmatrix}_A^X$$

or $$(xi_x - y) - (x_A\, i_A - y_A) = \frac{1}{EI}(A_x\bar{x}_x)$$

or $$xi_x - y - 0 + 0 = \frac{1}{EI}(A_x\bar{x}_x)$$

or $$y = x\left[i_A + \frac{A_x}{EI}\right] + \frac{A_x}{EI}\bar{x}_x$$

or $$y = xi_A + \frac{A_x}{EI}(x - \bar{x}_x) \quad \text{...(6.77a)}$$

or $$y = xi_A + \frac{1}{EI}\begin{bmatrix}\text{Moment of B.M. area} \\ \text{about the section XX}\end{bmatrix}_A^X \quad \text{...(6.67b)}$$

## 6.5.2 Simply Supported Beam with a Point Load not at the Centre

Fig. 6.13(*b*) shows the bending moment diagram for the beam shown in Fig. 6.13(*a*)

$$A = \frac{1}{2}l\left(\frac{-Wab}{l}\right) = -\frac{Wab}{2}$$

$$i_A = -\frac{A\,(l-\bar{x})}{EIl}$$

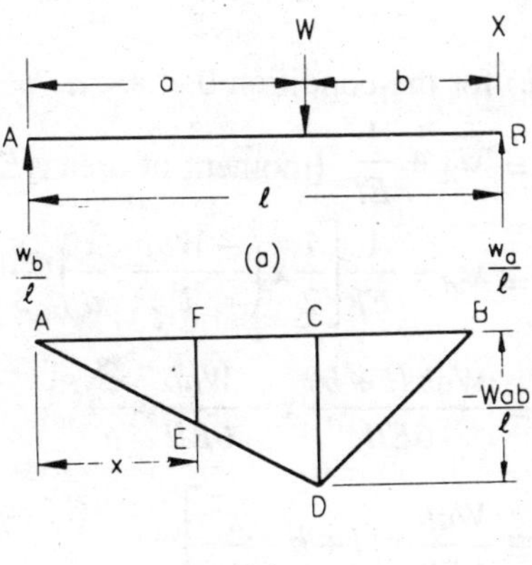


*Fig. 6.13*

or
$$i_A = -\frac{1}{EIl}\left[\begin{matrix}\text{Moment of B.M.}\\ \text{area about B}\end{matrix}\right]_A^B$$

$$= -\frac{1}{EIl}\left[\begin{matrix}\text{Sum of the moments of areas}\\ ACD \text{ and } CDB \text{ about } B\end{matrix}\right]_A^B$$

$$= -\frac{1}{EIl}\left[\frac{1}{2}a\left(\frac{-Wab}{l}\right)\left(\frac{a}{3}+b\right)+\frac{1}{2}b\left(\frac{-Wab}{l}\right)\left(\frac{2}{3}b\right)\right]$$

$$= \frac{Wab}{2EIl^2}\left[\frac{a^2}{3}+ab+\frac{2}{3}b^2\right]$$

$$= \frac{Wab}{6EIl^2}(a^2+3ab+2b^2)$$

$$= \frac{Wab}{6EIl^2}[(a+b)^2+b(a+b)]$$

*i.e.*
$$i_A = \frac{Wab}{6EIl}(l+b) \qquad ...(6.78)$$

Now
$$i_X = i_A + \frac{A_x}{EI}$$

$$= i_A + \frac{1}{EI}[\text{Area } AFE] \ [\text{Area } AFE]$$

$$= i_A + \frac{1}{EI}\left[\frac{1}{2}x\left(\frac{-Wab}{l}\right)\frac{x}{a}\right]$$

$$= \frac{Wab}{6EIl}(l+b) - \frac{Wab}{2EIl}\cdot\frac{x^2}{a}$$

*i.e.*,
$$i_x = \frac{Wab}{6EIl}\left[l+b-\frac{3x^2}{a}\right] \qquad ...(6.79)$$

From equation (6.77*b*) for the condition $0 < x < a$

$$y = xi_A + \frac{1}{EI}\text{ (moment of area } AEF \text{ about } F)$$

$$= xi_A + \frac{1}{EI}\left[\frac{1}{2}x\left(\frac{-Wab}{l}\cdot\frac{x}{a}\right)\cdot\frac{x}{3}\right]$$

$$= \frac{Wab\,(l+b)}{6EIl}x - \frac{Wab}{6EIl}\cdot\frac{x^3}{a}$$

or $$y = \frac{Wab}{6EIl}x\left[l + b - \frac{x^2}{a}\right] \quad ...(6.80)$$

If $x = a$, then from (6.80)

$$y_C = \frac{Wa^2b^2}{6EIl}(l + b - a)$$

or $$y_C = \frac{Wa^2b^2}{3EIl} \quad [\because l - a = b] \quad ...(6.81)$$

and from equation (6.79)

$$i_C = \frac{Wab}{6EIl}(l + b - 3a)$$

$$= -\frac{Wab\,(a-b)}{3EIl}$$

## Simply Supported Beam With a Central Point Load

This problem can be tackled independently exactly in a similar manner as has been done above. However using the equations (6.80) and (6.81) and letting

$$a = b = \frac{l}{2}$$

we see

For the range $0 < x < \frac{l}{2}$

$$y = \frac{W\left(\frac{l}{2}\right)\left(\frac{l}{2}\right)x}{6EIl}\cdot\left[l + \frac{l}{2} - \frac{2x^2}{l}\right]$$

or $$y = \frac{Wx}{48EI}(3l^2 - 4x^2) \quad ...6.82)$$

when $x = l/2$

$$y = \frac{W\left(\frac{l}{2}\right)}{48EI}\left[3l^2 - 4\left(\frac{l^2}{4}\right)\right] = \frac{Wl^3}{48EI} \quad ...(6.83)$$

## 6.5.3 Simply Supported Beam with a Uniformly Distributed Load

$$A = \frac{2}{3}l\left(-\frac{wl^2}{8}\right)$$

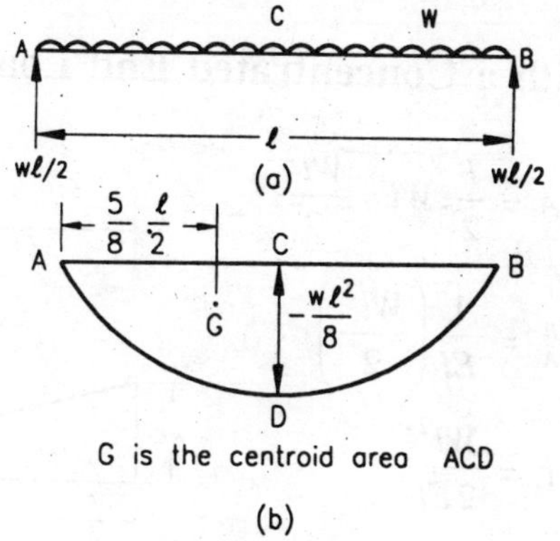

C W
A B
$l$
$wl/2$ (a) $wl/2$
$\frac{5}{8}\,\frac{l}{2}$
A C B
G $-\frac{wl^2}{8}$
D
G is the centroid area ACD

(b)

*Fig. 6.14*

*i.e.,* $$A = -\frac{wl^3}{12}$$

$$i_A = -\frac{1}{EI}\,.\ \text{(Moment of B.M. area about } B\text{)}$$

$$= -\frac{1}{EIl}\left(-\frac{wl^3}{12}\right)\frac{l}{2} = \frac{Wl^3}{24\,EI}$$

The value of $i_A$ can also be found in a much more convenient way in this case. Applying equation (6.71) for the point $A$ and $C$.

$$[i]_A^C = \frac{1}{EI}\,[\text{B.M. area}]_A^C$$

$$i_C - i_A = \frac{1}{EI}\left[\frac{1}{2}\cdot\frac{2}{3}l\left(-\frac{wl^2}{8}\right)\right]$$

*i.e.* $$i_A = \frac{Wl^3}{24EI} \qquad \text{(Since } i_C = 0\text{)}$$

Applying equation (6.72) for the points $A$ and $C$ and taking origin at $A$

$$[xi - y]_A^C = \frac{1}{EI}\left(\begin{matrix}\text{Moment of B.M. Area}\\ \text{about origin}\end{matrix}\right)_A^C$$

or $(x_C i_C - y_C) - (x_A\, i_A - y_A)$

$$= \frac{1}{EI}\left[\frac{1}{2}\cdot\frac{2}{3}l\left(-\frac{wl^2}{8}\right)\cdot\frac{5}{8}\left(\frac{l}{2}\right)\right]$$

or $$(0 - y_C) - (0 - 0) = \frac{5\,wl^4}{384\,EI}$$

or $$y_C = \frac{5\,wl^4}{384\,EI}$$

## 6.6.1 Cantilever with a Concentrated End Load

$$A = \frac{l}{2}.\,Wl = \frac{Wl^2}{2}$$

$$[i]_A^B = \frac{1}{EI}\left(\frac{Wl^2}{2}\right)$$

$$i_B - i_A = \frac{Wl^2}{2EI}$$

or $$i_B = \frac{Wl^2}{2EI}$$

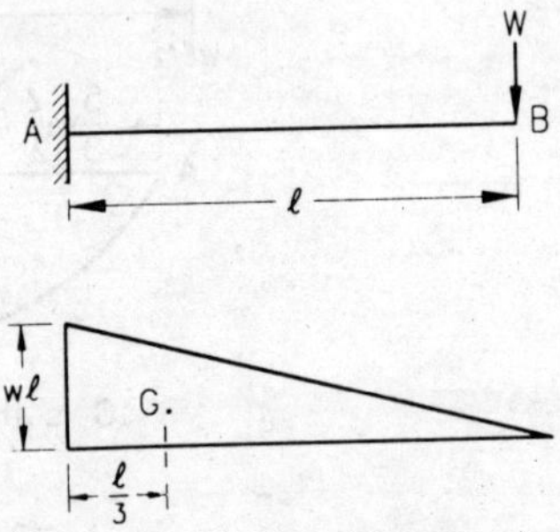

Fig. 6.15

Taking origin at $A$ and applying equation (6.72) for points $A$ and $B$

$$[xi - y]_A^B = \frac{1}{EI}\left[\begin{array}{c}\text{Moment of B.M. area}\\ \text{about origin}\end{array}\right]_A^B$$

or $$(x_B i_B - y_B) - (x_A i_A - y_A)$$

$$= \frac{1}{EI}\left(\frac{Wl^2}{2}.\frac{l}{3}\right)$$

or $$\left(l\frac{Wl^2}{2EI} - y_B\right) - (0 - 0)$$

$$= \frac{Wl^3}{6EI}$$

or $$y_B = \frac{Wl^3}{3EI}$$

## 6.6.2 Cantilever with Uniformly Distributed Load

$$[i]_A^B = \frac{1}{EI}\,[\text{B.M. area}]_A^B$$

$$i_B - i_A = \frac{1}{EI}\left[\frac{1}{3}\,\frac{wl^2}{2}.\,l\right]$$

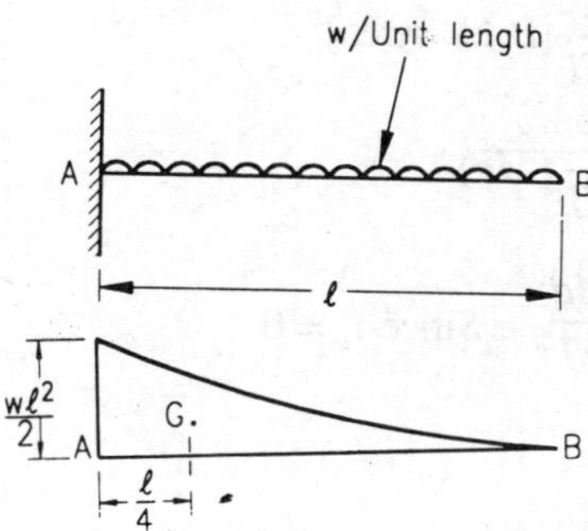


*Fig. 6.16*

i.e. $$i_B = \frac{wl^3}{6EI} \qquad \text{Since } i_A = 0$$

Taking origin at $A$ and applying equation (6.72)

$$[xi - y]_A^B = \left(\frac{1}{EI}\right)\left[\begin{array}{c}\text{Moment of } BM \text{ area}\\ \text{about origin}\end{array}\right]_A^B$$

or $$(x_B i_B - y_B) - (x_A i_A - y_A)$$

$$= \frac{1}{EI}\left(\frac{1}{3}\frac{wl^2}{2}.l\right)\left(\frac{1}{4}l\right)$$

or $$\left[l.\frac{wl^3}{6EI} - y_B\right] - (0 - 0)$$

$$= \frac{1}{EI}.\frac{wl^4}{24}$$

i.e. $$y_B = \frac{wl^4}{8EI}$$

### 6.6.3 Cantilever with a Couple

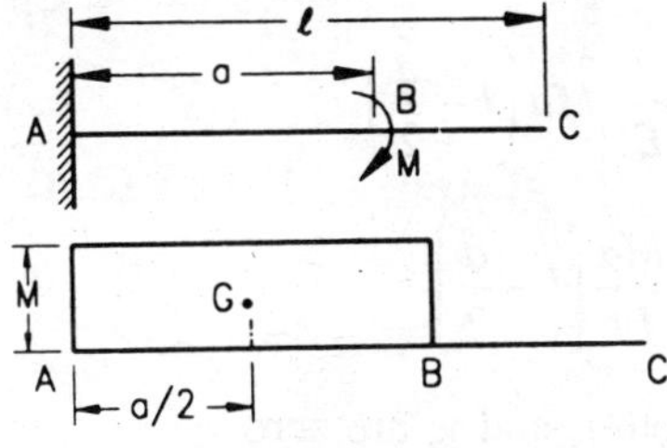


*Fig. 6.17*

$$[i]_A^B = \frac{1}{EI}[\text{B.M. Area}]_A^B$$

$$i_B - i_A = \frac{1}{EI}(Ma)$$

*i.e.* $$i_B = \frac{Ma}{EI} \quad \text{Since } i_A = 0$$

Also for points $A$ and $C$

$$i_c - i_A = \frac{Ma}{EI}$$

or $$i_c = \frac{Ma}{EI}$$

Taking origin at $B$ and applying equation (6.72)

$$[xi - y]_A^B = \frac{1}{EI}\begin{bmatrix}\text{Moment of B.M. Area} \\ \text{about origin}\end{bmatrix}_B^A$$

or $$(x_A i_A - y_A) - (x_B i_B - y_B)$$

$$= \frac{1}{EI} \times (Ma) \times \frac{a}{2}$$

or $$(0 - 0) - (0 - y_B) = \frac{Ma^2}{2EI}$$

*i.e.* $$y_B = \frac{Ma^2}{2EI}$$

For finding deflection at $C$, take origin at $C$ and apply equation (6.72)

$$[xi - y]_C^A = \frac{1}{EI}\begin{bmatrix}\text{Moment of B.M. Area} \\ \text{about origin}\end{bmatrix}_C^A$$

or $$(x_A i_A - y_A) - (x_C i_C - y_C)$$

$$= \frac{1}{EI}(Ma)\left[\frac{a}{2} + (l - a)\right]$$

or $$(0 - 0) - (0 - y_C) = \frac{1}{EI} Ma\left(l - \frac{a}{2}\right)$$

*i.e.* $$y_C = \frac{Ma}{EI}\left[l - \frac{a}{2}\right]$$

**Note.** For the point $A$ both $i_A$ and $y_A$ are zero.

## 6.7.1 Varying Load on a Cantilever

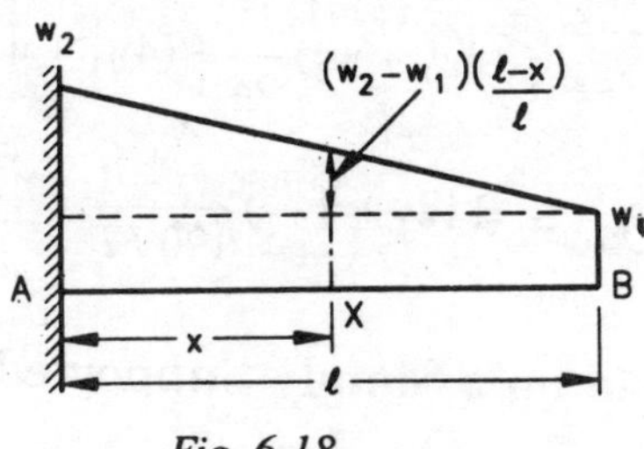


*Fig. 6.18*

Let the varying load increase from $w_1$ per unit length at the free end to $w_2$ per unit length at the fixed end.

$$M_x = \frac{w_1}{2}(l-x)^2 + \frac{(l-x)}{2}(w_2 - w_1)\frac{(l-x)}{l}\cdot\frac{(l-x)}{3}$$

Therefore

$$EL\frac{d^2y}{dx^2} = M_x = \frac{w_1}{2}(l-x)^2 + \frac{(w_2-w_1)}{6l}(l-x)^3$$

Integrating w.r.t. '$x$'

$$EI\frac{dy}{dx} = -\frac{w_1}{6}(l-x)^3 - \frac{(w_2-w_1)}{24l}(l-x)^4 + C_1 \quad ...(6.84)$$

Slope being zero at point $A$ for which $x = 0$, we have from equation (6.84)

$$0 = -\frac{w_1}{6}l^3 - \frac{(w_2-w_1)}{24l}l^4 + C_1$$

*i.e.*

$$C_1 = \frac{w_1 l^3}{8} + \frac{w_2 l^3}{24}$$

Substituting the value of $C_1$ in equation (6.84) and integrating once again

$$Ely = \frac{w_1}{24}(l-x)^4 + \frac{(w_2-w_1)}{120l}(l-x)^5 + \left(\frac{w_1}{8}l^3 + \frac{w_2 l^3}{24}\right)x + C_2 \quad ...(6.85)$$

But

$$y = 0 \quad x = 0$$

so that

$$C_2 = -\frac{w_1 l^4}{30} - \frac{w_2 l^4}{120}$$

Thus the deflection equation (6.85) reduces to

$$EI\,y = \frac{w_1}{24}(l-x)^4 + \frac{(w_2-w_1)}{120\,l}(l-x)^5 + \frac{l^3}{24}(3w_1 + w_2)x - \frac{l^4}{24}(4w_1 + w_2) \quad ...(6.85)$$

The maximum deflection will occur at $B$ ($x = l$)

i.e. $$EI\, y_{max} = (3w_1 + w_2)\frac{l^4}{24} - (4w_1 + w_2)\frac{l^4}{120}$$

i.e., $$y_{max} = (11w_1 + 4w_2)\frac{l^4}{120\, EI} \qquad ...(6.86)$$

## 6.7.2 Varying Load on a Simply Supported Beam

For the equilibrium of the beam, taking moments about $B$ (Fig. 6.19), we get

$$lR_A = (w_1 l)\frac{l}{2} + \frac{1}{2}l(w_2 - w_1)\frac{l}{3}$$

or $$R_A = \frac{w_1 l}{3} + \frac{w_2 l}{6} \qquad ...(6.87)$$

Therefore

$$EI\frac{d^2y}{dx^2} = M_X = -\left(\frac{w_1}{3}l + \frac{w_2 l}{6}\right)x + \frac{1}{2}w_1 x^2 + \left[\frac{x}{2}(w_2 - w_1)\frac{x}{l}\right]\frac{x}{3}$$

or $$EI\frac{d^2y}{dx^2} = -\frac{1}{6}(2w_1 + w_2)\,x + \frac{1}{2}w_1 x^2 + \frac{1}{6}(w_2 - w_1)\frac{x^3}{l}$$

Integrating w.r.t. '$x$' we get the equation for slope

$$EI\frac{dx}{dx} = -\frac{1}{12}(2w_1 + w_2)\,x^2 + \frac{w_1}{6}x^3 + \frac{(w_2 - w_1)}{24}\frac{x^4}{l} + C_1 \qquad ...(6.88)$$

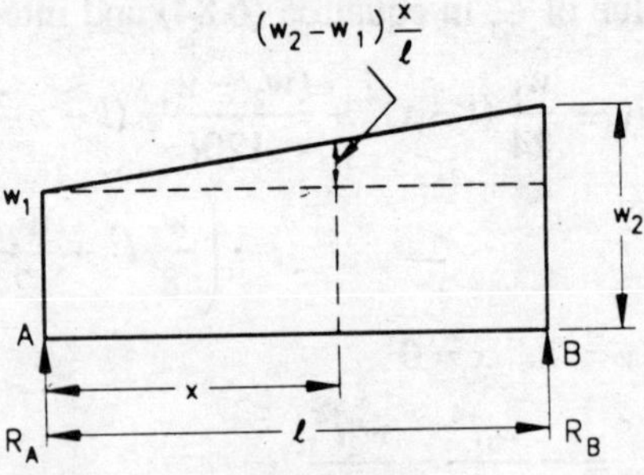


*Fig. 6.19*

Integrating again w.r.t. '$x$'

$$EI_y = -\frac{1}{36}(2w_1 + w_2)\,x^3 + \frac{w_1}{24}x^4 + \frac{(w_2 - w_1)}{24}\cdot\frac{x^5}{5l} + C_1 x + C_2 \qquad ...(6.89)$$

For evaluating the constants $C_1$ and $C_2$ we know that the deflection is zero at the supports. Therefore from equation (6.89) at $x = 0$ and $x = l$ we get

$$C_2 = 0$$

and $$C_1 = \frac{7l^3}{360}(5w_1 + w_2)$$

Thus substituting the values of $C_1$ and $C_2$ in equations (6.88) and (6.89), the slope and deflection equations are

$$EI\frac{dy}{dx} = -\frac{l}{12}(2w_1 + w_2)x^2 + \frac{w_1}{6}x^2 + \frac{(w_2 - w_1)}{24}\frac{x^4}{l} + \frac{7l^2}{360}(5w_1 + w_2) \quad ...(6.90)$$

$$EIy = -\frac{1}{36}(2w_1 + w_2)x^3 + \frac{w_1}{24}x^4 + \frac{(w_2 - w_1)}{120}\frac{x^5}{l} + \frac{7l^3}{360}(5w_1 + w_2)x \quad ...(6.91)$$

## 6.8 Overhanging Beams

Here again the Macaulay's method will be applied. The representative equation for the bending moment will be obtained by taking the section within the portion $CD$. Thus taking the origin at $A$.

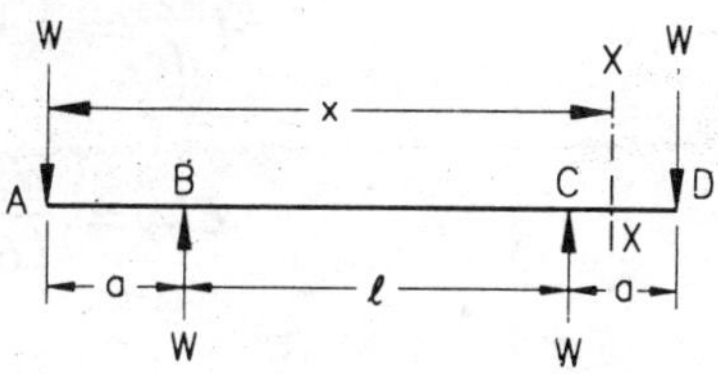


*Fig. 6.20*

$$EI\frac{d^2y}{dx^2} = Mx = W.x - [W(x-a)] - [W\{x - (l+a)\}] \quad ...(6.92)$$

Effect of last term to be neglected when $x < b$ where $b = (l + a)$ and effect of last and last but one term to be neglected when $x < a$.

Integrating equation (6.92) w.r.t. '$x$'

$$EI\frac{dy}{dx} = \frac{1}{2}Wx^2 + C_1 - \frac{W}{2}[(x-a)]^2 - \frac{W}{2}[x-b]^2 \quad ...(6.93)$$

Integrating again w.r.t. '$x$'

$$EI\,y = \frac{1}{6}Wx^3 + C_1x + C_2 - \frac{W}{6}[x-a]^3 - \frac{W}{6}[x-b]^3 \quad ...(6.94)$$

When $x = a$, $y = 0$ so that

$$aC_1 + C_2 = -\frac{Wa^3}{6} \quad ...(6.95)$$

Also when $x = b$, $y = 0$ so that

$$bC_1 + C_2 = -\frac{Wb^3}{6} + \frac{W}{6}[b - a]^3$$

or $$(l + a)\, C_1 + C_2 = -\frac{Wb^3}{6} + \frac{Wl^3}{6} \qquad \text{...(6.96)}$$

Subtracting equation (6.95) from equation (6.96)

$$lC_1 = -\frac{Wb^3}{6} + \frac{Wl^3}{6} + \frac{Wa^3}{6}$$

or $$C_1 = -\frac{W}{6l}(b^3 - l^3 - a^3) = -\frac{W}{2}a\,(a + l) \qquad \text{...(6.97)}$$

From equations (6.95) and (6.97)

$$C_2 = \frac{Wa^2}{6}(2a + 3l) \qquad \text{...(6.98)}$$

Substituting the values of $C_1$ and $C_2$ in equations (6.93) and (6.94) we get the slope and deflection equations as

$$EI\frac{dy}{dx} = \frac{1}{2}Wx^2 - \frac{W}{2}a(a + l) - \frac{W}{2}[x - a]^2 - \frac{W}{2}[x - b]^2 \qquad \text{...(6.99)}$$

$$EI\,y = \frac{Wx^3}{6} - \frac{W}{2}a\,(a + l)\,x + \frac{Wa^2}{6}(2a + 3l) - \frac{W}{6}[x - a]^3 - \frac{W}{6}[x - b]^3 \qquad \text{...(6.100)}$$

## 6.9 Propped Beams and Cantilevers

Prop is an additional support placed under a beam somewhere within the span. Evidently a portion of the load will be shared by the prop. If the prop does not yield due to the load on it, it is known as a rigid prop. But in actual practice prop will definitely yield a bit due to its elasticity. Such type of prop is known as an elastic prop.

A rigid prop which is at the same height as the original position of the unloaded beam, will not allow the beam to deflect at the prop point. It amounts to saying that the downward deflection of the beam due to load, with prop absent, is equal in magnitude of the upward deflection of beam due to the prop reaction, with load absent.

In case the position of the prop is lower than the original level of the beam (say by $d$) then the deflection removed by the prop is equal to the deflection due to load at prop point minus $d$.

**Example 6.1.** *A cantilever carrying a point load W at its free end is propped at its mid span by a rigid prop. If the level of the prop is same as that of the unloaded cantilever, find the load taken up by the prop. What is then the deflection of the free end?*

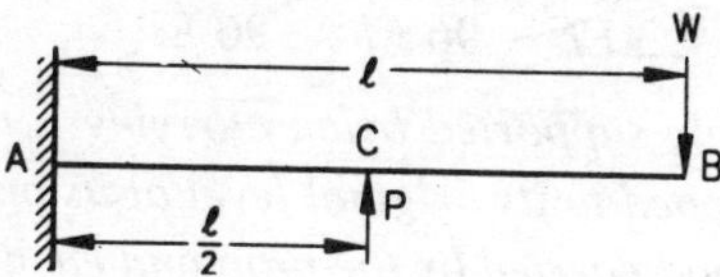


*Fig. 6.21*

**Solution.** Let the prop reaction be $P$. Then upward deflection of the point $C$ of the cantilever due to load $P$ is

$$\frac{1}{3}\frac{P(l/2)^3}{EI} = \frac{Pl^3}{24\,EI} \qquad ...(6.101)$$

The downward deflection of the point $C$ due to $W$ alone, is obtained with the help of equation (6.11)

$$y_c = \frac{W}{EI}\left[\frac{l}{2}\left(\frac{l}{2}\right)^2 - \frac{1}{6}\left(\frac{l}{2}\right)^3\right]$$

$$= \frac{5}{48}\frac{Wl^3}{EI} \qquad ...(6.102)$$

As the prop is rigid and at the same level as the beam, these two must be equal.

*i.e.*
$$\frac{Pl^3}{24\,EI} = \frac{5Wl^3}{48\,EI}$$

or
$$P = \frac{5}{2}W \qquad ...(6.103)$$

(*b*) Downward deflection of *B* due to *W* alone

$$\delta_w = \frac{Wl^3}{3\,EI} \qquad ...(6.104)$$

Upward deflection of *B* due to *P* alone from equation (6.16)

$$\delta_P = \frac{P\left(\frac{l}{2}\right)^3}{3EI} + \frac{P\left(\frac{l}{2}\right)^2}{2EI}\cdot\left(\frac{l}{2}\right)$$

$$= \frac{Pl^3}{24EI} - \frac{Pl^3}{16\,EI} = \frac{5\,Pl^3}{48\,EI} \qquad ...(6.105$$

Thereefore net downward deflection of *B*, from equations (6.104) and (6.105)

$$y_B = \frac{Wl^3}{3\,EI} - \frac{5Pl^3}{48\,EI}$$

$$= \frac{Wl^3}{3\,EI} - \frac{25\,Wl^3}{96\,EI} = \frac{7\,Wl^3}{96\,EI} \qquad \text{...(6.106)}$$

**Example 6.2.** *A simply supported beam carrying a uniformly distributed load over the whole span is propped to its original level at its mid-point with the help of a rigid prop. Calculate the load carried by the prop and each of the supports.*

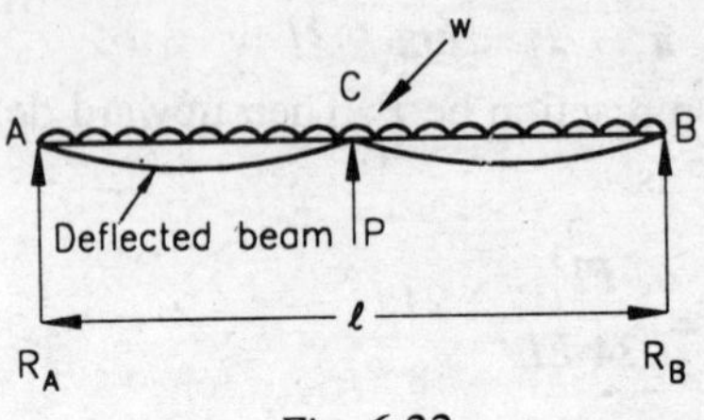


*Fig. 6.22*

**Solution.** Let the prop load be *P*. Then upward deflection of point *C* due to *P* alone

$$\delta_P = \frac{Pl^3}{48\,EI} \qquad \text{...(6.107)}$$

Also downward deflection of point *C* due to the given distributed load

$$\delta_P = \frac{5wl^4}{384\,EI} \qquad \text{...(6.108)}$$

Since these two are equal, we have

$$\frac{Pl^3}{48\,EI} = \frac{5wl^4}{384\,EI}$$

$$P = \frac{5wl}{8}$$

Also since $R_A = R_B$,

we must have $R_A = R_B = \frac{1}{2}(Wl - P)$

$$= \frac{1}{2}\left(wl - \frac{5}{8}wl\right) = \frac{3}{16}\,wl$$

## 6.9.1 Elastic or Sinking Props

**Example 6.3.** *Solve the Example 6.1 assuming the prop to be an elastic one having the stiffness of 's' (i.e. a force of s units is required to compress the prop through unit distance).*

**Solution.** (*a*) For the prop load *P* the downward deflection of the point *C* due to the elasticity of the prop is given by (*P*/*s*).

Then with the help of equations (6.101) and (6.102) we get

$$\frac{5}{48}\frac{Wl^3}{EI} - \frac{Pl^3}{24\,EI} = \frac{P}{s}$$

*i.e.*
$$P = \frac{5\,Wl^3\,s}{2\,(sl^3 + 24\,EI)} \qquad ...(6.109)$$

(*b*) From equations (6.104), (6.105) and (6.109) the net downward deflection of *B*, in this case is given by

$$y_B = \frac{Wl^3}{3\,EI} - \frac{5\,Pl^3}{48\,EI}$$

$$= \frac{Wl^3}{3\,EI} - \frac{5l^3}{48\,EI} \cdot \frac{5\,Wl^3 s}{2[sl^3 + 24\,EI]}$$

$$= \frac{Wl^3}{96\,EI}\left[\frac{768\,EI + 7\,sl^3}{24\,EI + sl^3}\right] \qquad ...(6.110)$$

**Example 6.4.** *Solve Example 6.2 assuming the prop to be elastic having the stiffness 's'.*

**Solution.** Downward deflection of the prop due to its elasticity

$$= (P/s)$$

Therefore from equations (6.107) and (6.108)

$$\frac{5wl^4}{384\,EI} - \frac{Pl^3}{48\,EI} = \frac{P}{s}$$

*i.e.*
$$P = \frac{5wl^4\,s}{8\,[48\,EI + sl^3)}$$

Then
$$R_A = R_B = \frac{1}{2}(wl - P)$$

$$= \frac{1}{2}wl\left[\frac{5sl^3}{8\,(48\,EI + sl^3)}\right]$$

$$= \frac{3}{16}wl\left[\frac{sl^3 + 128\,EI}{sl^3 + 48\,EI}\right]$$

**Example 6.5.** *A long flat 50 mm wide and 3.2 mm thick is lying on a flat horizontal plane. One end of the strip is now lifted 25 mm from the plane by a vertical force applied at the end. The strip is so long that the other end remains undisturbed. Calculate*

*(a) the force required to lift the end, (b) the maximum stress in the steel. Take the weight of steel as 7800 kg/m³. E = 205000 N/mm².*

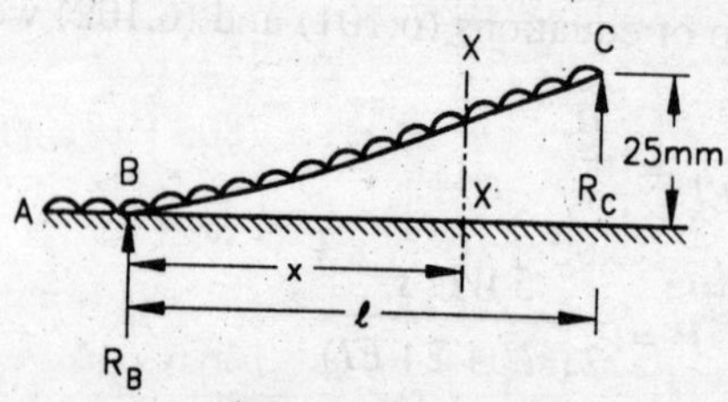


*Fig. 6.23*

**Solution.** For the portion *AB* of the strip the ground reaction just balances the weight of the strip and hence there is no bending of the portion *AB*. But at the point *B* where the strip leaves the plane there will be the reaction $R_B$ and the conditions are as though the length *AB* were cut. The slope at *B* is zero.

Choosing the origin at *B* and assuming the end *C* to be simply supported, the bending moment equation can be written

$$EI\frac{d^2y}{dx^2} = M_X = R_B x + \frac{wx^2}{2}$$

Now $\quad M_X = 0$ at $x = l$,

whence $\quad R_B = \dfrac{wl}{2}$

Integrating with respect to $x$

$$EI\frac{dy}{dx} = -R_B\frac{x^2}{2} + \frac{wx^3}{6}$$

(The constant of integration being zero as slope is zero at $x = 0$)

Integrating once again

$$EI\,y = -R_B\frac{x^3}{6} + \frac{wx^4}{24}$$

(Constant of integration is zero as $y = 0$ as $x = 0$)

Now at $\quad x = l,\ y = -25 \times 10^{-3}$ m $\quad$ (– *ve* as it is upward)

Therefore $-EI\,(25 \times 10^{-3})$

$$= -\left(\frac{wl}{2}\right)\left(\frac{l^3}{6}\right) + \frac{wl^4}{24}$$

or $\quad \dfrac{wl^4}{EI} = 0.6$ m

or $\quad l^4 = \left(\dfrac{0.6\,EI}{w}\right) m^4$

Now $$I = \frac{50 \times (3.2)^3}{12} = 136.6 \text{ mm}^4$$

$$= 136.6 \times 10^{-12} \text{ m}^4$$

$$w = \left(\frac{50}{1000} \times \frac{3.2}{1000} \times 1 \times 7800\right) 9.81 = 12.24 \text{ N/m}$$

$$E = 20500 \times 10^9 \text{ N/m}^2$$

Therefore $$l^4 = \frac{0.6 \times (205 \times 10^9) \times (136.6 \times 10^{-12})}{12.24}$$

$$= 1.37 \text{ m}^4$$

or $$l = 1.08 \text{ m}$$

Then $$R_B = \frac{wl}{2} = \frac{12.24 \times 1.08}{2} = 6.63 \text{ N}$$

For maximum bending moment

$$\frac{dM_x}{dx} = 0 = -R_B + wx$$

or $$x = \frac{R_B}{w} = \frac{wl}{2} \cdot \frac{1}{w} = \frac{l}{2}$$

Therefore $$M_{max} = -R_B\left(\frac{l}{2}\right) + \frac{w}{2}\left(\frac{l}{2}\right)^2$$

$$= -\frac{wl}{2} \cdot \frac{l}{2} + \frac{wl^2}{8}$$

$$= -\frac{wl^2}{8} = -\frac{12.24\,(1.08)^2}{8}$$

$$= -1.786 \text{ Nm}$$

Negative sign of $M_{max}$ indicates that the moment is sagging *i.e.* the upper fibres are in compression. Therefore maximum bending stress is given by

$$\sigma_b = \frac{1.786}{136.6 \times 10^{-12}}\left(\frac{3.2}{2} \times 10^{-3}\right)$$

$$= 2.09 \times 10^7 \text{ N/m}^2$$

$$= 20.9 \text{ MN/m}^2$$

**Example 6.6.** *A uniform beam ABCD of flexural rigidity EI, is simply supported at B and C as shown in figure 6.24. Calculate the deflection at the point M midway between B and C in terms of W, l and EI. Also find the deflection at A.*

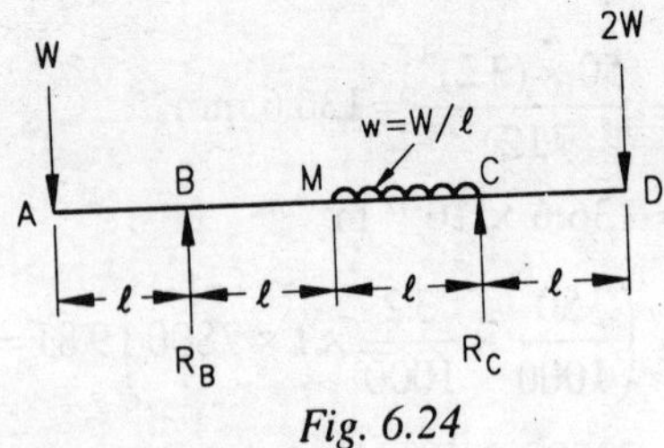


*Fig. 6.24*

**Solution.** Taking moments about $C$

$$W(3l) + W\frac{1}{2} = R_B\,(2l) + 2Wl$$

or $$R_B = \frac{3}{4}W$$

and then $$R_C = 4W - \frac{3W}{4} = \frac{13}{4}W$$

*Fig. 6.25*

Extend the given distributed load up to $D$ and applying a similar compensating load acting upwards over the portion $CD$ as shown in figure 6.25. Taking the origin at $A$ and using Macualay's method we have

$$EI\frac{d^2y}{dx^2} = Wx - R_B\,[x - l] + \frac{W}{2l}[x - 2l]^2 - R_C\,[x - 3l] - \frac{W}{2l}[x - 3l]^2$$

$$= Wx - \frac{3}{4}W[x - l] + \frac{W}{2l}[x - 2l]^2 - \frac{13}{4}W[x - 3l] - \frac{W}{2l}[x - 3l]^2$$

Integrating twice this equation w.r.t. '$x$'

$$EI\frac{dy}{dx} = \frac{Wx^2}{2} - \frac{3W}{8}[x - l]^2 + \frac{W}{6l}[x - 2l]^3 - \frac{13}{8}W[x - 3l]^2 - \frac{W}{6l}[x - 3l]^3 + A$$

$$EI\,y = \frac{Wx^3}{6} - \frac{3}{24}W[x-l]^2 + \frac{W}{24l}[x-2l]^4 + A$$

$$-\frac{13}{8}W[x-3l]^3 - \frac{W}{24l}[x-3l]^3 + Ax + B$$

Inserting the conditions that

$$y = 0 \text{ at both } x = \text{ and } x = 3l$$

we get two simultaneous equations in $A$ and $B$ whose solution is

$$A = -\frac{27}{16}Wl^2$$

and $$B = \frac{73}{48}Wl^3$$

Then the deflection equation reduces to

$$EI_y = \frac{Wx^3}{6} - \frac{W}{8}[x-l]^3 + \frac{W}{24l}[x-2l]^4$$

$$-\frac{13}{24}[x-3l]^3 + \frac{W}{24l}[x-3l]^4 - \frac{27}{16}Wl^2x + \frac{73}{48}Wl^3$$

When $x < 2l$, (third, fourth and fifth terms on R.H.S. Will be neglected)

Therefore $$EI\,y_M = \frac{W}{6}(8l^3) - \frac{Wl^3}{8} - \frac{27}{16}Wl^2\,(2l) + \frac{73}{48}Wl^3$$

or $$y_M = -\frac{31}{48}\frac{Wl^3}{EI}$$ (*i.e.* deflection is upwards at $M$)

Again when $x = 0$ (all terms with square brackets to be omitted)

$$EI\,y_A = \frac{73}{48}Wl^3$$

or $$y_A = \frac{73}{48}\frac{Wl^3}{EI}.$$

**Example 6.7.** *A vertical column 6 m high is fixed at the base and a clockwise moment of 1.4 kNm of is applied at the top of the column. A horizontal force P is applied to the column at a height of 3 m above the base so as to give a counter clockwise moment.*

*Determine the value of P so that the horizontal deflections at the top of the column and at the point of application of P shall be equal, (a) when the deflections are on the same side, (b) when the deflections are on opposite sides of the vertical line through the foot of the column.*

**Solution.** Choose the origin at $B$. Using the Macaulay's method we have

$$EI\frac{d^2y}{dx^2} = M - P\,[x - 3]$$

Effect of the term in square bracket to be neglected when $x < 3$.

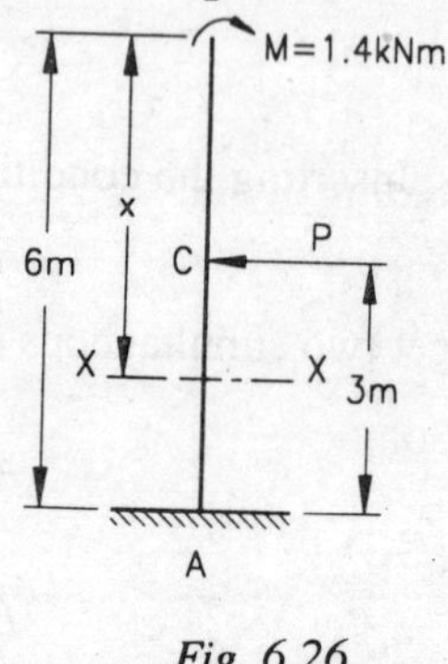

*Fig. 6.26*

Integrating w.r.t. '$x$'

$$EI\frac{dy}{dx} = Mx - \frac{P}{2}\,[x - 3]^2 + A$$

Now the slope is zero at $x = 6$, therefore

$$0 = 6M - \frac{P}{2}(9) + A$$

or $$A = (4.5\,P - 6M)$$

Integrating once again

$$EI\,y = \frac{Mx^2}{2} - \left(\frac{P}{2}\right)\frac{1}{3} \times [x - 3]^2 + (4.5\,P - 6M)\,x + B$$

When $x = 6$, $y = 0$

Therefore $$0 = \frac{M}{2} \times 36 - \frac{P}{6} \times 27 + (4.5\,P - 6M)\,6 + B$$

or $$B = (18M - 22.5P)$$

Thus the deflection equation reduces to

$$EI\,y = \frac{Mx^2}{2} - \frac{P}{6}[x - 3]^3 + (4.5\,P - 6M)\,x + (18M - 22.5P)$$

Therefore at $\mathbf{x} = 0$, neglecting the term with square brackets

$$EIy_B = (18M - 22.5P)$$

$$= (25.2 - 22.5P) \qquad \text{(as } M = 1.4 \text{ kNm)}$$

and at $x = 3$

$$EIy_C = \frac{9M}{2} + (4.5P - 6M)\,3 + (18M - 22.5P)$$

$$= (4.5\,M - 9P)$$

$$= (6.3 - 9P)$$

(*a*) Again when

$$y_B = y_C$$

we have $$25.2 - 22.5P = 6.3 - 9P$$

or $$P = \frac{18.9}{13.5} = 1.4 \text{ kN}$$

(*b*) When $y_B = -y_C$, we have

$$25.2 - 22.5P = 9P - 6.3$$

or $$P = \frac{31.5}{31.5} = 1.0 \text{ kN}$$

**Example 6.8.** *A horizontal cantilever, 1.5 m long, tapers in section from 200 mm deep by 75 mm wide at the fixed end to 75 mm square at the extreme end. It carries and end load of 2.7 kN. Calculate the deflection of the free end. E = 14 GN/m².*.

Solution. $$I_B = \frac{(75)^4}{12} = 2.64 \times 10^6 \text{ mm}^4$$

$$I_A = \frac{75 \times (200)^3}{12} = 50 \times 10^6 \text{ mm}^4$$

$$I_x = \left[I_B + (I_A - I_B)\frac{x}{l}\right]$$

or $$I_x = \left[(50 - 2.64)\frac{x}{l} + 2.64\right] \times 10^6 \text{ mm}^4$$

$$= \left[2.64 + 47.36\frac{x}{l}\right] \times 10^{-6} \text{ m}^4$$

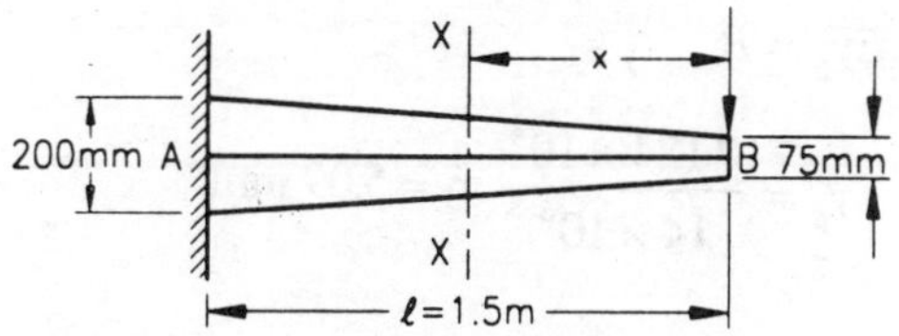

*Fig. 6.27*

Taking origin at *B*

$$EI_x\,(d^2y/dx^2) = Wx$$

or $$E\,(d^2y/dx^2) = \frac{Wxl}{(2.64l + 47.36x) \times 10^{-6}}$$

$$= \frac{(2.7 \times 1.5)\,10^6 x}{(21.64 \times 1.5 + 47.36x)}$$

or $$Ed^2y/dx^2 = 8.56 \times 10^4 - \frac{3.39 \times 10^5}{47.36x + 3.96}$$

Integrating w.r.t '*x*'

$$E\,dy/dx = 8.56 \times 10^4 x - \frac{(3.39 \times 10^5)}{47.36} \ln(47.36x + 3.96) + A$$

When $x = l = 1.5$, $dy/dx = 0$ ; then

$$0 = 8.56 \times 10^4 \times 1.5 - 7.16 \times 10^3$$

$$\ln(47.36 \times 1.5 + 3.96) + A$$

or $$A = 7.16 \times 10^3 \ln(75.0) - 12.84 \times 10^4$$

$$= 3.09 \times 10^4 - 12.84 \times 10^4 = -8.75 \times 10^4$$

Integrating again

$$E{\bullet}y = \frac{8.56 \times 10^4}{2} x^2 - \frac{7.16 \times 10^3 (47.36x + 3.96)}{47.36}$$

$$\times [\ln(47.36x + 3.96) - 1] + Ax + B$$

$$= 4.28 \times 10^4 x^2 - 1.515 \times 10^2 [47.36x + 3.96]$$

$$\times [\ln(47.36x + 3.96) - 1] - 8.75 \times 10^4 x + B$$

Now $y = 0$ at $x = 1.5$,

Therefore $$0 = 4.28 \times 10^4 \times (1.5)^2 - 1.515 \times 10^2 (47.36 \times 1.5 + 3.96)$$

$$\times [\ln(47.36 \times 1.5 + 3.96) - 1] - 8.75 \times 10^4 + B \times 1.5 + B]$$

or $$B = -9.64 \times 10^4 + 11350 [\ln(75) - 1] + 13.105 \times 10^4$$

$$= 3.465 \times 10^4 + 11350 \times 3.32$$

$$= 7.124 \times 10^4$$

Thus when x= 0, the deflection at the free end $B$ is

$$Ey_B = B$$

or $$y_B = \frac{7.124 \times 10^4}{14 \times 10^6} \text{ m} = 5.09 \text{ mm}$$

## 6.10. Fixed Beams

A fixed or built-in beam has both of its ends rigidly fixed so that the slope at the ends remains zero. Such a beam is also known as '*encastre beam*'. However it is assumed that the beam is not constrained horizontally. A possible arrangement is shown

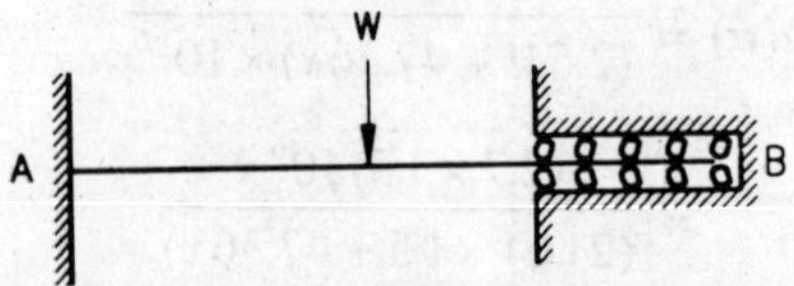


*Fig. 6.28*

in figure 6.28. One of the ends is held on rollers which allows the end $B$ to move longitudinally keeping the slope at $B$ as zero. In addition to the vertical reactions (for downward loads) the supports exert fixing couples at both the ends of the beam. These couples are anti-clockwise and clockwise at the ends $A$ and $B$ respectively and these must be included in the moment equations. The effect of these fixing moments is to reduce the deflection in general. The problems of fixed beams can be tackled by taking the beam to be simply supported with superimposed fixing moments at the ends, the two fixing moments being equal in the case of a symmetrically loaded beam.

## 6.11. Deflection of a Fixed Beam Carrying a Central Point Load

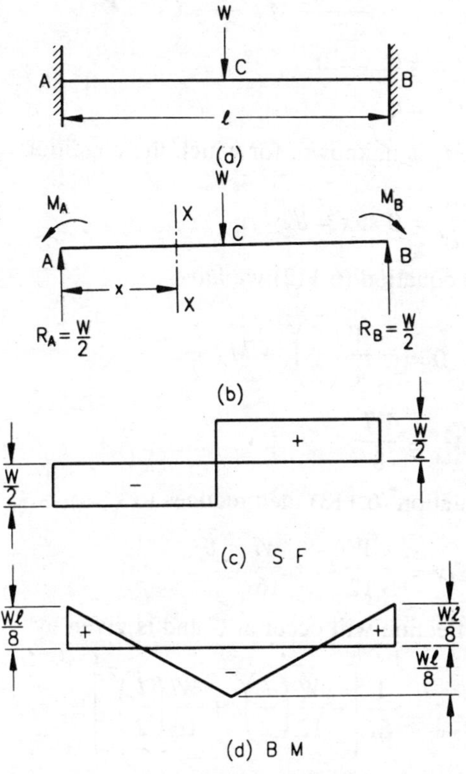


*Fig. 6.29*

This being a symmetrical case, the reactions as well as the end moments will be equal at the two ends

*i.e.* $M_A = M_B$

and $R_A = R_B = \dfrac{W}{2}$

Here Macaulay's method will be followed to evaluate the deflection etc.

Taking the origin at $A$ and for the range $0 < x < l/2$

$$EI\, d^2 y/dx^2 = M_x = -\frac{W}{2}x + M_A \qquad \text{...(6.111)}$$

Integrating equation (6.111)

$$EI\frac{dy}{dx} = -\frac{Wx^2}{4} + M_A x + C_1 \qquad \text{...(6.112)}$$

Since $dy/dx = 0$ at $x = 0$,

we get $C_1 = 0$

Integrating once again

$$EIy = -\frac{Wx^3}{12} + M_A\frac{x^2}{2} + C_2 \quad ...(6.113)$$

Also for $x = 0, y = 0$

so that $C_2 = 0$

Still the value of $M_A$ is unknown, for which the condition

$$\frac{dy}{dx} = 0 \text{ at } x = l/2$$

is applied. That is from equation (6.112) we have

$$0 = -\frac{W}{4}\left(\frac{l}{2}\right)^2 + M_A\frac{l}{2}$$

or

$$M_A = \frac{Wl}{8} \quad ...(6.114)$$

The deflection equation (6.113) then reduces to

$$EIy = -\frac{W}{12}x^3 + \frac{Wl}{16}x^2 \quad ...(6.115)$$

The maximum deflection will occur at $C$ and is given by

$$y_{max} = \frac{1}{EI}\left[-\frac{W}{12}\left(\frac{l}{2}\right)^3 + \frac{Wl}{16}\left(\frac{l}{2}\right)^2\right]$$

$$= \frac{Wl^3}{192\,EI} \quad ...(6.116)$$

Thus we see that the maximum deflection is exactly one fourth that of the same beam with ends simply supported.

The slope and deflection for the portion $CB$ may be found by symmetry. The shearing force diagram is shown in figure (6.29*c*). The bending moment diagram is drawn with the help of equation (6.111) and is shown in figure (6.29*d*).

## 6.12. Deflection of a Fixed Beam Carrying a Uniformly Distributed Load

Taking the origin at $A$, for the section XX

$$EI\,(d^2y/dx^2) = M_x = M_A - \frac{wl}{2}x + \frac{wl^2}{2} \quad ...(6.117)$$

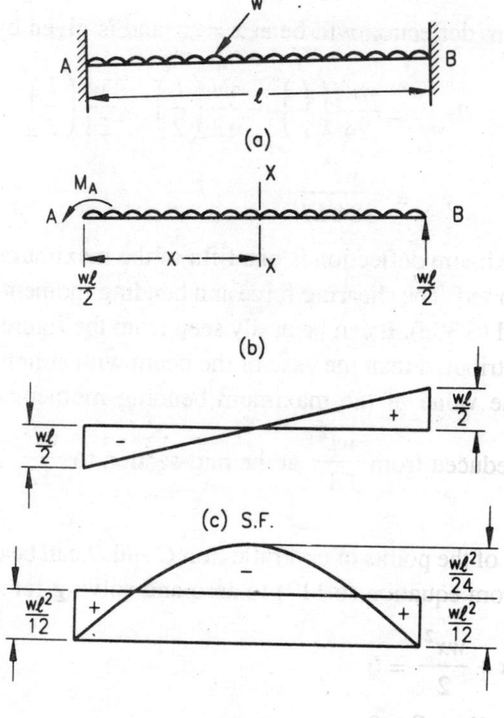

*Fig. 6.30*

Integrating (6.117)

$$EI\,(dy/dx) = M_A x - \frac{wlx^2}{4} + \frac{wx^3}{6} + C_1 \qquad ...(6.118)$$

Since $dy/dx = 0$ at $x = 0$, we get $C_1 = 0$

Integrating again

$$Ely = \frac{M_A x^2}{2} - \frac{wlx^3}{12} + \frac{wx^4}{24} + C_2 \qquad ...(6.119)$$

For $x = 0, y = 0$

so that $C_2 = 0$

For finding $M_A$ use the condition $dy/dx = 0$ at $x = l$. Therefore from equation (6.118)

$$0 = M_A l - \frac{wl^3}{4} + \frac{wl^3}{6} \qquad ...(6.120)$$

or $$M_A = \frac{wl^2}{12}$$

The deflection equation (6.119) then reduces to

$$Ely = \frac{wl^2}{24}x^2 - \frac{wlx^3}{12} + \frac{wx^4}{24} \qquad ...(6.121)$$

The maximum deflection will be at $x = \frac{l}{2}$ and is given by

$$EIy_{max} = \frac{wl^2}{24}\left(\frac{l}{2}\right)^2 - \frac{wl}{12}\left(\frac{l}{2}\right)^3 + \frac{w}{24}\left(\frac{l}{2}\right)^4$$

or $$y_{max} = \frac{wl^4}{384\,EI} \qquad ...(6.122)$$

Here the maximum deflection is one fifth of the maximum deflection if the ends were simply supported. The shearing force and bending moment diagrams are given in figures (6.30*c*) and (6.30*d*). It can be easily seen from the figure (6.30*d*) that the B.M. is more evenly distributed than the case of the beam with simply supported ends (Fig. 3.6). Moreover the value of the maximum bending moment to which the beam is subjected is also reduced from $\frac{wl^2}{8}$ at the mid-section to $\frac{wl^2}{12}$ at the ends by making the ends fixed.

The location of the points of contraflexion $C$ and $D$ can be calculated by equating the value of $M_x$ from equation (6.117) to zero and solving for $x$, **i.e.**

$$\frac{wl^2}{12} - \frac{wl}{2}x + \frac{wx^2}{2} = 0$$

or $$6x^2 - 6lx + l^2 = 0$$

or $$x = \frac{1}{12}\left[6l \pm \sqrt{36l^2 - 24l^2}\right]$$

*i.e.* $$x = 0.211l \text{ or } 0.789l \text{ i.e., } x = 0.211l \text{ or } 0.789l$$

In other words the points of contraflexion lie at a distance of $0.289l$ on either side of the mid-section.

## 6.13. Deflection of a Fixed Beam Due to a Point Load not at Mid-Section

Let $R_A$, and $R_B$ be the end reactions and $M_A$ and $M_B$ the fixing moments as in Fig. (6.31*a*). Taking the origin at $A$ and applying the Macaulay's method, the bending moment equation at any section $x$ is

$$M_x = M_A - R_A x + W\,[x - a]$$

with the condition that the terms in square brackets are to be omitted when $x < a$.

Thus the differential equation of flexure is

$$EI\frac{d^2y}{dx^2} = M_x = M_A - R_A x + W\,[x - a] \qquad ...(6.123)$$

Integrating equation (6.123)

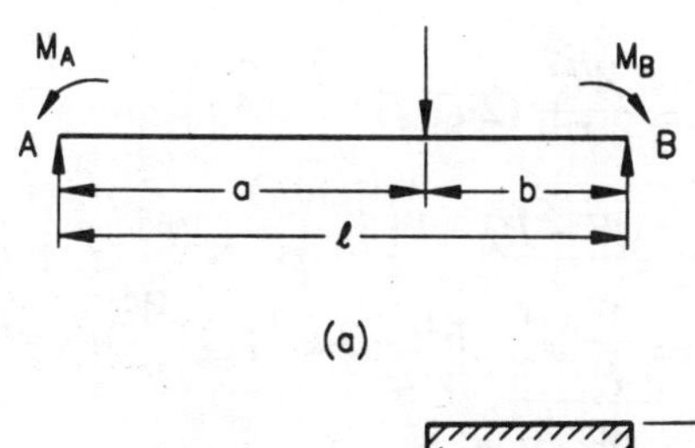


(a)

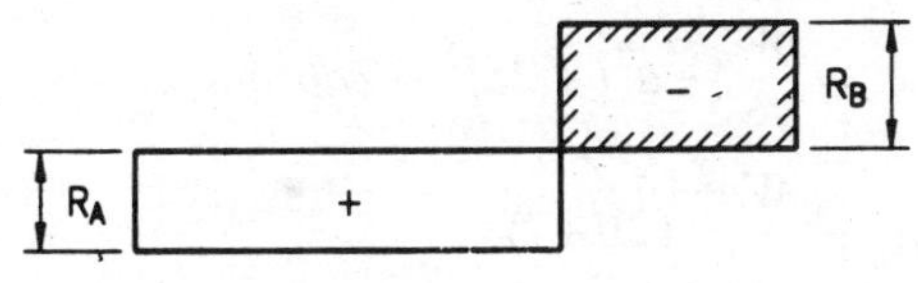


(b) S.F.

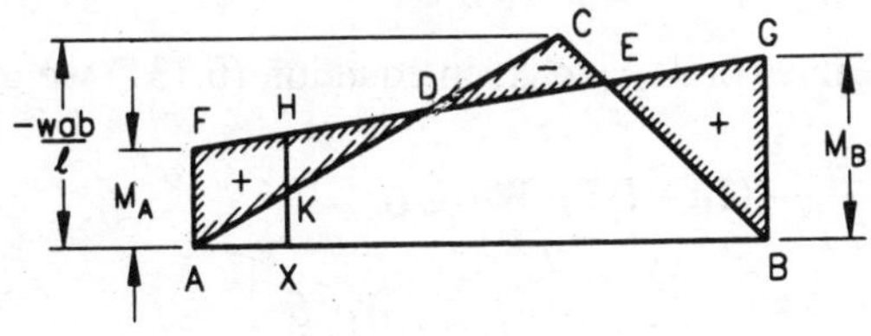


(C) B.M.

*Fig. 6.31*

$$EI\frac{dy}{dx} = M_X = M_A - R_A x + \frac{W[x-a]^2}{2} + C_1 \qquad ...(6.124)$$

Since the slope is zero at $x = 0$ and the effect of last term is to be neglected when $x < a$, the value of the constant

$$C_1 = 0$$

Integrating equation (6.124)

$$EIy = M_A\frac{x^2}{2} - R_A\frac{x^2}{2} + C_2 + \frac{W}{6}[x-a]^3 \qquad ...(6.125)$$

Again at $\quad x = 0, y = 0$

so that $\quad C_2 = 0,$

the last term being neglected.

Also at the end $B$ of the beam, the slope and deflection are both zero. Hence from equations (6.124) and (6.125) we have

$$0 = M_A l - R_A\frac{l^2}{2} + \frac{W}{2}[l-a]^2 \qquad ...(6.126)$$

and

$$0 = M_A\frac{l^2}{2} - R_A\frac{l^2}{6} + \frac{W}{6}[l-a]^3 \qquad ...(6.127)$$

Eliminating $R_A$ from the simultaneous equations (6.126) and (6.127) and keeping in mind that

$$l = (a+b),$$

we get

$$M_A = \frac{Wab^2}{l^2} \qquad ...(6.128)$$

Similarly the value of $R_A$ is given by

$$R_A = \frac{Wb^2}{l^3}(2a+l) \qquad \text{...(6.129)}$$

Then
$$R_B = W - R_A = W - \frac{Wb^2}{l^3}(2a+l)$$
$$= \frac{W}{l^3}[l^3 - b^2 l - 2ab^2] = \frac{W}{l^3}[l^3 - l(l-a)^2 - 2ab^2]$$
$$= \frac{W}{l^3}[-a^2 l + 2al^2 - 2ab^2]$$

or
$$R_B = \frac{Wa^2}{l^3}(2b+l) \qquad \text{...(6.130)}$$

For the equilibrium of the beam, taking moments about $B$.

$$M_B - M_A + R_A l - Wb = 0 \qquad \text{...(6.131)}$$

Substituting the values of $M_A$ and $R_A$ in equation (6.131) we get

$$M_B - \frac{Wab^2}{l^2} + \frac{Wb^2}{l^3}(2a+l)\,l - Wb = 0$$

Simplifying
$$M_B = \frac{Wa^2 b}{l^2} \qquad \text{...(6.132)*}$$

Maximum deflection can be found by first locating the section where the slope is zero. Evidently such a section will be between $A$ and $C$ if $a > b$.

Therefore substituting for $M_A$ and $R_A$ and putting $dy/dx) = 0$ we have from equation (6.124) omitting the terms in square brackets

$$0 = \frac{Wab^2}{l^2}x - \frac{Wb^2}{l^2}(2a+l)\frac{x^2}{2}$$

*i.e.* either
$$x = 0 \text{ or } x = 2al/(2a+l) \qquad \text{...(6.133)}$$

The latter value corresponds to the desired result, as the former one refers to minimum deflection which is zero at the end $A$.

Thus substituting the value of '$x$' from (6.133) in equation (6.125) we get

$$y_{max} = \frac{1}{EI}\left[\frac{Wab^2}{2l^2}\left(\frac{2al}{2a+l}\right)^2 - \frac{Wb^2}{6l^3}(2a+l)\left(\frac{2al}{2a+l}\right)^3\right]$$
$$= \frac{Wb^2}{6EI\,l^3}\left(\frac{2al}{2a+l}\right)^2[3al - 2al] = \frac{2Wa^3b^2}{3(2a+l)^2 EI}$$

or
$$y_{max} = \frac{2Wa^3b^2}{3(2a+l)^2 EI} \qquad \text{...(6.134)}$$

---

* **Note :** Expressions for $M_B$ and $R_B$ can easily be written straightway from equations (6.128) and (6.129) by interchanging '$a$' and '$b$'

Again from (6.133), the distance of the point of maximum deflection is

$$x = \frac{2al}{2a+l} = \frac{2l}{2+l/a} \quad ...(6.135)$$

For $a > b$, the value of $a$ will be between $\frac{l}{2}$ to $l$.

Evidently the expression (6.135) is maximum when

$$a = l.$$

In that event the maximum deflection will occur at a point $\frac{2}{3}l$ from $A$.

In other words the maximum deflection point always lies within $\left(\frac{2}{3}l - \frac{1}{2}l\right)$ $= l/6$ of the mix-section.

The deflection under the load can be found by substituting '$a$' for $x$ in equation (6.125).

$$y_C = \frac{1}{EI}\left[\frac{Wab^2}{l^2}\frac{a^2}{3} - \frac{Wb^2}{l^3}(2a+l)\frac{a^3}{6}\right] = \frac{Wa^3b^2}{6EIl^3}[3l - (2a+l)]$$

or $$y_C = \frac{Wa^3b^3}{3EIl^3} \quad ...(6.136)$$

Further knowing the values of $M_A$ and $R_A$ the bending moment at any section can be calculated with the help of equation(6.123) and B.M. diagram drawn. But a more convenient way of plotting the bending moment diagram is described below.

The B.M. diagram of the beam is drawn by assuming the ends to be simply supported (Triangle $ABC$ in figure 6.31$c$). Over this negative B.M. triangle is superimposed a positive trapezium made by joining the ends of two vertical lines $AF$ and $BG$ drawn to represent $M_A$ and $M_B$ respectively. Then the diagram $AFDCEBG$ is the resultant B.M. diagram. Bending moment at any section of the beam is given by the vertical height of the diagram *e.g.* at section $XX$. The proof is given below.

From equations (6.123), (6.128) and (6.129) the bending moment at section $XX$ is

$$M_x = \frac{Wab^2}{l^2} - \frac{Wb^2}{l^3}(2a+l)\,x \quad ...(6.137)$$

Also $$|HK| = |HX| - |KX|$$

$$= \left[M_A + (M_B - M_A)\frac{x}{l}\right] - \frac{Wab}{l}\cdot\frac{x}{a}$$

$$= \frac{Wab^2}{l^2} + \left(\frac{Wa^2b}{l^2} - \frac{Wab^2}{l^2}\right)\frac{x}{l} - \frac{Wbx}{l}$$

$$= \frac{Wab^2}{l^2} + \frac{Wbx}{l^3}(a^2 - ab - l^2)$$

$$= \frac{Wab^2}{l^2} + \frac{Wbx}{l^3}[(a-l)(a+l) - ab]$$

$$= \frac{Wab^2}{l^2} - \frac{Wb^2x}{l^2}(2a+1) \qquad ..(6.138)$$

The value as obtained in equation (6.138) is same as given in equation (6.137).

**Example 6.9.** *A beam of 6 m span is built-in horizontally at the ends and carries a uniformly distributed load of 12 kN/m on the right half of the beam together with a concentrated load of 60 kN at 2 m from the right hand end. Find the reactions and fixing moments. Draw the bending moment diagram. Also calculate the deflections at points 3 m and 5 m away from the left hand end if* $E = 2 \times 10^5$ *MN/m²* *and* $I = 0.5 \times 10^8$ *mm⁴.*

**Solution.** Take the origin at *A* and section *XX* as shown in figure (6.32 *a*). Using Macaulay's methods

$$EI\frac{d^2y}{dx^2} = M_A - R_A x - \frac{12}{2}[x-3]^2 + 60[x-4] \qquad ...(6.139)$$

Integrating equation (6.139)

$$EI\frac{dy}{dx} = M_A x - R_A\frac{x^2}{2} - 2[x-3]^3 - 30[x-4]^2 + C_1 \qquad ...(6.140)$$

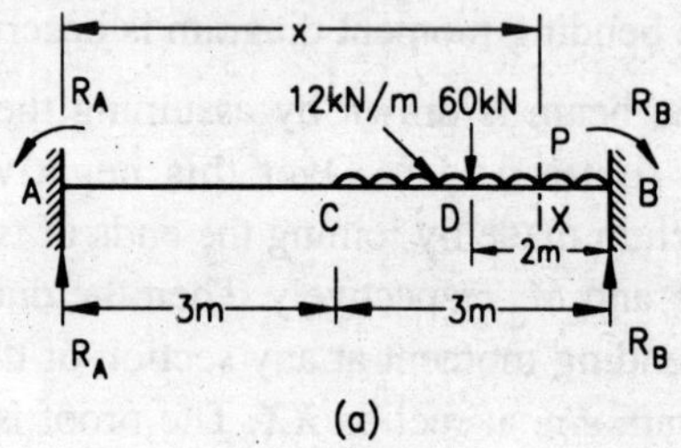

(a)

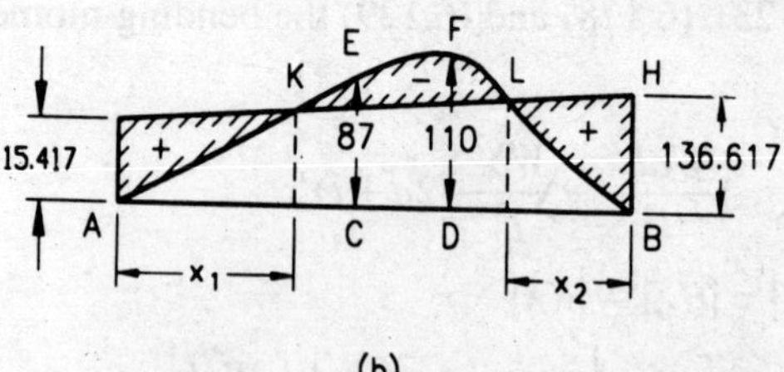

(b)

*Fig. 6.32*

When $\quad x = 0,\ dy/dx = 0$

Therefore $\quad C_1 = 0$

Integrating again

$$EIy = M_A \frac{x^2}{2} - R_A \frac{x^3}{6} - \frac{1}{2}[x-3]^4 + 10[x-4]^3 + C_2 \qquad ...(6.141)$$

Again when $x = 0, y = 0$

Therefore $C_2 = 0$

For the end $B$ we have the slope and deflection both zero. Therefore from equations (6.140) and (6.141) we have

$$M_A \times 6 - R_A \times \frac{6^2}{2} - 2 \times (3)^3 + 30(2)^2 = 0$$

or $$M_A - 3R_A = -11 \qquad ...(6.142)$$

and $$M_A \frac{(6)^2}{2} - R_A \frac{(6)^3}{6} - \frac{1}{2}(3)^4 + 10(2)^3 = 0$$

or $$36M_A - 72R_A = -79 \qquad ...(6.143)$$

Solving the simultaneous equations (6.142) and (6.143) we get

$$M_A = 15.417 \text{ kNm}$$

$$R_A = 8.8 \text{ kN}$$

But $R_A + R_B = 12 \times 3 + 60 = 96$

*i.e.* $R_B = 96 - 8.8 = 87.2$ kN

Taking moment about $B$

$$M_B - M_A + R_A \times 6 - \frac{12 \times 3^2}{2} - 60 \times 2 = 0$$

or $$M_B = 15.417 - 8.8 \times 6 + 6 \times 9 + 120$$

$$= 136.617 \text{ kNm}$$

## Bending Moment Diagram

For B.M. diagram assume the ends to be simply supported. In that case taking moment about $B$

$$6R'_A = 60 \times 2 + \frac{12 \times 3^2}{2}$$

*i.e.* $R'_A = 29$ kN and $R'_B = 60 + 36 - 29 = 67$ kN.

Then $M_C = -29 \times 3 = -87$ kNm

$$M_D = 29 \times 4 - \frac{12 \times 1^2}{2} = 110 \text{ kNm.}$$

The Bending Moment diagram for simply supported ends is shown in Fig. (6.32*b*) as *A E F B*. Superimpose the positive trapezium *A G H B* which is drawn by joining the ends $G$ and $H$ of the perpendiculars $AG$ and $BH$ such that $AG = M_A = 15.417$ and $BH = M_B = 136.617$.

The points of contraflexion $K$ and $L$ can be found by solving for $x_1$ and $x_2$ (Fig. 6.32*b*) by equating the *BM* to zero.

$$R'_A x_1 = M_A + (M_B - M_A)\frac{x_1}{6}$$

or $$29x_1 = 15.417 + (136.617 - 15.417)\ \frac{x_2}{6}$$

*i.e.* $$x_1 = 1.69 \text{ m}$$

and $$R'_B x_2 - \frac{12 \times x_2^2}{2} = M_B - (M_B - M_A)\frac{x_2}{6}$$

or $$67x_2 - 6x_2^2 + 136.617 - (136.617 - 15.417)\ \frac{x_2}{6}$$

or $$6x_2^2 - 87.2x_2 + 136.617 = 0$$

$$x_2 = \frac{1}{12}\left[87.2 \pm \sqrt{(87.2)^2 - 24 \times 136.617}\right]$$

$$= \frac{1}{12}(87.2 \pm 65.9)$$

or $$x_2 = 1.775 \text{ m or } 4.225 \text{ m from } A$$

(obviously latter value of $x_2$ is not admissible)

## Deflections

At $C$ for which $x = 3$. From equation (6.141)

$$EI\ y_C = \frac{15.417 \times (3)^2}{2} - 8.8 \times \frac{(3)^3}{6} = 23.5 \text{ kNm}^3$$

or $$y_C = \frac{23.5 \times 10^{12}}{(2 \times 10^5)(.5 \times 10^8)} = 2.35 \text{ mm}$$

At $x = 5$, from equation (6.141)

$$EIy_P = \frac{15.417 \times 5^2}{2} - \frac{8.8\,(5)^3}{6} - \frac{1}{2}(5-3)^4 + 10\,(5-4)^3$$

$$= 27.2 \text{ kNm}^3$$

Therefore $$y_P = \frac{27.2 \times 10^{12}}{(2 \times 10^5)(.5 \times 10^8)} = 2.72 \text{ mm.}$$

**Example 6.10.** *A girder of span 5 metres is built-in at each end and carries two point loads of 90 kN and 120 kN placed at 1 meter and 2 metres from the left support respectively. Find the reactions and fixing moments at the ends.*

*If E = 2 × 10⁵ MN/m² and I = 0.8 × 10⁸ mm⁴ find the deflections at the load points.*

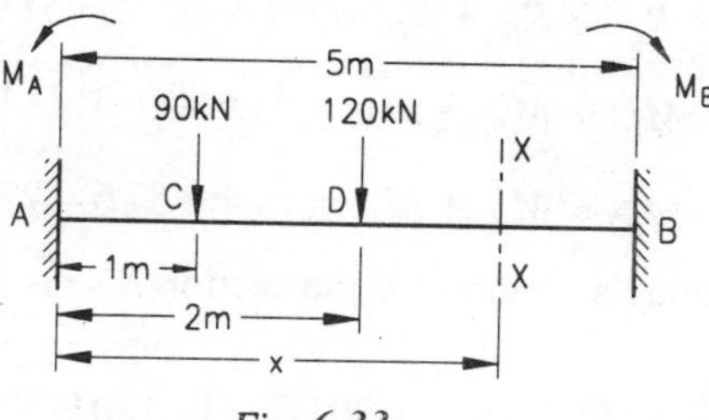


*Fig. 6.33*

**Solution.** (*i*) Reactions and end moments due to the load of 90 kN for which $a_1 = 1$ m and $b_1 = 4$ m

$$R_{A_1} = \frac{W_1 b_1^2}{l^3}(2a_1 + l)$$

$$= \frac{90 \times (4)^2}{(5)^2}(2 \times 1 + 5) = 80.5 \text{ kN}$$

$$R_{B_1} = 90 - 80.6 = 9.4 \text{ kN}$$

$$M_{A_1} = \frac{W_1 a_1 b_1^2}{l^2} = \frac{90 \times 1 \times (4)^2}{5^2} = 57.6 \text{ kNm}$$

$$M_{B_1} = \frac{W_1 a_1^2 b_1}{l^2} = \frac{90 \times (1)^2 \times (40}{5^2} = 14.4 \text{ kNm}$$

(*ii*) Reactions and end moments due to $W_2 = 120$ kN

Here $a_2 = 2$ ; $b_2 = 3$

Therefore $$R_{A_2} = \frac{W_2 B_2^2}{l^2}(2a_2 + l)$$

$$= \frac{120 \times (3)^2}{(5)^2} \times (2 \times 2 + 5)$$

$$= 77.76 \text{ kN}$$

and $$R_{B_2} = 120 - 77.76 = 42.24 \text{ kN}$$

$$M_{A_2} = \frac{W_2 a_2 b_2^2}{l^2} = \frac{120 \times 2 \times 2^2}{5^2} = 86.4 \text{ kNm}$$

$$M_{B_2} = \frac{W_2 b_2 a_2^2}{l^2} = \frac{120 \times 3 \times 2^2}{5^2} = 57.6 \text{ kNm}$$

Therefore resultant reactions and fixing moments are

$$R_A = R_{A_1} + R_{A_2} = 80.6 + 77.76 = 158.36 \text{ kN}$$

$$R_B = R_{B_1} + R_{B_2} = 9.4 + 42.24 = 51.64 \text{ kN}$$

$$M_A = M_{A_1} + M_{A_2} = 57.6 + 86.4 = 144.0 \text{ kNm}$$

$$M_B = M_{B_1} + M_{B_2} = 14.4 + 57.6 = 72.0 \text{ kNm}$$

(*b*) Using Macaulay's method for the section *XX*

$$EI\frac{d^2y}{dx^2} = M_A - R_A x + 90\,[x-1] + 120\,[x-2] \qquad \text{...(6.144)}$$

Integrating equation (6.144)

$$EI\frac{dy}{dx} = M_A x - R_A\frac{x^2}{2} + 45\,[x-1]^2 + 120\frac{[x-2]^2}{2} + C_1 \qquad \text{...(6.145)}$$

As $dy/dx = 0$ at $x = 0$ ; $C_1 = 0$

Integrating equation (6.145)

$$EIy = M_A\frac{x^2}{2} + R_A\frac{x^3}{6} + 15\,[x-1]^3 + 20[x-2]^3 + C_2 \qquad \text{...(6.146)}$$

Also $y = 0$ at $x = 0$ ; $C_2 = 0$.

Thus the deflections at *C* and *D* can be evaluated by substituting proper values of *x* in equation (6.146) along with the values of $M_A$ and $R_A$.

*i.e.*
$$EIy_C = 144 \times \frac{(1)^2}{2} - 158.36 \times \frac{(1)^3}{6}$$

$$= 45.44 \text{ kNm}^3$$

or
$$y_C = \frac{45.44 \times 10^{12}}{(2 \times 10^5) \times (.8 \times 10^8)} = 2.84 \text{ mm}$$

and
$$EIy_D = 144 \times \frac{(2)^2}{2} - 158.36 \times \frac{(2)^2}{6} + 15\,(2-1)^3$$

$$= 91.854 \text{ kNm}^3$$

or
$$y_D = \frac{91.854 \times 10^{12}}{(2 \times 10^5)\,(.8 \times 10^8)} = 5.741 \text{ mm}$$

**Example 6.11.** *A built-in beam of 6 metre span carries a uniformly distributed load of 2 kN/m run from left end to the mid-span. In addition it carries a point load of 3 kN at 4 metre from the left hand support. Find the deflection at the mid-span and under the point load if* $E = 2 \times 10^5$ *N/mm*$^2$ *and* $I = 10^7$ *mm*$^4$.

**Solution.** There are two methods to solve this problem by Macaulay's method. The easier way will be to assume the origin at *B* and proceed as in Example 6.10. The

Other method is to extend the distributed load over the whole span and to apply an equal and upward load over the portion *CB* as shown in figure (6.34 *b*) and choose the origin at *A*. This method is illustrated below.

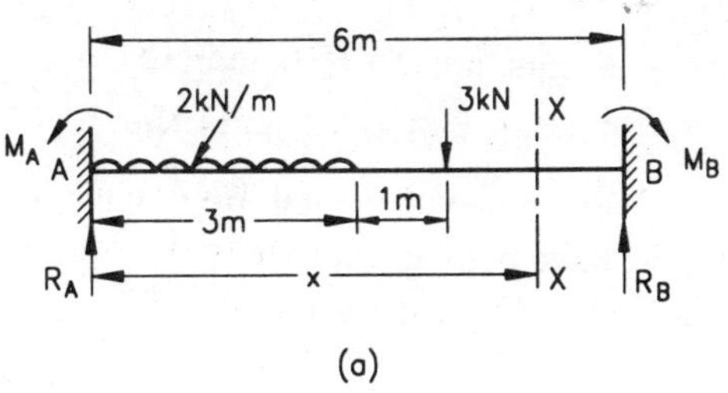


(a)

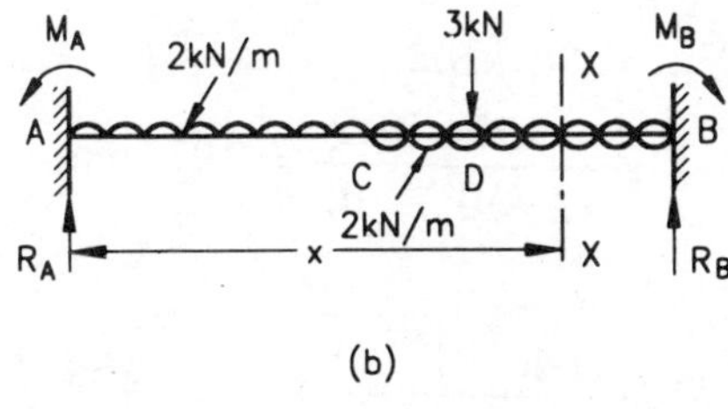


(b)

*Fig. 6.34*

Take a section *XX* at a distance x from the origin *A* . Then following Macaulay's method

$$EI\frac{d^2y}{dx^2} = M_A - R_A x + 2\frac{x^2}{2} - \frac{2}{2}[x-3]^2 + 3[x-4] \quad ...(6.147)$$

Integrating equation (6.147)

$$EI\frac{dy}{dx} = M_A x - \frac{1}{2}R_A x^2 + \frac{x^3}{3} - \frac{1}{3}[x-3]^3 + \frac{3}{2}[x-4]^2 + A \quad ...(6.148)$$

As the slope is zero at $x = 0$, the constant $A = 0$.

Integratng once again

$$EIy = M_A\frac{x^2}{2} - \frac{R_A}{6}x^3 + \frac{x^4}{12} - \frac{1}{12}[x-3]^4 + \frac{1}{2}[x-4]^3 + B$$

But $y = 0$ at $x = 0$ ; therefore $B = 0$

For evaluating $M_A$ and $R_A$ we employ the end conditions for the point *B* where $dy/dx = 0$ and $y = 0$. Thus from equations (6.148) and (6.149) we get

$$6M_A - \frac{1}{2}R_A(6)^2\,\frac{(6)^2}{3} - \frac{1}{3}[6-3]^3 + \frac{3}{2}[6-4]^2 = 0$$

or $$2M_A - 6R_A = -23 \quad ...(6.150)$$

and $$\frac{6^2}{2}M_A - \frac{1}{6}R_A(6)^3 + \frac{1}{12}(6)^4 - \frac{1}{12}(6-3)^4 + \frac{1}{2}(6-4)^3 = 0$$

or $$72M_A - 144R_A = -421 \qquad ...(6.151)$$

Solving the simultaneous equations (6.150) and (151) we get

$$R_A = 5.65 \text{ kN and } M_A = 5.46 \text{ kNm}$$

Therefore substituting the values of $R_A$ and $M_A$ in equation (6.149) along with the respective values of $x$ we have by neglecting both the square brackets when $x \leq 3$ and neglecting the last square bracket when $x \leq 4$.

$$EIy_C = 5.46 \times \frac{(3)^2}{2} - \frac{1}{6} \times 5.65 \times (3)^2 + \frac{1}{12}(3)^4$$

$$= 5.95 \text{ kNm}^3$$

or $$y_C = \frac{5.95 \times 10^{12}}{(2 \times 10^5) \times 10^7} = 2.975 \text{ mm}$$

Similarly when $x = 4$ m,

$$EIy_D = \frac{5.46 \times 4^2}{2} - \frac{1}{6} \times 5.65 \times 4^3 + \frac{1}{12}(4)^4 - \frac{1}{12}(4-3)^4$$

$$= 4713 \text{ kNm}^3$$

*i.e.* $$y_D = \frac{(4.713) \times 10^{12}}{(2 \times 10^5) \times 10^7} = 2.3565 \text{ mm}$$

## 6.14. Continuous Beams

A beam is called continuous when it is supported at more than two points. Figure (6.35 *a*) shows a continuous beam which is supported at n number of points, out of which only four have been shown, carrying a general type of load.

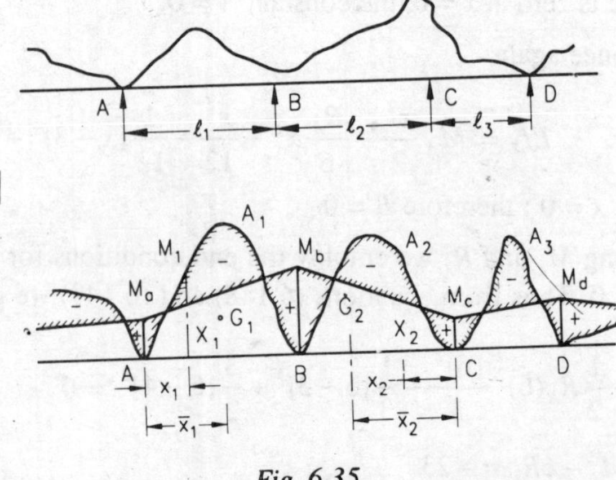


*Fig. 6.35*

## 6.14.1. Theorem of the Three Moments (Clapeyron's Theorem)

In a continuous beam when we fix our attention to a certain span, say *BC* here, the presence of the load over the adjacent spans *i.e.* *AB* and *CD* will offer fixing moments at the supports *B* and *C* respectively. The effect of these fixing moments at the points *B* and *C* of the span *BC* will be opposite to that of the load acting within this span. Similarly fixing moments will be there on each and every support point. In case extreme ends of the continuous beam ar simply supported the bending moment at these ends will be zero or else they will be equal to the fixing moments of the wall if they are fixed.

Let $A_1$, $A_2$ , $A_3$, ..... etc, be the areas of the negative bending moment diagrams, with centroids $G_1$, $G_2$, $G_3$,...... due to loads acting within the spans *AB, BC, CD,* ...... etc respectively if each of these spans of the continuous beam were individually simply supported at its ends free from the effect of the loads on the other spans and the effect of the presence of the walls if they are there at the extreme ends. Then $A_1$, $A_2$ , $A_3$, ..... spread on their respective spans, taken collectively are known as the "free bending moment "diagram of the continuous beam. If $M_a$ , $M_b$, $M_c$,...... etc. are the fixing moments at the supports *A*, *B*, *C*....... respectively, then following the method of article 6.13 the resultant bending moment diagram will be found by superimiposing the positive B.M. diagram due to fixing moments on the " free B.M. diagram " as shown in Fig. (6.35 *b*). The shaded area will then be the resultant B.M. diagram of the continuous beam. The bending moment at any section will be given by the vertical ordinate at the section.

Consider any two consecutive spans of the beam, say *AB* and *BC*. The distances of the centroids $G_1$, and $G_2$ of the "free B.M. areas" of these spans are $\bar{x}_1$ and $\bar{x}_2$ from *A* and *C* respectively as shown in figure (6.35 *b*). Let $I_1$ and $I_2$ be the uniform second moment of the areas of the cross-sections of the beam in the spans *AB* and *BC* respectively and *E* the Young's modulus of its material.

Taking origin at *A* measure distances towards right as positive for the span *AB*. Let the 'free' bending moment at any section $X_1$ at a distance $x_1$ from *A* be $M_2'$ which will be negative in sign. Then the deflection *y* at the section $X_1$ is given by the solution of the equation.

$$EI_1 \frac{d^2y}{dx_1^2} = M_a + \frac{x_1}{l_1}(M_b - M_a) + M_1' \qquad ...(6.152)$$

Multiplying equation (6.152) by $x_1$ and integrating the left hand side by parts for the limits o to $l_1$, we have

$$EI_1\left[\left(\frac{dy}{dx}\right)x_1 - y\right]_0^{l_1} = M_a \int_0^{l_1} x_1 dx_1 + \frac{1}{l_1}(M_b - M_a)\int_0^{l_1} x_1^2\, dx_1 + \int_0^{l_1} M_1'\, x_1 dx_1 \qquad ...(6.153)$$

In equation (6.153) $\frac{dy}{dx_1}$ represents the slope at any point $x_1$ within the span $AB$. Let it be represented by $i$ and its particular value at $x_1 = li$ be $i_{ba}$. That is, $i_{ba}$ is the slope of the beam at the point $B$ as seen in the direction $A$ towards $B$. Further $\int_0^{l_1} M_1' \, x_1 \, dx_1$ represents the moment of the "free $BM$ area" of the span $AB$ about $A$ and it must be equal to $A_1 \bar{x}_1$ where $\bar{x}_1$ is the distance of $G_1$ from $A$.

Thus the equation (6.153) reduces to

$$EI_1 \left[ \left\{ (i)_{x_1 = l_1} \times l_1 - y_b \right\} - \left\{ (i)_{x=0} \times 0.y_1 \right\} \right]$$

$$= M_a \frac{l_1^2}{2} + (M_b - M_1) \frac{l_1^2}{3} + A_1 \bar{x}_i$$

where $y_a$ and $y_b$ represent the respective sinkings of supports at $A$ and $B$.

Therefore,

$$EI_1 \, l_1 \, i_{ba} = M_a \frac{l_1^2}{2} + (M_b - M_a) \frac{l_1^2}{3} + A_1 \bar{x}_1 + EI_1 (y_b - y_a)$$

or

$$i_{ba} = \frac{1}{EI_1 l_1} \left[ \frac{l_1^2}{6} (M_a + 2M_b) + A_1 \bar{x}_1 \right] + \frac{1}{l_1} (y_b - y_a) \qquad \text{...(6.154)}$$

Similarly if for the span $BC$ the origin is taken at $C$ and distances are measured towards the left as positive and represented by $x_2$ as shown in figure (6.35 *b*), then the slope of the beam at $B$ as viewed from the direction $C$ towards $B$ will be given by $i_{ba}$ as

$$i_{bc} = \frac{1}{EI_2 l_2} \left[ \frac{l_2^2}{6} (M_c + 2M_b) + A_2 \bar{x}_2 \right] + \frac{1}{l_2} (y_2 - y_c) \qquad \text{...(6.155)}$$

where $I_2$ = Second moment of area of uniform section the beam in thespan $BC$

$l_2$ = span $BC$

$\bar{x}_2$ = Distance of $G_2$ from $C$

$y_c$ = Sinking of the support at $C$

Further $i_{ba}$ and $i_{bc}$ both represent the slope of the beam at $B$ but viewed from opposite directions.

Therefore $\quad i_{ba} = - i_{bc} \qquad$ ...(6.156)

Thus from equations (6.154) to (6.156) we get

$$\frac{1}{EI_1 l_1}\left[\frac{l_1^2}{6}(M_a + 2M_b) + A_1\bar{x}_1\right] + \frac{1}{l_1}(y_b - y_a)$$

$$= -\frac{1}{EI_2 l_2}\left[\frac{l_2^2}{6}(M_a + 2M_b) + A_2\bar{x}_2\right] + \frac{1}{l_2}(y_b - y_c)$$

or
$$M_a \frac{l_1}{I_1} + 2M_b\left[\frac{l_1}{I_1} + \frac{l_2}{I_2}\right] + M_c \frac{l_2}{I_2} + 6\left[\frac{A_1\bar{x}_1}{I_1 l_1} + \frac{A_2\bar{x}_2}{I_2 l_2}\right]$$

$$+ 6E\left[\frac{1}{l_1}(y_b - y_a) + \frac{1}{l_2}(y_b - y_c)\right] = 0 \qquad \text{...(6.157)}$$

Equation (6.157) is known as the "Clapeyron's theorem" or "Theorem of three moments". In case the section of the beam is uniform all along its length, then $I_1 = I_2 = \text{I}$.

The equation (6.157) reduces to

$$M_a l_1 + 2M_b(l_1 + l_2) + M_c l_2 + 6\left[\frac{A_1\bar{x}_1}{l_1} + \frac{A_2\bar{x}_2}{l_2}\right]$$

$$+ 6EI\left[\frac{1}{l_1}(y_b - y_a) + \frac{1}{l_2}(y_b - y_c)\right] \qquad \text{...(6.158)}$$

when the supports are rigid and they do not sink or else they all sink by equal amounts the equation (6.158) reduces to

$$M_a l_1 + 2M_b(l_1 + l_2) + M_c l_2 + 6\left[\frac{A_1\bar{x}_2}{l_1} + \frac{A_2\bar{x}_2}{l_2}\right] = 0 \qquad \text{...(6.159)}$$

Clapeyron's equations is used to evaluate the fixing moments at the supports, which are then superposed on the "free B.M. diagrams" of various spans to get the bending moment diagram of the entire beam. However since the Clapeyron's equation, for two consecutive spans, there are three unknown quantities viz. $M_a$, $M_b$ and $M_c$, we will have to form a set of three simultaneous equations to solve $M_a$, $M_b$ and $M_c$. The detailed procedure is given below.

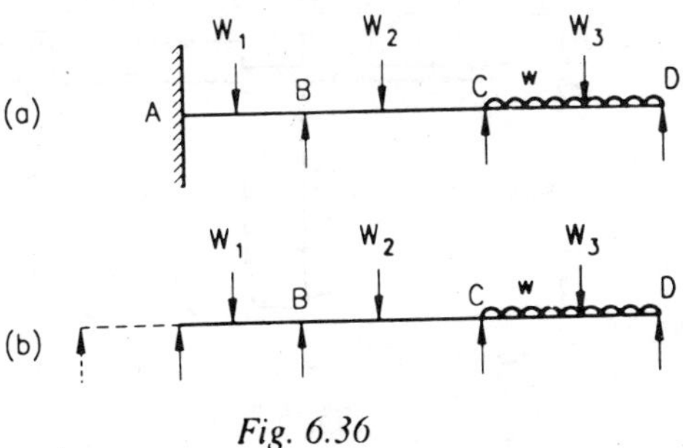


*Fig. 6.36*

Let there be a continuous beam *ABCD* as shown in Figure (6.36 *a*) for which the B.M. diagram is to be drawn. On the fixed end side the beam can be regarded as being continued beyond the fixed end *A* to *E* (Figure 6.36 *b*) where *EA* is imaginary span of zero length carrying no load. The Clapeyron's equation is then applied in succession to spans *EA*, *AB* ; *AB*, *BC* ; *BC*, *CD* to give three simultaneous equations for the three unknowns $M_a$, $M_b$ and $M_c$ to be evaluated. $M_d$ will be zero as the support at *D* is simple. In addition, equations can also be formed by equating moments at certain points to zero, for the equilibrium of the beam. This helps in evaluating the reactions at various supports.

**Example 6.12.** *A continuous beam, ABC, is fixed horizontally at C and supported (to the same level) at A and B. AB = 3m, BC = 5m. The loading on AB consists of a vertical concentrated load of 40 kN at 1 m from B, and on BC, of a vertical concentrated load of 100 kN at 1 m from B.*

*Give the dimensioned sketches of the B.M. and S.F. diagrams for the beam ABC.*

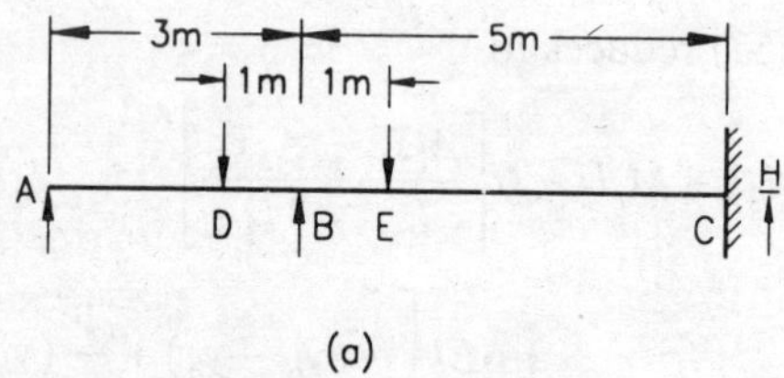


(a)

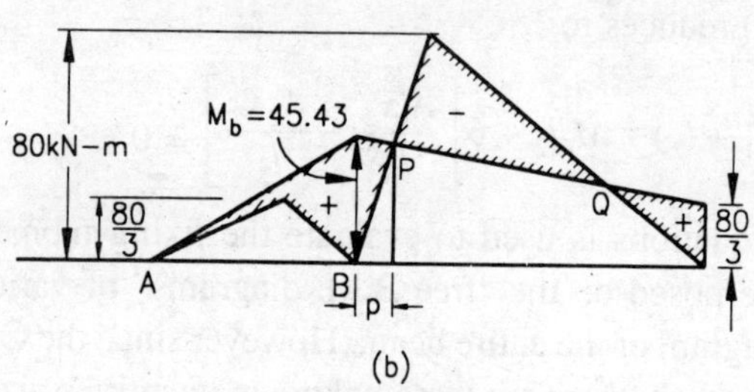


(b)

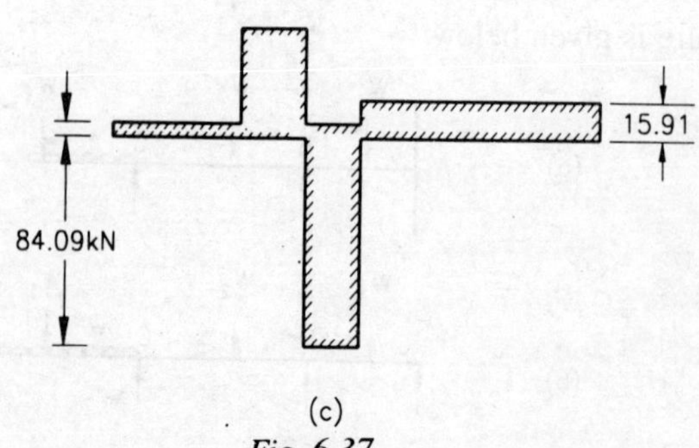


(c)

*Fig. 6.37*

**Solution**. The free bending moment diagrams for the spans $AB$ and $BC$ will be two triangles over the respective spans as bases, giving the negative bending moments at various sections. The alttiudes of these triangles will correspond to the 'free $B.M.$' at $D$ and $E$ respectively. These values of 'free $B.M.$' are

$$M'_a = \frac{40\ 2\times 1}{3} = \frac{80}{3}\ \text{kN / m}$$

$$M'_c = \frac{100\times 4\times 1}{5} = 80\ \text{mN-m}$$

Let the fixing moments at $B$ and $C$ be $M_b$ and $M_c$. There will be no fixing moment at $A$ as it is a simply supported end. Now extending the beam at the right hand end beyond the fixed support to $H$ as shown in the figure (6.37 $a$) such that the span $CH$ is of zero length of carries no load. Then applying the Clapeyron's equation (6.159) to the spans $AB$ and $EC$ ; $BC$ and $CH$ in turn, we get.

For spans $AB$ and $EC$

$$M_a\times 3+2M_b\,(3+5)+M_c\times 5+6\left[\frac{1}{3}\left\{\frac{1}{2}\times 2\times\left(-\frac{80}{3}\right)\left(\frac{2}{3}\times 3\right)+\frac{1}{2}\times 1\times\left(-\frac{80}{3}\right)\left(2+\frac{1}{3}\right)\right.\right.$$

$$\left.\left.+\frac{1}{5}\left\{\frac{1}{2}\times 1\times(-80)\left(\frac{1}{3}+4\right)+\frac{1}{2}\times 4\ (-80)\left(\frac{2}{3}\times 4\right)\right\}\right] = 0 \quad ...(6.160)$$

For spans $BC$ and $CH$

$$M_b\times 5+2M_c\,(5+0)+0+6\left[\frac{1}{5}\left\{\frac{1}{2}\times 1\times(-80)\frac{2}{3}+\frac{1}{2}\right.\right.$$

$$\left.\left.\times 4\times(-80)\times\left(1+\frac{4}{3}\right)\right\}+0\right] = 0 \quad ...(6.161)$$

Since $M_a = 0$, equations (6.160) to (6.161) respectively on simplification reduce to

$$16\,M_b + 5M_c = \frac{2560}{3} \quad ...(6.162)$$

$$5M_b + 10M_c = 480 \quad ...(6.163)$$

Solving the simultaneous equations (6.162) and (6.163) we get

$$M_b = 45.43\ \text{kN-m}$$

$$M_c = 25.3\ \text{kN}_0\text{m}$$

From figure (6.37 $b$) evidently one of the points of contraflexion $P$, will be between the points $B$ and $E$ . Let it be at a distance '$p$' from $B$ .

Then Free $BM$ at $P$ = $M_c + (M_b - M_c)\dfrac{(5-p)}{5}$

or $$\frac{80\,p}{1} = M_c + (M_b - M_C)\frac{(5-p)}{5}$$

or $$p = 0.54 \text{ m}$$

Similarly for the other point of contraflexion $Q$, let its distance from $C$ be $q$.

Then Free B.M. at Q = $M_c + (M_b - M_c)\dfrac{q}{5}$

or $$\frac{80}{4} \times q = 25.3 + (45.43 - 25.3)\ \frac{q}{5}$$

or $$q = 1.584 \text{ m}$$

The complete B.M. diagram is drawn by superposing the fixing moment diagram on the the 'free B.M.' diagram as shown in figure (6.37 *b*).

(*b*) For plotting the S.F. diagram, the reactions $R_a$, $R_b$ and $R_c$ will have to be evaluated.

Taking moments about $B$ (starting from left hand end)

$$-R_a \times 3 + 40 \times 1 = M_b$$

$$= 45.43$$

Therefore $$R_a = 1.81 \text{ kN.}$$

Similarly taking moments about $C$

$$-R_a \times 8 + 40 \times 6 - R_b \times 5 + 100 \times 4 = M_c$$

or $$14.48 + 240 - 5R_b + 400 = 25.3$$

or $$R_b = 125.9 \text{ kN}$$

Again for equilibrium of the beam

$$Rx + R_b + R_a = 40 + 100$$

or $$R_a = 140 - R_a - R_b$$

$$= 140 + 1.81 - 125.9$$

$$= 15.91 \text{ kN.}$$

The S.F. diagram can then be plotted as shown in Figure (6.37 *c*)

**Example 6.13.** *A continuous beam, 30 m long carries a uniformly distributed load of 50 kN/m all along its length. It is supported at four supports as shown in the figure 6.38. Draw the bending moment and shearing force diagrams for this beam. Determine the position of points of contraflexion.*

**Solution.** The "free bending moment diagram" is shown in figure (6.38 *a*). It consists of three parabolas, the maximum heights of these are $\dfrac{-50 \times 6^2}{8}$ (= – 225),

$\frac{-155 \times 18^2}{8}$ (=-2025), and (– 225) kN-m for the spans *AB, BC* and *CD* respectively. Since the extreme ends *A* and *D* of the beam are simply supported, the fixing moments $M_a$ and $M_d$ will both be zero. Further let $M_b$ and $M_c$ be the fixing moments at the points *B* and *C*.

Then applying the Clapeyron's equation to the spans *AB* and *BC* ; *BC* and *CD* separately we have

$$0 \times 6 + 2M_b\,(6 + 18) + M_c \times 18 + 6\left[\frac{1}{6}\left\{\frac{2}{3}(-225)\,6 \times \frac{6}{2}\right\} + \frac{1}{18}\left\{(-2025\,18 \times \frac{18}{2}\right\}\right] = 0$$

and $M_b \times 18 + 2M_c\,(18 + 6) + 0 \times 6 + 6\left[\frac{1}{18}\left\{\frac{2}{3}(-2025)\,18 \times \frac{18}{2}\right\} \frac{1}{6}\left\{\frac{2}{3}(-225)\,6 \times \frac{6}{2}\right\}\right] = 0$

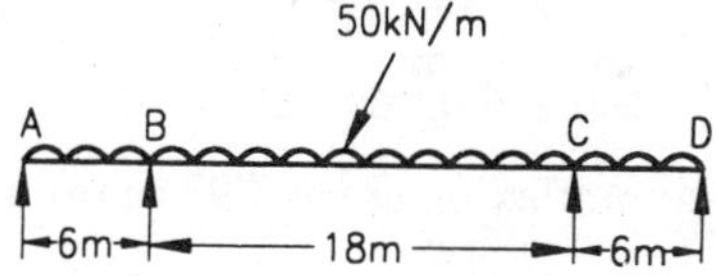


(a)

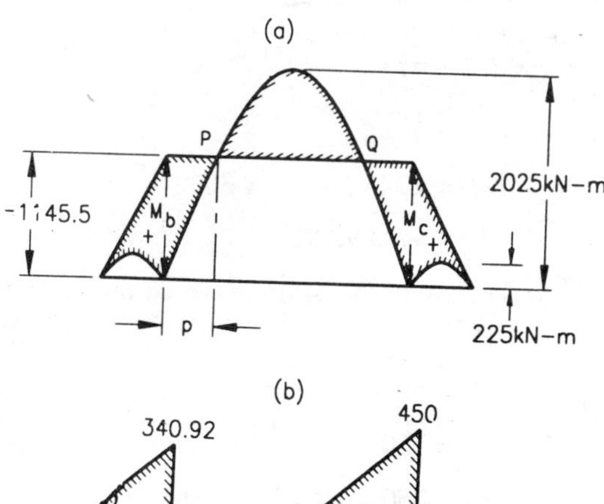


(b)

(c)

*Fig. 6.38*

These equations on simplification reduce to

$$8M_b + 3M_c = 12600 \qquad ...(6.164)$$

$$3M_b + 8M_c = 12600 \qquad ...(6.165)$$

From equations (6.164) and (6.165)

$$M_b = M_c + 1145.5 \text{ kN-m}$$

Thus superposing the fixing moment diagram on to the "free B.M." diagram we get the resultant B.M. diagram as shown in figure (6.38 *b*). The position of the points of contraflexion $P$ and $Q$ can be found by equating the resultant bending moments at $P$ and $Q$ to zero. Therefore if '$p$' is the distance of $P$ from $B$ , then

$$(\text{Free B.M. at } P) + (M_b) = 0$$

$$\frac{-50 \times 18}{2} \times p + \frac{50}{2} p^2 + 1145.5 = 0$$

or $$p^2 - 18p + 45.82 = 0$$

which gives $$p = 9 \pm \frac{1}{2}\sqrt{18^2 - 4 \times 45.82}$$

$$= 9 \pm 5.93 \text{ m}$$

Thus the points of contraflexion are at 5.93 m from the mid- section of the beam.

For shearing force diagram, the reaction at various supports will be required, for which taking moments about $B$ (starting from the left end)

$$-R_a \times 6 + \frac{50}{2}(6)^2 = M_b$$

$$= 1145.5$$

or $$-6R_a = 1145.5 - 25 \times 36$$

*i.e.* $$R_a = -40.92 \text{ kN}$$

Also for the equilibrium of the beam in which due to symmetry $R_e = R_b$ and $R_a = R_d$, we have

$$2R_a + 2R_b = 50 \times 30$$

or $$R_a + R_b = 750$$

or $$R_b = 750 - R_a = 750 + 40.92$$

$$= 790.92 \text{ kN.}$$

The shearing force diagram can then be drawn as shown in figure (6.38 *c*). Evidently maximum shearing force occurs just to the right of $B$ and left of $C$ and its value is 450 kN.

**Example 6.14.** *Draw the bending moment diagram of a uniform section ($I = 8.0 \times 10^{-6}$ m⁴) continuous beam ABC as shown in figure 6.39. The prop at B sinks by 2 mm below the common level of the supports at A and C. Take E = 200GN/m².*

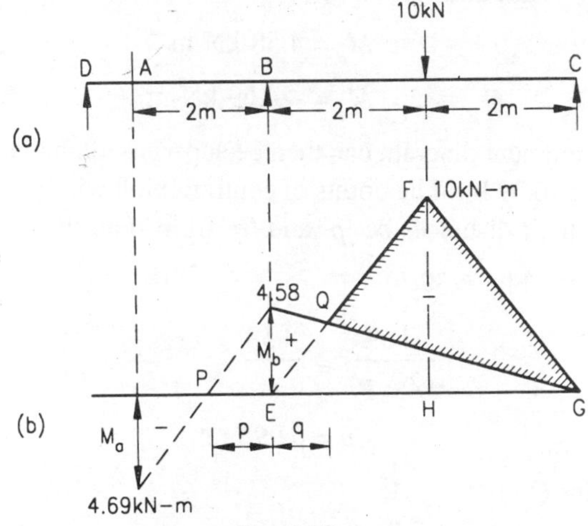


*Fig. 6.39*

**Solution.** As *C*, the right hand end of the beam, is simply supported $M_c = 0$. Further the end *A* of the beam is rigidly supported, so extend if left wards to the point *D* such that *DA* = 0 and the portion *DA* carries no load.

The 'free B.M. diagram' of the beam for the portion *BC* will be a triangle *EFG* of base equal to *BC* and height *FH* equal to $\dfrac{-10 \times 4}{4} = -10$ kN-m. For the portion *D* to *B* of the beam the free bending moment will be zero.

Here the prop at *B* sinks 2 mm below the common level of *A* and *C*. Therefore

$$(y_b - y_a) = (y_b - y_c) = 0.002 \text{ m}$$

Also $\quad (y_a - y_d) = 0 - 0 = 0$

Thus using Clapeyron's equation (6.158) for the spans *DA* and *AB* ; *AB* and *BC* in turn, we get,

For the spans *DA* and *AB*

$$0 + 2M_a\,(0 + 2) + M_b \times 2 + 6\,[0 + 0] + 6\left[\frac{200 \times 10^9}{10^3}\right] \times \left[0 + \frac{.002}{2}\right] = 0$$

For the spans *AB* and *BC*

$$M_a \times 2 + 2M_b\,(2 + 4) + 0 \times 4 + 6\left[0 + \frac{1}{4}\left\{(2) - (10)\left(\frac{2}{3} + 2\right) + \frac{1}{2} \times 2\,(-10\left(\frac{2}{3} \times 2\right)\right\}\right]$$

These equations on simplification reduce to

$$2M_a + M_b + 4.8 = 0 \quad ...(6.166)$$

$$2M_a + 12M_b + 45.6 = 0 \quad ...(6.167)$$

Equations (6.166) and (6.167) give

$$M_b = 4.58 \text{ kN-m}$$

$$M_a = -4.69 \text{ kN-m}$$

The fixing moment diagram can then be superposed on the "free B.M. diagram" as shown in figure (6.39 b ). The points of contraflexion will be given by the locations of $P$ and $Q$ . Let their distances be '$p$' and '$q$' from $B$ as shown in figure (6.39 $b$).

Then from the Fig. 6.39($b$)

$$\frac{4.69}{(2-p)} = \frac{4.58}{p}$$

or $$p = 0.998 \text{ m}$$

For the point $Q$

(Free B.M. at $Q$) + (Fixing B.M. at $Q$) = 0

or $$\left(\frac{FH}{2}\right)q + \frac{M_b}{4}(4-) = 0$$

or $$\frac{-10}{2}q + \frac{4.58}{4}(4-q) = 0$$

or $$q = 0.745 \text{ m.}$$

## PROBLEMS FOR PRACTICE

1. A cantilever of effective length L with a concentrated load $W$ at the free end is propped at a distance '$a$' from the fixed end to the same level as the fixed end. Find the load on the prop.

   Show that there is always a real points of inflexion and find its distance from the fixed end. $\left[\textbf{Ans. } \frac{W}{2a}(3L-a); \frac{a}{3}\right]$

2. A cantilever of uniform section has a length L and carries a uniformly distributed load of w per unit length on the whole length. Derive, from first principles, expressions for the slope and deflection at the free end of the cantilever. If the load is removed for the length L/4 measured from the free end, what will then be the slope and deflection at the free end ?

   $$\left[\textbf{Ans. } \frac{wL^3}{6\,EI}; \frac{wL^4}{8\,EI}; \frac{9wL^3}{128\,EI}; \frac{117}{2048}\frac{wL^4}{EI}\right]$$

3. A simply supported beam of length $L$ carries a load $W$ at distance a from one end, b from the other ($a > b$). Find the position and magnitude of the maximum deflection and show that the position is always within $L/13$, approximately, of the centre.

4. A beam $AB$ of length $l$ is rigidly fixed at the end $A$ and simply supported at the end $B$. It is acted upon by a bending couple $M$ in the vertical plane applied about an axis $l/4$ from the end $B$. Find the fixing moment at $A$ and the reaction at $B$. Draw the bending moment diagram.

[Hint : Treat it as a propped cantilever.] $\left[\textbf{Ans.}\ \frac{13}{32}M\,;\ \frac{45}{32}L\right]$

5. A uniform section beam $AB$ ($I = 2 \times 10^9\ \text{mm}^4$) of length 12 m is simply supported at its ends. It carries a load which reduces from 30 kN/m run at A to 20 kN/m run at $B$. Find the position and magnitude of maximum deflection. Take $E = 200$ GN/m$^2$. [**Ans.** 5.97 m from $A$ ; 16 mm]

6. A wooden beam of rectangular section 240 mm wide by 80 mm deep is 4 m long. It is simply supported a its ends and carries two concentrated loads of 1 kN each at 1.2 m m from each end. Neglecting the weight of the beam, find the deflection under the loads and at the mid section of the beam. $E = 14000$ kN/m$^2$.

[**Ans.** 12.65 mm ; 14.75 mm]

7. A uniform section beam of length $L$ is simply supported at its ends and carries a single concentrated load $W$ at a distance $L/3$ from one end. Working from fundamental beam theory, derive formulae for the deflection ($a$) under the load, ($b$) at the centre, ($c$) at the point of maximum deflection.

$$\left[\textbf{Ans.}\ \frac{0.1646\ WL^3}{EI}\,;\ 0.01775\frac{WL^3}{EI}\,;\ 0.01794\frac{WL^3}{EI}\right]$$

8. A beam of uniform section, $I = 4 \times 10^6\ \text{mm}^4$, is 1.8 m long and is supported horizontally at section 0.3 m from one end and 0.6 m from the other end. It carries a uniformly distributed load of 1.5 kN/m run. Calculate the deflection at the two ends relative to the supports. $E = 200000$ MN/m$^2$.

[**Ans.** 0.007595 mm ; 0.06455 mm)

9. A uniform beam of length $2L$ and simply supported at ends, is propped at its middle point. It carries two equal point loads, $W$, symmetrically placed at a distance $x$ on each side of the central support. Find the value of $x$ in terms of $L$, so that the reactions on the three supports are equal. Neglect the weight of the beam itself and assume that all three supports are at the same level.

[**Ans.** 0.774 $L$]

10. A beam of uniform flexural rigidity and length l is encastre at both ends. At a distance $a$ ($a < l/2$) form one end it carries a point load $W$ and an applied

moment $Wl$ which tends further to depress the centre of the beam. Show that when $a = l/\left(1+\sqrt{2}\right)$ one of the end fixing moments is zero.

11. A horizontal beam, built-in at each end, has a clear span of 4.5 m, and carries loads of 50 kN at 1.5 m and 70 kN at 2.5 m from its left-hand end. Calculate the fixing moments and the position and amount of the maximum bending moment.

12. A horizontal I-beam, rigidly built-in at the ends, and 7.5 m long, carries a total uniformly distributed load of 100 kN as well as a central concentrated load of 40 kN. If the bending stress is limited to 70 MN/m$^2$ and the deflection must not exceed 2.5 mm, find the depth of the section required. $E$ = 200 GN/m$^2$.

[**Ans.** 594 mm]

13. A 250 mm × 112.5 mm steel beam, $I = 47.6 \times 10^6$ mm$^4$, is used as a horizontal beam with fixed ends and a clear span of 3 m. Calculate from first principles the load which can be applied at one-third span if the bending stress is limited to 120 MN/m$^2$.

[**Ans.** 103 kN]

14. A built-in beam of 5 m span carries a uniformly distributed load of 30 kN/m extending from one support to the centre of the span. If the moment of inertia of the section is $200 \times 10^{-6}$ m$^4$, calculate (*a*) the end fixing moments, (*b*) the end reaction and (*c*) the position and magnitude of the maximum deflection. $E$ = 200 GN/m$^2$.

[**Ans.** 19.53 kNm ; 43 kNm ; 14.0 kN ; 61.0 kN ; 3.15 mm at 2.8 m from unloaded end]

15. A continuous beam *ABCD* is simply supported at four supports *A*, *B*, *C* and *D* which are at the same level and *AB* = 10 m, *BC* = 12 m, and *CD* = 8 m. It carries three point loads of 5 kN, 7 kN and 6 kN at distances of 7 m, 14 m and 24 m respectively from *A*. Find the bending moments at the inner supports, the reactions at the four supports and the position of the points of contraflexion. Draw the B.M. and S.F. diagrams.

[**Ans.** $M_b$ = 10.51 kN-m ; $m_c$ = 7.46 kN-m ; $R_a$ = 0.449 kN ; $R_b$ = 9.471 kN ; $R_c$ = 7.51 kN ; $R_d$ = 0.567 kN ; Points of contraflexion are at 7.9 m, 12.14 m, 18.5 m and 23.4 m from A]

16. Solve the Problem No. 9 by Clapeyron's method.

17. A continuous beam *ABCD* is simply supported at four supports *A*, *B*, *C* and *D* which are at the same level and *AB* = *BC* = *CD* = l. If the beam carries a uniformly distributed load w per unit length, find the bending moments and reactions at the supports.

[**Ans.** $M_a = M_d = 0$ ; $M_b = M_c = 0.1\ wl^2$ ; $R_a = R_d = 0.4\ wl$ ; $R_b = R_c = 1.1\ wl$]

18. Solve Problem No. 17, if the end *A* of the beam is firmly built-in.

$$\left[\textbf{Ans. } M_a = \frac{9}{104}wl^2;\ M_b\ \frac{1}{13}wl^2\ ;\ M_c\ \frac{11}{104}wl^2\ ;\ M_d = 0\right.$$

$$\left. R_a = \frac{53}{104}wl\ ;\ R_b = \frac{35}{26}wl\ ;\ R_c = \frac{59}{62}wl\ ;\ R_d = \frac{41}{104}wl\right]$$

19. Solve Problem No. 17, if both the ends $A$ and $D$ of the beam are fixed horizontally.

$$[\textbf{Ans. } M_a = M_b = M_c = M_d = \frac{wl^2}{12};\ R_a = R_d = \frac{wl}{2},\ R_b = R_c = w_l]$$

20. A beam of uniform section is 10 m long and is fixed horizontally at both ends. It is supported at the centre by a prop, level with both ends. It carries a uniformly distributed load of 60 kN/m extending from one fixing to the central prop. Calculate (*a*) the reaction on the prop, (*b*) the bending moment at the two ends, (*c*) the moment over the prop.

[**Ans.** 150 kN ; – 31.25 kN-m ; 156.25 kN-m ; 62.5 kN-m]

21. Solve the problem No. 20 if the prop sinks by 1 mm. Take I = $80 \times 10^{-6}$ m$^4$ and $E$ = 200 GN/m$^2$.

22. A uniform continuous beam *ABCD* is built-in at A and simply supported at *B*, *C* and *D*, 4, 10 and 13 m respectively from *A*. All the supports are at the same level. It carriesa point load of 40 kN, 3m from *A* and *a* uniformly distributedload of 5 kN/m on *BC*. Sketch the B.M. diagram, inserting principal numericl values.

[**And.** $M_a$ = 8.9 kN-m = 19.7 kN-m ; $M_c$ = 8.43 kN-m ; $M_d$ = 0]

23. A continuous beam ABCd of uniform section is simply supported at *A*, *B*, *C* and *D* (*AB* = *BC* = *CD* = 4 m). It carries two uniformly distrited loads one of 4000 kg/m spread over *AC* and the other of 2000 kg/m spread over the entire length *AD*. If the support *B* sinks by 2.5 mm below the level of the other supports find the moments at the supports. Take $I = 150 \times 10^{-6}$ m$^4$ and $E = 2 \times 106$ kg/cm$^2$.

Chapter–7

# Torsion of Circular Shafts

## 7.1. Torsion of Circular Shafts

When a circular shaft is acted upon by an axial torque and is not carrying any bending moment or axial thrust, its each and every cross-section is known to be under pure shear. The shaft can be assumed to be composed of a large number of thin discs, face of each disc having a tendency to slip with respect to that of the adjacent disc. Total twist of the free end of the shaft is then equal to the sum total of the individual twists of all these elementary discs. The shearing stress induced in the material over any cross-sectional plane produces the resisting moment equal and opposite to the applied torque. The complementary shear stress on the longitudinal planes will cause a distortion of the elements which were originally lying along the longitudinal direction.

In evaluating the stress due to twisting following assumptions are made :

1. The material is homogeneous and obeys Hooke's Law.
2. It is stressed within the limit of proportionality.
3. Plane cross-sections remain plane even after the application of torque.
4. All the radii in the cross-sectional plane remain straight and radial even after the application of torque.

Fig. 7.1 shows a hollow cylindrical shafts of inner and outer radii $r$ and $R$ respectively, fixed at one end and acted upon by a torque $T$ on the other end, distant $l$ from the fixed end. Figure 7.2 shows the free body diagram of an annular cylinder of radius $x$ and thickness $dx$. $AB$ is a line drawn on the surface of this cylinder in the axial direction. After the application of the twisting moment the line $AB$ becomes a helix with point B shifting to position C corresponding to the twist θ about $ZZ$ of

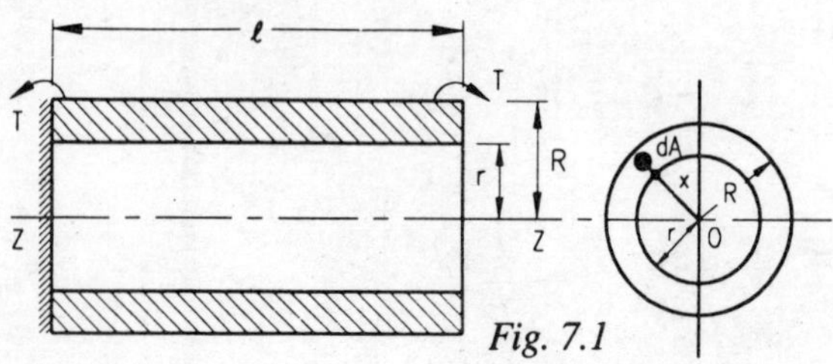

*Fig. 7.1*

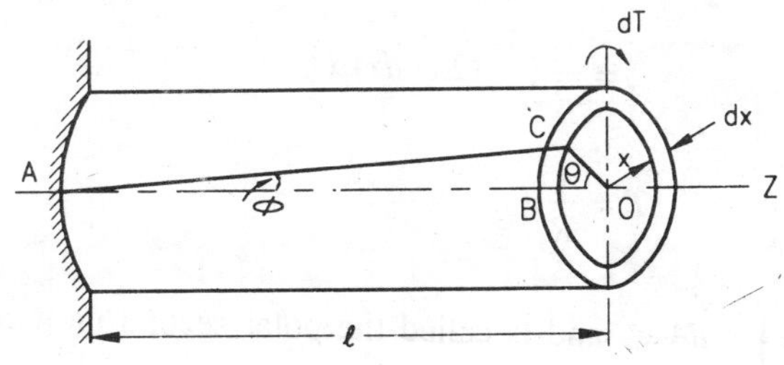


*Fig. 7.2*

the section at B with respect to the fixed section at *A*. The inclination of the helix *AC* with the *ZZ* axis will give the shear strain (refer to article 1.5). It may be seen that the shear strain, and hence the shear stress, on any of the cross-sectional planes of the annular cylinder along the length *AB*, will be same.

If the shear stress is within the elastic limit, the shear strain angle $\phi$ and the angle of twist $\theta$ due to torque *T*, will be small, then

$$BC = l\phi = x\theta$$

*i.e.*

$$\phi = \frac{x\theta}{l} \qquad ...(7.1)$$

Thus if *G* is the modulus of rigidity of the shaft material, then the shear stress $\tau_x$ on the sectional plane of the annular cylinder with be given by

$$\tau_x = \phi G$$

$$= Gx\,\frac{\theta}{l}$$

or

$$\frac{\tau_x}{x} = \frac{G\theta}{l} \qquad ...(7.2)$$

For any given shaft and given torque *G*, $\tau$ and *l* will be fixed ; *i.e.* the right hand side of equation (7.2) is constant. In other words the ratio $(\tau_x/x)$ is constant. That is the shear stress goes on increasing linearly with radius '*x*' and at the axis where $x = 0$, it is zero. Therefore if $\tau$ is the shear stress on the section at outer radius, *R*, it will be maximum and we will have

$$\frac{\tau_x}{x} = \frac{\tau}{R} = \frac{G\theta}{l} = \text{constant} \qquad ...(7.3)$$

Again if *dT* is the torque transmitted by the annular cylinder, then the total resisting moment of the cylinder, which is equal to *T* is given, by referring to figure 7.1, as

$$T = \int_{x=r}^{R} dT = \int_{x=r}^{R} (\tau_x \,.\, dA)\, x$$

$$= \int_{Y}^{R} \frac{\tau}{R} x^2\, dA$$

$$= \frac{\tau}{R}\int_r^R (2\pi x\, dx)\, x^2$$

or $$T = \frac{\tau}{R} J \qquad ...(7.4)$$

Where $J = \int_r^R dA\, x^2$ and is called the polar second moment of area of cross-section about ZZ (Fig. 7.1). and is given by

$$J = \int_r^R (2\pi x\, dx)\, x^2$$

$$= \frac{\pi}{2}(R^4 - r^4) = \frac{\pi}{32}(D^4 - d^4)$$

where $D = 2R$ and $d = 2r$.

Thus from equation (7.3) and (7.4), we have

$$\frac{T}{J} = \frac{\tau}{R} = \frac{\tau_x}{x} = \frac{G\theta}{l} \qquad ...(7.5)$$

For a solid shaft

$$J = \int_0^R (2\pi\, x dx)\, x^2 = \frac{\pi}{2}R^4 = \frac{\pi D^4}{32}$$

Therefore $$\tau = \frac{T}{J}\cdot\frac{D}{2} = \frac{16T}{\pi D^3} \qquad ...(7.6)$$

or $$T = \frac{\pi}{16}\tau D^3 \qquad ...(7.7)$$

## 7.2. Composite Shafts

A shaft which is composed of two or more materials rigidly fixed together so that they share the torque being transmitted, is called composite shaft (Fig. 7.3).

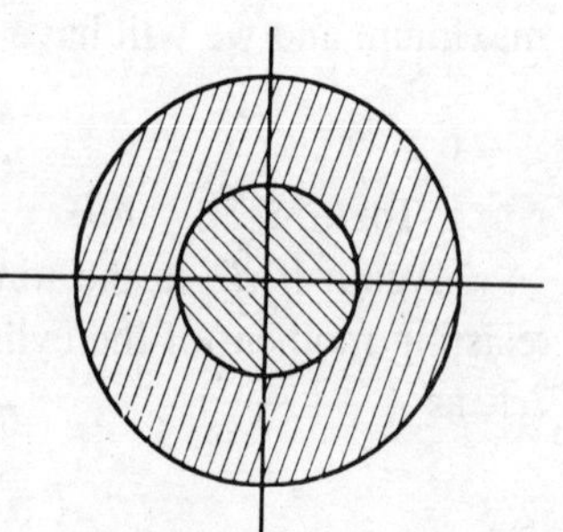

Fig. 7.3

Let $T_1$ and $T_2$ be the torques shared by the two portions respectively, then

$$T_1 + T_2 = T \qquad ...(7.8)$$

Also as the twist per unit length will be same for the two portions, we will have from equation (7.5)

$$\frac{\theta_1}{l} = \frac{\theta_2}{l_2}$$

That is
$$\frac{T_1}{J_1G_1} = \frac{T_2}{J_2G_2}$$

Or
$$\frac{T_1}{T_2} = \frac{J_1G_1}{J_2G_2} \quad ...(7.9)$$

Knowing total torque $T$, equations (7.8) and (7.9) can be solved to evaluate $T_1$ and $T_2$. Using equation (7.5) the shear stress can then be calculated at any point in the cross-section.

**Example 7.1.** *A horizontal bar of 60 mm diameter is rigidly fixed at each end, the fixings being 1200 mm apart. A rigid bracket is fixed at 800 mm from one end of the bar, at right angles to its axis and in the same horizontal plane.*

*Determine the maximum radius-arm of the bracket at which a vertical load of 1000 N can be suspended if the deflection of the load is not to exceed 0.1 mm.*

*Take E = 200 GN/m² and G = 80 GN/m².*

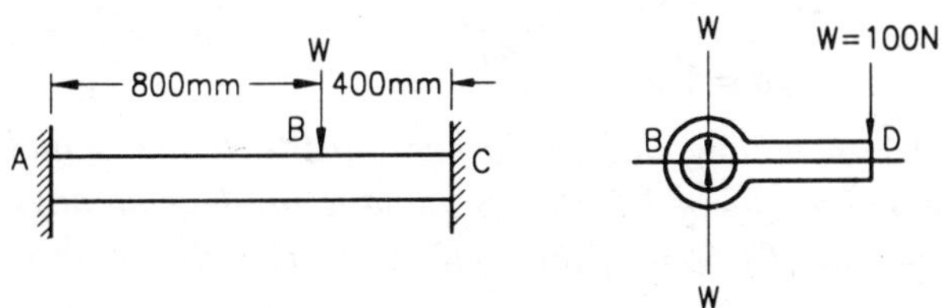


*Fig. 7.4*

**Solution.** Let a be the radius arm. The load at $D$ will give rise to a load $W$ in the vertical plane of the bar and a twisting moment equal to $(W \times a)$. The downward movement of $D$ will be equal to the downward deflection of the beam at $B$ due to load $W$ at $B$ plus the movement of $D$ due to rotation of the section $B$. $(\theta_{AB} = \theta_{BC})$

Now $\quad W = 1$ kN,

the deflection at $B$ from equation 6.136 is

$$y_B = \frac{1000 \times (800)^3 \times (400)^3}{3 \times (200 \times 10^3)\left[\frac{\pi}{64} \times 60^4\right] \times (1200)^3} \text{ mm } = 0.0496 \text{ mm}$$

**Torques $T_{AB}$ and $T_{BC}$ sustained by the portions $AB$ and $BC$**

$$T_{AB} + T_{BC} = W \times a = 1 \times a \quad ...(1)$$

$$\theta_{AB} = \theta_{BC}$$

$$\frac{T_{AB}\, l_{AB}}{JG} = \frac{T_{BC}\, l_{BC}}{JG}$$

or
$$T_{AB} \times 800 = T_{BC} \times 400$$

*i.e.*
$$T_{BC} = 2\, T_{AB} \quad ...(2)$$

From equations (1) and (2)

$$T_{AB} = \frac{a}{3}$$

and $$T_{BC} = \frac{2a}{3} \quad \text{...(3)}$$

Therefore

$$\theta_{AB} = \frac{T_{AB}\, l_{AB}}{JG} = \frac{a}{3} \times 800 \times \frac{32}{\pi \times (60)^4} \times \frac{1}{(80 \times 10^3)}$$

$$= 2.62 \times 10^{-6} \text{ a radians}$$

Thus the downward movement of end $D$, due to twisting

$$= \theta_{AB} \times a = 2.62 \times 10^{-6}\, a^2 \text{ mm}$$

Therefore

$2.62 \times 10^{-6}\, a^2 + .0496 = 0.1$

or $$a^2 = \frac{0.0504}{2.62 \times 10^{-6}}$$

or $$a = 138.5 \text{ mm}$$

**Example 7.2.** *A hollow shaft is 50 mm outside diameter and 30 mm internal diameter. An applied torque of 1.6 kNm is found to produce an angular twist of 0.4°, measured on a length of 0.2 m of the shaft. Calculate the value of the modulus of rigidity.*

*Calculate also the maximum power which could be transmitted by the shaft at 2000 r.p.m. if the maximum allowable shearing stress is 65 MN/m².*

**Solution.**

(*a*) $$\frac{T}{J} = \frac{G\theta}{l}$$

or $$G = \frac{Tl}{J\theta} = 1.6 \times .2 \times \frac{180}{0.4 \times \pi} \times \frac{32}{\pi\,(.05^4 - .03^4)}$$

$$= 86.4 \times 10^6 \text{ kN/m}^2$$

(*b*) $$\frac{T}{J} = \frac{\tau}{R}$$

or $$T = \frac{J\tau}{R}$$

$$= \frac{\pi}{32}(.05^4 - .03^4) \times \frac{65 \times 10^6 \times 2}{.05}$$

$$= 1.39 \times 10^3 \text{ Nm}$$

Therefore

$$\text{Power} = 1.39 \times 10^3 \times \frac{2\pi \times 2000}{60} W$$

$$= 292 \text{ kW}$$

**Example 7.3.** *The diameter ratio of a hollow shaft is 3 : 5. It is required to transmit 600 kW at 100 r.p.m., the maximum torque being 12% greater than the mean. If the twist in a length of 3 m is not to exceed 1° and the shearing stress is limited to 60 MN/m², calculate the minimum external diameter of the shaft satisfying these conditions. Take G = 80 × 10³ MN/m².*

**Solution.** $\frac{D}{d} = \frac{5}{3}$

or $d = 0.6\, D$

Now $J = \frac{\pi}{32}(D^4 - d^4)$

$$= \frac{\pi}{32} D^4 (1^4 - 0.6^4)$$

$$= 0.0855\, D^4$$

Mean torque $= \frac{\text{Power} \times 60}{2\pi N} = \frac{60 \times 1000 \times 60}{2\pi \times 100}$

$$= 5.72 \times 10^4 \text{ Nm}$$

Therefore $T = 1.12 \times 5.72 \times 10^4 = 6.41 \times 10^4$ Nm

**Case 1.** If the torsional rigidity were the limiting case, then

$$T = JG\,\theta/l$$

or $6.41 \times 10^4 = 0.855\, D^4 \times (80 \times 10^9) \times \frac{\pi}{180} \times \frac{1}{3}$

or $D^4 = 16.11 \times 10^{-4}$

*i.e.* $D = 0.2001$ m

For this value of $D$, the maximum shear stress induced will be

$$\tau_{max} = \frac{D}{2} \cdot \frac{T}{J} = \left(\frac{0.2001}{2}\right) \cdot \frac{0.41 \times 10^4}{0.0855 \times (0.2001^4)}$$

$$= 46.7 \times 10^6 \text{ N/m}^2$$

which is less than the allowable stress of 60 MN/m².

**Case 2.** If the shear stress is the limiting criterion, then

$$\frac{T}{J} = \frac{\tau}{R}$$

or $$\frac{6.41 \times 10^4}{(0.0855)\, D^4} = \frac{60 \times 10^6}{D} \times 2$$

or $$D^3 = \frac{6.41 \times 10^4}{0.0855 \times 120 \times 10^6} = 6.25 \times 10^{-3}$$

or $$D = 0.184 \text{ m}$$

For this value of $D$ the twist, over 3 m length is

$$= \frac{6.41 \times 10^4 \times 3}{0.0855 \times (.184)^4 \times 80 \times 10^9} \text{ radians}$$

$$= 0.467 \text{ degrees.}$$

This also is less than the allowable limit of 1°. Thus both the criterion are safe. Hence the least external diameter is 0.2001 m which will satisfy both the restrictions.

**Example 7.4.** *A hollow shaft of diameter ratio 0.6 is required to transmit 1500 kW at a speed of 2500 rpm. It is to be fitted with a flanged coupling having 8 bolts on a circle of diameter twice that of the shaft. If the allowable shear stresses in shaft and bolts are 70 MN/m² and 55 MN/m² respectively, find the shaft and bolt diameters.*

Solution.

**For shaft** $$d = 0.6\, D$$

$$J = \frac{\pi}{32}(D^4 - d^4) = 0.0855\, D^4$$

$$\text{Torque} = T = \frac{\text{Power} \times 60}{2\pi \text{N}} = \frac{(1500 \times 1000) \times 60}{2\pi \times 2500}$$

$$= 5740 \text{ Nm}$$

$$\frac{T}{J} = \frac{2\tau}{D}$$

or $$\frac{5740}{0.0855\, D^4} = 2 \times \frac{2 \times 70 \times 10^6}{D}$$

or $$D = 0.0782 \text{ m } i.e. \text{ } 78.2 \text{ mm}$$

$$d = 0.6 \times 78.2 = 46.9 \text{ mm.}$$

## Bolts

Shearing area of each bolt

$$= \frac{\pi}{4} D_b^2$$

Shearing force of each bolt

$$= \frac{\pi}{4} D_b^2 \times 55 \times 10^6 \text{ N}$$

Distance of bolt axis from shaft axis = $\frac{1}{2}(2D) = D$

Torque due to eight bolts

$$= \left(8 \times \frac{\pi}{4} D_b^2 \times 55 \times 10^6\right) D$$

$$= 110 \times 10^6 \pi D_b^2 = 5740$$

Thus $110 \times 10^6 \times \pi \times .0782 \times D_b^2 = 5740$

$\therefore$ $D_b = 0.01454 \text{ m} = 14.54 \text{ mm}.$

**Example 7.5.** *A solid alloy shaft 50 mm diameter is to be coupled in series with a hollow steel shaft of the same external diameter. Find the internal diameter of the steel shaft if the angle of twist per unit length is to be 75 per cent of that of the alloy shaft.*

*Determine the speed at which the shafts are to be driven to transmit 200 kW if the limits of shearing stresses are to be 55 and 80 MN/m² in the alloy and steel respectively. $G_{steel}$ = 2.2 time $G_{alloy}$.*

**Solution.** As the shafts are in series, the torque transmitted by each will be same

$$T_a = T_s \quad ...(1)$$

$$\frac{J_a G_a \theta_a}{l_a} = \frac{J_s G_s \theta_s}{l_s}$$

Therefore

$$J_s = J_a \left(\frac{G_a}{G_s}\right)\left(\frac{\theta_a}{\theta_s}\right) \qquad (l_a = l_s = 1)$$

or $$\frac{\pi}{32}(50^4 - d^4) = \frac{\pi}{32}(50)^4 \left(\frac{1}{2.2}\right)\left(\frac{1}{0.75}\right)$$

or $$d^4 = (50)^4 \left[1 - \frac{4}{6.6}\right]$$

*i.e.* $$d = 50\left(\frac{2.6}{6.6}\right)^{1/4} = 39.6 \text{ mm}$$

(*b*) **Case 1**. If the stress in alloy is the limiting case, then

$$\frac{T_a}{J_a} = \frac{2\tau_a}{D_a}$$

or $$T_a = \frac{2 \times 60 \times 10^6}{.050} = \frac{\pi}{32} \times (.050)^4$$

$$= \frac{200 \times 1000 \times 60}{2\pi N}$$

*i.e.* $$N = \frac{200 \times 1000 \times 60 \times 60}{2\pi^2 \times 160 \times 10^6 \times (.05)^3} = 976 \text{ rpm}.$$

At this speed stress in steel is found as follows :

$$T_s = \frac{200 \times 1000 \times 60}{2\pi \times 976}$$

$$= J_s \frac{2\tau_s}{D_s} = \frac{\pi}{3.2}(.05^4 = .0396^4) \times 2 \times \frac{\tau_s}{.05}$$

or $$\tau_s = \frac{200 \times 1000 \times 60 \times .05 \times 32}{2\pi^2 \times 9.76 \times 2(.05^4 \times .0396^4)}$$

$$= 132 \text{ MN/m}^2$$

which is more than allowable.

Thus the assumption of case 1 is not valid.

**Case 2.** If the stress in steel is the limiting case, then

$$\frac{T_s}{J_s} = \frac{2\tau_s}{D_s}$$

$$\tau_s = \frac{T_s D_s}{2I_s}$$

or $$80 \times 10^6 = \frac{200 \times 1000 \times 50}{2\pi N} \times \frac{.050 \times 32}{2 \times \pi (.05^4 \times .0396^4)}$$

or $$N = \frac{200 \times 1000 \times 60 \times .05 \times 32}{4\pi^2 (.05^4 - .396^4) \times 80 \times 10^6}$$

$$= 1620 \textit{ rpm}.$$

At this speed, the stress in alloy is given by

$$\frac{2\tau_a}{D_a} J_a = \frac{200 \times 1000 \times 60}{4\pi N}$$

or $$\tau_a = \frac{200 \times 1000 \times 60}{4\pi\, NJ_a} \times D_a$$

Since $\tau_a$ is inversely proportional to $N$, the stress at 1620 r.p.m. will be smaller than that at 976 r.p.m. Thus the assumption of case 2 holds good and the required working speed will be 1620 r.p.m.

## 7.3. Combined Torsion and Bending

In all the practical cases of shafts transmitting power, they will be subjected to bending moments due to gravity or inertia loads. Stress will be set up in the shaft due to bending moment, torque and shearing forces. But the effect of the stresses due to shearing forces is negligible, particularly as its maximum value occurs at the neutral axis (Chapter 4) where there is no bending stress. Evidently both direct and shear stresses will be set up at any point of the cross-section of the shaft. For design purposes, it will be essential to find the principal stresses along with the maximum shear stress at any particular point in the cross-section.

Thus if $\sigma_b$ is the greatest bending stress and $\tau$ the greatest shear stress on the cross-sectional plane, then

$$\sigma_b = \frac{32M}{\pi D^3} \quad \text{...(7.10)}$$

and $$\tau = \frac{16T}{\pi D^3} \quad \text{...(7.11)}$$

where, $M$ = Bending moment at the section.

$T$ = Torque being transmitted by the same section.

D = Diameter of the solid shaft.

Both $\sigma_b$ and $\tau$ occur together at the ends of the vertical diameter for the bending moment to be in the vertical plane. Therefore referring to the point diagram (Fig. 7.5 which is the top view) for the point near the end of the vertical diameter (line $BC$ and $AD$ are parallel to cross-sectional plane), we have the maximum principal stress as

$$\sigma_{max} = \frac{1}{2}[\sigma_b + \sqrt{\sigma_b^2 + 4\tau^2}]$$

$$= \frac{1}{2}\left[\frac{32M}{\pi D^3} + \sqrt{\left(\frac{32M}{\pi D^3}\right)^2 + 4\left(\frac{16T}{\pi D^3}\right)^2}\right]$$

or $$\sigma_{max} = \frac{16}{\pi D^3}[M + \sqrt{M^2 + T^2}] \quad \text{...(7.12)}$$

A, B, C, D, $\tau$, $\sigma_b$

*Fig. 7.5*

Suppose we had the pure bending moment *Me* acting on the shaft such that the magnitude of the bending stress induced by it would have been $\sigma_{max}$, then

$$\sigma_{max} = \frac{32M_e}{\pi D^3} \qquad ...(7.13)$$

Equating (7.12) and (7.13), we get

$$M_e = \frac{1}{2}[M + \sqrt{M^2 + T^2}] \qquad ...(7.14)$$

This $M_e$ is called the equivalent bending moment. It may be noted that the purpose of $M_e$ is just to calculate the magnitude of the maximum principal stress. It does not give the location of the plane on which the maximum principal stress acts, which actually is inclined at $\left[\frac{1}{2}\tan^{-1}(-2\tau/\sigma_b)\right]$ with direction of $\sigma_b$.

Again the magnitude of maximum shear stress is given by

$$\tau_{max} = \frac{1}{2}\sqrt{\sigma_b^2 + 4\tau^2} = \frac{1}{2}\sqrt{\left(\frac{32M}{\pi D^3}\right)^2 + 4\left(\frac{16T}{\pi D^3}\right)^2}$$

$$= \frac{16}{\pi D^3}\sqrt{M^2 + T^2} = \frac{16}{\pi D^3}T_e \text{ (from 7.10 \& 7.11)} \qquad ...(7.15)$$

Thus if $\tau_e$, called the equivalent torque, acting alone on the shaft, induces the maximum shear stress of magnitude $\tau_{max}$, then

$$\tau_{max} = \frac{16}{\pi D^3}T_e \qquad ...(7.16)$$

Therefore from equations (7.15) and (7.16)

$$\tau_e = \sqrt{M^2 + T^2} \qquad ...(7.17)$$

$\tau_e$ is also used to evaluate the magnitude of the maximum shear stress. It does not give the location of the planes of maximum shear stress which are at angles of 45° to the principal planes.

**Example 7.6.** *A solid steel shaft transmits 1000 kW at 100 r.p.m., the maximum torque being 50 per cent greater than the mean. The shaft is supported in ball bearings 1.5 m apart with a 7500 kg fly-wheel midway between bearings. Find the shaft diameter if the maximum tensile stress is 60 MN/m².*

Solution.

$$\text{Mean torque} = \frac{(1000 \times 1000) \times 60}{2\pi \times 100} = 95600 \text{ Nm}$$

$\therefore \quad T = 1.5 \times 9.56 \times 10^4 = 14.3 \times 10^4$ Nm

$$M = (7500 \times 9.81)\frac{1.5}{4} = 2.76 \times 10^6 \text{ Nm}$$

$$M_e = \frac{1}{2}[2.76 \times 10^4 + \{(14.3 \times 10^4)^2 + (2.76 \times 10^4)^2\}^{1/2}]$$

$$= 8.67 \times 10^4 \text{ Nm}$$

Now $\sigma_{max} = \dfrac{32\, M_e}{\pi D^2}$

or $D^3 = \dfrac{32 \times 8.67 \times 10^4}{\pi \times 60 \times 10^6} = .01475$

*i.e.* $D = .246$ m or 246 mm.

**Example 7.7.** *A solid phosphor-bronze shaft 50 mm diameter is rotating at 7.40 r.p.m. It is subjected to torsion only. An electrical resistance strain gauge mounted on the surface of the shaft with its axis at 45° to the shaft axis, gives the strain reading as 4.17 × 10⁴. Find the power being transmitted if the modulus of elasticity for bronze is 105000 MN/m². Take ν = 0.3.*

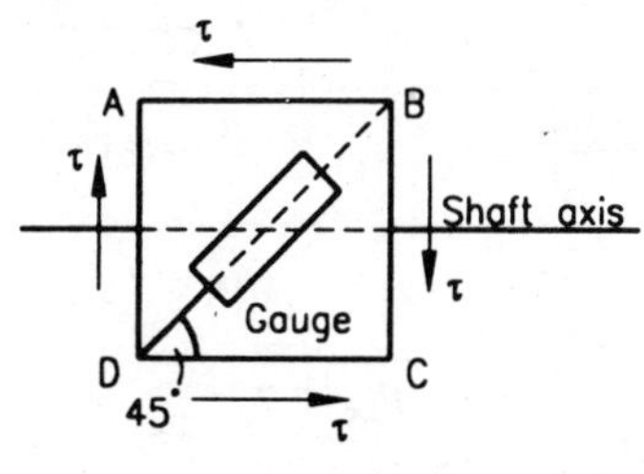


*Fig. 7.6*

**Solution.** If the maximum shear on the cross-sectional plane *BC* is τ then referring to Chapter II.

(*i*) Principal stress along *DB*

$$= \sigma_1 = -\frac{1}{2}\sqrt{4\tau^2}$$

$$= -\tau \text{ (i.e. comp.).}$$

(*ii*) Principal stress along *AC*

$$= \sigma_2 = \frac{1}{2}\sqrt{4\tau^2} = \tau \text{ (i.e. Tensile)}$$

Thus magnitude of the compressive strain along *DB* is

$$\varepsilon_{DB} = \frac{\tau}{E}(1 + \nu) + 4.17 \times 10^{-4}$$

*i.e.* $$\tau = \frac{4.17 \times 10^{-4} \times 105000}{(1 + 0.3)} = 33.7 \text{ MN/m}^2$$

Torque being transmitted, then, is

$$T = \frac{\pi}{16} \tau D^3 = \frac{\pi}{16} \times 33.7 \times (50)^3 \text{ Nmm} = 826 \text{ Nm}$$

Therefore $$\text{power} = \frac{826 \times 2\pi\ 740}{60} W$$

$$= 64000 \text{ W} = 64 \text{ kW}$$

**Example 7.8.** *A propeller shaft of a ship is 0.45 m diameter and it supports a propeller of mass 15 tonnes. The propeller can be considered as a load concentrated at the end of a cantilever of length 2 m. When the speed of the ship is 32 km/hr the propeller is revolving at 100 r.p.m. If the engine develops 15 MW, calculate the principal stress and maximum shear stress in the shaft. It may be assumed that the propulsive efficiency of the propeller is 85%.*

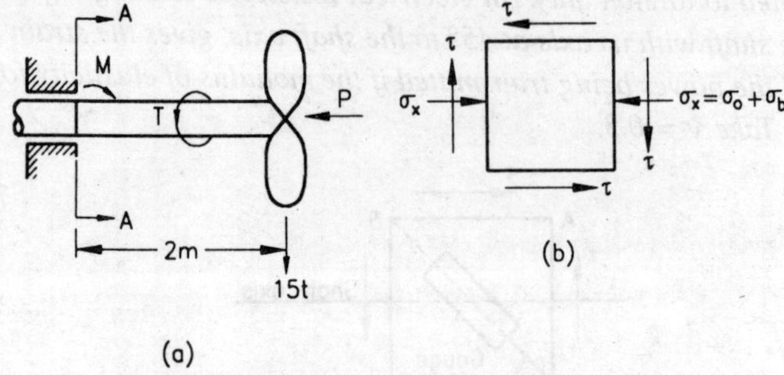


*Fig. 7.7*

**Solution.** At the bearing, Fig. 7.7(*a*),

Bending moment

$$M = (15000 \times 9.81) \times 2 \text{ Nm} = 0.2943 \text{ MN- m}$$

Torque $$T = \frac{\text{Power} \times 60}{2\pi \text{N}} = \frac{15 \times 10^6 \times 60}{2\pi \times 100} \text{ Nm}$$

$$= 1.433 \text{ MN-m}$$

If $P$ is the propulsive force and $v$ is the linear speed of the ship, then

$$\text{Engine Power} = \frac{Pv}{\text{efficiency}}$$

*i.e.* $$P = \frac{15 \times 10^6 \times 0.85 \times 3600}{32 \times 10^3} N = 1.435 \text{ MN}$$

Therefore direct stress on the cross-section $AA$,

$$\sigma_0 = \frac{P}{\text{Area}} = \frac{1.435 \times 4}{\pi \times (.45)^2} = 9.02 \text{ MN/m}^2 \text{ (comp.)}$$

Max. Bending stress due to M, on the cross section $AA$

$$\sigma_b = \frac{32M}{\pi D^3} = \frac{32 \times 0.2943}{\pi \times (.45)^3} = 32.9 \text{ MN/m}^2$$

(comp. at the lower most point)

Therefore $\sigma_x = 32.9 + 9.02 = 41.92 \text{ MN/m}^2$ (comp.)

Shear stress due to torque $T$,

$$\tau = \frac{16T}{\pi D^2}$$

$$= \frac{16 \times 1.433}{\pi \times (.45)^3} = 80 \text{ MN/m}^2$$

The point diagram for the bottom most point of the section $AA$ is given in Figure 7.7($b$). Thus the principal stresses are given by (Chapter 2)

$$\sigma_{1,2} = \frac{1}{2}\left[\sigma_x \pm \sqrt{\sigma_x^2 + 4\tau^2}\right]$$

$$= \frac{1}{2}\left[-41.92 \pm \sqrt{(41.92)^2 + 4 \times 80^2}\right]$$

$$= \frac{1}{2}[-41.92 \pm 165.48]$$

*i.e.* $\sigma_1 = 61.8 \text{ MN/m}^2$ (tensile)

and $\sigma_2 = -103.7 \text{ MN/m}^2$ (*i.e.* comp.

$$\tau_{max} = \frac{\sigma_1 - \sigma_2}{2} = \sqrt{\sigma_x^2 + 4\tau^2}$$

$$= \frac{61.8 + 103.7}{2} = 82.75 \text{ MN/m}^2$$

**Example 7.9.** *A horizontal shaft, securely fixed at each end, has a free length of 10 m. Viewed from one end of the shaft, axial couples of 30 kNm clockwise and 40 kNm counter clockwise act on the shaft at distance of 4 and 7 m respectively from the viewed end. Determine the end fixing couples in magnitude and direction and find the diameter of the solid shaft for a maximum shearing stress of 60 MN/m². Draw the torque distribution diagram for the shaft.*

*Also draw a diagram to show how a line, originally parallel to the axis and on the outer surface of the shaft will appear after the application of the couples and find the position along the shaft where the shaft suffers no angular twist.*

**Solution.** Let the reaction torque at the end $P$ be $T$ (clockwise) as seen from $P$ towards $S$ (Fig. 7.8).

Torque on $PQ = T$ kNm

Torque on $QR = T + 30$ kNm

Torque on $RS = T + 30 - 40 = T - 10$ kNm

For the above torques

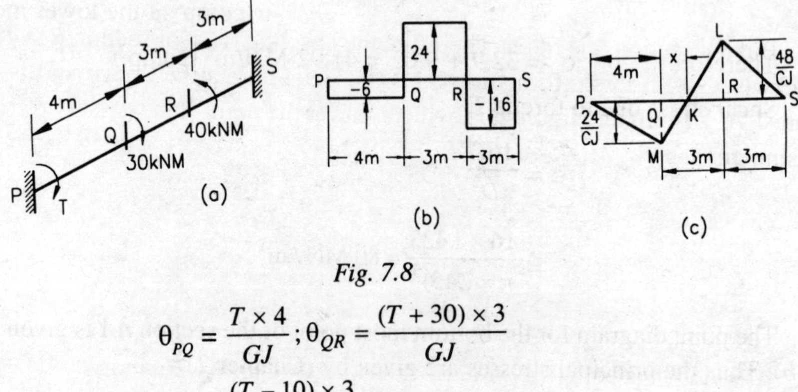


*Fig. 7.8*

$$\theta_{PQ} = \frac{T \times 4}{GJ}; \theta_{QR} = \frac{(T+30) \times 3}{GJ}$$

$$\theta_{RS} = \frac{(T-10) \times 3}{GJ}$$

But the ends being securely held, there is no relative twist between $P$ and $S$. Therefore

$$\theta_{PS} = \theta_{PQ} + \theta_{QR} + \theta_{RS} = 0$$

or $\quad 4T + 3\ (\mathbf{T} + 30) + 3\ (T - 10) = 0$

or $\quad T = -6$ kNm $\quad$ (*i.e.* anticlockwise as seen from $P$ to $S$)

Substituting the value of $T$, we have

$$T_{PQ} = -6,\ T_{QR} = 24 \text{ and } T_{RS} = -16 \text{ kNm}$$

These values enable us to draw the torque diagram of the shaft, shown in Fig. 7.8 *b*.

(*b*) From the torque diagram it is evident that the maximum torque occurs in the portion $QR$ and is equal to 24 kNm. Corresponding value of the maximum shear stress tau> can then be found.

$$T = \frac{\pi}{16} \tau D^3$$

or $\quad D^3 = \dfrac{16 \times (24 \times 10^3)}{\pi \times (60 \times 10^6)} \quad$ *i.e.* $D = 0.127$ m

(*c*) With respect to *P*, the twist at *Q* is

$$\theta_Q = \frac{-6 \times 4}{GJ} = \frac{-24}{GJ} \qquad \left(i.e. \frac{24}{GJ} \text{ anticlockwise}\right)$$

Similarly with respect to *R*, the twist of *S*

$$\theta_S = \frac{-16 \times 3}{GJ} = \frac{-48}{GJ}$$

But the section S is fixed. Therefore the absolute twist of section *R* is 48 over *GJ* (clockwise). With this information the twist diagram can be drawn and is shown in figure 7.7(*c*). From this figure it is evident that at the point *K* the twist is zero and from the similar triangles *MQK* and *KRL*.

$$\frac{x}{24} = \frac{3-x}{48} \quad i.e.\ x = 1 \text{ m}$$

## PROBLEMS FOR PRACTICE

1. A hollow steel shaft of 400 mm outside diameter is running at 120 r.p.m. and transmitting 9 MW power. Find the internal diameter and the maximum shearing stress induced if the twist over a length of 2 m is observed to be 0.45 degrees. Take $G = 80000$ MN/m$^2$. **[Ans.** 0.22 m ; 62.9 MN/m$^2$]

2. A torque of 1.6 kNm acting on a hollow shaft of 50 mm outside diameter and 30 mm internal diameter, is found to produce an angular twist of 0.4 degrees over a length of 0.2 m of the shaft. Find the value of the modulus of rigidity. Also determine the maximum power which could be transmitted by this shaft at 2000 r.p.m. if the maximum allowable shearing stress is 65000 MN/m$^2$.

   **[Ans.** 86000 MN/m$^2$ ; 292 kW]

3. A hollow steel shaft transmits 1600 kW at 180 r.p.m. If the internal and external diameters of the shaft are 125 mm and 200 mm respectively, find the shear stress at the outer and inner radii. Take $G = 84000$ MN/m$^2$.

   **[Ans.** (64.5 MN/m$^2$ ; 40.5 MN/m$^2$]

4. A flange coupling is required for a shaft transmitted 220 kW at 240 rev./min. The bolts are 16 mm diameter and are to be arranged on a circle of 250 mm diameter, the working shear stress in the both must not exceed 60 MN/m$^2$. Determine the number of bolts required and the actual stress realised.

   Briefly describe and explain how the power transmitted by the shaft may be measured. **[Ans.** 6 ; 59.4 MN/m$^2$]

5. A shaft runs at 300 r.p.m. and transmits power from a pulley A at one end to two pulleys, *B* and *C* ; each of which drives a machine in a workshop. The distance

between pulleys $A$ and $B$ is 3 m and that between $B$ and $C$ is 2.4 m (*i.e.* the total length of the shaft is 5.4 m). The shaft has a diameter of 50 mm between $A$ and $B$ and a diameter of 40 mm between $B$ and $C$. If the maximum permissible shear stress in the shaft is 80 MN/m$^2$, calculate the maximum power which may be supplied from each of the pulleys $B$ and $C$, assuming that both machines would be in operation at the same time. Also calculate the total angle of twist of one end of the shaft relative to the other when running on full load. $G$ = 8000 MN/m$^2$.

[**Ans.** 30.1 kW ; 31.6 kW ; 13.75°]

6. A steel shaft *ABCD* having a total length of 1.5 m is made of three lengths *AB*, *BC* and *CD*, each of 0.5 m long. *AB* and *BC* are solid, having diameters of 60 mm and 70 mm respectively, and *CD* is hollow having outside and inside diameters of 70 mm and 45 mm respectively. When an axial torque of 2 kN-m is transmitted from one end of the shaft to the other, the total angle of twist from $A$ to $D$ is 1.2°. Determine : (*a*) the maximum shearing stress in the shaft and state where it occurs, (*b*) the angle of twist for each of the three lengths *AB*, *BC* and *CD*, (*c*) the modulus of rigidity of the material.

[**Ans.** 49.4 MN/m$^2$ in *AB* ; 0.548° ; 0.296° ; 0.356° ; 82.2 *GN*/m$^2$]

7. A solid cylindrical shaft 50 mm diameter and 1.5 m long, is passed through the centre of a hollow cylindrical shaft of the same material 1.5 m long 55 mm and 75 inner and outer diameters. The ends of the two shafts are rigidly joined, with the shafts concentric. The composite shaft so formed is used to transmit 375 kW at a speed of 600 rev./min. Find the maximum and minimum stress in the two shafts.

[**Ans.** 79.3 MN/m$^2$ ; 58.1 MN/m$^2$ ; 52.9 MN/m$^2$]

8. A steel bar 19 mm diameter is encased in a closely fitting brass tube of 32 mm external diameter, securely fixed together at the ends. The compound bar is subjected to a torque of 520 Nm and the angle of twist measured on a gauge length of 250 mm is found to be 1.8°. If $G$ for steel is assumed to 80 GN/m$^2$ calculate $G$ for the brass.

   Find also the maximum shearing stresses in the two materials.

   [**Ans.** 34.6 GN/m$^2$ ; 95.5 MN/m$^2$ ; 69.5 MN/m$^2$]

9. A shaft is required for an engine which indicates 746 kW power at 80 r.p.m. The maximum twisting moment on the shaft is 1.8 × the mean twisting moment. The main bearings are 4500 mm apart, and the shaft carries a flywheel mid-way between bearings. The mass of the flywheel is 90000 kg. The bending moment due to the flywheel is additional to that due to steam pressure, the latter of which is 0.8 × the mean twisting moment. Find the diameter of the shaft to satisfy the condition that the maximum tensile stress in the material is not to exceed 62.78 MN/m$^2$. [**Ans.** 558 mm]

Chapter–8

# Struts and Columns

## 8.1 Definition

A strut is a structural member which is in compression. A vertical strut is known as a column or a pillar or a stanchion. Struts are further classified as slender and stocky struts. A slender strut is long in comparison to its lateral dimensions while a stocky one is short in comparison with its lateral dimensions. In the case of slender struts the failure is usually due to the lateral deflection and is known as buckling. The buckling is caused by the inherent eccentricity of loading and the crookedness of the strut. Evidently at each an every section of the strut the load will have eccentricity and the eccenticity on each plane in general will not be in the same vertical plane passing through the load line. However in, one in the million cases, it may happen that the eccentricities of loading at various sections taken along the length of the strut may lie in one vertical plane and that too the amount of eccenticities are such that of the points of intersections of the load lines with various sections from a loop of the shape as shown in the figure 8.1, then this will be the worst possible case of eccenticities and it will need the least posible axial load to cause buckling, provided further the value of the second moment of area of the section is least for bending in this plane. Bending moment due to load and these defects will be balanced by the restoring moment offered by the elasticity of the material of the strut. The load may then be made to increase till a stage comes when a very slight increase in it will overweigh the resisting moment and cause the strut to buckle all of a sudden. The value of the load at this stage, known as critical equilibrium, is called the buckling or crippling load of the strut. In the ideal case of perfect axial loading the problem is one of determining the critical load which causes elastic instability. This consists in finding, for what load the strut, if slightly displaced in the lateral direction in the form of fig 8.1, which is the worst possible assumed case of defect, though they are rare, will be rendered to be in neutral equilibrium under the action of the load and the elastic restoring forces of the strut. In other words if the load is slightly more than this critical value then the small lateral displacement of shape as in fig. 8.1 will make the bending moment due to the load to overweigh the restoring moment of elastic forces and the strut will go on bending further causing the bending moment to increase still further till the strut buckles. In case the load is smaller in value than the critical and a slight lateral displacement is given to the strut, it will come back to its original position (may be after a little bit of vibrations).

Further for any strut the buckling load depends upon the end conditions, whether fixed or free. A fixed end is one which is so clamped as to constrain the direction of the strut at that point and a free end means one which by being hinged or pivoted is free to take up any angular position due to the bending of the strut.

## 8.2 Euler's Column Theory

This theory refers to the struts which are very long as compared to the lateral dimensions. Following assumptions are made :

1. The strut or column is very long.
2. The strut is originally perfectly straight and homogeneous.
3. The compressive load is perfectly axially applied.
4. The stresses in the strut are within the elastic limit.
5. The value of Young's modulus of the material in tension as well as compression is same
6. The fibres are free to expand or contract without any constraints from the adjoining fibres.
7. Plane sections remain plane even during the slight bent position caused by the small lateral displacement given by an external agency.

**Case 1. Strut with both ends pinned.** Taking the origin at the mid-section, if the lateral displacement is given in the shape as shown in the Fig 8.1 which is the worst combination of defects at various sections of the strut and $P$ is the buclking load then for neutral equilibrium bending moment at section $XX$ will be equal to the restoring moment due to the elasticity. Thus if $y$ is the deflection at a distance $x$ and $I$ the second moment of area of the section about the centroidal bending axis, then for neutral equilibrium we have

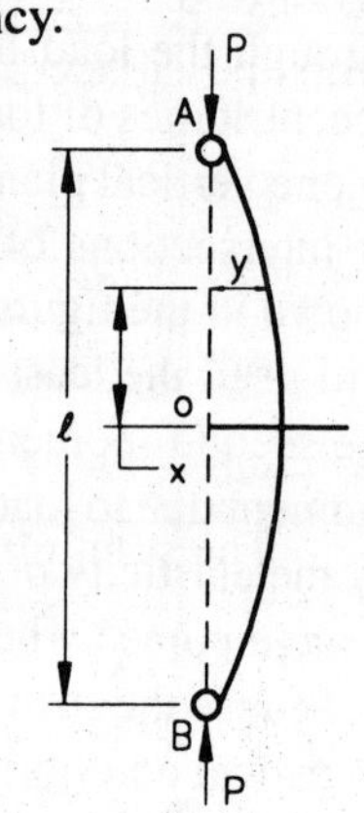

*Fig. 8.1*

$$EI\frac{d^2y}{dx^2} = M = -Py$$

or $$\frac{d^2y}{dx^2} + m^2y = 0 \qquad ...(8.1)$$

where $$m^2 = P/(EI) \qquad ...(8.2)$$

The solution of differential equation (8.1) is

$$y = A\cos mx + B\sin mx \qquad ...(8.3a)$$

and $$dy/dx = -mA\sin mx + mB\cos mx \qquad ...(8.3b)$$

For the end condition $dy/dx = 0$ at $x = 0$, we get $B = 0$; and for the condition $y = 0$ at $x = \frac{l}{2}$ we have from equation (8.3$a$).

$$0 = A\cos\frac{ml}{2}$$

Therefore either $A = 0$ or $\cos\frac{ml}{2} = 0$. But $A$ cannot be zero because then $y = 0$ for all values of $x$, *i.e.* a situation of no buckling.

Thus $$\cos\frac{ml}{2} = 0$$

or $$\frac{ml}{2} = \frac{\pi}{2}, \frac{3\pi}{2} \text{ etc.} \qquad ...(8.4)$$

from which the smallest value of $P$ is obtained by taking first value from equation 8.4 (this refers to buckling in one loop as shown in Fig. 8.1) and substituting in equation (8.2)

$$\frac{P}{EI}\frac{l^2}{4} = \left(\frac{\pi}{2}\right)^2$$

or $$P = \frac{\pi^2 EI}{l^2} \qquad ...(8.5)$$

Case 2. Strut with one end fixed and one end free, Fig. 8.2.

Taking the origin at $A$. The differential equation for the buckling load $P$ in this case is given by (for the assumed displacement by external agency compatible with worst possible combination of eccentricities as explained above)

$$EI\frac{d^2y}{dx^2} = P\,(a - y)$$

or $$EI\frac{d^2y}{dx^2} + m^2y = m^2a \qquad ...(8.6)$$

where $$m^2 = P/(EI) \qquad ...(8.7)$$

The solution of the equation (8.6) is

$$y = A\cos mx + B\sin mx + a \qquad ...(8.8a)$$

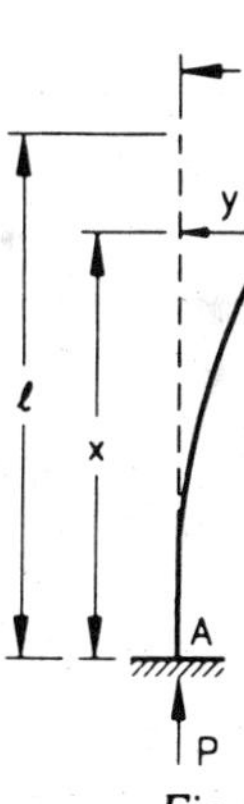

and $$\frac{dy}{dx} = -mA\sin mx + mB\cos mx \qquad ...(8.8b)$$

Fig. 8.2

when $x = 0, y = 0$ ; $\therefore A = -a$

when $x = 0, \frac{dy}{dx} = 0$;

$\therefore$ $B = 0$

Thus equation (8.8$a$) reduces to

$$y = a - a\cos mx \qquad ...(8.9)$$

Also at $x = l$, $y = a$, therefore from equation (8.9), we have

$$\cos ml = 0$$

or $$ml = \pi/2$$

from which
$$P = \frac{\pi^2 EI}{4l^2} \quad ...(8.10)$$

Case 3. Strut with both ends fixed, (Fig. 8.3)

Taking the origin at the centre, we have in this case, if $M$ is the fixing moment. Further let $P$ be the buckling load and the external agency displaces the centre line of the shirt as shown in fig. 8.3 which is the worst possible assumed combination of eccentricities and all lying in one vertical plane along which $I$ of section is least.

$$EI\frac{d^2y}{dx^2} = -Py + M$$

or
$$\frac{d^2y}{dx^2} + m^2y = \frac{M}{EI} \quad ...(8.11)$$

Complementary function of equation (8.11) is

$$y = A\cos mx + B\sin mx$$

The particular integral is given by

$$P.I = \frac{1}{(D^2+m^2)}\left(\frac{M}{EI}\right)$$

$$= \frac{1}{m^2}\left(1+\frac{D^2}{m^2}\right)^{-1}\left(\frac{M}{EI}\right)$$

$$= \frac{1}{m^2}\left[1-\frac{D^2}{m^2}+\ldots\right]\left(\frac{M}{EI}\right)$$

$$= \frac{1}{m^2}\cdot\frac{M}{EI} = \frac{EI}{P}\cdot\frac{M}{EI} = \frac{M}{P}$$

Thus the complete solution of equation (8.11) is

$$y = A\cos mx + B\sin mx + \frac{M}{P} \quad ...(8.12b)$$

Fig. 8.3

and then
$$\frac{dy}{dx} = -mA\sin mx + mB\cos mx$$

When $x = 0$, $dy/dx = 0$ ; $B = 0$

When $x = \frac{l}{2}$, $dy/dx = 0$ ; then from (8.12*b*)

$$0 = -mA\sin\frac{ml}{2}$$

or
$$\sin\frac{ml}{2} = 0$$

*i.e.*
$$\frac{ml}{2} = \pi$$
(taking the least no-trivial value)

from which
$$P = \frac{4\pi^2 EI}{l^2} \quad ...(8.13)$$

**Case 4. Strut with one end fixed and the other guided.**

Since, in this case, the lateral movement at $B$ is restricted, the hinges at $B$ will exert a horizontal force $F$ (say) to keep the end $B$ of the strut on the vertical axis. Further the Product $F \times l$ will balance the fixing moment $M$ at the base.

Here again assuming P to be buckling load and external agency gives lateral displacment as in fig. 8.4 which corresponds to the worst combination of defects at various sections that too in the plane of bending where $I$ is least.

Taking the origin at $A$, the buckling load $P$ will be given by

$$EI\frac{d^2y}{dx^2} = -Py + F(l-x)$$

or
$$\frac{d^2y}{dx^2} + m^2y = \frac{F}{EI}(l-x) \qquad ...(8.14)$$

Fig. 8.4

Complementary function of equation (8.4) is again

$$y = A\cos mx + B\sin mx$$

and
$$\text{P.I.} = \frac{1}{(D^2+m^2)}\left[\frac{F}{EI}(l-x)\right]$$

$$= \frac{1}{m^2}\left(1+\frac{D^2}{m^2}\right)^{-1}\left[\frac{F}{EI}(l-x)\right]$$

$$= \frac{1}{m^2}\cdot\frac{F}{EI}\left(1-\frac{D^2}{m^2}+...\right)[l-x] = \frac{EI}{P}\cdot\frac{F}{EI}(l-x)$$

*i.e.*
$$\text{P.I.} = \frac{F}{P}(l-x)$$

Thus the complete solution of equation (8.14) is

$$y = A\cos mx + B\sin mx + \frac{F}{P}(l-x) \qquad ...(8.15a)$$

and
$$\frac{dy}{dx} = -mA\sin mx + mB\cos mx - \frac{F}{P} \qquad ...(8.15b)$$

when $x = 0$, $y = 0$ giving

$$A = -\frac{Fl}{P}$$

when $x = 0$, $\frac{dy}{dx} = 0$ giving

$$B = \frac{F}{mP}$$

Therefore from (8.15$a$)

$$y = \frac{F}{P}\left[-l\cos mx + \frac{1}{m}\sin mx + (l-x)\right] \qquad ...(8.16)$$

Again when $x = l$, $y = 0$, therefore from equation (8.16)

$$0 = \frac{F}{P}\left[-l\cos ml + \frac{1}{m}\sin ml\right]$$

or $$\tan ml = ml$$

The least solution of this is

$$ml = 4.5 \text{ roman radians}$$

from which $$P = (4.5)\frac{EI}{l^2} \simeq \frac{2\pi^2 EI}{l_e^2} \quad \text{...(8.17)}$$

## 8.2.1 Equivalent Hinged Length

Euler's crippling load derived in equations (8.5), (8.10), (8.13) and (8.17) can be expressed by a single equation in the form.

$$P = \frac{\pi^2 EI}{l_e^2} \quad \text{...(8.18)}$$

Where $l_e$ is known as the equivalent hinged length and its value is given below for various end conditions.

(*a*) Both ends pinned $l_e = l$

(*b*) One end fixed and other free $l_e = 2l$

(*c*) Both ends fixed $l_e = \frac{l}{2}$

(*d*) One end fixed and other pinned $l_e = \frac{l}{\sqrt{2}}$

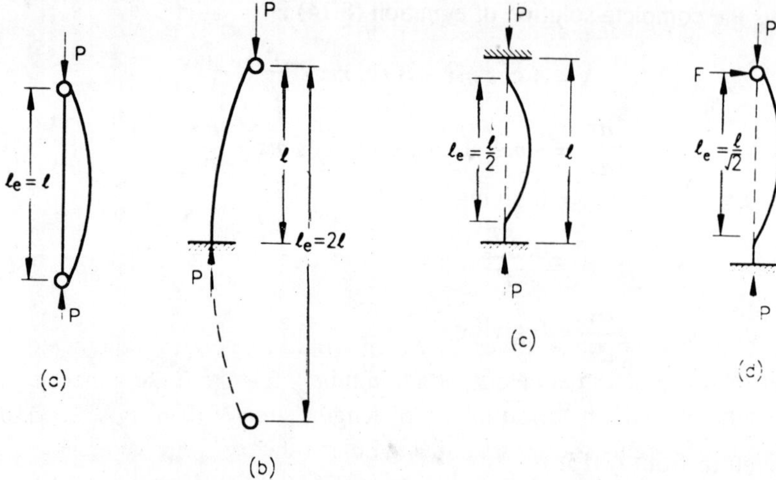


*Fig. 8.5*

These values are rather apparent from the Fig. 8.5 which gives an easy way to remember the values of $l_e$ in terms of the length of the strut.

## 8.3 Slenderness Ratio and its Effect on Euler's Theory

If $k$ is the least radius of gyration and $A$ the area of cross-section of the strut, from equation (8.81), Euler's crippling load is given by

$$P = \frac{\pi^2 EI}{l_e^2}$$

$$= \frac{\pi^2 EA\,k^2}{l_e^2}$$

or $$P = \frac{\pi^2\ EA}{(l_e / k)^2} \qquad ...(8.19a)$$

or $$\sigma_E = \frac{P}{A} = \frac{\pi^2 E}{(l_e / k)^2} \qquad ...(8.19b)$$

Where $\sigma_E$ is called the Euler's crippling stress. The ratio $l_e/k$ is known as the slenderness ratio. Evidently when $l_e/k$ is very small the value of $\sigma_E$ will be very large. Thus as the slenderness ratio is reduced a stage may come when the corresponding value of $\sigma_E$ may become just equal to the compressive yield stress, $\sigma_C$, of the material. Thus below this value of the slenderness ratio the Euler's buckling theory will not apply. This point is called the validity limit for the Euler's theory.

For example, for a mild steel strut, freely hinged at both ends and taking $E$ = 200000 MN/m$^2$ and the yield stress as 325 MN/m$^2$, the shortest length, to which the equation (8.18*a*) or (8.19*b*) could be applied, would be such that

$$325 = \frac{\pi^2 \times 200000}{(l_e / k)^2}$$

or $$l_e/k = \left(\frac{200000}{325}\right)^{1/2} \simeq 78.2 \qquad ...(8.20)$$

That is, length is about 80 times, $k$, which would be about 20 diameters for a solid circular section and 28 diameters for a thin tube. Thus the Euler's theory would give good results only when the struts are of lengths greater than those mentioned above. For struts of lengths shorter than these, Euler's theory is not applicable and if used will evidently give much too high values of the buckling load. In practice such shorter or medium-length struts are, however, very common. For such cases empirical formulae are used.

## 8.4 Rankine's Empirical Formula

As discussed in the previous article, Euler's equation (8.18) is applicable for very long struts for which let the crippling load be represented by $P_e$. Similarly for very short struts the failure will be by crushing and let the crushing load be giving by $P_C$. Then for any length '$l$' and cross-sectional area '$A$' Rankine suggested the expression for crippling load known after him as Rankine's crippling load and represented by $P_R$, as

$$\frac{1}{P_R} = \frac{1}{P_e} + \frac{1}{P_C} \qquad \text{...(8.21)}$$

For very long struts $P_e$ will be very small and thus $1/P_C << 1/P_e$. That is, from equation (8.21) $P_R \simeq P_e$. Similarly for very short struts $P_e$ will be very large, rendering $1/P_C >> 1/P_e$. That is $P_R \simeq P_C$ from equation (8.21). Further since the change in the value of the crippling load with increasing length, for a constant $A$, must be a continuous change, the equation (8.21) is taken to be reasonable.

Equation (8.21) can be re-written as

$$\frac{1}{P_R} = \frac{P_C + P_e}{P_C P_e}$$

or

$$P_R = \frac{P_C}{(1 + P_C / P_e)}$$

$$= \frac{A\sigma_C}{1 + \dfrac{A\sigma_C l_e^2}{\pi^2 EI}}$$

$$= \frac{A\sigma_C}{1 + \dfrac{\sigma_C}{\pi^2 E}(l_e / k)^2}$$

or

$$P_R = \frac{A\sigma_C}{1 + \sigma(l_e / k)^2}$$

Where $\alpha = \sigma_C/(\pi^2 E)$, a constant for a given material.

If $\alpha_R$ is the corresponding mean intensity of compressive stress on the cross-section of the strut, then

$$\alpha_R = \frac{P_R}{A} = \frac{\sigma_C}{1 + \sigma(l_e / k)^2} \qquad \text{...(8.22}b\text{)}$$

Based on extensive experimentation, the usual accepted values of the Rankine's constants $\alpha_C$ and $\alpha$, to be used in Rankine's formula (8.22) are given in the following table.

| Material | $\alpha_C$ kgf/cm$^2$ | $\alpha_C$ MN/m$^2$ | $\alpha$ for pinned ends |
|---|---|---|---|
| Cast iron | 5683 | 557 | 1/1600 |
| Wrought Iron | 2336 | 229 | 1/9000 |
| Mild Steel | 3315 | 325 | 1/7500 |
| Medium Carbon Steel | 4736 | 465 | 1/5000 |
| Timber | 473.6 | 46.5 | 1/2000 |

## 8.5 Other Empirical Formulae

A large number of experiments to determine the ultimate, strength of struts have been conducted under various conditions and several empirical formulae have been devised to suit these results. As expected the experimental results have been consistent and in agreement with empirical formulae where the conditions of loading and end fixings have approached nearly to the ideal. But such conditions are not realised in practical struts as used in machines and structures which deviate from the ideal for want of (*i*) straightness, (*ii*) homogenity of material, (*iii*) concentricity of the load, (*iv*) frictionlessness of the pin-joints. For these reasons the test results obtained for struts under working conditions show great variations and no formula, empirical or otherwise, can accurately predict the load at which failure will take place in a given case. Thus for the purposes of design the accuracy of any empirical formula is generally about the same as any other. Evidently the formula simplest in form is the best to use. The constants in any formula are usually found after experimentation and are valid only for a short range of values of $l/k$.

Based on the considerations outlined above, following empirical formulae are very commonly used.

(*i*) Empirical straight line formula :

For its simplicity the formula as given below is very common in U.S.A. The buckling load $P$ is given as a linear function of the slenderness ratio $l/k$.

$$P = \sigma_c A\left[1 - n\frac{l_e}{k}\right] \qquad \text{...(8.23)}$$

Where $A$, $\alpha_c$, $l_e$ and $k$ represent the same quantities as in Rankine's formula and $n$ is a constant. Typical values for mild-steel struts are $\alpha_C = 100$ MN/m$^2$, $n = 1/200$ for pinned ends and 1/250 for fixed ends. For structural steel $\alpha_C = 140$ MN/m$^2$, $n = 0.0054$ for pinned ends and 0.0038 for fixed ends.

(*ii*) Johnson's Parabolic formula :

$$P = \alpha_C A\ [1 - b\ (l_e/k)^2] \qquad \text{...(8.24)}$$

Here again $P$, $\alpha_c$, $A$, $l_e$ and $k$ represent the same quantities as in Rankine's formula and $b$ is a constant. For mild steel $\alpha_c$ = 290 MN/mm² and $b$ = 0.00003 for pinned ends and 0.00002 for fixed ends.

## 8.6. Eccentrically Loaded Struts

**(*a*) Pinned at both ends :**

Fig. 8.6 shows a pinned end strut with the load $P$ applied at each end with eccentricity $e$. In such a strut there is no buckling load rather the strut starts bending immediately after the load is applied and the deflection at any point is dependent upon the load.

Taking the origin at $A$, if $y$ is the deflection at the distance $x$, then

$$M_x = EI\frac{d^2y}{dx^2}$$

$$= -P(y+e)$$

Fig. 8.6

or
$$\frac{d^2y}{dx^2} + m^2y = -m^2e \qquad ...(8.25)$$

where
$$m^2 = P/EI \qquad ...(8.26)$$

Complementary function of equation (8.25) is

$$y = A\cos mx + B\sin mx \qquad ...(8.27)$$

The particular integral is given by

$$\text{P.I.} = \frac{1}{(D^2+m^2)}[-m^2e]$$

$$= \frac{1}{m^2}\left(1+\frac{D^2}{m^2}\right)^{-1}[-m^2e]$$

$$= \frac{1}{m^2}\left(1-\frac{D^2}{m^2}+...\right)[-m^2e]$$

$$= \frac{1}{m^2}.(-m^2e) = -e \qquad ...(8.28)$$

Thus the complete solution of equation (8.25) is

$$y = A\cos mx + B\sin mx - e \qquad ...(8.29)$$

## End Conditions

(*i*) For $x = 0$, $y = 0$ ; therefore $A = e$ ...(8.30)

Thus from equations (8.29) and (8.30)

$$\frac{dy}{dx} = -\,mA \sin mx + mB \cos mx \qquad \text{...(8.31)}$$

(*ii*) Therefore from equation (8.31), since at $x = \frac{l}{2}$, $dy/dx = 0$, we have

$$0 = -\,me \sin \frac{ml}{2} + mB \cos \frac{ml}{2}$$

or

$$B = e \tan \frac{ml}{2} \qquad \text{...(8.32)}$$

Substituting values of $A$ and $B$ from equations (8.30) and (8.32) in equation (8.29), we get the deflection equation

$$y = e \cos mx - e \tan \frac{ml}{2} \sin mx - e \qquad \text{...(8.33)}$$

Maximum deflection will be at $x = l/2$, therefore from equation (8.33)

$$y_{max} = e \cos \frac{ml}{2} - e \tan \frac{ml}{2} \sin \frac{ml}{2} - e$$

$$= e\left(\cos^2 \frac{ml}{2} + \sin^2 \frac{ml}{2}\right) \sec \frac{ml}{2} - e$$

or

$$y_{max} = e\left(\sec \frac{ml}{2} - 1\right) \qquad \text{...(8.34)}$$

Maximum bending moment M sub {ma x}> is then given by

$$M_{max} = P\,(y_{max} + e)$$

$$= eP \sec \frac{ml}{2} \qquad \text{(from 8.34)}$$

Corresponding maximum bending stress, if c is the maximum distance of the fibre from centroidal bending axis of the section (This will be compressive on the concave side of the bent strut) is then

$$(\sigma_b)_{max} = \left(eP \sec \frac{ml}{2}\right) \frac{c}{I}$$

$$= \frac{ecP}{Ak^2} \sec \frac{ml}{2}$$

Where $k$ is the radius of gyration of the section. In addition there will be the direct compressive stress over the whole section. Thus the resultant maximum compressive stress in the strut material

$$\sigma_{max} = (\sigma_b)_{max} + \frac{P}{A}$$

$$= e\frac{Pc}{Ak^2}\sec\frac{ml}{2} + \frac{P}{A}$$

or
$$\sigma_{max} = \frac{P}{A}\left[1 + \frac{ec}{k^2}\sec\frac{ml}{2}\right]$$

$$= \frac{P}{A}\left[1 + \frac{ec}{k^2}\sec\left(\sqrt{\frac{P}{EI}}\frac{l}{2}\right)\right] \quad ...(8.35)$$

Equation (8.35) is solved graphically for $P$. Further Professor Perry pointed out that

$$\sec\left(\sqrt{\frac{P}{EI}}\frac{l}{2}\right) \text{ or } \sec\left(\frac{\pi}{2}\sqrt{\frac{P}{P_e}}\right)$$

where $P_e$ is the Euler's crippling load, is approximately equal to $\frac{1.2\,P_e}{(P_e - P)}$. Using this approximation equation (8.35) reduces to a simple form

$$\sigma_{max} \simeq \frac{P}{A}\left[1 + 1.2\frac{ec}{k^2}.\frac{P_e}{P_e - P}\right] \quad ...(8.36)$$

**(*b*) One end fixed and the other free :**

Fig. 8.7 shows a strut which is fixed at $B$ and free at $A$ but loaded eccentrically at $C$. Taking the origin at fixed end $B$ we have for any section $XX$, the bending moment equation

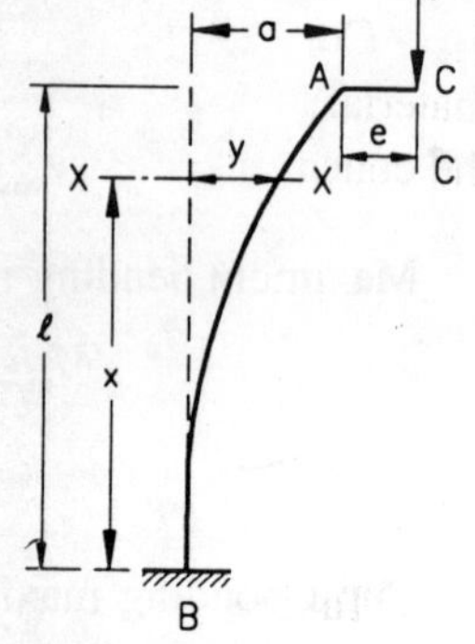

*Fig. 8.7*

$$M = EI\frac{d^2y}{dx^2}$$

$$= P\,(a + e - y)$$

or
$$\frac{d^2y}{dx^2} + \frac{P}{EI}y = \frac{P}{EI}(a + e) \quad ...(8.37)$$

The solution of differential equation (8.37) is

$$y = A\cos mx + B\sin mx + (a + e) \quad ...(8.38)$$

where
$$m^2 = P/EI \quad ...(8.39)$$

## End Conditions :

(*i*) $x = 0$, $y = 0$. Therefore from equation (8.38)

$$A = -(a + e)$$

(*ii*) At $x = 0$, $dy/dx = 0$. But from equation (8.38)

$$\frac{dy}{dx} = -mA \sin mx + mB \cos mx$$

i.e.

$$0 = 0 + mB$$

or

$$B = 0 \text{ as } m \neq 0 \qquad \text{...(8.41)}$$

Therefore the complete solution is

$$y = -(a + e) \cos mx + (a + e) \qquad \text{...(8.42)}$$

Also at $x = l$, $y = a$, therefore from (8.42)

$$a = -(a + e) \cos ml + (a + e)$$

or

$$e = (a + e) \cos ml$$

or

$$(a + e) = e \sec\left(l\sqrt{\frac{P}{EI}}\right) \qquad \text{...(8.43)}$$

Now maximum bending moment is at the fixed end

$$M_{max} = P(a + e)$$

$$= e\,P \sec\left(l\sqrt{P/EI}\right) \qquad \text{...(8.44)}$$

Corresponding maximum bending stress, if $c$ is the maximum distance of the material from the centroidal bending axis of the section (this will be compressive on the concave side of the bent strut), is given by

$$(\sigma_b)_{max} = \frac{(M_{max})}{I} \times c$$

$$= \frac{ceP}{I} \sec\left(l\sqrt{\frac{P}{EI}}\right)$$

Taking into account the direct compressive stress, the resultant maximum compressive stress is

$$\sigma_{max} = (\sigma_b)_{max} + \frac{P}{A}$$

$$= \frac{ce\,P}{I} \sec\left(l\sqrt{P/EI}\right) + \frac{P}{A}$$

$$= \frac{P}{A}\left[1 + \frac{ce}{k^2} \sec\left(l\sqrt{P/EI}\right)\right] \qquad \text{...(8.45)}$$

Comparing equations (8.35) and (8.45) we see that the general expression in terms of equivalent hinged length could be written as

$$\sigma_{max} = \frac{P}{A}\left[1 + \frac{ce}{k^2}\sec\left(\frac{l_e}{2}\sqrt{\frac{P}{EI}}\right)\right] \quad ...(8.46)$$

## 8.7 Struts with Initial Curvature

Fig. 8.8 shows a pin jointed strut *AB* of length *l* with initial curvature such that the maximum crookedness, e is assumed at the mid section. We assume an equation to the curve as

$$y_0 = e \sin \frac{\pi x}{l} \quad ...(8.47)$$

(As a matter of fact any other equation such as a parabola or a circle could as well be assumed, conforming to the end condition *i.e.* at $x = 0$ and $l$, $y_0 = 0$ and at $x = l/2$, $y_0 = e$. It can be shown that the results will not be much different).

Due to the application of axial load *P* let the strut take the shape *ADB* such that there is additional deflection *y* at section *XX*. Then the bending equation can be written as

$$EI\frac{d^2y}{dx^2} = -P(y_0 + y)$$

or
$$\frac{d^2y}{dx^2} + \frac{P}{EI}y = -\frac{P}{EI}y_0 \quad ...(8.48)$$

or
$$\frac{d^2y}{dx^2} + m^2y = -\frac{P}{EI}y_0$$

where
$$m^2 = P/EI \quad ...(8.49)$$

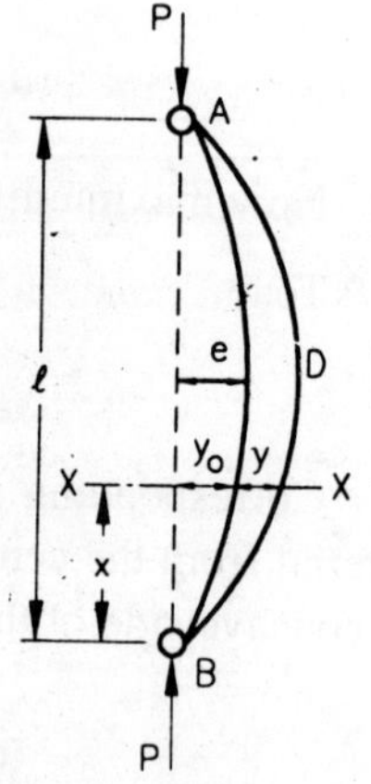


*Fig. 8.8*

Complementary function of equation (8.48) is

$$y = A\cos mx + B\sin mx$$

and the particular integral is given by

$$\text{P.I.} = \left[\frac{1}{D^2 + m^2}\right]\left[-m^2 e \sin\frac{\pi x}{l}\right]$$

$$= \frac{-m^2 e \sin\dfrac{\pi x}{l}}{-\dfrac{\pi^2}{l^2} + m^2}$$

Thus the complete solution of equation (8.48) is

$$y = A\cos mx + B\sin mx + \frac{m^2 e \sin\dfrac{\pi x}{l}}{\dfrac{\pi^2}{l^2} - m^2} \quad ...(8.50)$$

## End Conditions :

($i$) At $x = 0$, $y = 0$ ; therefore from equation (8.50) $A = 0$

($ii$) At $x = l$, $y = 0$ for which we get

$$0 = B \sin ml + \frac{m^2 e \sin \pi}{\frac{\pi^2}{l^2} - m^2}$$

or $$0 = B \sin ml$$

*i.e.* either $B = 0$ or $\sin ml = 0$. The latter actually corresponds to the Euler's solution for a pin jointed strut. Therefore $\sin ml \neq 0$. Thus

$$B = 0$$

In the light of these, the solution of equation (8.50) is

$$y = \frac{m^2 e \sin \frac{\pi x}{l}}{\frac{x^2}{l^2} - m^2} \qquad ...(8.51)$$

Total eccentricity at any section $XX$

$$= y_0 + y$$

$$= e \sin \frac{\pi x}{l} + \left(m^2 e \sin \frac{\pi x}{l}\right) \Big/ \left[\frac{\pi^2}{l^2} = m^2\right]$$

$$= \left[e \sin \frac{\pi x}{l}\right]\left[1 + \frac{P/(EI)}{\frac{p^2}{l^2} - \frac{P}{EI}}\right]$$

$$= \left[e \sin \frac{\pi x}{l}\right]\left[1 + \frac{P}{\frac{\pi^2 EI}{l^2} - P}\right]$$

$$= \left(e \sin \frac{\pi x}{l}\right)\left(1 + \frac{P}{P_e - P}\right)$$

$$= \left(e \sin \frac{\pi x}{l}\right)\left(\frac{P}{P_e - P}\right)$$

Where $$P_e = \frac{\pi^2 EI}{l^2}$$

Thus Max. bending moment which occurs at $x = l/2$, is given by

$$M_{max} = P\left(e \sin \frac{\pi}{l} \cdot \frac{l}{2}\right)\left(\frac{P_e}{P_e - P}\right)$$

$$= \frac{e\,P\,P_e}{P_e - P}$$

Corresponding maximum bending stress, if $c$ is the maximum distance of the fibre from the centroidal bending axis, is then given by

$$(\sigma_b)_{max} = \left(\frac{eP\,P_e}{P_e - P}\right)\frac{c}{I}$$

Resultant compressive stress

$$\sigma_{max} = (\sigma_b)_{max} + \frac{P}{A}$$

$$= \frac{ec\,PP_e}{(Ak^2)(P_e - P)} + \frac{P}{A}$$

$$\sigma_{max} = \frac{P}{A}\left[\frac{ec}{k^2}\left(\frac{P_e}{P_e - P}\right) + 1\right] \quad \text{...(8.52)}$$

## 8.8 Strut with Transverse Loads

**(a) Point load at the centre :**

Taking origin at $A$ for any section $XX$ ( $0 < x < l/2$)

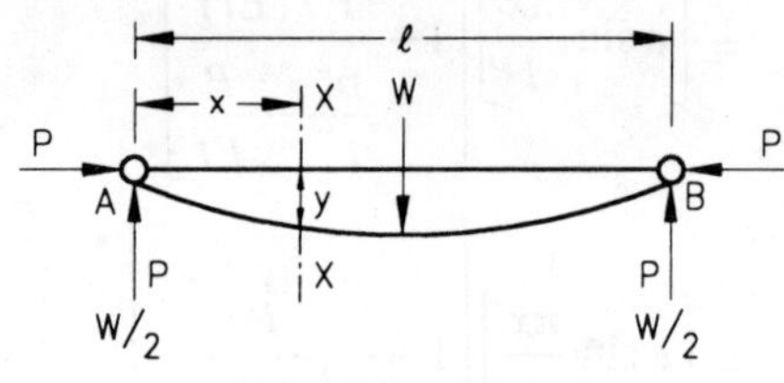


*Fig. 8.9*

$$EI\frac{d^2y}{dx^2} = -Py - \frac{W}{2}x$$

or

$$\frac{d^2y}{dx^2} + \frac{P}{EI}y = -\frac{W}{2EI}x \quad \text{...(8.53)}$$

If $m^2 = \frac{P}{EI}$, the solution of equation (8.53) is

$$y = A\cos mx + B\sin mx - \frac{Wx}{2P} \quad \text{...(8.54)}$$

## End Conditions :

(*i*) At $x = 0$, $y = 0$, therefore $A = 0$

Also $$\frac{dy}{dx} = m\,B\cos mx - \frac{Wx}{2P}$$

At $x = \frac{l}{2}, \frac{dy}{dx} = 0$, therefore

$$0 = m\,B\cos\frac{ml}{2} - \frac{W}{2P}$$

or $$B = \frac{W}{2P}\frac{1}{m}\sec\frac{ml}{2}$$

Therefore complete solution of equation (8.53) is

$$y = \frac{W}{2Pm}\sec\frac{ml}{2}\sin mx - \frac{Wx}{2P} \quad ...(8.55)$$

Therefore $$M_x = P.\,y + \frac{Wx}{2}$$

$$= \left(\frac{W}{2}\frac{1}{m}\sec\frac{ml}{2}\sin mx - \frac{Wx}{2}\right) + \frac{Wx}{2}$$

or $$M_x = \frac{W}{2}\frac{1}{m}\sec\frac{ml}{2}\sin mx$$

At $x = \frac{l}{2}$, we have

$$M_{max} = \frac{W}{2}\frac{1}{m}\sec\frac{ml}{2}\sin\frac{ml}{2}$$

$$= \frac{W}{2}\frac{1}{m}\tan\frac{ml}{2}$$

Therefore $$(\sigma_b)_{max} = \frac{M_{max}}{I}.\,c$$

$$= \frac{W}{2}\frac{c}{Im}\tan\frac{ml}{2}$$

Total maximum compressive stress

$$\sigma_{max} = \frac{P}{A} + (\sigma_b)_{max}$$

or $$\sigma_{max} = \frac{P}{A} + \frac{W}{2l}\frac{c}{m}\tan\frac{ml}{2}$$

$$= \frac{P}{A}\left[1 + W\frac{c}{2k^2}\sqrt{\frac{EI}{P}}\tan\left(\frac{1}{2}\sqrt{\frac{P}{EI}}\right)\right] \quad ...(8.56)$$

(*b*) **Uniformly distributed load :**

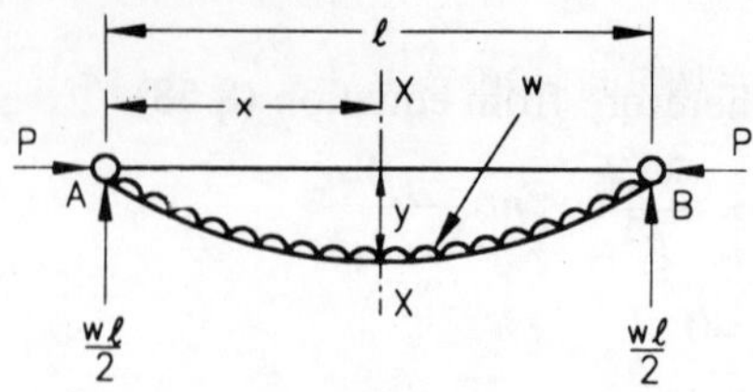


*Fig. 8.10*

Taking origin at A, for any section *XX*

$$EI\frac{d^2y}{dx^2} = -Py - \frac{wl}{2}x + \frac{wx^2}{2}$$

or $$\frac{d^2y}{dx^2} + \frac{P}{EI}y = \frac{-wl}{2EI}x + \frac{wx^2}{2EI} \quad \text{...(8.57)}$$

For the solution of equation (8.57) complementary function is

$$y = A \sin mx + B \cos mx$$

Where $$m^2 = P/EI$$

and roman $$\text{P.I.} = \left[\frac{1}{D^2 + m^2}\right]\left[\frac{-wl}{2EI}x + \frac{wx^2}{2EI}\right]$$

$$= \frac{1}{m^2}\left[1 - \frac{D^2}{m^2} + \ldots\right]\left[\frac{-wl}{2EI}x + \frac{wx^2}{2EI}\right]$$

$$= \frac{1}{m^2}\left[\left(\frac{-wlx}{2EI} + \frac{wx^2}{2EI}\right) - \frac{1}{m^2}\left(0 + \frac{w}{2EI}.2\right)\right]$$

$$= \frac{EI}{P}\left[\left(\frac{-wlx}{2EI} + \frac{wx^2}{2EI}\right) - \frac{EI}{P}\frac{w}{EI}\right]$$

$$= \frac{-wl}{2P}x + \frac{wx^2}{2P} - \frac{wEI}{P^2}$$

Thus the complete solution of equation (8.57) is

$$y = A \sin mx + B \cos mx - \frac{wlx}{2P} + \frac{wx^2}{2P} - \frac{wEI}{P^2} \quad \text{...(8.58)}$$

$$\therefore \quad \frac{dy}{dx} = m\,A \cos mx - B\,m \sin mx - \frac{wl}{2P} + \frac{wx}{P} \quad \text{...(8.59)}$$

## End conditions :

(*i*) At $x = 0$, $y = 0$ ; therefore from equation (8.58)

$$B = \frac{wEI}{P^2}$$

(*ii*) At $x = \frac{l}{2}$, $dy/dx = 0$

Therefore from equation (8.59), are get

$$0 = mA\cos\frac{ml}{2} - \frac{wEI}{P^2} m\sin\frac{ml}{2} - \frac{wl}{2P} + \frac{wl}{2P}$$

or

$$A = \frac{wEI}{P^2}\tan\left(\frac{ml}{2}\right)$$

Thus

$$y = \frac{wEI}{P^2}\tan\frac{ml}{2}\sin mx + \frac{wEI}{P^2}\cos mx - \frac{wlx}{2P} + \frac{wx^2}{2P} - \frac{wEI}{P^2} \quad ...(8.60)$$

At $x = l/2$

$$y_{max} = \frac{wEI}{P^2}\tan\frac{ml}{2}\sin\frac{ml}{2} + \frac{wEI}{P^2}\cos\frac{ml}{2} - \frac{wl^2}{4P} + \frac{wl^2}{8P} - \frac{wEI}{P^2}$$

$$= \frac{wEI}{P^2}\left[\sec\frac{ml}{2}\right] - \frac{wl^2}{8P} - \frac{wEI}{P^2}$$

or

$$y_{max} = \frac{wEI}{P^2}\left[\sec\frac{ml}{2} - 1\right] - \frac{wl^2}{8P} \quad ...(8.61)$$

Thus the maximum bending moment, which will occur at mid section is

$$M_{max} = Py_{max} + \frac{wl}{2}\cdot\frac{l}{2} - \frac{wl}{2}\cdot\frac{l}{4}$$

$$= P\cdot\frac{wEI}{P^2}\left[\sec\frac{ml}{2} - 1\right] - P\cdot\frac{wl^2}{8P} + \frac{wl^2}{4} - \frac{wl^2}{8}$$

$$= \frac{wEI}{P}\left[\sec\frac{ml}{2} - 1\right]$$

Corresponding maximum bending stress

$$(\sigma_b)_{max} = \frac{M_{max}}{I}\cdot c$$

$$= \frac{wEIc}{PI}\left(\sec\frac{ml}{2} - 1\right)$$

Total compressive stress

$$\sigma_{max} = \frac{P}{A} + (\sigma_b)_{max}$$

$$= \frac{P}{A} + \frac{wEc}{P}\left[\sec\frac{l}{2}\sqrt{P/EI} - 1\right] \qquad ...(8.62)$$

**Example 8.1.** *A straight cylindrical bar is 16 mm diameter and 1.2 m long. It is freely supported at its two ends in a horizontal position and loaded at the centre with a concentrated load of 90 N. The central deflection is found to be 5 mm.*

*If placed vertical and loaded along its axis, what load would cause it to buckle ? (use Euler's formula). What is the ratio of the maximum stresses in the two cases ?*

**Case I.** The central deflection for simply supported beam is given by

$$y = \frac{WL^3}{48\,EI}$$

*i.e.*

$$EI = \frac{90 \times (1200)^2}{48 \times 5}$$

$$M_{max} = \frac{WL}{4}$$

$$= \frac{90 \times 1200}{4} = 2700 \text{ N} - \text{mm}$$

$$(\sigma_b)\text{max} = \frac{2700 \times 64}{\pi \times (16)^4}\,\frac{16}{2} = 67.1 \text{ N/mm}^2$$

Case II.

$$P_e = \frac{\pi^2\,EI}{l^2}$$

$$= \frac{\pi^2 \times 90\,(1200)^2}{(1200)^2 \times 48 \times 5} = 4400 \text{ N}$$

$$\sigma_e = \frac{P_e}{A} = \frac{4440 \times 4}{\pi\,(16)^2} = 22 \text{ N/mm}^2$$

Therefore ratio $\dfrac{(\sigma_b)_{max}}{\sigma_e} = \dfrac{67.1}{22} = 3.04$ .

**Example 8.2.** (*a*) *From some tests on steel struts with end fixed in position but free in direction, two of the results obtained were :*

| *Test No.* | *1* | *2* |
|---|---|---|
| *Ratio of slenderness* | *70* | *170* |
| *Average Stress of failure ($MN/m^2$)* | *200* | *69* |

*Assuming these two values are in agreement with the Rankine formula, find the two constants in the formula.*

*(b) If a steel bar of rectangular section 60 mm by 20 mm and of length 1.25 m is used as a strut with both ends fixed in position and direction, find the safe load using the constants derived in (a) and employing a factor of safety of 4.*

**Solution.**

(*a*) $$\frac{P_R}{A} = \frac{\sigma_c}{1+\alpha\,(l/k)^2}$$ (The ends are pinned in part a)

For test No. 1

$$200 = \frac{\sigma_c}{1+\alpha\,(70)^2} \qquad ...(A)$$

For test No. 2

$$69 = \frac{\sigma_c}{1+\alpha\,(170)^2} \qquad ...(B)$$

Dividing *A* and *B*

$$\frac{200}{69} = \frac{1+\alpha\,(170)^2}{1+\alpha\,(70)^2}$$

$$200 + 200 \times 70^2\,\alpha = 69 + 69 \times 170^2\,\alpha$$

or $$\alpha = \frac{131}{69 \times 170^2 - 200 \times 70^2} = \frac{1}{7750}$$

Then from *A*

$$\sigma_c = 200\left(1+\frac{1}{7750}\times 70^2\right) = 326.4\text{MN/m}^2$$

(*b*) For fixed ends

$$P_R = \frac{A\sigma_c}{1+\alpha\left(\frac{l}{2k}\right)^2} \qquad (\text{as } l_e = l/2)$$

Now least $$k^2 = \frac{I_{\text{least}}}{A} = \frac{60\,(20)^3}{12} \times \frac{1}{60\times 20}\ \text{mm}^2$$

$$= \frac{1}{3 \times 10^4} \text{m}^2$$

Therefore safe load $P$ is given by

$$P = \frac{P_R}{A} = \frac{1}{4} \times \frac{(.06 \times .02)\,326.4}{\left[1 + \frac{(1.25)^2 \times 3 \times 10^4}{4 \times 7750}\right]} = 38900 \text{ N.}$$

**Example 8.3.** *What are the fundamental assumptions in deriving the Euler formula for the buckling load of a strut ? Calculate this buckling load for a strut of T-section, the flange width being 100 mm, overall depth 80 mm, and with both flange and stem 10 mm thick. The strut is 3 m long and hinged at both ends. E = 200 GN/m².*

**Solution.** Taking moments of areas about top of flange where '$h$' is the distance of centroid $G$, from top of flange, we get

$$h = \frac{100 \times 10 \times 5 + 70 \times 10 \times 15}{1000 + 700}$$

$$= 21.47 \text{ mm}$$

$$I_{xx} = \frac{100 \times 10^3}{12} + 1000\,(21.47 - 5)^2$$

$$+ \frac{70^3 \times 10}{12} + 700\,(58.53 - 35)^2$$

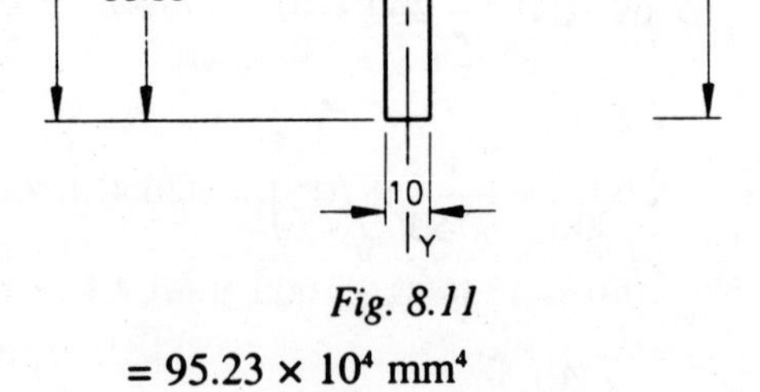

*Fig. 8.11*

$$= 95.23 \times 10^4 \text{ mm}^4$$

$$I_{yy} = \frac{10 \times (100)^3}{12} + \frac{70 \times 10^3}{12}$$

$$= 83.88 \times 10^4 \text{ mm}^4$$

$$= 83.88 \times 10^{-8} \text{ m}^4$$

Thus $I_{yy} < I_{xx}$

Now $$P_e = \frac{\pi^2 EI}{l_e^2}$$

$$= \frac{\pi^2 \times 100 \times 10^9 \times 83.88 \times 10^{-8}}{(3)^2}$$

$$= 184000 \text{ N.}$$

**Example 8.4.** *A hollow circular steel strut with its ends position fixed has a length of 2.4 m, its external diameter being 100 mm and its internal diameter 87.5 mm. Before loading, the strut is bent with a maximum deviation of 4.5 mm. Assuming that the centre line of the strut is sinusoidal, determine the maximum stress due to a central compressive end load of 10 kN. E = 200 GN/m².*

**Solution.** From equation (8.52), the max. stress is

$$\sigma_{max} = \frac{P}{A}\left[\frac{ec}{k^2}\left(\frac{P_e}{P_e - P}\right) + 1\right]$$

Now $$k^2 = \frac{D^2 + d^2}{16} = \frac{100^2 + 87.5^2}{16}$$

$$= 1104.38 \text{ mm}^2$$

$$A = \frac{\pi}{4}(100^2 - 87.5^2)$$

$$= 1840 \text{ mm}^2$$

$$P_e = \frac{\pi^2 EI}{l^2}$$

$$= \frac{\pi^2 \times 200 \times 10^9}{(2.4)^2}\left(\frac{1840}{10^6} \times \frac{1104.38}{10^6}\right) \text{N}$$

$$= 697 \text{ kN}$$

$$\sigma_{max} = \frac{10000}{.001840}\left[\frac{4.5}{1000} \times \frac{50}{1000} \times \frac{10^6}{1104.38} \times \left(\frac{697}{697 - 10}\right) + 1\right]$$

$$= \frac{10000}{.001840} \times 1.206 \text{ N/m}^2$$

$$= 6.55 \text{ MN/m}^2.$$

**Example 8.5.** *A locomotive coupling rod, 2.5 m between centres, is of symmetrical I-section : the outside dimensions are : 40 mm wide by 120 mm deep, web 14 mm thick* *and flanges each 20 mm thick. Find the maximum compressive stress in the rod due to an end thrust of 150 kN and a uniform transverse load of 5 kN/m. E = 200 GN/m².*

*Assume the bending about x-axis and ends pin jointed.*

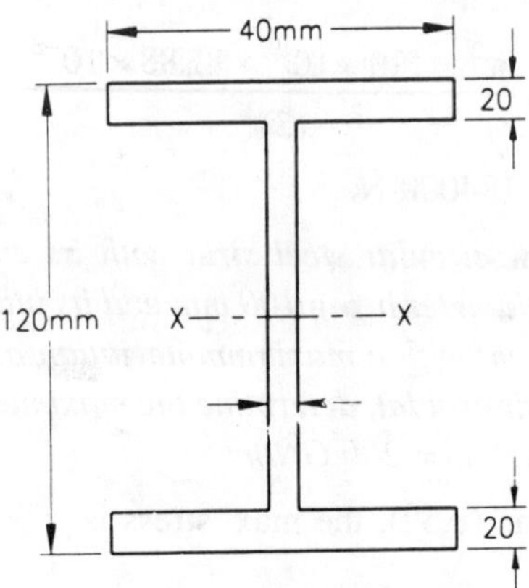


*Fig. 8.12*

**Solution.**

$$I_{xx} = \frac{40 \times 120^3}{12} - 2 \times \frac{13 \times 80^3}{12}$$

$$= 576 \times 10^4 - 111 \times 10^4$$

$$= 465 \times 10^4 \text{ mm}^4$$

From equation (8.62)

$$\sigma_{max} = \frac{P}{A} + \frac{wcE}{P}\left[\sec\frac{l}{2}\sqrt{\frac{P}{EI}} - 1\right]$$

Now $\sec\frac{1}{2}\sqrt{\frac{P}{EI}} = \sec\left[\frac{2.5}{2}\sqrt{\frac{150 \times 1000}{200 \times 10^9 \times 465 \times 10^{-8}}}\right]$

$$= \sec\,[0.502] = 1.17$$

$$\therefore \quad \sigma_{max} = \frac{150 \times 10^3}{2720 \times 10^{-6}} + \frac{5000 \times 200 \times 10^9 \times .060}{150 \times 10^3}(1.17 - 1)$$

$$= 5.52 \times 10^7 + 6.8 \times 10^7 \text{ N/m}^2$$

$$= 123.2 \text{ MN/m}^2.$$

**Example 8.6.** *A tubular steel strut 12.3 m long having outer and inner diameters of 38 mm and 33 mm, is loaded through pin joints at both ends. It is gradually loaded axially until it buckles. Assuming Euler's formula to apply in this case, estimate the maximum central deflection before the steel passes the yield point at 376 N/mm². Take E = 200 GN/m².*

**Solution.** $$I = \frac{\pi}{64}(38^4 - 33^4) = 4.41 \times 10^4 \text{ mm}^4$$

$$P_e = \frac{\pi^2 EI}{l_e^2} = \frac{\pi^2 \times 200 \times 10^9 \times 4.41 \times 10^{-3}}{(12.3)^3}$$

$$= 575 \text{ N}$$

Direct stress $$\sigma_e = \frac{P_e}{A} = \frac{575 \times 4}{\pi (38^2 - 33)^2}$$

$$= 206 \text{ N/mm}^2$$

Therefore $$\sigma_b = 376 - 206 = 170 \text{ N/mm}^2$$

$\therefore$ $$M = P_e \cdot y = \frac{I\sigma_b}{c}$$

*i.e.* $$y = \frac{I\sigma_b}{cP_e} = \frac{4.41 \times 10^4 \times 170}{19 \times 575}$$

$$= 687 \text{ mm}$$

## PROBLEMS FOR PRACTICE

1. A straight bar of alloy 1 m long and 12.5 mm × 5 mm in section is mounted in a strut testing machine and loaded axially till it buckles. Assuming the Euler formula for pinned ends to apply, estimate the maximum central deflection before the material attains its yield point, at 280 MN/m$^2$, $E$ = 75 GN/m$^2$. [**Ans.** 0.15 m]

2. Derive the Euler formula for a strut hinged at both ends. State clearly what you regard as the three most important assumptions made in deriving this formula.

   A straight cylindrical bar is 16 mm diameter and 1.2 m long. It is freely supported at its two ends in a horizontal position and loaded at the centre with a concentrated load of 90 N. The central deflection is found to be 5 mm.

   If placed vertical and loaded along its axis, what load would cause it to buckle ? What is the ratio of the maximum stresses in the two cases ? Assume ends to be hinged. [**Ans.** 4.44 kN ; 3.04]

3. A cast iron column 200 mm external diameter is 20 mm thick and has a length of 4.5 m. Assuming it can be treated as rigidly fixed at each end, calculate the safe load by Rankine's formula, using the following empirical constants.

   $\sigma_c$ = 550 MN/m$^2$, $\alpha$ (for hinged ends) = 1/1600, factor of safety = 4.

   [**Ans.** 879 kN]

4. A hollow cast iron column with fixed ends supports an axial load of 1 MN. If the column is 4.5 m long and has an external diameter of 250 mm, find the thickness of metal required. Use the Rankine formula, taking a constant of 1/6400 for pinned ends and a working stress of 80 MN/m$^2$. [**Ans.** 28 mm]

5. In the experimental determination of the buckling loads for 12.5 mm diameter mild steel pin-ended struts of various lengths, two of the values obtained were : (*i*) length 0.5 m, load 9.25 kN, (*ii*) length 0.2 m, load 25 kN.

(*a*) Make the necessary calculations and then state whether either of the values conforms with the Euler formula for the critical load.

(*b*) Assuming that both values are in agreement with the Rankine formula, find the two constants for this formula.

Take $E$ as 200 GN/m$^2$. **[Ans.** 301.5 MN/m$^2$ ; 1/8540]

6. Compare the crippling loads given by Euler's and Rankine's formulae for a tubular steel strut 2.25 m long, having outer and inner diameters of 38 mm and 33 mm respectively, loaded through pin joints at both ends. Take the yield stress as 325 MN/m$^2$, the Rankine constant (for pinned ends) = 1/7500 and $E$ = 200 GN/m$^2$. For what length of strut of this cross-section does the Euler formula cease to apply ? **[Ans.** 17.24 kN ; 17.2 kN ; 0.98 m]

7. A strut 3 long is constructed of steel tube 75 mm outside diameter and 3 mm thick. The ends are pin jointed, but the end load of 50 kN is applied eccentrically through a line parallel and 2.5 mm away from the axis of the strut, which is initially straight. Find the deflection and the maximum stress at the centre of the length. E = 200 GN/m$^2$. **[Ans.** 3.38 mm ; 98.7 MN/m$^2$]

8. A steel bar 25 mm diameter and 1.8 m long is tested as a pin-jointed strut. Calculate the crippling load if the bar is initially straight and is centrally loaded. Find the load which will produce the yield stress of 300 MN/m$^2$ (3300 kg/cm$^2$) in this bar, if it is centrally loaded, but has initial curvature with an eccentricity of 9 mm at the centre of its length. $E$ = 200 GN/m$^2$. **[Ans.** 11.7 kN ; 9.4 kN]

9. A vertical strut of uniform section is fixed rigidly at the base and carries a vertical load $W$ at the top acting with an eccentricity $e$. In addition there is a horizontal force at the top, $H$, acting so as to produce bending in the same plane as W and tending to increase the deflection. Obtain a formula for the maximum bending moment.

If the column is a tube 50 mm outside diameter and 44 mm inside diameter of free length 1.5 m and if $W$ = 9 kN acting with an eccentricity of 25 mm, find $H$ to produce a maximum stress of 275 MN/m$^2$, $E$ = 200 GN/m.

$$\left[\textbf{Ans. } We \sec ml + \frac{H}{m} \tan ml \,;\, 419 \text{ N}\right]$$

10. A rolled steel joist 300 mm by 125 mm and 6 m long is used as a strut with hinged ends. It carries an axial load of 300 kN together with a lateral load of 16 kN/m uniformly distributed along one flange over the entire length. Determine the maximum stress produced. $I = 86 \times 10^{-6}$ m$^4$ ; $A = 5.89 \times 10^{-3}$ m$^2$ ; $E$ = 200 GN/m$^2$. **[Ans.** 185.6 MN/m$^2$]

Chapter–9

# Thin Shells

**Thin Shells**. Vessels used for storing fluid under pressure are called shells. These are usually cylindrical or spherical in shape. Cylindrical shells with hemispherical ends are not uncommon. A shell with wall thickness small as compared to the internal diameter ($d/t \geq 20$ ) is called a thin shell. When empty the pressure inside and outside the shell is atmospheric and the shell material is said to be under zero stress. In other words, the atmospheric pressure is taken as the reference pressure.

## 9.1 Cylindrical Shells

When the shell is subjected to an internal pressure $p$ (gauge pressure) this will tend to burst the shell. The stress induced on the longitudinal section *XX* (Fig 9.1) due to the force trying to separate the upper and lower halves is termed as the circumferential or hoop stress ($\sigma_h$) which is tensile in nature. On the other hand the stress on the circumferential section *YY* due to force tending to separate the left and right hand portions of the cylinder is called the longitudinal stress ($\sigma_l$).

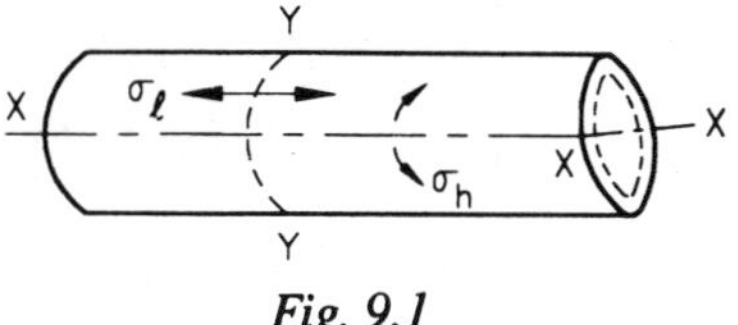


*Fig. 9.1*

The longitudinal and circumferential stresses are calculated on the assumption that the thickness is small in comparison to diameter so that the stress variation along the thickness is ignored and it is taken as constant. In addition it is also assumed that the ends give no support to the sides. Let the internal diameter of the shell be d , the wall thickness t and internal pressure $p$. Then the longitudinal and circumferential stresses are determined as follows :

### 9.1.1 Longitudinal Stress

Considering the equilibrium of the left and right hand halves created by the section *YY* (Fig. 9.2), if $P_l$ is the force on the cover due to interval pressure, we have

$$P_l = \sigma_l \, \pi dt = \frac{\pi d^2}{4} P$$

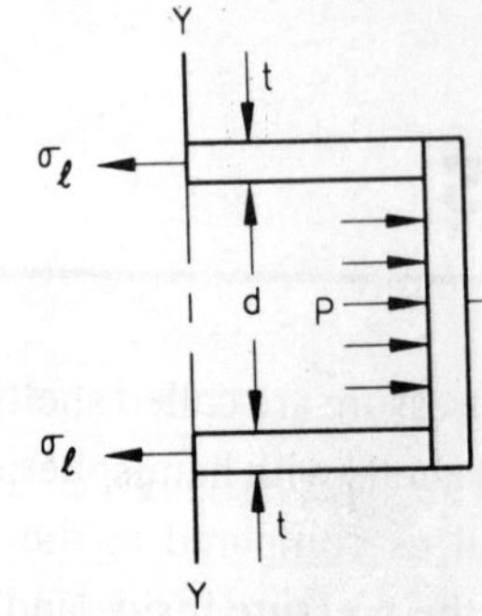

Fig. 9.2

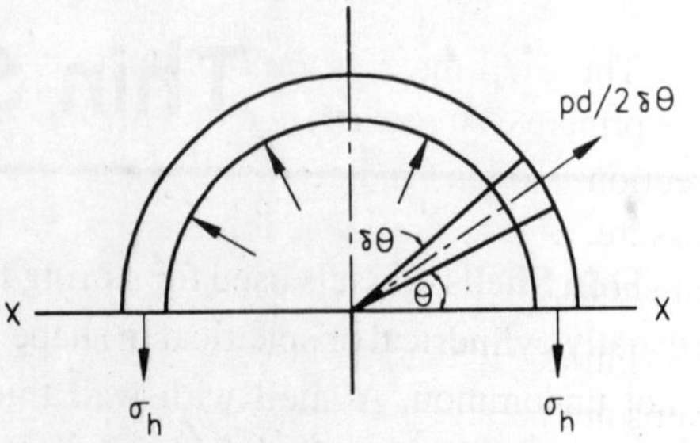

Fig. 9.3

or $$\sigma_l = \frac{pd}{4t} \quad ...(9.1)$$

### 9.1.2 Hoop or Circumferential Stress

This can be found by considering the equilibrium of the top and bottom portions of the cylinder obtained by taking the section along the longitudinal plane (Fig. 9.3). For the equilibrium the resultant vertical upward force due to pressure $p$ on the semicylindrical surface will be equal to the resisting force $2\,\sigma_h\, tl$ which acts on two rectangular areas $t \times l$ of the shell of length $l$ and thickness t . The upward resultant force due to p can be found by first considering an element of length $\left(\frac{d}{2} d\theta\right)$ at an angle $\theta$ with respect to the $x$-axis. Thus resultant vertical force $V$ is

$$V = \int_0^{\pi} \left[\left(\frac{d}{2}\, d\theta\right) l.p\right] \sin\theta$$

$$= \frac{1}{2}(pld)\left[-\cos\theta\right]_0^{\pi} = lpd$$

Thus $$V = lpd$$

or $$V = p \times \left[\begin{array}{l}\text{Projected area of the curved surface on} \\ \text{horizongal plane passing through the} \\ \text{shell axis}\end{array}\right] \quad ...(9.2)$$

Therefore for the equilibrium of the upper half of the shell

$$2\,\sigma_h\, tl = lpd$$

or $$\sigma_h = \frac{pd}{2t} \quad ...(9.3)$$

Evidently from equations (9.1) and (9.3)

$$\sigma_h = 2\,\sigma_l \qquad \text{...(9.4)}$$

### 9.1.3 Longitudinal and Diametral Strains

The shell material on the inner surface of the shell will be under the action of three principal stresses (*i*) $\sigma_l$ along the longitudinal axis, (*ii*) $\sigma_h$ along the circumferential direction and (*iii*) radial compressive stress which is numerically equal to the fluid pressure. The last of these three is generally negligible in comparison to $\sigma_h$ and $\sigma_l$ for thin shells.

Further if ν is the Poisson's ratio of the material and *E* its Young's modulus the strains are then given by

(*a*) Longitudinal strain

$$\sigma_l = \frac{\sigma_l}{E} - \nu\frac{\sigma_h}{E}$$

$$= \frac{1}{E}\left[\frac{pd}{4t} - \nu\frac{pd}{2t}\right]$$

or

$$\varepsilon_l = \frac{pd}{4tE}[1 - 2\nu] \qquad \text{...(9.5)}$$

(*b*) Circumferential strain

$$\varepsilon_h = \frac{1}{E}[\sigma_h - \nu\sigma_l]$$

$$= \frac{1}{E}\left[\frac{pd}{2t} - \nu\frac{pd}{4t}\right]$$

or

$$\varepsilon_h = \frac{pd}{4tE}(2 - \nu) \qquad \text{...(9.6)}$$

The circumference, *c*, is in direct proportion to the internal diameter

$$c = \pi d \qquad \text{...(9.7}a\text{)}$$

Differentiating equation (9.7*a*)

$$\delta c = \pi \delta d \qquad \text{...(9.7}b\text{)}$$

$$\frac{\delta c}{c} = \frac{\delta d}{d}$$

Therefore from equations (9.7)

$$\frac{\delta c}{c} = \frac{\delta d}{d} \qquad \text{...(9.8)}$$

Evidently from equation (9.8) the diametral and circumferential strains are equal. Thus from equations (9.6) and (9.8)

$$\varepsilon_h = \frac{\delta c}{c} = \frac{\delta d}{d} = \frac{pd}{4tE}(2 - \nu) \qquad \text{...(9.9)}$$

### 9.1.4 Increase in the Volume of Shell Due to Internal Pressure

$$\text{Volume} = V = \frac{\pi}{4} d^2 l \qquad \text{...(9.10)}$$

If the diameter increases by δd and length by $\delta l$, the corresponding increase in volume $\delta V$ is obtained by differentiating equation (9.10).

$$\delta V = \frac{\pi}{4}[(2d\,\delta d)\,l + d^2\,\delta l] \qquad \text{...(9.11)}$$

Dividing equation (9.11) by (9.10)

$$\frac{\delta V}{V} = 2\frac{\delta d}{d} + \frac{\delta l}{l} \qquad \text{...(9.12)}$$

That is volumetric strain = twice the diametrial strain + longitudinal strain

Therefore substituting equations (9.5) and (9.9) in equation (9.12)

$$\frac{\delta V}{V} = 2\frac{pd}{4tE}(2 - \nu) + \frac{pd}{4tE}(1 - 2\nu)$$

or

$$\frac{\delta V}{V} = \frac{pd}{4tE}(5 - 4\nu) \qquad \text{...(9.13)}$$

The change in the volume can then be found by multiplying the volumetric strain with volume of the shell.

## 9.2 Spherical Shell

Let d be the internal diameter and $t$ the thickness of a thin spherical shell Fig. 9.4 which is subjected to an internal pressure $p$. The shell being symmetrical will have

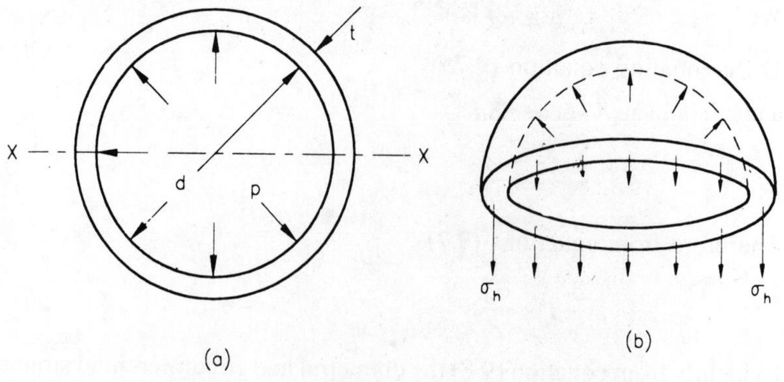


*Fig. 9.4*

equal tendency to burst in all directions. In other words the hoop or circumfrential stress in the material will be same in all directions. Take the section along the plane *XX*, the upper hemisphere Fig. 9.4(*b*) will be in equilibrium under the action of the resisting force due to the lower half portion and the upward bursting force due to the pressure $p$ acting on the internal hemispherical surface.

Consider a thin spherical shell subjected to an internal pressure as shown in Fig. 9.4.

Let $p$ = Intensity of internal pressure,

$d$ = Diameter of the shell and

$t$ = Thickness of the shell.

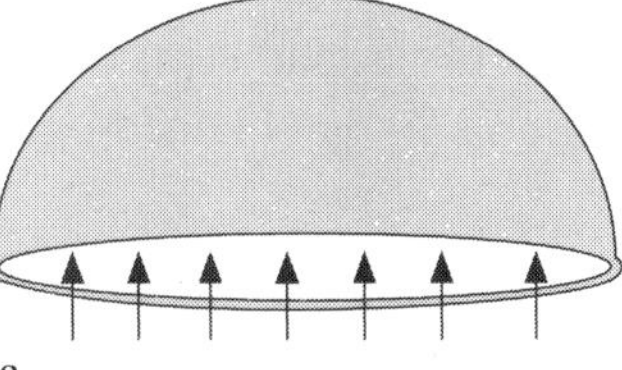

*Fig. 9.4* Spherical Shell

As a result of this internal pressure, the shell is likely to be torn away along the centre of the sphere. Therefore, total pressure acting along the centre of the sphere,

$$P = \text{Intensity of internal pressure} \times \text{Area}$$

$$= p \times \frac{\pi}{4} \times d^2$$

and stress in the shell material,

$$\sigma = \frac{\text{Total pressure}}{\text{Resisting section}} = \frac{p \times \frac{\pi}{4} \times d^2}{\pi d \times t} = \frac{pd}{4t}$$

**Note.** If $\eta$ is the efficiency of the riveted joints of the spherical shell, then stress,

$$\sigma = \frac{pd}{4t\eta}$$

### 9.2.1 Diametral Strain and Increase in Volume

Consider a thin spherical shell subjected to an internal pressure.

Let $d$ = Diameter of the shell,

$p$ = Intensity of internal pressure,

$t$ = Thickness of the shell and

$\nu$ = Poisson's ratio.

We have already discussed in the last article that the stress in a spherical shell,

$$\sigma = \frac{pd}{4t}$$

and strain in any one direction,

$$\varepsilon = \frac{\sigma}{E} - \frac{\nu\sigma}{E} \qquad ...(\because \sigma_1 = \sigma_2 = \sigma)$$

$$= \frac{pd}{4tE} - \frac{\nu pd}{4tE} = \frac{pd}{4tE}(1-\nu)$$

$\therefore$ Change in diameter,

$$\delta d = \varepsilon . d = \frac{pd}{4tE}(1-\nu)d$$

$$= \frac{pd^2}{4tE}(1-\nu)$$

We also know that original volume of the sphere,

$$V = \frac{\pi}{6} \times (d)^3$$

and final volume due to pressure,

$$V + \delta V = \frac{\pi}{6} \times (d + \delta d)^3$$

where $(d + \delta d)$ = Final diameter of the shell.

∴ Volumetric strain,

$$\frac{\delta V}{V} = \frac{(V + \delta V) - V}{V}$$

$$= \frac{\frac{\pi}{6}(d + \delta d)^3 - \frac{\pi}{6} \times d^3}{\frac{\pi}{6} \times d^3}$$

$$= \frac{d^3 + (3d^2\,\delta d) - d^3}{d^3}$$

...(Ignoring second and higher power of $\delta$d)

$$= \frac{3.\delta d}{d} = 3\varepsilon$$

and

$$\delta V = V.\, 3\varepsilon$$

$$= \frac{\pi}{6} d^3 \times 3 \times \frac{pd}{4tE}(1 - \nu)$$

$$= \frac{\pi d^4}{8tE}(1 - \nu)$$

**Example 9.1.** *A cylindrical compressed air drum is 2 m in diameter with plates 12.5 mm thick. The efficiencies of the longitudinal and circumferential joints are respectively 85 and 45%. If the tensile stress in the plating is to be limited 100 MN/l², find the maximum safe air pressure.*

**Solution.** For longitudinal joint, the efficiency of 85% will increase the hoop stress to $\delta_h/\eta_l$ where $\eta_l$ is the efficiency of the longitudinal joint.

Thus if p is the internal pressure

$$\frac{pd}{2t\eta_l} = 100$$

or

$$p = \frac{100 \times 2 \times 12.5 \times 0.85}{2 \times 1000}$$

$$= 1.063 \text{ N/mm}^2 \text{ or } MN/\text{m}^2$$

For the circumferential joint, the efficiency of 45% will increase the longitudinal stress to $\frac{\sigma_l}{\eta_h}$ where $\eta_h$ is the efficiency of the circumferential joint.

Thus if the internal pressure is $p$

$$\frac{pd}{4t\,\eta_h} = 100$$

$$p = \frac{100 \times 4 \times 12.5 \times 0.45}{2 \times 1000}$$

$$= 1.125 \text{ N/mm}^2 \text{ or MN/m}^2$$

Evidently the limiting pressure is decided on the basis of hoop stress.

Thus $P = 1.063 \text{ MN/m}^2$.

**Example 9.2.** *A spherical vessel of 0.75 m diameter is made from 11 mm steel plate and is just filled with water at atmospheric pressure. When an additional 0.000885 m³ of water is pumped in the pressure rises to 5.8 MN/m² (gauge). Find the bulk modulus of water. For steel take E = 200 GN/m² and Poisson's ratio = 0.286.*

**Solution.** The additional quantity of water pumped in accounts for the change in volume of the shell as well as the compression of the water in it. Thus if $K_w$ is the bulk modulus of water,

$$0.000885 \; m^3 = \delta V_{\text{shell}} + \delta V_{\text{water}}$$

Now $$\delta V_{\text{shell}} = \frac{3pd}{4tE}(1-\nu) \times \frac{\pi}{6} d^3$$

$$= \frac{3 \times 5.8 \times 10^6 \times 9.75)^2 \times .714 \times \pi \times 10^3}{4 \times 11 \times 200 \times 10^9 \times 6}$$

$$= 0.000234 \text{ m}^3$$

and $$\delta V_{\text{water}} = \frac{pV}{K_w} = \frac{p}{K_w}\frac{\pi}{6}d^3$$

or $$0.000885 - 0.000234 = \frac{p}{K_w}\frac{\pi}{6} \times (0.75)^3$$

or $$K_w = \frac{\pi}{6} \times (0.75)^3 \times \frac{p}{0.000651}$$

$$= \frac{\pi \times (0.75)^3}{6\,(0.000651)} \times 5.8 \times 10^6$$

$$= 1.27 \times 10^9 \text{ N/m}^2$$

$$= 1.97 \text{GN/m}^2$$

**Example 9.3.** *The ends of a thin cylinder, 150 mm internal diameter and wall thickness 2.5 mm, are closed by rigid plates and it is then filled with a liquid. When an axial compressive force of 28 kN is applied to the cylinder, the pressure of the liquid rises by 60 kN/m². Find the bulk modulus of liquid. Take E = 200 GN/f² and ν = 0.3.*

**Solution.** Since no liquid is pumped in or is allowed to escape the compressive volumetric strain of the liquid will be equal to the compressive volumetric strain of the shell. Thus if $P$ be the axial force on the shell which raises the pressure of liquid to $p$, then

$$\sigma_l = \frac{P}{\pi dt} - \frac{pd}{4t}$$

$$= \frac{28 \times 10^3}{\pi \times .150 \times .0025} - \frac{60 \times 10^3 \times .150}{4 \times .0025}$$

$$= 21.7 \times 10^6 \text{ N/m}^2 \text{ (Comp.)}$$

$$\varepsilon_h = \frac{pd}{2t}$$

$$= \frac{60 \times 10^3 \times .150}{2 \times .0025}$$

$$= 1.8 \times 10^6 \text{ N/m}^2 \text{ (Tensile)}$$

$$\varepsilon_l = \frac{1}{E}(\sigma_l - \nu\sigma_h)$$

$$= \frac{21.7 \times 10^6}{200 \times 10^9} - \frac{0.3 \times 1.8 \times 10^6}{200 \times 10^9}$$

or $$\varepsilon_l = -111.2 \times 10^{-6} \text{ (Comp.)}$$

$$\varepsilon_h = \frac{\sigma_h}{E} + \nu\frac{\sigma_l}{E}$$

$$= \frac{1.8 \times 10^6}{200 \times 10^9} + \frac{0.3 \times 21.7 \times 10^6}{200 \times 10^9}$$

$$= 41.5 \times 10^{-6} \text{ (tensile)}$$

Also the diametral strain = Circumferential strain

$$= 41.5 \times 10^{-6} \text{ (tensile)}$$

Therefore volumetric strain of shell

$$\varepsilon_V = \varepsilon_l + 2\varepsilon_h$$

$$= -111.2 \times 10^{-6} + 2 \times 41.5 \times 10^{-6}$$

$$= -28.2 \times 10^{-6} \text{ (Comp.)}$$

But the compressive volumetric strain of liquid

$$(\varepsilon_v)_{\text{liq}} = \frac{p}{K} = \frac{60 \times 10^3}{K}$$

Thus equating these two, we get

$$K = \frac{60 \times 10^3}{28.2 \times 10^{-6}}$$

$$= 2.13 \times 10^9 \text{ N/m}^2 \text{ or } 2.13 \text{ GN/m}^2.$$

**Example 9.4.** *A solid cylindrical test piece, 75 mm long and 50 mm diameter, is enclosed within a hollow pressure vessel. With the test piece in the vessel, 20 × 10³ mm³ of oil are required just to fill the pressure vessel. Measurement shows that a further 50 mm³ of oil have to be pumped into the vessel to raise the oil pressure to 7 MN/m².*

*The experiment is repeated using the same pressure vessel and oil, but without the test piece inside the vessel. This time, after initially filling the pressure vessel, a further 364 mm³ of oil are needed to raise the pressure to 7 MN/m².*

*The test piece is made of Aluminium, for which Young's modulus and Poisson's ratio are 70 GN/m² and 0.3 respectively. Find the bulk modulus of oil.*

**Solution.** Volume of aluminium test piece

$$= \frac{\pi}{4} \times 50^2 \times 75 = 147.4 \times 10^3 \text{ mm}^3$$

Therefore volume of vessel

$$= 20 \times 10^3 + 147.4 \times 10^3$$

$$= 167.4 \times 10^3 \text{ mm}^3$$

Also $$K_{Al} = \frac{E}{3(1-2\nu)} = \frac{70}{3(1-0.6)}$$

$$= 58.3 \text{ GN/m}^2$$

In the first test the additional volume of oil pumped in is absorbed in expanding the pressure vessel, compressing the aluminium test piece and compressing the oil itself. Thus if $\delta V$ is the increase in the volume of the vessel for pressure rise to 7 MN/m², we have

$$\delta V + \frac{7 \times 10^6 \times 147.4 \times 10^3}{58.3 \times 10^9} + \frac{7 \times 10^6 \times 20 \times 10^3}{K_{oil}} = 50 \qquad \text{...}(A)$$

Similarly in the second test when the aluminium test piece is not there in the vessel, again at 7 MN/m², we have

$$\delta V + \frac{7 \times 10^6 \times 167.4 \times 10^3}{K_{oil}} = 364 \qquad \text{...}(B)$$

Subtracting ($A$) from ($B$)

$$\frac{7 \times 147.4 \times 10^9}{K_{oil}} + \frac{7 \times 147.4}{58.3} = 314$$

or $$K_{oil} = 3.48 \times 10^9 \text{ N/m}^2$$

$$= 3.48 \text{ GN/m}^2$$

## PROBLEMS FOR PRACTICE

1. A closed cylindrical vessel with plane ends is made of steel plate 3 mm thick, the internal dimensions being : length 0.6 m and diameter 0.25 m. Determine first the longitudinal and hoop stresses in the cylindrical shell due to an internal pressure of 2.8 MN/m$^2$, ignoring any stiffening effects due to the ends, and then calculate the increase in ($a$) length ($b$) diameter and ($c$) volume of the vessel. For steel take $E$ = 200 GN/m$^2$. Poisson's ratio = 0.286.

   [**Ans.** 0.075 mm increase ; 0.125 mm increase ; and 33.2 × 10$^3$ mm$^3$ increase]

2. A copper tube of 50 mm internal diameter, 1.2 m long and 1.25 mm thick, has closed ends and is filled with water under pressure. Neglecting any distortion of the end plates, determine the alteration of pressure when an additional volume of 3 × 10$^{-6}$ m$^3$ of water is pumped into the tube. Modulus of elasticity for copper = 100 GN/m$^2$. Poisson's ratio = 0.3. Bulk Modulus for water = 2 GN/m$^2$.

   [**Ans.** 1.445 MN/m2]

3. A spherical shell of 1.5 m internal diameter has wall made from 10 mm thick steel plate. It is to be hydraulically tested and is just filled with water. Find the volume of the additional water to be pumped into it so as to raise the pressure to 10 MN/m$^2$ value. Take the bulk modulus of water as 2GN/m$^2$ and for steel $E$ 200 GN/m$^2$, $\nu$ = 0.286). [**Ans.** 0.016 m$^3$]

4. The ends of long thin-walled tube are closed by rigid flat plates. The vessel is then subjected to internal fluid pressure under the following conditions :

   ($a$) Free movement of the ends being allowed.

   ($b$) The ends being rigidly stayed together.

   If Poisson's ratio is 0.25, find the ratio of the resulting increase in volume of the vessel in the above two cases. [**Ans.** 16 : 15]

5. The internal diameter of a thin straight metal tube full of water is 300 mm and its thickness is 3 mm. The ends are closed with rigid end plates and an axial compressive load L is applied to it. If the rise in pressure of the water is observed to be 51 kN/m$^2$, find the load $L$. Neglect bending in the tube due to end effects.

For the metal, $E = 140$ GN/m$^2$ and Poisson's ratio 0.35. Bulk modulus for water = 2.05 GN/m$^2$. **[Ans.** 10.6 kN]

6. A steel tube having outside and inside diameters of 45 mm and 35 mm is firmly plugged at both ends leaving an internal length of 250 mm between the flat ends of the plugs. The plugs are designed so that water can be admitted to the inner space and also so that an axial pull can be applied to the tube. If the tube is subjected to an axial pull of 40 kN and in addition is filled with water at a gauge pressure of 1.7 MN/m$^2$, find the volume of water which will escape from the tube if the axial load is removed and inner space opened to atmosphere.

   $K$ for water = 2.1 GN/m$^2$,

   $E = 200$ GN/m$^2$, Poisson's ratio = .286. **[Ans.** 400 mm$^3$]

7. Two thin shells, one spherical and the other cylindrical, have same diameter, wall thickness and volume. If they are subjected to the same internal fluid pressure find the ratio of change of their volumes.

   Take Poisson's ratio as ν. $\left[\textbf{Ans. } \dfrac{3(1-\nu)}{(5-4\nu)}\right]$

## Chapter–10

# Thick Cylinders and Spherical Shells

## 10.1 Lame's Theory

Cylindrical vessels, containing fluid under pressure and whose wall thickness is not small ($t \geq d/20$) are classified as thick cylinders. Unlike thin shells the radial stress in the wall thickness is not negligible, rather it varies from the inner surface where it is equal to the magnitude of the fluid pressure to the outer surface where usually it is equal to zero if exposed to the atmosphere. Further the circumferential stress also varies along the thickness. The variation in the radial as well as circumferential stresses across the thickness are obtained with the help of Lame's theory. The assumptions made in this theory are : (1) The material is homogeneous and isotropic, (2) Plane sections perpendicular to the longitudinal axis of the cylinder remain plane after the application of internal pressure, (3) The material is stressed within the elastic limit and (4) All the fibres of the material are free to expand or contract independently without being constrained by the adjacent fibres.

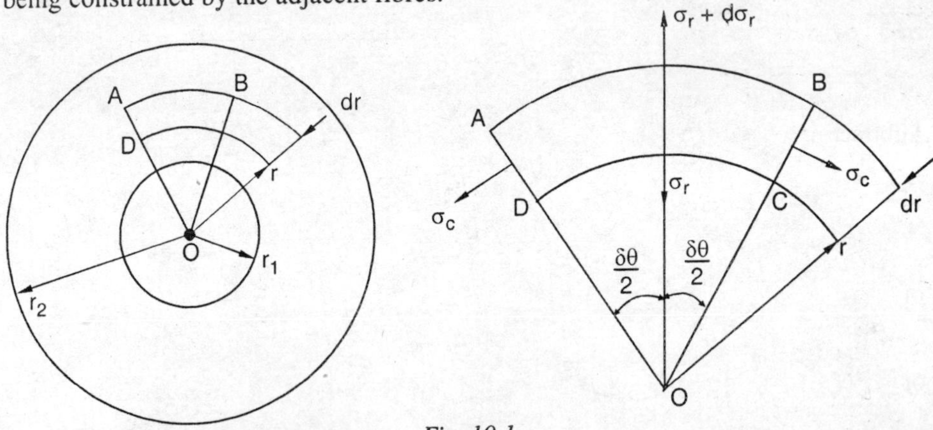


*Fig. 10.1*

Fig. 10.1 (*a*), *ABCD* is an element out of the annular cylinder of radius *r* and radial thickness dr. The element *ABCD* subtends an alone $\delta\theta$ at the centre *O*. fig. 10.1 (*b*) shows the free body diagram of the element *ABCD* on which act the hoop or circumferential stress $\sigma_C$ and radial stresses $\sigma_r$ and $(\sigma_r + d\sigma_r)$ at radii *r* and $(r + dr)$

respectively. These stresses have been assumed tensile. In case their value comes out under certain situation, negative it will mean that stress is compressive.

Considering the equilibrium of the clement *ABCD* and resolving the forces due to these stresses along the radial direction, we get for unit length of the cylinder.

$$(\sigma_r + d\sigma_r)(r + dr)\,\delta_\theta - \sigma_r r d_\theta - 2\sigma_C dr \sin\frac{\delta\theta}{2} = 0$$

Neglecting the product of small quantities and setting $\sin\frac{\delta\theta}{2}$ equal to $\frac{\delta\theta}{2}$, it reduces to

$$\sigma_C - \sigma_r = r\frac{d\sigma_r}{dr} \qquad ...(10.1)$$

From the second assumption it implies that this longitudinal strain at each and every point of the cross-section of the cylinder will be equal. That is the longitudinal strain $\delta_l$ at any radius $r$ will be constant. Also the stress in the longitudinal direction is constant.

Therefore

$$\varepsilon_l = \frac{\sigma_l}{E} - \frac{\nu\sigma_C}{E} - \frac{\nu\sigma_r}{E} = \text{constant}$$

or $$-\frac{\nu}{E}(\sigma_C + \sigma_r) = \left(\varepsilon_l - \frac{\sigma_l}{E}\right) = \text{constant}$$

or $$\frac{\nu}{E}(\sigma_C + \sigma_r) = \text{constant} = 2a \text{ (say)}$$

or $$\sigma_C + \sigma_r = 2a \qquad ...(10.2)$$

Subtracting equation (10.1) from equation (10.2)

$$2\sigma_r = 2a - \frac{rd\sigma_e}{d_r}$$

or $$\frac{r\,d\sigma_r}{dr} = 2a - 2\sigma_r$$

or $$\frac{d\sigma_r}{(a - \sigma_r)} = \frac{2dr}{r} \qquad ...(10.3)$$

Integrating equation (10.3)

$$2\ln(r) = \frac{\ln(a - \sigma_r)}{(-1)} + \ln(b)$$

where '$b$' is a constant of integration

or $$\ln(a - \sigma_r) = \ln(\mathrm{b}) - 2\ln(r)$$

$$= \ln\left(\frac{b}{r^2}\right)$$

or $$a - \sigma_r = \frac{b}{r^2}$$

or $$\sigma_r = a - \frac{b}{r^2} \qquad ...(10.4)$$

Then from equation (10.2) and (10.4) we get

$$\sigma_C = a + \frac{b}{r^2} \qquad ...(10.5)$$

**Case I : When inside pressure $p_1$ and outside pressure $p_2$ are given.**

The two constants '*a*' and '*b*' can be determined by applying the end conditions; *viz.* (*i*) at $r = r_1$, $\sigma_r = -p_1$ (*ii*) at $r = r_2$, $\sigma_r = -p_2$

Therefore from equation (10.4)

$$-p_1 = a - \frac{b}{r_1^2} \qquad ...(10.6)$$

$$-p_2 = a - \frac{b}{r_2^2} \qquad ...(10.7)$$

Evidently from equations (10.6) and (10.7)

$$p_1 - p_2 = \frac{b}{r_1^2} - \frac{b}{r_2^2}$$

$$= \frac{b\,(r_2^2 - r_1^2)}{r_1^2 r_2^2}$$

or $$b = \frac{r_1^2 r_2^2\,(p_1 - p_2)}{(r_2^2 - r_1^2)} \qquad ...(10.8a)$$

and from (10.6) $$a = \frac{b}{r_1^2} - p_1$$

$$= \frac{r_1^2\, r_2^2\,(p_1 - p_2)}{r_1^2\,(r_2^2 - r_1^2)} - p_1$$

$$= \frac{r_2^2 p_1 - r_2^2 P_2 - p_1 r_2^2 + p_1 r_1^2}{(r_2^2 - r_1^2)}$$

or $$a = \frac{p_1 r_1^2 - p_2 r_2^2}{(r_2^2 - r_1^2)} \qquad ...(10.8\ b)$$

Substituting values of *a* and *b* from equations (10.8*a*) and (10.8*b*) in equation (10.4) and (10.5) we get

$$\sigma_C = \frac{p_1 r_1^2 - p_2 r_2^2}{(r_2^2 - r_1^2)} + \frac{r_1^2 r_2^2 (p_1 - p_2)}{r^2 (r_2^2 - r_1^2)}$$

or
$$\sigma_C = \frac{1}{(r_2^2 - r_1^2)} \left[ p_1 r_1^2 - P_2 r_2^2 + \frac{r_1^2 r_2^2}{r^2} (p_1 - p_2) \right] \quad ...(10.9)$$

and
$$\sigma_r = \frac{p_1 r_1^2 - p_2 r_2^2}{(r_2^2 - r_1^2)} - \frac{r_1^2 r_2^2 (p_1 - p_2)}{r^2 (r_2^2 - r_1^2)}$$

or
$$\sigma_r = \frac{1}{(r_2^2 - r_1^2)} \left[ p_1 r_1^2 - p_2 r_2^2 - \frac{r_2^2 r_1^2 (p_1 - p_2)}{r^2} \right] \quad ...(10.10)$$

**Cast II :** When there is only internal pressure and the outer surface of the cylinder is exposed to atmospheric pressure, then the two end conditions become

$$\sigma_r = -p_2 = 0 \quad \text{at } r = r_2$$

$$\sigma_r = -p_1 \quad \text{at } r = r_1$$

Then the corresponding values of the constants '$a$' and '$b$' are obtained by equating $p_2 = 0$ in equations (10.8$a$) and (10.8$b$).

That is
$$a = \frac{p_1 r_1^2}{(r_2^2 - r_1^2)} \quad ...(10.11)$$

and
$$b = \frac{p_1 r_1^2 r_2^2}{(r_2^2 - r_1^2)} \quad ...(10.12)$$

Similarly the corresponding values of $\sigma_r$ and $\sigma_r$ are obtained by putting $p_2 = 0$ in equations (10.9) and (10.10) as

$$\sigma_r = \frac{p_1 r_1^2}{(r_2^2 - r_1^2)} - \frac{r_1^2 r_2^2 p_1}{r^2 (r_2^2 - r_1^2)}$$

$$= \frac{p_1 r_1^2}{(r_2^2 - r_1^2)} \left[ 1 - \frac{r_2^2}{r^2} \right] \quad ...(10.13)$$

and
$$\sigma_C = \frac{1}{(r_2^2 - r_1^2)} \left[ p_1 r_1^2 + \frac{r_1^2 r_2^2 p_1}{r^2} \right]$$

$$= \frac{p_1 r_1^2}{(r_2^2 - r_1^2)} \left[ 1 + \frac{r_2^2}{r^2} \right] \quad ...(10.14)$$

The distribution of stresses $\sigma_r$ and $\sigma_C$ can then he plotted with the help of equations (10.13) and (10.14) as shown in fig. (10.2). Evidently the maximum values of both $\sigma_r$ and $\sigma_C$ occur at the inner surface.

At $r = r_1$ $\quad (\sigma_r)_{max} = \dfrac{p_1 r_1^2}{(r_2^2 - r_1^2)}\left[1 - \dfrac{r_2^2}{r_1^2}\right] = -p_1$ ...(10.13*b*)

And at $r = r_1$

$$(\sigma_C)_{max} = \frac{p_1 r_1^2}{(r_2^2 - r_1^2)}\left[1 + \frac{r_2^2}{r_1^2}\right] = \frac{p_1(r_2^2 + r_1^2)}{(r_2^2 - r_1^2)} \quad \text{...(10.14a)}$$

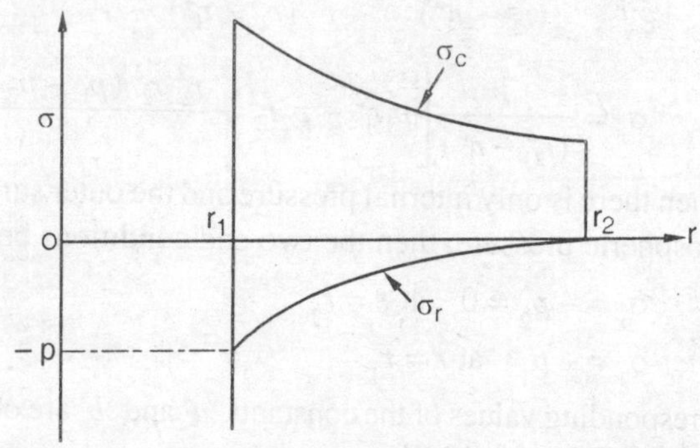


*Fig. 10.2*

Further for the closed ends of the cylinder the longitudinal stress is

$$\sigma_l = \frac{p_1 \pi r_1^2}{\pi(r_2^2 - r_1^2)}$$

$$= \frac{p_1 r_1^2}{(r_2^2 - r_1^2)} = a \qquad \text{(from 10.11)} \qquad \text{...(10.14b)}$$

**Case III :** When there is only external pressure and the internal surface of the cylinder is exposed to atmospheric pressure then the two end conditions are

$$\sigma_r = -p_1 = 0 \quad \text{at } r = r_1$$

and $\qquad \sigma_r = -p_2 \quad \text{at } r = r_2$

Then the corresponding values of the constants '*a*' and '*b*' are obtained by equating $p_1 = 0$ in equations (10.8*a*) and (10.8*b*)

That is $\qquad a = \dfrac{-p_2 r_2^2}{(r_2^2 - r_1^2)}$

and $\qquad b = \dfrac{p_2 r_1^2 r_2^2}{(r_2^2 - r_1^2)}$

Similarly the corresponding values of $\sigma_C$ and $\sigma_r$ are obtained by putting $p_1 = 0$ in equations (10.9) and (10.10)

as $\qquad \sigma_C = \dfrac{1}{(r_2^2 - r_1^2)}\left[-p_2 r_2^2 - \dfrac{p_2 r_1^2 r_2^2}{r^2}\right]$

$$= \frac{-p_2 r_2^2}{(r_2^2 - r_1^2)}\left[1 + \frac{r_1^2}{r^2}\right] \qquad ...(10.15a)$$

and $$\sigma_r = \frac{1}{(r_2^2 - r_1^2)}\left[-p_2 r_2^2 + \frac{p_2 r_1^2 r_2^2}{r^2}\right]$$

$$= \frac{p_2 r_2^2}{(r_2^2 - r_1^2)}\left[\frac{r_1^2}{r^2} - 1\right]$$

or $$\sigma_r = \frac{-p_2 r_2^2}{(r_2^2 - r_1^2)}\left(1 - \frac{r_1^2}{r^2}\right) \qquad ...(10.15b)$$

**Cast IV : Solid shaft subjected to external pressure $p_2$.**

The values of $\sigma_c$ and $\sigma_r$ are given by equations (10.4) and (10.5). Since $\sigma_r$ and $\sigma_C$ can not be infinite at $r = 0$, it follows that constant '$b$' must be zero so that the second term on the right hand side of equations (10.4) and (10.5) vanish and then it will result in

$$\sigma_r = a$$

and $$\sigma_C = a$$

But at $r = r_2 \quad \sigma_r = -p_2$

Therefore $$\sigma_r = -p_2 = a \qquad ...(10.16C)$$

or $$a = -p_2$$

which in turn yields

$$\sigma_C = a = -p_2 \qquad ...(10.15d)$$

In other words $\sigma_C = \sigma_r = a = -p_2$ = constant.

Thus the radial and hoop stresses are equal and constant anywhere with the solid shaft subjected to radial pressure and these stresses are both compressive in nature.

## 10.2 Comparison with Thin Cylinder Theory

For a cylinder with internal pressure only, we have from equations (10.5) and (10.9) the maximum value of $\sigma_C$ occurring at $r = r_1$ and then we have from equation (10.14$a$)

$$(\sigma_C)_{max} = p_1\left[\frac{r_2^2 + r_1^2}{r_2^2 - r_1^2}\right] \qquad ...(10.16)$$

Further let $t = r_2 - r_1$ and $d = 2r_1$, then equation (10.16) reduces to

$$(\sigma_C)_{max} = \frac{\left(\frac{d^2}{2} + td + t^2\right)}{t(d+t)} p_1 = \frac{\left[\frac{d^2}{2} + t(d+t)\right]}{t(d+t)} p_1 = \left[1 + \frac{d^2}{2t\,(d+t)}\right] p_1$$

or $$\frac{(\sigma_C)_{max}}{p_1} = \left[1 + \frac{1}{2k(1+k)}\right] \quad \text{...(10.17)}$$

where $t/d = k$

As proved earlier equation (9.31) for a thin shell where $\sigma_C$ is assumed to be constant across the thickness

$$\sigma_C = \frac{p_1 d}{2t}$$

or $$\frac{\sigma_C}{p_1} = \frac{1}{2k} \quad \text{...(10.18)}$$

The relative values of $(\sigma_C/p_1)$ as calculated from equations (10.17) and (10.18) for various values to $k$ are shown in fig. 10.3.

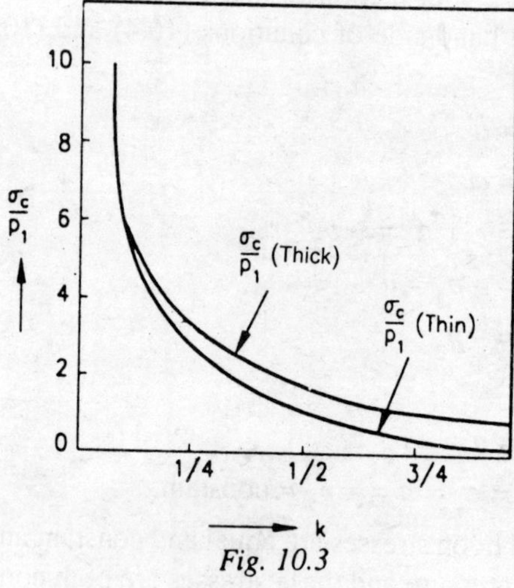


*Fig. 10.3*

The error in the value of the non dimensional stress $\sigma_l / p_1$ will be of the order of 5% when $k = 0.05$ (*i.e.*, when $d = 20t$) and that is why it is customary to use thin shell theory where $d \geq 20t$. It should be noted that if mean diameter is used in the thin cylinder formula the error is practically eliminated.

## 10.3 Longitudinal and Shear Stress

For a cylinder with closed ends the tensile longitudinal stress, $\sigma_l$, which is assumed to be constant over the cross-section, is given by the force acting on the end cover due to internal pressure divided by the area of cross-section of the cylinder.

Thus, $$\sigma_l = \frac{\pi r_1^2 \, p_1}{\pi(r_2^2 - r_1^2)} = \frac{r_1^2 p_1}{(r_2^2 - r_1^2)} \quad \text{...(10.19)}$$

Further in the absence of any torque acting on the cylinder, $\sigma_C$, $\sigma_r$ and $\sigma_l$ will all be the principal stresses. Out of these three $\sigma_r$ is compressive while $\sigma_C$ and $\sigma_l$ are both tensile. Moreover $\sigma_C$ is greater than $\sigma_l$. In other word they can be arranged in the following algebraical order

$$\sigma_C > \sigma_l > \sigma_r \cdot$$

Thus the maximum shear stress, which is half the maximum difference between the principal stresses (equation 2.43) is, then

$$\tau_{\max} = \frac{1}{2}(\sigma_C - \sigma_r)$$

Substituting values of $\sigma_r$ and $\sigma_C$ from equations (10.4) and (10.5), it reduces to

$$\tau_{\max} = \frac{1}{2}\left[\left(a + \frac{b}{r^2}\right) - \left(a - \frac{b}{r^2}\right)\right] = \frac{1}{2}\left(\frac{2b}{r^2}\right)$$

or

$$\tau_{\max} = \frac{b}{r^2} \quad ..(10.20)$$

The absolute maximum value of $\tau_{\max}$ will be on the inner surface, but it will act on the planes which are inclined to the longitudinal axis at 45°.

## 10.4 Compound or Shrunk Cylinders

From Fig. 10.2, it is evident that the hoop stress falls off appreciably as the radius increases. In other words the material of the outer layers of the cylinder is not economically used. The make the stress distribution more uniform the cylinder is made in two parts, one part being shrunk on the other in which the outer diameter of the inner tube is slightly more than the inner diameter of the outer tube. That is the common diameter of the two tubes has interference fit. By this process the material of the inner cylinder tube is subjected to compressive stresses and that in the outer tube is put to tensile stresses. In addition when this shrunk cylinder is subjected to internal pressure only (*i.e.*, the external pressure $p_2$, on the shrunk cylinder being zero), it will induce a tensile hoop stress in both the tubes. These stresses will be superposed on the shrinkage stresses in the respective tube and the resultant stress which is the algebraic sum of two sets of stresses in the two tubes, will be determined. Specially when the two cylinders are of the same material the stress distribution due to internal fluid pressure alone will be obtained with the help of equations (10.4), (10.5), (10.11) and (10.12) by treating the assembly of two shrunk cylinder, as one cylinder. Then the shrinkage stresses in each cylinder are calculated from the knowledge of radial pressure at the common surface of the two cylinders. The radial pressure at the common surface is related to the shrinkage allowance or the diametral interference at the common surface, before the cylinders are shrunk. If the interference allowernce at the common surface is known then the radial pressure at the common surface can be calculated which in turn will enable to determine the shrinkage stresses in both the tubes.

Thus if $p$ is the pressure at the common surface, then treating the two cylinders separately, once outer cylinder experiencing internal pressure $p$ and secondly the inner cylinder acted upon by pressure $p$ at its outer surface the radial and hoop stresses in both the cylinders can be calculated with the help of equations (10.4) and (10.5). It may be noted that the constants of integrations '$a$' ad '$b$' will be different for these two cylinders.

Due to common pressure $p$, let $\sigma_{C_2}$ be the compressive hoop stress at the outer surface of the inner cylinder and $\sigma_{C_2'}$ be the tensile hoop stress at the inner surface of the outer cylinder, then due to shrinkage the outer diameter of the inner cylinder will be reduced by

$$(\Delta d_2)_i = (d_2)_i\, \varepsilon_2$$

where $d_2$ is the nominal diameter of the common surface and $\varepsilon_2$ in the circumferential strain (or diameterial strain) at the common surface of the inner cylinder Further let $E_1$ and $\nu_1$ be the youngs modulus and poisons' ratio of inner cylinder material

Then $$(\Delta d_2)_i = (d_2)_i \left[\frac{-\sigma'_{C_2}}{E_1} + \frac{\nu_1 p}{E_1}\right]$$

$$= \frac{-d_2}{E_1}(\sigma'_{C_2} - \nu_1 p) \qquad \text{...(10.21)}$$

Negative sign indicates that the outer diameter of the inner cylinder will reduce after shrinkage.

Similarly if $\sigma'_{C_2}$ is the tensile hoop stress on the inner surface of the outer cylinder due to pressure $p$ at the common surface the circumferential strain (or the diametral strain) at the inner diameter of the outer cylinder is

$$\sigma_{C_2} = \left(\frac{\sigma'_{C_2}}{E_2} + \frac{\nu_2 p}{E_2}\right)$$

and $$(\Delta d_2)_0 = (d_2)_0\, \varepsilon'_2 = \frac{d_2}{E_2}(\sigma'_{C_2} + \nu_2 p) \qquad \text{...(10.22)}$$

where $E_2$ and $\nu_2$ are youngs modulus and poisson's ratio of outer cylinder material.

Then the interference allowance of the two cylinders at the common surface is

$$(\Delta d_2)_0 - (\Delta d_2)_C = \frac{d_2}{E_2}\left(\sigma'_{C_2} + \nu_2 p\right) + \frac{d_2}{E_1}\left(\sigma_{C_2} - \nu_1 p\right) \qquad \text{...(10.23)}$$

In case the material of both the cylinders is same than

$$E_1 = E_2 = E \text{ (say)}$$

and $$\nu_1 = \nu_2 = \nu \text{ (say)}$$

and interference allowance

$$= \frac{d_2}{E}\left(\sigma'_{C_2} + \sigma_{C_2}\right) \qquad \text{...(10.24)}$$

## 10.4.1 Evaluation of Stresses

(*a*) Shrinkage stress : The constant '*a*' and '*b*' (equations 10.4 and 10.5) can be formed by applying boundary conditions

(*i*) For miner tube $\sigma_r = 0$ at $r = r_1$

$\sigma_r = -p$ at $r = r_2$

(*ii*) for outer tube $\sigma_r = 0$ at $r = r_3$

$\sigma_r = -p$ at $r = r_2$

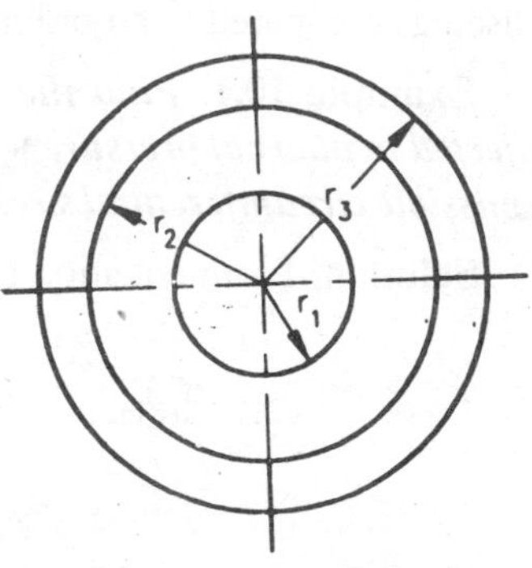


*Fig. 10.4*

Knowing the values of these constants the variations in $\sigma_r$ and $\sigma_c$ due to internal pressure $p_1$ are obtained by treating it as one cylinder of internal radius $r_1$ and outer radius $r_3$. The boundary conditions will then be

$\sigma_r = -p_1$ at $r = r_1$

$\sigma_r = 0$ at $r = r_3$

The stresses due $\sigma_r$ and $\sigma_c$ at any radius *r* internal pressure $p_1$ are then calculated with the help of equation (10.13) and (10.14) in which $r_2$ in replaced by $r_3$

*i.e.*

$$\sigma_r = \frac{p_1 r_1^2}{(r_3^2 - r_1^2)}\left[1 - \frac{r_3^2}{r^2}\right] \qquad ...(10.25)$$

and

$$\sigma_c = \frac{p_1 r_1^2}{(r_3^2 - r_1^2)}\left(1 + \frac{r_3^2}{r^2}\right) \qquad ...(10.26)$$

Stresses due to shrinkage and due to internal pression $p_1$ are then combined algebraically as shown in fig. 10.5 evidently the resultant stress distribution is more evenly

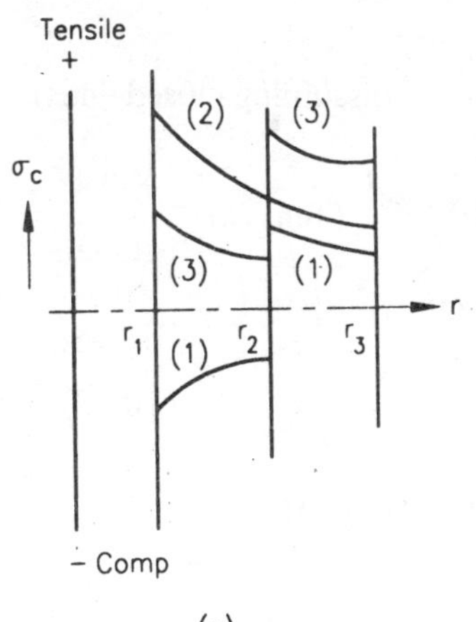


(a)

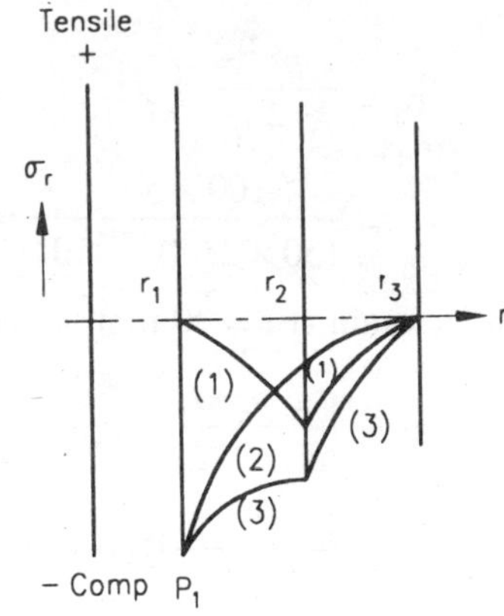


(b)

*1. Stress due to Shrinkage.*

*2. Stress due to internal pressure* $p_I$.

*3. Resultant Stress.*

*Fig. 10.5*

spread over the material and the maximum resultant circumferential stress in the shrunk cylinder is less than for a homogeneous cylinder of the same cross-section. In other words keeping the maximum stress at the same level, thinner compound cylinder can be used as compared to an ordinary one.

**Example 10.1.** ***Find the ratio of thickness to internal diameter for a tube subjected to internal pressure when the pressure is 6.8 of the value of the maximum permissible circumferential stress.***

**Solution.** From equation (10.4), at $r = r_1$

$$(\sigma_C)_{max} = p_1 \left(\frac{r_2^2 + r_1^2}{r_2^2 - r_1^2}\right)$$

or
$$\frac{8}{5} = \frac{\left(\frac{r_2}{r_1}\right)^2 + 1}{\left(\frac{r_2}{r_1}\right)^2 - 1}$$

from which
$$\frac{r_2}{r_1} = 2.08$$

or
$$r_2 = 2.08 r_1$$

∴
$$t = r_2 - r_1 = 1.08\, r_1$$

so that
$$\frac{t}{d} = \frac{1.08 r_1}{2 r_1} = 0.54$$

(*b*) At $r$ = 50 mm $\sigma_r = -p = -$ 100 MN/m$^2$ (comp.) and $\sigma_c$ = (5/8) × 100 = 160 MN/m$^2$ (tensile)

Further
$$\sigma_l = \frac{p r_1^2}{r_2^2 - r_1^2} \quad \text{(assuming closed ends)}$$

$$= \frac{100 \times 50^2}{(50 \times 2.08)^2 - 50^2} = 30 \text{MN/m}^2 \quad \text{(Tensile)}$$

The circumferential strain at r = 30 mm, is then given by

$$\varepsilon_c = \frac{\sigma_c}{E} - \nu\frac{\sigma_r}{E} - \nu\frac{\sigma_l}{E}$$

$$= \frac{1}{200 \times 10^9}[160 + 0.286\,(100 - 30)] \times 10^6$$

$$= 0.9001 \times 10^3$$

Since $\varepsilon_c = \varepsilon_d$, the increase in the internal diameter is given by

$$\delta d_i = 100 \times 0.9001 \times 10^{-3} = 0.09001 \text{ mm}$$

**Example 10.2.** *Find the change in the wall thickness of the tube in example 10.1.*

**Solution.** At $r = r_2 = 2.08 \times 50 = 104$ mm

$$\sigma_r = 0$$

and $$\sigma_c + \sigma_r = 2a = (\sigma_c)_{r1} + (\sigma_r)_{r1}$$

(since $\sigma_c + \sigma_r$ = constant from equation 10.2)

or $$(\sigma_c - 0) = 160 - 100$$

*i.e.,* $$\sigma_c = 160 - 100 = 60 \text{ MN/m}^2 \text{ (Tensile)}$$

Therefore $$\varepsilon_c = 1/E[\sigma_c - \nu\sigma_l - \nu\sigma_r]$$

$$= \frac{1}{200 \times 10^9}[60 - 0.286(30) - 0] \times 10^6$$

$$= 0.2556 \times 10^{-3}$$

Since $\varepsilon_d = \sigma_c$, we have increase in external diameter

$$(\delta d)_{r_2} = 2r_2\varepsilon_c$$

$$= 2\,(2.08 \times 50) \times .2556 \times 10^{-3}$$

$$= 0.05316 \text{ mm}$$

Therefore decrease in wall thickness

$$= \frac{1}{2}(.09001 - 0.05316)$$

$$= 0.01852$$

**Example 10.3.** *A steel cylinder is 1 m inside diameter and is to be designed for an internal pressure of 8 MN/m$^2$. Calculate the thickness if the maximum shearing stress is not to exceed 35 MN/m$^2$. Calculate the increase in volume, due to working pressure, if the cylinder is 6m long with closed ends. Neglect any constraints due to ends. E = 200 GN/m2, ν = 1/3.*

**Solution.** At $r = r_1$ $\sigma_r = -8$ MPa (Given)

Now $$\sigma_c = \frac{b}{r^2} + a$$

$$\sigma_r = -\frac{b}{r^2} + a$$

Therefore $$\tau_{max} = \frac{1}{2}[\sigma_c - \sigma_r)] = \frac{b}{r^2}$$

Absolute maximum shearing stress will act on the internal surface

That is $$35 = \frac{b}{r_1^2} = \frac{b}{(0.5)^2}$$

or $$b = 8.75 \text{ MN}$$

Also at the inner surface

$$\sigma_r = -8 = a - \frac{b}{r_1^2} = -35 + a$$

*i.e.*, $$a = 27 \text{ MN/m}^2$$

Further at $r = r_2$

$$(\sigma_r)_{r=r_2} = 0 = \frac{-b}{r_2^2} + a$$

or $$\frac{b}{r_2^2} = a$$

*i.e.*, $$r_2 = \sqrt{\frac{b}{a}} = \sqrt{\frac{8.75}{27}} = 0.57 \text{ m}$$

Then the wall thickness

$$= r_2 - r_1 = 0.57 - 0.5 = 0.07 \text{ m} = 70 \text{ mm}$$

At $r = r_1$ $$\sigma_c = \frac{b}{r_1^2} + a = 35 + 27$$

$$= 62 \text{ MN/m}^2 \text{ (Tensile)}$$

$$\sigma_l = \frac{pr_1^2}{r_2^2 - r_1^2}$$

$$= \frac{8 \times (0.5)^2}{(0.57)^2 - (0.5)^2} = 26.7 \text{ MN/m}^2$$

Therefore $$(\varepsilon_d)_{r_1} = \frac{1}{E}\left[\sigma_C - \nu\{\sigma_r + \sigma_l\}\right]$$

$$= \frac{1}{200 \times 10^9}\left[62 - (-8 + 26.7)\frac{1}{3}\right] \times 10^6$$

$$= 0.27885 \times 10–3$$

Also $$(\varepsilon_l)_{r_1} = 1/E[\sigma_l + \nu(\sigma_r - \sigma_c)]$$

$$= \frac{1}{200 \times 10^9}\left[26.7 + \frac{1}{3}(8 - 62)\right] \times 10^6$$

$$= 0.0435 \times 10^{-3}$$

Therefore volumetric strain

$$\frac{\delta V}{V} = 2(\varepsilon_d)_{r_1} + (\varepsilon_l)_{r_1}$$

$$= 2 \times .27885 \times 10^{-3} + 0.0435 \times 10^{-3}$$

$$= 0.6012 \times 10^{-3} \times \text{(internal volume)}$$

Then $\delta V = 0.6012 \times 10^{-3} (\pi/4 \times 1^2 \times 6)$

$$= 0.002835 \text{ m}^3. \text{ (increase)}$$

**Example 10.4.** *On the outer surface of a closed thick cylinder of diameter ratio 2.5 were fixed strain gauges to measure the longitudinal and circumferential strains. At an internal pressure of 230 MN/m² these strains were recorded as 9.18 × 10–5 and 36.9 × 10–5 respectively. Determine the value of the Young's modulus, modulus of rigidity and poison's ratio.*

**Solution.**

At $r = r_2$

$$\sigma_r = 0 = \frac{-b}{r_2^2} + a$$

*i.e.,* $a = \dfrac{b}{r_2^2}$

or $b = a\,r_2^2$

At $r = r_1$ $\sigma_r = -230 = -\dfrac{b}{r_1^2} + a$

or $230 = \dfrac{ar_2^2}{r_1^2} - a$

$$= a\,(2.5^2 - 1)$$

*i.e.,* $a = \dfrac{230}{5.25} = 43.8 \text{ MN/m}^2$

Again at $r = r_2$

$$\sigma_c = \frac{b}{r_2^2} + a = 2a = 87.6 \text{ MN / m}^2 \text{ (Tensile)}$$

$$\sigma_l = \frac{pr_1^2}{r_2^2 - r_1^2}$$

$$= \frac{p}{\left(\dfrac{r_2}{r_1}\right) - 1}$$

$$= \frac{230}{5.25} = 43.8 \text{ MN/m}^2 \qquad \text{(Tensile)}$$

$$\sigma_r = 0 \text{ (Comp.)}$$

Therefore $$\varepsilon_c = \frac{1}{E}[\sigma_c - \nu(\sigma_l + \sigma_r)] \times 10^6$$

or $$36.9 \times 10^{-5} = \frac{1}{E}[87.6 - \nu \times 43.8] \times 10^6 \quad ...(A)$$

Similarly $$\varepsilon_l = \frac{1}{E}[\sigma_l - \nu(\sigma_c + \sigma_r)] \times 10^6$$

or $$9.18 \times 10^{-5} = \frac{1}{E}[43.8 - \nu \times 87.6] \times 10^6 \quad ...(B)$$

Solving equations (*A*) and (*B*)

$$\nu = 0.287,\ E = 203 \text{ GN/m}^2$$

Then $$G = \frac{E}{2(1+\nu)} = \frac{203}{2(1+0.287)}$$

$$= 78.8 \text{ MN/m}^2.$$

**Example 10.5.** *A bronze liner of 50 mm external diameter is to be shrunk on a steel rod of 40 mm diameter. Calculate the maximum radial pressure between liner and rod if the maximum stress in the liner is limited to 120 MN/m², also the difference between the liner bore and shaft diameter before shrinking E for steel and bronze are 200 GN/m² and 90 GN/m² respectively. Passion's ratio is 0.3 for both steel and bronze.*

**Solution.** (*a*) Let the normal pressure between shaft and the linear be *p*. Then for the bronze linear, we haveing $r_1$ inner radius and $r_2$ outer radius, we have

$$(\sigma_e)_{r_2} = 0 = \frac{-b}{r_2^2} + a$$

or $$a = \frac{b}{r_2^2}$$

$$(\sigma_c)_{r_1} = 120 = \frac{b}{r_1^2} + a = \frac{b}{r_1^2} + \frac{b}{r_2^2}$$

$$= b\left[\frac{r_2^2 + r_1^2}{r_1^2\, r_2^2}\right]$$

Therefore $$b = \frac{120 \times (20)^2 \times (25)^2}{(25)^2 + (20)^2} = \frac{12}{41} \times 10^5$$

and $$a = \frac{12}{41} \times 10^5 \times \frac{1}{(25)^2}$$

$$= \frac{960}{41} \text{ N/mm}^2 \text{ or MN/m}^2 \text{ or MPa}$$

Also $$(\sigma_r)_{r_1} = -p = \frac{-b}{r_1^2} + a$$

$$= -\frac{12}{41} \times \frac{10^5}{(20)^2} + \frac{960}{41}$$

or $$-p = 26.35 \text{ MN/m}^2$$

or $$p = 26.35 \text{ MN/m}^2$$

(*b*) For the steel shaft :

$$\sigma_r = a_2 = -p \qquad \text{(Comp.)}$$

$\therefore$ $$a_2 = -p$$

and $$\sigma_c = a_2 = -p = -26.35 \qquad (\textit{i.e.}, \text{ Comp})$$

Therefore diametral strain in the steel shaft

$$\varepsilon'_c = \frac{1}{E}[\sigma_c - \nu\sigma_r]$$

$$= \frac{1}{200 \times 10^9}[-26.35 + 0.3 \times 26.35] \times 10^6$$

$$= -9.2225 \times 10^{-5}$$

Therefore decrease in diameter of shaft

$$(\Delta d_2)_s = 9.2225 \times 10–5 \times 40 = 0.003689 \text{ mm}$$

For the bronze liner at the inner surface :

Diametral strain $$\varepsilon''_c = \frac{1}{E}[(\sigma_c)_{r_1} - \nu(\sigma_r)_{r_1}]$$

$$= \frac{1}{90 \times 10^9}[120 + 0.3 \times 26.35] \times 10$$

$$= 1.42 \times 13^{-3}$$

$\therefore$ Increase in the inner diameter of liner

$$(\Delta d_2)_l = 1.42 \times 10^{-3} \times 40$$

$$= 0.0568 \text{ mm}$$

Therefore difference in shaft diameter and inner diameter of liner before shrinking

$$\delta d_2 = .0568 + .003689 = .060489 \text{ mm.}$$

**Example 10.6.** *A solid plug gauge of steel has a diameter of 25.005 mm, and is forced into a ring gauge of the same material, which measures 25 mm inside diameter and 50 mm outside diameter. Its axial length is 20 mm; what is the maximum stress in the ring, and what force is required to slide the plug, assuming the coefficient of friction is 0.3 ? Take E = 200 GN/m² and v = 0.287.*

**Solution.** Let the normal radial pressure at the common surface be $p$ N/mm$^2$.

(*i*) For the plug gauge :

$$\sigma_r = a_1 = -\text{p} \qquad \text{(Comp.)}$$

or $$\sigma_r = -p$$

and $$\sigma_c = a_1 = -p \qquad (\textit{i.e.}, \text{ comp})$$

Therefore diameteral strain in the plug gauge

$$\varepsilon_c' = \frac{1}{E}[\sigma_c - \nu\sigma_r]$$

$$= \frac{1}{E}[-p + \nu p)$$

$$= -\frac{p}{E}(1-\nu)$$

$\therefore$ Decrease in the diameter of plug

$$= \frac{p}{E}(1-\nu) \times 25 \text{ mm}$$

(*ii*) For the steel ring gauge

$$(\sigma_r)_{r_2} = 0 = \frac{-b}{r_2^2} - a_2 \text{ or } a_2 = \frac{b_2}{r_2^2}$$

$$(\sigma_r)_{r_1} = -p = \frac{-b_2}{r_1^2} + \frac{b_2}{r_2^2}$$

$$= -b_2 \frac{r_2^2 - r_1^2}{r_1^2 r_2^2}$$

or $$b_2 = \frac{p\, r_1^2 r_2^2}{r_2^2 - r_1^2}$$

$$= p \,.\, \frac{(12.5)^2 (20)^2}{(20)^2 - (12.5)^2} = 256.5\, p$$

$$(\sigma_c)_{r_1} = \frac{b_2}{r_1^2} + a_2$$

But $$a_2 = \frac{b_2}{r_2^2}$$

$\therefore$ $$a_2 = \frac{256.5\, p}{(20)^2} = 0.642\, p$$

Therefore $$(\sigma_r)_{r_1} = \frac{256.5\, p}{(20)^2} + 0.642\, p$$

$$= 2.284 \text{ p.}$$

Therefore diametral strain at the inner surface of the ring

$$\varepsilon_c'' = \frac{1}{E}[(\sigma_c)_{r_1} - v(\sigma_r)_{r_1}]$$

$$= \frac{1}{E}[2.284\, p + vp]$$

$$= \frac{p}{E}[2.284 + v]$$

Therefore Increase in the inner diameter of ring

$$= 25\frac{p}{E}[2.284 + v]\text{ mm}$$

Thus equating the shrinkage allowance to 0.005, we get

$$0.005 = \frac{25p}{200 \times 10^3}(1 - 0.286) + \frac{25p}{200 \times 10^3}(2.284 + 0.286)$$

or $$= \frac{.005 \times 200 \times 10^3}{25 \times 3.284} = p$$

or $$p = 12.18 \text{ N/mm}^2 \text{ or MN/m}^2$$

Thus maximum stress in the ring

$$= (\sigma_c)_{r_1}$$

$$= 2.284 \times 12.18$$

$$= 27.8 \text{ MN/m}^2 \text{ or N/mm}^2$$

Force required to slide the plug

= Frictional force = (Normal reaction due to $p$) × $\mu$

$$= (2\pi r_1\, lp)\,\mu = 2\pi \times 12.5 \times 20 \times 12.18 \times 0.3 N$$

= 5.47 kN.

**Example 10.7.** *Find the torque required to rotate the plug with respect to the ring in Example 10.6.*

**Solution.** Tangential frictional force = $2\pi r_1 lp\mu$

Therefore torque required = $(2\pi r_1 l\, p\, \mu)\, r_1$

$$= 2\pi \times (12.5)^2 \times 20 \times 20 \times 12.18 \times .3 \text{ N/mm}$$

= 7.0 Nm.

**Example 10.8.** *One steel cylinder is shrunk on to another, to form a compound cylinder having an inside diameter of 100 mm, an outside diameter of 200 mm and a diameter of 150 mm at the surface in contact. If shrinkage produces a radial pressure $p_0$ at the surfaces in contact, after which the compound cylinder is subjected to an internal* *pressure $p_1$, find the ratio of $p_0$ to $p_1$, so that the maximum hoop stresses in the two cylinders shall be the same.*

**Solution.** (*i*) Stresses due to shrinkage only

For the inner cylinder :

$$\sigma_r = 0 \text{ at } r = 500 \text{ mm},$$

so that $$0 = \frac{-b_1}{(50)^2} + a_1$$

or $$b_1 = 2500\, a_1$$

Also $\sigma_r = -p_0$ at $r = 75$ mm so that

$$-p_0 = \frac{-b_1}{(75)^2} - a_1 = -\frac{25000}{(75)^2} - a_1$$

or $$p_0 = \frac{4}{9}a_1 - a_1$$

or $$a_1 = \frac{9}{5}p_0$$

Therefore $$b_1 = -4500\, p_0$$

Then $$(\sigma_c)_{max} = \frac{b_1}{(50)^2} + a_1$$

$$= \left[\frac{-4500}{(50)^2} - \frac{9}{5}\right] p_0 = -\frac{18}{5}p_0 \quad \text{(Comp.)}$$

**For the outer cylinder :**

Since $\sigma_r = 0$ at $r = 100$ mm, therefore

$$0 = \frac{-b_2}{(100)^2} + a_2$$

or $$b_2 = (100)^2 a_2$$

Also $\sigma_r = -p_0$ at $r = 75$ mm, so that

$$-p_0 = \frac{-b_2}{(75)^2} + a_2$$

or $$p_0 = \frac{(100)^2 a_2}{(75)^2} - a_2$$

or $$a_2 = \frac{9}{7}p_0$$

$$\therefore \quad b_2 = (100)^2 \times \frac{9}{7}p_0$$

Then $$(\sigma_c)_{max} = \frac{b_2}{(75)_2} + a_2$$

$$= \frac{100^2}{(75)^2} \times \frac{9}{7} p_0 + \frac{9}{7} p_0$$

$$= \frac{25}{7} p_0 \quad \text{(Tensile)}$$

***(ii)* Stresses due to internal pressure only : – (Treat the compound cylinder as one unit)**

$$\sigma_r = \text{at } r = 100 \text{ m}$$

So that $$0 = \frac{-b_3}{(100)^2} + a_3$$

or $$b_3 = (100)^2 a_3$$

$$\sigma_r = -p_1 \text{ at } r = 50, \text{ so that}$$

$$-p_1 = \frac{-b_3}{(50)^2} + a_3$$

or $$p_1 = \frac{(100)^2}{(50)^2} \times a_3 - a_3 = 3a_3$$

Therefore $$a_3 = \frac{1}{3} p_1$$

Also $$b_3 = (100)^2 a_3$$

or $$b_3 = 100^2 \frac{p_1}{3}$$

Then $$(\sigma_c)_{r=50} = \frac{b_3}{(50)^2} + a_3$$

$$= \frac{100^2}{50^2} \cdot \frac{p_1}{3} + \frac{p_1}{3} = \frac{5}{3} p_1 \quad \text{(Tensile)}$$

and $$(\sigma_c)_{r=75} = \frac{b_3}{(75)^2} + a_3$$

$$= \frac{100^2}{75^2} \frac{p_1}{3} + \frac{p_1}{3} = \frac{25}{27} p_1 \quad \text{(Tensile)}$$

***(iii)* Combined effect of both shrinking and internal pressure :**

Max. resultant tensile hoop stress in inner cylinder

$$(\sigma_c)_{max_i} = \left(\frac{5}{3} p_1 - \frac{18}{5} p_0\right)$$

Max. resultant tensile hoop stress in outer cylinder

$$(\sigma_{c\max})_0 = \left(\frac{25}{27}p_1 + \frac{25}{7}p_0\right)$$

Equating these two

$$\frac{5}{3}p_1 - \frac{18}{5}p_0 = \frac{25}{27}p_1 + \frac{25}{7}p_0$$

or $$\frac{20}{27}p_1 = \frac{251}{35}p_0$$

or $$\frac{p_0}{p_1} = 0.103$$

## 10.4.1 Shrinkage Stress : Derivation of Formula

Let the common surface pressure be "$p$". Then for the inner cylinder the end conditions are :

At $r = r_1$ $\sigma_r = 0$

At $r = r_2$ $\sigma_r = -p$

Then from equation (10.4)

$$0 = a_1 - \frac{b_1}{r_1^2}$$

or $$a_1 = \frac{b_1}{r_1^2}$$

and $$-p = a_1 - \frac{b_1}{r_2^2}$$

$$= \frac{b_1}{r_1^2} - \frac{b_1}{r_2^2}$$

$$= \frac{b_1(r_2^2 - r_1^2)}{r_1^2 r_2^2}$$

or $$b_1 = \frac{-pr_1^2 r_2^2}{(r_2^2 - r_1^2)}$$

and then $$a_1 = \frac{-pr_2^2}{(r_2^2 - r_1^2)}$$

Then at $r = r_2$ for inner cylinder

$$\left(\sigma_{c_i}\right)_{r_2} = a_1 + \frac{b_1}{r_2^2}$$

$$= \frac{-pr_2^2}{(r_2^2 - r_1^2)} - \frac{pr_1^2}{(r_2^2 - r_1^2)}$$

$$= \frac{-p(r_2^2 + r_1^2)}{(r_2^2 - r_1^2)} \qquad ...(10.27)$$

$$\left(\sigma_{c_i}\right)_{r_2} = a_1 - \frac{b_1}{r_2^2}$$

$$= \frac{-pr_2^2}{(r_2^2 - r_1^2)} + \frac{pr_1^2}{(r_2^2 - r_2^2)} = -p \qquad ...(10.28)$$

Therefore circumferential strain in the inner cylinder at $r = r_2$ will be as under if $E_i$ is youngs modulus of material of inner cylinder and $v_i$ is its poisson's ratio

$$\left(\varepsilon_{c_i}\right)_{r_2} = \frac{1}{E_i}\left(\sigma_{c_i}\right)_{r_2} - \frac{v_i}{E_i}\left(\sigma_{r_i}\right)_{r_2}$$

$$= \frac{1}{E_i}\left[\frac{-p(r_2^2 + r_1^2)}{(r_2^2 - r_1^2)}\right] + \left[\frac{pv_i}{E_i}\right]$$

$$= \frac{-p}{E_i}\left[\frac{r_2^2 + r_1^2}{r_2^2 - r_1^2} - v_i\right]$$

Therefore $\left(\dfrac{\delta D_2}{D_2}\right)_i = \left(\varepsilon_{c_i}\right)_{r_2}$

or $$(\delta D_2)_i = \frac{D_2 p}{E_i}\left[-\frac{r_2^2 + r_1^2}{r_2^2 - r_1^2}\right] + \left[\frac{pv_i}{E_i}\right] D_2$$

$$= \frac{pD_2}{E_i}\left[\frac{-r_2^2 + r_1^2}{r_2^2 - r_1^2} + v_i\right] \qquad ..(10.29)$$

For the outer cylinder the boundary conditions are

At $r = r_2$ $\sigma_r = -p$

At $r = r_3$ $\sigma_r = 0$

Therefore from equation (10.4)

$$0 = a_2 - \frac{b_2}{r_3^2}$$

or $$a_2 = \frac{b_2}{r_3^2}$$

and $$-p = a_2 - \frac{b_2}{r_2^2} = \frac{b_2}{r_3^2} - \frac{b_2}{r_2^2} = \frac{b_2(r_2^2 - r_3^2)}{r_3^2 r_2^2}$$

or $$b_2 = \frac{r_2^2 r_3^2 p}{(r_3^2 - r_2^2)}$$

Then $$a_2 = \frac{pr_2^2}{(r_3^2 - r_2^2)}$$

Therefore the radial and circumferential stresses at $r = r_2$ in the outer cylinder are

$$\left(\sigma_{r_0}\right)_{r_2} = -p$$

$$\left(\sigma_{c_0}\right)_{r_2} = a_2 + \frac{b_2}{r_2^2} = \frac{pr_2^2}{(r_3^2 - r_2^2)} + \frac{pr_3^2}{(r_3^2 - r_2^2)}$$

$$= \frac{p(r_2^2 + r_3^2)}{(r_3^2 - r_2^2)} \qquad \text{...(10.30)}$$

Thus the circumferential strain in the enter cylinder at the common surface where $r = r_2$ is

$$\left(\varepsilon_{c_0}\right)_{r_2} = \frac{1}{E_0}\left(\varepsilon_{c_0}\right)_{r_2} - \frac{v_0}{E_0}\left(\varepsilon_{r_0}\right)_{r_2}$$

$$= \frac{1}{E_0}\left[\frac{p(r_2^2 + r_3^2)}{(r_3^2 - r_2^2)} + \frac{v_0 p}{E_0}\right]$$

$$= \frac{p}{E_0}\left[\frac{(r_2^2 + r_3^2)}{(r_3^2 - r_2^2)} + v_0\right] = \left(\frac{\delta D_2}{D_2}\right)_{0_{r_2}}$$

Thus if $E_0$ and $v_0$ are youngs modulus and passion's ratio of the material of outer cylinder, then

$$\left(\delta D_2\right)_{0r_2} = \frac{D_2 p}{E_0}\left[\frac{r_2^2 + r_3^2}{r_3^2 - r_2^2} + v_0\right] \qquad \text{...(10.31)}$$

Therefore the interference allowance on the common diameter for unlike materials in then given by

$$(\Delta D_2)_{\text{unlike}} = \frac{D_2 p}{E_0}\left[\frac{r_2^2 + r_3^2}{r_3^2 - r_2^2} + v_0\right] - \frac{pD_2}{E_i}\left[-\frac{r_2^2 + r_1^2}{r_2^2 - r_1^2} + v_i\right]$$

$$= \frac{D_2 p}{E_0}\left[\frac{r_3^2 + r_2^2}{r_3^2 - r_2^2} + v_0\right] + \frac{pD_2}{E_i}\left[\frac{r_2^2 + r_1^2}{r_2^2 - r_1^2} - v_i\right] \qquad \text{...(10.32)}$$

Equation (10.32) for alike materials of the two cylinders where $E = E_i = E_0$ and $v = v_i = v_0$, reduces to

$$(\Delta D_2)_{\text{alike}} = \frac{D_2 p}{E}\left[\frac{r_3^2 + r_2^2}{r_3^2 - r_2^2} + \frac{r_2^2 + r_1^2}{r_2^2 - r_1^2}\right] \quad ...(10.33)$$

Corresponding shrinkage stresses in the two cylinders are then calculated as follows circumferential stress in inner cylinder any radius $r$ is given by

$$\left(\sigma_{c_i}\right)_r = a_1 \frac{b_1}{r^2}$$

$$= \frac{-pr_2^2}{(r_2^2 - r_1^2)} - \frac{pr_1^2 r_2^2}{(r_2^2 - r_1^2)\,r^2}$$

or

$$\left(\sigma_{c_i}\right)_r = \frac{-pr_2^2}{(r_2^2 - r_1^2)}\left[1 + \frac{r_1^2}{r^2}\right] \quad ...(10.34)$$

and

$$\left(\sigma_{c_i}\right)_r = a_1 - \frac{b_1}{r^2}$$

$$= \frac{-pr_2^2}{(r_2^2 - r_1^2)} + \frac{pr_1^2 r_2^2}{(r_2^2 - r_1^2)\,r^2}$$

$$= \frac{-pr_2^2}{(r_2^2 - r_1^2)}\left[1 - \frac{r_1^2}{r^2}\right] = -p\left[\frac{r_2^2 + r_1^2}{r_2^2 - r_1^2}\right] \quad ...(10.35)$$

Thus for $r = r_1$, we have from equation (10.34)

$$(\sigma_{c_i})_{r=r_1} = \frac{-pr_2^2}{(r_2^2 - r_1^2)}\left[1 + \frac{r_1^2}{r_1^2}\right]$$

$$= \frac{-2pr_2^2}{(r_2^2 - r_1^2)} \quad ...(10.36)$$

Again from equation (10.34) at $r = r_2$ we have

$$\left(\sigma_{c_i}\right)_{r=r_2} = \frac{-pr_2^2}{(r_2^2 - r_1^2)}\left[1 + \frac{r_1^2}{r_2^2}\right] = -p\left[\frac{r_2^2 + r_1^2}{r^2 - r_1^2}\right] \quad ...(10.37)$$

And circumferential stress at any radius $r$ in the outer cylinder is given below

$$\left(\sigma_{c_0}\right)_r = a_2 + \frac{b_2}{r^2} = \frac{pr_2^2}{(r_3^2 - r_2^2)} + \frac{pr_2^2 r_3^2}{(r_3^2 - r_2^2)} \times \frac{1}{r^2}$$

or

$$(\sigma_{c_0})_r = \frac{pr_2^2}{(r_3^2 - r_2^2)}\left[1 + \frac{r_3^2}{r^2}\right] \quad ...(10.38)$$

and

$$(\sigma_{r_0})_r = a_2 - \frac{b_2}{r^2}$$

$$= \frac{pr_2^2}{(r_3^2 - r_2^2)} - \frac{pr_2^2\, r_3^2}{(r_3^2 - r_2^2)} \frac{1}{r^2}$$

$$= \frac{pr_2^2}{(r_3^2 - r_2^2)}\left[1 - \frac{r_3^2}{r^2}\right] \quad ...(10.39)$$

Thus at $r = r_2$ we have from equation (10.38) the circumferential stress at radius $r_2$ in the outer cylinder

$$\left(\sigma_{c0}\right)_{r=r_2} = \frac{pr_2^2}{(r_3^2 - r_2^2)}\left[1 + \frac{r_3^2}{r_2^2}\right]$$

$$= \frac{p\,(r_3^3 + r_2^2)}{(r_3^2 - r_2^2)} \quad ...(10.40)$$

Again from equation (10.38) the circumferential stress in the enter cylinder at radius = $r_3$

$$\left(\sigma_{c0}\right)_{r=r_3} = \frac{pr_2^2}{(r_3^2 - r_2^2)}\left[1 + \frac{r_3^2}{r_3^2}\right]$$

$$= \frac{2pr_2^2}{(r_3^2 - r_2^2)} \quad ...(10.41)$$

**Example 10.9.** *A compound cylinder is made by shrinking one cylinder on to another of the same material. The outside diameter of the compound cylinder is 300 mm and the inside diameter is 150 mm. The diameter at the common surface is 250 mm. Calculate the shrinkage allowance on the common diameter if the pressure to be generated at the common surface is to be 12 MPa. Then estimate the minimum temperature rise to which the outer cylinder must be subjected the outer cylinder must so that is can easily be slipped over the other cylinder. Take*

*Co efficient of linear expansion* $\alpha = 0.6 \times 10^{-5}/°C$

*Young's modules* $E = 200$ *GPa*

*(b) If now the shrink cylinder is subjected to internal fluid pressure of 15 MPa, find the resultant stresses in the circumferential direction at the inner and outer radii of the compound cylinder.*

**Solution.** **(*a*) Shrinking Stress**

Here $r_1 = \dfrac{150}{2} = 75$ mm

$$r_2 = \frac{250}{2} = 125 \text{ mm}$$

$$r_3 = \frac{300}{2} = 150 \text{ mm}$$

$$E = 200 \times 10^3 \text{ MPa}$$

Common surface pressure = $p$ = 12 MPa

As the material of the two cylinder, which are shrunk over each other, is same we have from equation (10.33) the shrinkage allowance on common diameter $D_2$.

*i.e.,*
$$(\Delta D_2)_{\text{alike}} = \frac{pD_2}{E}\left[\frac{r_3^2 + r_2^2}{r_3^2 - r_2^2} + \frac{r_2^2 + r_1^2}{r_2^2 - r_1^2}\right]$$

$$= \frac{12 \times 250}{200 \times 10^3}\left[\frac{150^2 + 125^2}{150^2 - 125^2} + \frac{125^2 + 75^2}{125^2 - 75^2}\right]$$

$$= \frac{12 \times 250}{200 \times 10^3}\,(2.12 + 5.455)$$

$$= 0.115 \text{ mm}$$

It $t$ = temperature rise, then

$$(\Delta D_2)_{\text{alike}} = D_2 \alpha t$$

or
$$0.115 = 250\,(0.6 \times 10^{-5})t$$

Therefore
$$t = \frac{0.115 \times 10^5}{250 \times 0.6} = 76.66°\text{C}$$

**Shrinkage Stress for Inner cylinder**

At $r = r_1$ we have from equation 10.36

$$(\sigma_{c_i})_{r_1} = \frac{-2pr_2^2}{(r_2^2 - r_1^2)}$$

$$= \frac{-2 \times 12 \times (125)^2}{(125^2 - 75^2)}$$

$$= -37.5 \text{ MPa } (\textit{i.e.}, \text{ Comp.})$$

From equation (10.37) we have at $r = r_2$

$$\left(\sigma_{c_i}\right)_{r_2} = \frac{-p(r_2^2 + r_1^2)}{(r_2^2 - r_1^2)}$$

$$= \frac{-12(125^2 + 75^2)}{(125^2 - 75^2)} = \frac{-12 \times 21250}{10000}$$

$$= -25.5 \text{ MPa } (\textit{i.e.}, \text{ Comp.})$$

Similarly from equation (10.40) at $r = r_2$ we have at the common surface in the outer cylinder,

$$\left(\sigma_{c0}\right)_{r=r_2} = \frac{p(r_3^2 + r_2^2)}{(r_3^2 - r_2^2)}$$

$$= \frac{12(150^2 + 125^2)}{(150^2 - 125^2)} = \frac{12 \times 38125}{6875} = 66.55 \text{ MPa}$$

Again from equation (10.41) we have the circumferential stress in outer cylinder at $r = r_3$

$$\left(\sigma_{c_0}\right)_{r=r_3} = \frac{2pr_2^2}{(r_3^2 - r_2^2)} = \frac{2 \times 12 \times (125)^2}{(150^2 - 125^2)} = 54.54 \text{ MPa}$$

**(*b*) Stresses due to internal pressure of 15 MPa acting on compound cylinder**

From equation (10.13) in which $r_2$ is replaced by $r_3$ we have for composite cylinder, where $p_1$, = 15 MPa, at any radius $r$

$$\left(\sigma_{r_c}\right)_r = \frac{p_1 r_1^2}{(r_3^2 - r_1^2)}\left(1 - \frac{r_3^2}{r^2}\right)$$

and
$$\left(\sigma_{c_c}\right)_r = \frac{p_1 r_1^2}{(r_3^2 - r_1^2)}\left(1 + \frac{r_3^2}{r^2}\right)$$

where $\left(\sigma_{c_c}\right)_r$ represents the circumferential stress in the composite cylinder due to internal pressure at any radius $r$. Similarly $\left(\sigma_{r_c}\right)_r$ is the radial stress in it at any $r$.

Then substituting values of various quantities we get

$$\left(\sigma_{C_C}\right)_{r=r_1} = \frac{15 \times 75^2}{(150^2 - 75^2)}\left[1 + \frac{150^2}{75^2}\right] = 25 \text{ MPa}$$

and
$$\left(\sigma r_c\right)_{r=r_3} = \frac{15 \times 75^2}{(150^2 - 75^2)}\left[1 + \frac{r_3^2}{r_2^2}\right] = \frac{15 \times 75^2 \times 2}{(150^2 - 75^2)} = 10 \text{ MPa}$$

Thus the resultant circumferential stresses at inner and outer surfaces of composite cylinder are

$$\text{Resultant } (\sigma_c)_{r=r1} = \left(\sigma_{c_c}\right)_{r_1} + \left(\sigma_{c_i}\right)_{r_1}$$
$$= 25 - 37.5 = -12.5 \text{ MPa} \qquad (\textit{i.e.}, \text{ compressive})$$

$$\text{Resultant } (\sigma_c)_{r=r3} = \left(\sigma_{c_c}\right)_{r_3} + \left(\sigma_{c_0}\right)_{r_3}$$
$$= 10 + 53.54 = 64.54 \text{ MPa} \qquad (\text{Tensile})$$

**Example 10.10.** *A high tensile steel tyre 20 mm thick, is shrunk on a cast-iron rim having 480 mm out side diameter and 60 mm thick. Find the inside diameter of the steel tyre of the nearest hundredth of a mm, if after shrinking on, the tyre exerts a radial pressure of 50 MPa on cast iron rim. Take E for steel = 210 GPa and E for cast iron = 100 GPa. Poissious radio for both = 0.25.* *(Engineering Services)*

**Solution.** Inner dia of inner CI cylinder = 480 – 120 = 360 mm

Outside dia of inner CI cylinder = 480 mm

Outside dia of steel tyre = 480 + 40 = 520 mm

Therefore for the compound cylinder we have

$$r_1 = \frac{360}{2} = 180 \text{ mm}$$

$$r_2 = \frac{480}{2} = 240 \text{ mm} \qquad \text{(radius of common surface)}$$

$$r_3 = \frac{520}{2} = 260 \text{ mm}$$

$$p = 50 \text{ MPa}$$

Since the materials of the two cylinders forming the compound cylinder are different we have for this problem, from equation (10.32) the value of interference allowance on the common diameter $\Delta D_2$

$$\Delta D_2 = \frac{D_2 p}{E_0}\left[\frac{r_3^2 + r_2^2}{r_3^2 - r_2^2} + v_0\right] + \frac{pD_2}{E_i}\left[\frac{r_2^2 + r_1^2}{r_2^2 - r_1^2} - v_i\right]$$

where $E_0$ = Youngs modulus of material of outer cylinder

$v_0$ = poissous ratio of material of outer cylinder

$E_i$ = Youngs modulus of material of inner cylinder

$v_i$ = poissous ratio of material of inner cylinder

Here $E_0 = 210 \times 10^3$ MPa and $v_0 = 00.25$

$E_i = 100 \times 10^3$ MPa and $v_i = 0.25$

Therefore

$$\Delta D_2 = \frac{480 \times 50}{210 \times 10^3}\left[\frac{260^2 + 240^2}{260^2 - 240^2} + 0.25\right] + \frac{480 \times 50}{100 \times 10^3}\left[\frac{240^2 + 180^2}{240^2 - 180^2} - 0.25\right]$$

$$= 0.1143\,[12.52 + 0.25] + 0.24\,[3.57 - 0.25]$$

$$= 1.4596 + 0.7968$$

$$= 2.2564 \text{ mm}$$

Therefore inside dia of steel tyre

$$= 480 - 2.2564 = 477.74 \text{ mm}$$

**Example 10.11.** *A steel hoop of 200mm outer and 130 mm inner diameters, is shrunk on a hollow steel cylinder of 80 mm inner diameter, the pressure of shrinkage being 20 MPa.*

*When subjected to internal fluid pressure of 70 MPa, what will be :*

*(a) the greatest circumferential stress induced in the cylinder*

*(b) the radial pressure between the cylinder and the hoop*

*(c) the greatest circumferential stress in hoop*

*Assume the stresses induced are within proportional limit.*

*(Engineering services)*

**Solution.** Given data

$$r_1 = \frac{80}{2} = 40 \text{ mm}$$

$$r_2 = \frac{130}{2} = 65 \text{ mm}$$

$$r_3 = \frac{200}{2} = 100 \text{ mm}$$

$$p = 20 \text{ MPa}$$

**(i) Stresses due to shrinkage :**

In the hoop the stress in the circumferential direction at $r = r_3$ given by equation (10.41)

$$\left(\sigma_{c0}\right)_{r_3} = \frac{2pr_2^2}{(r_3^2 - r_2^2)}$$

$$= \frac{2 \times 20 \times 65^2}{(100)^2 - (65)^2} = 29{,}264 \text{ MPa}$$

From equation (10.40)

$$\left(\sigma_{c0}\right)_{r_2} = \frac{p\,(r_3^2 + r_2^2)}{(r_3^2 - r_2^2)}$$

$$= \frac{20 \times (100^2 + 65)^2}{(100^2 - 65^2)}$$

$$= 49.264 \text{ MPa}$$

For stresses in the cylinder we have

From eqns. (10.36) and (10.37)

$$\left(\sigma_{c_i}\right)_{r_1} = \frac{-2pr_2^2}{(r_2^2 - r_1^2)}$$

$$= \frac{-2 \times 20 \times 65^2}{(65^2 - 4^2)}$$

$$= -64.381 \text{ MPa}$$

$$\left(\sigma_{c_i}\right)_{r_2} = \frac{-p(r_2^2 + r_1^2)}{(r_2^2 - r_1^2)} = \frac{-20 \times (65^2 + 40^2)}{(65^2 - 40^2)} = -44.381 \text{ MPa}$$

***(ii)* Stresses due to internal fluid pressure of 70 MPa contained in the compound cylinder**

From equiation (10.14) we have for stresses in the circumferential direction due to $p_1$ = 70 MPa at $r = r_1$ and $r = r_3$

$$\left(\sigma_{c_c}\right)_{r_3} = \frac{2p_1r_2^2}{(r_3^3 - r_1^2)} = \frac{2 \times 70 \times (40)^2}{(100^2 - 40^2)}$$

$$= \frac{140 \times 40^2}{(140 \times 36)} = 26.667 \text{ MPa}$$

And $$\left(\sigma_{cC}\right)_{r=r_2} = \frac{p_1r_1^2}{(r_3^2 - r_1^2)}\left[1 + \frac{r_3^2}{r_2^2}\right]$$

$$= \frac{70 \times 40^2}{(100^2 - 40^2)}\left(1 + \frac{100^2}{65^2}\right) = 44.892 \text{ MPa}$$

$$\left(\sigma_{c_c}\right)_{r_1} = \frac{p_1\left(r_1^2 + r_3^2\right)}{\left(r_3^2 - r_1^2\right)} = \frac{70 \times (40^2 + 100^2)}{(100^2 - 40^2)} = 96.667 \text{ MPa}$$

From equation (10.13) we have at $r = r_2$ for the compound cylinder with inner radius $r_1$ and outer radius $r_3$

$$\left(\sigma_{r_c}\right)_{r_2} = \frac{p_1r_1^2}{(r_2^2 - r_1^2)}\left[1 - \frac{r_3^2}{r_2^2}\right]$$

$$= \frac{70 \times 40^2}{(100^2 - 40^2)}\left[1 - \frac{(100)^2}{(65)^2}\right]$$

$$= \frac{70 \times 40^2}{140 \times 60}(-1.36686)$$

$$= -18.225 \text{ MPa } (i.e., \text{ Comp.})$$

(*a*) Therefore the greatest circumferential stress in cylinder will be at $r = r_1$ and it will be tin resultant of the stresses due to fluid pressure and shrinking.

*i.e.* $$\left(\sigma_{c_i}\right)_{max} = \left(\sigma_{c_c}\right)_{r_1} + \left(\sigma_{c_i}\right)_{r_1}$$

$$= 96.667 - 64.381 = 32.286 \text{ MPa (Tensile)}$$

(*b*) The greatest circumferential stress in hoop will be at $r = r_1$ and will be the resultant of stresses induced due to fluid pressure and shrinkage

*i.e.,* $$\left(\sigma_{c_0}\right)_{max} = \left(\sigma_{c_0}\right)_{r_2} + \left(\sigma_{c_c}\right)_{r_2}$$

$$= 49.264 + 44.892 = 94.156 \text{ MPa}$$

(*c*) The resultant radial pressure at $r = r_2$

$$= \text{Radial stress due to } p_1 + \text{shrinking pressure}$$

$$= \left(\sigma_{r_c}\right)_{r_2} + p = 18.225 + 20$$

$$= 38.225 \text{ MPa (Comp.)}$$

## 10.5 Autofrettage

In the article 10.4 the theory of shrunk cylinders was developed with the sole purpose of having stress distribution uniform as far as possible. In this process the outer cylinder is shrunk on to the inner cylinder in which the outer diameter of the inner cylinder is slightly bigger than the inner diameter of the outer cylinder *i.e.* these two diameters have interference fit. In so doing the inner cylinder is subjected to radial compression and the outer to tension. This results in lessening the weight of the shrunk cylinder. This is particularly helpful in the case of field guns, when they have to be carried to the hilly region during war. The shrinking process is done either by heating the outer cylinder or by cooling the inner cylinder to such an extent that in the case of heating the outer cylinder the inner diameter of outer cylinder becomes more than the outer diameter of the inner cylinder and then the outer cylinder is slipped on the inner cylinder. In the case of cooling of the inner cylinder the cooling is carried out till the outer diameter of the inner cylinder reduces sufficiently so that its outer diameter is smaller than the inner diameter of the outer cylinder and then outer cylinder is slipped on the inner cylinder and the assembly is then brought to the room temperature. In the light of the theoretical analysis followed by the calculations for stress-distribution in shrunk cylinder shown in the previous article it appears to be very fascinating. But in actual practice the practical problems encountered during the manufacture of long barrel field gums are too many. Even through it is possible theoretically to calculate the shrinkage allowance on the common diameter of the two cylinders but in practice in case where the length of the barrel of the field gums in large, it will be next to impossible to keep uniform interference fit allowance all along the length of the two cylinders. The outer surface of the inner cylinder will be made by turning process on precision lathe, while the inner diameter of the outer cylinder will be machined on a boring machine. The manufacturing tolerance on the inside diameter of the outer cylinder will be next to impossible to maintain. Thus on this account alone, the common pressure at the mating surface of the shrunk gun barrel will not be same all along the length of the barrel. This will result in nonuniform stress distribution along the length of the barrel, which in turn will give rise to stress concentration leading to failure.

Secondly if we resort to the process of heating the outer cylinder we need a heating chamber which should be more that twice the length of the cylinders or field guns, whose length is of the order of 4 to 5m. In addition, due to heating, warpage of the outer cylinder will take place which in turn will result in uneven stress distribution. Similarly if we adopt the process of cooling the inner cylinder by immercing it in liquified nitrogen there may not be warpage but the requirement of very large cooling chamber will make the process very costly. In addition this will be a very slow process.

In order to overcome above difficulties a new process called the AUTOFRETTAGE has been developed. In this process only one cylinder is taken whose ends are closed with detachable covers of special design so that water or oil under high pressure can be pumped into the cylinder with the help of a reciprocating pump. To start with the cylinder is just filled with water and then additional predetermined quantity of water is pumped into the cylinder under high pressure. Due to this internal fluid pressure the maximum circumferential stress which is equal to the yield stress of the cylinder material is generated on the inner surface of the cylinder. Pressure in increased further to pre determined value so that the plasticity spreads to a certain layer who radius is more than the inner radius and less than the outside radius of the cylinder. This fluid pressure is maintained for about 24 hours so that the stresses developed in the material get stabilised. Then the fluid pressure is released and the fluid is drained out. By so doing, the inner layers which had earlier been stressed beyond yield point will partially recover the strain but a certain amount of permanent deformation, depending upon the level of stress to which it had been subjected, will be left. On the contrary all those layers where the stress level, to which they ware subjected was lower than yield stress will tend to regain back their original position or they will try to recover the complete deformation to which they had been subjected. But due to the presence of permanently deformed inner layers their inward movement or their recovery will be checked. In other words they will exert pressure on inner layers and as a reaction they will be pushed out by the inner layers. However when the equilibrium is achieved it will end up in a cylinder similar to shrunk cylinder. Since the stress strain curve of the material is monotonicallly increasing upto the ultimate point or instability point, each layer of the cylinder, which is in plastic range, will have different stress. Therefore actually there will be infinite number of shrunk cylinders. However for the purpose of making a beginner to understand the process a simplifying assumption is made to the effect that

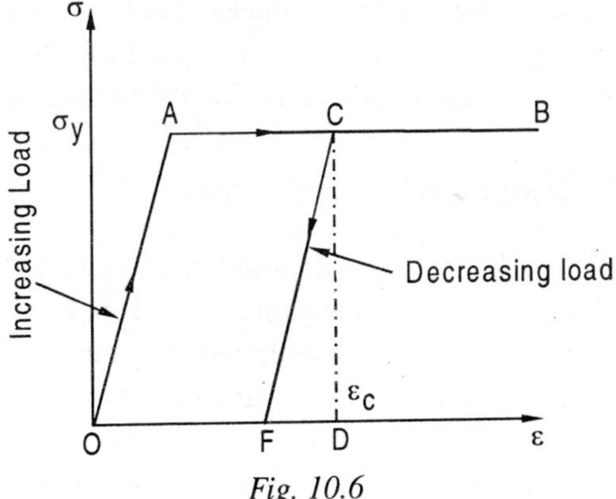


*Fig. 10.6*

the material of the cylinder is elastic-perfectly plastic, that is it does not strain harden. Its stress-strain curve is of the shape as shown in fig. 10.6. For cylinder of such materials if the pressure of water is raised then the strain on the layers adjacent to the inner radius of the cylinder will go on increasing but the stress level will remain at $\sigma_y$. However there will be a limit upto which the water pressure can be raised and at this value of $p_w$ the corresponding total strain at $r = r_1$ will become equal to $\varepsilon_{ult}$ corresponding to ultimate stress point or instability point.

Let at any water pressure $p_w$, the layer at radius $r_2$ experiences the strain $\varepsilon_c$ corresponding to point $C$ on the stress-strain curve. When the water pressure is removed, the material will follow the path $CF$ which is parallel to the elastic portion of the stress strain curve (This is called the Baushinger Effect). The total strain $\varepsilon_c$ at point $C$ consists of two components as explained below

$$\begin{aligned} \varepsilon_c &= OD \\ &= OF + FD \\ &= \varepsilon_p + \varepsilon_e \end{aligned} \qquad ...(10.42)$$

The component $FD$ is called the elastic component of total strain which is recovered when the load is removed. This is also called the spring back. The other portion $OF$, which is known as plastic component is not recovered and forms part of the permanent change in shape. Thus the material at radius $r_2$ is stained permanently upto level of $\varepsilon_p$. Corresponding to the strain $\varepsilon_p$ the interference fit at diameter $D_2 (= 2r_2)$ will be obtained by multiplying $\varepsilon_p$ will $D_2$

*i.e.* $\Delta D_2 = D_2\varepsilon_p$

= Interference fit of thus created equivalent shrunk cylinder at common surface of diameter $D_2$

Then knowing the interference fit in the equivalent shrunk cylinder of radius $r_1$ at inner surface and $r_3$ of outer surface with common surface radius $r_2$, the residual stresses in the two equivalent shrunk cylinders can be determined with the help of equation (10.33), (10.35) and (10.38).

However, the mathematical model for the evaluation of water pressure to which the inner surface of the given thick cylinder is to be subjected, so that it could be autofrettaged is developed by the author in the following paragraphs.

## 10.5.1(*a*) Simple Theory of Autofrettage

Mostly the pressure vessels which are required to sustain high internal pressure are made of variety of steels. The stress-strain curve of these steels is of the general shape as given in fig. (1.2*a*), where the plastic range of the curve upto ultimate load point is known as strain-hardening portion. However here we make a simplifying assumption that the material of the cylinder is elastic perfectly plastic and the corresponding stress-strain curve is of the shape as shown in fig. 10.6, it does not exhibit strain hardening.

To start with, the water pressure in the cylinder is raised gradually from zero value, the stresses both in circumferential and radial directions start rising.

Thus for a given cylinder having inner radius $r_1$ and outer radius $r_3$ when subjected to water pressure of $p_{w_1}$, the radial and circumferential stresses can be evaluated with the help of equations (10.13) and (10.14) but the stresses are less than or at the most equal to $\sigma_y$, the yield stress.

The radial and circumferential stresses at and radius $r$ are then

$$\sigma_r = \frac{p_{w_1} r_1^2}{(r_3^2 - r_1^2)}\left(1 - \frac{r_3^2}{r^2}\right) \quad ...(10.43)$$

and

$$\sigma_c = \frac{p_{w_1} r_1^2}{(r_3^2 - r_1^2)}\left(1 + \frac{r_3^2}{r^2}\right) \quad ...(10.44)$$

Thus the value of the maximum tensile stress, which occurs at the inner surface of radius $r_1$, will be

$$(\sigma_c)_{max} = (\sigma_c)_{r=r_1}$$

$$= \frac{p_{w1} r_1^2}{(r_3^2 - r_1^2)}\left(1 + \frac{r_3^1}{r_1^2}\right)$$

$$= p_{w_1}\left(\frac{r_3^2 + r_1^2}{r_3^2 - r_1^2}\right) \quad ...(10.44a)$$

or

$$(\sigma_c)_{max} = (\sigma_c)_{r_1} = p_{w_1}\left[\frac{r_3^2 + r_1^2}{r_3^2 - r_1^2}\right] \quad ...(10.44b)$$

As the water pressure is further raised gradually a stage will arrive when the maximum circumferential tensile stress at the inner surface of the cylinder will become equal to $\sigma_y$ which is the yield stress of the cylinder material. This stage when $(\sigma_c)_{max}$ becomes equal to $\sigma_y$ is called the "FIRST YIELD STAGE"

**That is at first yield stage**

$$(\sigma_c)_{max} = \sigma_y \quad ...(10.45)$$

Therefore from equations (10.44*a*) and (10.45) we get

$$p_{w_1}\left[\frac{r_3^2 + r_1^2}{r_3^2 - r_1^2}\right] = \sigma_y$$

or

$$p_{w_1} = \sigma_y\left[\frac{r_3^2 - r_1^2}{r_3^2 + r_1^2}\right] \quad ...(10.46a)$$

or

$$\overline{p}_{w_1} = \frac{c_3^2 - 1}{c_3^2 + 1}$$

where $\overline{p}_{w_1}$ which is called the non diamensionalised water pressure at first yield and is defined as

$$\overline{p}_{w_1} = \frac{p_{w_1}}{\sigma_y} \qquad ...(10.47)$$

Similarly $c$, $c_2$ and $c_3$ are the non dimensionalised radii, defined as below

$$c = \frac{r}{r_1} \qquad ...(10.48a)$$

$$c_2 = \frac{r_2}{r_1} \qquad ...(10.48b)$$

and
$$c_3 = \frac{r_3}{r_1} \qquad ...(10.48c)$$

The water pressure required to intiate yielding *i.e.,* $p_{w_1}$ is calculated with the help of equation (10.46) if the values of $\sigma_y$, $r_2$, and $r_3$ are known.

The water pressure is then gradually raised further beyond $p_{w_1}$. In so doing the yielding of the cylinder material will spread radially outwards. However in the yielding material the circumferented stress, for the assumed type of material, will remain constant at $\sigma_y$ level, but the circumferential strain will go on increasing along with the spread of yielding. In other words in the yielded material even through the circumferential stress will be $\sigma_y$ at all the radii within the yielded zone but the circumferential strain will go on increasing as we move radially out wards from the inner radius $r_1$ to say radius $r_2$ upto which the yielded zone extends. However the maximum circumferential strain at $r = r_1$ will occur and the maximum value of the water pressure will generate the total circumferential strain equal to strain at ultimate point or instability point of the material. Then if $\varepsilon_{ult}$ is the strain in uniaxial tension test on the material, then

$$\varepsilon_{ult} = \frac{\sigma_c}{E} + \frac{\nu_p}{E}(p_w)_{max}$$

where $\nu_p$ = Poissons ratio of the material in the plastic range*

$= 0.5$

or
$$\varepsilon_{ult} = \frac{\sigma_y}{E} + \frac{0.5}{E}(p_w)_{max}$$

$$= \frac{\sigma_y}{E}\left(1 + 0.5\frac{(p_w)_{max}}{\sigma_y}\right) \qquad ...(10.48)$$

Thus if
$$\overline{\varepsilon}_{ult} = \frac{\varepsilon_{ult}}{\varepsilon_e}$$

where
$$\varepsilon_e = \frac{\sigma_y}{E} = \text{strain at elastic limit point}$$

---

* **Note :** *For proof see miscellaneous problem 20.9 in chapter 20.*

Then $$\varepsilon_{ult} = \varepsilon_e\,[1 + 0.5\,(\bar{p}_w)_{\max})$$

or $$\bar{\varepsilon}_{ult} = \frac{\varepsilon_{ult}}{\varepsilon_e} = 1 + 0.5\,(\bar{p}_w)_{\max}$$

or $$(\bar{p}_w)_{\max} = 2(\bar{\varepsilon}_{ult} - 1) \qquad ...(10.50a)$$

Thus the autofrettaging pressure $\bar{p}_w$ to which the thick cylinder is subjected lies within the range

$$\bar{p}_{w_1} < \bar{p}_w < (\bar{p}_w)_{\max} \qquad ...(10.50)$$

In the following paragraphs, theory has been developed to determine the radius $r_2$ upto which the material has yielded when the water pressure is raised to the level of $p_w$ where $\bar{p}_{w_1} < \bar{p}_w < (\bar{p}_w)_{\max}$ as given in equation (10.50*b*). Thus the inner zone stretching from radius $r_1$ to radius $r_2$ is like a cylinder having internal pressure $p_w$ at radius $r_1$ and the circumferential stress equal to $\sigma_y$ at all the radii from $r_1$ to $r_2$. In figure (10.7) is shown the free body diagram of a small circular element, contained in the plastically deformed zone stretching from radius $r_1$ to $r_2$; which lies between the radii *r* and $(r + dr)$ and subtends an angle $d\theta$ at the centre. Let $\sigma_r$ be the radial stress at radius r and $(\sigma_r + d\sigma_r)$ at radius $(r + dr)$ Then for the static equilibrium of this element *ABCD* we get by resolving the forces along the radially out ward direction per unit length of cylinder.

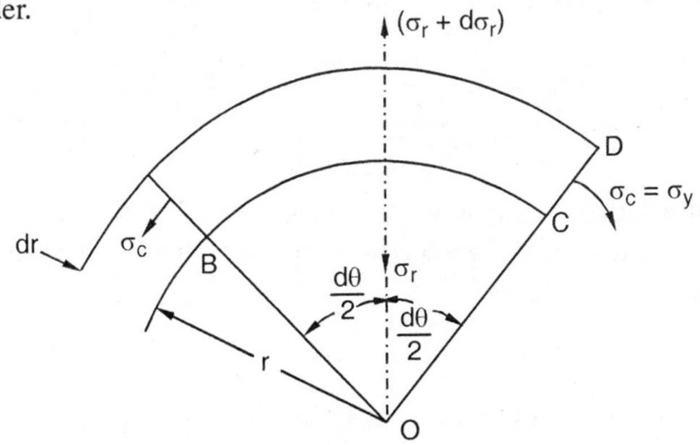

*Fig. 10.7*

$$(\sigma_r + d\sigma_r)\,(r + dr)\,d\theta - \sigma_r r d\theta - 2\sigma_y dr \sin\frac{d\theta}{2} = 0$$

Then neglecting the product of two small quantities and equating $\sin\frac{d\theta}{2} \approx \frac{d\theta}{2}$, this equation reduces to

$$(r\sigma_r + rd\sigma_r + \sigma_r dr)\,d\theta - r\sigma_r d\theta - \sigma_y dr d\theta = 0$$

or $$rd\sigma_r + \sigma_r dr - \sigma_y dr = 0$$

or $$(\sigma_r - \sigma_y) = -r\frac{d\sigma_r}{dr}$$

or $$\frac{d_r}{r} = \frac{d\sigma_r}{(-\sigma_r + \sigma_y)} \qquad ...(10.51)$$

Integrating equation (10.51) we get

$$\ln (r) = -\ln (-\sigma_r + \sigma_y) + \ln A$$

where $A$ is a constant of integration

or $$\ln (r) = \ln \left( \frac{A}{-\sigma_r + \sigma_y} \right)$$

or $$r = \frac{A}{-\sigma_r + \sigma_y}$$

or $$\sigma_y - \sigma_r = \frac{A}{r}$$

or $$\sigma_r = \sigma_y - \frac{A}{r} \qquad ...(10.52)$$

For evaluating the constant A, we insert the boundary condition which in

At $r = r_1$ $\quad \sigma_r = -p_w$

Therefore from (10.52)

$$-p_w = \sigma_y - \frac{A}{r_1}$$

or $$A = r_1 (\sigma_y + p_w) \qquad ...(10.53)$$

Thus substituting the value of A from equation (10.53) into equation (10.52) we get the final equation for the variation of $\sigma_r$ with $r$ within the yielded zone of the material

$$\sigma_r = \sigma_y - \frac{r_1}{r}(\sigma_y + p_w) \qquad ...(10.54a)$$

Dividing equation (10.54*a*) by $\sigma_y$ we get the radial pressure in non dimensionalized form as

$$\bar{\sigma}_r = 1 - \frac{r_1}{r}(1 + \bar{p}_w) \qquad ...(10.54b)$$

Thus if $p_2$ is the radial pressure at the common surface of diameter $D_2$ $(= 2r_2)$ between the plastic or yielded zone and the elastic zone beyond radius $r_2$ stretching upto radius $r_3$, then from equation (10.54*a*) we get

$$\text{At } r = r_2 \quad (\sigma_r)_{r_2} = -p_2 = \sigma_y - \frac{r_1}{r_2}(\sigma_y + p_w)$$

*i.e.,* $$p_2 = -\sigma_y + \frac{r_1}{r_2}(\sigma_y + p_w) \qquad ...(10.55a)$$

and in non dimentionalised from equation (10.55*a*) is re-written as given below

$$\bar{p}_2 = \frac{p_2}{\sigma_y}$$

$$\bar{p}_w = \frac{p_w}{\sigma_y}$$

and $$c_2 = \frac{r_2}{r_1}$$

Since at water pressure $p_w$ the cylinder material is divided into two distinct zones: one stretching from radius $r_1$ to radius $r_2$ which is yielded and other from $r_2$ to $r_3$ which is still elastic. Thus these zones behave like two different cylinders. Let the stresses in the yielded zone be designated with a suffix '$i$' and that in the elastic zone with suffix '0'

Then for yielded zone we have from equation (10.55)

$$(\sigma_{r_i})_{r_2} = -p_2 = \sigma_y - \frac{r_1}{r_2}(\sigma_y + p_w)$$

$$p_2 = -\sigma_y + \frac{r_1}{r_2}(\sigma_y + p_w) \qquad \text{...(10.55}a\text{)}$$

and $$\bar{p}_2 = -1 + \frac{1}{c_2}(1 + \bar{p}_w) \qquad \text{...(10.55}b\text{)}$$

Further $$\left(\sigma_{c_i}\right)_r = \left(\sigma_{c_i}\right)_{r_1} = \left(\sigma_{c_i}\right)_{r_2} = \sigma_y$$

because everywhere in the inner zone the circumferential stress is $\sigma_y$.

Therefore circumferential strain at radius $r_2$ is given by

$$\left(\varepsilon_{c_i}\right)_{r_2} = \frac{1}{E}\left(\sigma_{c_i}\right)_{r_2} - \frac{\nu_p}{E}\left(\sigma_{r_i}\right)_{r_2}$$

$$= \frac{1}{E}\sigma_y - \frac{\nu_p}{E}(-p)$$

Therefore substituting values from equation (10.55$a$) we get

or $$\left(\varepsilon_{c_i}\right)_{r_2} = \frac{1}{E}\sigma_y + \frac{\nu_p}{E}\left[-\sigma_y + \frac{r_1}{r_2}(\sigma_y + p_w)\right] \qquad \text{...(10.56}a\text{)}$$

$$= \frac{\sigma_y}{E} + \frac{\nu_p}{E}\sigma_y\left[-1 + \frac{1}{c_2}(1 + \bar{p}_w)\right] \qquad \text{...(10.56}b\text{)}$$

After analysing the stresses in the inner cylinder where the material is in plastic range we now proceed to analyse the stresses in the outer cylinder which extends from radius $r_2$ to radius $r_3$. The stresses in this cylinder are within the elastic limit and this constitutes a thick cylinder having internal pressure $p_2$ at radius $r_2$ and outer pressure zero at $r = r_3$. Thus using the equations (10.13) and (10.14) are have for this cylinder

$$\left(\sigma_{c0}\right)_r = \frac{p_2 r_2^2}{(r_3^2 - r_2^2)}\left[1 + \frac{r_3^2}{r^2}\right] \quad \text{...(10.57}a\text{)}$$

and

$$\left(\sigma_{r0}\right)_r = \frac{p_2 r_2^2}{(r_3^2 - r_2^2)}\left[1 - \frac{r_3^2}{r^2}\right] \quad \text{...(10.57}b\text{)}$$

Therefore at $r = r_2$

$$\left(\sigma_{r0}\right)_{r_2} = \frac{p_2 r_2^2}{(r_3^2 - r_2^2)}\left[1 - \frac{r_3^2}{r_2^2}\right] = -p_2$$

and

$$\left(\sigma_{c0}\right)_{r_2} = \frac{p_2 r_2^2}{(r_3^2 - r_2^2)}\left[1 + \frac{r_3^2}{r_2^2}\right]$$

$$= p_2\left(\frac{r_3^2 + r_2^2}{r_3^2 - r_2^2}\right) \quad \text{...(10.58}a\text{)}$$

or

$$= p_2\left(\frac{c_3^2 + c_2^2}{c_3^2 - c_2^2}\right) \quad \text{...(10.58}b\text{)}$$

where

$$c_3 = \frac{r_3}{r_1} \text{ and } c_2 = \frac{r_2}{r_1}$$

However the circumferential stress at $r = r_2$ is $\sigma_y$. Therefore substituting $\sigma_y$ for $\left(\sigma_{c0}\right)_{r_2}$ in equation (10.58) we get a relationship between $p_2$ the radial pressure at the common surface, and radius $r_2$ (or $c_2$) as

$$\sigma_y = p_2 \frac{(c_3^2 + c_2^2)}{(c_3^2 - c_2^2)}$$

or

$$\frac{p_2}{\sigma_y} = \frac{(c_3^2 - c_2^2)}{(c_3^2 + c_2^2)} = \overline{p}_2 \quad \text{...(10.59)}$$

where $\overline{p}_2$ is the non-dimensionalised pressure $= \dfrac{p_2}{\sigma_y}$

Substituting the value of $\overline{p}_2$ from equation (10.55$b$) in equation (10.58) we get the equation the solution of which will give the value of $c_2$ (or $r_2$) in terms of $\overline{p}_w$ the non-dimensionalised water pressure

$$\frac{c_3^2 - c_2^2}{c_3^2 + c_2^2} = -1 + \frac{1}{c_2}(1 + \overline{p}_w)$$

or $$c_3^2 - c_2^2 = -(c_3^2 + c_2^2) + \frac{(c_3^2 + c_2^2)}{c_2}(1 + \bar{p}_w)$$

$$2c_3^2 c_2 = (c_3^2 + c_2^2)(1 + \bar{p}_w)$$

or $$2c_3^2 c_2 = c_3^2 + c_2^2 + c_3^2 \bar{p}_w + c_2^2 \bar{p}_w$$

or $$c_2^2(1 + \bar{p}_w) - 2c_3^2 c_2 + c_3^2(1 + \bar{p}_w) = 0 \quad ...(10.60)$$

Equation (10.59) is quadratic equation in $c_2$ and its solution gives the value of $c_2$ as

$$c_2 = \frac{2c_3^2 \pm \left[4c_3^4 - 4c_3^2(1 + \bar{p}_w)^2\right]^{\frac{1}{2}}}{2(1 + \bar{p}_w)} \quad ...(10.61)$$

The roots given by expression (10.61) for $c_2$ will be real if

$$c_3 \geq 1 + \bar{p}_w \quad ...(10.62a)$$

or $$\bar{p}_w \leq c_3 - 1 \quad ...(10.62b)$$

In the limiting case when $c_3 = 1 + \bar{p}_w$, then from equation (10.61) $c_2 = c_3$ which will mean that $\bar{p}_w$ is such that the entire cylinder material from radius $r_1$ to radius $r_3$ has been rendered plastic and autofrettaging will not be possible. Thus we conclude from equation (10.62*b*) that $\bar{p}_w$ must be less than $(c_3 - 1)$, which apparently fixes the upper limit of water pressure. However, as already discussed earlier that the upper most limit of pressure is given by equation (10.50) as $(\bar{p}_w)_{max}$. The minimum limit for $p_w$ will be the one which initiates the yielding of the material *i.e.*, at first yielding point and will be equal to $\bar{p}_{w_1}$ as given by equation (10.46*b*). Thus we conclude

$$\bar{p}_{w_1} < \bar{p}_w < (c_3 - 1) < (\bar{p}_w)_{max} \quad ...(10.63)$$

Further as we have concluded above from relationship (10.62) that $c_3 > (1 + \bar{p}_w)$ along with the fact that $c_2 < c_3$, it is evident from equation (10.61) that only the negative root will be valid in the numeator of equation (10.61)

That is $$c_2 = \frac{c_3^2 - c_3\left[c_3^2 - (1 + \bar{p}_w)^2\right]^{\frac{1}{2}}}{(1 + \bar{p}_w)} \quad ...(10.64)$$

It is re-iterated that all the equations derived so far in the article 10.5 apply when the water pressure of intensity $\bar{p}_w$ is retained in the cylinder. However after the water after the water pressure is released when the stresses have stabilised it will result in the creation of residual stresses due to interaction of the elastically deformed zone trying to come back to it original position but being opposed by the one which had been plastically deformed. When the interaction is over and equilibrium has reached, it will be equivalent to two shrunk cylinders with common surface of diameter $D_2$.

## 10.5.1(*b*) Equivalent Shrunk Cylinders

As explained above, when the water pressure is released by draining out water the interaction between the two zones ensues. The two zones act as if they are two cylinders outer shrunk onto the inner. The outer layer of the inner cylinder at radius $r_2$ has been strained beyond elastic limit. Thus when it is forced by the outer zone to return it will do so partially to the extent that only elastic portion of the total strain will be recovered and the plastic portion will stay thereby creating a situation equivalent to two shrunk cylinder with interference fit at diameter $D_2$ (= $2r_2$) equal to the plastic component of the total strain.

That is "interference fit" at common dia $D_2$ of the equivalent shrunk cylinders $\Delta D_2$, is given by

$$\Delta D_2 = D_2 \text{ [plastic component of total circumferential strain at } r_2]$$

$$= D_2\left[\left(\varepsilon_{c_i}\right)_{r_2} - \left(\varepsilon_e\right)_{r_2}\right]$$

where $\left(\varepsilon_e\right)_{r_2}$ = Elastic component of circumferential strain at $r_2$

$$= \frac{\sigma_y}{E}$$

and $\left(\varepsilon_{c_i}\right)_{r_2}$ = circumferential strain at $r = r_2$

Therefore $\Delta D_2$ = Interference fit

$$= D_2\left[\left(\varepsilon_{c_i}\right)_{r_2} - \frac{\sigma_y}{E}\right]$$

Substituting value of $\left(\varepsilon_{c_i}\right)_{r_2}$ from equation (10.56) we get

$$\Delta \mathbf{D}_2 = D_2 \frac{\nu_p}{E}\sigma_y\left[-1 + \frac{1}{c_2}(1 + \bar{p}_w)\right] \qquad \text{...(10.65)}$$

where $\nu_p$ = Poissons ratio of the material in plastic zone = 0.5

$E$ = Youngs modulus of elasticity

Thus after the water pressure is released the resultant material is an assembly of equivalent shrunk cylinders having inner radius $r_1$ and outer radius $r_3$ with $r_2$ as common surface radius with interference fit of $\Delta D_2$ at common diameter of $D_2$ ( = $2r_2$).

### 10.5.2 Radial Pressure at the Common Surface of Equivalent Shrunk Cylinders and the Associated Residual Stresses

After having determined the interference fit at the common surface of radius $r_2$ we proceed to find the radial pressure at the common surface of the equivalent shrunk cylinders which are the result of subjecting original thick cylinder of internal and outer radii $r_1$ and $r_3$ respectively to high water pressure $p_w$ for sufficient length of time and then releasing the water.

Let $p_2$ be the radial pressure at the common surface of radius $r_2$ of the two equivalent shrunk cylinders. The interference fit on common diameter is $\Delta D_2$ as given in equation (10.65) and this fit generates common pressure $p_2$ at $r = r_2$ as given by equation (10.33) which is reproduced below as (10.66)

$$\Delta D_2 = \frac{D_2 p_2}{E}\left[\frac{r_3^2 + r_2^2}{r_3^2 - r_2^2} + \frac{r_2^2 + r_1^2}{r_2^2 - r_1^2}\right] \qquad ...(10.66a)$$

or

$$\Delta D_2 = \frac{D_2 p_2}{E}\left[\frac{c_3^2 + c_2^2}{c_3^2 - c_2^2} + \frac{c_2^2 + 1}{c_2^2 - 1}\right] \qquad ...(10.66b)$$

Thus equating the values of $\Delta D_2$ from equations (10.65) and (10.66) we get

$$\frac{D_2 \sigma_y}{2E}\left[-1 + \frac{1}{c_2}(1 + \overline{p}_w)\right] = \frac{D_2 p_2}{E}\left[\frac{c_3^2 + c_2^2}{c_3^2 - c_2^2} + \frac{c_2^2 + 1}{c_2^2 - 1}\right]$$

or

$$\frac{1}{2c_2}\left[1 + \overline{p}_w - c_2\right] = \frac{\overline{p}_2\left[c_3^2\,c_2^2 + c_2^4 - (c_3^2 + c_2^2) + c_2^2\,c_3^2 - c_2^4 + c_3^2 - c_2^2\right]}{(c_3^2 - c_2^2)(c_2^2 - 1)}$$

or

$$\frac{1}{2c_2}\left[1 + \overline{p}_w - c_2\right] = \frac{\overline{p}_2\,(2c_3^2\,c_2^2 - 2c_2^2)}{(c_3^2 - c_2^2)(c_2^2 - 1)}$$

or

$$\overline{p}_2 = \frac{(1 + \overline{p}_w - c_2)\left(c_3^2 - c_2^2\right)\left(c_2^2 - 1\right)}{4c_2^3\left(c_3^2 - 1\right)} \qquad ...(10.67)$$

where

$$\overline{p}_2 = \frac{p_2}{\sigma_y} \qquad ...(10.68)$$

Knowing the values of $v_p$, $c_2$, $c_3$ and $\overline{p}_w$, $\overline{p}_2$ (or $p_2$) is calculated. This value of $p_2$ will then be used to find the residual stresses in the two equivalent shrunk cylinders with the help of equations (10.34), (10.35), (10.36), (10.37), (10.38) and (10.39) where in the value of $p_2$ will be inserted in place of $p$.

The residual stresses in equivalent shrunk cylinders are then calculated as follows

**For inner cylinder :** At any radius r the circumferential stress is

$$\left(\sigma_{c_i}\right)_r = \frac{-p_2 r_2^2}{\left(r_2^2 - r_1^2\right)}\left(1 + \frac{r_1^2}{r^2}\right) \qquad \text{...(10.69)}$$

$$\left(\sigma_{r_i}\right)_r = \frac{-p_2 r_2^2}{(r_2^2 - r_1^2)}\left(1 - \frac{r_1^2}{r^2}\right) \qquad \text{...(10.70)}$$

Therefore at the inner surface of the inner cylinder we get from equation (10.68)

$$\left(\sigma_{c_1}\right)_{r_1} = \frac{-2p_2\, r_2^2}{(r_2^2 - r_1^2)}$$

$$= \frac{-2p_2\, c_2^2}{(c_2^2 - 1)} \qquad (10.71b)$$

For outer cylinder, similarly, we get from equation (10.38) and (10.39), by replacing $p$ with $p_2$

$$\left(\sigma_{c0}\right)_r = \frac{p_2 r_2^2}{(r_3^2 - r_2^2)}\left(1 + \frac{r_3^2}{r^2}\right) \qquad \text{...(10.72)}$$

$$\left(\sigma_{r0}\right)_r = \frac{p_2 r_2^2}{(r_3^2 - r_2^2)}\left(1 - \frac{r_3^2}{r^2}\right) \qquad \text{...(10.73)}$$

Therefore at the inner surface of the outer cylinder *i.e.* at $r = r_2$ we have from equation (10.72)

$$\left(\sigma_{c0}\right)_{r_2} = \frac{p_2 r_2^2}{(r_3^2 - r_2^2)}\left(1 + \frac{r_3^2}{r_2^2}\right)$$

$$= \frac{p_2(r_3^2 + r_2^2)}{(r_3^2 - r_2^2)} \qquad \text{...(10.74a)}$$

Substituting the value of $p_2$ as found with the help of equation (10.67) in equations (10.67) through (10.74), the values of the residual stresses in the antofrettaged cylinder can be determined.

### 10.5.3 Resultant Stress In Autofrettaged Cylinder due to New Internal Pressure

Let $p_n$ be the new internal pressure to which the autofrettaged cylinder is subjected. The radial and circumferential stress developed due to $p_n$ can be calculated with the help of equation (10.13) and (10.14) where inner radius will be taken as $r_1$ and outer radius $r_3$ and treating the entire cylinder as one unit.

Let the stresses, both circumferential and radial be represented with symbol having prime. Then from equation (10.13) and (10.14) we have

$$(\sigma'_r)_r = \frac{p_n r_1^2}{(r_3^2 - r_1^2)}\left(1 - \frac{r_3^2}{r^2}\right) \quad ...(10.75)$$

and

$$(\sigma'_c)_r = \frac{p_n r_1^2}{(r_3^2 - r_1^2)}\left(1 + \frac{r_3^2}{r^2}\right) \quad ...(10.76)$$

Therefore at $r = r_1$

$$(\sigma'_c)_{r_1} = \frac{p_n r_1^2}{(r_3^2 - r_1^2)}\left(1 + \frac{r_3^2}{r_1^2}\right)$$

$$= \frac{p_n r_1^2}{(r_3^2 - r_1^2)} \frac{(r_1^2 + r_2^2)}{r_1^2} \quad ...(10.77a)$$

or

$$(\sigma'_c)_{r_1} = \frac{p_n (c_3^2 + 1)}{(c_3^2 - 1)} \quad ...(10.77b)$$

Similarly at $r = r_2$

$$(\sigma'_c)_{r2} = \frac{p_n r_1^2}{(r_3^2 - r_1^2)}\left(1 + \frac{r_3^2}{r_2^2}\right)$$

$$= p_n \frac{r_1^2}{r_2^2}\left(\frac{r_3^2 + r_2^2}{(r_3^2 - r_1^2)}\right) \quad ...(10.78a)$$

or

$$(\sigma'_c)_{r2} = \frac{p_n}{c_2^2}\left(\frac{c_3^2 + c_2^2}{c_3^2 - 1}\right) \quad ...(10.78b)$$

The resultant stresses at these radii are then determined by taking the algebraic sum of residual stresses and the stresses induced due to $p_n$.

Thus the resultant circumferential stress at $r = r_2$ is obtained from equations (10.71) and (10.77)

$$(\sigma_{cR})_{r_1} = (\sigma_{c_i})_{r_1} + (\sigma'_c)_{r_1}$$

$$= \frac{-2p_2 c_2^2}{(c_2^2 - 1)} + \frac{p_n (c_3^2 + 1)}{(c_3^2 - 1)} \quad ...(10.79)$$

Similarly the resultant circumferential stress at $r = r_2$ is obtained from equations (10.73) and (10.78*b*).

$$(\sigma_{cR})_{r_2} = (\sigma_{c0})_{r_2} + (\sigma'_c)_{r_2}$$

$$= \frac{p_2 (c_3^2 + c_2^2)}{(c_3^2 - c_2^2)} + \frac{p_n}{c_2^2} \frac{(c_3^2 + c_2^2)}{(c_3^2 - 1)} \quad ...(10.80)$$

**Example 10.12.** A steel cylinder inside diameter 80 mm aŋd outside diameter 240 mm is autofrettaged with high pressure water of intersity 360 MPa. If the material of cylinder is assumed as elastic-perfectly plastic having yield stress $\sigma_y$ = 240 MPa, $E$ = 200 GPa and poissou's ratio = 0.25, find

The dimensions of the equivalent shrunk

(*a*) Cylinders

(*b*) The radial pressure at the common diameter of equivalent shrink cylinders and the interference fit at the common diameter.

(*c*) Residual stresses at the inner and outer diameters of the euqavalent shrunk cylinders after the high pressure water is released.

(*d*) The new internal pressure $p_n$ which this antofrettaged cylinder can sustain without the stress crossing the yield stress $\sigma_y$ any where in the material.

(*e*) The ratio of $p_n$ to $p_{w_1}$ and hence the percentage increase in pressure bearing capacity due to autofrettage

**Solution.** Given data

$r_1$ = 40 and $r_3$ = 120 mm

$$E = 200 \times 10^3 \text{ MPa}$$

$$\sigma_y = 240 \text{ MPa}$$

$$\nu = 0.25 \text{ within elastic limit point but in plastic range} = 0.5$$

$$p_w = 360 \text{ MPa}$$

(*a*) Let $r_2$ be the radius of the common surface of the equivalent shrunk cylinder

then
$$c_3 = \frac{r_3}{r_1} = \frac{120}{40} = 3$$

$$c_2 = \frac{r_2}{r_1}$$

From equation (10.63)

$$c_2 = \frac{c_3^2 - c_3\left[c_3^2 - (1+\bar{p}_w)^2\right]^{\frac{1}{2}}}{(1+\bar{p}_w)}$$

where
$$\bar{p}_w = \frac{p_w}{\sigma_y} = \frac{360}{240} = 1.5$$

Therefore
$$c_2 = \frac{9 - 3\,[9-(1+1.5)^2]^{\frac{1}{2}}}{(1+1.5)}$$

$$= \frac{9 - 3[9 - (6.25)^2]^{\frac{1}{2}}}{(2.5)}$$

$$= \frac{9 - 3 \times 1.6583}{2.5} = 1.6$$

Therefore $r_2 = c_2 r_1 = 1.6 \times 40 = 64$ mm

and $$D_2 = 2r_2 = 2 \times 64 = 128 \text{ mm}$$

The diameteral interference fit between the two equivalent shrunk cylinders is given by equation (10.64) as

$$\Delta D_2 = D_2 \frac{v}{E} \sigma_y \left[ -1 \frac{1}{c_2} (1 + \bar{p}_w) \right]$$

$$= \frac{128.0.5}{20 \times 10^3} \times 240 \left[ -1 + \frac{1}{1.6} (1 + 1.5) \right]$$

$$= 0.0432 \text{ mm}$$

Thus the dimensions of two equivalent shrunk cylinders are

$$r_1 = 40 \text{ mm}, r_2 = 64 \text{ mm}$$
$$r_2 = 64 \text{ mm}, r_3 = 120 \text{ mm}$$

with interference fit at $D_2 = 0.0432$ mm

(*b*) The radial pressure at common surface of these shrunk cylinders is given by equation (10.66)

$$(\bar{p}_2) = \frac{(1 + \bar{p}_w - c_2)(c_3^2 - c_2^2)(c_2^2 - 1)}{4C_2^3 (c_3^2 - 1)}$$

$$= \frac{(1 + 1.5 - 1.6)(9 - 1.6^2)(1.6^2 - 1)}{4(1.6)^3 (9 - 1)}$$

$$= \frac{0.9 \times 6.444 \times 1.56}{4 \times (1.6)^3 \times 8} = 0.069026$$

Therefore $$p_2 = \bar{p}_2 \times \sigma_y$$

$$= 240 \times 0.069026 = 16.566 \text{ MPa}$$

(*c*) Residual stresses after antofrettage : For inner cylinder, the circumferential stress at radius $r_1$ is given by (10.70*b*)

$$\left(\sigma_{c_i}\right)_{r_1} = \frac{-2p_2 c_2^2}{\left(c_2^2 - 1\right)}$$

$$= \frac{-2 \times 16.566 \times (1.6)^2}{(1.6^2 - 1)}$$

$$= -54.37 \text{ MPa}$$

Similarly at $r = r_2$ we have form equation (10.68) for inner cylinder

$$\left(\sigma_{c_i}\right)_{r_2} = \frac{-p_2\left(r_2^2 + r_1^2\right)}{\left(r_2^2 - r_1^2\right)}$$

$$= \frac{-p_2\left(c_2^2 + 1\right)}{\left(c_2^2 - 1\right)}$$

$$= \frac{-16.566\,(1.6^2 + 1)}{(1.6^2 - 1)}$$

$$= -37.79 \text{ mPa}$$

For outer cylinder, we have from equation (10.37) the circumferential stress at radius $r_2$

$$\left(\sigma_{c_0}\right)_{r_1} = \frac{p_2\left(c_3^2 + c_2^2\right)}{\left(c_3^2 - c_2^2\right)}$$

$$= \frac{16.566 \times (9 + 1.6)^2}{(9 - 1.6^2)}$$

$$= \frac{16.566 \times 11.56}{6.44}$$

$$= 29.7 \text{ MPa}$$

For outer cylinder at $r = r_3$ we have from equation (10.72)

$$\left(\sigma_{c_0}\right)_{r_3} = \frac{2p_2 r_2^2}{\left(r_3^2 - r_2^2\right)}$$

$$= \frac{2p_2 c_2^2}{\left(c_3^2 - c_2^2\right)}$$

$$= \frac{2 \times 16.55596 \times (1.6)^2}{(9 - 1.6^2)}$$

$$= 13.16258 \text{ MPa}$$

(*d*) Let $p_n$ be the new internal pressure acting on the inner surface of autofrettaged cylinder of radii $r_1$ = 40 mm and $r_3$ = 120 mm

Then from equation (10.76) we, the circumferential stress $\sigma'_c$ induced by $p_n$ at $r_1$

$$(\sigma'_c)_{r_1} = \frac{p_n\left(c_3^2 + 1\right)}{\left(c_3^2 - 1\right)}$$

$$= p_n\left(\frac{9+1}{9-1}\right)$$

$$= 1.25\ p_n$$

Similarly at $r = r_2$

$$(\sigma'_c)_{r_2} = \frac{p_n\left(c_3^2 + c_2^2\right)}{c_2^2\left(c_3^2 - 1\right)}$$

$$= \frac{p_n\,(9 + 1.6^2)}{(1.6)^2\ (9 - 1)}$$

$$= \frac{p_n\ 11.56}{(2.56)\,(8)}$$

$$= 0.5644\ p_n$$

And at $r = r_3$

$$(\sigma'_c)_{r_3} = \frac{2p_n\, r_1^2}{\left(r_3^2 - r_1^2\right)}$$

$$= \frac{2p_n}{\left(c_3^2 - 1\right)}$$

$$= \frac{2p_n}{8}$$

$$= 0.25\, p_n$$

There are three options for stress to reach $\sigma_y$ value *i.e.*, at $r_1$, $r_2$, $r_3$

(*i*) Thus if the resultant stress reaches the limiting value of $\sigma_y$ at $r = r_1$, then

$$\left(\sigma_{R_c}\right)_{r_1} = 240 = \left(\sigma_{c_i}\right)_{r_1} + \left(c_c'\right)_{r_1}$$

$$= -54.33752 + 1.25\, p_n$$

or $$p_n = \frac{240 + 54.33750}{1.25}$$

$$= 235.47 \text{ MPa}$$

(*ii*) It the resultant stress reaches the limiting value of $\sigma_y$ at $r = r_2$, of the outer cylinder, then

$$\left(\sigma_{R_c}\right)_{r_2} = 240\left(\sigma_{c_o}\right)_{r_2} + \left(\sigma_c'\right)_{r_2}$$

$$= 29.718 + 0.5644\, p_n$$

or $$p_n = \frac{240 - 29.718}{0.5644}$$

$$= 372.576 \text{ MPa}$$

(*iii*) It the resultant stress reaches the limiting value of $\sigma_y$ at $r = r_3$ of the outer cylinder, then

$$\left(\sigma_{R_c}\right)_{r_3} = \left(\sigma_{c0}\right)_{r_3} + \left(\sigma_c'\right)_{r_3}$$

$$240 = 13.16258 + 0.25\, p_n$$

or $$p_n = \frac{240 - 13.16258}{0.25}$$

$$= 907.3497 \text{ MPa}$$

The least of the three values of $p_n$ so obtained will be the allowable internal pressure which the autofrettaged cylinder can sustain without yielding

That is $$p_n = 335.47 \text{ MPa}$$

(*e*) The water pressure required to initiate yielding in original cylinder *i.e.*, $\overline{p}_{w_1}$ is given by equation (10.46)

That $$\overline{p}_{w_1} = \frac{c_3^2 - 1}{c_3^2 + 1}$$

$$= \frac{9-1}{9+1} = \frac{8}{19}$$

$$= 0.8$$

or $$p_{w_1} = \bar{p}_{w_1} \sigma_y$$

$$= 0.8 \times 240 = 192 \text{ MPa}$$

Therefore the ratio

$$\frac{p_n}{p_{w_1}} = \frac{235.47}{192}$$

$$= 1.2264$$

or percentage increase in pressure bearing capacity after autofrettage

$$= \frac{p_n - p_{w_1}}{p_{w_1}} \times 100$$

$$= (1.2264 - 1) \times 100$$

$$= 22.64\%$$

**Example 10.13.** *An artillary gun with 120 mm bore end 180 mm wall thickness is made of steel haing yield stress 250 MPa, Youngs modulus = 200 GPa and Poissons ratio 0.25. In order to have maximum firing range the gun ponder pressure is such that it just produces the circumferential stress at inner radius equal to yield stress of the material. With the purpose of increasing the firing range, the method of autofrellage is to be invoked by using water at pressure of 275 MPa. Assume the material to be elastic-perfectly prastic. Find the % increase in firing range.*

**Solution.** Given data

(*a*) Given $$\sigma_y = 250 \text{ MPa}$$

$$r_1 = \frac{120}{2} = 60 \text{ mm}$$

$$r_3 = 60 + 180 = 240 \text{ mm}$$

$$c_3 = \frac{r_3}{r_1} = \frac{240}{60} = 4$$

If $$p_{w_1} = \text{Water pressure at first yield}$$

$$= \text{Capacity without autofrettage}$$

From equation (10.46)

$$\bar{p}_{w_1} = \frac{c_3^2 - 1}{c_3^2 + 1}$$

$$= \frac{16-1}{16+1}$$

$$= \frac{15}{17} = 0.88235$$

Therefore $$p_{w_1} = \bar{p}_{w1}\sigma_y$$

$$= 0.88235 \times 250$$

$$= 220.588 \text{ MPa}$$

Let $r_2$ be the radius of the common surface in the equivalent shrunk cylinders after autofrettage due to internal pressure of intersity of 275 MPa.

Then if $c_2 = \dfrac{r_2}{r_1}$ we have from equation (10.64)

Therefore $$c_2 = \frac{c_3^2 - c_3\left[c_3^2 - (1+\bar{p}_w)^2\right]^{\frac{1}{2}}}{(1+\bar{p}_w)}$$

where $$\bar{p}_w = \frac{p_w}{\sigma_y} = \frac{275}{250} = 1.1$$

Therefore $$c_2 = \frac{16 - 4\,[16 - (1+1.1)^2\,]^{\frac{1}{2}}}{(1+1.1)}$$

$$= \frac{16 - 4\,[16 - 4.41]^{\frac{1}{2}}}{2.1}$$

$$= \frac{16 - 4 \times 3.4044}{2.1}$$

$$= \frac{16 - 13.6176}{2.1}$$

$$= 1.13446$$

Therefore $$r_2 = r_1 c_2 = 60 \times 68.06755$$

$$= 68.06755 \text{ mm}$$

or $$D_2 = 2r_2 = 2 \times 68.06755$$
$$= 136.1351 \text{ mm}$$

Therefore interference fit at common $D_2$ of the equivalent shrunk cylinder is given by equation (10.65) as

$$\Delta D_2 = \frac{D_2}{2E}\sigma_y\left[-1 + \frac{1}{c_2}(1 + \bar{p}_w)\right]$$

$$= \frac{136.1351}{2 \times 200 \times 10^3} \times 250\left[-1 + \frac{1}{1.3446}(1 + 1.1)\right]$$

$$= [-1 + 1.85100] = 0.07242 \text{ mm}$$

(*h*) If $p_2$ is the normal pressure at the common dia $D_2$ of the equivalent shrunk cylinders, then from equation (10.67).

$$\bar{p}_2 = \frac{p_2}{\sigma_y}$$

$$= \frac{v(1 + \bar{p}_w - c_2)\left(c_3^2 - c_2^2\right)\left(c_2^2 - 1\right)}{2c_2^3\left(c_3^2 - 1\right)}$$

Now for plastically deforming material

$$v = 0.5$$

$$\therefore \quad \bar{p}_2 = \frac{(1 + 1.1 - 1.13446)(16 - 1.13446^2)(1.13446^2 - 1)}{4(1.13446)^2(16 - 1)}$$

$$= \frac{(0.96554)(14.7130)(0.286999)}{4(1.460049)15}$$

$$= 0.04654$$

or $$p_2 = \bar{p}_2 \times \sigma_y$$

$$.04654 \times 250 = 11.6352 \text{ MPa}$$

(*c*) Therefore residual stresses after autofrettage are as follows

For inner equivalent cylinder at radius $r_1$ as given in equation (10.71*b*)

$$\left(\sigma_{c_i}\right)_{r_1} = \frac{-2p_2c_2^2}{\left(c_2^2 - 1\right)}$$

$$= \frac{-2 \times 11.6352 \times (1.13446)^2}{(1.13446^2 - 1)}$$

$$= \frac{.1.6352 \times 1.286999}{0.286999}$$

$$= -\ 104.356 \text{ MPa}$$

Similarly at $r = r_2$ we have from equation (10.69) for inner cylinder

$$\left(\sigma_{c_i}\right)_{r_2} = \frac{-\,p_2\left(r_2^2 + r_1^2\right)}{\left(r_2^2 - r_1^2\right)}$$

$$= \frac{-\,p_2\left(c_2^2 + 1\right)}{\left(c_2^2 - 1\right)}$$

$$= \frac{-\,11.6352 \times (1.13446^2 + 1)}{(1.13446^2 - 1)}$$

$$= \frac{-\,11.6352 \times 2.286999}{0.286999}$$

$$= 92.7170 \text{ MPa}$$

For outer equivalent cylinder, we have from equation (10.74) the circumferential stress at $r = r_2$

$$\left(\sigma_{c0}\right)_{r_2} = \frac{p_2\left(c_3^2 + c_2^2\right)}{\left(c_3^2 - c_2^2\right)}$$

$$= \frac{11.6352\,(4^2 + 1.13446^2)}{(16 - 1.13446^2)}$$

$$= \frac{11.6352\,(4^2 + 1.13446^2)}{(16 - 1.13446^2)}$$

$$= \frac{11.6352 \times 17.286999}{14.713}$$

$$= 13.6707 \text{ MPa}$$

Further for the outer cylinder at $r = r_3$

We have from equation (10.73)

$$\left(\sigma_{c0}\right)_{r_3} = \frac{2p_2r_2^2}{\left(r_3^2 - r_2^2\right)}$$

$$= \frac{2p_2c_2^2}{\left(c_3^2 - c_2^2\right)}$$

$$= \frac{2 \times 11.6352 \times 1.13446^2}{(16 - 1.13446^2)}$$

$$= \frac{11.6352 \times 1.286999}{14.713}$$

$$= 2.0355 \text{ MPa}$$

(*d*) Let $\overline{p}_n \left( = \frac{p_n}{\sigma_y} \right)$ be the new internal pressure which the autofrettaged cylinder can withstand without the maximum circumferential stress anywhere going beyond $\sigma_y$. Therefore from equation (10.77).

$$(\sigma_c')_{r_1} = \frac{p_n\left(c_3^2 + 1\right)}{\left(c_3^2 - 1\right)}$$

$$= p_n \frac{(16+1)}{(16-1)}$$

$$= \frac{17}{15} p_n$$

And the residual stress at $r = r_1$ as calculated earlier (eq. 10.71*b*)

$$\left(\sigma_{c_i}\right)_{r_1} = \frac{-2p_2\, c_2^2}{\left(c_2^2 - 1\right)}$$

$$= -\ 104.356 \text{ mPa}$$

Therefore resultant stress at $r = r_1$

$$\left(\sigma_{R_c}\right)_{r_1} = \left(\sigma_{c_i}\right)_{r_1} + \left(\sigma_c'\right)_{r_1}$$

$$= 250$$

or $$-\,104.356 + \frac{17p_n}{15} = 250$$

or $$p_n = \frac{354.356 \times 15}{17}$$

$$= 312.667 \text{ MPa}$$

Therefore $\dfrac{p_n}{p_{w_1}} = \dfrac{312.667}{220.588}$

$= 1.4174$ *i.e.*, 41.74% or increase

## 10.6 Stresses in thick Spherical Shells

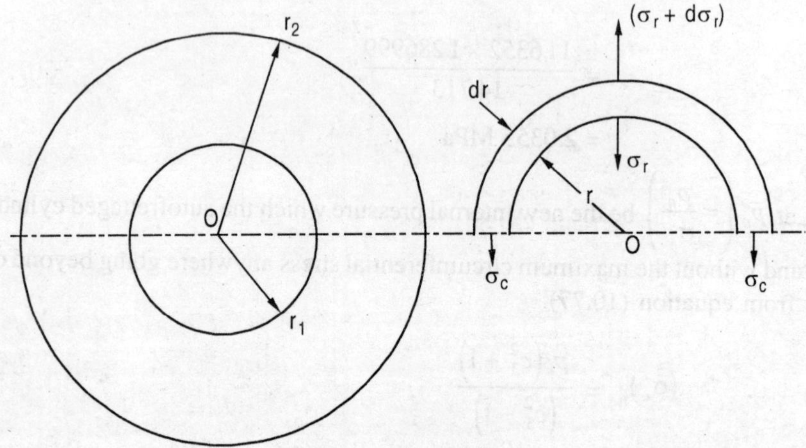


*Fig. 10.8*

In fig. 10.8*a* is shown a thick spherical shell of inner and outer radii $r_1$ and $r_2$ respectively. Let the internal pressure be $p_1$ and at outer surface $p_2$. In figure (10.8*b*) is shown an elemental hemisphere sphere of radii $r$ and $(r + dr)$. The circumferential stresses along two directions will be $\sigma_c$ and the radial stress $(\sigma_r)$ at radius r and $(\sigma_r + d\sigma_r)$ at radius $r$ $(r + dr)$.

Then considering the static equilibrium of this elemental hemisphere and resolving the forces in the radialy outward direction we get

$$(\sigma_r + d\sigma_r)\,\pi(r + dr)^2 - \sigma_r\pi(r)^2 - 2(\pi r dr)\sigma_c = 0$$

$$(\sigma_r + d\sigma_r)\,(r^2 + 2rdr + dr^2) - \sigma_r r^2 = 2rdr\sigma_c$$

$$r^2\sigma_r + r^2 d\sigma_r + 2r\sigma_r dr - r^2\sigma_r = 2r\sigma_c dr$$

(neglecting the product of two small quantities)

or $$r^2 d\sigma_r + 2r\sigma_r dr = 2r\sigma_c dr$$

or $$\sigma_c = \sigma_r + \frac{rd\sigma_r}{2dr} \qquad \text{...(10.81)}$$

Differentiating equation (10.81) with respect to $r$ we get

$$\frac{d\sigma_c}{dr} = \frac{d\sigma_r}{dr} + \frac{r}{2}\frac{d^2\sigma_r}{dr^2} + \frac{1}{2}\frac{d\sigma_r}{dr}$$

$$= \frac{3}{2}\frac{d\sigma_r}{dr} + \frac{r}{2}\frac{d^2\sigma_r}{dr^2} \qquad \text{...(10.82)}$$

Due to the a action of fluid pressures let us assume that the radius $r$ of the elemental sphere increases by $u$ and the radius $(r + dr)$ increases by $(u + du)$.

The the circumferential strain at radius $r$ is

$$\varepsilon_c = \frac{2\pi(r+u) - 2\pi r}{2\pi r} = \frac{u}{r} \qquad \text{...(10.83)}$$

and the radial strain

$$\varepsilon_r = \frac{\left(u + \frac{\partial u}{\partial r}dr\right) - u}{dr} = \frac{du}{dr} \qquad \text{...(10.84)}$$

Also

$$\varepsilon_c = \frac{\sigma_c}{E} - \frac{v}{E}(\sigma_r + \sigma_c) \qquad \text{...(10.85)}$$

Therefore from (10.83) and (10.85) we get

$$\frac{u}{r} = \frac{1}{E}[\sigma_c - v(\sigma_r + \sigma_c)] \qquad \text{...(10.86}a\text{)}$$

or

$$u = \frac{r}{E}[\sigma_c - v(\sigma_r + \sigma_c)] \qquad \text{...(10.86}b\text{)}$$

Differentiating equation (10.86*b*) with respect to '*r*'

$$\frac{du}{dr} = \frac{r}{E}\left[\frac{d\sigma_c}{dr} - v\left(\frac{d\sigma_r}{dr} + \frac{d\sigma_c}{dr}\right)\right] + \frac{1}{E}[\sigma_c - v(\sigma_r + \sigma_c)]$$

$$= \frac{r}{E}\left[(1-v)\frac{d\sigma_c}{dr} - v\frac{d\sigma_r}{dr}\right] + \frac{1}{E}[(1-v)\,\sigma_c - v\sigma_r] \qquad \text{...(10.87)}$$

Again radial strain in terms of stresses is given by

$$\varepsilon_r = \frac{\sigma_r}{E} - \frac{v}{E}(\sigma_c + \sigma_c)$$

$$= \frac{1}{E}[\sigma_r - 2v\sigma_c] \qquad \text{...(10.88)}$$

Equation (10.84) and (10.88) give

$$\frac{du}{dr} = \frac{1}{E}[\sigma_r - 2v\sigma_c] \qquad \text{...(10.89)}$$

From equations (10.87) and (10.89) we get

$$\frac{1}{E}[\sigma_r - 2v\sigma_c] = \frac{r}{E}\left[(1-v)\frac{d\sigma_c}{dr} - v\frac{d\sigma_r}{dr}\right] + \frac{1}{E}[(1-v)\,\sigma_c - v\sigma_r]$$

Substituting values of $\sigma_c$ and $\frac{d\sigma_c}{dr}$ from equation (10.81) and (10.82) we get

$$\frac{1}{E}\left[\sigma_r - 2v\left\{\sigma_r + \frac{rd\sigma_r}{2dr}\right\}\right] = \frac{r}{E}\left[(1-v)\left\{\frac{3}{2}\frac{d\sigma_r}{dr} + \frac{r}{2}\frac{d^2\sigma_r}{dr^2}\right\} - v\frac{d\sigma_r}{dr}\right] + \frac{1}{E}\left[(1-v)\left\{\sigma_r + \frac{rd\sigma_r}{2dr}\right\} - v\sigma_r\right]$$

or

$$\sigma_r - 2v\left[\sigma_r + \frac{rd\sigma_r}{2dr}\right] = (1-v)\left[\frac{3}{2}\frac{rd\sigma_r}{dr} + \frac{r^2}{2}\frac{d^2\sigma_r}{dr^2}\right] - rv\frac{d\sigma_r}{dr} + (1-v)\left[\sigma_r + \frac{r}{2}\frac{dr_r}{dr}\right] - v\sigma_r$$

On simplification is reduces to

$$u\frac{d\sigma_r}{dr} + r\frac{d^2\sigma_r}{dr^2} = 0 \qquad \text{...(10.90)}$$

Let $$x = \frac{d\sigma_r}{dr} \qquad \text{...(10.91)}$$

Then $$\frac{d^2\sigma_r}{dr^2} = \frac{dx}{dr}$$

Then equation (10.90) reduces to

$$\frac{rdx}{dr} + 4x = 0$$

or $$\frac{dx}{x} = -4\frac{dr}{r}$$

Integrating, $\ln(x) = -4\ln(r) + \ln(6B)$

where 6b is a constant

or $$\ln x + 4\ln r = \ln(6B)$$

$$\ln(xr^4) = \ln(6B)$$

*i.e.,* $$xr^4 = 6B$$

$$x = \frac{6B}{r^4}$$

or $$\frac{d\sigma_r}{dr} = \frac{6B}{r^4} \qquad \text{(From 10.91)}$$

or $$\frac{r^4 d\sigma_r}{dr} = 6B$$

$$d\sigma_r = (6B)\frac{dr}{r^4}$$

Again integrating $\sigma_r = -(6B)\frac{1}{r^3}\frac{1}{(3)} + A$

where A is another constant of integrating

or $$\sigma_r = -\frac{2B}{r^3} + A$$

*i.e.*, $$\sigma_r = A - \frac{2B}{r^3} \quad ...(10.93)$$

Therefore substituting (10.92) and (10.93) in equation (10.81) we get

$$\sigma_c = A - \frac{2B}{r^3} + \frac{r}{2}\left(\frac{6B}{r^4}\right)$$

$$= A - \frac{2B}{r^3} + \frac{3B}{r^3}$$

or $$\sigma_c = A + \frac{B}{r^3} \quad ...(10.94)$$

Re writing these equations we have for thick spherical shell

$$\sigma_c = A + \frac{B}{r^3} \quad ...(10.94)$$

$$\sigma_r = A - \frac{2B}{r^3} \quad ...(10.95)$$

The boundary conditions can then be applied to these equations to find the values of constants $A$ and $B$.

### 10.6.1 Thick Spherical Shell with Internal Pressure Only

It internal pressure is $p$, and out side pressure is zero the boundary conditions will then be

At $r = r_1$ $\sigma_r = -p_1$

At $r = r_2$ $\sigma_r = 0$

That is $$-p_1 = A - \frac{2B}{r_1^3}$$

and $$0 = A - \frac{2B}{r_2^3}$$

*i.e.*, $$A = \frac{2B}{r_2^3}$$

or $$-p_1 = \frac{2B}{r_2^3} - \frac{2B}{r_1^3}$$

$$= 2B\left[\frac{r_1^3 - r_2^3}{r_1^3\, r_2^3}\right]$$

*i.e.,* $$B = \frac{r_1^3\, r_2^3\, p_1}{2\left(r_2^3 - r_1^3\right)}$$

and $$A = \frac{2}{r_2^3}\,\frac{r_1^3 r_2^3 p_1}{2\left(r_2^3 - r_1^3\right)} = \frac{r_1^3\, p_1}{\left(r_2^3 - r_1^3\right)}$$

Then corresponding expressions for $\sigma_c$ and $\sigma_r$ become

$$\sigma_c = \frac{p_1 r_1^3}{\left(r_2^3 - r_1^3\right)} + \frac{p_1 r_1^3 r_2^3}{2\left(r_2^3 - r_1^3\right) r^3} \quad ...(10.96)$$

$$\sigma_c = \frac{p_1 r_1^3}{\left(r_2^3 - r_1^3\right)} - \frac{r_1^3 r_2^3 p_1}{\left(r_2^2 - r_1^3\right) r^3} \quad ...(10.97)$$

Evidently from equation (10.96)

At $r = r_1$ $\sigma_c$ will be maximum

$$\therefore \quad (\sigma_c)_{max} = \frac{p_1 r_1^3}{\left(r_2^3 - r_1^3\right)} + \frac{p_1 r_2^3\, r_1^3}{2\left(r_2^3 - r_1^3\right) r_1^3}$$

$$= \frac{p_1 r_1^3}{\left(r_2^3 - r_1^3\right)} + \frac{p_1 r_2^3}{2\left(r_2^3 - r_1^3\right)}$$

or $$(\sigma_c)_{max} = \frac{p_1}{2\left(r_2^3 - r_1^3\right)}\left(2r_1^3 + r_2^3\right) \quad ...(10.98)$$

Since $\sigma_r$ is compressive, it will also be maximum in magnition at minimum radius is at $r = r_1$ [from 10.97)]

At $r = r_1$ $$(\sigma_r)_{max} = \frac{p_1 r_1^3}{\left(r_2^3 - r_1^3\right)} - \frac{p_1 r_2^3}{\left(r_2^2 - r_1^3\right)}$$

$$= -p_1$$

### 10.6.2 Thick Spherical Shell with both internal and external Pressures

Here the boundary conditions will be

At $r = r_1$, $\sigma_r = -p_1$

And at $r = r_2$, $\sigma_r = -p_2$

Then we have from equation (10.95)

$$-p_1 = A - \frac{2B}{r_1^3} \qquad ...(10.99)$$

and
$$-p_2 = A - \frac{2B}{2r_2^3} \qquad ...(10.100)$$

Substracting (10.99) from (10.100)

$$p_1 - p_2 = 2B\left[\frac{1}{r_1^3} - \frac{1}{r_2^3}\right]$$

$$= \left(\frac{r_2^3 - r_1^3}{r_1^2\, r_2^3}\right)(2B)$$

or
$$B = \frac{r_1^3 r_2^3\,(p_1 - p_2)}{2\left(r_2^3 - r_1^3\right)} \qquad ...(10.101)$$

Then from (10.99) and (10.101)

$$A = -p_1 + \frac{2}{r_1^3}\left[\frac{r_1^3 r_2^3\,(p_1 - p_2)}{2\left(r_2^3 - r_1^3\right)}\right]$$

$$= \frac{-p_1\left(r_2^3 - r_1^3\right) + r_2^3\,(p_1 - p_2)}{\left(r_2^3 - r_1^3\right)}$$

or
$$A = \frac{p_1 r_1^3 - p_2 r_2^3}{r_2^3 - r_1^3} \qquad ...(10.102)$$

Substituting these values of $A$ and $B$ in equations (10.94) and (10.95) we get

$$\sigma_c = \frac{p_1 r_1^3 - p_2 r_2^3}{\left(r_2^3 - r_1^3\right)} + \frac{r_1^3 r_2^3\,(p_1 - p_2)}{2r^3\left(r_2^3 - r_1^3\right)} \qquad ...(10.103)$$

$$\sigma_r = \frac{p_1 r_1^3 - p_2 r_2^3}{\left(r_2^3 - r_1^3\right)} - \frac{r_1^3 r_2^3\,(p_1 - p_2)}{r^3\left(r_2^3 - r_1^3\right)} \qquad ...(10.104)$$

Maximum $\sigma_c$ and $\sigma_r$ will occur at $B = r_1$

Therefore
$$(\sigma_c)_{\max} = \frac{p_1 r_1^3 - p_2 r_2^3}{\left(r_2^3 - r_1^3\right)} + \frac{r_2^3\,(p_1 - p_2)}{2\left(r_2^3 - r_1^3\right)} \qquad ...(10.105)$$

and
$$(\sigma_r)_{\max} = \frac{p_1 r_1^3 - p_2 r_2^3}{\left(r_2^3 - r_1^3\right)} - \frac{r_2^3\,(p_1 - p_2)}{\left(r_2^3 - r_1^3\right)} \qquad ...(10.106)$$

Also at $r = r_1$, the maximum shear stress will be equal to half of the difference between these two stresses as they are principal stresses

*i.e.*,
$$\tau_{max} = \frac{1}{2}\left[(\sigma_c)_{max} - (\sigma_r)_{max}\right]$$

$$= \frac{1}{2}\left[\frac{r_2^3(p_1 - p_2)}{2\left(r_2^3 - r_1^3\right)} + \frac{r_2^3(p_1 - p_2)}{\left(r_2^3 - r_1^3\right)}\right]$$

$$= \frac{3}{4}\frac{r_2^3(p_1 - p_2)}{\left(r_2^2 - r_1^3\right)} \qquad \text{...(10.107)}$$

**Example 10.14.** *A thick spherical sheet of steel is used to store a fluid at 10 MPa pressure. If its inside diameter is 500 mm and outside diameter 700 mm, find the maximum hoop and radial stresses induced in the steel. Also find the change in volume of the shell. Take E = 200 GPa and v = 0.25.*

**Solution.**

Here $r_1 = 250$ mm, $r_2 = 350$ mm

$p_1 = 10$ MPa and $p_2 = 0$

From equation (10.98)

$$(\sigma_c)_{max} = \frac{p_3\left(2r_1^3 + r_2^3\right)}{2\left(r_2^3 - r_1^3\right)}$$

$$= \frac{10\left(2 \times 250^3 + 350^3\right)}{2[(350)^2 - (250)^3]}$$

$$= \frac{5(2 \times 15.621 \times 10^6 + 42.875 \times 10^6}{2(42.875 - 15.625)\,10^6}$$

$$= \frac{5[74.125 \times 10^6}{2(27.25 \times 10^6}$$

$$= 6.8005 \text{ MPa (Tensile)}$$

And at $r = r_1$, we have

$$(\sigma_r)_{max} = -p_1$$
$$= -10 \text{ MPa}$$

Now Diametral strain = circumferential strain

$$= \frac{\sigma_c}{E} - \frac{v}{E}(\sigma_r + \sigma_c)$$

$$= \frac{6.8005}{200 \times 10^3} - \frac{0.25}{200 \times 10^3}(-10 + 6.68005)$$

$$= 0.034 \times 10^{-3} + .003999 \times 10^{-3}$$

$$= .03799 \times 10^{-3}$$

Therefore $\frac{\delta V}{V}$ = volumetric strain

$$= 3\left(\frac{\delta D}{D}\right)$$

$$= 3 \times 0.037999 \times 10^{-3}$$

Then $\delta V$ = charge in volume

= (vol strain) × volume

$$= (3 \times .037999 \times 10^{-3}) \times \frac{\pi D^3}{6}$$

$$= 3 \times .037999 \times 10^{-3} \times \frac{\pi}{6} \times (5000)^3$$

$$= 7.4611 \times 10^3 \text{ mm}^3$$

$$= 7461.1 \text{ mm}^3.$$

## PROBLEMS FOR PRACTICE

1. A steel pipe 100 mm outside diameter and 50 mm inside diameter is subjected to an internal pressure of 14 MN/m$^2$ and an external pressure of 5.5 MN/m$^2$. Plot a curve showing the distribution of tangential stress across the wall of the pipe. Prove any formula used. [**Ans.** Max. Stress = 8.67 MN/m$^2$]

2. Find the ratio of thickness to internal diameter for a tube subjected to internal pressure when the ratio of internal pressure to the greatest circumferential stress is 0.5.

   Find the alteration in thickness of metal in such a tube, 0.2 m internal diameter, when the internal pressure is 7.5 MN/m$^2$ $E$ = 200 GN/m$^2$ and $\nu$ = 0.3.

   [**Ans.** 0.366 ; – 0.0253 mm]

3. A steel tube is 18 mm internal diameter and 3 mm thick. One end is closed and the other is screwed into a pressure vessel. The projecting length is 300 mm. Neglecting any constraints due to the ends, calculate the safe internal pressure if the allowable stress is not to exceed 150 MN/m$^2$. Calculate the increase in the internal volume under this pressure. $E$ = 200 GN/m$^2$ $\nu$ = 1/3.5.

   [**Ans.** 42 MN/m$^2$, 120.6 mm$^3$]

4. A thick steel tube with closed ends, of inside and outside diameters 50 mm and 70 mm respectively, contains oil at a pressure of 1 MN/m$^2$. The oil is allowed to escape until the pressure in the tube has fallen to 0.75 MN/m$^2$. Find how much oil has been released per metre length of tube, if bending due to end effects is negligible.

   $E$ for steel = 200 GN/m$^2$. K for oil = 2.75 GN/m$^2$, $\nu$ for steel = 0.25.

   **[Ans.** 195 mm$^3$]

5. A thick walled cylinder, 0.2 m inside diameter, is to contain fluid at a pressure of 50 MN/m$^2$. Find the necessary thickness if the maximum shearing stress is not to exceed 100 MN/m$^2$. What will then be the greatest and least values of the hoop stress in the material ?

   If the inner surface becomes corroded and the cylinder has to re- bored, by how much can the inside diameter be increased without raising more than 5 per cent the maximum shearing stress induced by the same internal pressure.

   **[Ans.** 150 MN/m$^2$, 100 MN/m$^2$, 4.6 mm]

6. A steel rod, 0.06 m diameter, is forced into a bronze casing having an outside diameter of 0.1 m and thereby produces a hoop tension at the outer circumference of the casing of 35 MN/m$^2$. Determine (*a*) the radial pressure between the rod and the casing, and (*b*) the rise in temperature which would just eliminate the force fit.

   For steel $E$ = 200 GN/m$^2$, $\nu = 0.28$, $\alpha = 12 \times 10^{-6}$/degree *C*. For bronze $E$ = 110 GN/m$^2$, $\nu = 0.33$, $\alpha = 19 \times 10^{-6}$/degree *C*. **[Ans.** 31.11 MN/m$^2$ ; 115° C]

7. A steel cylindrical plug of 125 mm diameter is forced into a steel sleeve of 200 mm external diameter and 100 mm long. If the greatest circumferential stress in the sleeve is 100 MN/m$^2$. Find the torque required to turn the plug in the sleeve assuming the coefficient of friction between the plug and the sleeve is 0.2.

   **[Ans.** 21.5 kNm]

8. In problem 7 find the axial force to drive the plug into the sleeve.

   **[Ans.** 34.4 kN]

9. A gun metal cylinder, 150 mm external diameter and 99.925 mm internal diameter is forced onto a steel cylinder 100 mm external diameter and 50 mm internal diameter. Calculate the maximum resulting stresses in the steel and gun metal.

   $E$ for steel = 200 GN/m$^2$ $E$ for gun-metal = 100 GN/m$^2$ Poisson's ratio = 0.35 for both metals. **[Ans.** 55.4 MN/m$^2$ ; 54.1 MN/m$^2$]

10. A steel cylinder 200 mm external diameter and 150 mm internal diameter has another cylinder 250 mm external diameter shrunk on to it. If the maximum tensile stress induced in the outer cylinder is 80 MN/m$^2$ find the radial compressive stress between the cylinders.

Determine the circumferential stresses at inner and outer diameter of both cylinders and show by means of a diagram, how these stresses vary with radius. Calculate the necessary shrinkage allowance at the common surface. $E = 200$ GN/m$^2$.

[**Ans.** 17.6 ; 80 ; 62.5 ; 80.5 ; 62.8 MN/m$^2$ ; 0.1428 mm]

11. A steel tube of 80 mm outside diameter is to be shrunk on another steel tube of 40 mm inside diameter and 60 outside diameter. Calculate the shrinkage alowance if in the compound tube the final maximum stress in each tube is same, when it is subjected to an internal pressure of 50 MN/m$^2$. Also find the value of this stress and draw the stress distribution diagram. Take E = 207 GN/N$^2$.

[**Ans.** 0.0092 mm ; 64.7 MN/m$^2$]

12. A steel cylinder of outside diameter 300 mm and inside diameter 250 mm is shrunk onto one having diameters 250 mm and 200 mm, the interference fit being such that under an internal pressure $p$ the inner tensile stress in both cylinders = 84 MN/m$^2$.

Find the initial difference in the nominal 250 mm diameters, and the value of $p$ if $E = 1200$ GN/m$^2$. [**Ans.** 0.0273 mm ; 36.9 MN/m$^2$]

Chapter–11

# Springs

## 11.1 Closed Coil Helical Springs with Axial Load

A coil spring is formed by bending a wire, around a mandril, in the form of a helix. If the coils are in close contact with each other, the spring thus formed is called a closed coil spring. Evidently the helix angle, which is defined as the angle which the centre line of the wire makes with the plane normal to the axis of the spring, will be very small in the case of the closed coil spring. The actual value of the helix angle is given by $\tan^{-1}(p/\pi D)$ where $p$ is called the pitch which is defined as the distance between the similar points on the adjacent coils of the spring and $D$ is called the mean coil diameter which is the diameter of an imaginary cylinder which contains the centre line of the wire that has been wound to form a spring. In the case of closed coil springs the pitch will be equal to $d$, the diameter of the spring wire. When the pitch is more than $d$ the spring is called an open coil helical spring. Evidently the closed coil spring will not be able to take up any compressive load, rather it can take only axial pull. However in actual practice when the helix angle is not very large the springs are designed as closed coil spring even though the pitch may be more than $d$ and such springs can sustain both axial tensile as well as compressive loads.

Fig. 11.1 shows a coil spring under the action of an axial pull $W$. When the helix angle $\alpha$ is small, *i.e.* in the case of a closed coil spring, each section of the spring wire may be assumed to be under the action of a torque ($=W \times D/2$) only, the effects of bending and direct shear being negligible.

Thus if

$l$ = length of the wire.

$n$ = number of coils.

$\delta$ = deflection of the spring or the axial movement of one end of the spring with respect to the other end.

$G$ = Modulus of rigidity.

Then $l \simeq n \times$ length of wire in each coil

$$= \pi D n \qquad \text{...(11.1)}$$

Now, shear stress induced in the wire due to the torque $T$ ($=WD/2$) is given by $\tau$ from equation (7.7) as

$$\tau = \frac{WD/2}{\left(\dfrac{\pi}{16}d^3\right)}$$

*i.e.*
$$\tau = \frac{8\,WD}{\pi\,d^3} \qquad \text{...(11.2)}$$

Also the twist of one end of the wire relative to the other end of the wire is given by equation (7.6)

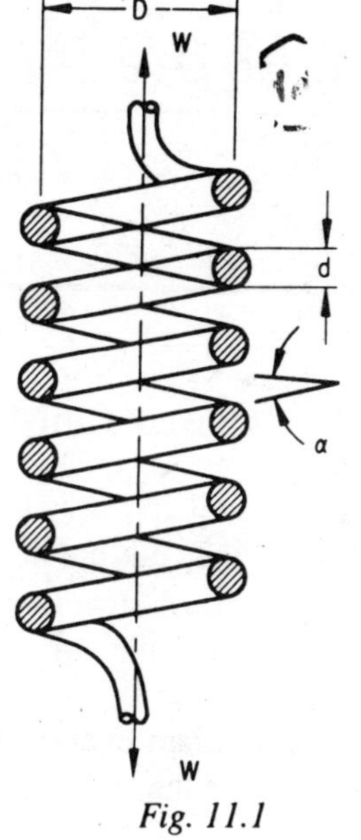


*Fig. 11.1*

$$\tau = \frac{Tl}{GJ} = \frac{\dfrac{WD}{2} \times \pi Dn}{D \times \dfrac{\pi d^2}{32}}$$

$$= \frac{16\,WnD^2}{Gd^4} \qquad \text{...(11.3)}$$

Therefore axial movement of one end with respect to the other end of the spring.

$$\delta = \theta \times \text{radius of spring coil}$$

$$= \frac{16\,WnD^2}{Gd^2} \times \frac{D}{2}$$

or
$$\theta = \frac{8\,WnD^3}{Gd^4} \qquad \text{...(11.4)}$$

Evidently the deflection of any spring is directly proportional to the load. Thus the load extension curve of a closed coil spring will be a straight line as shown in Fig. 11.2. For any load $W$ and the corresponding deflection $\delta$, the work done on the spring will be given by area $OAB$ under the curve $OA$. Thus the work done, which will be stored in the spring as strain energy is given by

$$U = \frac{1}{2}W\delta$$

$$= \frac{1}{2}W \cdot \frac{8WnD^3}{Gd^4} \qquad \text{(from 11.4)}$$

$$= \frac{4nD^3}{Gd^4}\left[\frac{\pi d^3 \tau}{8D}\right]^2 \qquad \text{(from eqn. 11.2)}$$

$$= \frac{\tau^2}{4G}\left[\pi Dn \cdot \frac{\pi d^2}{4}\right]$$

or $$U = \frac{\tau^2}{4G} \times \text{Volume of wire} \qquad ...(11.5)$$

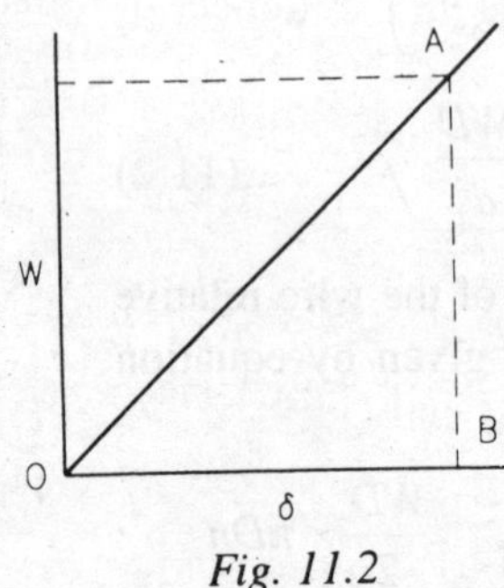


*Fig. 11.2*

The strain energy per unit volume stored in the spring wire is known as the resilience of the spring.

Stiffness of a coil spring is defined as the load required to deflect it through unit distance. Thus from equation 11.4, the stiffness, $s$, is given by

$$s = \frac{Gd^4}{8nD^3} \qquad ...(11.6)$$

Since $G$, $d$, $n$ and $D$ are constants for any particular spring, its stiffness will also be constant and this will be equal to the slope of the line $OA$ in Fig. 11.2.

## 11.2 Closed Coil Helical Spring with Axial Couple

A closed coil helical spring is shown in Fig. 11.3 under the action of an axial couple $M$. The axial couple will exert a bending moment (= $M$) on each and every section of the wire. If the couple tends to increase the number of turns of the spring it is called the bending moment and is reckoned positive. If its sense is such that it tends to reduce the number of turns it is called the *unbending moment* and reckoned negative. The additional bending stresses due to positive moment are tensile on the outer fibres of the coil wire. In case of the unbending moment the bending stresses induced due to $M$ will be compressive on the outer fibres of the coil wire.

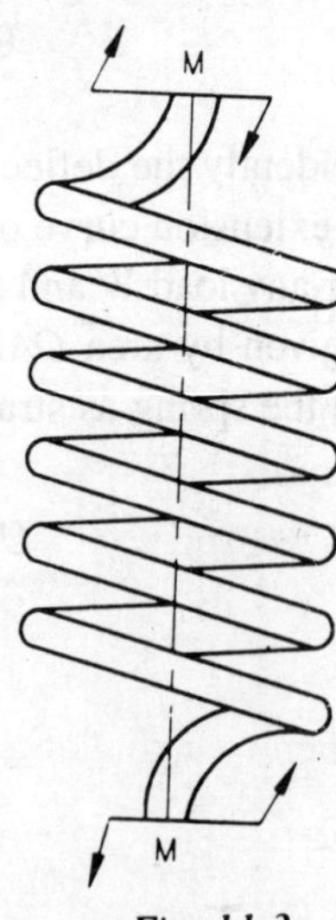


*Fig. 11.3*

Therefore maximum bending stress is given by

$$\sigma_b = \frac{M \times \frac{d}{2}}{\frac{\pi}{64} d^4} = \frac{32M}{\pi d^3} \quad \text{...(11.7)}$$

Under the action of $M$, let the radius of curvature of the coils change from $R$ to $R'$ and the corresponding increase in the number of turns from $n$ to $n'$.

Then approximately

$$M = EI\left(\frac{1}{R'} - \frac{1}{R}\right) \quad \text{...(11.8)}$$

Also length of wire $= l = 2\pi Rn = 2\pi R'n'$ ...(11.9)

Thus from equations (11.8) and (11.9)

$$M = EI \times \frac{2\pi}{l}(n' - n) \quad \text{...(11.10)}$$

But $2\,\pi(n' - n)$ is the rotation of the free end of the spring, about the spring axis, with respect to the fixed end. If it is denoted by $\phi$, then from equation (11.10)

$$M = \frac{EI\phi}{l}$$

or

$$\phi = \frac{Ml}{EI} \quad \text{...(11.11a)}$$

It can also be written in another form

$$\phi = \frac{M \times \pi Dn}{E \times \frac{\pi}{64} d^4}$$

or

$$\phi = \frac{64\,MDn}{Ed^4} \quad \text{...(11.11b)}$$

## Second Method

The rotation of the free end of the spring can also be found from energy considerations. The strain energy $U$ (Chaper 12) of the wire of the spring due to the bending moment $M$ is given by

$$U = \frac{M^2 l}{2EI} \quad \text{...(11.12)}$$

Therefore equating the work done by $M$ to the strain energy $U$, we get

$$\frac{1}{2} M\,\phi = \frac{M^2 l}{2EI}$$

or $$\phi = \frac{Ml}{El}$$ ...(11.13)

## 11.3 Open Coil Spring Under the Action of Axial Load

Fig. 11.4 shows the section of a coil of an open coil spring with a vertical plane *ABCD*, *AD* coinciding with the axis of the spring. After the section with this plane, only upper portion of the coil is shown. In the figure is also shown the annular cylinder which contains the axis of the spring wire. Evidently the diameter of this cylinder will be equal to the mean coil diameter of the spring. The edge *BC* of the plane *ABCD* lies on the surface of this cylinder. The intersection of the wire at point *P*, with the plane *ABCD* will be an ellipse. The horizontal tangent $X_1' PX_1$ to the annular cylinder at the point *P* will be normal to the plane *ABCD*. The tangent $X'_2 \ PX_2$ to the helix of the wire at *P* will lie in the vertical plane through *P* and will be inclined to $X'_1 \ PX_1$ at an angle α equal to the helix angle of the spring.

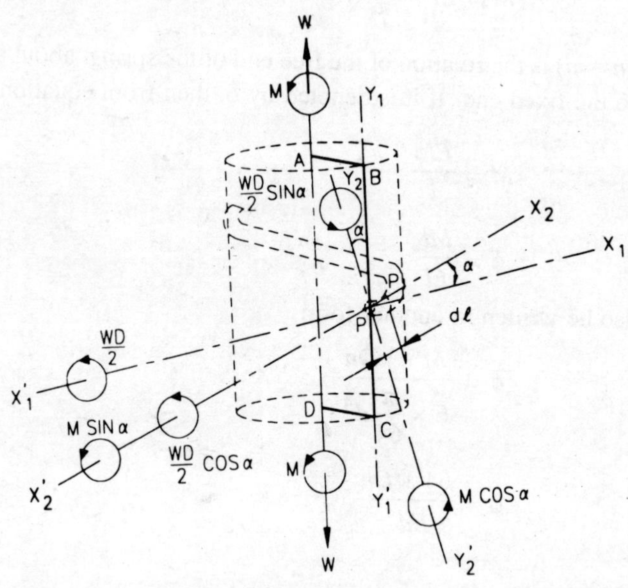


*Fig. 11.4*

The downward load *W* applied at the lower end of the spring can be replaced by an equal load *W* at *P* and an anticlockwise couple *WD*/2 acting on the wire along the axis $X'_1 \ PX_1$. The load *W* at *P* can then be resolved into *W* sin α and *W* cos α components, normal to and along the cross-sectional plane of the wire at $P_1$ respectively. These in turn will induce uniform direct stress of $\frac{4W \sin \alpha}{\pi d^2}$ and shear stresses due to transverse shearing force *W* cos α (article 5.1.2), over the cross-sectional plane of wire at *P*. However the magnitude of these shear stresses is very small and they will be neglected in comparison to the stresses induced by the couple. Further the anticlockwise couple $WD/2$ along $X'_1\ PX_1$ can be resolved along two perpendicular axes $X'_2\ PX_2$ and $Y_2\ PY'_2$. This will result in an anticlockwise twisting moment $\frac{WD}{2}$ (cos α) along $X'_2$ $PX_2$ and an unbending moment $\frac{WD}{2}$ (sin α) along $Y_2\ PY'_2$ .

(*a*) **Effect of twisting moment $\frac{WD}{2}$ cos α :**

This twisting moment acting along the wire axis will give rise to shear stresses. The maximum shear stress induced in the cross- sectional plane normal to the axis of the wire is given by (Chapter VII).

$$\tau = \frac{\left(\frac{WD}{2}\cos\alpha\right)}{\left(\frac{\pi d^4}{32}\right)} \cdot \left(\frac{d}{2}\right)$$

*i.e.*

$$\tau = \frac{8WD\cos\alpha}{\pi d^3} \qquad \text{...(11.14)}$$

In addition, this twisting moment will twist the wire, about its axis. If $d\theta'$ is the anticlockwise twist of the section at $P$ of the wire with respect to the section $P'$ of the wire distant $dl$ from $P$ and $G$ the modulus of rigidity of wire, then

$$d\theta' = \frac{\left(\frac{WD}{2}\cos\alpha\right)dl}{\frac{\pi d^4}{32} \times G}$$

*i.e.*

$$d\theta' = \frac{16WD\cos\alpha\ dl}{\pi d^4\ G} \qquad \text{...(11.15)}$$

The total twist of the free end of the spring wire about the wire axis with respect to the upper end of the spring will then be obtained by integrating equation (11.5) because at each and every point of the wire axis the twisting moment is same.

Thus

$$\theta' = \frac{16WD\cos\alpha}{\pi d^4\ G} \int_0^l dl$$

$$= \frac{16\ WDl\cos\alpha}{\pi d^4\ G}$$

Its component about $X'_1\ PX_1$

$$\theta_1 = \frac{16\ WDl}{\pi d^4\ G}\cos^2\alpha \qquad \text{(anticlockwise) ...(11.16)}$$

and component about $Y'_1 PY_1$

$$\phi_1 = \frac{16WDl}{\pi d^4 G} \cos\alpha \sin\alpha \qquad \text{(anticlockwise)} \quad ...(11.17)$$

or $\phi_1$ will be clockwise about $Y_1PY'_1$.

(*b*) **Effect of the unbending moment $\frac{WD}{2}$ sin α :**

This anticlockwise moment with axis $Y_2PY'_2$ will give rise to bending stresses on the wire cross-section, which will be compressive on the outer fibres and tensile on the fibres on the inner side of the coil. The magnitude of the maximum bending stress is given by

$$\sigma_b = \frac{\frac{WD}{2}\sin\alpha}{\frac{\pi d^4}{64}} \times \frac{d}{2} = \frac{16\, WD \sin\alpha}{\pi d^3} \quad ...(11.18)$$

In addition, this moment will rotate anticlockwise the wire section at $P$ with respect to that at $P'$ about $Y_2PY'_2$ and will be obtained with the help of equation (11.11*a*) as

$$\left(\frac{WD}{2}\sin\alpha\right)\frac{dl}{EI} \qquad \text{(anticlockwise)}$$

The component of this about $Y_1PY'_1$ is :

$$\left(\frac{WD}{2}\sin\alpha\right)\frac{dl}{EI} \times \cos\alpha \qquad \text{(anticlockwise)} \quad ...(11.19)$$

Further for various equal-lengths (= *dl*) spread over the entire length, the rotation about *Y*-axis is same, therefore integrating equation (11.19), we get the rotation of the free end of the spring with respect to the upper end, about the spring axis as seen along $Y_1\ PY'_1$

$$\phi_2 = \int_0^l \left(\frac{WD\sin\alpha\cos\alpha}{2EI}\right) dl$$

*i.e.*

$$\phi_2 = \frac{32\, WDl}{\pi E d^4} \sin\alpha\cos\alpha \qquad \text{(anticlockwise)} \quad ...(11.20)$$

Similarly the component of rotation about the horizontal axis $X'_1\ PX_1$

$$\theta_2 = \int_0^l \frac{WD\sin^2\alpha}{2EI}\, dl$$

$$\theta_2 = \frac{32\, WDl\sin^2\alpha}{\pi E d^4} \qquad \text{(anticlockwise)} \quad ...(11.21)$$

Thus from equations (11.16) and (11.21) total rotation of the lower end of the spring about horizontal axis $X'_1PX$

$$\theta = \theta_1 + \theta_2$$

$$= \frac{16\,WDl\cos^2\alpha}{\pi d^4 G} + \frac{16\,WDl\sin^2\alpha}{\pi d^4 E}$$

$$= \frac{16\,WDl}{\pi d^4}\left(\frac{\cos^2\alpha}{G} + \frac{2\sin^2\alpha}{E}\right)$$

Therefore the axial deflection $\delta$ under the axial load $W$ is

$$\delta = \frac{D}{2}\theta = \frac{8\,WD^2 l}{\pi d^4}\left(\frac{\cos^2\alpha}{G} + \frac{2\sin^2\alpha}{E}\right) \quad ...(11.22)$$

But $\qquad l = \pi\,Dn\sec\alpha \qquad ...(11.23)$

Thus from equations (11.22) and (11.23)

$$\delta = \frac{8\,WD^3\,n\sec\alpha}{d^4}\left(\frac{\cos^2\alpha}{G} + \frac{2\sin^2\alpha}{E}\right) \quad ...(11.24)$$

Again from equations (11.17) and (11.20) the total clockwise rotation of the free end about the spring axis as seen along $Y_1PY'_1$

$$\phi = \frac{16\,WDl}{\pi G d^4}\cos\alpha\sin\alpha - \frac{32\,WDl}{\pi E d^4}\sin\alpha\cos\alpha$$

or
$$\phi = \frac{16\,WDl}{\pi d^4}(\sin\alpha\cos\alpha)\left(\frac{1}{G} - \frac{2}{E}\right) \quad ...(11.25a)$$

Also substituting the value of $l(= \pi\,Dn\sec\alpha)$, we have

$$\phi = \frac{16\,WD^2 n}{d^4}\sin\alpha\left(\frac{1}{G} - \frac{2}{E}\right) \quad ...(11.25b)$$

The bending stress $\sigma_b$ found in equation (11.18) is accompanied by the shear stress $\tau$ (equation 11.14). The maximum principal stress and the absolute maximum shear stress in the wire are then given by

$$\sigma = \frac{1}{2}\sigma_b + \frac{1}{2}\sqrt{\sigma_b^2 + 4\tau^2}$$

$$\tau_{max} = \frac{1}{2}\sqrt{\sigma_b^2 + 4\tau^2}$$

Substituting values of $\sigma_b$ and $\tau$ from equations (11.18) and (11.14)

$$\sigma = \frac{8\,WD}{\pi d^3}(\sin\alpha + 1) \quad ...(11.26)$$

$$\tau_{max} = \frac{8\,WD}{\pi d^3} \qquad \text{...(11.27)}$$

The value of $\sigma_{max}$ is same as found in closed coil spring (equation 11.2).

## 11.4 Open Coil Spring Under the Action of Axial Couple

In the Fig. 11.4 is also shown an axial couple *M* acting on the spring. The sign of M is taken positive if it tends to increase the curvature of the wire and negative if it reduces the curvature.

At the point *P* the couple *M*, which has its axis parallel to $Y'_1\,PY_1$ can be resolved along two perpendicular axes $X'_2PX_2$ and $Y'_2\,PY_2$ into ($M \sin \alpha$) and ($M \cos \alpha$) respectively. The effect of these components is considered below separately.

(*a*) **Effect of $M \sin \alpha$ :**

The couple which is anticlockwise as seen along $X'_2PX_2$ will twist the wire about its axis and induce shear stress in it. The maximum shear stress in the cross-sectional plane of the wire is given by (Chapter VII).

$$\tau = \frac{(M \sin \alpha)}{\left(\dfrac{\pi d^4}{32}\right)}\left(\frac{d}{2}\right)$$

*i.e.*

$$\tau = \frac{16\,M \sin \alpha}{\pi d^3} \qquad \text{...(11.28)}$$

If $d\theta'$ is the anticlockwise twist of the section *P* with respect to the section *P′* of the wire distant *dl* from *P* and *G* the modulus of rigidity of the wire material, then

$$d\theta' = \frac{M \sin \alpha}{\left(\dfrac{\pi d^4}{32}\right) \times G}\,dl$$

or

$$d\theta' = \frac{32\,M \sin \alpha}{\pi d^4\,G}\,dl \qquad \text{...(11.29)}$$

The total twist of the free end of the wire with respect to the upper end will then be given by

$$\theta' = \int_0^l \frac{32M \sin \alpha}{\pi d^4 G}\,dl = \frac{32M \sin \alpha}{\pi d^4 G}$$

Its component about $X'_1PX_1$ is

$$\theta_1 = \frac{32Ml}{\pi d^4 G} \sin \alpha \cos \alpha \qquad \text{(anticlockwise) ...(11.30)}$$

and component about $Y'_1\ YP_1$ is

$$\phi_1 = \frac{32Ml}{\pi d^4 G} \qquad \text{(anticlockwise) ...(11.31)}$$

(*b*) **Effect of bending moment M cos α :**

This anticlockwise moment with axis $Y'_2PY_2$ will give rise to bending stresses on the wire-cross-section. The nature of the stress at any point will be opposite to that induced by un-bending couple left $\left(\frac{WD}{2}\sin\alpha\right)$ of the previous article. The magnitude of the maximum bending stress on the cross-section is then

$$\sigma_b = \frac{M\cos\alpha}{\left(\frac{\pi d^4}{64}\right)} \times \frac{d}{2} = \frac{32M\cos\alpha}{\pi d^3} \qquad \text{...(11.32)}$$

Further this moment will rotate anticlockwise the wire section at $P$ with respect to that at $P'$ about $Y'_2\ PY_2$ and will be obtained with the help of equation (11.11a) as :

$$(M\cos\alpha)\frac{dl}{EI}$$

Its component about $Y'_1PY_1$ is :

$$(M\cos\alpha)\frac{dl}{EI} \qquad \text{(anticlockwise) ...(11.32)}$$

Again for various-equal lengths (= $dl$) spread over the entire length, the rotation about $Y$-axis is same everywhere, therefore integrating equation (11.32) we get the rotation of the free end of the spring with respect to the upper end, about the spring axis as seen along $Y'_1PY_1$

$$\phi_2 = \int_0^l \frac{M\cos^2\alpha}{EI}\,dl$$

*i.e.*

$$\phi_2 = \frac{Ml\cos^2\alpha}{EI} \qquad \text{(anticlock) ...(11.33)}$$

Similarly the component of rotation about the horizontal axis $X_1PX'_1$

$$\theta_2 = \int_0^l \frac{M\cos\alpha\sin\alpha}{EI}\,dl$$

or

$$\theta_2 = \frac{64\,Ml}{\pi d^3 E}\cos\alpha\sin\alpha \qquad \text{(anticlockwise) ...(11.34)}$$

*i.e.*, $\theta_2$ will be clockwise if seen along $X'_1PX_1$.

Thus from equations (11.30) and (11.34), total anticlockwise rotation of the lower end of the spring about horizontal axis $X'_1PX_1$

$$\theta = \frac{32\,Ml}{\pi d^4 G}\sin a\cos a - \frac{64\,Ml}{\pi d^4 E}\sin\alpha\cos\alpha$$

$$= \frac{32\,Ml}{\pi d^4 G}\sin\alpha\left(\frac{1}{G}-\frac{2}{E}\right)$$

Therefore the axial deflection $\delta$ due to the action of $M$ is

$$\delta = \frac{D}{2}.\theta = \frac{16\,MlD}{\pi d^4}\sin\alpha\cos\alpha\left(\frac{1}{G}-\frac{2}{E}\right)$$

$$= \frac{16\,MD^2 n}{d^4}\sin\alpha - \left(\frac{1}{G}-\frac{2}{E}\right) \qquad ...(11.35)$$

Also from equations (11.31) and (11.33) the total anticlockwise rotation of the free end about the spring axis as seen along $Y'_1PY_1$ (*i.e.*, the axis of the applied couple) is

$$\phi = \phi_1 + \phi_2$$

$$= \frac{32\,Ml}{\pi d^2 G}\sin^2\alpha + \frac{64\,Ml}{\pi d^4 E}\cos^2\alpha$$

*i.e.*,

$$\phi = \frac{32\,Ml}{\pi d^4}\left(\frac{\sin^2\alpha}{G}+\frac{2\cos^2\alpha}{E}\right) \qquad ...(11.36a)$$

or

$$\phi = \frac{32\,MDn\sec\alpha}{\pi d^4}\left(\frac{\sin^2\alpha}{G}+\frac{2\cos^2\alpha}{E}\right) \qquad ...(11.36)$$

As in article (11.3) the absolute maximum shear stress and the maximum principal stress are given by

$$\tau_{max} = \frac{1}{2}\sqrt{\sigma_b^2 + 4\tau^2}$$

$$\sigma = \frac{1}{2}\sigma_b + \frac{1}{2}\sqrt{\sigma_b^2 + 4\tau^2}$$

Substituting values of $\sigma_b$ and $\tau$ from equations (11.32) and (11.28), we get

$$\tau_{max} = \frac{16\,M}{\pi d^3} \qquad ...(11.37)$$

and

$$\sigma = \frac{16\,M}{\pi d^3}(1+\cos\alpha) \qquad ...(11.38)$$

## 11.5 Open Coil Spring Under the Combined Action of Load and Couple

Let an axial load $W$ and couple $M$ be acting on the spring as shown in the Fig. 11.4. The axial deflection of the free end of the spring will be obtained by the algebraic addition of equations (11.24) and (11.35) as

$$\delta = \frac{8\,WD^3 n \sec\alpha}{d^4}\left(\frac{\cos^2\alpha}{G} + \frac{2\sin^2\alpha}{E}\right) + \frac{16\,MD^2}{d^4}\sin\alpha\left(\frac{1}{G} - \frac{2}{E}\right) \quad ...(11.39)$$

Similarly from equations (11.25*b*) and (11.36*b*) the rotation of the free end of the spring about the spring axis is

$$\phi = \frac{16\,WD^2 n}{d^4}\sin\alpha\left(\frac{1}{G} - \frac{2}{E}\right) + \frac{32\,MDn}{d^4}\sec\alpha\left(\frac{\sin^2\alpha}{G} + \frac{2\cos^2\alpha}{E}\right) \quad ...(11.40)$$

## 11.6 Carriage, Leaf or Laminated Springs

The carriage or leaf spring is made by placing circularly bent spring steel strips or plates of same radius one over the other as shown in the Fig. 11.5. Each plate is free

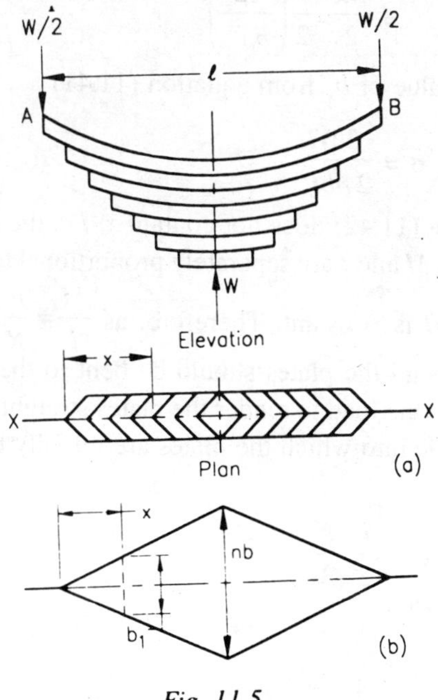


*Fig. 11.5*

to slide relative to the adjacent plates as the spring deflects. Both the ends of each plate are tapered to provide a uniform change in effective width between centre and the ends. The taper of any plate starts where that of lower plate ends. Further if the plates were cut along *XX* and placed side by side, they would form a diamond-shaped plate as shown in Fig. 11.5(*b*).

Let width of each plate = $b$

Thickness of each plate = $t$

Load on the spring = $W$

Number of plates = $n$

The width of the equivalent plate Fig. 11.5(*b*) at the centre will be nb and the width $b_1$ at a section distant x from one end will be given by

$$b_1 = \frac{x}{l/2}(nb) = \frac{2\,nbx}{l} \qquad \text{...(11.41)}$$

Further the bending moment at this section will be

$$M = \frac{Wx}{2}$$

Thus the bending stress at the skin will be

$$\sigma = \frac{M}{I}\cdot\left(\frac{t}{2}\right) \qquad \text{...(11.45)}$$

$$= \frac{Wx}{2}\cdot\frac{t}{2}\left(\frac{12}{b_1 t^3}\right)$$

Substituting the value of $b_1$ from equation (11.41)

$$\sigma = \frac{3\,Wl}{2\,nbt^2}$$

Evidently equation (11.42) does not contain *x*, *i.e.* the skin stress is same at all sections. Moreover both *M* and *I* are separately proportional to *x*, *i.e.* *M* is proportional to *I*. In other words *M/I* is constant. Therefore, as $\frac{M}{I} = \frac{E}{R}$, we conclude that *R* is constant. In other words all the plates should be bent to the same radius. The *proof load*, which is defined as the load to render the plates straight, can then be found from the geometry of the circle into which the plates are initially bent.

From Fig. (11.6)

$$y\,(2R - y) = \frac{l}{2}\cdot\frac{l}{2}$$

or approximately if $R >> y$

$$2yR = \frac{l^2}{4}$$

i.e. $$y = \frac{l^2}{8R} \quad ...(11.43)$$

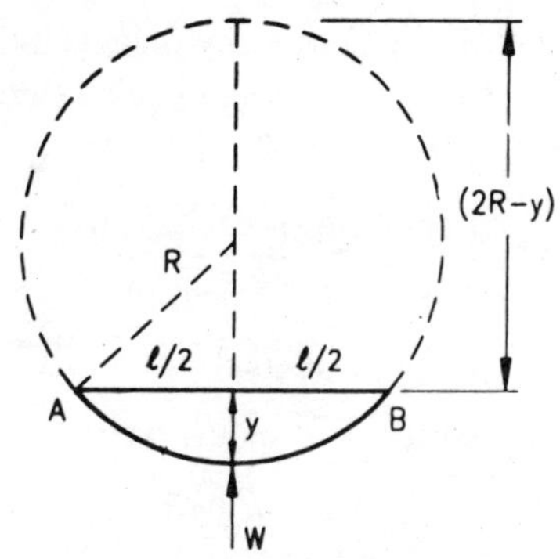

*Fig. 11.6*

Substituting the value of $R$ in terms of $M$, $I$ and $E$

$$y = \frac{l^2}{8} \cdot \frac{M}{EI}$$

$$= \frac{l^2}{8} \cdot \frac{Wx}{2} \cdot \frac{12t}{Eb_1 t^3}$$

$$= \frac{l^2}{8} \cdot \frac{Wx}{2} \cdot \frac{12l}{Et^3\,(2nbx)}$$

or $$y = \frac{3\,Wl^3}{8\,Enbt^3} \quad ...(11.44a)$$

or $$W = \frac{8\,yEnbt^3}{3l^3} \quad ...(11.44b)$$

Equation (11.44*b*) will give the proof load when the dip *y*, which will be the corresponding deflection of the spring, is known. Alternatively for any value of *W* (less than the proof load) the corresponding deflection can be found from equation (11.44*a*)

The strain energy stored in the spring is given by

$$U = \frac{1}{2} Wy = \frac{1}{2} . W . \frac{3Wl^3}{8\,Enbt^3}$$

or $$U = \frac{3}{16} \frac{W^2 l^3}{Enbt^3} \quad ...(11.45a)$$

This may be written in another form

$$U = \left(\frac{3\,Wl}{2\,nbt^2}\right)^2 \left(\frac{nbtl}{12\,E}\right)$$

$$= \frac{\sigma^2}{6E} \cdot \frac{nb}{2} . tl$$

$$= \frac{\sigma^2}{6E} \times \text{(volume of spring)} \quad ...(11.45b)$$

## 11.7 Flat Spiral Spring

Fig. 11.7 shows a flat spiral spring, made of rectangular steel strip of thickness *t* and width *b*. The winding torque *T* is applied to the spindle *A* while winding the spiral spring. The other end of the spring is hinged at *B* where the reaction of the pin on the

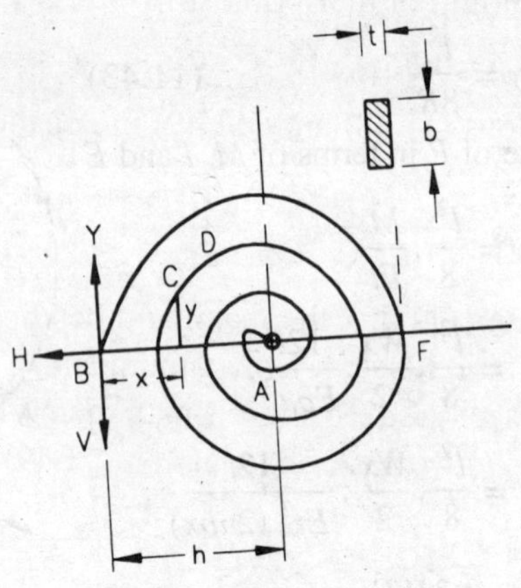


*Fig. 11.7*

spring can be resolved into two components $H$ and $V$ along and perpendicular to the line $BA$. Let $CD$ be a small length $dl$ with co-ordinates of the point $C$, with respect to the origin at $B$ as $x$ and $y$.

Bending moment $M$ at the element (taken positive if increases curvature) will be given by

$$M = V.x - H.y \qquad ...(11.46)$$

Due to this bending moment the rotation ($d\phi$) of the end $D$ with respect to the end $C$ of the element $dl$ will be

$$d\phi = \frac{Mdl}{EI}$$

or $$d\phi = \frac{(xV - yH)}{EI}dl \qquad \text{(From 11.46)}$$

Integrating this for the entire length of the strip, the rotation of the end $A$ with respect to the hinged end $B$ will be

$$d\phi = \frac{V}{EI}\int x\,dl - \frac{H}{EI}\int ydl \qquad ...(11.47)$$

If the centroid of the spring is assumed to be at the centre of the spindle A, the first moment of the length $l$ about the line $AB$ will be zero.

*i.e.* $$\int y\,dl = 0 \qquad ...(11.48a)$$

and moment of the length $l$ about $yy$ line gives

$$\int xdl = lh \qquad ...(11.48b)$$

where $l$ is the length of the spring strip and $h$ the distance between points $A$ and $B$.

Also for the equilibrium of the spring strip

$$T = Vh \qquad \text{...(11.49)}$$

Substituting equations (11.48) and (11.49) in equation (11.47), we get

$$\phi = \frac{(T/h)}{EI} . (lh) = \frac{Tl}{EI} \qquad \text{...(11.50)}$$

Further from equation (11.46), the maximum bending moment will be obtained on such a section of the strip where x is maximum and y is minimum. Evidently the point *F* is such a point where *y* is zero and *x* is maximum. Again assuming the distance *BF* to be approximately 2*h*, we get from equation (11.49)

$$M_{max} = M_F = \frac{T}{h} . (2h) = 2T \qquad \text{...(11.51)}$$

Corresponding maximum bending stress on the cross-section of the strip is

$$\sigma_{max} = \frac{2T}{I} . \frac{t}{2} = \frac{12\,Tt}{bt^3} = \frac{12T}{bt^2} \qquad \text{...(11.52a)}$$

or

$$T = \frac{1}{12} bt^2 \sigma_{max} \qquad \text{...(11.52 )}$$

Knowing the maximum allowable stress $\sigma_{max}$, the value of the applied torque *T* can be found from the equation (11.52*b*).

The energy stored corresponding to the torque *T* can be calculated as

$$U = \frac{1}{2} T\phi$$

$$= \frac{1}{2} T . \frac{Tl}{EI}$$

$$= \frac{1}{2} \frac{T^2 l\, 12}{Ebt^2} = \frac{6T^2 l}{Ebt^3} \qquad \text{...(11.53)}$$

The maximum energy which can be stored, when the stress reaches its maximum value will be obtained by substituting the value of *T* from equation (11.52*b*) in equation (11.53)

$$U_{max} = \frac{6l}{Ebt^3} \left( \frac{bt^2 \sigma_{max}}{12} \right)^2$$

$$= \frac{(\sigma_{max})^2}{24E} (btl)$$

$$= \frac{\sigma_{max}^2}{24E} \text{(volume)} \qquad \text{...(11.54)}$$

Energy per unit volume = Resilience of spring

$$= \frac{\sigma^2_{max}}{24E} \qquad ...(11.55)$$

## 11.8 Composite Springs

When two or more helical springs are connected to each other to take up the applied load, they form what is called a compound or composite spring. When they are in series as shown in Fig. 11.8(*a*) the load sustained by each is equal to the applied load W and the total deflection of this will be equal to the sum of the individual deflections of the springs. Thus if $\delta_1$ and $\delta_2$ are the individual deflections and $s_1$ and $s_2$ the stiffnesses of the respective springs then the deflection $\delta$ and stiffness $s$ of the composite series spring are calculated as follows

$$W_1 = W_2 = W \qquad ...(11.56)$$

$$\delta = \delta_1 + \delta_2 \qquad ...(11.57)$$

or
$$\frac{W}{s} = \frac{W_1}{s_1} + \frac{W_2}{s_2}$$

or
$$\frac{1}{s} = \frac{1}{s_1} + \frac{1}{s_2} \qquad ...(11.58)$$

However when the springs of equal free lengths placed in parallel as shown in Fig. 11.8(*b*) they share the applied load W in such a way that their individual deflations are equal to that of the composite spring.



*Fig. 11.8*

Then
$$\delta_1 = \delta_2 = \delta \qquad ...(11.59)$$

$$W_1 + W_2 = W \qquad ...(11.60)$$

or
$$\delta_1 s_1 + \delta_2 s_2 = \delta s$$

or
$$s_1 + s_2 = s \qquad ...(11.61)$$

**Example 11.1.** *A closed-coil helical spring is to have a stiffness of 70 kN/m and to exert a force of 2.25 kN. If the mean diameter of the coils is to be 90 mm, and the working stress 230 MN/m², find the required number of coils and the diameter of the steel rod from which the spring should be made. Take the modulus of rigidity as 80 GN/m².*

**Solution.** From equation (11.2)

$$\tau = \frac{8WD}{\pi d^3}$$

or
$$d^3 = \frac{8 \times 2250 \times 0.09}{\pi \times 230 \times 10^6} = 2.24 \times 10^{-6}$$

*i.e.* $d = 1.308 \times 10^{-2}$ m = 13.08 mm

Also from equation (11.6), we have

$$s = \frac{Gd^4}{8nD^3}$$

or
$$70 \times 1000 = \frac{80 \times 10^9\ (1.308 \times 10^{-2})^4}{8n\ (.09)^3}$$

or
$$n = 5.81.$$

**Example 11.2.** *In a compound helical spring the inner spring is arranged within and concentric with the outer one but is 10 mm shorter. The outer spring has 10 coils of mean diameter 25 mm and the diameter of the wire is 3 mm. Find the stiffness of the inner spring if the axial load of 150 N causes the outer one to compress 20 mm. If the radial clearance between the two springs is to be 1.6 mm, find the diameter of the wire of the inner spring when it has 8 coils. G = 80 GN/m² for both spring.*

**Solution.** When the outer spring deflects through 20 mm, the corresponding deflection of the inner spring will be 10 mm. If the loads shared by them are $W_0$ and $W_i$, then as the springs are in parallel

$$W_0 + W_i = W = 150 \text{ N}$$

For the outer spring from equation (11.6)

$$s_0 = \frac{Gd_0^4}{8\,n_0 D_0^3} = \frac{80 \times 10^3 \times 3^4}{8 \times 10 \times (25)^3} = 5.59 \text{ N/ mm}$$

$$W_0 = \delta_0 s_0 = 20 \times 5.69 = 103.8 \text{ N}$$

$$W_i = W - W_0 = 150 - 103.8 = 46.2 \text{ N}$$

Now
$$D_i = D_0 - d_0 - d_i - 2 \times 1.6$$
$$= 25 - 3 - d_i - 3.2$$
$$= (18.8 - d_i) \text{ mm}$$

$$\delta_i = \frac{8\,W_i n_i D_i^3}{Gd_i^4}$$

$$10 = \frac{8 \times 46.2 \times 8 \times (18.8 - di)^3}{(80 \times 10^3) \times d_i^4}$$

which gives
$$d_i = 2.04 \text{ mm}$$

Also
$$s_i = \frac{W_i}{\delta_i} = \frac{46.2}{10} = 4.62 \text{ N / mm}$$
$$= 4.62 \text{ kN/m}.$$

**Example 11.3.** *Derive an expression for the axial extension of an open-coiled helical spring produced by an axial twisting couple.*

*An open-coiled helical spring made of 10 mm diameter rod has six free coils of 100 mm mean diameter. The ends of the spring are fastened to two discs kept 0.75 m apart, which is the free length of the spring. Calculate the force on the discs, acting along the axis of the spring, when one disc is rotated through 10° to coil the spring, E = 200 GN/m² and G = 80 GN/m².*

**Solution.** Pitch $= \dfrac{0.75}{6} = 0.125$ m

$$\tan\alpha = \frac{.125}{0.1\,\pi} = 0.3979$$

or $\alpha = 21°\ 42'$

$$l = \pi\, Dn \sec$$

$$= \pi \times .10 \times 6 \sec 21°\ 42° = 203 \text{ m}$$

Also $\dfrac{D^2 n \sin\alpha}{d^4} = \dfrac{(0.1)^2 \times 6 \times \sin 21°42'}{(0.01)^4} = 22.18 \times 10^5$

As the discs are kept at the same distance the net deflection of the spring will be zero and the spring will be under the action of both axial load and torque. Therefore from equation (11.39), we have

$$\delta = \frac{8WD^3 n \sec\alpha}{d^4}\left[\frac{\cos^2\alpha}{G} + \frac{2\sin^2\alpha}{E}\right] + \frac{16MD^2 n \sin\alpha}{d^4}\left(\frac{1}{G} - \frac{2}{E}\right) = 0$$

$$\frac{W(0.1)^3 \times 6 \times \sec 21°\ 42'}{(.01)^4}\left[\frac{0.9291^2}{80 \times 10^9} + \frac{2\,(.3697)^2}{200 \times 10^9}\right] + 16\,M \times 22.18 \times 10^5 \times \left[\frac{1}{80 \times 10^9} - \frac{2}{200 \times 10^9}\right] = 0$$

from which $M = -\,0.7078\ W$

Again from equation (11.40)

$$\phi = \frac{16WD^2}{d^4}\sin\alpha\left(\frac{1}{G} - \frac{2}{E}\right) + \frac{32\,MDn}{d^4}\sec\alpha\left(\frac{\sin^2\alpha}{G} + \frac{2\cos^2\alpha}{E}\right) = \frac{10\pi}{180}$$

*i.e.*

$$\frac{16W\,(0.1)^2 \times 6}{(.01)^4}(.3697)\left[\frac{1}{80 \times 10^9} - \frac{2}{200 \times 10^9}\right] + \frac{32\,(-\,0.7078\ W) \times .1 \times 6}{(.01)^4 \times .9291}\left[\frac{.3697^2}{80 \times 10^9} + \frac{2 \times .9291^2}{200 \times 10^9}\right] = \frac{10\pi}{180}$$

from which $W = -11.607$ N (*i.e.* compressive force). The reaction on disc will be 11.6 N.

**Example 11.4.** *A quarter-elliptic or cantilever leaf spring of length l with n plates each of breadth b and thickness t is so shaped that circular bending occurs. Derive (a) expression for the maximum deflection due to an end load W.*

*If the maximum allowable bending stress is 500 MN/m² and the spring has 5 plates each 70 mm wide by 10 mm thick with the length of the largest plate 0.4 m, find (b) the maximum value for the load W and (c) the deflection at the end. Assuming that the spring just straightens under these conditions, determine the initial radius of curvature of the plates. For steel E = 200 GN/m².*

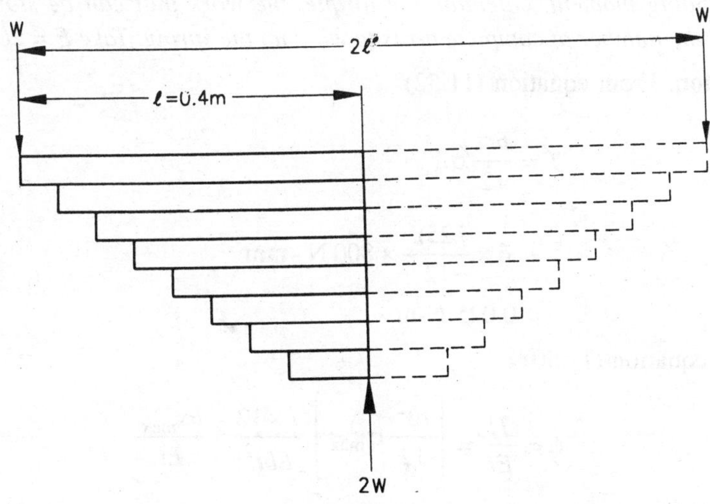


*Fig. 11.9*

**Solution.** The quarter-elliptic spring is shown in Fig. 11.9. The end deflection and stress are the same as in an ordinary semi-elliptic leaf spring Fig. 11.5 of span 2l and carrying a central load 2*W*. That is from equation (11.42)

$$\sigma = \frac{3(2W)(2l)}{2nbt^2} = \frac{6Wl}{nbt^2} \qquad ...(A)$$

or

$$W = \frac{nbt^2\sigma}{6l} = \frac{5 \times 70 \times 10^2 \times 400}{6 \times 400} \text{N}$$

$$= 5.833 \text{ kN.}$$

Again from equation (11.44*a*) we have for the quarter elliptic leaf spring

$$y = \frac{3(2W)(2l)^3}{8Enbt^3} = \frac{6Wl^3}{Enbt^4} \qquad ...(B)$$

$$= \frac{6 \times 5833 \times (400)^3}{(200 \times 10^3) \times 5 \times 70 \times (10)^3}$$

$$= 32 \text{ mm}$$

For the radius of curvature for the cantilever leaf spring we have from equation (11.43).

$$R = \frac{(2l)^2}{8y} = \frac{l^2}{2y} \qquad \text{...(C)}$$

$$= \frac{(400)^2}{2 \times 32} \text{ mm} = 2.5 \text{ m.}$$

**Example 11.5**. *A flat spiral spring is 6 mm wide and 0.25 mm thick, the length being 2.5 m. Assuming the maximum stress of 800 MN/m² to occur at the point of greatest bending moment, calculate the torque, the work that can be stored in the spring, and the number of complete turns to wind up the spring. Take E = 200 GN/m².*

**Solution.** From equation (11.52)

$$T = \frac{bt^2}{12}\sigma_{max}$$

$$= 6 \times \frac{(.25)^2}{12} \times 800 \text{ N - mm}$$

$$= 0.025 \text{ N-m}$$

From equation (11.50)

$$\phi = \frac{Tl}{EI} = \left[\frac{bt^2}{12}\sigma_{max}\right]\frac{l \times 12}{Ebt^2} = \frac{l\sigma_{max}}{Et}$$

$$= \frac{bt^2}{12}\sigma_{max}\frac{l \times 12}{Ebt^2} = \frac{l\sigma_{max}}{Et}$$

$$= \frac{2.5 \times 800 \times 10^6}{.(200 \times 10^9)(0.25 \times 10^{-3})} = 40 \text{ rad}$$

$$U_{max} = \frac{1}{2}T\phi = \frac{1}{2} \times 0.025 \times 40$$

$$= 0.5\ J.$$

## PROBLEMS FOR PRACTICE

1. A close-coiled helical spring is to have a stiffness of 70 kN/m and to exert a force of 2.25 kN. If the mean diameter of the coils is to be 90 mm, and the working stress 230 MN/m², find the required number of coils and the diameter of the steel rod from which the spring should be made. Take the modulus of rigidity as 80 GN/m². **[Ans.** 6.58 ; 81 ; 13.08 mm]

2. A closed coil helical spring made of round steel wire is required just to get over a rod 30 mm diameter and to carry an axial load of 120 N without causing the deflection to exceed 20 mm. The maximum allowable shearing stress is 200 MN/m$^2$ and G for the steel is 80 GN/m$^2$. Find the diameter of the wire, the mean diameter of the coil and the number of turns.

[**Ans.** 3.72 mm ; 33.72 mm ; 8.52]

3. A closed coil spring is to have a stiffness of 1 kN/m of compression, a maximum load of 50 N and a maximum shearing stress of 120 MN/m$^2$. The solid length of the spring (*i.e.* when the coils are touching) is to be 45 mm.

   Find the diameter of the wire, the mean diameter of the coils and the number of coils required. $G = 50$ GN/m$^2$. [**Ans.** 0.0034 m ; 0.0436 m ; 12.54]

4. A composite spring has two close-coiled helical springs connected in series ; each spring has 12 coils at a mean diameter of 25 mm. Find the diameter of wire in one of the spring if the diameter of the wire in the other springs is 2.5 mm and the stiffness of the composite spring is 700 N/m.

   Estimate the greatest load that can be carried by the composite spring and the corresponding extension for a maximum shearing stress of 180 MN/m$^2$ $G = 80$ GN/m$^2$. [**Ans.** 2.11 mm ; 26.55 N ; 37.9 mm]

5. The spring load against which a valve is opened is provided by an inner helical spring arranged within and concentric with an outer helical spring. Both springs are of steel, close-coiled, and the free length of the inner spring is 6 mm longer than that of the outer spring. The outer spring has 12 coils of mean diameter 25 mm, diameter of wire 3 mm and an initial compression of 5 mm when the valve is closed. Find the stiffness of the inner spring if the greatest force required to open the valve 10 mm is 150 N.

   If the radial clearance between the springs is approximately 1.5 mm, find the diameter of the wire of the inner spring if it has 10 coils. $G = 80$ GN/m$^2$.

   [**Ans.** 4057 N/m ; 2.08 mm]

6. An open coiled spring of 125 mm mean diameter has 10 coils of 12 mm diameter wire, at a slope of 30° to the horizontal when the coil axis is vertical. Find expressions for the longitudinal extension and the rotation for the joint application of an axial load $W$ and an axial torque $T$. Hence find the axial load and torque necessary to extend the spring 5 mm if rotation is prevented, indicating whether torque tends to wind up or upwind spring. $E = 200$ GN/m$^2$ ; $G = 80$ GN/m$^2$.

   [**Ans.** 48.9 N ; 0.312 Nm ; unwinding]

7. A helical spring has a mean diameter of 65 mm, and consists of 12 turns of 6 mm diameter wire. Initially the inclination of the turns is 7°. Calculate the axial and

angular deformations caused by an axial '*winding up*' couple of 9 Nm. Explain carefully any approximations made. $E$ = 200 GN/m$^2$ $G$ = 80 GN/m$^2$.

**[Ans.** 1.72 mm ; 1.754 rad.**]**

8. A laminated spring, made of 12 steel plates, is 0.9 m long. The maximum central load is 7.2 kN. If the maximum allowable stress in steel is 230 MN/m$^2$ and the maximum deflection approximately 38 mm, calculate the width and thickness of the plates E = 200 GN/m$^2$. **[Ans.** 93.8 mm ; 6.13 rad.**]**

9. Deduce an expression for the resilience of a loaded carriage spring, the maximum bending stress being given. A carriage spring 1.35 m long has leaves of 100 × 12.5 mm section. The maximum bending stress is 150 MN/m$^2$ and the spring must absorb 125 $J$ when straightened. Calculate the number of leaves and their initial curvature. $E$ = 200 GN/m$^2$. **[Ans.** 8 ; 8.43 m**]**

10. A flat spiral spring is made of steel 12 mm wide and 0.5 mm thick. The end at the greatest radius is attached to a fixed point and the other end to the spindle. The length of the steel is 6 m.

    **Determine :** (*a*) the maximum turning moment which can be applied to the spindle if the stress in the strip is not to exceed 550 MN/m$^2$ ; (*b*) the number of turns required to be given to the spindle ; (c) the energy then stored in the spring $E$ = 200 GN/m$^2$. **[Ans.** 0.1375 Nm ; 5.26 rev ; 2.27 $J$**]**

Chapter–12

# Strain Energy and Deflection Due to Shear and Bending

## 12.1 Strain Energy and Resilience

When a bar is acted upon by a tensile or compressive load there is a change in the length and the point of application of the load moves against the resistance offered by the material which gets strained under the action of the load. In so doing the work is done on the bar. If the bar had been loaded within the elastic limit of the material the work done on the bar will be recovered when the load is removed. In other words, the work done on the bar, while loading, is stored in the bar as strain energy which can be recovered wholly or partially depending upon whether the bar was loaded up to or beyond its elastic limit. The strain energy stored per unit volume is called the resilience of the body. Further the '*Proof Resilience*' is the strain energy per unit volume stored in the body at its elastic limit. Strain energy is always a positive quantity and, being work units, will be expressed in Nm (or Joules).

If the material is uniformly stressed everywhere the strain energy per unit volume will be constant. However if the stress varies from point to point the strain energy per unit volume will vary from point to point and will depend upon the stress distribution and shape of the body.

## 12.2 Strain Energy in Tension and Compression

### 12.2.1 Gradually Applied Load

A gradually applied load $P$ is one which increases from zero to its current value $P$ in infinitely large number of very small steps. In actual practice if the strain rate is less than $2 \times 10^{-3}$/sec. it is called a gradually applied or static load.

For a gradually applied tensile load the load-extension (or load-contraction for compressive load) curve for a bar of uniform section will be a straight line up to the limit of proportionality point Fig. 12.1. Corresponding to the load $P$ the stretch will be $x$ and the work done will be represented by the shaded area. Thus the strain energy $U$ is given by

$$U = \frac{1}{2} Px \qquad ...(12.1)$$

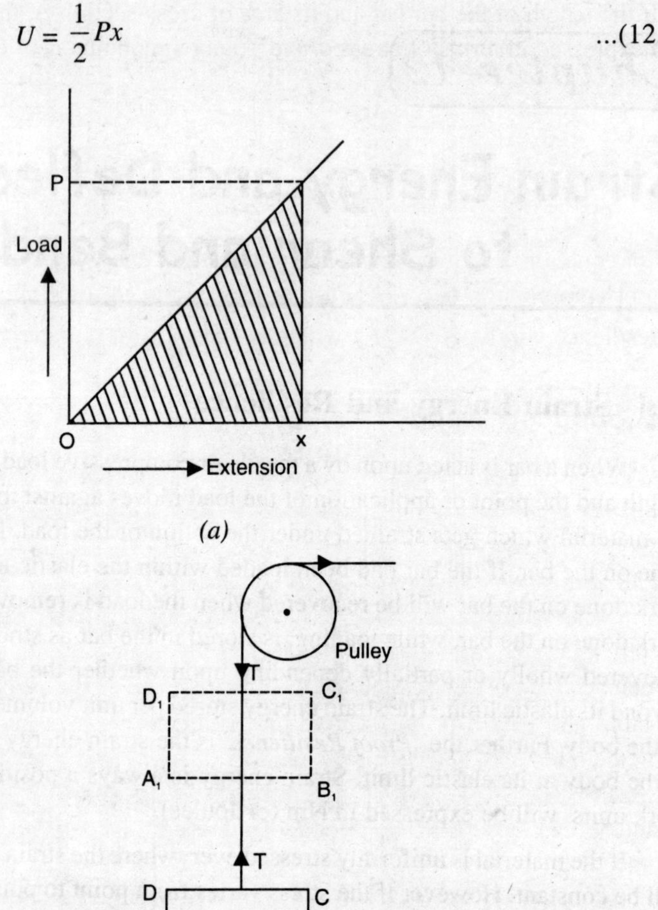


*(b)*

*(i)* *Bottom of body AB, just touching table and*

$$R = 0$$

$$T = Mg$$

*Rope snapped, then sudden load on table*

*(ii)* *Rope tight but carries tension T, (< Mg)*

$$R + T = Mg$$

*(iii)* *Rope just loose so that R = Mgand Tension = 0 gradually applied load.*

*(iv)* *Bottom of body, AB at height h above table rope snapped → shock on impact load.*

*Fig. 12.1*

If the length of the bar is $l$ and its area of cross-section $A$, then the strain energy can be expressed in terms of stress $\sigma$ and Young's modulus $E$ of the material

$$U = (A\sigma)\frac{1}{2}\left(\frac{\sigma}{E}l\right) = \frac{\sigma^2}{2E}(Al)$$

or $$U = \frac{\sigma^2}{2E}(\text{Volume}) \qquad ...(12.2)$$

For equation (12.2) the resilience or the strain energy per unit volume in simple tension or compression is $\sigma^2/2E$. Further if $\sigma_e$ is the elastic limit of the material, the proof resilience will be $\sigma_e^2/(2E)$ which is constant for a material.

### 12.2.2 Suddenly Applied Load

A suddenly applied load $P_1$ is one which is applied in one step of value $P_1$, that is, it remains constant during the period the stretch or contraction attains the final value. Consider a body $ABCD$ of mass $M$ hanging from the end of a rope as shown in Fig. 12.1$b$ and is being lowered slowly by releasing the rope with the help of a worm and worm wheel arrangement attached to the other end of the rope. When the body is completely in air and not touching the table, the tension $T$ in the rope will be equal to $Mg$. When the body is lowered slowly its lower face $AB$ will come close to the top face of the table and it will be just above the table surface at a height $h \rightrightarrows 0$ but not touching the table then again $T = Mg$ and the reaction, $R$ of the table on body will be euqal to zero. But when the rope is further released slowly, the reaction $R$ will start increasing and the tension $T$ will simultaneously start decreasing at the same rate and after some time R will become equal Mg and $T$ will reduce to zero and the rope will become loose. Evidently the load on the table will be equal to $R$ which in turn will be equal to Mg.

Thus a load of Mg has been applied which has increased from zero to current value Mg gradually and hence it will be called gradually applied load. Secondly when the bottom of the body was just above the table at zero height but $T = Mg$ and $R = 0$ at that time had we cut the rope, the etire load $Mg$ would have come on table in one go. It would then be called a suddenly applied load.

Thirdly when the body was hanging in the air and at height '$h$' above the table, if we cut the rope, the body will fall and come in contact with the table with certain velocity. We call this a shock or impact load. In the light of above, in the case of a suddenly applied load, $P_1$, the strain energy is given by

$$U = P_1 x \qquad ...(12.3)$$

If the deformation $x$ is same as in the gradually-applied-load case the bar will experience the same strain and will therefore be under the action of same stress which in turn will result in the same strain energies in the two cases. Then comparing equations (12.1) and (12.3)

$$\frac{1}{2}Px = P_1 x$$

or $$P = 2P_1 \qquad ...(12.4)$$

From equation (12.4) it is observed that in order to produce the same strain and same stress a gradually applied load should be twice the value of the suddenly applied load. Alternatively a suddenly applied load will induce stress in the bar which is twice the one induced by same load on the bar but applied gradually. It may be noted that in the case of the suddenly applied load the strain and stress considered are instantaneous maximum ones obtained immediately after the application of the sudden load. As a matter of fact in the case of suddenly applied load the system is set into free damped vibrations and the stress in the bar changes in magnitude and nature from tension to compression to tension and so on. The maximum value of the stress on each side progressively goes on diminishing till it ultimately settles down to a value equal to that due to gradually applied load* when the vibrations die out.

### 12.2.3 Impact Load

An impact load is one which strikes the body with certain velocity as is the case of a falling mass *M* through height h (Fig. 12.2) above the collar attached to the lower end of the rod *AB* of length *l*.

It is assumed that the mass *M* falls freely without any loss of energy in impact and produces an instantaneous extension *x* and due to it instantaneous stress σ is induced in this rod. Then the loss of potential energy of the mass will be stored in the form of strain energy in the rod.

$$Mg\,(h + x) = \frac{\sigma^2}{2E}(Al) \qquad ...(12.5)$$

or $$Mg\left(h + \frac{\sigma l}{E}\right) = \frac{\sigma^2}{2E}(Al)$$

or $$\sigma^2 - \sigma\left(\frac{2Mg}{A}\right) - \frac{2\,MghE}{Al} = 0 \qquad ...(12.6)$$



*Fig. 12.2*

Taking only positive root of equation (12.6)

$$\sigma = \frac{Mg}{A}\left[1 + \sqrt{1 + \frac{2AEh}{Mgl}}\right] \qquad ...(12.7)$$

*Note : The reader is urged to refer to any good book on vibrations.

Case I.

When $h = O$, the impact velocity will be zero and the load will be a suddenly applied load. Then the equation (12.7) gives $\sigma = \frac{Mg}{A} = 2\frac{P}{A}$, , which is same as obtained in article 12.2.2.

**Case II.**

When $x$ is very small as compared to h, then from equation (12.5)

$$Mgh = \frac{\sigma^2}{2E}(Al)$$

or

$$\sigma = \sqrt{\frac{2\,Mgh\,E}{Al}} \qquad ...(12.8)$$

**Example 12.1.** *A round bar of steel 1 m long is 20 mm diameter for half its length and 25 mm diameter over the rest. A gradually applied axial load of 1 kN acts on it. Find the total strain energy and strain energy per unit volume. E = 200 GN/m².*

**Solution.**

$$A_1 = \frac{\pi}{4}(0.02)^2 = \pi \times 10^{-4}\ \text{m}^2$$

$$\sigma_1 = \frac{1000}{\pi \times 10^{-4}} = 3.18 \times 10^6\ \text{N/m}^2$$

$$A_2 = \frac{\pi}{4}(.025)^2 = 4.91 \times 10^{-4}\ \text{m}^2$$

$$\sigma_2 = \frac{1000}{4.91 \times 10^{-4}} = 2.035 \times 10^6\ \text{N/m}^2$$

$$\text{Total strain energy} = \frac{\sigma_1^2}{2E}(A_1 l_1) + \frac{\sigma_2^2}{2E}(A_2 l_2) = \frac{l_1}{2E}[\sigma_1(\sigma_1 A_1) + \sigma_2 A_2)]$$

$$= \frac{0.5}{2D}(P\sigma_1 + P\sigma_2)$$

[Because $P = \sigma_1 A_1 = \sigma_2 A_2 = 1000$ N and $l_1 + l_2 = .5$]

$$= \frac{100 \times 0.5}{2 \times 200 \times 10^9}(3.18 \times 10^6 + 2.035 \times 10^6) = 0.006515\ J$$

Total volume $= A_1 l_1 + A_2 l_2 = 0.5\ (\pi \times 10^{-4} + 4.91 \times 10^{-4})$

$= 4.025 \times 10^{-4}\ \text{m}^3$

Therefore average strain energy per unit volume

$$U/V = \frac{0.006515}{4.025 \times 10^{-4}} = 16.2\ \text{J/m}^3.$$

**Example 12.2.** *A wagon, of mass 5 t, is attached to a steel wire rope, the other end of which is wound round a brake drum. The wagon is descending a slope at a uniform speed of 5 km/h and the rope is taut. By the action of the brake the drum is suddenly stopped and the wagon is brought to rest. If the length of the rope between the wagon and the drum at that instant was 250 m and its cross-sectional area is 300 mm², find the maximum stress in the rope. E = 200 GN/m².*

**Solution.** Strain energy of rope = Kinetic energy of wagon

$$\frac{\sigma^2}{2E}Al = \frac{1}{2}Mv^2$$

or
$$\sigma^2 = 5000 \times \left(\frac{5000}{3600}\right)^2 \times \frac{200 \times 10^9}{(300 \times 10^{-6}) \times 250}$$

or
$$\sigma^2 = \frac{(10)^{10}}{36\sqrt{3}}$$

i.e.
$$\sigma = 160 \times 10^6 \text{ N/m}^2 = 160 \text{ MN/m}^2$$

**Example 12.3.** *A bar of certain material, 40 mm diameter and 1.2 m long, has a collar securely fitted to one end. It is suspended vertically with the collar at the lower end and a mass of 2000 kg is gradually lowered on to the collar, producing an extension in the bar of 0.25 mm. Find the height from which this load could be dropped on to the collar if the maximum tensile stress in the bar is to be 100 MN/m².*

**Solution.** In the first case when the load is applied gradually

$$\sigma_1 = \frac{(2000 \times 9.81) \times 4}{\pi \times (0.040)^2} = 15.6 \times 10^6 \text{ N/m}^2$$

and
$$\varepsilon = \frac{0.25 \times 10^{-3}}{1.2}$$

Therefore,
$$E = \frac{\sigma_1}{\varepsilon} = \frac{15.6 \times 10^6 \times 1.2}{(0.25 \times 10^{-3})} = 75.0 \times 10^9 \text{ N/m}^2$$

For the second case $\sigma_2 = 100 \times 10^6$ N/m²
Corresponding extension

$$x = \frac{(100 \times 10^6)}{75 \times 10^9} \times 1.2 = 1.6 \times 10^{-3} \text{ m}$$

If $h$ is the height through which the mass has to be dropped, then from equation (12.5), we have

$$(2000 \times 9.81)\,(h + 1.6 \times 10^{-3}) = \frac{(100 \times 10^6)}{75 \times 10^{-3}} \times 1.2$$

from which $h$ = 3.52 mm.

## 12.3 Strain Energy Due to Simple Shear

Referring to Fig. 12.2(*a*) if *a*, *b* and *t* are the length, breadth and thickness of the rectangular block under the action of gradually applied shearing force *P* due to which the movement of the face $AB$ is $AA'$. Then the work done, whicn is stored in the block as shear strain energy is given by the expression given below, where $\phi$ is the shear strain angle

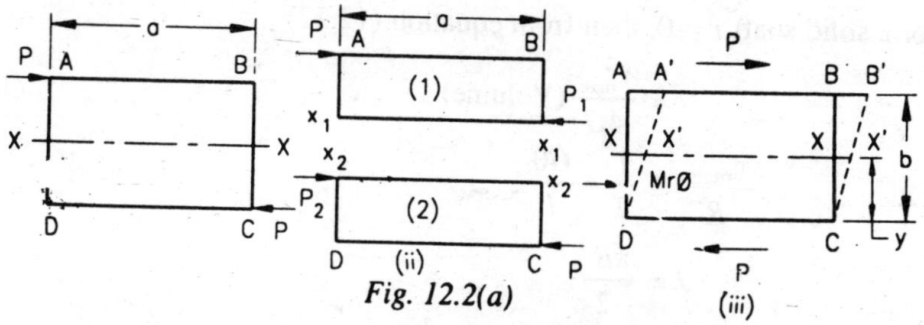

Fig. 12.2(a)

$$U = \frac{1}{2} P(AA') = \frac{1}{2}(\tau . a . t)(b\,\phi)$$

$$= \frac{1}{2}\tau\phi\,(\text{volume}) = \frac{1}{2}\tau\left(\frac{\tau}{G}\right)\text{volume}$$

*i.e.*
$$U = \frac{\tau^2}{2G}(\text{volume}) \quad ...(12.9)$$

## 12.4 Strain Energy Due to Torsion

When a shaft is under the action of a torque, the shear stress over the cross-section is not constant and is directly proportional to the radius. Let the shear stress at radius $x$ be $\tau$, which is given in terms of the maximum shear stress $\tau_{max}$ at the outer radius $R$ of the shaft, as

$$\tau = \frac{x}{R}\tau_{max} \quad ...(12.10)$$

Taking the shear stress $\tau$ constant for the thickness $dx$ of the elementary cylinder of thickness $dx$, the shear strain energy of this annular cylinder of length $l$ is given by the equation 12.9

$$\delta U = \frac{1}{2G}\left(\frac{x}{R}\tau_{max}\right)^2 d\gamma \quad ...(12.11)$$

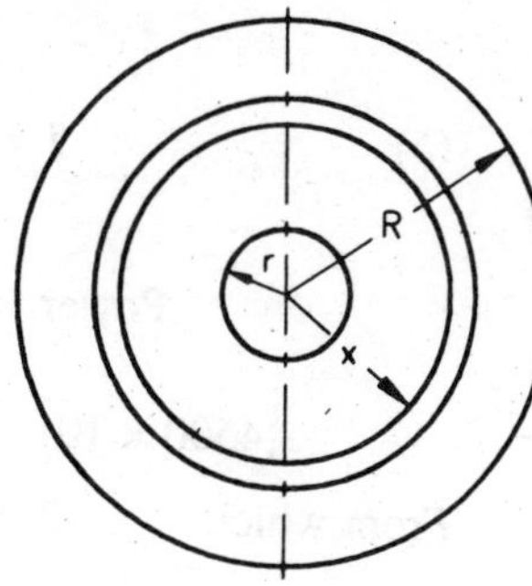

Fig. 12.3

Therefore, integrating equation (12.11), the total shear strain energy of the hollow shaft is

As $\quad dV = (2\pi x dx)l$

$$\therefore \quad U = \int_r^R \frac{1}{2G}\frac{x^2}{R^2}\tau_{max}^2\,(2\pi\, xldx) = \frac{\pi l}{R^2 G}\tau_{max}^2 \int_r^R x^3\,dx$$

or $$U = \frac{\pi\tau_{max}^2}{4G R^2}(R^4 - r^4)l = \frac{\tau_{max}^2}{4G R^2}(R^2 + r^2)\{\pi(R^2 - r^2)l\}$$

*i.e.* $$U = \frac{\tau_{max}^2}{4 GR^2}(R^4 + r^4) \text{ Volume} \qquad \text{...(12.12)}$$

For a solid shaft $r = 0$, then from equation (12.12)

$$U = \frac{\tau_{max}^2}{4G}(\text{Volume}) \qquad \text{...(12.13}a\text{)}$$

But $$\frac{\tau_{max}}{R} = \frac{T}{J} = \frac{G\theta}{l}$$

and $$J = \frac{\pi R^4}{2}$$

$$\text{vol} = \pi R^2 l$$

Substituting these in equation (12.13) we get

$$U = \frac{1}{2}T\theta = \frac{T^2 l}{2GJ} \qquad \text{...(12.13}b\text{)}$$

**Example 12.4.** *A hollow shaft having the external diameter, twice the internal diameter, subjected to a pure torque, attains a maximum shears stress* $\tau$. *Show that the strain energy stored per unit volume of the shaft is* $5\tau^2/16\,G$.

*Such a shaft is required to transmit 4500 kW at 110 rev/min. with uniform torque, the maximum stress not exceeding 70 MN/m².*

*Calculate the shaft diameter and the energy stored per m³ when transmitting this power.* $G = 83$ *GN/m²*.

**Solution.** (*a*) For the hollow shaft from equation (12.12)

$$U/\text{Volume} = \frac{\tau^2}{4GR^2}(R^2 + r^2) = \frac{\tau^2}{4GR^2}\left(R^2 + \frac{R^2}{4}\right) = \frac{5\tau^2}{16\,G}$$

(*b*) $$T = J\left(\frac{2\tau_{max}}{D}\right) = \frac{\pi}{32}\left(D^4 - \frac{D^4}{16}\right)\frac{2}{D}(70 \times 10^6) \text{ Nm}$$

$$\text{Power} = 4500 \times 10^3 = T \times 2\pi \times \frac{110}{60}$$

or $$4500 \times 10^3 = \frac{\pi}{32} \times \frac{15}{16}D^4 \cdot \frac{2}{D} \times 70 \times 10^6 \times \frac{2\pi \times 110}{60}$$

From which $D = 0.312$ m or 312 mm

$$\therefore \quad d = \frac{312}{2} = 156 \text{ mm}$$

(*c*) $$U/\text{Volume} = \frac{5}{16}\frac{\tau^2}{G} = \frac{5}{16} \times \frac{(70 \times 10^6)^2}{83 \times 10^3} \text{ J/m}^3 = 18.44 \text{ kJ/m}^3.$$

## 12.5 Strain Energy Due to Bending

Fig. 12.4. represents a beam of uniform section with certain end conditions such that the bending moment varies along its length. Let the bending moment at a distance $x$ from the chosen origin $A$ be $M$ and it is assumed that it remains constant over a small

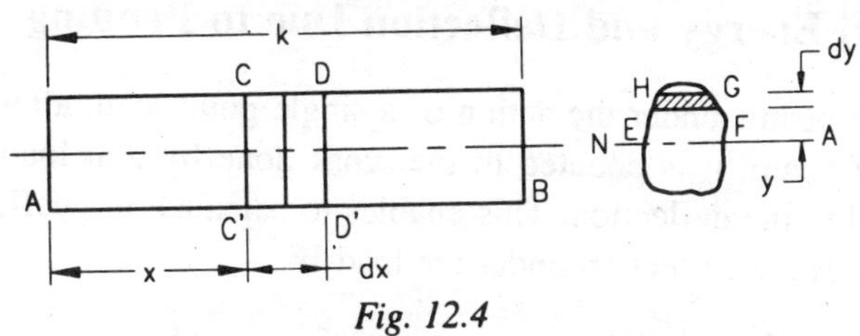


*Fig. 12.4*

length dx of the beam. Similarly the bending stress σ normal to the cross-section at the height $y$ from the neutral axis is assumed to be constant for a small distance dy. Thus for a small volume of material bounded between the sections $C$ and $D$ over the length $dx$ and between planes $EF$ and $HG$ the normal stress σ can be taken as constant.

Therefore if $E$ is the modulus of elasticity, the strain energy stored in this elemental volume of area $EFGH$ (= $dA$) is

$$dU = \frac{\sigma^2}{2E} \text{ (volume of strip)}$$

or

$$dU = \frac{\sigma^2}{2E}(dA\,dx) \qquad \text{...(12.14)}$$

Since the stress varies over the section with $y$ and in addition being dependent on $M$ it varies with x, the total strain energy of the beam will be found by integrating equation (12.14) twice ; once along the depth of the section and secondly along the length of the beam. That is

$$U = \int_0^l \left[\int_{c'}^{C} \frac{\sigma^2}{2E}\,dA\right] dx$$

$$= \int_0^l \left[\int_{c'}^{C} \frac{1}{2E}\left(\frac{M}{I}y\right)^2 dA\right] dx \qquad \text{(from equation 4.8)}$$

$$= \int_0^l \frac{M^2}{2EI^2}\left[\int_{c'}^{C} y^2 dA\right] dx$$

$$= \int_0^l \frac{M^2}{2EI^2} I\,dx \qquad \text{as } \left(\int_{c'}^{c} y^2\,dA = I\right)$$

i.e.

$$U = \int_o^l \frac{M^2}{2EI}\,dx \qquad \text{...(12.15a)}$$

For any given load and end conditions, $M$ can be expressed in terms of $x$ and then the total strain energy can be evaluated with the help of equation (12.15).

However if $M$ is constant over the length l

$$U = M^2 l/(2EI) \qquad \text{...(12.15}b\text{)}$$

## 12.5.1 Strain Energy and Deflection Due to Bending

In cases of beams under the action of a single point load, after calculating the strain energy of beam, it is equated to the work done by that load for its gradual movement equal to the deflection. This enables to calculate the deflection under the load. That is if δ is the deflection under the load $W$

$$\frac{1}{2} W\delta = U$$

or
$$\delta = \frac{2U}{W} \qquad \text{...(12.16)}$$

**Example 12.5.** *A beam of uniform section is simply supported at its ends and is acted upon by a load W at a distance 'a' from one end and 'b' from the other. Find the total strain energy of the beam in terms of the modulus of elasticity, the second moment of area of the section I and the length l of the beam. Hence calculate the deflection under the load.*

**Solution.** For any section $xx$ lying between $A$ and $C$ *i.e.* for $o < x < a$.

$$M = -\frac{Wb}{l}x$$

and for $a < x < l$.

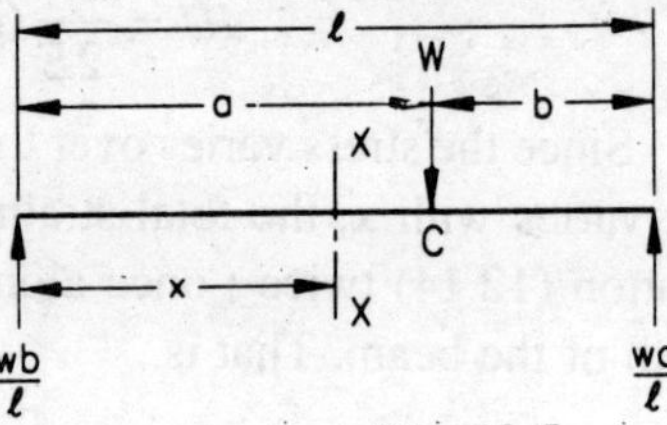

$$M = -\frac{Wa}{l}(l - x)$$

Fig. 12.5

$$U = U_{AC} + U_{BC}$$

$$= \int_0^a \frac{1}{2EI}\frac{W^2b^2}{l^2}x^2\,dx + \int_a^l \frac{1}{2EI}\frac{W^2a^2}{l^2}(l-x)^2\,dx$$

$$= \frac{W^2b^2b^2}{2EIl^2}\left(\frac{a^3}{3}\right) + \frac{W^2a^2}{2EI\,l^2}\left[\frac{(l-x)^2}{-3}\right]_a^l$$

or
$$U = \frac{W^2a^2b^2}{6EI\,l^2}[a \_ (l-a)] = \frac{W^2a^2b^2}{6EI\,l} \qquad \text{...(12.17)}$$

If δ is the deflection under the load $W$, then equating this to $\frac{1}{2}\delta W$, we get

$$\left(\frac{1}{2} W\delta\right) = \frac{W^2 a^2 b^2}{6EI\, l}$$

or
$$\delta = \frac{Wa^2 b^2}{3EI\, l}$$

**Example 12.6.** *Find the expression for the strain energy of a simply supported beam carrying a load W spread over its entire length uniformly.*

**Solution.** At any section *XX*

$$M = -\frac{wl}{2}x + \frac{wx^2}{2}$$

$$U = \int_0^l \frac{M^2}{2EI} dx$$

$$= \frac{1}{2EI}\int \left(\frac{-wl}{2}x + \frac{wx^2}{2}\right)^2 dx$$

$$= \frac{w^2}{8EI}\int_0^l (l^2x^2 + x^4 - 2lx^3)dx$$

$$= \frac{w^2}{8EI}\left[\frac{l^2x^3}{3} + \frac{x^5}{5} - \frac{2lx^4}{4}\right]_0^l$$

$$= \frac{w^2 l^5}{240\, EI} = \frac{W^2 l^3}{240\, EI} \qquad [(\because W = wl)] \qquad ...(12.18)$$

*Fig. 12.6*

## 12.6 Deflection of Beams Due to Impact Load

Let a mass *M* drop from height h at a point on a beam. It is required to find the instantaneous deflection under the load. Let $W_e$ be the equivalent weight which when applied gradually at the same point will produce the same deflection y. Since in these two cases the deflection is same, the strain energies in the two cases will be equal. Further since the strain energy of the beam is equal to the work done by the individual loads in the two cases, we have

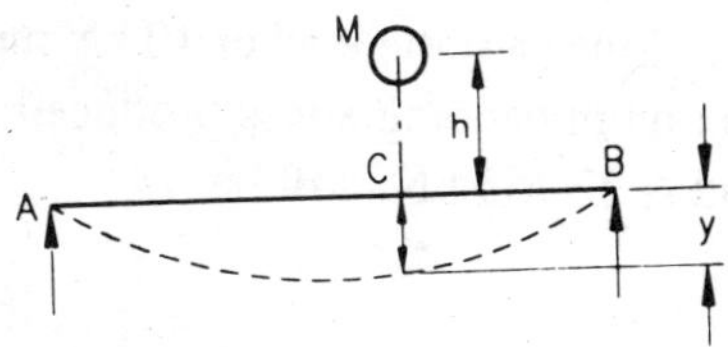


*Fig. 12.7*

$$Mg\,(h + y) = \frac{1}{2}W_e y \qquad ...(12.19)$$

Further depending on the end conditions the deflection y can be found in terms of $W_e$ (Chapter VI) and in general y is proportional to $W_e$

$$y = aW_e \qquad ...(12.20)$$

Where '*a*' is constant which depends upon the beam and its end conditions.

For equations (12.19) and (12.20)

$$Mg\ (h + aW_e) = \frac{1}{2} aW_e^2 \qquad ...(12.21)$$

The solution of the quadratic equation (12.21) gives the value of $W_e$. Then from equation (12.20) the value of y can be evaluated.

**Example 12.7.** *A static concentrated load of 10 kN applied to a simply supported beam at mid-span, produces deflection of 6 mm and a maximum bending stress of 25 MN/m². Calculate the maximum value of the momentary stress produced when a mass of 500 kg is allowed to fall through a height of 18 mm on to the beam at the middle of the span.*

**Solution.** Let $W_e$ N be the static load equivalent to the given impact load.

Further since 10 kN static load produces a deflection of 6 mm, then $W_e$ will produce a deflection δ such that

$$\delta = \frac{6}{1000} \times W_e = 0.006\ W_e \text{ mm}$$

Thus equating the work done by the equivalent static load and the given moving load, we get

$$Mg\ (h + \delta) = \frac{1}{2} \delta W_e$$

or $$500 \times 9.81 \left( .018 + \frac{1}{1000} \times .0006\ W_e \right) = \frac{1}{2} W_e \times \frac{0.0006}{1000} W_e$$

or $$W_e^2\ 9810\ W_e - 2943 \times 10^5 = 0$$

From which $W_e$ = 22755 N or 22.755 kN

Since a static load of 10 kN induces a maximum bending stress of 25 kN/m², the maximum bending stress produced in the impact case for which the equivalent static load is 22.755 kN, will be

$$\sigma_{max} = 25 \times \frac{22.755}{10} = 56.89 \text{ Mn / m}^2$$

## 12.7 Strain Energy of a Closed Coil Spring

**Case I :** Spring under the action of an axial load *W* :

The strain energy due to axial load *W* has already been calculated in equation (11.5).

$$U = \frac{1}{2}W\delta = \frac{\tau^2}{4G} \text{ (Volume of wire)} \qquad ...(12.22)$$

**Case II :** Spring under the action of axial couple $M$ :

Due to the axial couple, the rotation of the free end with respect to the fixed end is given by equation (11.11a) as

$$\phi = \frac{Ml}{EI}$$

Therefore strain energy

$$U = \frac{1}{2}M\phi \text{ or } U = \frac{M^2 l}{2EI} \qquad ...(12.23a)$$

Also from equation (11.11*b*)

$$\phi = \frac{64\, M\, D_n}{E\, d^4}$$

and from equation (11.7), $M$ can be expressed in terms of the maximum bending stress $\sigma_b$ in the spring wire as

$$M = \frac{\pi d^3}{32}\sigma_b$$

Thus the expression (12.23*a*) for strain energy reduces to

$$U = \frac{1}{2}\left(\frac{\pi d^3}{32}\sigma_b\right)\left[\frac{64 Dn}{Ed^4}\left(\frac{\pi d^3}{32}\sigma_b\right)\right]$$

$$= \frac{\sigma_b^2}{8E}(\pi Dn)\left(\frac{\pi d^2}{4}\right)$$

$$= \frac{\sigma_b^2}{8E} \text{ (Volume of wire)} \qquad ...(12.23b)$$

## 12.8 Strain Energy of an Open Coil Helical Spring

### 12.8.1

**Case I :** Due to axial load $W$

As proved in article 11.3, the load $W$ will give rise to

(*i*) Twisting moment on wire $-\dfrac{WD}{2}\cos\alpha$

(*ii*) Bending moment on wire $= -\dfrac{WD}{2}\sin\alpha$ (unbending)

Thus with the help of equations (12.13*b*) and (12.15*b*), the total strain energy due to these components can be written as

$$U = \frac{1}{2}W\delta = \left(\frac{WD}{2}\cos\alpha\right)^2 \frac{l}{2GJ} + \left(\frac{WD}{2}\sin\alpha\right)^2 \frac{l}{2EI}$$

$$= \frac{W^2D^2l}{8}\left[\frac{\cos^2\alpha}{GJ} + \frac{\sin^2\alpha}{EI}\right] \qquad ...(12.24)$$

Equation (12.24) also gives the value of deflection as

$$\delta = \frac{WD^2l}{4}\left[\frac{\cos^2\alpha}{GJ} + \frac{\sin^2\alpha}{EI}\right] \qquad ...(12.25)$$

### 12.8.2

**Case II :** Due to axial couple *M*.

A proved in article 11.4, *M* will give rise to

(*i*) Twisting moment on wire = $M \sin\alpha$

(*ii*) Bending moment on wire = $M \cos\alpha$

Thus with the help of equations (12.13*b*) and (12.25*b*), the total strain energy due to these components can be written as

$$U = \frac{1}{2}M\phi = \frac{(M\sin a)^2\, l}{2GJ} + \frac{(M\cos\alpha)^2 l}{2EI}$$

$$= \frac{M^2 l}{2}\left[\frac{\sin^2\alpha}{GJ} + \frac{\cos^2\alpha}{EI}\right] \qquad ...(12.26)$$

Equation (12.26) also gives the value of the rotation $\phi$ of free end of the spring with respect to the fixed end as

$$\phi = Ml\left[\frac{\sin^2\alpha}{GJ} + \frac{\cos^2\alpha}{EI}\right] \qquad ...(12.27)$$

### 12.8.3

**Case III :** Under the combined action of axial load *W* and axial couple *M* :

When the load *W* and couple *M* are applied simultaneously, then referring to article 11.3.

(*a*) Total twisting moment on wire = $\frac{WD}{2}\cos\alpha + M\sin\alpha$

(*b*) Total bending moment on wire = $-\frac{WD}{2}\sin\alpha + M\cos\alpha$

$$\therefore \quad U = \frac{1}{2}\left[\frac{WD}{2}\cos\alpha + \mathrm{M}\sin\alpha\right]^2 \frac{l}{GJ} + \frac{1}{2}\left[-\frac{WD}{2}\sin\alpha + M\cos\alpha\right]^2 \frac{l}{EI} \qquad \ldots(12.28)$$

### 12.8.4 Strain Energy of a Leaf Spring

As proved in equation (11.45*b*).

$$U = \frac{\sigma^2}{6E} \text{ (Volume of spring)}$$

### 12.8.5 Strain Energy of Flat Spiral Spring

From equation (11.54)

$$U = \frac{\sigma_{max}^2}{24E} \text{ (Volume).}$$

## 12.9 Strain Energy Due to a Tri-axial Stress System

Let a triaxial stress system wth $\sigma_1$, $\sigma_2$ and $\sigma_3$ as principal stresses be acting on a rectangular parallepiped of dimensions $l_1 \times l_2 \times l_3$ as shown in Fig. 12.8.

The strains in the respective three directions are

$$\varepsilon_1 = \frac{1}{E}[\sigma_1 - \nu(\sigma_2 + \sigma_3)] \qquad \ldots(12.29a)$$

$$\varepsilon_2 = \frac{1}{E}[\sigma_2 - \nu(\sigma_3 + \sigma_1)] \qquad \ldots(12.29b)$$

$$\varepsilon_2 = \frac{1}{E}[\sigma_3 - \nu(\sigma_1 + \sigma_2)] \qquad \ldots(12.29c)$$

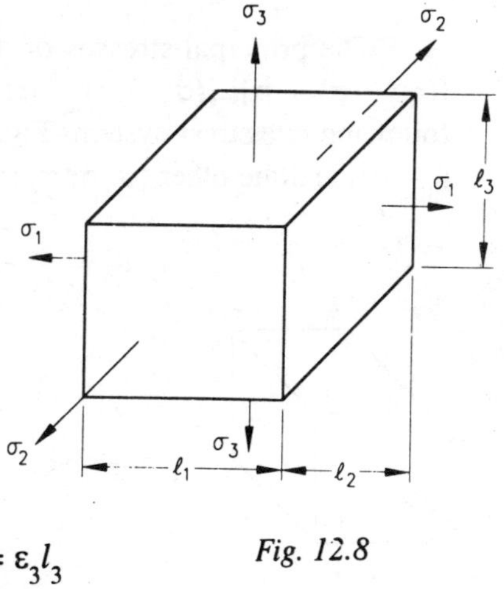

Fig. 12.8

Where $E$ is the Young's modulus and $\nu$ the Poisson's ratio. The extensions in three directions are

$$x_1 = \varepsilon_1 l_1 \; ; x_2 = \varepsilon_2 l_2 \text{ and } x_3 = \varepsilon_3 l_3$$

Thus the work done by each of the forces due to the stresses $\sigma_1$, $\sigma_2$ and $\sigma_3$ will be

$\frac{1}{2}(\sigma_1 l_2 l_3)\, x_1, \frac{1}{2}(\sigma_2 l_3 l_1)\, x_4,$ and $\frac{1}{2}(\sigma_3 l_1 l_2)\, x_3$ respectively.

Then the total strain energy of the block which is equal to the work done on it will be

$$U = \frac{1}{2}(\sigma_1 l_2 l_3)\,\varepsilon_1 l_1 + \frac{1}{2}(\sigma_2 l_3 l_1)\,\varepsilon_2 l_2 + \frac{1}{2}(\sigma_3 l_1 l_2)\,\varepsilon_3 l_3$$

$$= \frac{1}{2} l_1 l_2 l_3 \left[\sigma_1\left\{\frac{\sigma_1}{E} - \frac{\nu}{E}(\sigma_2 + \sigma_3)\right\} + \sigma_2\left\{\frac{\sigma_2}{E} - \frac{\nu}{E}(\sigma_3 + \sigma_1)\right\} + \sigma_3\left(\frac{\sigma_3}{E} - \frac{\nu}{E}(\sigma_1 + \sigma_2)\right)\right]$$

$$= \frac{1}{2E}[(\sigma_1^2 + \sigma_2^2 + \sigma_3^2) - 2\nu(\sigma_1\sigma_2 + \sigma_3\sigma_1)] \times \text{Volume}$$

Therefore energy per unit volume

$$U/\text{volume} = \frac{1}{2E}\left[(\sigma_1^2 + \sigma_2^2 + \sigma_3^2) - 2\nu(\sigma_1\sigma_2 + \sigma_2\sigma_3 + \sigma_3\sigma_1)\right] \quad ...(12.30)$$

## 12.9.1 For the Special Case

When $\sigma_1 = \sigma_2 = \sigma_3 = \sigma$ ; then equation (12.30) reduces to

$$U/\text{Volume} = \frac{3\sigma^2}{2E}(1 - 2\nu) \quad ...(12.31a)$$

$$= \frac{\sigma^2}{2K} \quad [\because E = 3K(1 - 2\nu)] \quad ...(12.31b)$$

## 12.9.2 Shear Strain Energy Due to a Tri-axial Stress System

The principal stresses of the given stress system Fig. 12.8 can be split into $[(\sigma_1 - \sigma) + \sigma]$, $[(\sigma_2 - \sigma) + \sigma]$ and $[(\sigma_3 - \sigma) + \sigma]$ respectively to give rise the following two stress systems Fig. 12.9 : one of principal stresses $([\sigma_1 - \sigma), (\sigma_2 - \sigma), (_3 - \sigma)]$ and the other $[\sigma, \sigma, \sigma]$ superimposed on each other.

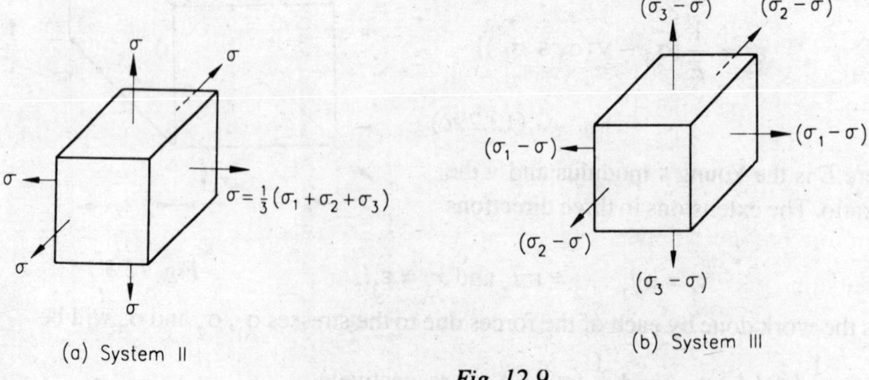


*Fig. 12.9*

Further let $\sigma = \frac{1}{3}(\sigma_1 + \sigma_2 + \sigma_3)$ ...(12.32)

For the given system Fig. 12.8 : from equation (12.29) volumetric strain

$$[\delta V/V]_I = \varepsilon_1 + \varepsilon_2 + \varepsilon_3 \quad \text{(Substituting values from eq. (12.29)}$$

$$= \frac{(\sigma_1 + \sigma_2 + \sigma_3)}{E}(1 - 2\nu) \quad ...(12.33)$$

In the system III Fig. 12.9(*b*) :

The volumetric strain can be found with the help of equation (12.33) by replacing $\sigma_1$, $\sigma_2$ and $\sigma_3$ respectively by $(\sigma_1 - \sigma)$, $(\sigma_2 - \sigma)$ and $(\sigma_3 - \sigma)$ to give

$$\text{Volumetric strain} = \left(\frac{\delta V}{V}\right)_{III} = \frac{1}{E}(1 - 2\nu)\,[(\sigma_1 - \sigma) + (\sigma_2 - \sigma) + (\sigma_3 - \sigma)]$$

$$= \frac{(1 - 2\nu)}{E}(\sigma_1 + \sigma_2 + \sigma_3 - 3\sigma)$$

$$= 0 \quad \text{(from 12.32)} \quad ...(12.34)$$

In the system II Fig. 12.9(*a*) :

The volumetric strain

$$= \left[\frac{\delta V}{V}\right]_{II} = \frac{(\sigma + \sigma + \sigma)}{E}(1 - 2\nu) = \frac{3\sigma}{E}(1 - 2\nu)$$

$$= \frac{(\sigma_1 + \sigma_2 + \sigma_3)}{E}(1 - 2\nu) = \left[\frac{\delta V}{V}\right] \quad \text{(form 12.32)} \quad ...(12.35)$$

= Volumetric strain of given system. [from eq. (12.33)]

Evidently from equations (12.33) through (12.35) the volumetric strain of the given system is same as the volumetric strain of system II. Moreover there is no shear strain energy in system No. II, as in this system due to equal values of principal stresses there will be no shear stress (equation 2.2).

Thus the volumetric strain energy of the given system will be equal to the strain energy of the system number II and its shear strain energy will be equal to the strain energy of system number III. Further in system III there will be no volumetric strain energy as $\left[\frac{\delta V}{V}\right]_{III}$ is zero from equation (12.34).

Therefore per unit volume

Shear strain energy of given system

= Total strain energy of given system

– total strain energy of system II. ...(12.36)

But from equations (12.31*a*) and (12.32)

Volumetric strain energy per unit volume

$$= \text{Strain energy per unit volume of system II.}$$

$$= \frac{3}{2E}(1-2\nu)\left(\frac{\sigma_1+\sigma_2+\sigma_3}{3}\right)^2$$

$$= \frac{(1-2\nu)}{6E}(\sigma_1+\sigma_2+\sigma_3)^2 \qquad ...(12.37)$$

Substituting equations (12.30) and (12.37) in equation (12.36), we get

Shear strain energy per unit volume

$$= \frac{1}{2E}\left[(\sigma_1^2+\sigma_2^2+\sigma_3^2)-2\nu(\sigma_1\sigma_2+\sigma_2\sigma_3+\sigma_3\sigma_1)\right]$$

$$-\frac{(1-2\nu)}{6E}\left[(\sigma_1^2+\sigma_2^2+\sigma_3^2)-2(\sigma_1\sigma_2+\sigma_2\sigma_3+\sigma_3\sigma_1)\right]$$

$$= \frac{1}{6E}\left[(\sigma_1^2+\sigma_2^2+\sigma_3^2)\ (2+2\nu) - (2+2\nu)\times(\sigma_1\sigma_2+\sigma_2\sigma_3+\sigma_3\sigma_1)\right]$$

$$= \frac{(1+\nu)}{6E}\left[2(\sigma_1^2+\sigma_2^2+\sigma_3^2)-2(\sigma_1\sigma_2+\sigma_2\sigma_3+\sigma_3\sigma_1)\right]$$

$$= \frac{1}{12G}\left[(\sigma_1-\sigma_2)^2+(\sigma_2-\sigma_3)^2+(\sigma_3-\sigma_1)^2\right] \qquad ...(12.38)$$

as $E = 2G(1+\nu)$.

## 12.10 Strain Energy and Deflection Due to Shear

In Chapter 4 the deflection of beams was calculated taking into account the effect of bending moments only. In fact due to the change of bending moment, there will be

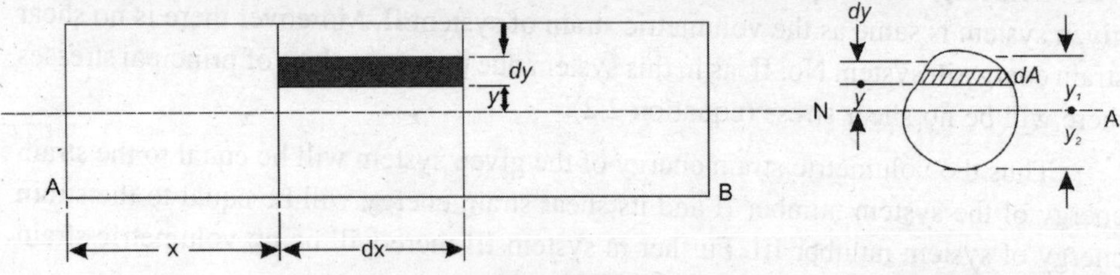


*Fig. 12.10*

shearing force (=$dM/dx$) on each and every section of the beam. Due to this the transverse sections will slip will respect to the adjacent sections and will give rise to what is called the deflection due to shear. To evaluate this the help of the strain energy method is taken. As in equation (12.9) the strain energy due to shear is given by $(\tau^2/2G)\times$ volume. However $\tau$ varies over the section of the beam. For an elemental area $dA$ at a distance $y_0$ from neutral axis and for the length $dx$ of the beam, the shear stress $\tau$ can be taken to be constant Fig. 12.10.

Then for the whole beam $AB$, the shear strain energy is given by

$$U_s = \int_0^l \int_{-y_1}^{y_2} \frac{\tau^2}{2G} dA\, dx \qquad ...(12.39)$$

The integration can be done for particular shape of the section over which the variation of $\tau$ is known.

For the rectangular section, equation (5.5) gives the distribution of $\tau$ in terms of '$b$' and '$d$', the width and depth of the section respectively

i.e.,
$$\tau = \frac{6F}{bd^3}\left(\frac{d^2}{4} - y_0^2\right)$$

Where $F$ = Shearing force over the section

Therefore
$$U_s = \int_0^l \int_{-d/2}^{d/2} \frac{1}{2G} \frac{36F^2}{b^2d^6}\left[\frac{d^4}{16} + y_0^4 - \frac{d^2y_0^2}{2}\right](b\, d\, y_0)\, dx$$

$$= \frac{18}{Gb^2d^6}\int_0^l F^2\left[\frac{d^4y_0}{16} + \frac{y_0^5}{5} - \frac{d^2y_0^3}{6}\right]_{-d/2}^{d/2} dx$$

$$= \frac{18}{Gbd}\int_0^l F^2\left(\frac{1}{16} - \frac{1}{24} + \frac{1}{80}\right) dx$$

or
$$U_s = \frac{3}{5Gbd}\int_0^l F^2\, dx \qquad ...(12.40)$$

Depending upon the type of loading and end conditions of the beam the value of F as function of $x$ can be substituted in equation (12.40) and integration done.

## 12.10.1

### Case I : Cantilever with Point Load $W$ at the Free End

In this case the shearing force throughout the length of the cantilever will be constant and equal to $W$.

That is $F = W$

Therefore from equation (12.40)

$$U_s = \frac{3}{5Gbd}\int_0^l W^2 dx = \frac{3\,W^2 l}{6Gbd} \qquad ...(12.41)$$

But the work done due to $W$, if the deflection due to shear under the load is $\delta_s$, must be $\frac{1}{2}W\delta_s$..

Therefore

$$\frac{1}{2}W\delta_S = \frac{3W^2 l}{6Gbd}$$

or $$\delta_S = \frac{6Wl}{5Gbd} \quad ...(12.42)$$

## 12.10.2

### Case II : Cantilever with uniformly distributed load.

At the section *XX* distant x from A (Fig. 12.11) the load (*w dx*) on a small length *dx* can be considered to be a point load which will produce a deflection due to shear from equation (12.42) = 6(*wdx*)*x*/(5*Gbd*) at *x* as well as at any point towards the right of section *XX* as shown in Fig. 12.11.

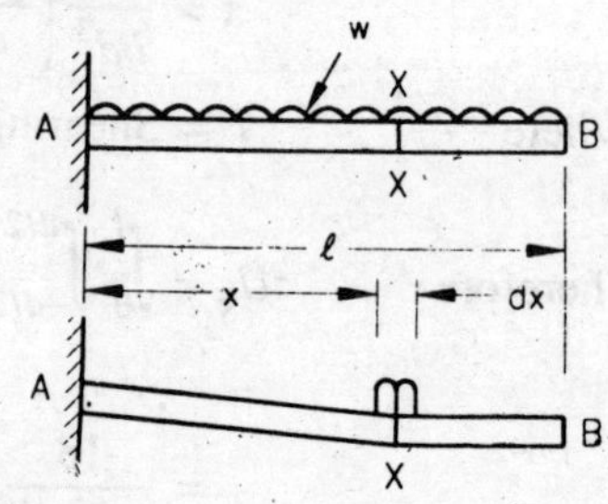

*Fig. 12.11*

Hence the total deflection at the free end

$$(\delta_S)_B = \int_0^l \frac{6\,(wdx)\,x}{5Dbd} = \frac{3wl^2}{5Gbd} \quad ...(12.43)$$

## 12.10.3

### Case III : Simply supported beam with central load *W* :

The magnitude of shearing force over *BC* and *AC* will be $\frac{W}{2}$. Therefore for the entire length

$$F^2 = \frac{W^2}{4}.$$

Then from equation (12.40)

$$U_S = \frac{3}{5Gbd}\int_0^l \frac{W^2}{4}\,dx$$

or $$U_S = \frac{3W^2 l}{20\,Gbd} \quad ...(12.44)$$

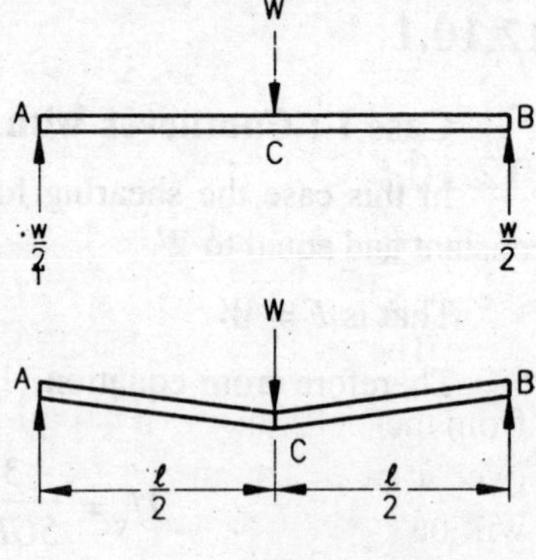

*Fig. 12.12*

Equating this to $\frac{1}{2}W\delta_S$, we get

$$\delta_S = \frac{3Wl}{10\,Gbd} \quad ...(12.45)$$

## Alternative method :

As the shearing force is constant in magnitude throughout the length, the deflection curve due to shear will be as shown in Fig. (12.12). Evidently the deflection due to shear of the portions *AC* and *BC* can be considered separately as if they were cantilevers of lengths $\frac{l}{2}$ with load $\frac{W}{2}$ at the free end. In that case, also, the deflection due to shear from equation (12.42) will be

$$\delta_S = \frac{6\left(\frac{W}{2}\right)\left(\frac{l}{2}\right)}{5\,Gbd} = \frac{3Wl}{10\,Gbd}$$

Which is same as found above in equation (12.45).

## 12.10.4

### Case IV : Simple supported beam with load not at centre :

In this case the magnitude of shearing force over *AC* will be $\left(\frac{Wl_2}{l}\right)$ and that over $CB_1$, $\left(\frac{Wl_1}{l}\right)$. The deflection curve due to shear will be shown in Fig. 12.13. Treating either of the portions *AC* or *CB* as cantilever, the deflection due to shear at *C* will be obtained from equation (12.42) by replacing *W* with the end reaction and *l* by the length of the corresponding portion.

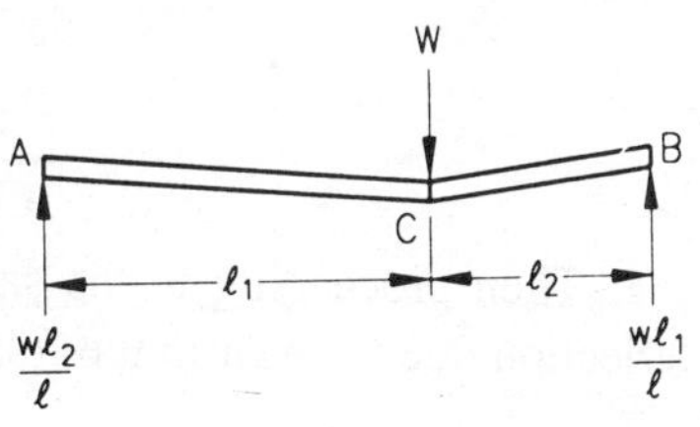


*Fig. 12.13*

*i.e.*
$$(\delta_s)_e = 6\frac{\left(\frac{Wl_2}{l}\right)l_1}{5\,Gbd} = \frac{6Wl_1\,l_2}{5\,Gbdl} \qquad ...(12.46)$$

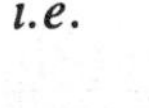

## 12.10.5

### Case V : Simply supported beam with uniformly distributed load :

The deflection at section *XX* distant $x\left[\leq\frac{1}{2}\right]$ from the end A due to a point load ($w\,dx$) spread over a small length $dx$, from equation (12.46) will be

$$= \frac{6\,(wdx)\,x\,(l-x)}{5\,Gbdl}$$

By proportion the deflection at the centre C, due to (wdx) will be

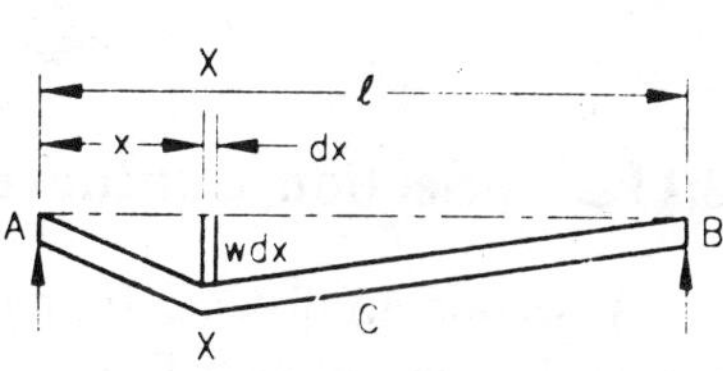


*Fig. 12.14*

$$= \frac{6\,(wdx)\,x\,(l-x)}{5\,Gbdl} \cdot \frac{(l/2)}{(l-x)} = \frac{3\,wx\,dx}{5\,Gbd}$$

Thus the total deflection at $C$ due to shear

$$(\delta_s)_C = 2\int_0^l \frac{3\,wx\,dx}{5\,Gbd} = \frac{3\,wl^2}{20\,Gbd} \qquad ...(12.47)$$

## 12.11 Deflection of I-section Beams Due to Shear

In the case of I-section beams it is customary to treat the shearing force as uniformly distributed over the web and neglect the shearing stress over the flanges.

Thus if $b$ and $d$ are the width and depth respectively of the web of the I-section, then

$$\tau = \frac{F}{bd} \qquad ...(12.48)$$

In that event for a beam of length $l$

$$U_s = \int_0^l \frac{\tau^2}{2G}(bd)\,(dx) = \int_0^l \frac{F^2}{2Gbd}\,dx$$

*i.e.*

$$U_s = \frac{1}{2Gbd}\int_0^l F^2\,dx \qquad ...(12.49)$$

Then proceeding in the same manner as in the article 12.10 we can find the deflection due to shear in the following cases of I-section beams.

### 12.11.1 I-section Cantilever with Load at Free End

$$F = W$$

Therefore from equation (12.49)

$$U_s = \frac{1}{2Gbd}\int_0^l W^2 dx = \frac{W^2 l}{2Gbd} \qquad ...(12.50)$$

Then $$\frac{1}{2}W\delta_s = \frac{W^2 l}{2Gbd}$$

*i.e.*

$$\delta_s = \frac{Wl}{Gbd} \qquad ...(12.51)$$

### 12.11.2 I-Section Cantilever with Uniformly Distributed Load

At section $XX$ distant $x$ from $A$ (Fig. 12.11) deflection due to shear due to a point load $(wdx)$ spread over small length $dx$ will be $[(wdx)x/Gbd]$. Therefore total deflection at free end is

$$\delta_s = \int_0^l \frac{(wdx)x}{Gbd} = \frac{Wl^2}{2Gbd} \qquad ...(12.52)$$

### 12.11.3 I-section Beam Simply Supported at Ends with Load W at the mid Section

As in article 12.10.3

$$F = \pm W/2$$

Then from equation (12.51) and replacing $l$ by $l/2$ and $W$ by end reactions $W/2$, we have

$$(\delta_s)_C = \frac{(W/2)(l/2)}{Gbd} = \frac{Wl}{4\,Gbd} \qquad ...(12.53)$$

### 12.11.4 I-section Beam Simply Supported at Ends with Load not at Centre

Referring to Fig. 12.13 and applying equation (12.51) to the two portions of the beam *i.e.* replacing l by the length of one portion and $W$ by the end reaction on that side, we get

$$(\delta_s)_C = \frac{\left(\frac{Wl_2}{l}\right)l_1}{Gbd} = \frac{Wl_1l_1}{Gbdl} \qquad ...(12.54)$$

### 12.11.5 I-section Beam Simply Supported at Ends with Distributed Load

Again referring to Fig. 12.14 and treating ($wdx$) as a point load when $dx$ is a small length at a distance $x$ $\left[\le \frac{1}{2}\right]$,, from $A$, from equation (12.54) the deflection at $XX$ due to ($wdx$) will be

$$= \frac{w(dx)\,x\,(l-x)}{(Gbdl)}$$

By proportion the deflection at the centre $C$, due to (wxdx) will be

$$= \frac{(wdx)\,x\,.(l-x)}{Gbdl}\cdot\left(\frac{l/2}{l-x}\right)$$

$$= \frac{wx\,dx}{2Gbd}$$

Thus total deflection at $C$ due to shear

$$(\delta_s)_C = 2\int_0^{l/2} \frac{wx\,dx}{2Gbd} = \frac{wl^2}{8\,Gbd} \quad ...(12.55)$$

## 12.12 Castigliano's Theorem

**Statement :** If $U$ is the total strain energy of any structure due to the application of external gradually applied loads $W_1$, $W_2$ ..... $W_n$ at points $P_1$, $P_2$ .... $P_n$ respectively in the directions $P_1X_1$, $P_2X_2$, ..... , $P_nX_n$ and due to gradually applied couples $M_1$, $M_2$,...., $M_m$, at points $Q_1$, $Q_2$, $Q_3$ .... $Q_m$ respectively then the deflections at the points $P_1$, $P_2$,..... $P_n$ in the directions $P_1X_1$, $P_2X_2$,..... $P_nX_n$ are $\partial U/\partial W_1$, $\partial U/\partial W_2$ ......... $\partial_U/\partial W_n$, respectively and the angular rotations of the couples are $\partial U/\partial M_1$, $\partial U/\partial M_2$ ........ $\partial U/\partial M_m$ at the respective points of application,

*Proof for concentrated loads*

Let the displacements or deflections at the points $P_1$, $P_2$ .... $P_n$ in the direction of respective gradually applied loads be $x_1$, $x_2$, $x_3$ ..... $x_n$ and the rotations due to the respective gradually applied couples $M_1$, $M_2$ ..... $M_m$ at their points of application be $\phi_1$, $\phi_2$, ..... $\phi_m$. The total strain energy $U$ is given by

$$U = \left[\frac{1}{2}W_1x_1 + \frac{1}{2}W_2x_2 + ... + \frac{1}{2}W_nx_n\right]$$

$$+\left[\frac{1}{2}M_1\phi_1 + \frac{1}{2}M_2\,\phi_2 + ...\frac{1}{2}M_m\phi_m\right] \quad ...(12.56)$$

Now let the load $W_1$ alone be increased by $\delta W_1$, gradually, then the points $P_1$, $P_2$, $P_3$, ..... $P_n$ will deflect further in the respective directions of loads by amounts $\delta x_1$, $\delta x_2$.... $\delta x_n$ along with the rotations $\delta\phi_1$, $\delta\phi_2$, ....., $\delta\phi_m$ of the points of application of the respective couples. Evidently as the loads $P_2$, $P_3$, .... $P_n$ and couples $M_1$, $M_2$, ..... $M_m$, do not change and continue to act at their respective 'points of application during the time this change in $W_1$ occurs, the increase in external work will then be given by $\delta U$ as

$$\delta U = \left[\left(W_1 + \frac{1}{2}\delta W_1\right)\delta x_1 + W_2\delta x_2 + ... + W_n\delta x_n\right]$$

$$+[M_1\delta\phi_1 + M_2\delta\phi_2 + ...\, M_m\,\delta\phi_m]$$

Neglecting the product of two small quantities it reduces to

$$\delta U = [W_1\delta x_1 + W_2\delta x_2 + ...\, W_n\,\delta x_n]$$

$$+[M_1\delta\phi_1 + M_2\delta\phi_2 + ... + M_m\delta\phi_m] \quad ...(12.57)$$

Again if the loads $(W_1 + \delta W_1)$, $W_2$, $W_2$, .... $W_n$ and couples $M_1$, $M_2$, $M_3$ ..... $M_m$ were all applied gradually starting from zero, the total strain energy would have beem same *i.e.* $(U + \delta U)$ and the corresponding deflections and rotations would be

$(x_1 + \delta x_1), (x_2 + \delta x_2), ..... (x_n + \delta x_n)$ and $(\phi_1 + \delta\phi_1), (\phi_2 + \delta\phi_2), ...... (\delta_m + \delta\phi_m)$. The total strain energy can also be expressed as

$$(U + \delta U) = \left[\frac{1}{2}(W_1 + \delta W_1)(x_1 + \delta x_1) + \frac{1}{2} W_2 (x_2 + \delta x_2) + ... \right.$$
$$\left. \frac{1}{2} + W_n (x_n + \delta x_n)\right] + \left[\frac{1}{2} M_1 (\phi_1 + \delta\phi_1) + \frac{1}{2} M_2 (\phi_2 + \delta\phi_2) + ... \right.$$
$$\left. + \frac{1}{2} M_m (\phi_m + \delta\phi_m)\right] \quad ...(12.58)$$

Subtracting equation (12.56) from equation (12.58) and neglecting the product of two small quantities, we get

$$\delta U = \frac{1}{2} x_1 \delta W_1 + \left[\frac{1}{2} W_1 \delta x_1 + \frac{1}{2} W_2 \delta x_2 + ... \frac{1}{2} W_n \delta x_n\right]$$
$$+ \left[\frac{1}{2} M_1 \delta\phi_1 + \frac{1}{2} M_2 \delta\phi_2 + ... + \frac{1}{2} M_m \delta\phi_m\right]$$

or
$$2\delta U = x_1 \delta W_1 + [W_1 \delta x_1 + W_2 dx_2 + ... + W_n \delta x_n]$$
$$+ [M_1 \delta\phi_1 + M_2 \delta\phi_2 + M_m \delta\phi_m] \quad ...(12.59)$$

Subtracting equation (12.57) from equation (12.59)

$$\delta_U = x_1 \delta W_1$$

or
$$\frac{\delta U}{\delta W_1} = x_1$$

x sub 1>

and in the limiting case when $\delta W_1 \to 0$, we get

$$\frac{\partial U}{\partial W_1} = x_1 \quad ...(12.60)$$

Similarly
$$\frac{\partial U}{\partial W_2} = x_2, \frac{\partial U}{\partial W_3} = x_3, ... \frac{\partial U}{\partial W_n} = x_n \quad ...(12.61)$$

Proceeding in the same way the proof can be extended for the couples to yield

$$\frac{\partial U}{\partial M_1} = \phi_1, ... \frac{\partial U}{\partial M_2} = \phi_2, ... \frac{\partial U}{\partial M_m} = \phi_m \quad ...(12.62)$$

Castigliano's theorem can conveniently be applied to find the deflection of beams due to shearing or bending if we know the total strain energy due to shearing forces or bending moments as the case may be. It can also be used for finding the deflections of curved beams, springs and other similar problems. However following points must be kept in view :

(*i*) Treat all the loads and moments as variables and carry out the partial differentiation. Then afterwards substitute the numerical values of various loads and couples.

(*ii*) In case the deflection or rotation is to be found at a point of the structure where there is no load or couple acting, a dummy load W or dummy moment M may be assumed to act at that point and given a value zero at the end. That is

$$x = \left(\frac{\partial U}{\partial W}\right)_{W=0}$$

and
$$\phi = \left(\frac{\partial U}{\partial M}\right)_{M=0}$$

**Example 12.8.** *A simply supported beam of rectangular section b × d is of length l and carries a uniformly distributed load w over its entire length. Find the deflection due to shear at a point C distant $l_1$ and $l_2$ from the ends respectively. Hence derive the expression for the deflection at the mid section.*

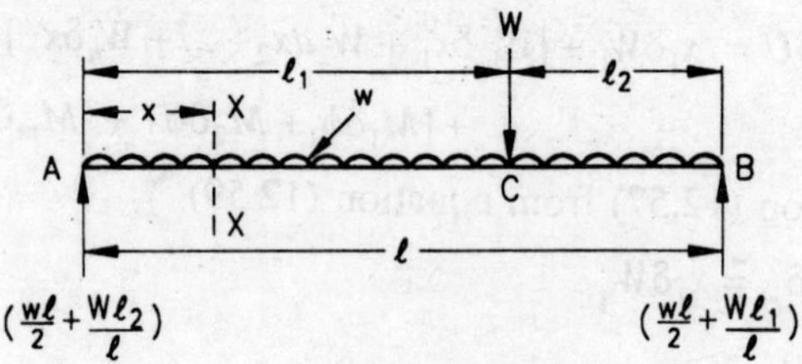


*Fig. 12.15*

**Solution.** (*a*) Assume a point load *W* to be acting at the point *C*. Then the magnitude of shearing force at a section *XX*, distant *x* from *A* is given as

For $o < x < l_1$ $\quad F = \left[\frac{wl}{2} + \frac{Wl_2}{l} - wx\right]$

For $l_1 < x < l$ $\quad F = \left[\frac{wl}{2} + \frac{Wl_2}{l} - wx - W\right]$

Therefore from equation (12.40), the strain energy of the beam due to shearing is

$$U_s = \frac{3}{5Gbd}\left[\int_o^{l_1}\left(\frac{wl}{2} + \frac{Wl_2}{l} - wx\right)^2 dx + \int_{l_1}^{l}\left(\frac{wl}{2} + \frac{Wl_2}{l} - wx - W\right)^2 dx\right]$$

or $$\frac{5Gbd}{3}U_s = \frac{1}{-3w}\left[\left(\frac{wl}{2}+\frac{Wl_2}{l}-wl_1\right)^3-\left(\frac{wl}{2}+\frac{Wl_2}{l}\right)^3\right]$$

$$+1 \quad \frac{1}{-3w}\left[\left(\frac{wl}{2}+\frac{Wl_2}{l}-wl-W\right)^3-\left(\frac{wl}{2}+\frac{Wl_2}{l}-wl_1-W\right)^3\right]$$

Differentiating partially with respect to $W$

$$5\,wGbd\left(\frac{\partial U_S}{\partial W}\right) = -3\left(\frac{wl}{2}+\frac{Wl_2}{l}-wl\right)^2\frac{l_2}{l}$$

$$+3\left(\frac{wl}{2}+\frac{Wl_2}{l}\right)^2\frac{l_2}{l}-3\left(\frac{wl}{2}-\frac{Wl_2}{l}-wl-W\right)^2$$

$$\times\left(\frac{l_2}{l}-1\right)+3\left(\frac{wl}{2}+\frac{Wl_2}{l}-wl_1-W\right)^2\left(\frac{l_2}{l}-1\right)$$

Then putting W = 0, we get

$$5\,wGbd\left(\frac{\partial U_S}{\partial W}\right)_{W=0} = -3\left(\frac{wl}{2}-wl_1\right)^2\frac{l_2}{2}+3\left(\frac{wl}{2}\right)^2\frac{l^2}{l}$$

$$-3\left(\frac{wl}{2}-wl\right)^2\left(\frac{l_2}{l}-1\right)+3\left(\frac{wl}{2}-wl_1\right)^2\left(\frac{l_2}{2}-1\right)$$

or $$5\,wGbd\,(\delta_S)_C = 3\left(\frac{l_2}{l}\right)\left[\left(\frac{wl}{2}-\frac{wl}{2}-wl_1\right)\left(\frac{wl}{2}-\frac{wl}{2}+wl_1\right)\right]$$

$$+3\left(\frac{l_2}{l}-1\right)\left[\left(\frac{wl}{2}-wl_2+\frac{wl}{2}-wl\right)\left(\frac{wl}{2}-wl_1-\frac{wl}{2}+wl\right)\right]$$

$$= 3\frac{l_2}{l}(wl_1)(wl-wl_1)+3\left(\frac{l_2}{l}-1\right)(-wl_1)(wl-wl_1)$$

$$= \frac{3w^2l_1l_2}{l}(l-l_1)+3\left(\frac{l-l_2}{l}\right)w^2\,l_1\,(l-l_1)$$

$$= \frac{3w^2l_1l_2^2}{l}+\frac{3w^2l_1^2l_2}{l} \qquad (\because l_1+l_2=l)$$

$$= \frac{3w^2l_1l_2}{l}(l_2+l_1) = 3w^2l_1l_2$$

or $$(\delta_S)_C = \frac{3wl_1l_2}{5\,Gbd}$$

(*b*) When $l_1 = l_2 = l/2$ *i.e.*, when $C$ is the mid point then

$$(\delta_S)_C = \frac{3}{5}\frac{w\,(l/2)^2}{Gbd} = \frac{3wl^2}{20\,Gbd}$$

which is same as in equation (1.47).

**Example 12.9.** *A rectangular section (b × d) cantilever of length l carries uniformly distributed load spread from free end to the mid section of the cantilever. Find the deflection due to shear at the free end with the help of Castigliano's theorem.*

*(Punjab University)*

**Solution.** Assume a point load $W$ acting at the point $C$. The shearing force at any section $XX$ distant $x$ from $C$ is as follows :

For $o < x < l/2$ $\quad F = (W + wx)$

For $l/2 < x < l$ $\quad F = \left(W + \frac{wl}{2}\right)$

Therefore from equation (12.40), the strain energy of cantilever due to shearing is

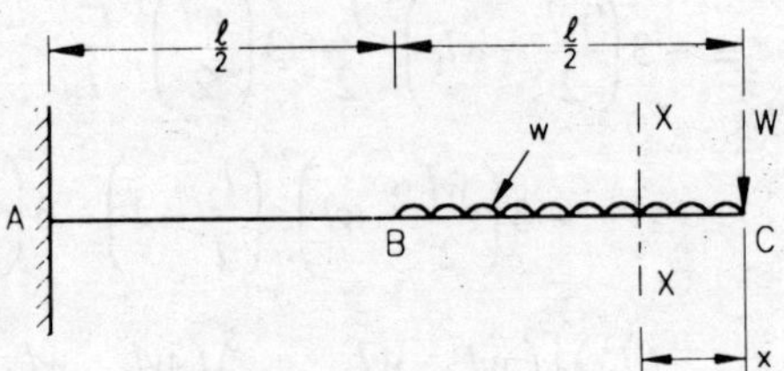


*Fig. 12.16*

$$U_S = \frac{3}{5Gbd}\left[\int_0^{\frac{l}{2}} (W + wx)^2\,dx + \int_{l/2}^{i} \left(W + \frac{wl}{2}\right)^2 dx\right]$$

or
$$\frac{5Gbd}{3}U_S = \left[\frac{(W + wx)^3}{3w}\right]_0^{\frac{l}{2}} + \left(W + \frac{wl}{2}\right)^2 [d]_{l/2}^{l}$$

$$= \frac{1}{3w}\left[\left(W + \frac{wl}{2}\right)^3 - W^3\right] + \left(W + \frac{wl}{2}\right)^2 \frac{l}{2}$$

Differentiating $U_S$ with respect to $W$ partially

$$\frac{5Gbd}{3}\left(\frac{\partial U_S}{\partial W}\right) = \frac{1}{3w}\left[3\left(W + \frac{wl}{2}\right)^2 - 3W^2\right] + 2\left(W + \frac{wl}{2}\right)\frac{l}{2}$$

Then equating $W = 0$, we get

$$\frac{5Gbd}{3}\left(\frac{\partial U_S}{\partial W}\right)_{W=0} = \frac{1}{3w}\left[\frac{3w^2l^2}{4}\right]+\frac{wl^2}{2}$$

$$= \frac{wl^2}{4}+\frac{wl^2}{2} = \frac{3}{4}wl^2$$

*i.e.*
$$(\delta_S)_C = \left(\frac{\partial U_S}{\partial W}\right)_{W=0} = \frac{9wl^2}{20\,Gbd}$$

**Example 12.10.** *In example 12.9 find the deflection at the free end due to bending with the help of Castigliano's theorem.*

**Solution.** Here also we assume a point load $W$ at the point where the deflection due to bending is to be found. The bending moment at any section $XX$, distant $x$ from the free end Fig. 12.16 is given as follows :

For $0 < x < \frac{l}{2}$
$$M = \left(Wx + \frac{wx^2}{2}\right)$$

For $l/2 < x < l$
$$M = Wx + \frac{wl}{2}\left(x - \frac{l}{4}\right)$$

$$= \left(Wx + \frac{wl}{2}x - \frac{wl^2}{8}\right)$$

Therefore from equation (12.15*a*), the strain energy of the cantilever due to bending is

$$U = \frac{1}{2EI}\left[\int_o^{\frac{l}{2}}\left(Wx + \frac{wx^2}{2}\right)^2 dx + \int_{l/2}^{l}\left(Wx + \frac{wlx}{2} - \frac{wl^2}{8}\right)^2 dx\right]$$

or
$$2EIU = \int_o^{\frac{l}{2}}\left(W^2x^2 + Wwx^2 + \frac{w^3x^4}{4}\right)dx$$

$$+\int_{\frac{l}{2}}^{l}\left(W^2x^2 + \frac{w^2l^2x}{4} + \frac{w^2l^4}{64} + Wwl^2x - \frac{Wwl^2x}{4} - \frac{w^3l^3}{8}x\right)dx$$

$$= \left[\frac{W^2x^3}{3} + \frac{Wwx^4}{4} + \frac{w^2x^5}{20}\right]_o^{\frac{l}{2}}$$

$$+\left[\frac{W^2x^3}{3} + \frac{w^2l^2x^3}{12} + \frac{w^2l^4}{64}x + \frac{Wwlx^3}{3} - \frac{Wwl^2x^2}{8} - \frac{w^2l^3x^2}{16}\right]_{\frac{l}{2}}^{l}$$

$$= \left[\frac{W^2l^3}{24} + \frac{Wwl^4}{64} + \frac{w^2l^5}{640}\right] + \left[\frac{W^2}{3}\left(l^3 - \frac{l^3}{8}\right)\right.$$

$$+ \frac{w^2l^2}{12}\left(l^3 - \frac{l^3}{8}\right) + \frac{w^2l^4}{64}\left(l - \frac{l}{2}\right) + \frac{Wwl}{3}\left(l^3 - \frac{l^3}{8}\right)$$

$$\left. - \frac{Wwl^2}{8}\left(l^2 - \frac{l^2}{4}\right) - \frac{w^2l^3}{16}\left(l^2 - \frac{l^2}{4}\right)\right]$$

$$= \left[\frac{W^2l^2}{24} + \frac{Wwl^4}{64} + \frac{w^2l^5}{64}\right]$$

$$+ \left[\frac{7}{24}W^2l^3 + \frac{7}{96}w^2l^5 + \frac{w^2l^5}{128} + \frac{7}{24}Wwl^4 - \frac{3}{32}Wwl^4 - \frac{3}{64}w^2l^5\right]$$

Differentiating partially with respect to W, we get

$$2EI\frac{\partial U}{\partial W} = \frac{Wl^3}{12} + \frac{wl^4}{64} + \frac{7}{12}Wl^3 + \frac{7}{24}wl^4 - \frac{3}{32}wl^4$$

Then equating $W = 0$, we have

$$2EI\left(\frac{\partial U}{\partial W}\right)_{W=0} = \frac{wl^4}{64} - \frac{7}{24}wl^4 - \frac{3}{32}wl^4$$

$$= \frac{41}{192}wl^4$$

or $$\delta_C = \left(\frac{\partial U}{\partial W}\right)_{W=0} = \frac{41wl^4}{384\,EI}$$

**Example 12.11.** *Find the axial deflection and the rotation of the free end of an open coil spring with the help of Castigliano's theorem in the following cases :*

*(a) When only an axial load W acts*

*(b) When only an axial moment M acts.*

**Solution.** (*a*) From equation (12.28) the total strain energy of the spring under the action of both $W$ and $M$ is

$$U = \left(\frac{WD}{2}\cos\alpha - M\sin\alpha\right)^2\frac{l}{2JG} + \left(-\frac{WD}{2}\sin\alpha + M\cos\alpha\right)^2\frac{l}{2EI}$$

*(a) For axial load only, M = 0*

$\therefore$ $$\delta = (\partial U/\partial W)_{M=0}$$

$$= \frac{1}{2JG}.2\left(\frac{WD}{2}\cos\alpha\right)\left(\frac{D}{2}\cos\alpha\right)l + \frac{1}{2EI} \times 2\left(-\frac{WD}{2}\sin\alpha\right)\left(-\frac{D}{2}\sin\alpha\right)l$$

$$= \frac{WD^2 l}{4}\left[\frac{\cos^2\alpha}{GJ} + \frac{\sin^2\alpha}{EI}\right]$$

$$\phi = (\partial U / \partial M)_{M=0}$$

$$= \frac{1}{2JG}.2\left(\frac{WD}{2}\cos\alpha\right)(\sin\alpha)\,l + \frac{1}{2EI} \times 2\left(-\frac{WD}{2}\sin\alpha\right)(\cos\alpha)\,l$$

$$= \frac{WDl}{2}\sin\alpha\cos\alpha\left(\frac{1}{GJ} - \frac{1}{EI}\right)$$

$$= \frac{WDl}{4}\sin 2\alpha\left(\frac{1}{GJ} - \frac{1}{EI}\right)$$

*(b) For axial moment only, W = 0*

$\therefore$ $$\delta = (\partial U/\partial W)_{W=0}$$

$$= \frac{1}{2JG}.2(M\sin\alpha)\left(\frac{D}{2}\cos\alpha\right)l + \frac{1}{2EI} \times 2(M\cos\alpha)\left(-\frac{D}{2}\sin\alpha\right)l$$

$$= \frac{MDl}{2}\sin\alpha\cos\alpha\left[\frac{1}{GJ} - \frac{1}{EI}\right]$$

$$= \frac{MDl}{4}\sin 2\alpha\left(\frac{1}{GJ} - \frac{1}{EI}\right)$$

$$\phi = (\partial U/\partial M)_{W=0}$$

$$= \frac{1}{2JG}2(M\sin\alpha)\sin\alpha\,.\,l + \frac{1}{2EI}2(M\cos\alpha)\cos\alpha\,.\,l$$

$$= Ml\left[\frac{\sin^2\alpha}{GJ} + \frac{\cos^2\alpha}{EI}\right]$$

**Example 12.12.** *A Steel bar of 12 mm diameter and 3 m long is fixed at the top and hangs vertically. At the lower end is fixed a disc on which is placed a helical spring of stiffness 2000 N/mm. A moving mass of 20 kg is allowed to fall freely on the spring through a distance of 50 mm before coming in contact with the top of the spring. Find the instantaneous maximum tensile stress induced in the rod. Take E for rod as 200 GN/m².*

**Solution.** Let the instantaneous maximum deflections of the spring and rod be $\delta$ and $x$ respectively and the corresponding maximum stress induced in the rod due to impact be $\sigma$ N/m². Let $W_e$ be the equivalent static load which when applied on the disc through the spring gradually will deflect the spring and the rod through the same amounts viz., $\delta$ and $x$ respectively and also induce the same stress $\sigma$ in the rod as is being done by the falling mass $M$.

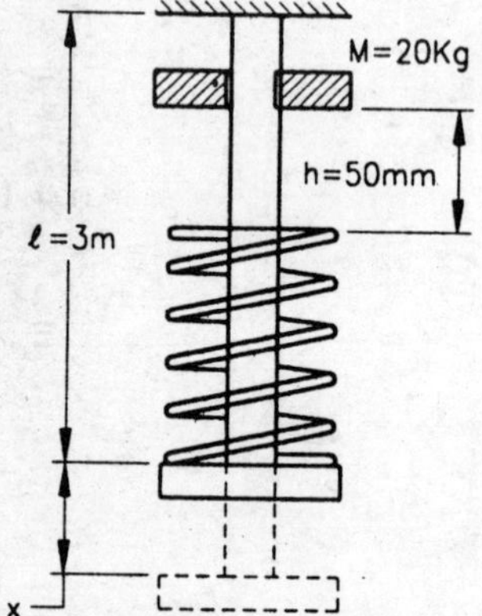


*Fig. 12.17*

$$W_e = \frac{\pi}{4}\left(\frac{12}{1000}\right)^2 \sigma = 113.1 \times 10^{-6}\ \sigma\text{N}$$

$$x = \frac{\sigma}{E} \times l = \frac{3\sigma}{200 \times 10^9}$$

$$= 1.5 \times 10^{-11}\ \sigma\text{m}$$

$$\delta = \frac{W_e}{s} = \frac{113.1 \times 10^{-6}\ \sigma}{2000 \times 1000} = 56.55 \times 10^{-12}\ \sigma\ m$$

Strain energy of spring

$$= \frac{1}{2} W_e \delta$$

$$= \frac{1}{2} \times 113.1 \times 10^{-6}\ \sigma \times 56.55 \times 10^{-12}\ \sigma$$

$$= 3198 \times 10^{-18}\ \sigma^2\ \text{Nm}$$

Strain energy of rod

$$= \frac{\sigma^2}{2E}\ \text{volume}$$

$$= \frac{\sigma^2}{2 \times 200 \times 10^9} \frac{3 \times \pi}{4} \times \left(\frac{12}{1000}\right)^2$$

$$= 84.82 \times 10^{-17}\ \sigma^2\ \text{Nm}$$

Loss of potential energy of falling mass

$$= Mg\ (h + \delta + x)$$

$$= 20 \times 9.81\ [50 \times 10^{-3} + 56.55 \times 10^{-12}\ \sigma + 1.5 \times 10^{-11}\ \delta]$$
$$= 9.81 + 11.1245 \times 10^{-9}\ \delta)\ \text{Nm}$$

Thus equating the strain energy to the loss of instantaneous potential energy, we get

$$3198 \times 10^{-18}\ \sigma^2 + 84.82 \times 10^{-17}\ \sigma^2 = 9.81 + 11.1245 \times 10^{-9}\ \sigma$$

or
$$4.04\sigma^2 - 11.1245 \times 10^6\ \sigma - 9.81 \times 10^{15} = 0$$

From which

$$\sigma = 49.1 \times 10^6\ \text{N/m}^2$$

or
$$= 49.1\ \text{MN/m}^2 \text{ or MPa.}$$

## 12.13 Maxwell's Reciprocal Theorem

**Statement :** When an elastic body is acted upon by two systems of gradually applied point loads $(P_i)_{i\,=\,1,\,n}$ and $(Q_j)_{j\,=\,1\text{ to }m}$ then the work done by the first system of loads $(P_j)_{j\,=\,1\text{ to }m}$ due to respective displacements of the load points of caused by the second system of loads equals the work done by the second system of loads $(Q_j)_{j\,=\,1\text{ to }m}$ due to respective displacements of loads points of second system caused by the first system of loads.

**Proof :** Let the two systems of gradually applied loads $(P_i)_{i\,=\,1\text{ to }n}$ and $(Q_j)_{j\,=\,1\text{ to }m}$ be acting on an elastic body.

Further let the following notations for various quantities as mentioned against each are used.

$(\delta_i)_{i\,=\,1\text{ to }n}$ = Deflections or displacements of the $n$ points of applications of respective loads $(P_i)_{i\,=\,i\,to\,n}$ of the first system of forces.

$(\delta_j)_{j\,=\,1\text{ to }m}$ = Deflections or displacements of $m$ number of points when the loads $(Q_j)_{j\,=\,1\text{ to }m}$ of the second system of forces are acting.

$\left[(\delta_i)_{II}\right]_{i=1\,\text{to}\,n}$ = Additional Deflection of the $n$ points of application of $n$ forces of first system due to the additional action of forces of second system

$\left[(\delta_j)_I\right]_{i=1\,\text{to}\,m}$ = Additional deflections of respective m point of application of $m$ forces of second system due to application of $n$ forces of first system.

$U_I$ = Total strain energy or work done due to n number of forces of first system of forces $(P_i)_{i\,=\,1\text{ to }n}$ acting alone.

$U_{II}$ = Total strain energy or work dine due to $m$ number of second system of forces $(Q_j)_{j\,=\,1\text{ to }m}$ acting alone.

$U$ = Total strain energy or work done on the body when both the systems $(P_i)_{i\,=\,1\text{ to }n}$ and $(Q_j)_{j\,=\,1\text{ to }m}$ are acting on the body.

$U_{I,\,II}$ = Work down on the body due to the application of second system of forces $(Q_j)_{j=1 \text{ to } m}$ when the forces of first system $(P_i)_{i=1 \text{ to } n}$ are already acting due to additional deflection of $n$ points where forces of first system are acting.

$U_{II,\,I}$ = Work done on the body due to the application of n forces of first system $(P_i)_{i=1 \text{ to } n}$ when m forces of second system are already acting due to the defections of the $m$ points where forces of second system are acting.

$\left[(\delta_i)_{II}\right]_{i=1 \, to \, n}$ = Deflection of respective $n$ points of application of forces of first system due to additional application of second system of forces.

$\left[(\delta_j)_I\right]_{j=1 \, to \, m}$ = Deflection of respective $m$ points of application of forces of second system due to the additional application of first system of forces.

The systems of forces $(P_i)_{i=1 \text{ to } m}$ and $(Q_j)_{j=1 \text{ to } m}$ can be applied on the body in two different ways. In the first case the first system of n forces $(P_i)_{i=n}$ is applied first which will deflect the point of application by amounts $(\sigma_i)_{i=1 \text{ to } n}$ and the corresponding work done on the body or the system energy will be given by

$$U_I = \sum_{i=1}^{n} \frac{1}{2} P_i \delta_i \qquad \text{...(12.63)}$$

After the first system, of m forces have done work, second system of $m$ forces are applied. In so doing in the duration all the points of the elastic body are further defecting the $n$ points were $n$ forces of first system are acting will also deflect further but in this duration on these $n$ points the respective loads $(P_i)_{m=1}$ to n will continue to act. Thus the additional work done due to the second system of $m$ forces will be as given below :

$$U_{I,\,II} = U_{II} + \sum_{i=1}^{n} (P_i)_I \, (\delta_i)_{II} \qquad \text{...(12.64)}$$

Thus the total strain energy due to the application of two systems in steps. ∪ is then given by adding equation (12.63) and (12.64)

$$U = U_I + U_{II} + \sum_{i=1}^{n} (P_i)_I \, (\delta_i)_{II} \qquad \text{...(12.65)}$$

Secondly the $m$ forces of system number two arc applied first and after they have done the work on the body, then the $n$ forces of system number one are applied. Due to this additional deflections of the $m$ points when m individual forces are already acting additional work will be done. Thus the total strain energy of the system will be given by

$$(U_A)_{II, I} = U_I + \sum_{j=1}^{m} (Q_j)_{II} (\delta_j)_I \qquad \text{...(12.66)}$$

Therefore the total stream energy due to the applications of two system in steps $U$ is

$$U = U_{II} + (U_A)_{II, I}$$

$$= U_{II} + U_I + \sum_{j=1}^{m} (Q_j)_{II} (\delta_j)_I \qquad \text{...(12.67)}$$

Thus from equation (12.65) and (12.67) we get

$$U_I + U_{II} + \sum_{i=1}^{n} (P_i)_I (\delta_i)_{II} = U_{II} + U_I + \sum_{j=1}^{m} (Q_j)_{II} (\delta_j)_I$$

or
$$\sum_{i=1}^{n} (P_i)_I (\delta_i)_{II} = \sum_{j=1}^{m} (Q_j)_{II} (\delta_j)_I$$

or
$$U_{I, II} = U_{II, I} \qquad \text{...(12.68)}$$

Thus, the work done by the first system of loads due to the deflections or displacements caused by the second system of loads is equal to the work done by the second system of loads due to the deflection or displacements caused by the first system of loads.

**Example 12.13.** *A cantilever AB of length l carries a load $W_1$ at the free end B and another load $W_2$ at a point C distance of which is 'a' the free end. Show that the Maxwells theorem holds good for the points B and C.*

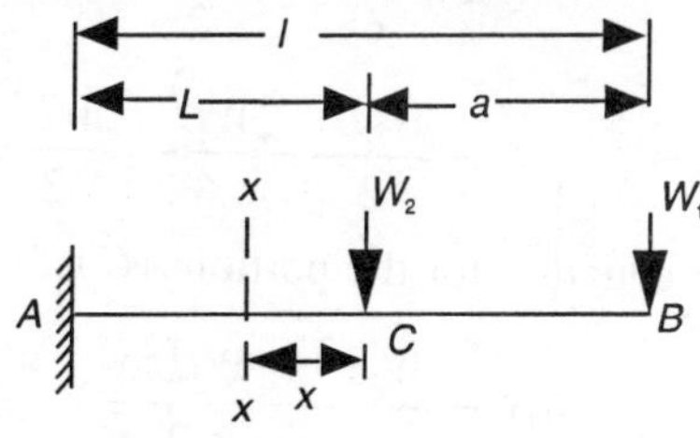


*Fig. 12.18*

**Solution.** Let $W_1$ be the system $I$ and $W_2$ system $II$. In other words $W_1$ is applied first and then $W_2$.

Therefore we consider $W_2$ acting alone and find out the deflection due to $W_2$ at the point $B$. For this we take the point $C$ as origin and take a section $XX$ at a distance $x$ from $C$.

Then
$$EI \frac{d^2 y}{dx^2} = M_x = W_2 x$$

Integrating tit gives

$$EI\frac{dy}{dx} = W_2\frac{x^2}{2} + A$$

But at $x = L \quad \frac{dy}{dx} = 0$

$$\therefore \quad 0 = W_2\frac{L^2}{2} + A \quad \text{or } A = -\frac{W_2L^2}{2}$$

Therefore slope equation becomes

$$EI\frac{dy}{dx} = W_2\frac{x^2}{2} - \frac{W_2L^2}{2} \qquad ...(12.69)$$

Thus at x = 0 *i.e.,* at point C

$$EI\left(\frac{dy}{dx}\right)_c = \frac{-W_2L^2}{2}$$

$$\left[\left(\frac{dy}{dx}\right)_c\right]_{W_2} = -\frac{W_2L^2}{2EI}$$

Integrading equation 12.66, we have

$$EIy = \frac{W_2x^3}{6} - \frac{W_2L^2}{2}x + B \qquad ...(12.70)$$

But $y = 0$ at $x = L$

$$\therefore \quad 0 = \frac{W_2L^3}{6} - \frac{W_2L^3}{2} + B$$

or

$$B = \frac{W_2L^3}{2} - \frac{WL^3}{6} = \frac{WL^3}{3}$$

Therefore deflection equation for the portion $AC$ is

$$EIy = \frac{W_2x^3}{6} - \frac{W_2L^2x}{2} + \frac{W_2L^3}{3} \qquad ...(12.71)$$

At $x = 0$

or

$$(y_c)_{W_2} = \frac{W_2L^3}{3EI}$$

But

$$(y_B)_{W_2} = (y_c)_{W_2} + \left[\left(\frac{dy}{dx}\right)_c\right]_{W_2} \times a$$

$$(y_B)_{W_2} = \frac{W_2(l-a)^3}{3ED} + \frac{W_2\,(l-a)^2}{2EI}a \qquad (\because L = l - a)$$

$$= \frac{W_2 (l-a)^2}{6EI}[2(l-a)+3a]$$

$$\therefore \quad U_{I,\,II} = \frac{W_1 W_2}{6EI}(l-a)^2\,(2l+a) \qquad ...(12.72)$$

Again considering the case when $W_2$ is applied first and then $W_1$. Here we write the diflection equation when $W_1$ is actng alone and find out $(y_c)_{W_1}$ for this in equation (12.69) and (12.71) we reeplace $L$ by $l$ and $W_2$ by $W_1$ we get the requisite equations provided we side by side measure distances from $B$ as origin.

That slope equation is

$$EI\frac{dy}{dx} = W_1\frac{x^2}{2} - \frac{W_1 l^2}{2} \qquad ...(12.73)$$

and diflection equation

$$EIy = \frac{W_1 x^3}{6} - W_1\frac{l^2 x}{2} + \frac{W_1 l^3}{3} \qquad ...(12.74)$$

Thus putting $x = a$, we get

$$EI(y_c) = \frac{W_1 a^3}{6} - W_1\frac{l^2 a}{2} + \frac{W_1 l^3}{3} \qquad ...(12.75)$$

$$= \frac{W_1}{6}(a^3 - 3l^2 a + 2l^3)$$

$$= \frac{W_1}{6}\left[(a^3 - l^2 a) - 2l^2 a + 2l^3\right]$$

$$= \frac{W_1}{6}\left[a(a^2 - l^2) - 2l^2\,(a-l)\right]$$

$$= \frac{W_1}{6}\left[(a-l)\,(a+l)\,a - 2l^2\,(a-l)\right]$$

$$= \frac{W_1}{6}\left[2l^2 - a(a+l)\right]\,(l-a)$$

$$= \frac{W_1}{6}(l-a)\,(l^2 - a^2 + l^2 - al)$$

$$= \frac{W_1}{6}(l-a)\left[(l-a)\,(l+a) + l(l-a)\right]$$

or $$EIy_c = \frac{W_1}{6}(l-a)^2\,(2l+a) \qquad ...(12.76a)$$

i.e., $$(y_c)_{W_1} = \frac{W_1}{6EI}(l-a)^2\,(2l+a) \qquad ...(12.76b)$$

$\therefore$ $$U_{II,\,I} = W_2(y_c)_{W_1} \qquad ...(12.77)$$

Therefore $$U_{II,\,I} = \frac{W_1W_2}{6EI}(l-a)^2\,(2l+a) = U_{I,\,II} \qquad ...(12.78)$$

It shows that the Maxwells thereom holds good.

**Example 12.14.** *A cantilever AB of length l carries a point load W at the free end and a distributed load w per unit length on its entire length. Show by calculations that Maxwell's theorem holds good for the free end B.*

**Solution.** Let first system be load *W* and second system u/d load *w*. Then for the second system acting alone the deflection equation is given as in equation (6.23).

Taking origin at *A* and measuring distance from *A* we have from equation (6.23), two deflection at any section *XY* distant *x* from *A* is given by *y*

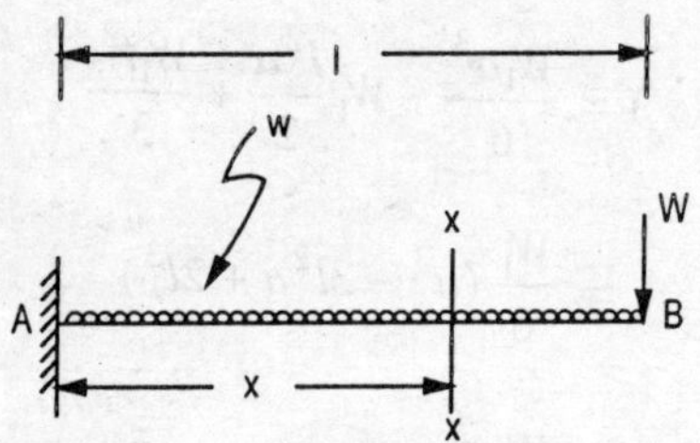


*Fig. 12.19*

$$EIy = \frac{w}{24}(l-x)^4 + \frac{wl^3}{6}x - \frac{wl^4}{24} \qquad ...(12.79)$$

Similarly the deflection at *B* is (from eqn. 6.24)

$$(y_B)_{II} = \frac{wl^4}{8EI} \qquad ...(12.80)$$

Therefore $$U_{II,\,I} = W(y_B)_{II} = \frac{Wwl^4}{8EI} \qquad ...(12.80)$$

Now consider the u/d load *w* as first system and the point load *W* as the second system which is superimposed on the *u/d* load.

When the system, of only one point load $W$ at and $B$ is super imposed on $u/d$ load the work is done by each elemental weight ($wdx$) acting on small length $dx$, situated at a distance $x$. The deflection of each such small load ($wdx$) is given by equation (6.11)

$$EI(y)_W = W\left(\frac{lx^2}{2} - \frac{x^3}{6}\right) \quad ...(12.81)$$

*i.e.*,

$$dU_{I,\,II} = (y)_W(wdx)$$

or

$$U_{I,\,II} = \int_{x=0}^{\ell} dU_{II,\,I}$$

$$= \int_{x=0}^{\ell} (y)_W (wdx) \quad ...(12.82)$$

Therefore from (12.81) and (12.82)

$$U_{I,\,II} = \int_0^{\ell} \frac{W}{EI}\left(\frac{lx^2}{2} - \frac{x^3}{6}\right) wdx$$

$$= \frac{wW}{EI}\left[\frac{lx^3}{6} - \frac{x^4}{24}\right]_0^l$$

$$= \frac{wW}{EI}\left(\frac{lx^3}{6} - \frac{x^4}{24}\right) = \frac{wWl^4}{8EI} \quad ...(12.83)$$

Thus from equation (12.83) and (12.80) we see that

$$U_{\text{I, II}} = U_{II,\,I}$$

which shows that Maxwell's theorem holds good.

**Example 12.15.** *A simply supported beam of length l is supported at its ends and carries a uniformly distributed load w per unit length along with a concentrated load W at the mid span. Show that the Maxwell's reciprocal theorem holds good for the mid section.*

**Solution.** Let the point load $W$ at $C$ be considered as system I.

Then if the distributed load is assumed to be system II. Then due to distributed load alone the deflection of mid point $C$ is given by equation (6.42).

$$(y_c)_w = \frac{5}{384}\frac{wl^4}{EI} \quad ...(12.86)$$

When the system II of distributed load is superimposed work is done by the point load $W$ at $C$ for its movement through distance $(y_c)_w$. The work done is then given by

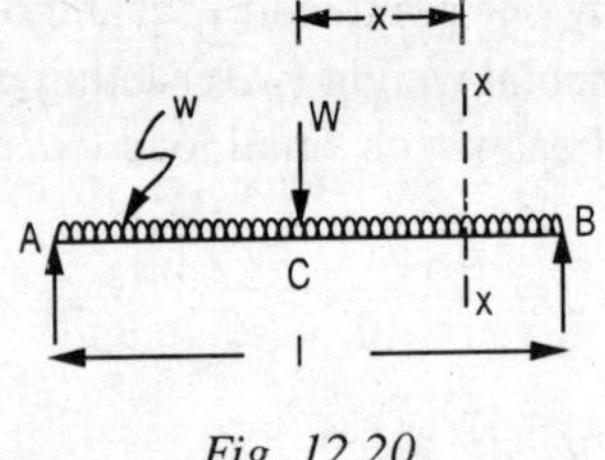


*Fig. 12.20*

$$U_{I,\,II} = W(y_c)_w$$

$$= \frac{5W\,wl^4}{384\,EI} \qquad ...(12.87)$$

Again consider the system of *u/d* load *w* is applied first & then the system of point load W is applied next.

Taking origin at *C* and considering half of the beam *AC* since the other half being symmetrical will have similar deflections, then from equation (6.37) the deflection at section *x* due to point load *W* is given by

$$EI\,(y)_W = \frac{Wl^3}{48} - \frac{W}{2}\left(\frac{lx^2}{4} - \frac{x^3}{6}\right) \qquad ...(12.88)$$

Now the elemential point load (*wdx*) of u/d load system situated at a distance *x*, will deflect though distance $(y)_W$ to result in the work done given by

$$dU_{II,\,I} = (y)_W\,(wdx)$$

Integrating for the entire length

$$U_{II,\,I} = 2\int_0^{l/2} (y)_w\,(wdx)$$

$$= 2\int_0^{l/2} \frac{w}{EI}\left[\frac{Wl^3}{48} - \frac{W}{2}\left(\frac{lx^2}{4} - \frac{x^3}{6}\right)\right] dx$$

$$= \frac{2wW}{48EI}\left[xl^3 - 6l\left(\frac{x^3}{3}\right) + 4\left(\frac{x^4}{4}\right)\right]_0^{\frac{l}{2}}$$

or $$U_{II,\,I} = \frac{wW}{24EI}\left[\frac{l^4}{2} - 2l\left(\frac{l^3}{8}\right) + \frac{l^4}{16}\right]$$

$$= \frac{wWl^4}{384EI}(8 - 4 + 1)$$

or $$U_{II,\,I} = \frac{5}{384}\frac{Wwl^4}{EI} \qquad \text{...(12.89)}$$

From equation (12.87) and (12.89) we see that

$$U_{I,\,II} = U_{II,\,I}$$

Thus the validity of Maxwell's reciprocal theorem for the point $C$ is proved.

**Example 12.16.** *A contilener AB of length l carries a point load W at the free end B. The other system of load consists of a u/d load w spread over the portion. AC when C is at a distance $\frac{l}{2}$ from the fixed end. Verify Maxwell theorem for these two system of loads.*

**Solution.** Let the point load $W$ be the first system and $u/d$ load w the second system.

Due to second system of load *i.e.*, $u/d$ load $w$ the slope and deflection of the cantilever at the point C is obtained with the help of equation (6.28) in which '$a$' is to be replaced by $\frac{l}{2}$.

*i.e.*, $$(y_c)_{II} = \frac{w\left(\frac{l}{2}\right)^4}{8EI} = \frac{wl^4}{128EI}$$

and $$\left[\left(\frac{dy}{dx}\right)_c\right]_{II} = \frac{w\left(\frac{l}{2}\right)^3}{6EI} = \frac{wl^3}{48\,EI}$$

Fig. 12.21

Then $$= (y_C)_{II} + \left(\frac{l}{2}\right)\left(\left(\frac{dy}{dx}\right)_C\right)_{II} \qquad \text{...(12.90)}$$

or $$(y_B)_{II} = \frac{wl^4}{128\,EI} + \frac{l}{2} \times \frac{wl^3}{48\,EI}$$

$$= \frac{wl^4}{384\,EI}(3 + 4)$$

$$= \frac{7\,wl^4}{384\,EI}$$

Therefore $$U_{I,\,II} = \frac{7}{384}\frac{wWl^4}{EI} \qquad \text{...(12.91)}$$

Again let the *u/d* load *w* be the first system and the point load *W* be second system.

Then the deflection equation due *W* for the camtilever can be written with the help of equation (6.11) where at a distance *x* from the fixed end the deflection is

$$EI(y)_I = W\left[\frac{lx^2}{2} - \frac{x^3}{6}\right] \qquad ...(12.92)$$

The wock done due to the deflection of a small load (*wdx*) at distance *x* from the fixed end *A* is

$$(\delta U_{II,\ I}) = (y)_I\, wdx$$

Therefore $$U_{II,\ I} = \int_0^{l/2} w \frac{W}{EI}\left[\frac{lx^2}{2} - \frac{x^3}{6}\right] dx$$

or $$U_{II,\ I} = \frac{wW}{EI}\left[l\left(\frac{x^2}{6}\right) - \frac{x^4}{x}\right]_0^{l/2}$$

$$= \frac{wW}{EI}\left[\frac{l^4}{48} - \frac{l^4}{384}\right]$$

$$= \frac{7wW\,l^4}{384\,ED} \qquad ...(12.97)$$

thus from equation (12.92) and (12.93)

$$U_{I,\ II} = U_{II,\ I'}$$

## PROBLEMS FOR PRACTICE

1. Two bars *A* and *B*, each 0.5 m long are subjected to the same axial blow. *A* is 50 mm diameter over 0.3 m length and 25 mm diameter over 0.2 m length and *B* is 50 mm diameter over 0.2 m length and 25 mm diameter over 0.3 m length. If the material of both the rods is same find the ratio of the maximum stress in A to the maximum stress in *B*. **[Ans.** $\sqrt{14/11}$ ]
2. A bar 10 mm diameter and 3 m long is fixed at the top and hangs vertically. At the lower end is fixed a disc and on this disc a mass of 10 kg. is allowed to fall freely through a distance of 50 mm. Calculate the maximum stress induced in the rod, stating any assumptions made and proving any formula used, $E$ = 200 GN/m$^2$. **[Ans.** 92.5 MN/m$^2$]
3. In the bar and disc arrangement of Problem 2, it is intended to reduce the maximum induced stress in the bar to 50 MN/m$^2$ by providing a coil spring over the disc as a bumper so that the given mass falls through a height of 50 mm before coming

in contact with the upper end of the spring. Find the necessary stiffness of the spring and its instantaneous maximum compression due to the impact.

[**Ans.** 2090 N/mm ; 1.88 mm].

4. A close coiled helical spring consists of 5 mm diameter wire made up into 10 coils of mean diameter 50 mm. A mass of 3 kg. falls from a height of 25 mm before it reaches the top of the spring and compresses it. Calculate the stiffness of the spring, the maximum deflection and shear stress due to falling mass if the modulus of rigidity $G$ = 80 GN/m$^2$. [**Ans.** 5 k N/m ; 24 mm ; 122 MN/m$^2$]

5. A vertical rod, 3 m long, 25 mm in diameter, fixed at the top end, is provided with a collar at the bottom end. A helical spring of mean diameter 240 mm, consisting of 5 coils of 40 mm diameter steel, is mounted on the collar. A sliding mass of 550 kg. is dropped down the rod on the spring. Find the height, measured from the top of the uncompressed spring, from which the weight should be dropped to produce an instantaneous stress of 70 MN/m$^2$ in the rod.

   Find also the maximum shearing stress in the spring. Take $E$ for the rod as 70 GN/m$^2$ and $G$ for the spring as 80 GN/m$^2$. Assume the spring closecoiled, but not quite closed up tight by the action of the falling weight.

   [**Ans.** 223 mm, 328.5 MN/m$^2$]

6. A mass of 20 kg. is dropped from a height of 100 mm on to the centre of a carriage spring which is simply supported at its ends. The spring has 10 steel plates each 50 mm wide and 6 mm thick, the longest plate being 0.75 m. Calculate the maximum instantaneous stress in the plates and the initial radius of curvature of the spring if the impact just flattens the spring. $E$ = 200 GN/m$^2$.

   [**Ans.** 158 MN/m$^2$]

7. Derive the formula for the strain energy stored per unit volume in a circular section rod subjected to pure torsion.

   A closely coiled helical spring of circular section is required to absorb 270 J of energy without the stress exceeding 350 MN/m$^2$. If the steel has a density of 7800 kg./m$^3$, calculate the mass of the spring.

   If the maximum load is 1.8 kN and the mean diameter of the coils is 100 mm, calculate the diameter of the rod. Take G = 75 GN/m$^2$.

   [**Ans.** 5.16 kg ; 10.94 mm]

8. Derive an expression for the strain energy due to bending stored in a beam of length $l$, simply supported at ends carrying a uniformly distributed load $w$ over its middle half portion. [**Ans.** $w^2l^5/(480\ EI)$]

9. A cantilever of uniform rectangular cross-section 50 mm wide × 100 mm deep of length 1 m carries a variable distributed load whose intensity varies linearly, from zero at the free end to 10 kN/m run at the fixed end. Find the strain energy

stored in the beam due to : (*a*) bending (*b*) shearing. Take $E = 200$ GN/m$^2$ and $G = 80$ GN/m$^2$. **[Ans.** 0.00222 ; 0.00757]

10. A hollow shaft subject to a pure torque, attains a maximum shearing stress $\tau$. Given that the strain energy stored per unit volume is $\tau^2/3G$, calculate the ratio of shaft diameters.

    Determine the actual diameters for such a shaft required to transmit 3600 kW at 110 r.p.m. with uniform torque when the energy stored is 20 kJ/m$^3$ of material. $G = 80$ GN/m$^2$. **[Ans.** $\sqrt{3}:1$ ; 298.2 mm ; 172 m]

11. A hollow shaft having the external diameter twice the internal diameter, subjected to a pure torque, attains a maximum shear stress $\tau$. Show that the strain energy stored per unit volume of the shaft is $5\tau^2/16G$.

    Such a shaft is required to transmit 4500 kW at 110 r.p.m. with uniform torque, the maximum stress not exceeding 70 MN/m$^2$. Calculate the shaft diameters and the energy stored per m$^3$ when transmitting this power. $G = 83$ GN/m$^2$.

    **[Ans.** 312 mm ; 156 mm ; 1844 kJ./m$^3$]

12. A straight rod of length $l$ and diameter $d$, is subjected to a twisting moment $T$ about the axis of the rod. A similar rod is wound into the form of a close coiled spring and subjected to a twisting moment $T$, about the axis of the spring. Compare the strain energies in the two cases. **[Ans.** $E : 2G$]

13. In question No. 8 if the section of the beam is $b \times d$ and uniform throughout the length find by Castigliano's theorem the deflection at the mid-section (*i*) due to the bending (*ii*) due to shearing. $\left[\textbf{Ans. } \dfrac{19wl^4}{3072\,EI} ; \dfrac{9}{160}\dfrac{wl^2}{Gbd}\right]$

14. In question No. 9, find by Castigliano's theorem the deflection at the free end due to (*i*) bending (*ii*) shearing. **[Ans.** 0.4 mm ; 0.005 mm]

## Chapter–13

# Theories of Failure

## 13.1 Various Theories of Failure

In a simple tension test a material is said to have failed elastically when the stress crosses the elastic limit of the material, $\sigma_e$. However in a complex stress system elastic failure does not necessarily occur when the greatest principal stress reaches the value of the elastic limit of the material, rather the presence of other two lesser principal stresses cause the material to fail when the value of the maximum principal stress may be much less than $\sigma_e$. Thus when a material is simultaneously acted upon by a complex system of stresses, *i.e.* when it is subjected to three principal stresses, it becomes important to know how the elastic failure will occure. Various theories have been put forward on elastic failure of material under complex system of steady stresses by different people and these theories are associated with their names. It may further be pointed out that these theories relate to such stress, often called the static stress and not necessarily to either suddenly applied loads or to stresses which fluctuate cyclically and are applied repeatedly.

Whatever the system of stress applied to a body, it can always be resolved into three principal stresses (Chapter 18), $\sigma_1$ , $\sigma_2$ and $\sigma_3$ say. In the discussion that is to follow it will be assumed that algebraically $\sigma_1 > \sigma_2 > \sigma_3$.

(*i*) *Maximum Principal Stress Theory (Rankine's Theory)*

Based on the observations of experimental work on brittle metals Rankine proposed the theory that elastic failure occurs when the greatest principal stress reaches the elastic limit value in simple tension test irrespective of the value of other two principal stresses, *i.e.* when

$$\sigma_1 = \sigma_e \qquad ...(13.1)$$

The theory appears to be approximately correct for ordinary cast irons and brittle metals.

(*ii*) *Maximum Principal Strain Theory (St. Venant's Theory)*

According to this theory the elastic failure occurs when the greatest principal strain reaches the strain at elastic limit point in the simple tension test.

Now from equation (2.53), the greatest principal strain when $\sigma_1 > \sigma_2 > \sigma_3 >$ is

$$\varepsilon_1 = \frac{1}{E}[\sigma_1 - \nu(\sigma_2 + \sigma_3)]$$

and in the simple tension test, at elastic limit the strain is $\varepsilon_e/E$. Thus according to this theory the elastic failure will occur when

$$\frac{1}{E}[\sigma_1 - \nu(\sigma_2 + \sigma_3)] = \frac{\sigma_e}{E}$$

or $$\sigma_1 - \nu(\sigma_2 + \sigma_3) = \sigma_e \qquad \text{...(13.2)}$$

Evidently if $\sigma_1$, $\sigma_2$ and $\sigma_3$ are all of the same nature, then from this theory $\sigma_1 > \sigma_e$ but this is not supported by experiments and the theory does not find general support. It overestimates the elastic strength of ductile metals.

*(iii) Maximum Shear Stress Theory (Guest's or Tresca's Theory)*

According to this theory the elastic failure occurs when the greatest shear stress reaches the value of maximum shear stress at the elastic limit in the simple tension test, *i.e.* when

$$\frac{1}{2}(\sigma_1 - \sigma_3) = \frac{1}{2}\sigma_e$$

or $$\sigma_1 - \sigma_3 = \sigma_e \qquad \text{...(13.3)}$$

This theory gives good correlation with results of experiments on ductile materials. In case of any of the three principal stresses is compressive then it must be taken as $\sigma_3$ and its proper sign taken in equation (13.3). That is, the maximum stress difference is to be equated to $\sigma_e$. In the case of two dimentional tensile stress system, third stress must be taken as zero and then the maximum stress difference calculated to equate it to $\sigma_e$.

*(iv) Strain Energy Theory (Beltrami's or Haigh's Theory)*

According to this theory the elastic failure occurs when the energy per unit volume in the strained material reaches the value of the strain energy per unit volume at the elastic limit point in the simple tension test. In other words, according to this theory the maximum energy which a body can store without deforming plastically is constant for that material irrespective of the manner of loading.

With the help of equation (12.3) the mathematical statement of the theory can be written as

$$\frac{1}{2E}\left[(\sigma_1^2 + \sigma_2^2 + \sigma_3^2) - 2\nu(\sigma_1\sigma_2 + \sigma_2\sigma_3 + \sigma_3\sigma_1)\right] = \frac{\sigma_e^2}{2E}$$

or $$(\sigma_1^2 + \sigma_2^2 + \sigma_3^2) - 2\nu(\sigma_1\sigma_2 + \sigma_2\sigma_3 + \sigma_3\sigma_1) = \sigma_e^2 \qquad \text{...(13.4)}$$

Beltrami and Haigh conducted extensive experiments on thick cylinders of ductile metals and the results were in good agreement with this theory. However this theory breaks down for a case when

$$\sigma_1 = \sigma_2 = \sigma_3 = -\sigma$$

as in that case the failure is predicted when

$$\sigma = \sigma_e / \sqrt{3(1-2\nu)}$$

but in fact with this type of loading *i.e.* when there is uniform pressure all round (which is called the hydrostatic pressure) no failure occurs.

(*v*) *Shear Strain Energy Theory (Von Mises Theory) :*

According to this theory the elastic failure occurs where the shear strain energy per unit volume in the stressed material reaches a value equal to the shear strain energy per unit volume at the elastic limit point in the simple tension test.

Now if $U_S$ is the shear strain energy per unit volume of the stressed material, then from equation (12.38)

$$U_S = \frac{1}{12G}\left[(\sigma_1 - \sigma_2)^2 (\sigma_2 - \sigma_3)^2 + (\sigma_3 - \sigma_1)^2\right]$$

But for the simple tension test at the elastic limit point where there is only one principal stress *i.e.* $\sigma_e$, we have the shear strain energy per unit volume given by $U'_s$ as

$$U'_s = \frac{1}{12G}[(\sigma_e - 0)^2 + 0^2 + (0 - \sigma_e)^2]$$

$$= \frac{\sigma_e^2}{6G}$$

Thus for this theory, equating $U_S$ and $U'_S$ we get

$$(\sigma_1 - \sigma_2)^2 + (\sigma_2 - \sigma_3)^2 + (\sigma_3 - \sigma_1)^2 = 2\sigma_e^2 \qquad ...(13.5)$$

This theory is actually a refinement of the one due to Beltrami and is regarded as the one to which conform most of the ductile materials under the actions of various types of loading.

## 13.1.1 Two Dimensional Stress Cases :

Substituting zero for $\sigma_3$ the equations for the above-mentioned five theories reduce to :

(*i*) Maximum Principal Stress Theory :

$$\sigma_1 = \sigma_e \qquad ...(13.6)$$

(*ii*) Maximum Principal Strain Theory :

$$\sigma_1 - \nu\sigma_2 = \sigma_e \qquad ...(13.7)$$

(*iii*) Maximum Shear Stress Theory :

(*a*) For like tensile stresses

$$\sigma_1 > \sigma_2 > 0$$

$$\sigma_1 - 0 = \sigma_e$$

*i.e.*

$$\sigma_1 = \sigma_e \qquad ...(13.8a)$$

(*b*) For unlike stresses $\sigma_1 > 0 > \sigma_2$ ($\sigma_1$ is Tensile and $\sigma_2$ compressive)

$$\sigma_1 - \sigma_2 = \sigma_e \qquad ...(13.8b)$$

and when $\sigma_2$ is tensile and $\sigma_1$ compressive *i.e.*

$$\sigma_2 > 0 > \sigma_1$$

$$\sigma_2 - \sigma_1 = \sigma_e \qquad ...(13.8c)$$

(*iv*) Strain Energy Theory :

$$\sigma_1^2 + \sigma_2^2 - 2\nu\sigma_1\sigma_2 = \sigma_e^2 \qquad ...(13.9)$$

(*v*) Shear Strain Energy Theory :

$$(\sigma_1 - \sigma_2)^2 + (\sigma_2 - 0)^2 + (0 - \sigma_1{}^2) = 2\,\sigma_e^2$$

or

$$\sigma_1^2 + \sigma_2^2 - \sigma_1\,\sigma_2 = \sigma_e^2 \qquad ...(13.10)$$

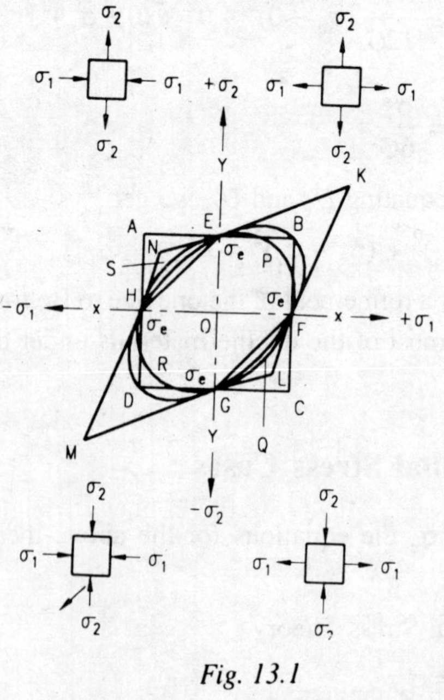

*Fig. 13.1*

The results of equations (13.6) through (13.10) are plotted in Fig. 13.1. Along the $X$-axis is plotted $\sigma_1$ and along $Y$-axis $\sigma_2$. The sign convention being : tensile $\sigma_1$ and $\sigma_2$ are taken to the right and upwards while compressive $\sigma_1$ and $\sigma_2$ to the left and downward respectively.

(*i*) In the case of maximum principal stress theory the limitations within which $\sigma_1$ and $\sigma_2$ must lie for non-failure, is the square $ABCD$, where

$$OF = OE = OM = OG = \sigma_e.$$

(*ii*) For the maximum strain theory the boundaries are straight lines as under

(*a*) When $\sigma_1$ is tensile and $\sigma_1 > \sigma_2$

$$\sigma_1 - \nu\sigma_2 = \sigma_e$$

It is represented by $KFL$

(*b*) When $\sigma_2$ is tensile and $\sigma_2 > \sigma_1$

$$\sigma_2 - \nu\sigma_1 = \sigma_e$$

It is represented by $KEN$

(*c*) When $\sigma_1$ is compressive and $\sigma_1 > \sigma_2$

$$\sigma_1 - \nu\sigma_2 = \sigma_e$$

It is represented by $NHM$

(*d*) When $\sigma_2$ is compressive and $\sigma_2 > \sigma_1$

$$\sigma_2 - \nu\sigma_1 = \sigma_e$$

It is represented by $LGM$

(*iii*) In the case of maximum shear stress theory we have for like stresses

$$\sigma_1 = \sigma_e \qquad (\text{for } \sigma_1 > \sigma_2)$$

and

$$\sigma_2 = \sigma_e \qquad (\text{for } \sigma_2 > \sigma_1)$$

That is in the first and third quadrants the boundary lines coincide with $FBE$ and $HDG$ respectively.

However for unlike stresses (*i.e.* for second and fourth quadrants)

$$\sigma_1 - \sigma_2 = \sigma_e \qquad (\sigma_1 > 0 > \sigma_2)$$

and

$$\sigma_2 - \sigma_1 = \sigma_e \qquad (\sigma_2 > 0 > \sigma_1)$$

These equations are represented by straight lines $FG$ and $EH$.

(*iv*) The strain energy theory is represented by an ellipse $PFQGRSE$ with its axes inclined at 45° to $XX$ and $YY$.

(*v*) The shear strain energy theory is likewise represented by an other ellipse $EBFGDHE$ with its axes also inclined at 45° to $XX$ and $YY$.

From the Fig. 13.1 it is evident that the results given by the maximum strain theory specially for the like stresses are very much away from those predicted by other theories. Secondly, the boundary representing the maximum principal stress theory for unlike stress is still farther away from other theories but in the case of like stresses is nevertheless very much near the other theories rather it coincides with the maximum shear stress theory. Thirdly the boundaries representing the maximum shear stress theory and the last two theories connected with strain energies are generally similar.

**Example 13.1.** *At a point in a steel member the major principal stress is 200 MN/m², and the minor principal stress is compressive. If the tensile yield point of the steel is 250 MN/m², find the value of the minor principal stress at which yielding will commence, according to each of the following criteria of failure : (a) maximum shearing stress ; (b) maximum total strain energy ; (c) maximum shear strain energy. Poisson's ratio = 0.28.*

*Prove any formula used for (b) and (c).*

**Solution.** (*a*) For unlike principal stresses we have from equation (13.8*b*)

$$\sigma_1 - \sigma_2 = \sigma_e$$

*i.e.*

$$\sigma_2 = \sigma_1 - \sigma_e = 200 - 250$$

$$= 50 \text{ M/m}^2 \text{ (Comp.)}$$

(*b*) From equation 13.9

$$\sigma_1^2 + \sigma_2^2 - 2\nu\,\sigma_1\sigma_2 = \sigma_e^2$$

or

$$(200)^2 + \sigma_2^2 - 2 \times .28 \times 200\,\sigma_2 = (250)^2$$

or

$$\sigma_2^2 - 112\sigma_2 - 22500 = 0$$

*i.e.*

$$\sigma_2 = \frac{1}{2}\left[112 - \sqrt{(112)^2 + 4 \times 22500}\right]$$

(only – ve sign is taken as $\sigma_2$ is comp.)

$$= 104.5 \text{ MN/m}^2 \text{ (Comp.)}$$

(*c*) From equation (13.10)

$$\sigma_1^2 + \sigma_2^2 - \sigma_1\sigma_2 = \sigma_e^2$$

or

$$(200)^2 + \sigma_2^2 - 200\,\sigma_2 = (250)^2$$

or

$$\sigma_2^2 - 200\sigma_2 - 22500 = 0$$

or

$$\sigma_2 = \frac{1}{2}\left[200 - \sqrt{(200)^2 + 4 \times 22500}\right]$$

$$= 80.1 \text{ MN/m}^2 \text{ (Comp.)}$$

**Example 13.2.** *A shaft of 100 mm diameter is subjected to a bending moment of 5 kNm. Find the value of the maximum torque which can be applied to the shaft for each of the following conditions : (a) maximum direct stress not to exceed 120 MN/m²: (b) maximum shearing stress not to exceed 60 MN/m² ; (c) maximum shear strain energy per unit volume not to exceed that induced by simple shear stress of 80 MN/m².*

**Solution.** Let $\sigma_b$ be the maximum bending stress induced due to *B.M.* then

$$\sigma_b = \frac{32M}{\pi d^3} = \frac{32 \times 5000}{\pi \times (0.1)^3}$$

$$= 5.1 \times 10^7 \text{ N/m}^2 \text{ or Pa}$$

Let the torque be $T$ Nm corresponding to which the maximum shear stress be $\tau$. Then

$$\tau = \frac{16T}{\pi d^3} = \frac{16}{\pi \times (0.1)^3} T = 5.1 \times 10^3\ T\ \ N/m^2$$

The two principal stresses are

$$\sigma_{1,2} = \frac{1}{2}\sigma_b \pm \sqrt{\left(\frac{\sigma_b}{2}\right)^2 + \tau^2}$$

$$= 2.55 \times 10^7 \pm \sqrt{(2.55 \times 10^7)^2 + (5.1 \times 10^3)^2\, T^2}$$

*i.e.*

$$\sigma_1 = 2.55 \times 10^7 \pm \sqrt{6.5 \times 10^{14} + 2.6 \times 10^7\, T^2}\,\text{N/m}^2$$

and

$$\sigma_2 = 2.55 \times 10^7 - \sqrt{6.5 \times 10^{14} + 2.6 \times 10^7 T^2}\ \text{N/m}^2$$

(*a*) Maximum direct stress

$$\sigma_1 = 120 \text{ MN/m}^2$$

$$\therefore \quad 120 \times 10^6 = 2.55 \times 10^7 + \sqrt{6.5 \times 10^{14} + 2.6 \times 10^7\, T^2}$$

or

$$(94.5 \times 10^6)^2 = 6.5 \times 10^{14} + 2.6 \times 10^7\, \text{T}^2$$

or

$$T^2 = \frac{(89.3 - 6.5)10^{14}}{2.6 \times 10^7} = \frac{82.8}{2.6} \times 10^7$$

$$T = 1.785 \times 10^4 \text{ N-m or } 17.85 \text{ kNm.}$$

(*b*)

$$\tau_{max} = 60 \times 10^6$$

$$= \sqrt{\left(\frac{\sigma_b}{2}\right)^2 + \tau^2} = \sqrt{6.5 \times 10^{14} + 2.6 \times 10^7\, T^2}$$

or

$$3600 \times 10^{12} = 6.5 \times 10^{14} + 2.6 \times 10^7\, \text{T}^2$$

$$T^2 = \frac{29.5 \times 10^{14}}{2.6 \times 10^7}$$

*i.e.* $T = 1.065 \times 10^4$ Nm or 10.65 kNm

(*c*) In the present case shear strain energy per unit volume is given by

$$U_s/\text{vol} = \frac{1}{12G}[(\sigma_1 - \sigma_2)^2 + (\sigma_2 - 0)^2 + (0 - \sigma_1)^2]$$

$$= \frac{1}{12G}\left[(\sigma_1 - \sigma_2)^2 + \left\{(\sigma_1 + \sigma_2)^2 - 2\sigma_1 \sigma_2\right\}\right]$$

Now $(\sigma_1 - \sigma_2) = 2\tau_{max} = 2\sqrt{\left(\frac{\sigma_b}{2}\right)^2 + \tau^2}$

and $(\sigma_1 + \sigma_2) = \sigma_b$

$$(\sigma_1\sigma_2) = \left(\frac{\sigma_b}{2} + \tau_{max}\right) \times \left(\frac{\sigma_b}{2} - \tau_{max}\right) = \left(\frac{\sigma_b^2}{4} - \tau_{max}^2\right)$$

$$= \frac{\sigma_b^2}{4} - \left(\frac{\sigma_b^2}{4} + \tau^2\right) = -\tau^2$$

$$\therefore \quad U_s/\text{vol} = \frac{1}{12G}\left[\left\{2\sqrt{\left(\frac{1}{2}\sigma_b\right)^2 + \tau^2}\right\}^2 + (\sigma_1 + \sigma_2)^2 - 2\sigma_1\sigma_2\right]$$

$$= \frac{1}{12G}\left[4\left\{\left(\frac{1}{2}\sigma_b\right)^2 + \tau^2\right\} + (\sigma_b)^2 - 2\left\{\left(\frac{\sigma_b^2}{2^2}\right) - \left(\frac{\sigma_b^2}{2^2} + \tau^2\right)\right\}\right]$$

$$= \frac{1}{12G}[2\sigma_b^2 + 6\tau^2]$$

Also the shear strain energy per unit volume for pure shear case

$$= \frac{1}{2G}(80 \times 10^6)^2$$

Equating these two and substituting values of $\sigma_b$ and $\tau$

$$= \frac{1}{12G}[2(5.1 \times 10^7)^2 + 6\,(5.1 \times 10^3\ T)^2]$$

$$= \frac{1}{2G}(80 \times 10^6)^2$$

or $15.6 \times 10^7\ T^2 = 384 \times 10^{14} - 52 \times 10^{14}$

$$T^2 = \frac{332 \times 10^{14}}{15.6 \times 10^7}$$

*i.e.* $T = 1.46 \times 10^4$ Nm or 14.6 kN-m

**Example 13.3.** *A bending moment M applied to a solid shaft carries a maximum direct stress σ at elastic failure. Determine the numerical relationships between M and a twisting moment T which, acting alone on the shaft, will produce elastic failure, according to each of the following theories of failure : (a) maximum principal stress ; (b) maximum principal strain ; (c) maximum strain energy ; (d) maximum shear stress; (e) shear strain energy. Poisson's ratio = 0.30.*

**Solution.** The stress at elastic limit is given as and is related to $M$ as

$$\sigma = \frac{32M}{\pi d^3}$$

When the torque $T$ is acting alone on the shaft, then the shear stress on the cross-sectional plane is

$$\tau = \frac{16T}{\pi d^3}$$

Moreover in this pure shear case the two principal stresses $\sigma_1$ and $\sigma_2$ will be (Chapter 2)

$$\sigma_1 = -\sigma_2 = \frac{16T}{\pi d^3}$$

Thus for this case the three principal stresses in descending order are

$$\frac{16T}{\pi d^3},\ 0,\ \frac{-16T}{\pi d^3}$$

(*a*) For the maximum principal stress theory

$$\sigma_1 = \sigma$$

*i.e.* $$\frac{16T}{\pi d^3} = \sigma = \frac{32M}{\pi d^3}$$

or $$T = 2M.$$

(*b*) Maximum principal strain theory : from equation (13.7)

$$\sigma_1 - \nu\sigma_2 = \sigma$$

or $$\frac{16T}{\pi d^3} - 0.3\left(\frac{-16T}{\pi d^3}\right) = \frac{32M}{\pi d^3}$$

or $$\frac{16T}{\pi d^2} \times 1.3 = \frac{32M}{\pi d^3}$$

*i.e.* $$T = \frac{2}{1.3} M = 1.54\ M$$

(*c*) Maximum strain energy theory : From equation (13.9)

$$\sigma_1^2 + \sigma_2^2 - 2\nu\sigma_1\sigma_2 = \sigma^2$$

or $$2\sigma_1^2 + 2\nu\sigma_1^2 = \sigma^2 \qquad (\because\ \sigma_2 = -\sigma_1)$$

or $$2 \times \left(\frac{16T}{\pi d^3}\right)^2 (1 + \nu) = M^2\left(\frac{32T}{\pi d^3}\right)^2$$

or $$2T^2\ (1 + .3) = 4\ M^2$$

*i.e.* $$T = \sqrt{\frac{4}{2.6}}\,M = 1.24M.$$

(*d*) Maximum shear stress theory :

As the principal stresses are unlike in nature, we have from equation (13.8*b*)

$$\sigma_1 - \sigma_2 = \sigma$$

or $$\frac{16T}{\pi d^3} + \frac{16T}{\pi d^3} = \frac{32M}{\pi d^3}$$

or $$T = M.$$

(*e*) Shear strain energy theory : From equation (13.10)

$$\sigma_1^2 + \sigma_2^2 - \sigma_1\sigma_2 = \sigma^2$$

or $$2\sigma_1^2 + \sigma_1^2 = \sigma^2$$

or $$3\left(\frac{16T}{\pi d^3}\right)^2 = \left(\frac{32M}{\pi d^3}\right)^2$$

or $$T = \sqrt{\frac{4}{3}}\,M = 1.155\,M.$$

*Example 13.4. A cast iron cylinder has outside and inside diameters of 200 mm and 125 mm. If the ultimate tensile strength of the cast iron is 150 MN/m², find, according to each of the following theories, the internal pressure which would cause rupture. (a) Maximum principal stress theory, (b) Maximum strain theory and (c) Maximum strain energy theory.*

*Assume no longitudinal stress in the cylinder. $\nu = 0.25$.*

**Solution.** If the internal pressure is $p_1$, then from equations (10.12) and (10.13) the maximum radial and circumferential stresses are given at inner radius $r_1$ as

$$\sigma_r = p_1 \text{ (Com.)}$$

*i.e.* $$\sigma_2 = \sigma_r = -p_1$$

and $$\sigma_c = \frac{p_1\,(r_2^2 + r_1^2)}{(r_2^2 - r_1^2)} \text{ Tensile}$$

*i.e.* $$\sigma_1 = \sigma_c = p_1\left(\frac{d_2^2 + d_1^2}{d_2^2 - d_1^2}\right) = \frac{p_1\,(200^2 + 125^2)}{(200^2 - 125^2)}$$

$$= 2.29\,p_1 \text{ (Tensile)}$$

(*a*) Maximum principal stress theory :

$$\sigma_1 = 2.29\,p_1 = 150$$

or $$p_1 = \frac{150}{2.29} = 65.7 \text{ MN/m}^2$$

(*b*) Maximum strain theory : From equation (13.7)

$$\sigma_1 - \nu\sigma_2 = \sigma$$

or $$2.29\, p_1 - .25\,(-\,p_1) = 150$$

or $$2.54\, p_1 = 150$$

*i.e.* $$p_1 = \frac{150}{2.54} = 59.2 \text{ MN / m}^2$$

(*c*) Maximum strain energy theory : From equation (13.9)

$$\sigma_1^2 - \sigma_2^2 - 2\nu\sigma_1\sigma_2 = 150$$

or $$(2.29\, p_1)^2 + (-\,p_1)^2 - 2 \times .25\,(2.29\, p_1)\,(-\,p_1) = (150)^2$$

or $$7.41\, p_1^{\;2} = (150)^2$$

*i.e.* $$p_1 = 55.2 \text{ MN/m}^2.$$

**Example 13.5**. *In a two-dimensional stress system the direct stresses on two mutually perpendicular planes are σ and 120 MN/m². In addition these planes carry a shear stress of 40 MN/m². Find the value of σ at which the shear strain energy is least. If the failure occurs at this value of the shear strain energy, estimate the elastic limit of the material in simple tension. Take the factor of safety on elastic limit as 3.*

**Solution.** The two principal stresses are given by

$$\sigma_{1,2} = \frac{120 + \sigma}{2} \pm \sqrt{\left(\frac{120 - \sigma}{2}\right)^2 + 40^2}$$

*i.e.* $$\sigma_1 = \left(60 + \frac{\sigma}{2}\right) + \tau_{max}$$

and $$\sigma_2 = \left(60 + \frac{\sigma}{2}\right) - \tau_{max}$$

where $$\tau_{max} = \sqrt{\left(60 - \frac{\sigma}{2}\right)^2 + 40^2}$$

and $$\sigma_1^2 + \sigma_2^2 = 2\left[\left(60 + \frac{\sigma}{2}\right)^2 + \tau_{max}^2\right]$$

Now the shear strain energy per unit volume $U$ is given by

$$U = \frac{1}{12G}\left[(\sigma_1 - \sigma_2)^2 + (\sigma_2 - 0)^2 + (0 - \sigma_1)^2\right]$$

$$= \frac{1}{12G}\left[(2\tau_{max})^2 + (\sigma_1^2 + \sigma_2^2)\right]$$

$$= \frac{1}{12G}\left[4\tau_{max}^2 + 2\left\{\left(60 + \frac{\sigma}{2}\right)^2 + \tau_{max}^2\right\}\right]$$

$$= \frac{1}{6G}\left[3\tau_{max}^2 + \left(60 + \frac{\sigma}{2}\right)^2\right]$$

$$= \frac{1}{6G}\left[3\left\{\left(60 - \frac{\sigma}{2}\right)^2 + 40^2\right\} + \left(60 + \frac{\sigma}{2}\right)^2\right]$$

*i.e.* $$U = \frac{1}{6G}[\sigma^2 - 120\sigma + 19200]$$

For minimum $U$,

$$\frac{dU}{d\sigma} = 0$$

*i.e.* $$2\sigma - 120 = 0$$

or $$\sigma = 60 \text{ MN/m}^2$$

Then $$\tau_{max} = \sqrt{(60 - 30)^2 + 40^2} = 50$$

$$\sigma_1 = 90 + 50 = 140$$

$$\sigma_2 = 90 - 50 = 40$$

(*b*) For failure at this level of shear strain energy, we have from equation (13.10)

$$\sigma_1^2 + \sigma_2^2 - \sigma_1\sigma_2 = \left(\frac{1}{3}\sigma_e\right)^2$$

or $$(140)^2 + (40)^2 - 140 \times 40 = \frac{\sigma_e^2}{9}$$

or $$\frac{\sigma_e^2}{9} = 15600$$

i.e. $$\sigma_e = 374.7 \text{ MN/m}^2.$$

## PROBLEMS FOR PRACTICE

1. At a point in stressed material the direct stresses on two perpendicular planes are 140 MN/m$^2$ tension and 90 MN/m$^2$ compression respectively, and the shearing stress on these planes is $\tau$. The yield stress for the material is 250 MN/m$^2$. Find the value of $\tau$ at which failure may be expected, according to each of the following theories of failure :

(*a*) maximum principal stress theory :

(*b*) maximum shearing stress theory ;

(*c*) maximum shear strain energy theory. [**Ans.** 193.4 ; 49 ; 86 $MN/m^2$]

2. A torque of 1.5 kNm is transmitted by a cylindrical tube 100 mm external diameter and of uniform thickness 2.5 mm. If the elastic limit of the material under simple tension is 250 $MN/m^2$, calculate the factor of safety when the criterion of failure is

   (*a*) maximum shearing stress,

   (*b*) maximum shear strain energy. [**Ans.** 3.04 ; 3.51]

3. A circular shaft 0.1 m diameter is subjected to combined bending and twisting moments, the bending moment being three times the twisting moment. If the direct tensile yield point of the material is 350 $MN/m^2$ and the factor of safety on yield is to be 4, calculate the allowable twisting moment by the following three theories of elastic failure :

   (*a*) maximum principal stress theory ;

   (*b*) maximum shearing stress theory ;

   (*c*) shear strain energy theory. [**Ans.** 2790 ; 2715 ; 2755 Nm]

4. A direct tension test on a specimen of steel gave elastic breakdown at 300 $MN/m^2$. A shaft made of this material is 50 mm diameter. Determine according to the following theories the torque required to produce elastic breakdown when the shaft also carries a bending moment of 2.5 KNm :

   (*a*) Maximum principal stress theory,

   (*b*) Maximum total strain energy theory. Poisson's ratio = 0.3.

   [**Ans.** 4.166 KNm ; 3.35 kNm]

5. Explain clearly what is meant by the theory of elastic failure

   (*a*) by maximum principal stress,

   (*b*) by maximum shear strain energy.

   A steel drum 0.6 m diameter is required to hold gas under a pressure of 3.5 $MN/m^2$. Calculate the thickness according to both the above hypothesis if the material used has a tensile working strength of 70 $MN/m^2$. [**Ans.** 15 mm ; 13 mm]

## Chapter–14

# Bending of Curved Bars

## 14.1 Bars of Initial Large Radius of Curvature

In Chapter 4 the bending equation was derived for the simple bending case where along with other fundamental assumptions the beam was taken to be initially straight. Theoretically the bending equation thus derived cannot be applied to the beams which are initially not straight. However this may be applied, without much impairing the exactness of the results, to cases where the radius of curvature of the central longitudinal axis is initially large in comparison with the dimensions of the section of beam, but in this case *M*/*EI* will represent the change in the value of the curvature. Thus if $R_1$ is the initial radius of curvature of the longitudinal axis of the beam $R_2$ its final radius of curvature after the action of bending moment *M* (reckoned position if it increases the curvature) then equations (4.1) and (4.2) are modified as

$$\varepsilon = y\left(\frac{1}{R_2} - \frac{1}{R_1}\right) \qquad \text{...(14.1}a\text{)}$$

$$\sigma = E\varepsilon = Ey\left(\frac{1}{R_2} - \frac{1}{R_1}\right) \qquad \text{...(14.1}b\text{)}$$

and

$$\frac{M}{I} = \frac{\sigma}{y} = E\left(\frac{1}{R_2} - \frac{1}{R_1}\right) \qquad \text{...(14.2}a\text{)}$$

or

$$\frac{M}{EI} = \left(\frac{1}{R_2} - \frac{1}{R_1}\right) \qquad \text{...(14.2}b\text{)}$$

Evidently when $R_1 \to \infty$ (*i.e.* when the beam is initially straight equation (14.2*b*) reduces to

$$\frac{M}{EI} = \frac{1}{R_2}$$

which is same as equation (4.8).

According to this theory the variation of stress on the section will be linear with the distance of the fibre from the neutral axis.

In actual practice in cases like hooks, links and rings generally the dimensions of the cross-section are not very small in comparison with either the radius of curvature or the length of the bar. The treatment of such cases is given in the following articles.

## 14.2 Bars of Initial Small Radius of Curvature

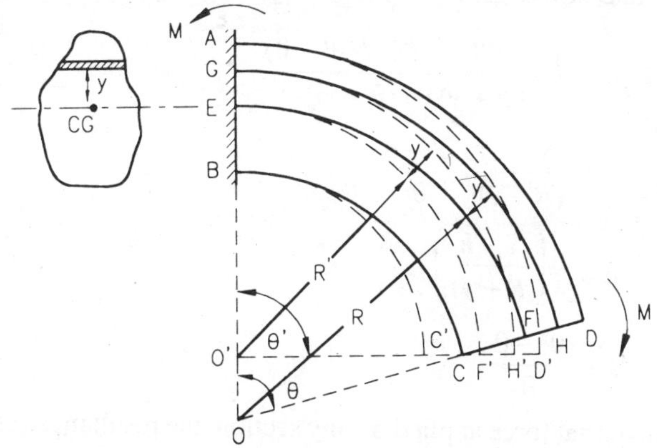


*Fig. 14.1*

Let $ABCD$ be a curved bar with initial radius of curvature of its centroidal axis $EF$ as $R$. The two sections $AB$ and $CD$ subtend a small angle $\theta$ at the centre of curvature 0. Due to the application of a bending moment $M$ (reckoned positive if it increases the curvature and negative otherwise) the radius of curvature of the centroidal axis decreases to $R'$ to occupy the position $EF'$. The angle $\theta$ changes to $\theta'$ along with the shift of centre of curvature to 0′.

For finding the strain and stress normal to the section, consider the fibre $GH$ at a distance $y$ from the centroidal axis. The distance y is reckoned as positive if measured outwards away from the centre of curvature, and negative inwards towards it.

Following assumptions are made in developing the theory :

(*i*) The distance y of any layer $GH$ from the centroidal surface $EF$ remains same even after the application of the moment $M$.

(*ii*) Plane sections remain plane during bending.

(*iii*) The fibres are free to expand or contract without any constraining effect from the adjacent fibres.

(*iv*) The value of the Young's modulus $E$, is same in tension and compression.

The strain along $GH$ at a distance y from the centroidal axis, is given by

$$\varepsilon = \frac{GH' - GH}{GH} = \frac{(R' + y)\,\theta' - (R + y)\theta}{(R + y)\,\theta}$$

$$= \frac{y(\theta' - \theta) + (R'\theta' - R\theta)}{(R + y)\,\theta} \qquad ...(14.3)$$

If neutral axis lies at a distance h from the centroidal surface $EF$, then its location can be found by substituting zero for the strain ε in equation (14.3) and normal equating $y$ to $h$

*i.e.*
$$h = -\frac{(R'\theta' - R\theta)}{(\theta' - y)\theta} \qquad ..(14.4)$$

Equations (14.3) and (14.4) give

$$\varepsilon = \frac{y + (R'\theta' - R\theta)/(\theta' - \theta)}{(R + y)\,\theta/(\theta' - \theta)}$$

$$= \frac{(y - h)}{(R + y)} \cdot \frac{\theta' - \theta}{\theta}$$

or
$$\varepsilon = k\left(\frac{y - h}{R + y)}\right) \qquad ...(14.5)$$

where
$$k = \frac{\theta' - \theta}{\theta} \qquad ...(14.6)$$

Since there is no external force applied on any section, the resultant stress normal to any section will be zero. That is for any section if σ is the stress at the distance $y$ from the centroidal axis. we have

$$\int \sigma\, dA = 0$$

or
$$\int (\varepsilon E)\, dA = 0 \quad \text{(Assumption } iii)$$

or
$$E \int k\left(\frac{y - h}{R + y}\right) dA = 0 \qquad ...(14.7)$$

or
$$\int \frac{y dA}{R + y} - h \int \frac{dA}{R + y} = 0$$

or
$$h = \frac{\int \frac{y}{R + y}\, dA}{\int \frac{1}{R + y}\, dA} = \frac{\int \left(1 - \frac{R}{R + y}\right) dA}{\int \frac{1}{R + y}\, dA}$$

or
$$h = \frac{A - A'}{A'/R} = -\frac{R(A' - A)}{A'} \qquad ...(14.8)$$

Where
$$A' = \int \frac{R}{R + y}\, dA \qquad ...(14.9)$$

The quantity $A'$ is called the modified or derived area and its value will depend upon the shape of the area. Thus the equations (14.8) and (14.9) will give the location of the neutral axis.

Further for any section equating the external moment M to the resisting moment of the section, we have

$$M = \int (\sigma\, dA)\, y$$

$$= \int \varepsilon\, Ey\, dy$$

$$= \int \frac{k\,(y-h)}{(R+y)}\, Ey\, dA$$

$$= kE \int \frac{y\,(y-h)}{(R+y)}\, dA$$

$$= kE \int (y-h)\left\{1 - \frac{R}{R+y}\right\} dA$$

$$= kE \int \left\{y - h - R\frac{(y-h)}{R+y}\right\} dA$$

or $$M = kE\left[\int yd\,A - hA - R\int \frac{y-h}{R+h}\, dA\right] \qquad ...(14.10)$$

Now $\int ydA = 0$ as it represents the first moment of the area about centroid.

Also from (14.7) the last term on the right hand side of equation (14.8) is zero.

Thus the equation (14.10) reduces to

$$M = kE\,(0 - hA - 0)$$

or $$k = \frac{-M}{E\,Ah} \qquad ...(14.11)$$

Substituting the value of the constant $k$ from equation (14.11) in equation (14.5) we get the value of the strain $\varepsilon$ and in turn the value of the longitudinal stress at the distance y from the centroidal surface as

$$\sigma = E\varepsilon = E\left(\frac{-M}{EAh}\right)\left(\frac{y-h}{R+h}\right)$$

or $$\sigma = \frac{M}{-hA}\left(\frac{y-h}{y+R}\right) \qquad ...(14.12)$$

The value of $\sigma$ can be expressed in another standard form by substituting the value of h from equation (14.8) in equation (14.12).

$$\sigma = \frac{M}{R\dfrac{(A'-A)}{A'}A}\left[\frac{y + R(A'-A)/A'}{y+R}\right]$$

$$= \frac{M}{R(A'-A)}\left[\frac{y+R-R\,RA/A'}{y+R}\right]\left(\frac{A'}{A}\right)$$

*i.e.*
$$\sigma = \frac{M}{R(A'-A)}\left[\frac{A'}{A}-\frac{R}{y+R}\right] \quad ...(14.13)$$

## 14.1.1 Values of bold A′ For various Sections

(*a*) **Rectangular Section :** Fig. 14.2 shows the rectangular section with centre of curvature *O* lying on *YY* axis. *XX* axis is the centroidal bending axis. Then from equation (14.9), we have

$$A' = \int_{-d/2}^{d/2} \frac{R}{R+y}\,dA$$

$$= \int_{-d/2}^{d/2} \frac{R\,bdy}{R+y}$$

$$= bR\,[\text{In}\,(R+y)]_{-d/2}^{d/2}$$

$$= bR\ln\left(\frac{R+(d/2)}{R-(d/2)}\right)$$

*i.e.*,
$$A' = bR\ln\left(\frac{2R+d}{2R-d}\right) \quad ...(14.14)$$

Fig. 14.2

## 14.2.2

**(*b*) Circular Section**

Referring to Figure 14.3

$$A' = \int_{-r}^{r} \frac{RdA}{R+y}$$

From Fig. 14.3

$$y = r\sin\theta\,;$$

$$dy = r\cos\theta\ d\theta$$

$$b = 2\,r\cos\theta$$

Therefore

$$A' = R\int_{-\pi/2}^{\pi/2} \frac{(2r\cos\theta)\,(r\cos\theta\,d\theta)}{R+r\sin\theta}$$

$$= 2r^2R\int_{-\pi/2}^{\pi/2} \frac{\cos^2\theta\,d\theta}{r\sin\theta+R}$$

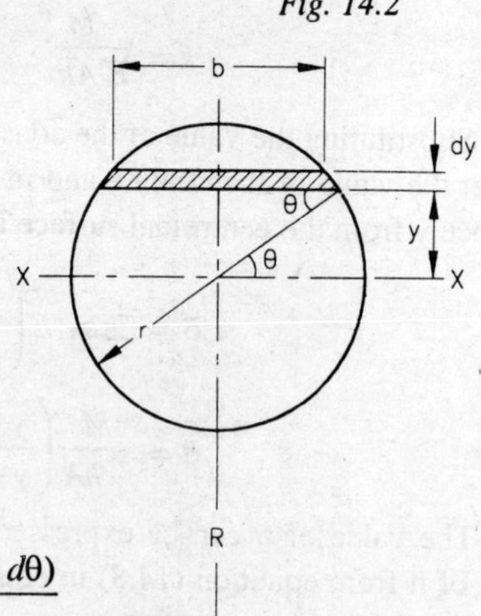


Fig. 14.3

$$= 2r^2 R \int_{-\pi/2}^{\pi/2} \frac{-\sin^2\theta + 1}{r\sin\theta + R}\, d\theta$$

Dividing numerator by denominator, we get

$$A' = 2r^2 R \int_{-\pi/2}^{\pi/2} \left[\frac{-\sin\theta}{r} + \frac{R}{r^2}\frac{(r^2 - R^2)}{(r\sin\theta + R)}\right] d\theta$$

$$= 2r^2 R\left[\frac{1}{r}\left\{\cos\frac{\pi}{2} - \cos\left(\frac{-\pi}{2}\right)\right\} + \frac{R}{r^2}\left(\frac{\pi}{2} + \frac{\pi}{2}\right)\right.$$

$$\left. + \frac{r^2 - R^2}{r^2}\int_{-\pi/2}^{\pi/2} \frac{d\theta}{(r\sin\theta + R)}\right]$$

$$= 2r^2 R\left[0 + \frac{\pi R}{r^2} + \frac{r^2 - R^2}{r^2}\int_{-\pi/2}^{\pi/2}\frac{d\theta}{r\sin\theta + R}\right]$$

*i.e.,*
$$A' = 2\pi R^2 - 2R(R^2 - r^2)\int_{-\pi/2}^{\pi/2}\frac{d\theta}{r\sin\theta + R} \qquad ...(14.15)$$

The last term in equation (14.15) contains a well known integral where $R > r$, its value is

$$\int_{-\pi/2}^{\pi/2}\frac{d\theta}{R + r\sin\theta} = \left[\frac{2}{\sqrt{R^2 - r^2}}\tan^{-1}\left\{\frac{R\tan\frac{\theta}{2} + r}{\sqrt{R^2 - r^2}}\right\}\right]_{-\pi/2}^{\pi/2}$$

$$= \frac{2}{\sqrt{R^2 - r^2}}\left[\tan^{-1}\left(\frac{R + r}{\sqrt{R^2 - r^2}}\right) - \tan^{-1}\left(\frac{-R + r}{\sqrt{R^2 - r^2}}\right)\right]$$

$$= \frac{2}{\sqrt{R^2 - r^2}}[\alpha + \beta]$$

Where $\alpha = \tan^{-1}\left(\dfrac{R + r}{\sqrt{R^2 - r^2}}\right)$ and $\beta = \tan^{-1}\left(\dfrac{R - r}{\sqrt{R^2 - r^2}}\right)$

So that, $\tan(\alpha + \beta) = (\tan\alpha + \tan\beta)/(1 - \tan\alpha\tan\beta)$

$$= \frac{\dfrac{R + r}{(R^2 - r^2)^{1/2}} + \dfrac{R - r}{(R^2 - r^2)^{1/2}}}{1 - \dfrac{(R^2 - r^2)}{(R^2 - r^2)}} = \infty$$

*i.e.*, $\alpha + \beta = \pi/2$

Therefore $\displaystyle\int_{-\pi/2}^{\pi/2} \frac{d\theta}{R + r\sin\theta} = \frac{2}{\sqrt{R^2 - r^2}} \cdot \frac{\pi}{2} = \frac{\pi}{\sqrt{R^2 - r^2}}$

Thus substituting this value in equation (14.15)

$$A' = 2\pi R^2 - 2R(R^2 - r^2) \times \frac{\pi}{\sqrt{R^2 - r^2}}$$

*i.e.* $$A' = 2\pi R\left[R - \sqrt{R^2 - r^2}\right] \qquad ...(14.16)$$

## 14.1.2 Stress Distribution in Rectangular Section

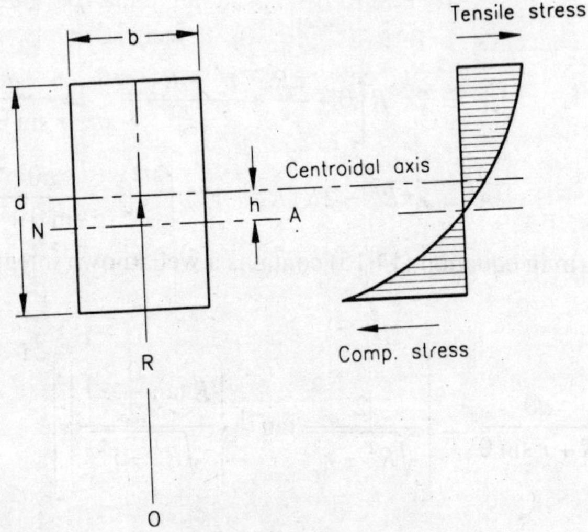


*Fig. 14.4*

Substituting the value of $A'$ from equation (14.12) in equation (14.8)

$$h = -R\frac{\left[bR\ln\left(\dfrac{2R+d}{2R-d}\right) - bd\right]}{bR\ln\left(\dfrac{2R+d}{2R-d}\right)}$$

*i.e.* $$h = -\frac{R\ln\left(\dfrac{2R+d}{2R-d}\right) - d}{\ln\left(\dfrac{2R+d}{2R-d}\right)} \qquad ...(14.17)$$

Again from equations (14.13) and (14.14)

$$\sigma = \frac{M\left\{\dfrac{bR}{bd}\ln\left(\dfrac{2R+d}{2R-d}\right)-\dfrac{R}{y+R}\right\}}{R\left\{bR\ln\left(\dfrac{2R+d}{2R+d}\right)-bd\right\}}$$

$$\sigma = \frac{M\left[\dfrac{R}{d}\ln\left(\dfrac{2R+d}{2R-d}\right)-\dfrac{R}{y+R}\right]}{bR\left\{R\ln\left(\dfrac{2R+d}{2R-d}\right)-d\right\}} \qquad \text{...(14.18)}$$

Substituting different values of y in equation (14.18) the stress distribution curve can be plotted as in Fig. 14.4. It must be kept in mind that for points which lie, in this case, above the centroidal axis $y$ will be positive and for points which lie below the centroidal axis *i.e.*, towards the centre of curvature $O$ the sign of $y$ will be taken as negative.

## 14.3 Stresses in Crane Hooks

Crane hooks form part of the family of curved beams and the equation (14.13) can be applied to find the bending stresses which are then combined with the direct

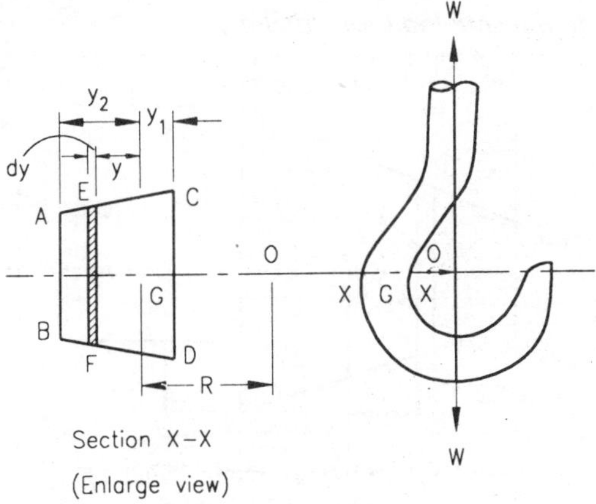


*Fig. 14.5*

stress. Usually the section of the crane hooks is trapezoidal. Let $O$ be the centre of curvature of the trapezoidal section $XX$ with $R$ as its radius of curvature. The distance of the centroid $G$ of the section from the line of action of the vertical load $W$ is '$e$'. For the section $XX$, the load $W$ is nothing but an eccentric load with eccentricity '$e$'. This will be equivalent to a tensile load $W$ at centroid $G$ of the section along with a bending moment $M$ (magnitude = $We$), causing the curvature to decrease. That is in addition the bending stresses $\sigma_b$ due to $M$, which will be compressive on fibres away from the centre of curvature and tensile on fibres nearer to the centre of curvature, there will be direct tensile stress $\sigma_0$ due to load $W$ at $G$ at each and every point of the section.

Thus the resultant stress at any fibre along $EF$ which is at a distance $y$ from $G$, is given by

$$\sigma_r = \sigma_b + \sigma_0$$

or
$$\sigma_r = \frac{(-We)}{R(A'-A)}\left[\frac{A'}{A} - \frac{R}{y+R}\right] + \frac{W}{A} \qquad \text{...(14.19)}$$

Evidently the maximum tensile stress will occur at points along $CD$ and compressive at points along $AB$. Thus if $y_1$ and $y_2$ are respectively the distances of $CD$ and $AB$ from $G$, then

$$\sigma_{CD} = \frac{-We}{R(A'-A)}\left[\frac{A'}{A} - \frac{R}{R-y_1}\right] + \frac{W}{A} \qquad \text{...(14.20a)}$$

$$\sigma_{AB} = \frac{-We}{R(A'-A)}\left[\frac{A'}{A} - \frac{R}{R+y_2}\right] + \frac{W}{A} \qquad \text{...(14.20b)}$$

**Example 14.1.** *The principal section of a hook is a symmetrical trapezium as shown in Fig. 14.6. The centre of curvature of the centroidal axis, at the principal section, is in the plane of the section and is 68 mm from the inside of it. The load line passes 56 mm from the inner side of the section. If the maximum allowable tensile stress is 115 MN/m², estimate the safe load for this hook.*

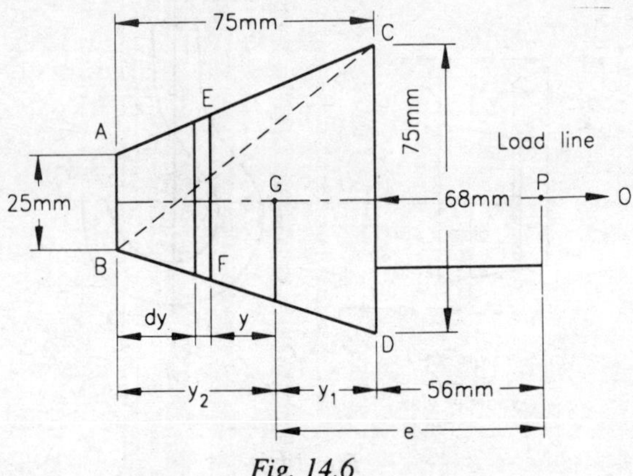


*Fig. 14.6*

**Solution.** If $y_1$ is the distance of the centroid $G$ of the section from $CD$, then taking moments of the areas of the triangles $ABC$ and $BCD$ about $CD$.

$$y_1 = \frac{\left(\frac{1}{2}\times 25\times 75\right)\left(\frac{2}{3}\times 75\right) + \left(\frac{1}{2}\times 75\times 75\right)\left(\frac{1}{3}\times 75\right)}{\left(\frac{1}{2}\times 25\times 75\right) + \left(\frac{1}{2}\times 75\times 75\right)}$$

$$= 31.25 \text{ mm}$$

$$\therefore \quad y_2 = 75 - 31.25 = 43.75 \text{ mm}$$

$$A = \frac{1}{2}(25 + 75) \times 75 = 3750 \text{ mm}^2$$

$$EF = 75 - \frac{50}{75} \times (y_1 + y) = 75 - \frac{2}{3}(31.25 + y)$$

$$= (54.17 - 0.667y)$$

$$R = OG = 68 + y_1 = 68 + 31.25 = 99.25 \times \text{mm}$$

$$e = PG = 56 + y_1 = 56 + 31.25 = 87.25 \times \text{mm}$$

$$A' = \int_{-y_1}^{y_2} \frac{R\,dA}{R + y} = R \int_{-y_1}^{y_2} \frac{(EF)\,dy}{R + y}$$

$$= 99.25 \int_{-31.25}^{43.75} \frac{(54.17 - 0.667\,y)}{99.25 + y}\,dy$$

or

$$A' = 99.25 \times 54.17\,[\ln(99.25 + y)]_{31.25}^{43.75}$$

$$- 99.25 \times 0.667 \int_{-31.25}^{43.75} \frac{y}{99.25 + y}\,dy$$

$$= 5370\,[\ln 143.0 - \ln 68.0] - 66.20\,[y - \ln(99.25 + y)]_{-31.25}^{43.75}$$

$$= 5436.20 \ln\left(\frac{143}{68}\right) - 66.20\,(43.75 + 31.25)$$

$$= 4020 - 49.65 = 3970.35 \text{ mm}^2$$

Let the allowable load on the hook be $W$ Newton.

Then $M = -We = -87.25$ W N-mm

The bending moment here is taken as negative as it tends to reduce the curvature of the hook. The maximum tensile stress will occur at points along $DC$ for which $y = -y_1 = -31.25 \times$ mm.

Therefore from equation (14.17), the resultant stress along $CD$ is

$$\sigma_{CD} = \frac{-87.25W}{(99.25)(2970.35 - 3750)}\left[\frac{3970.35}{3750} - \frac{99.25}{99.25 - 31.25}\right] + \frac{W}{3750}$$

$$= \frac{-87.25\,W}{99.25 \times 220.35} \times (1.06 - 1.455) + \frac{W}{3750} \text{ N/mm}^2 \text{ or MN/m}^2$$

$$= 1.58 \times 10^{-4}\,W + 2.67 \times 10^{-4}\,W = 4.25 \times 10^{-4}\,W\,MN/\text{m}^2$$

Equating this to the allowable stress, we get

$$W = \frac{115}{4.25 \times 10^{-4}} = 2.7 \times 10^5 \text{ N}.$$

### 14.3.1 Stresses in Rings

Fig. 14.7 shows a ring under the action of an axial load $W$. Any radial section $PQ$ which is inclined at an angle $\theta$ with the load line, will carry direct pull (or thrust for comp. $W$), shearing force and a bending moment. The direct pull (or thrust) and the shearing force can be found by drawing the free body diagram of the element $PQDC$ as shown in Fig. 14.7(*b*). Evidently the load on the sections $CD$ and $EF$ which are normal to the load line will be $W/2$. For the equilibrium, the resultant force on the section $PQ$ will also be $W/2$ in the vertical direction as shown in Fig. 14.7(*b*), which in turn will give rise to $\left(\frac{W}{2}\sin\theta\right)$ and $\left(\frac{W}{2}\cos\theta\right)$ as direct pull and shearing force respectively on the section $PQ$. Due to the latter the intensity of the shearing stress at

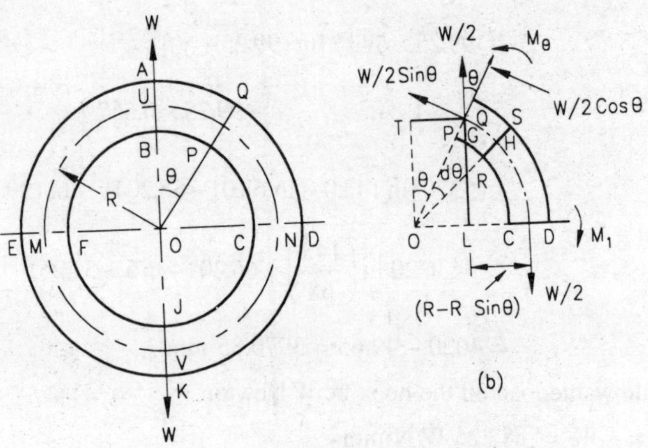


*Fig. 14.7*

the inner and outer edges of the section will be zero. Thus while calculating the stresses at the inner and outer edges of various radial sections the effect of ($W/2 \cos\theta$) will be ignored and only the direct tensile stress due ($W/2 \sin\theta$) will be taken into account.

Again if $M_1$ and $M_\theta$ are the bending moments (reckoned positive if they tend to increase the curvature) on sections $CD$ and $PQ$, then for equilibrium of the element $PQDC$ taking moments about the centroid $G$ of the section $PQ$, we have

$$M_\theta = M_1 + \frac{W}{2}(R - R\sin\theta)$$

or $$M_\theta = M_1 + \frac{1}{2} WR(1 - \sin\theta) \qquad ...(14.21)$$

Evidently $M_\theta$ cannot be evaluated in the absence of the value of $M_1$. For evaluating $M_\theta$ or $M_1$, it is assumed, for simplicity, that the formulae meant for straight beams are applicable here. Strictly speaking it should not be so, but it has been observed that error involved in the evaluation of bending moments by neglecting the curvature is not very large. Thus the rotation of the plane $RS$ with respect to the plane $PQ$ due to bending moment $M_\theta$ is approximately, given by

$$d\phi = \frac{M_\theta}{EI}(GH)$$

or $$d\phi = \frac{M_\theta}{EI}(Rd\theta) \qquad ...(14.22)$$

The total rotation of the plane $CD$ or $EF$ with respect to $AB$ will be obtained by integrating equation (14.22) for the limits of $\theta$ from zero to $\pi/2$. But due to symmetry the sections $AB$ and $CD$ will remain at right angles even after the application of load and there will be no change in their inclination.

*i.e.*, $$\int_0^{\pi/2} \frac{M_\theta}{EI}(Rd\theta) = 0$$

or $$\int_0^{\pi/2}\left[M_1 + \frac{1}{2}WR(1 - \sin\theta\right]\frac{R}{EI}d\theta = 0$$

or $$\frac{\pi}{2}M_1 + \frac{1}{2}WR[\theta + \cos\theta]_0^{\pi/2} = 0$$

or $$\pi M_1 + WR\left[\frac{\pi}{2} - 1\right] = 0$$

*i.e.*, $$M_1 = WR\left[\frac{1}{\pi} - \frac{1}{2}\right] \qquad ...(14.23)$$

Substituting the value of $M_1$ from equation (14.23) into equation (14.21) it gives the variation of the bending moment with $\theta$.

$$M_\theta = WR\left(\frac{1}{\pi} - \frac{1}{2}\right) + \frac{1}{2}WR(1 - \sin\theta)$$

or $$M_\theta = WR\left(\frac{1}{\pi} - \frac{1}{2}\sin\theta\right) \qquad ...(14.24)$$

Evidently $M_\theta = 0$ at $\theta = \theta_1 = \sin^{-1}\left(\frac{2}{\pi}\right)$ The variation of $M_\theta$ with $\theta$ in the case of the ring is usually plotted radially with the circle of radius $R$ as the base line. The positive values of the bending moment are plotted radially outward and negative values radially inwards as shown in Fig. 14.8(*a*). The distribution of bending stress at any

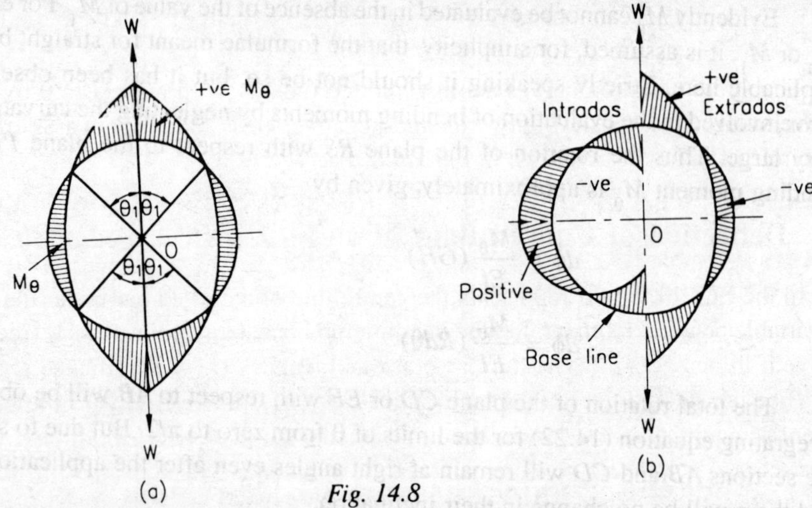

*Fig. 14.8*

section *PQ*, can be found by substituting the value of $M_\theta$ from equation (14.24) into equation (14.13). In addition to the bending stress, there is the direct tensile stress over the section of magnitude $\left(\dfrac{W}{2A}\sin\theta\right)$ as explained earlier.

Thus the resultant stress over any point of the section *PQ* is given by $\sigma_r$ as

$$\sigma_r = \frac{M_\theta}{R(A'-A)}\left[\frac{A'}{A}-\frac{R}{R+y}\right]+\frac{W}{2A}\sin\theta$$

or

$$\sigma_r = \frac{W\left(\dfrac{1}{\pi}-\dfrac{\sin\theta}{2}\right)}{(A'-A)}\left[\frac{A'}{A}-\frac{R}{R+y}\right]+\frac{W}{2A}\sin\theta \qquad \text{...(14.25)}$$

**Resultant Stress at Inside (Intrados) and outside (Extrados) fibres**

On the inner surface of the ring which is also called the *intrados* stress can be found by replacing *y* in equation (14.25) with (– *r*) where *r* is the radius of the bar stock out of which ring is made. Similarly for the outer surface which is known as extrados the resultant stress is calculated by replacing *y* with (+ *r*) in equation (14.25).

Thus for Intrados

$$(\sigma_r)_i = \frac{W\left(\dfrac{1}{\pi}-\dfrac{\sin\theta}{2}\right)}{(A'-A)}\left[\frac{A'}{A}-\frac{R}{R-r}\right]+\frac{W}{2A}\sin\theta \qquad \text{...(14.26)}$$

and for Extrados

$$\left(\sigma_r\right)_c = \frac{W\left(\dfrac{1}{\pi} - \dfrac{\sin\theta}{2}\right)}{(A' - A)}\left[\frac{A'}{A} - \frac{R}{R+r}\right] + \frac{W}{2A}\sin\theta \qquad ...(14.27)$$

The stress variation curves are also plotted radially with respect to θ by taking positive values (*i.e.*, tensile) outwards and negative stress radially inwards from the base circle of radius $R$ as shown in Fig. 14.8(*b*).

## 14.4 Deflection of Curved Bars

In the Fig. 14.9 *APB* represents the centre line of a curved bar under the action of a variable bending moment. Let the bending moment at any point $P$ be $M$ (reckoned positive if it increases the curvature). Let a small length $PQ$ ( $= dl$) situated about the point $P$ be considered to find the rotation of $Q$ with respect to $P$ due to the bending moment $M$ and rest of the curved bar is assumed rigid. Due to the rotation d $\phi$ of $Q$ with respect to $P$ let the portion $PB$ shift to the position $PC$. The displacement $BC$ of the point $B$ can be resolved into two components (*i*) $BD$ perpendicular to the chord $AB$ and (*ii*) $DC$ along $BA$. Evidently due to $M$ at the point $P$, the deformation of the chord $AB$ is $DC$ along the chord. Since the displacement of $B$ along the chord is towards $A$, $DC$ will be reckoned negative (signifying that $A$ and $B$ are coming closer).

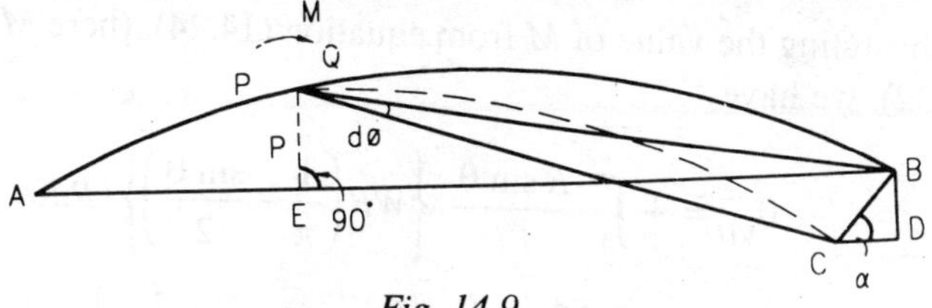


*Fig. 14.9*

Thus the deflection $\Delta\delta_{BA}$ of $B$, along $AB$ is given as

$$\Delta\delta_{BA} = -DC$$
$$= -BC\cos\alpha$$
$$= -(PB\ \delta\phi)\cos\alpha \qquad ...(14.28)$$

where $d\phi = \angle CPB$. The portions $AP$ and $QB$ being considered rigid, $d\phi$ will also represent the rotation of the end $Q$ with respect to $P$ of the element $PQ$.

As $CD$ and $EB$ are parallel, $\angle EBC = \angle BCD = \alpha$.

Further since $CB$ is perpendicular to $PB$ (for small rotation $d\phi$) $\angle PBE = (\pi/2 - \alpha)$. Therefore in the right angled triangle $PBE$, $\angle EPB = \alpha$. Equation (14.28), then reduces to

$$\Delta\delta_{BA} = -PB\,(d\phi)\frac{PE}{PB} = -p\,(d\phi) \qquad ...(14.29)$$

Where $p$ is the perpendicular distance of the point $P$ from the chord $AB$.

Again for the small length dl under the action of bending moment $M$, we have

$$d\phi = \frac{M}{EI}dl \quad \text{...(14.30)}$$

Substituting the value of $d\phi$ from equation (14.30) in equation (14.29), we get

$$\Delta\delta_{BA} = -p\frac{M}{EI}dl \quad \text{...(14.31)}$$

Therefore the total deflection of $B$ with respect to $A$ along the chord $AB$ is obtained by integrating equation (14.31)

$$\Delta\delta_{BA} = -\int_A^B p\frac{M}{EI}dl \quad \text{...(14.32)}$$

### 14.4.1 Deflection of rings

**(a) Deflection along load line**

Equation (14.32) can be applied to find the deflection of the ring under the action of an axial load $W$ Fig. 14.7. In this case $GT$ will be the perpendicular distance of the point $G$ from the chord $UV$ (here diameter) along which deflection is to be evaluated

*i.e.* $p = GT = R\sin\theta$

Thus substituting the value of $M$ from equation (14.24), (here $M = M\theta$) and $p$ in equation (14.32), we have

$$\theta_{VU} = -\int_0^\pi \frac{R\sin\theta}{EI}.\left\{WR\left(\frac{1}{\pi}-\frac{\sin\theta}{2}\right)\right\}(Rd\theta)$$

$$= -\frac{WR^3}{EI}\int_0^\pi\left\{\frac{\sin\theta}{\pi}-\frac{1}{4}(1-\cos 2\theta)\right\}d\theta$$

$$= -\frac{WR^3}{EI}\left[\frac{-\cos\theta}{\pi}-\frac{\theta}{4}+\frac{\sin 2\theta}{8}\right]_0^\pi$$

$$= -\frac{WR^3}{EI}\left[\frac{1}{\pi}(-1-1)-\frac{\pi}{4}\right]$$

*i.e.*
$$\delta_{VU} = \frac{WR^3}{EI}\left(\frac{\pi}{4}-\frac{2}{\pi}\right) \quad \text{...(14.33}a\text{)}$$

$$= 0.149\frac{WR^3}{EI} \quad \text{...(14.33}b\text{)}$$

From equation (14.33) it is evident that the deflection of the ring along the load line is directly proportional to the load applied. This property of the ring is actually responsible for its wide use in the form of proving rings deployed for the calibration purposes.

(*b*) **Deflection along the direction perpendicular to load line :**

Equation (14.32) can also be used to find the change in the diameter *MN* (or the movement of the point *N* relative to that of *M*) due to the load *W* as shown in Fig. 14.7.

In this case, *GL* represents the distance *p*. Thus from equations (14.32) and (14.24).

$$\delta_{NM} = -\int_M^N (GL)\frac{M_\theta}{EI}(R\,d\theta)$$

$$= -\int_M^N \left(\frac{R\cos\theta}{EI}\right)\left\{WR\left(\frac{1}{\pi}-\frac{1}{2}\,SIN\,\theta\right)\right\} R\,d\theta$$

$$= -\frac{WR^3}{EI}\times\int_o^{\pi/2}\left(\frac{\cos\theta}{\pi}-\frac{1}{4}\sin 2\theta\right)d\theta$$

$$= -\frac{2WR^3}{EI}\left[\frac{\sin\theta}{\pi}+\frac{\cos 2\,\theta}{8}\right]_o^{\pi/2}$$

or $$\delta_{NM} = -\frac{2\,WR^3}{EI}\left(\frac{1}{\pi}-\frac{1}{4}\right) = -\,0.137\frac{WR^3}{EI} \qquad ...(14.34)$$

The negative sign for the deflection $\delta_{NM}$ indicates that the points *N* and *M* are moving towards each other; in other words the diameter *MN* decreases.

## 14.5 Deflection of rings by Castigliano's theorem

(*a*) **Deflection along the load line**

The deflection of the ring along the load line can very easily be found with the help of Castigliano's theorem. The total strain energy of the ring due to bending will be four times that of the portion *ABCD* (Fig. 14.7).

$$U = 4\int_0^{\pi/2}\frac{1}{EI}(M_\theta)^2\,dl$$

$$= \frac{2}{EI}\int_0^{\pi/2} W^2R^2\left(\frac{1}{\pi}-\frac{\sin\theta}{2}\right)^2 R d\theta$$

$$= \frac{2W^2R^3}{EI}\int_0^{\pi/2}\left(\frac{1}{\pi^2}+\frac{1}{4}\sin^2\theta-\frac{1}{\pi}\sin\theta\right)d\theta$$

$$= \frac{2W^2R^3}{EI}\int_0^{\pi/2}\left[\frac{1}{\pi^2}+\frac{1}{8}(1-\cos 2\,\theta)-\frac{1}{\pi}\sin\theta\right]de$$

$$= \frac{2W^2R^3}{EI}\left[\left(\frac{1}{\pi^2}+\frac{1}{8}\right)\theta-\frac{1}{8}\sin 2\theta+\frac{1}{\pi}\cos\theta\right]_0^{\pi/2}$$

$$= \frac{2W^2R^3}{EI}\left[\left(\frac{1}{\pi}+\frac{1}{8}\right)\frac{\pi}{2} - 0 + \frac{1}{\pi}(0-1)\right]$$

i.e. $$U = \frac{W^2R^3}{EI}\left(\frac{\pi}{8}-\frac{1}{\pi}\right) \qquad \text{...(14.35)}$$

Therefore $$\delta_{VU} = \frac{\delta U}{\delta W} = \frac{WR^3}{EI}\left(\frac{\pi}{4}-\frac{2}{\pi}\right)$$

Which is same as found in equation (14.33).

(*b*) **Deflection along the direction perpendicular to the load line :**

The deflection of the ring along perpendicular direction to the load line can also be found with the help of Castigliano's theorem, by first assuming an additional load $W_1$ on the ring in the perpendicular direction to the load $W$. The additional bending moment over any section $PQ$ Fig. 14.7 due to tensile load $W_1$ along $MN$ can easily be found from equation (14.24) by replacing $W$ with $W_1$ and $\theta$ with $(\pi/2 - \theta)$.

That is, additional bending moment due to $W_1$

$$= W_1R\left[\frac{1}{\pi}-\frac{1}{2}\sin\left(\frac{\pi}{2}-\theta\right)\right]$$

$$= W_1R\left[\frac{1}{\pi}-\frac{1}{2}\cos\theta\right]$$

Therefore the total bending moment on and section $PQ$

= sum of bending moments due to $W$ and $W_1$

i.e., $$M_\theta = WR\left(\frac{1}{\pi}-\frac{1}{2}\sin\theta\right) + W_1R\left(\frac{1}{\pi}-\frac{1}{2}\cos\theta\right) \qquad \text{...(14.36)}$$

Therefore as there are four symmetrical segments like $ABCD$ in the ring, the total strain energy due to bending is given by

$$U = 4\int_0^{\pi/2}\frac{M_\theta^2}{2EI}dl$$

$$= \frac{2}{EI}\int_o^{\pi/2}\left[WR\left(\frac{1}{\pi}-\frac{1}{2}\sin\theta\right) + W_1R\left(\frac{1}{\pi}-\frac{1}{2}\cos\theta\right)\right]^2 Rd\theta$$

$$= \frac{2R^3}{EI}\int_0^{\pi/2}\left[W^2\left(\frac{1}{\pi}-\frac{1}{2}\sin\theta\right)^2 + W_1^2\left(\frac{1}{2}-\frac{1}{2}\cos\theta\right)^2\right.$$

$$\left. + 2WW_1\left(\frac{1}{\pi}-\frac{1}{2}\sin\theta\right)\left(\frac{1}{\pi}-\frac{1}{2}\cos\theta\right)\right]d\theta \qquad \text{...(14.37)}$$

Differentiating equation (14.37) partially with respect to $W_1$

$$\frac{\partial U}{\partial W_1} = \frac{2R^3}{EI}\int_0^{\pi/2}\left[0 + 2W_1\left(\frac{1}{\pi} - \frac{\cos\theta}{2}\right)^2 + 2W\left(\frac{1}{\pi} - \frac{\sin\theta}{2}\right)\left(\frac{1}{\pi} - \frac{\cos\theta}{2}\right)\right]d\theta \quad ...(14.38)$$

Substituting zero for $W_1$ in equation (14.38), we get lateral deflection of the ring.

$$\delta_{NM} = \left(\frac{\partial U}{\partial W_1}\right)_{W_1=0} = \frac{2R^3}{EI}\int_0^{\pi/2} 2W\left(\frac{1}{\pi} - \frac{\sin\theta}{2}\right)\left(\frac{1}{\pi} - \frac{\cos\theta}{2}\right)d\theta$$

$$= \frac{4WR^3}{EI}\int_0^{\pi/2}\left[\frac{1}{\pi^2} - \frac{\sin\theta}{2\pi} - \frac{\cos\theta}{2\pi} + \frac{1}{8}\sin 2\theta\right]d\theta$$

$$= \frac{4WR^3}{EI}\left[\frac{1}{\pi^2}.\theta + \frac{1}{2\pi}\cos\theta - \frac{1}{2\pi}\sin\theta - \frac{1}{16}\cos 2\theta\right]_0^{\pi/2}$$

$$= \frac{4WR^3}{EI}\left[\frac{1}{\pi^2}.\frac{\pi}{2} + \frac{1}{2\pi}(0-1) - \frac{1}{2\pi}(1-0) - \frac{1}{16}(-1-1)\right]$$

*i.e.* $$\delta_{NM} = -\frac{2WR^3}{EI}\left(\frac{1}{\pi} - \frac{1}{4}\right)$$

Which is same as found in equation (14.34).

**Example 14.2.** *A ring of mean diameter of 100 mm is made of round steel of 20 mm diameter. A pull of 5000 N acts on the ring. Calculate the greatest tensile and compressive stresses. Also estimate the increase in the diameter of the ring along the load line and the decrease in the diameter in the lateral direction. Take $E = 200\ GN/m^2$.*

**Solutioin.** $R = 50 \times$ mm ; $r = 10 \times$ mm

$A = \pi r^2 = 100\,\pi \times \text{mm}^2$

$$A' = 2\pi R\left(R - \sqrt{R^2 - r^2}\right) = 2\pi \times 50\left[50 - \sqrt{50^2 - 10^2}\right]$$

$$= 100\,\pi \times 1.0102$$

$$= 101.02\,\pi\ \text{mm}^2$$

For Intrados : Stresses are obtained from equation (14.26), for intrados

At $\theta = 0$

$$\sigma_r = \frac{500\,(1/\pi - 0)}{(101.02\,\pi - 100\pi)}\left[\frac{101.02\pi}{100\pi} - \frac{50}{50-10}\right] + 0$$

$$= \frac{5000}{\pi^2 (1.02)} (1.0102 - 1.25)$$

$$= -119 \times \text{N/mm}^2 \text{ or } 119 \text{ MN/m}^2 \text{ (Comp.)}$$

At $\theta = \frac{\pi}{2}$ ;

$$\sigma_r = \frac{5000\left(\frac{1}{\pi} - \frac{1}{2}\right)}{(101.02\pi - 100\pi)}\left[\frac{101.02\pi}{100\pi} - \frac{50}{50-10}\right] + \frac{5000}{2 \times 100\pi}$$

$$= \frac{-500 \times .182}{1.02\,\pi}(1.0102 - 1.25) + \frac{25}{\pi}$$

$$= 68.2 + 7.96 = 76.16 \times \text{N/mm}^2$$

or $76.16 \text{ MN/m}^2$ (Tensile)

For Extrados : Stresses for extrados are obtained from equation (14.27)

At $\theta = 0$

$$\sigma_r = \frac{5000\left(\frac{1}{\pi} - 0\right)}{(101.02\pi - 100\pi)}\left[\frac{101.02\pi}{100\pi} - \frac{50}{50+10}\right] + 0$$

$$= \frac{5000}{1.02\,\pi^2}(1.0102 - 0.8333)$$

*i.e.* $\sigma_r = 88 \times \text{N/mm}^2$ or $88 \text{ MN/m}^2$ (Tensile)

At $\theta = \frac{\pi}{2}$

$$\sigma_r = \frac{5000\left(\frac{1}{\pi} - \frac{1}{2}\right)}{(101.02\,\pi - 100\,\pi)}\left[\frac{101.02\,\pi}{100\,\pi} - \frac{50}{50+10}\right] + \frac{5000}{2 \times 100\,\pi}$$

$$= \frac{-5000 \times 0.182}{1.02\,\pi}(1.0102 - 0.8333) + \frac{25}{\pi}$$

$$= \frac{-5000 \times .182 \times .1769}{1.02\,\pi} + 7.96$$

$$= -50.25 + 7.96 = -42.29 \times \text{MN/m}^2 \text{ (Comp.)}$$

Thus the greatest tensile stress occurs on extrados at $\theta = 0$ and is equal to 88.0 $\text{MN/m}^2$. The greatest compressive stress is 119 $\text{MN/m}^2$ and occurs on Intrados at $\theta = 0°$, with the load line.

(*b*) The increase in the ring diameter along the load line is given by equation (14.33)

$$\delta_{VU} = \frac{WR^2}{EI}\left(\frac{\pi}{2} - \frac{2}{\pi}\right) = 0.149\frac{WR^3}{EI}$$

$$= \frac{0.149 \times 5000 \times (50)^3 \times 64}{(200 \times 10^9 \times 10^{-6}) \times 20^4}$$

$$= 0.0594 \text{ mm.}$$

Similarly the decrease in the lateral diameter is given by equation (14.34)

$$\delta_{NM} = 0.137\frac{WR^3}{EI}$$

$$= \frac{0.137 \times 5000 \times (50)^3 \times 64}{(200 \times 10^9 \times 10^{-6}) \times 20^4} = 0.0546 \text{ mm}$$

**Example 14.3**. *If the ring of example 14.2 has a saw cut along the horizontal diameter on one side, find the intensities of greatest tensile and compressive stresses induced in the ring due to the same load. Also find the deflection along the load line.*

**Solution.** There will be no bending stress in this case towards the left of load line. On the right hand side the bending moment over any section $PQ$ will be ($-W \times R \sin\theta$), tending to decrease curvature. In addition there will be direct pull of $W \sin\theta$ and direct shearing force of $W \cos\theta$ over the section.

Thus the direct stress over any section

$$\sigma_0 = \frac{W \sin\theta}{A} \text{ (Tensile)}$$

The bending stress due to bending moment ($-WR \sin\theta$), is given by the equation (14.11)

$$\sigma_b = \frac{-(WR\sin\theta)}{R(A' - A)}\left[\frac{A'}{A} - \frac{R}{y + R}\right]$$

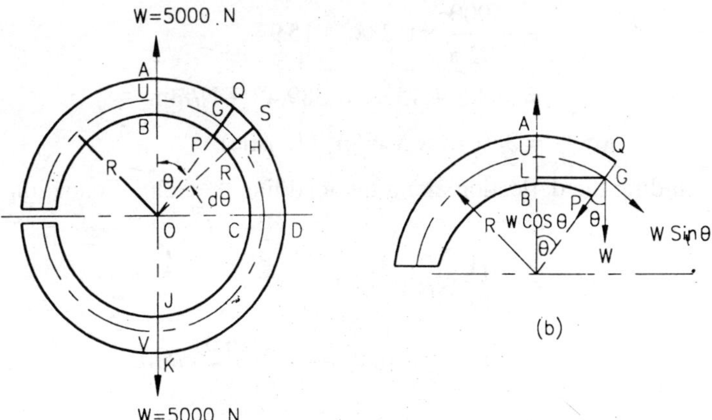


*Fig. 14.10*

and
$$\sigma_r = \sigma_b + \sigma_0$$
$$= -\frac{WR\sin\theta}{R(A'-A)}\left[\frac{A'}{A}-\frac{R}{y+R}\right]+\frac{W\sin\theta}{A}$$

Evidently for maximum values of $\sigma_r$,
$$\theta = 90°$$

Thus on Extrados, at $\theta = 90°$ *i.e.* at point $D$.
$$(\sigma_r)_D = \frac{-W}{(A'-A)}\left[\frac{A'}{A}-\frac{R}{r+R}\right]+\frac{W}{A}$$

*i.e.*
$$(\sigma_r)_D = \frac{-5000}{1.02\,\pi - 100\,\pi}\left[\frac{101.02\,\pi}{100\,\pi}-\frac{50}{50+10}\right]+\frac{5000}{100\,\pi}$$
$$= \frac{-5000}{1.02\,\pi}(1.0102 - 0.8333) + 15.92$$
$$= \frac{-5000 \times 0.1769}{1.02\,\pi} + 15.92$$
$$= -276.0 + 15.92 = -260.08 \times \text{N/mm}^2$$

or
$$(\sigma_r)_D = 260.08 \times \text{MN/m}^2 \text{ (Comp.)}$$

On intrados, at $\theta = 90°$ *i.e.* at point $C$.
$$(\sigma_r)_C = -\frac{W}{(A'-A)}\left[\frac{A'}{A}-\frac{R}{R-r}\right]+\frac{W}{A}$$
$$= \frac{-5000}{1.02\pi}\left(1.0102 - \frac{50}{40}\right)+\frac{5000}{100\,\pi}$$
$$= \frac{5000}{1.02\,\pi}\times 0.2388 + 15.92$$
$$= 373.5 + 15.92 = 389.42 \times \text{N/mm}^2$$

or
$$(\sigma_r)_C = 389.42 \times \text{MN/m}^2 \text{ (Tensile)}$$

(*b*) For finding the deflection along the load line the help of equation (14.32) is taken
$$(\sigma_{UV}) = -\int_U^V (GL)\frac{M}{EI}\,dl$$
$$= -\int_{\theta/0}^{\pi} (R\sin\theta)\frac{(-WR\sin\theta)}{EI}\,Rd\theta$$
$$= -\frac{WR^3}{2EI}\int_0^{\pi}(1-\cos 2\theta)\,d\theta$$

$$= -\frac{WR^3}{2EI}[(\pi - 0) - (0 - 0)] = \frac{\pi WR^3}{2EI}$$

$$= \frac{\pi \times 5000 \times (50)^3 \times 64}{2 \times (200 \times 10^9 \times 10^{-6})\, \pi \times (20)^4}$$

$$= 0.626 \times \text{mm}$$

Alternatively the axial deflection along the load line can be found with the help of Castigliano's theorem. It may be borne in mind that the portion on the left of $UV$ will not carry any bending moment.

$$\text{Strain energy} = U = \int_{\theta=0}^{\pi} \frac{M^2}{2EI}\, dl$$

$$= \frac{1}{EI}\int_0^{\pi} (-\,WR \sin\theta)^2\, R\, d\theta$$

$$= \frac{W^2R^3}{2EI}\int_0^{\pi} \frac{1}{2}(1 - \cos 2\theta)\, d\theta$$

i.e. $$U = \frac{\pi W^2 R^3}{4EI}$$

Therefore $$\delta_{UV} = \frac{\partial U}{\partial W} = \frac{\pi WR^3}{2EI}\text{, same as found above.}$$

## 14.6 Stresses in Simple Chain Link

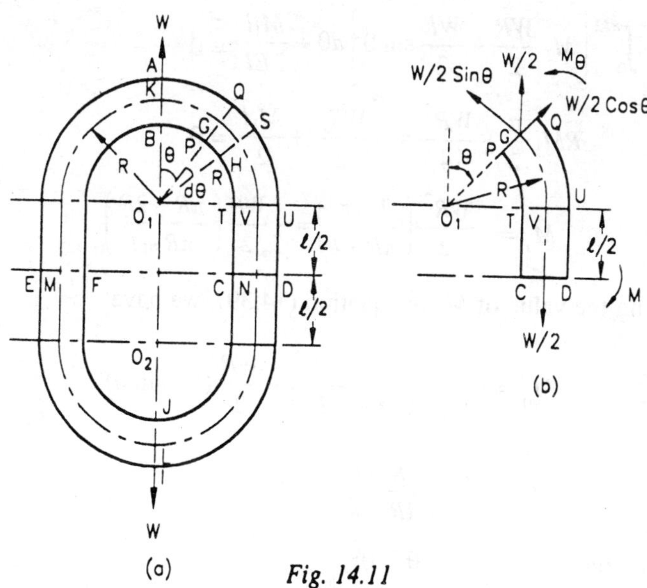


*Fig. 14.11*

Fig. 14.11 shows the simplest form of a chain link with straight sides connecting the semicircular ends of radius $R$, under the action of an axial load $W$.

The vertical pull $W/2$ on any section $PQ$ can be resolved into two components : t/2 sin θ normal and $W/2 \cos\theta$ along the plane. As in the case of a ring the effect of the shearing force (W/2 cos θ) on the section $PQ$ will be zero at points $P$ and $Q$. Further an approximate theory of stress calculation by considering the elemental portion $GH$ straight, as done in the case of ring, will be developed to calculate the stresses at various points of the chain link.

The bending moment, $M_1$, on the straight portion $TUDC$ being parallel to the load line, will remain constant. Considering the equilibrium of the element $PQDC$ we find by taking moment about $G$, the moment $M_\theta$ (reckoned positive if it tends to increase the curvature) over the section $PQ$

$$M_\theta = M_1 + \frac{W}{2}(R - R\sin\theta)$$

or
$$M_\theta = M_1 + \frac{WR}{2}(1 - \sin\theta) \quad ...(14.39)$$

For finding $M_1$, we find the total rotation of the section $CD$ with respect to $AB$ and equate to zero as these sections remain normal to each other even after the application of load.

*i.e.*
$$\int_{\theta=0}^{\pi/2} \frac{M_\theta}{EI}(GH) + \frac{M_1}{EI}\cdot\left(\frac{l}{2}\right) = 0$$

or
$$\int_{\theta=0}^{\pi/2} \frac{1}{EI}\left[M_1 + \frac{WR}{2}(1 - \sin\theta)\right] Rd\theta + \frac{M_1 l}{2EI} = 0$$

or
$$\frac{R}{EI}\int_0^{\pi/2}\left[M_1 \frac{WR}{2} - \frac{WR}{2}\sin\theta\right] d\theta + \frac{M_1 l}{2EI} = 0$$

or
$$RM_1\frac{\pi}{2} + \frac{WR^2}{4}\pi - \frac{WR^2}{2} + \frac{M_1 l}{2} = 0$$

'*i.e.*
$$M_1 = \frac{WR^2}{2}\left(\frac{2-\pi}{\pi R + l}\right) = -\frac{WR}{2}\left(\frac{\pi R - 2R}{\pi R + l}\right) \quad ...(14.40)$$

Substituting the value of $M_1$ in equation (14.39), we have

$$M_\theta = -\frac{WR}{2}\left(\frac{\pi R - 2R}{\pi R + l}\right) + \frac{WR}{2}(1 - \sin\theta)$$

$$M_\theta = \frac{WR}{2}\left[\frac{2R + l}{\pi R + l} - \sin\theta\right] \quad ...(14.41)$$

From equation (14.41) when $\theta = 0$,

$$M_\theta = \frac{WR}{2}\left(\frac{2R+l}{\pi R+l}\right)$$

When $\theta = \sin^{-1}\left(\frac{2R+l}{\pi R+l}\right) = \theta_1$ (say), $M_{\theta_1} = 0$

When $\theta = 90°$, $M_{90} = M_1$ (equation 14.40)

The bending moment diagram here also is obtained by plotting the values, as obtained from equation (14.41), radially (positive outwards and negative inwards) with the centre line *KNLM* as the base line. The diagram is shown in Fig. 14.14(*a*).

The bending stress over any point of the curved portions can be found with the help of equation (14.13). The resultant stress can then be found by algebraically adding to it the direct tensile stress of magnitude $\frac{W}{2A}\sin\theta$.

*i.e.*
$$\sigma_r = \frac{WR}{2}\left[\frac{2R+l}{\pi R+l} - \sin\theta\right]\frac{1}{R(A'-A)}\left[\frac{A'}{A} - \frac{R}{R+y}\right] + \frac{W}{2A}\sin\theta$$

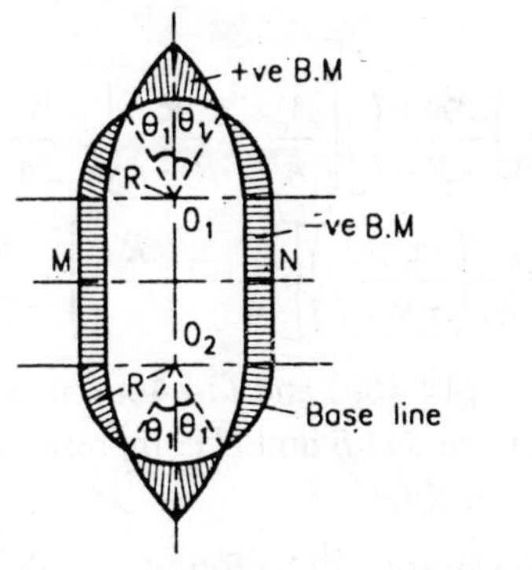


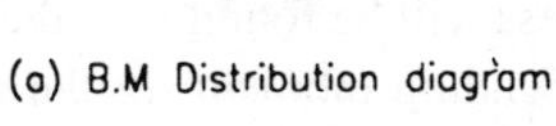
(a) B.M Distribution diagram

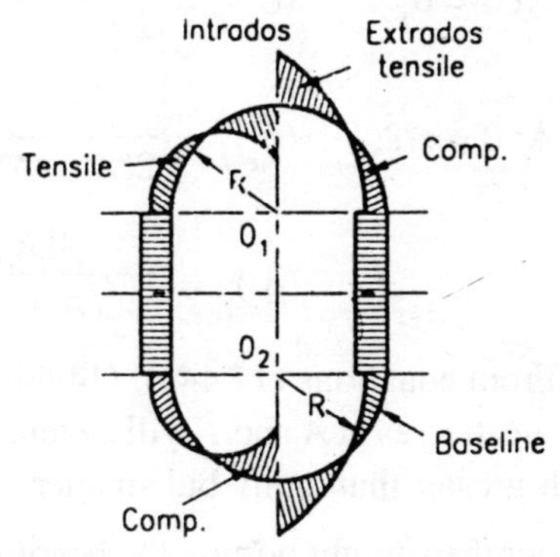

(b) Stress distribution diagram.

*Fig. 14.12*

or
$$\sigma_r = \frac{W}{2(A'-A)}\left[\frac{2R+l}{\pi R+l} - \sin\theta\right]\left[\frac{A'}{A} - \frac{R}{R+y}\right] + \frac{W}{2A}\sin\theta \quad ...(14.42)$$

Thus if *r* is the radius of the bar stock of which the chain link is made, we have the following expressions for the extrados and the intrados by replacing *y* with (+ *r*) and (– r) respectively.

(*i*) **For extrados :** From equation (14.42)

$$\sigma_r = \frac{W}{2(A'-A)}\left(\frac{2R+l}{\pi R+l} - \sin\theta\right)\left(\frac{A'}{A} - \frac{R}{R+r}\right) + \frac{W}{2A}\sin\theta \quad ...(14.43)$$

at $\theta = 0$, $$(\theta_r)_A = \frac{W}{2(A'-A)}\left(\frac{2R+l}{pR+l}\right)\left[\frac{A'}{A}\frac{R}{R+r}\right] \quad ...(14.24a)$$

At $\theta = \theta_1$, $$\theta_r = \frac{W}{2A}\sin\theta_1 \quad ...(14.44b)$$

At $\theta = \pi/2$, $$(\sigma_r)_U = \frac{W}{2(A'-A)}\left(\frac{2R+l}{\pi R+l}-1\right)\left[\frac{A'}{A}-\frac{R}{R+r}\right]+\frac{W}{2A}$$

or $$(\sigma_r)_U = \frac{WR}{2(A'-A)}\left(\frac{\pi-2}{\pi R+l}\right)\left[\frac{A'}{A}-\frac{A}{R+r}\right]+\frac{W}{2A} \quad ...(14.44c)$$

***(iii)* For Intrados :** From equation 14.42

$$\sigma_r = \frac{W}{2(A'-A)}\left[\frac{2R+l}{\pi R+l}\sin\theta\right]\left[\frac{A'}{A}-\frac{R}{R-r}\right]+\frac{W}{2A}\sin\theta \quad ...(14.45)$$

At $\theta = 0$, $$(\sigma_r)_B = \frac{W}{2(A'-A)}\left(\frac{2R+l}{\pi R+l}\right)\left[\frac{A'}{A}-\frac{R}{R-r}\right] \quad ...(14.46a)$$

At $\theta = \theta_1$ $$(\sigma_r)_T = \frac{W}{2A}\sin\theta_1 \quad ...(14.46b)$$

At $\theta = \pi/2$, $$(\sigma_r)_T = \frac{W}{2(A'-A)}\left[\frac{2R+l}{\pi R+l}\right]\left[\frac{A'}{A}-\frac{R}{R-r}\right]+\frac{W}{2A}$$

or $$(\sigma_r)_T = -\frac{WR}{2(A'-A)}\left(\frac{\pi-2}{\pi R+l}\right)\left[\frac{A'}{A}-\frac{R}{R-r}\right]+\frac{W}{2A} \quad ...(14.46c)$$

From equations (14.44*a*), (14.44*c*), (14.46*a*) and (14.46*c*), it is evident thai the resultant stresses at $A$ and $T$ will be tensile and at $B$ and $U$ compressive because $A'/A$ is though greater than unity but smaller than $R/(R-r)$.

For the straight portion the bending stress will be found with the help of "theory of simple bending." (equation 4.8). As the bending moment for the straight portion, $M_1$, is negative, the bending stress along the inner edge *i.e.*, $CT$ will be tensile and on $UD$ compressive. To this will be added the direct tensile stress of magnitude $W/2A$.

$$(\sigma_r)_{UD} = \frac{WR}{2}\left(\frac{\pi R-2R}{\pi R-l}\right).\frac{r}{I}+\frac{W}{2A} \text{ (tnesile)} \quad ...(14.47a)$$

$$(\sigma_r)_{UD} = -\frac{WR}{2}\left(\frac{\pi R-2R}{\pi R+l}\right)\frac{r}{I}+\frac{W}{2A} \quad ...(14.47b)$$

$(\sigma_r)_{CT}$ will be tensile and $(\sigma_r)_{UD}$ being negative will be compressive . As done in the case of a ring, the stress distribution diagram for a chain link is also drawn radially with tensile stresses plotted outwards and compressive stresses inwards, with respect to centroidal axis as the reference line. The stresses as calculated from equations (14.43) to (14.47) are shown in Fig. 14.14(*b*).

## 14.7 Deflection of the Simple Chain link

Here also the equation (14.32) can be used to find the deflection of the simple chain link due to bending. Thus due to bending

$$\delta_{KL} = -\int_K^L \frac{pM}{EI}\,dl = -2\int_K^N \frac{pM}{EI}\,dl$$

$$= -2\left[\int_K^V \frac{(R\sin\theta)}{EI} M_\theta\, dl + \int_V^N \frac{R}{EI}. M_1\, dl\right]$$

Now $M_1$ is constant over the portion *VN* and its value is given in equation (14.40). Therefore

$$\delta_{KL} = -\frac{2R}{EI}\int_0^{\pi/2} (\sin\theta)\frac{WR}{2}\left[\frac{2R+l}{\pi R+l} - \sin\theta\right] Rd\theta - \frac{2RM_1}{EI}\left(\frac{l}{2}\right)$$

$$= \frac{WR^3}{EI}\int_0^{\pi/2}\left[\left(\frac{2R+l}{\pi R+l}\right)\sin\theta - \frac{1}{2}(1-\cos 2\theta)\right] d\theta$$

$$-\frac{Rl}{EI}\left[-\frac{WR}{2}\left(\frac{\pi R-2R}{\pi R+l}\right)\right]$$

$$= -\frac{WR^3}{EI}\left[\frac{2R+l}{\pi R+l}(-\cos\theta) - \frac{\theta}{2} + \frac{\sin 2\theta}{4}\right]_0^{\pi/2} + \frac{WR^3 l}{2EI}\left(\frac{\pi-2}{\pi R+l}\right)$$

$$= -\frac{WR^3}{EI}\left[\frac{2R+l}{\pi R+l} - \frac{\pi}{4}\right] + \frac{WR^3 l}{2EI}\left[\frac{\pi-2}{\pi R+l}\right]$$

or

$$\delta_{KL} = \frac{WR^3}{4EI}\left[\frac{(\pi^2-8)R + (3\pi-8)l}{(\pi R+l)}\right] \qquad ...(14.48)$$

In addition to the deflection due to bending the ends of the link will move apart due to the stretch in the straight portions caused by the direct load *W*/2 carried by each by the straight portions. The additional deflection due to this will be $\left(\frac{Wl}{2AE}\right)$.

Thus the total deflection will be

$$\delta = \delta_{KL} + \frac{Wl}{2AE}$$

$$= \frac{WR^3}{4WI}\left[\frac{(\pi^2-8)R + (3\pi-8)l}{(\pi R+l)}\right] + \frac{Wl}{2AE} \qquad ...(14.49)$$

In equation (14.47), $Wl/(2AE)$ is usually neglected being small. However when the length $l$ of the straight portions is very large this term may not be small to be neglected.

**Second Method :**

Alternatively the deflection $\delta_{KL}$ due to bending may be found by the application of Castigliano's theorem.

Total strain energy of the link will be four times that of the portion *ABCD*.

Therefore

$$U = 4\left[\int_0^{\pi/2} \frac{M_\theta^2}{2EI} dl + \frac{M_1^2}{2EI}\left(\frac{l}{2}\right)\right]$$

$$= \frac{l}{EI}\left[-\frac{WR}{2}\left(\frac{\pi R - 2R)}{\pi R + l}\right)\right]^2 + \frac{2}{EI}\int_0^{\pi/2}\left[\frac{WR}{2}\left(\frac{2R + l}{\pi R + l} - \sin\theta\right)\right]^2 Rd\theta$$

$$= \frac{W^2R^4 l}{4\,EI}\left(\frac{\pi - 2}{\pi R + l}\right)^2 + \frac{W^2R^3}{2\,EI}$$

$$\times \int_0^{\pi/2}\left[\left(\frac{2R + l}{\pi R + l}\right)^2 - 2\left(\frac{2R + l}{\pi R + l}\right)\sin\theta + \frac{1}{2}(1 - \cos 2\theta)\right] d\theta$$

$$= \frac{W^2R^4 l}{4\,EI}\left(\frac{\pi - 2}{\pi R + l}\right)^2 + \frac{W^2R^3}{2\,EI}\left[\frac{\pi}{2}\left(\frac{2R + l}{\pi R + l}\right)^2 - 2\left(\frac{2R + l}{pR + l}\right) + \frac{\pi}{4}\right]$$

Therefore

$$\delta_{KL} = \frac{\partial U}{\partial W} = \frac{WR^4 l}{2EI}\frac{(\pi^2 + 4 - 4\pi)}{(\pi r + l)^2} + \frac{WR^3}{4EI\,(\pi R + l)^2}$$

$$\times [(\pi^3 - 8\pi)\,R^2 + Rl\,(2\pi^2 - 16) + (3\pi - 8)\,l^2]$$

$$= \frac{WR^3}{4\,EI\,(\pi R + l)^2}[(\pi^3 - 8\pi)\,R^2 + (4\pi^2 - 8\pi - 8)\,Rl + (3\pi - 8)\,l^2]$$

$$= \frac{WR^3}{4\,EI\,(\pi R + l)}[(\pi^2 - 8)\,R + (3\pi - 8)\,l]$$

which is same as found in equation (14.48).

**Example 14.4.** *Fig. 14.13 shows a flat ring of steel 25 mm wide by 6 mm thick, loaded with a central load of 600 N. Calculate the maximum bending moment in the ring. Sketch the B.M. diagram and find the position of the points of inflexion.*

*Taking E as 200 GN/m², find the vertical deflection produced by the load of 600 N.*

**Solution.** For the bending moment diagram two sections will be taken : one at the straight portion Fig. 14.13(*b*) normal to the centre line and the other radial again normal to the centre line Fig. 14.13(*c*).

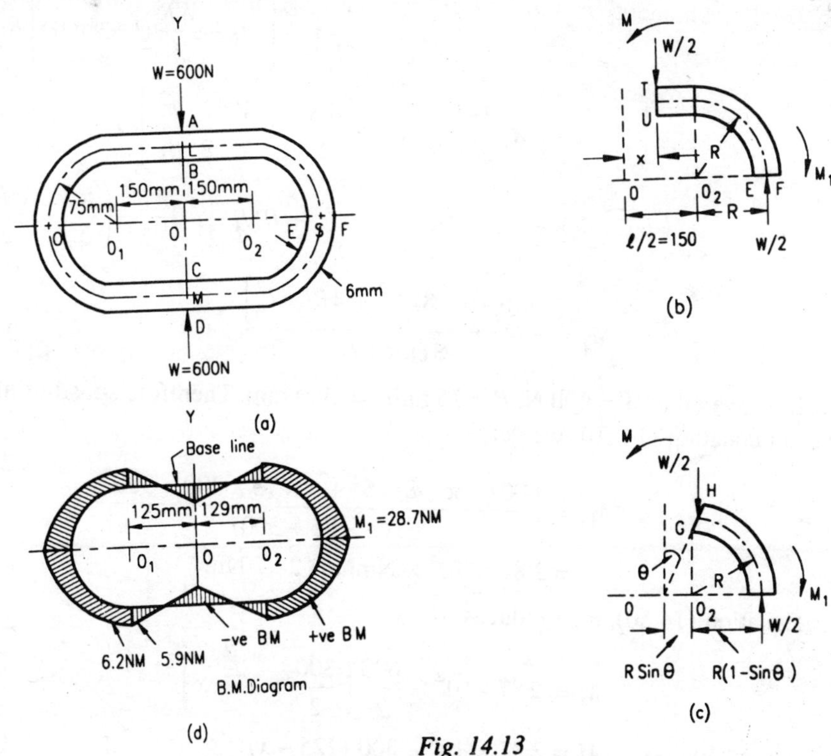


*Fig. 14.13*

Let the *B.M.* at any section *TU* at a distance *x* from the load line be *M* (reckoned positive if the centre of curvatue due to *M* is on the inner side of link). Then from Fig. 14.13(*b*), for the equilibrium of the portion *TUEF*, we have, assuming $M_1$ to be the moment on *EF*

$$M = M_1 - \frac{W}{2}\left(\frac{l}{2} + R - x\right) \qquad ...(14.50)$$

Similarly if $M_\theta$ is the *B.M.* (Reckoned positive if it increases the curvature) at any radial section *GH* Fig. 14.13(*c*) we have for the equilibrium of the element *GHEF*

$$M_\theta = M_1 - \frac{WR}{2}(1 - \sin\theta) \qquad ...(14.51)$$

Due to symmetry, the sections *AB* and *EF* of the link remain at right angles even after the application of load, that is, the total rotation of the section *EF* with respect to *AB* will be zero.

Thus $$\int_0^{l/1} \frac{M}{EI} dx + \int_0^{\pi/2} \frac{M_\theta}{EI} (Rd\theta) = 0$$

Substituting values of M and $M_\theta$, we get

$$\frac{1}{EI}\int_0^{l/2}\left[M_1 - \frac{W}{2}\left(\frac{l}{2} + R - x\right)\right] dx$$

$$+ \frac{1}{EI}\int_0^{\pi/2}\left[M_1 - \frac{WR}{2}(1 - \sin\theta)\right] Rd\theta = 0$$

or $$\left[M_1 \frac{l}{2} - \frac{Wl}{4}.\frac{l}{2} - \frac{WR}{2}\frac{l}{2}.\frac{W}{2}.\frac{l}{2}.\frac{l^2}{4}\right]$$

$$+ RM_1 \frac{\pi}{2} - \frac{WR^2}{2}\left(\frac{\pi}{2}\right) + \frac{WR^2}{2} = 0$$

or $$M_1 = \frac{W\left[(4\pi - 8) R^2 + 4Rl + l^2\right]}{8(\pi R + l)} \qquad ...(14.52)$$

It is given that $W = 600$ N, $R = 75$ mm, $l = 300$ mm. Therefore substituting these values in equation (14.50), we get

$$M_1 = \frac{600\,[(4\pi - 8)\,75^2 + 4 \times 75 \times 300 + 300^2}{8(\pi \times 75 + 300)}$$

$$= 2.87 \times 10^4 \times \text{Nmm} = 28.7 \text{ Nm}$$

Equation (14.50), then reduces to

$$M = 2.87 \times 10^4 - \frac{600}{2}\left(\frac{300}{2} + 75 - x\right)$$

or $$M = 2.87 \times 10^4 - 300\,(225 - x) \qquad ...(14.53)$$

Therefore $$(M)_{x=0} = 2.87 \times 10^4 - 300 \times 255 = -3.88 \times 10^4 \times \text{Nmm}$$

$$= -38.8 \times \text{Nm}$$

$\therefore$ $$(M)_{x=\frac{l}{2}=150} = 2.87 \times 10^4 - 300\,(225 - 150)$$

$$= 0.59 \times 10^4 \times \text{Nmm} = 5.9 \times \text{Nm}$$

Evidently the point of inflexion will be obtained by equating

$$M = 0$$

*i.e.* $$0 = 2.87 \times 10^4 - 300\,(225 - x), \text{ hence } x = 129 \text{ mm.}$$

Similarly equation (14.51) reduces to

$$M_\theta = 2.87 \times 10^4 - \frac{600}{2} \times 75\,(1 - \sin\theta)$$

or
$$M_\theta = 2.87 \times 10^4 - 22500\,(1 - \sin\theta) \qquad ...(14.54)$$
$$(M)_\theta = 0 = 2.87 \times 10^4 - 2.25 \times 10^4$$
$$= 0.62 \times 10^4 \text{ Nmm} = 6.2 \text{ Nm}$$
$$(M)_{\theta = \pi/2} = 2.87 \times 10^4 \text{ Nmm} = 28.7 \text{ Nm}$$

Bending moment diagram is plotted with the help of equation (14.53) and (14.54) as shown in Fig. 14.13(*d*).

For finding the deflection along the load line, the help of equation (14.32) is taken

$$\delta_{LM} = -2\int_0^{l/2} \frac{pM}{EI}\,dx - 2\int_0^{\pi/2} \frac{pM_\theta}{EI}\,Rd\theta$$

*i.e.*,
$$\delta_{LM} = \frac{-2}{EI}\int_0^{l/2} x\left[M_1 - \frac{W}{2}\left(\frac{l}{2} + R - x\right)\right]dx$$
$$-\frac{2}{EI}\int_0^{l/2}\left(\frac{l}{2} + R\sin\theta\right)\left[M_1 - \frac{W}{2}R.(1 - \sin\theta)\right]Rd\theta$$
$$= -\frac{2}{EI}\left[\left(M_1 - \frac{Wl}{4} - \frac{WR}{2}\right)\frac{l^2}{8} + \frac{W}{2}\frac{1}{3}\frac{l^3}{8}\right]$$
$$-\frac{2R}{EI}\int_0^{l/2}\frac{l}{2}\left[M_1 - \frac{W}{2}R(1 - \sin\theta)\right]d\theta$$
$$-\frac{2R^2}{EI}\int_0^{l/2}\left[\left(M_1 - \frac{WR}{2}\right)\sin\theta + \frac{WR}{2} + \sin^2\theta\right]d\theta$$
$$= \frac{-2}{EI}\left(\frac{l^2}{8}\right)\left[M_1 - \frac{Wl}{4} - \frac{WR}{2} + \frac{Wl}{6}\right] - \frac{Rl}{El}$$
$$\times\left[\left(M_1 - \frac{WR}{2}\right)(\theta) + \frac{WR}{2}(-\cos\theta)\right]_0^{\pi/2}$$
$$-\frac{2R^2}{EI}\left[\left(M_1 - \frac{WR}{2}\right)(-\cos\theta) + \frac{WR}{4}(\theta) - \frac{WR}{8}\sin 2\theta\right]_0^{\pi/2}$$

or
$$\delta_{LM} = -\frac{l^2}{48\,EI}(12M_1 - 6WR - Wl) - \frac{Rl}{EI}$$
$$\times\left[\left(M_1 - \frac{WR}{2}\right)\frac{\pi}{2} + \frac{WR}{2}\right] - \frac{2R^2}{EI}\left(M_1 - \frac{WR}{2} + \frac{\pi WR}{8}\right) \qquad ...(14.55)$$

Substituting values of various quantities

$$\delta_{LM} = \frac{-(300)^2 \times 12\,(12 \times 2.87 \times 10^4 - 6 \times 600 \times 75 - 600 \times 300)}{48 \times 200 \times 10^3 \times 25 \times 6^3}$$

$$- \frac{75 \times 300 \times 12}{200 \times 10^3 \times 25 \times 6^3}\left[\left(2.87 \times 10^4 - \frac{600 \times 75}{2}\right) \times \frac{\pi}{2} + \frac{600 \times 75}{2}\right]$$

$$- \frac{2 \times 75^2 \times 12}{200 \times 10^3 \times 25 \times 6^3}\left[2.87 \times 10^4 - \frac{600 \times 75}{2} + \frac{\pi}{8} \times 600 \times 75\right]$$

$$= 2.20 - 2.99 - 8.06 = -8.85 \times \text{mm}$$

(Negative sign indicates that the points $L$ and $M$ move towards each other).

**Example 14.5.** *A steel rod of radius $r_1$ is bent into a shape shown in Figure (14.14) one end being rigidly clamped, the other end being constrained to move along the slide. Determine the deflection of the slider when acted upon by a force P.*

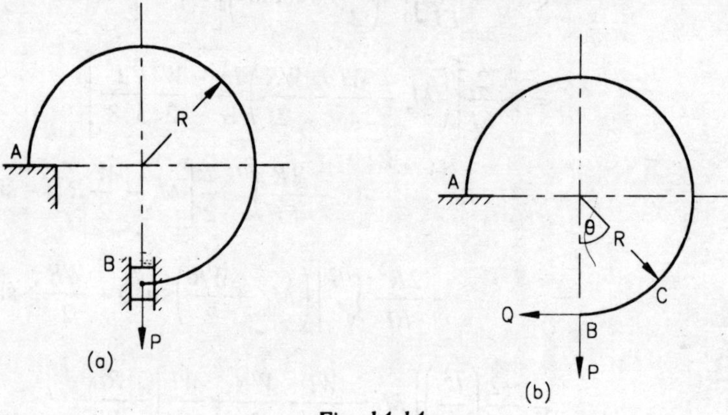


*Fig. 14.14*

**Solution.** Let the reaction due to the guides at $B$ be $Q$ as shown in Fig. 14.14(*b*). The bending moment $M$ due to both $P$ and $Q$ about any section $C$ is given as

$$M = P\,R \sin\theta - QR\,(1 - \cos\theta)$$

Then using Castigliano's theorem, the deflection of $B$ along $P$ is given by

$$\theta_B = \frac{\partial U}{\partial P}$$

$$= \frac{\partial}{\partial p}\left[\int_o^l \frac{M^2}{2\,EI}\,dl\right]$$

$$= \frac{1}{EI}\left[\int_o^l \frac{\partial M}{\partial P}\,dl\right]$$

$$= \frac{1}{EI}\int_o^{3\pi/2} [PR\sin\theta - QR(1-\cos\theta)][R\sin\theta].\,Rd\theta$$

$$= \frac{R^3}{EI}\int_o^{\frac{3\pi}{2}} [P\sin^2\theta - Q(\sin\theta - \sin\theta\cos\theta)]\,d\theta$$

$$= \frac{R^3}{EI}\int_o^{3\pi/2}\left[\frac{P}{2}(1-\cos 2\theta) - Q\sin\theta + \frac{Q}{2}\sin 2\theta\right]d\theta$$

$$= \frac{R^3}{EI}\left[\frac{P}{2}\theta - \frac{P}{4}\sin 2\theta + Q\sin\theta - \frac{Q}{4}\cos 2\theta\right]_0^{3\pi/2}$$

$$= \frac{R^3}{EI}\left[\frac{P}{2}\frac{3\pi}{2} - 0 + Q(0-1) - \frac{Q}{4}(-1-1)\right]$$

*i.e.,*
$$\delta_B = \frac{R^3}{EI}\left[\frac{3\pi}{4}p - \frac{Q}{2}\right] \quad ...(14.56)$$

The reaction $Q$ is evaluated by the condition that the deflection of $B$ along $Q$ is zero.

*i.e.*
$$\frac{\partial U}{\partial Q} = 0$$

or
$$\frac{1}{EI}\int_0^l M\frac{\partial M}{\partial Q}dl = 0$$

or
$$\frac{1}{EI}\int_o^{3\pi/2}[PR\sin\theta - QR(1-\sin\theta)][-R(1-\cos\theta)]\,Rd\theta = 0$$

or
$$\frac{R^3}{EI}\int_0^{3\pi/2}[P\sin\theta - P\sin\theta.\cos\theta) - Q(1 - 2\cos\theta + \cos^2\theta)]\,d\theta = 0$$

or
$$P\int_o^{3\pi/2}\left[\sin\theta - \frac{1}{2}\sin 2\theta\right]d\theta = Q\int_o^{3\pi/2}\left[1 - 2\cos\theta + \frac{1}{2}(1+\cos 2\theta\right]d\theta$$

or
$$P\left[-\cos\theta + \frac{\cos 2\theta}{4}\right]_0^{3\pi/2} = Q\left[\theta - 2\sin\theta + \frac{\theta}{2} + \frac{\sin 2\theta}{4}\right]_0^{3\pi/2}$$

or
$$P\left(1 - \frac{2}{4}\right) = Q\left[\frac{3\pi}{2} - 2(-1) + \frac{3\pi}{2}\right]$$

or
$$\frac{P}{2} = Q\left(\frac{9\pi}{4} + 2\right)$$

or
$$Q = \frac{2P}{9\pi + 8} \quad ...(14.57)$$

From equations (14.56) and (14.57), we get

$$\delta_B = \frac{R^3}{EI}\left[\frac{3\pi}{4}P - \frac{P}{9\pi+8}\right]$$

$$= \frac{PR^3}{EI}\left[\frac{3\pi}{4} - \frac{1}{9\pi+8}\right] \quad ...(14.58)$$

Substituting the value of $I$, the final result is

$$\delta_B = \frac{4PR^3}{\pi r^4 E}\left[\frac{3\pi}{4} - \frac{1}{9\pi+8}\right] \quad ...(14.59)$$

**Example 14.6.** *A circular steel wire of diameter d is bent as shown in Fig. 14.15. It is rigidly held at the end A so that the centre line of the wire lies in the horizontal plane. Calculate the vertical deflection at the free end B when a vertical load W is placed there.*

**Solution.** The wire $AB$ is in the horizontal plane and the load $W$ at end $B$ is vertical. Evidently any section $C$ of the wire situated at angle $\theta$ from $OB$, will experience a bending moment $M$ (= $W \times BS$) and twisting moment $T$ (= $W \times CS$). The strain energy of the wire can then be found, the partial differentiation of which with respect to $W$ will give the deflection of the free end $B$.

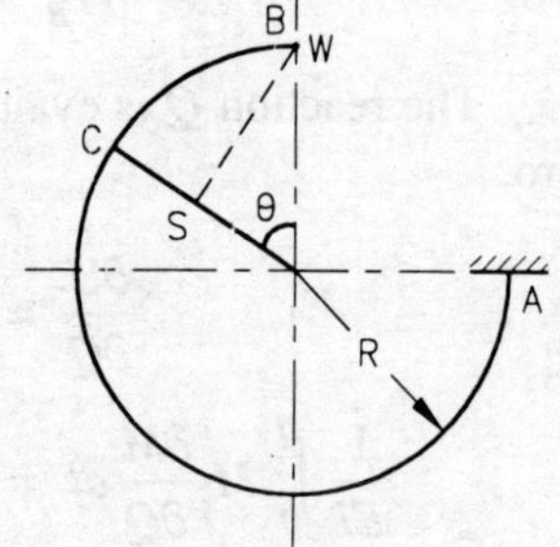


Fig. 14.15

$$U = \int_0^l \frac{M^2}{2EI}dl + \int_0^l \frac{T^2}{2GJ}dl$$

$$\delta_B = \frac{\partial U}{\partial W} = \frac{1}{2EI}\int_0^l \left(2M\frac{\partial M}{\partial W}\right)dl + \frac{1}{2GJ}\int_0^l 2T\frac{\partial T}{\partial W}dl \quad ...(14.60)$$

Now $\quad M = W(BS) = WR\sin\theta\ ;\ \frac{\partial M}{\partial W} = R\sin\theta \quad ...(14.61a)$

and $\quad T = W(CS) = WR(1-\cos\theta)\ ;\ \frac{\partial T}{\partial W} = R(1-\cos\theta) \quad ...(14.61b)$

From equations (14.60) and (14.61)

$$\therefore \quad \delta_B = \frac{1}{EI}\int_0^l \left(M\frac{\partial M}{\partial W}\right)dl + \frac{1}{GJ}\int_0^l \left(T\frac{\partial T}{\partial W}\right)dl$$

or

$$\delta_B = \frac{1}{EI}\int_0^{3\pi/2} (WR\sin\theta)(R\sin\theta)\,Rd\theta$$

$$+ \frac{1}{GJ}\int_0^{3\pi/2} (1 - 2\cos\theta + \cos^2\theta)\,d\theta$$

$$= \frac{WR^3}{2EI}\int_0^{3\pi/2}(1-2\cos\theta)\,d\theta + \frac{WR^3}{GJ}\int_0^{3\pi/2}(1-2\cos\theta+\cos^2\theta)\,d\theta$$

$$= \frac{WR^3}{2EI}\int_0^{3\pi/2}\left[\theta-\frac{\sin 2\theta}{2}\right]_0^{3\pi/2} + \frac{WR^3}{GJ} \times \int_0^{3\pi/2}\left[1-2\cos\theta+\frac{1}{2}(1+\cos 2\theta)\right]$$

$$= \frac{WR^3}{2EI}\left[\frac{3\pi}{2}-0\right] + \frac{WR^3}{GJ}\left[\frac{3}{2}(\theta)-2\sin\theta+\frac{\sin 2\theta}{4}\right]_0^{3\pi/2}$$

$$= \frac{3\pi\, WR^3}{4\,EI} + \frac{WR^3}{GJ}\left[\frac{9\pi}{4}-2(-1)+0\right]$$

$$= \frac{3\pi\, WR^3}{4\,EI} + \frac{WR^3}{GJ}\left(\frac{9\pi}{4}+2\right)$$

$$= \frac{3\pi}{4}\,\frac{WR^3\times 64}{E\pi d^4} + \frac{WR^3\times 32}{G\pi d^4}\left(\frac{9\pi}{2}+2\right)$$

*i.e.,*

$$\delta_B = \frac{16WR^3}{d^4}\left[\frac{3}{E}+\frac{2}{G}\left(\frac{9}{4}+\frac{2}{\pi}\right)\right] \qquad \text{...(14.62)}$$

## PROBLEMS FOR PRACTICE

1. A rectangular bar 50 mm wide by 75 mm deep is bent into a circular form with the plane of the arc parallel to its depth, the radius of curvature being 100 mm. If the bar is now subjected to a bending moment of 37.5 Nm tending to decrease its curvature, draw the stress distribution diagram for the section.

2. A round bar of steel 38 mm diameter is bent into a curve of mean radius 31.7 mm. If a bending moment of 4.6 Nm tending to increase the curvature, acts on the bar find the intensities of maximum tensile and compressive stresses.

   [**Ans.** 0.564 MN/m$^2$ (Tensile) ; 1.605 MN/m$^2$ (comp).]

3. A crane hook has a symmetrical trapezoidal section 56 mm deep, the inner and outer widths being 50 mm and 25 mm respectively. Estimate the extreme intensities of stress when the hook carries a load of 12.5 KN, the load line passing 50 mm from the inside edge of the section and the centre of curvature being in the load line. [**Ans.** 61.2 MN/m$^2$ tensile ; 36 MN/m$^2$ comp.]

4. A steel ring of rectangular cross-section 7.5 mm wide by 5 mm thick has mean diameter of 300 mm. A narrow radial saw cut is made and tangential separating forces of 5 N each are applied at the cut in the plane of the ring. Determine the additional separation due to these forces. $E = 200 \times \text{GN/m}^2$ **[Ans.** 4.52 mm]

5. A steel ring, having a narrow radial saw cut, is made of a flat steel strip 20 mm × 3 mm deep with the depth parallel to the radius. The mean diameter of the ring is 200 mm. A pull of 37.5 N acts along the diameter perpendicular to the diameter passing through the point of saw cut. Find the increase in the opening at the point of saw cut. Take $E = 200\ \text{GN / m}^2$. **[Ans.** 14.9 mm]

   **Hint :** Assume a pull **Q** at the saw cut point. The increase in opening will then be

   $$\left(\frac{\partial U}{\partial Q}\right)_{Q=0}.$$

6. In question number 5, find the intensities of maximum tensile and compressive stresses.

7. A proving ring is 250 mm diameter, 40 mm wide and 6 mm thick. The maximum stress permitted is 550 MN/m$^2$. Find the load to cause this stress and the load to give 6 mm deflection of the ring in the direction of the loading.

   Prove any formulae used which are particular to this problem.

   $E = 200\ \text{GN/m}^2$ **[Ans.** 3.325 kN ; 2.98 kN]

8. A ring of mean diameter of 250 mm is made of round steel of 20 mm diameter. A pull of 4 kN acts on the ring. Draw the bending moment diagram for the ring. Calculate the greatest tensile and compressive stresses. Also estimate the increase in the mean diameter of the ring along the load line and the decrease in the diameter in the lateral direction. Take $E = 200\ \text{GN/m}^2$.

   **[Ans.** 33.6 MN/m$^2$ tensile ; 30.9 MN/m$^2$ comp ; 2.96 mm ; 2.72 mm]

9. A chain link as shown in Fig. 14.11, is made of round steel rod of 6 mm diameter. If $R = 25$ mm and $l = 45$ mm, calculate the ratio of the maximum tensile stress at the section where load is applied to that at the section half way along the straight portion. **[Ans.** 2.68]

10. A steel bar 65 mm diameter is bent to the shape shown in Fig. 14.16 and the lower end is firmly fixed in the ground in a vertical position. A load of 1 kN is applied at the free end. Calculate the vertical deflection of the free end. $E = 200\ \text{GN/m}^2$. **[Ans.** 12.95 mm]

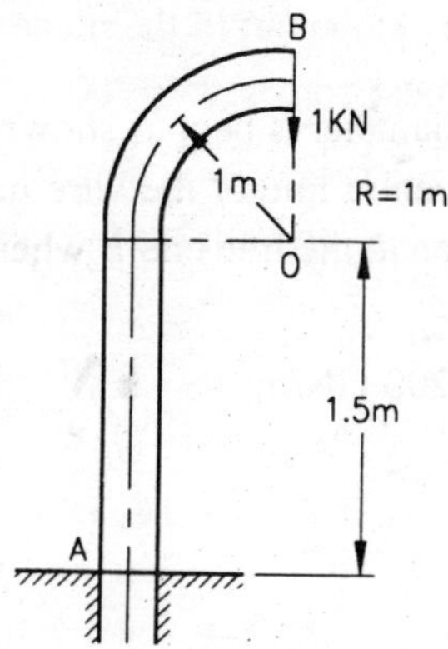


*Fig. 14.16*

11. A steel spring is shown in Fig. 14.17. It is formed in a circular arc subtending an angle of 270° at the centre. The lower end is rigidly fixed and a vertical force of 5 N is applied at the free end.

    If the section of the spring is 12 mm wide and 3 mm thick, calculate vertical and horizontal displacements at the free end.

    $E = 200$ GN/m$^2$ **[Ans.** 3.65 mm ; 0.195 mm]

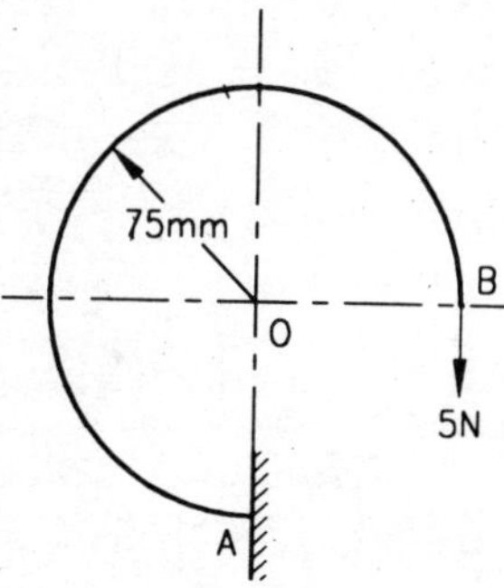


*Fig. 14.17*

12. Fig. 14.14 shows a steel rod. 10 mm diameter, with one end firmly fixed to a horizontal table. The remainder of the rod is bent into the form of three quarters of a circle and the free end of the rod is constrained by guides to move in a vertical direction. If the mean radius to which the rod is bent is 150 mm, determine

the vertical deflection of the free end when a 100 kN load is gradually applied there.

$E = 200$ GN/m$^2$ **[Ans.** 8 mm]

13. A circular steel wire 10 mm diameter is bent as shown in Fig. 14.15. It is rigidly held at the end $A$ so that the centre line of the wire lies in the horizontal plane. Calculate the vertical deflection at the free end $B$ when a vertical load of 20 $N$ is placed there. Take

$R = 150 \times$ mm, $E = 200$ GN/m$^2$

and $G = 80$ GN m$^2$ **[Ans.** 9.42 mm]

## Chapter–15

# Stresses Due to Rotation

## 15.1 Rotating Ring

Let a ring of mean radius $r$ be rotating about an axis through its centre of gravity '$O$' and perpendicular to its plane of rotation with an angular speed of $\omega$ radians per second as shown in Fig. 15.1.

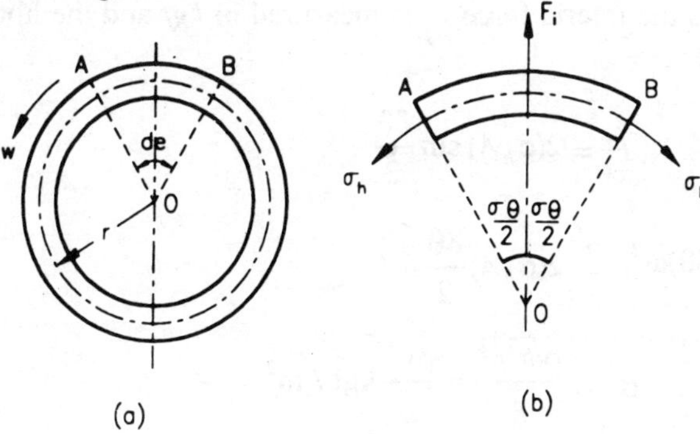


*Fig. 15.1*

Due to rotation each and every element of the ring, like $AB$, will experience an intertia or centrifugal force $F_i$ which will tend to expand the ring radially outwards. This in turn will induce the circumferential or hoop stress $\sigma_h$ in the ring which will be tensile in nature. In order to evaluate this stress following assumptions are made.

(*i*) The cross-sectional dimensions of the ring are small as compared to its mean radius.

(*ii*) The hoop stress on the cross-sectional area '$A$' of the ring is uniform.

(*iii*) The constraining effect of the spokes is neglected. If $\rho$ is the density of the material in kg/m$^3$, the inetria force $F_i$ in Newtons on a small element $AB$ which subtends an angle $\delta\theta$ at the centre, will be equal to the product of mass of the element and the centripetal acceleration which is equal to $\omega^2 r$ but directed in the radially outward direction. Thus considering the equilibrium of the element $AB$ and resolving the forces along $F_i$ Fig. 15.1(*b*), we get

$$F_i = 2(\sigma_h A) \sin \frac{\delta\theta}{2}$$

Since $\delta\theta$ is small,

$$\sin \frac{\delta\theta}{2} \simeq \frac{\delta\theta}{2}$$

Therefore $$\rho(Ar\delta\theta)\omega^2 r \simeq 2(r_h A)\frac{\delta\theta}{2}$$

or $$\theta_h = \rho\omega^2 r^2 = \rho v^2 \text{ N/m}^2 \quad \text{...(15.1)}$$

Where $v = r\,\omega$ = The mean peripheral speed.

If $\sigma_a$ is the allowable stress for any material, then limiting peripheral speed for that material can be evaluated from equation (15.1) by substituting $\sigma_a$ for $\sigma_h$.

In M.K. Units the interia force $F_j$ is measured in *kgf* and the hoop stress, $\sigma_h$ in kgf/m$^2$, in that case

$$F_i = 2(\sigma_h A) \sin \frac{\delta\theta}{2}$$

or $$\frac{\rho}{g}(Ar\delta\theta)\omega^2 r \simeq 2(\sigma_h A)\frac{\delta\theta}{2}$$

or $$\sigma_h = \frac{\rho\omega^2 r^2}{g} = \frac{\rho v^2}{g} \text{kgf / m}^2 \quad \text{...(15.2)}$$

**Example 15.1.** *Calculate the hoop stress in a thin rim, 0.6 m mean diameter revolving about its axis at 800 r.p.m. steel weighs 7700 kg/m$^3$.*

**Solution.**

$$\omega = \frac{2\pi \times 800}{60} - 83.7758 \text{ rad / sec.}$$

Then $$\sigma_h = \rho\omega^2 r^2 = 7700\,(83.7758)^2 \left(\frac{0.6}{2}\right)^2 \text{ N / m}^2$$

$$= 4.864 \text{ MN/m}^2.$$

**Example 15.2.** *For a cast iron flywheel find the limiting peripheral speed, if the allowable stress in C.I. is 6.6 MN/m$^2$ and it weighs 7212 kg/m$^3$.*

**Solution.** From equation (15.1)

$$v = \sqrt{\frac{\sigma_h}{\rho}} = \sqrt{\frac{6.6 \times 10^6}{7212}} \text{ m/sec.}$$

$$= 30.25 \text{ m/sec.}$$

## 15.2 Thin Rotating Disc

Let a thin circular disc be rotating at an angular speed ω rad/sec about its central axis, normal to the plane of rotation. It is assumed that the thickness '*t*' of the disc is uniform and very small as compared to the radius. Since the flat faces normal to *Z*-axis are free there will be no shear stress on them. The flat faces for disc will be principal planes of zero principal stresses. Evidently the radial and hoop stresses on any element Fig. 15.2(*b*) will also be principal stresses.

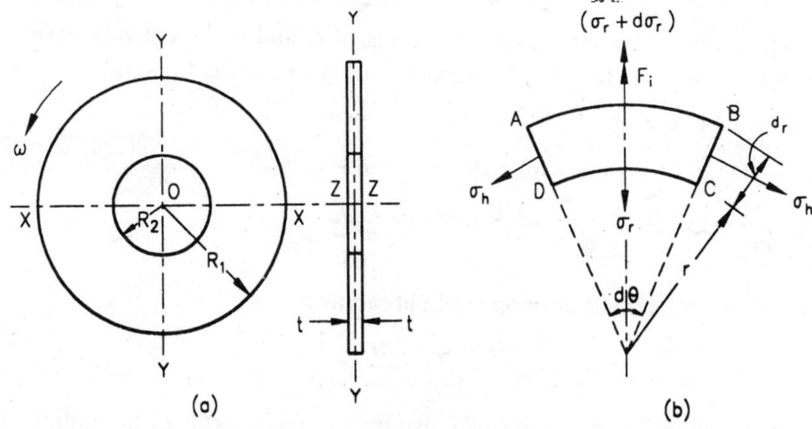


*Fig. 15.2*

In Fig. 15.2(*b*) is shown the free body diagram of an element *ABCD* such that the radius of the face *DC* is *r* and that of *AB* (*r* + *dr*) and the element is bounded within radial lines *AO* and *BO* subtending a small angle *d*θ at centre *O*. The hoop and radial stresses at radius r are represented by $\sigma_h$ and $\sigma_r$. N/m$^2$ (They are reckoned positive when tensile). Let the radial stress at radius (*r* + *dr*) be ($\sigma_r + d\sigma_r$). Let ρ be the density of the material in kg/m$^3$, then $F_i$ N, the centrifugal or inertia force on the element *ABCD* due to rotation will be equal to the product of the mass of the element and the centripetal acceleration in magnitude.

Then considering the equilibrium of the element and resolving the forces along the radial line, we have

$$F_i + (\sigma_r + d\sigma_r)\,(r + dr)\,d\theta\,.\,t - \sigma_r\,(rd\theta t) - 2\sigma_h\,(dr.t)\sin\frac{d\theta}{2} = 0$$

Neglecting the product of two small quantities and substituting the value of $F_i$ and also letting $\sin\frac{d\theta}{2} \simeq \frac{d\theta}{2}$, we get

$$(\rho r\,.\,d\theta\,.\,dr.t)\,\omega^2 r + [\sigma_r\,dr + rd\,\sigma_r - \sigma_h dr]\,t.d\theta = o$$

or

$$\sigma_h = \rho\omega^2 r^2 + \sigma_r + r\frac{d\sigma_r}{dr} \quad ...(15.3)$$

## Strain considerations

When the disc starts rotating, due to the centrifugal force on each and every element it will tend to expand. Let the radius of the arc $CD$ increase from $r$ to $(r + u)$ and that of $AB$ from $(r + dr)$ to $(r + dr + u + du)$.

Then the radial strain is given by

$$\sigma_r = \frac{D'A' - DA}{DA}$$

Where $D'$ and $A'$ are the new positions of $D$ and $A$ respectively. Evidently the radial movement of $D$ (= $DD'$) is $u$ and that of $A$ (= $AA'$) is $(u + du)$.

Thus $D'A' = DA + (AA' - DD')$

$$= dr + (u + du) - u = dr + du$$

Then $$\varepsilon_r = \frac{(dr + du) - dr}{dr} = \frac{du}{dr} \qquad \text{...(15.4)}$$

The circumferential or hoop strain at radius $r$

$$\varepsilon_h = \frac{2\pi(r + u) - 2\pi r}{2\pi r} = \frac{u}{r} \qquad \text{...(15.5)}$$

The radial and hoop strains can also be written in terms of the radial and hoop stresses and the Poisson's ratio (Chapter 2)

$$\varepsilon_r = \frac{du}{dr} = \frac{1}{E}(\sigma_r - \nu\sigma_h) \qquad \text{...(15.6)}$$

and $$\varepsilon_r = \frac{u}{r} = \frac{1}{E}(\sigma_h - \nu\sigma_r) \qquad \text{...(15.7)}$$

Solving the simultaneous equations (15.6) and (15.7)

$$\sigma_r = \frac{E}{1 - \nu^2}\left(\frac{\nu u}{r} + \frac{du}{dr}\right) \qquad \text{...(15.8)}$$

$$\sigma_h = \frac{E}{1 - \nu^2}\left(\frac{u}{r} + \nu\frac{du}{dr}\right) \qquad \text{...(15.9)}$$

Substituting equations (15.8) and (15.9) in equation (15.3)

$$r\frac{d^2u}{dr^2} + \frac{du}{dr} - \frac{u}{r} = -\rho\frac{\omega^2}{E}(1 - \nu^2)\,r^2 \qquad \text{...(15.10)}$$

or $$\frac{d^2u}{dr^2} + \frac{1}{r}\frac{du}{dr} - \frac{u}{r^2} = -\rho\frac{\omega^2}{E}(1 - \nu^2)\,r \qquad \text{...(15.11)}$$

For the solution of the differential equation (15.11) its complementary function is

$$\frac{d^2u}{dr^2} + \frac{1}{r}\frac{du}{dr} - \frac{u}{r^2} = 0$$

or
$$\frac{d^2u}{dr} + \frac{d}{dr}\left(\frac{u}{r}\right) = 0$$

Which on integration gives

$$\frac{du}{dr} + \frac{u}{r} = 2A \quad \text{...(15.12)}$$

Where $2A$ is a constant of integration.

Equation (15.12) can be re-written as

$$r\frac{du}{dr} + u = 2Ar$$

or
$$\frac{d}{dr}(ur) = 2Ar$$

Again on integration, we get

$$ur = Ar^2 + B$$

or
$$\frac{u}{r} = A + \frac{B}{r^2} \quad \text{...(15.13)}$$

Where $B$ is another constant.

From equations (15.12) and (15.13), we have

$$\frac{du}{dr} = A - \frac{B}{r^2} \quad \text{...(15.14)}$$

For finding the particular integral of equation (15.11) assume

$$u = cr^3 \quad \text{...(15.15)}$$

where c is a constant

or
$$\frac{u}{r} = cr^2 \quad \text{...(15.16)}$$

Also by differentiating equation (15.15)

$$\frac{du}{dr} = 3cr^2 \quad \text{...(15.17)}$$

and
$$\frac{d^2u}{dr^2} = 6cr \quad \text{...(15.18)}$$

Then substituting equations (15.15) through (15.18) in equation (15.11), we get the value of $c$ as

$$6cr + \frac{1}{r}3cr^2 - cr = -\rho\frac{\omega^2}{E}(1 - \nu^2)r$$

or
$$c = -\frac{\rho\omega^2}{8E}(1-\nu^2) \quad ...(15.19)$$

Thus from equations (15.13), (15.14), (15.16), (15.17) and (15.19) the complete solution is

$$\frac{u}{r} = A + \frac{B}{r^2} - \frac{\rho\omega^2}{8E}(1-\nu^2)\, r^2 \quad ...(15.20)$$

$$\frac{du}{dr} = A - \frac{B}{r^2} - \frac{3\rho\omega^2}{8E}(1-\nu^2)\, r^2 \quad ...(15.21)$$

Inserting the equations (15.20) and (15.21) in equations (15.8) and (15.9), we get the radial and hoop stresses as

$$\sigma_r = \frac{E}{(1-\nu^2)}\left[(1+\nu)A - (1-\nu)\frac{B}{r^2} - (3+\nu)\frac{\rho\omega^2}{8E}(1-\nu^2)r^2\right] \quad ...(15.22)$$

$$\sigma_h = \frac{E}{(1-\nu^2)}\left[(1+\nu)A + (1-\nu)\frac{B}{r^2} - (1+3\nu)\frac{\rho\omega^2}{8E}(1-\nu^2)r^2\right] \quad ...(15.23)$$

The values of the constants $A$ and $B$ will depend upon the end conditions.

### 15.2.1 Case I. Hollow Disc

Let $R_1$ be the outer radius of the disc and $R_2$ the radius of the central hole. Then evidently the radial stresses at these radii will be zero. These conditions will help to evaluate the constants of integration $A$ and $B$.

Thus from equation (15.22)

$$(1+\nu)A - (1-\nu)\frac{B}{R_1^2} = (3+\nu)\frac{\rho\omega^2}{8E}(1-\nu^2)R_1^2 \quad ...(15.24)$$

and
$$(1+\nu)A - (1-\nu)\frac{B}{R_2^2} = (3+\nu)\frac{\rho\omega^2}{8E}(1-\nu^2)R_2^2 \quad ...(15.25)$$

Solving the simultaneous equations (15.24) and (15.25), we get

$$A = (3+\nu)(1-\nu)\frac{\rho\omega^2}{8E}(R_1^2 + R_2^2) \quad ...(15.26)$$

$$B = (3+\nu)(1+\nu)\frac{\rho\omega^2}{8E}R_1^2 R_2^2 \quad ...(15.27)$$

Inserting these values in equations (15.22) and (15.23), we get the values of radial and hoop stresses in a hollow disc

$$\sigma_r = \frac{\rho\omega^2}{8}(3+\nu)\left[R_1^2 + R_2^2 - \frac{R_1^2 R_2^2}{r^2} - r^2\right] \quad ...(15.28)$$

$$\sigma_h = \frac{\rho\omega^2}{8}\left[(3+\nu)(R_1^2 + R_2^2) + \frac{R_1^2 R_2^2}{r^2} - (1-3\nu)\, r^2\right] \qquad ...(15.30)$$

*Variation of* $\sigma_r$ : Value of $\sigma_r$ at both $r = R_1$ and $r = R_2$ is zero. Moreover it will be tensile at all values of $r$ between $R_2$ and $R_1$. Further for its maximum value $\frac{d\sigma_r}{dr} = 0$. Thus from equation (15.28) for maximum $\sigma_r$, we have

$$\frac{\rho\omega^2}{8}(3+\nu)\left[\frac{2R_1^2 R_2^2}{r^3} - 2r\right] = 0$$

*i.e.*

$$r = \sqrt{R_1 R_2} \qquad ...(15.31)$$

at which radius from equation (15.29), the maximum radial tensile stress is

$$(\sigma_r)_{max} = \frac{\rho\omega^2}{8}(3+\nu)(R_1 - R_2)^2 \qquad ...(15.32)$$

*Variation of* $\sigma_h$ : From equation (15.30) it is evident that the value of $\sigma_h$ is always positive and goes on decreasing with the increase of $r$. Maximum $\sigma_h$ will occur at $r = R_2$ and its value will then be

$$(\sigma_h)_{max} = \frac{\rho\omega^2}{4}\left[(3+\nu) R_1^2 + (1-\nu) R_2^2\right] \qquad ...(15.33)$$

As a special case when $R_2$ approaches $R_1$ equation (15.33) reduces to equation (15.1), for the rotating ring.

### 15.2.2

**Hollow disc with a pin hole at the centre :**

Here $R_2 \to 0$, the from equations (15.32) and (15.33)

$$(\sigma_r)_{max} = \frac{\rho\omega^2}{8}(3+\nu) R_1^2 \qquad ...(15.34)$$

$$(\sigma_h)_{max} = \frac{\rho\omega^2}{4}(3+\nu) R_1^2 \qquad ...(15.35)$$

The variation of $\sigma_r$ and $\sigma_h$ is shown in Fig. 15.3(*a*).

### 15.2.3 Case II. Solid Disc

The end conditions for this case are (*i*) At $r = R_1$, $\sigma_r = 0$ and (*ii*) at $r = 0$, $u = 0$. Then the second condition with the help of equation (15.20) gives $B = 0$. This can also be viewed in another way, that is if $B$ is not zero then at $r = 0$ both $\sigma_r$ and $\sigma_h$ tend to become infinite [equations (15.22) and (15.23),] even at very low speeds. This is not possible. The direct conclusion then is that B must be zero for a solid disc.

Accordingly with the help of first condition we have from equation (15.22) for a solid disc

$$A = \frac{\rho\omega^2}{8E}(3+\nu)(1-\nu)R_1^2 \quad \text{...(15.36}a\text{)}$$

$$B = 0 \quad \text{...(15.36}b\text{)}$$

Thus from equations (15.22), (15.23) and (15.36)

$$\sigma_r = \frac{\rho\omega^2}{8}(3+\nu)(R_1^2 - r^2) \quad \text{...(15.37)}$$

$$\sigma_h = \frac{\rho\omega^2}{8}\left[(3+\nu)R_1^2 - (1+3\nu)r^2\right] \quad \text{...(15.38)}$$

*Variation of* $\sigma_r$ *in solid disc* : Evidently from equation (15.37) $\sigma_r$, being positive, is always tensile and goes on decreasing with increase of $r$ and its maximum value will occur at centre.

$$(\sigma_r)_{max} = \frac{\rho\omega^2}{8}(3+\nu)R_1^2 \quad \text{...(15.39)}$$

*Variation of in* ($\sigma_h$) *solid disc* : From equation (5.38) the value of $\sigma_h$ is maximum when $r = 0$.

That is $$(\sigma_h)\text{max} = \frac{\rho\omega^2}{8}(3+\nu)R_1^2 \quad \text{...(15.40)}$$

This value is same as that of $(\sigma_r)_{max}$ given by equation (15.39). However if we compare it with equation (15.35) it is seen that it is just half in value of the maximum hoop stress induced in disc with pin hole at the centre. It is due to this reason that small holes (or as a matter of fact small blow holes in cast grinding wheels) are avoided at the centre of rotating discs.

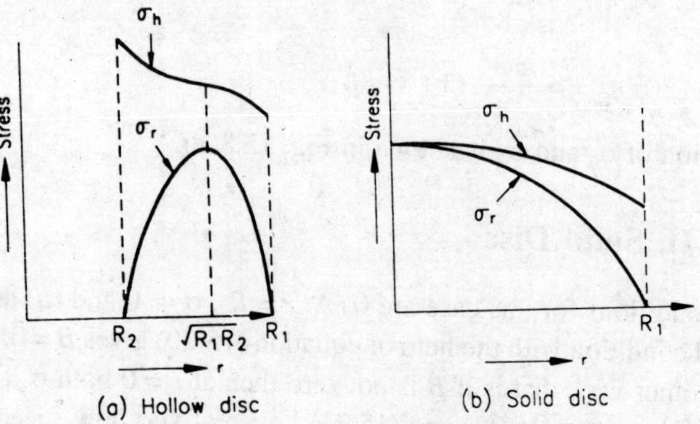


*Fig. 15.3*

The variation of $\sigma_r$ and $\sigma_h$ for solid disc is shown in Fig. 15.3(*b*)

Note about units : In the equations derived above if $\rho$ is in kg/m$^3$, $\omega$ in radians per sec, radius in metres then the stress will be in N/m$^2$.

However if these expressions are to be used in M.K.S. system $\rho$ in kg/m$^2$ has to be divided by '$g$' taken in m/sec$^2$ units to give stress in kg/m$^2$ units.

## 15.3 Rotating Cylinders

The analysis is similar to that of a thin disc, the major difference being that the length of the axis is large as compared to the radius. The problem is tackled by making following simple assumptions.

(*i*) Plane cross-sections remain plane even at high speeds of rotation. Strictly speeking this is true for sections far away from the ends.

(*ii*) On an element lying at the central cross-sectional plane of the cylinder, due to symmetry, there will be no shear stress and thus the radial, circumferential and longitudinal stresses will all be principal stresses. It will be assumed that it is nearly true at all sections except near the ends.

Let $\sigma_r$, $\sigma_h$ and $\sigma_l$ be the radial, hoop and longitudinal stresses at any element of a section of the cylinder at radius $r$ Fig. 15.2(*b*). They are all reckoned positive when tensile. Then proceeding as in article 15.2 radial equilibrium and the strain equations in terms of stresses and Poisson's ratio are

$$\sigma_h = \rho\omega^2 r^2 + \sigma_r + r\frac{d\sigma_r}{dr} \quad ...(15.41)$$

$$\varepsilon_h = \frac{u}{r} = \frac{1}{E}[\sigma_h - (\sigma_r + \sigma_1)\,\nu] \quad ...(15.42)$$

$$\varepsilon_r = \frac{du}{dr} = \frac{1}{E}[\sigma_r - (\sigma_l + \sigma_h)\,\nu] \quad ...(15.43)$$

Where u represents the radial displacement of the point at radius $r$ as in the case of thin disc. In addition the strain along the axial direction will be given by

$$\varepsilon_l = \frac{1}{E}[\sigma_1 - (\sigma_h + \sigma_r)\nu] \quad ...(15.44a)$$

or

$$\sigma_l = E\sigma_l + (\sigma_r + \sigma_h)\,\nu \quad ...(15.44b)$$

Further from the assumption $\varepsilon_l$ must be constant for all values of $r$.

Substituting the value of $\sigma_l$, as calculated from equation (15.44*b*) in equations (15.42) and (15.43) we can express $\sigma_h$ and $\sigma_r$ in terms of strains

$$\sigma_h = \frac{E}{(1-2\nu)(1+\nu)}\left[(1-\nu)\frac{u}{r} + \nu\frac{du}{dr} + \nu\varepsilon_l\right] \quad ...(15.45)$$

$$\sigma_r = \frac{E}{(1-2\nu)(1-\nu)}\left[\nu\frac{u}{r} + (1-\nu)\frac{du}{dr} + \nu\varepsilon_l\right] \quad ...(15.46)$$

Inserting these values of $\sigma_h$ and $\sigma_r$ in equation (15.41), we get

$$\frac{d^2u}{dr^2}+\frac{1}{r}\frac{du}{dr}-\frac{u}{r^2}=-\frac{\rho\omega^2}{E}\frac{(1+\nu)(1-2\nu)}{(1-\nu)}r \quad ...(15.47)$$

Proceeding exactly in the same manner as in the case of thin disc, the complete solution of the differential equation (15.47) is

$$\frac{u}{r}=A+\frac{B}{r^2}-\frac{\rho\omega^2}{8E}\frac{(1+\nu)(1-2\nu)}{(1-\nu)}r^2 \quad ...(15.48)$$

$$\frac{du}{dr}=A-\frac{B}{r^2}-\frac{\rho\omega^2}{8E}\frac{(1+\nu)(1-2\nu)}{(1-\nu)}r^2 \quad ...(15.49)$$

Where $A$ and $B$ are constants of integration to be evaluated with the help of end conditions.

### 15.3.1 Case I. Hollow Cylinder

As in the case of thin disc, the end conditions are $\sigma_r = 0$ at both $r = R_1$ and $R_2$. Substituting these conditions along with equations (15.48) and (15.49) in equation (15.46) we get, after solving the two simultaneous equations so obtained, the values of the constants $A$ and $B$ as

$$A=\frac{\rho\omega^2}{8E}\frac{(1+\nu)(1-2\nu)(3-2\nu)}{(1-\nu)}(R_1^2+R_2^2)-\nu\varepsilon_l \quad ...(15.50)$$

$$B=\frac{\rho\omega^2}{8E}\frac{(1+\nu)(3-2\nu)}{(1-\nu)}R_1^2R_2^2 \quad ...(15.51)$$

*Evaluation of constant* $\varepsilon_l$ : If we take a section at the middle of the cylinder, then as the two portions so obtained will have no motion parallel to the axis, the resultant force on the section in the axial direction will be zero :

$$\int_{R_1}^{R_2}\sigma_1\,(2\pi r dr)=0 \quad ...(15.52)$$

Now from equations (15.44*b*), (15.45), (15.46), (15.48) and (15.49), we get

$$\sigma_l=E\varepsilon_l+\frac{\nu E}{(1-2\nu)(1-\nu)}\left[2A-\frac{\rho\omega^2}{2E}\frac{(1-\nu)(1-2\nu)}{(1-\nu)}r^2+2\nu\varepsilon_l\right] \quad ...(15.53)$$

Inserting the value of $A$ from equation (15.50) in equation (15.53) and then substituting the value of $\sigma_l$ so obtained in equation (15.52) we get on integrating

$$\frac{\rho\omega^2}{4}\left[\frac{(3-2\nu)^2}{1-\nu}\frac{\nu(R_1^4-R_2^4)}{2}-\frac{(R_1^4-R_2^4)^2}{2(1-\nu)}\right]+\frac{1}{2}E\varepsilon_l\,(R_1^2-R_2^2)=0$$

or
$$\varepsilon_l = -\frac{\rho\omega^2}{2E}(R_1^2 + R_2^2) \qquad ...(15.54)$$

Therefore from equations (15.50) and (15.54), the value of $A$ is

$$A = \frac{\rho\omega^2}{8E}\frac{(3-5\nu)}{(1-\nu)}(R_1^2 + R_2^2) \qquad ...(15.55)$$

Thus substituting the values of $A$ and $B$ in equations (15.48) and (15.49), we get the desired expressions for the circumferential and radial strains as

$$\frac{u}{r} = \left(\frac{\rho\omega^2}{8E}\right)\left[\frac{(3-5\nu)}{1-\nu}(R_1^2 + R_2^2) + \frac{(1-\nu)(3-2\nu)\,R_1^2\,R_2^2}{(1-\nu)\,r^2} - \frac{(1+\nu)(3-2\nu)}{(1-\nu)}r^2\right] \qquad ...(15.56)$$

$$\frac{du}{dr} = \frac{\rho\omega^2}{8E}\left[\frac{(3-5\nu)}{(1-\nu)}(R_1^2 + R_2^2) - \frac{(1-\nu)(3-2\nu)}{(1-\nu)}\frac{R_1^2 R_2^2}{r^2} - \frac{3(1+\nu)(1-2\nu)}{(1-\nu)}r^2\right] \qquad ...(15.57)$$

Inserting the values of strains from equations (15.56) and (15.57) in equations (15.45) and (15.46), we get the expressions for hoop and radial stresses as

$$\sigma_h = \frac{\rho\omega^2}{8}\left(\frac{1}{1-\nu}\right)\left[(3-2\nu)\left(R_1^2 + R_2^2 + \frac{R_1^2\,R_2^2}{r^2}\right) - (1+2\nu)r^2\right] \qquad ...(15.58)$$

$$\sigma_r = \frac{\rho\omega^2}{8}\frac{(3-2\nu)}{(1-\nu)}\left[R_1^2 + R_2^2 - \frac{R_1^2 R_2^2}{r^2} - r^2\right] \qquad ...(15.59)$$

Similarly from equations (15.53) through (15.55), we have

$$\sigma_l = \frac{\rho\omega^2}{8}\left(\frac{\nu}{1-\nu}\right)(R_1^2 + R_2^2 - 2r^2) \qquad ...(15.60)$$

*Variation of hoop stress* : Evidently from equation (15.58) the hoop stress is always positive and hence tensile and goes on decreasing with increasing $r$. The greatest value of $\sigma_h$ occurs at $r = R_2$ *i.e.*

*i.e.*
$$(\sigma_h)_{max} = \frac{\rho\omega^2}{4}\left(\frac{1}{1-\nu}\right)\left[3-2\nu)\,R_1^2 + (1-2\nu)\,R_2^2\right] \qquad ...(15.61)$$

The variation of $\sigma_h$ is shown in Fig. 15.4.

Again as in thin disc when there is a pin hole at the centre for which $R_2 \simeq 0$, equation (15.61) reduces to

$$(\sigma_h)_{max} = \frac{\rho\omega^2}{4} \frac{(3-2\nu)}{(1-\nu)} R_1^2 \qquad ...(15.62)$$

Further when $R_2$ approaches $R_1$, $(\sigma_h)_{max}$ from equation (15.61) attains the value $\rho\omega^2 R_1^{\ 2}$, the value which was obtained for a thin rim (article 15.1).

Variation of radial stress : As in thin disc, the radial stress is never negative and is zero at both $r = R_1$ and $R_2$. Further since the equation (15.59) is exactly similar to that for the thin disc except for the constant multiplier, the maximum value of $\sigma_r$ will occur at $r = \sqrt{R_1 R_2}$ for which

$$(\sigma_h)_{max} = \frac{\rho\omega^2}{8} \frac{(3-2\nu)}{(1-\nu)} (R_1 - R_2)^2 \qquad ...(15.63)$$

The variation of $\sigma_r$ is shown in Fig. 15.4.

*Variation of axial stress* : It is evident from equation (15.60) that $\sigma_l$ is maximum when $r$ is least, that is $(\sigma_l)_{max}$ will occur at $r = R_2$ and will go on decreasing continuously as r increases. Thus

$$(\sigma_l)_{max} = \frac{\rho\omega^2}{4} \left(\frac{\nu}{1-\nu}\right) (R_1^2 - R_2^2) \qquad ...(15.64)$$

Further $\sigma_l$ will be zero when

$$2r^2 = R_1^2 + R_2^2$$

or

$$r = \sqrt{\frac{R_1^2 + R_2^2}{2}} \qquad ...(15.65)$$

For all values of $r$ which are more than the value given by equation (15.65) and less than $R_1$, the axial stress, being negative, will be compressive. The maximum compressive axial stress will occur at $r = R_1$

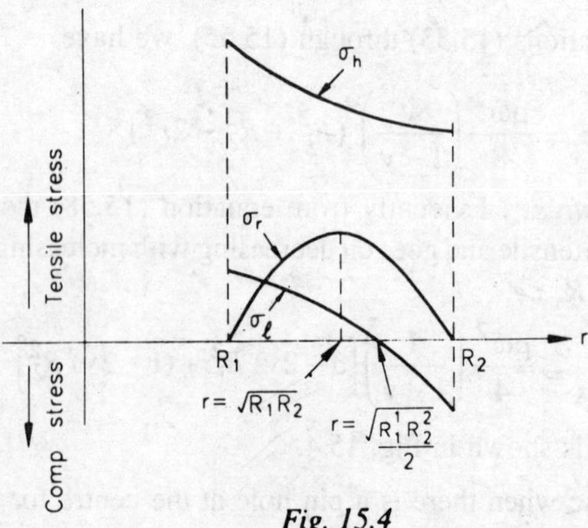


*Fig. 15.4*

$$(\sigma_l)_{r=R_1} = \frac{\rho\omega^2}{8}(R_1^2 - R_2^2) \text{ (Comp.)} \quad ...(15.66)$$

The variation of $\sigma_l$ is shown in Fig. 15.4.

### 15.3.2 Case II. Solid Cylinder

As in the case of solid thin disc, here again the end conditions are (*i*) $\sigma_r = 0$ at $r = R_l$ and (*ii*) $u = 0$ at $r = 0$ for which from equations (15.48) and (15.49)

$$B = 0 \quad ...(15.67)$$

Further from equation (15.46) in the light of first end condition and equation (15.67)

$$A = \frac{\rho\omega^2}{8E}\frac{(1-\nu)(1-2\nu)(3-2\nu)}{(1-\nu)}R_1^2 - \nu\varepsilon_l \quad ...(15.68)$$

Again as in the hollow cylinder, here also the total axial force on any section will be zero, for which

$$\int_o^{R_1} \sigma_l\,(2\pi r dr) = 0 \quad ...(15.69)$$

Substituting equations (15.44*b*), (15.45) and (15.46) in equation (15.69), we find

$$\varepsilon_l = -\frac{\rho\omega^2}{2E}\nu R_1^2 \quad ...(15.70)$$

From equation (15.68) and (15.70)

$$A = \frac{\rho\omega^2}{8E}\frac{3-5\nu}{(1-\nu)}R_1^2 \quad ...(15.71)$$

Thus inserting the values of constants *A* and *B* in equations (15.48) and (15.49) and in turn putting these values in equations (15.45) and (15.46), we get

$$\sigma_h = \frac{\rho\omega^2}{8}\left(\frac{1}{1-\nu}\right)\left[(3-2\nu)R_1^2 - (1-2\nu)r^2\right] \quad ...(15.72)$$

$$\sigma_r = \frac{\rho\omega^2}{8}\frac{(3-2\nu)}{(1-\nu)}(R_1^2 - r^2) \quad ...(15.73)$$

Similarly equation (15.44*b*) reduces to

$$\sigma_l = \frac{\rho\omega^2}{4}\frac{r}{(1-\nu)}(R_1^2 - 2r^2) \quad ...(15.74)$$

*Variation of hoop stress* : Here also the hoop stress is always positive and is maximum at $r = 0$ from where it goes on reducing as r increases.

$$(\sigma_h)_{\max} = \frac{\rho\omega^2}{8}\left(\frac{3-2\nu}{1-\nu}\right)R_1^2 \qquad ...(15.75)$$

It is interesting to note that this value is exactly half of that for a hollow cylinder with a pin hole along the axis (compare it with equation 15.62). The variation of $\sigma_h$ is shown in Fig. 15.5.

*Variation of radial stress* : From equation (15.73) it is evident that $\sigma_r$ is maximum at

$$r = 0$$

$$(\sigma_h)_{\max} = \frac{\rho\omega^2}{8}\left(\frac{3-2\nu}{1-\nu}\right)R_1^2 \qquad ...(15.76)$$

which is same in magnitude as $(\sigma_h)_{\max}$. Further $\sigma_r$, being positive, is always tensile. The variation is shown in Fig. 15.5.

*Variation of longitudinal stress* : The maximum value of $\sigma_l$ occurs at $r = 0$ and its value from equation (15.74) is

$$(\sigma_l)_{\max} = \frac{\rho\omega^2}{4}\left(\frac{\nu}{1-\nu}\right)R_1^2 \qquad ...(15.77)$$

Further it is zero, at $2r^2 = R_1^2$

or $$r = \frac{R_1}{\sqrt{2}}$$

For the range $0 < R_1/\sqrt{2} < R_1$ the stress being negative is compressive.

At $$r = R_1;$$

$$(\sigma_l)_{r = R1} = \frac{\rho\omega^2}{4}\left(\frac{\nu}{1-\nu}\right)R_1^2 \text{ (Comp.)} \qquad ...(15.78)$$

The variation in $\sigma_l$ is shown in Fig. 15.5.

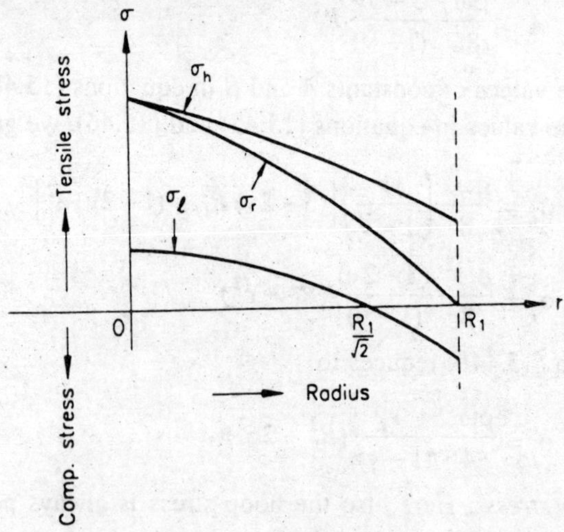


*Fig. 15.5*

**Example 15.2.** *A turbine rotor, 0.4 m external diameter and 0.2 m internal diameter, is revolving at 1000 r.p.m. Taking the weight of rotor as 7700 kg/m³ and Poisson's ratio 0.3, find the maximum hoop and radial stresses assuming (i) rotor to be a thin disc (ii) rotor to be a long cylinder. And calculate the percentage error in assuming it to be a thin disc.*

**Solution.** (*i*) **For thin disc**

$$\omega = \frac{2\pi N}{60} = \frac{2\pi}{60}(1000)$$

$$= 104.72 \text{ rad/sec}$$

Therefore from equations (15.32) and (15.33)

$$(\sigma_r)_{\max} = \frac{\rho\omega^2}{8}(3+\nu)(R_1 - R_2)^2$$

$$= \frac{7700 \times (104.72)^2}{8}(3.3)(0.1)^2 \text{ N/m}^2$$

$$= 0.3475 \text{ NM/m}^2$$

$$(\sigma_h)_{max} = \frac{\rho\omega^2}{4}[(3+\nu)R_1^2 + (1-\nu)R_2^2]$$

$$= \frac{700 \times (104.72)^2}{4}[3.3(0.2)^2 + 0.7(0.1)^2] \text{ N/m}^2$$

$$= 2.93 \text{ MN/m}^2.$$

(*ii*) **For long cylinder**

From equations (15.63) and (15.62) we have

$$(\sigma_h)_{max} = \frac{\rho\omega^2}{4}\left(\frac{3-2\nu)}{1-\nu}\right)(R_1 - R_2)^2$$

$$= \frac{7700 \times (104.72)^2}{8}\left(\frac{2.4}{0.7}\right)(0.1)^2 \text{ N/m}^2$$

$$= 0.3619 \text{ MN/m}^2$$

$$(\sigma_h)_{max} = \frac{\rho\omega^2}{4(1-\nu)}[3-2\nu)R_1^2 + (1-2\nu)R_2^2]$$

$$= \frac{7700 \times (104.72)^2}{4 \times 0.7}[2.4(0.2)^2 + 0.4(0.1)^2] \text{ N/m}^2$$

$$= 3.0157 \text{ MN/m}^2.$$

(*iii*) Error in $(\sigma_r)_{max} = \frac{0.3619 - 0.3475}{.3619} \times 100$

$$= 3.97 \%$$

$$\text{Error in} \quad (\sigma_r)_{max} = \frac{3.0157 - 2.93}{3.0157} \times 100$$
$$= 2.7\ \%$$

**Example 15.3.** *Solve the Example 15.2 if the rotor were either a solid disc or solid cylinder.*

**Solution.** (*i*) **For thin solid disc**. From equations (15.39) and (15.40), we have

$$(\sigma_r)_{\max} = (\sigma_h)_{\max} = \frac{\rho\omega^2}{8}(3+\nu)\,R_1^2$$

or

$$(\sigma_r)_{\max} = (\sigma_h)_{\max} = \frac{7700 \times (104.72)^2}{8}(3.3)\,(0.2)^2\ \text{N/m}^2$$
$$= 1.39\ \text{MN/m}^2$$

(*ii*) **For long solid cylinder**. From equations (15.76) and (15.75)

$$(\sigma_r)_{\max} = (\sigma_h)_{\max} = \frac{\rho\omega^2}{8}\left(\frac{3-2\nu}{1-\nu}\right)R_1^2$$
$$= \frac{700 \times (104.72)^2}{8} = \left(\frac{24}{0.7}\right)(0.2)^2\ \text{N/m}^2$$
$$= 1.4475\ \text{MN/m}^2$$

(*iii*) Error in $(\sigma_r)_{\max}$ = Error in $(\sigma_h)_{\max}$

$$= \frac{1.4475 - 1.39}{1.4475} \times 100 = 3.97\%$$

### 15.3.3 Comparison of Disc and Cylinder Formulae

From Examples (15.2) and (15.3) it is evident that the maximum error involved in using the thin disc formulae is of the order of 4%. For engineering design purposes this error is not much. On the other hand the thin disc formulae are easier to apply. Thus for convenience of calculations the designers often apply thin disc formulae even to long rotating cylinders.

## 15.4 Disc of Uniform Strength

In article (15.2) it has been seen that radial and hoop stresses in a rotating disc of uniform thickness go on changing with change in radius. The aim of the present article is to find the profile of the disc, which will give the variation of the thickness (parallel to the axis of rotation) with its radius such that the stresses induced in both the radial and circumferential directions are same and further they are same everywhere in the disc, that is it does not change with the change in radius. Fig. 15.6(*a*) shows the elevation of half of such a disc and Fig. 15.6(*b*) shows the free body diagram of an

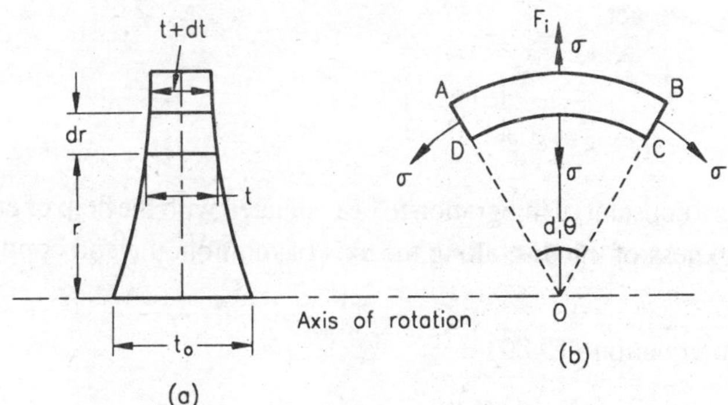


*Fig. 15.6*

element *ABCD* of the disc which subtends an angle $d\theta$ at the centre *O*. Further the element is bounded by surfaces *DC* and *AB* whose radii are *r* and $(r + dr)$ respectively with the corresponding axial thinknesses as *t* and $(t + dt)$. Let the uniform stress in the radial and circumferential directions be $\sigma$ as shown. $F_i$ is inertia force on the element and is equal to the product of the mass of the element and the centripetal acceleration due to the angular speed $\omega$ rad/sec. Thus considering the equilibrium of the element *ABCD* and resolving the forces along the radial direction we get.

$$F_i + \sigma(r + dr)(t + dt)\, d\theta - \sigma rt\, d\theta - 2\sigma\, t\, dr \sin\frac{d\theta}{2} = 0$$

Neglecting the product of two small quantities and equating

$$\sin\frac{d\theta}{2} \simeq \frac{d\theta}{2}$$

for small angle $d\theta$ we get from this

$$F_i + \sigma r\, dt\, d\theta = 0$$

or $\quad \rho(r\,.d\,\theta\,.\,dr.t)\,\omega^2\, r + \sigma r\, dt\, d\theta = 0$

or
$$\frac{dt}{dr} + \left(\frac{\rho\omega^2}{\sigma} r\right) t = 0$$

The solution of this differential equation is obtained by multiplying with the integrating factor *I* where

$$I = e^{\int \frac{\rho\omega^2}{\sigma} r dr} = e^{\frac{\rho\omega^2}{2\sigma} r^2}$$

The equation (15.79), then reduces to

$$I\frac{dt}{dt} + I\left(\frac{\rho\omega^2}{\sigma} r\right) t = 0$$

or
$$\frac{d}{dr}\left[t\, e^{\frac{\rho\omega^2}{2\sigma} r^2}\right] = 0$$

Integrating, we get

$$t = A\,e^{\left[-\frac{\rho\omega^2}{2\sigma}r^2\right]} \qquad ...(15.80)$$

Where $A$ is a constant of integration to be evaluated with the help of end conditions. Thus if $t_0$ is thickness of the disc along the axis of rotation, the end condition will be $t = t_0$ at $r = 0$.

Then from equation (15.80)

$$t_0 = Ae^{\circ} = A$$

That is the constant A in the equation (15.80) represents the thickness of the disc along the axes of rotation. Evidently from equation (15.80), the thickness goes on reducing as $r$ increases and is maximum along the axes of rotation.

**Example 15.4.** *A grinding wheel is 300 mm diameter with the bore at the centre 25 mm diameter. If the thickness of the wheel at the outer radius is 25 mm, what should be the thickness at the bore diameter for a uniform allowable stress of 10 MN/m² at 2800 r.p.m. ? Take density of the wheel material as 2700 kg/m³.*

**Solution.** $r_1 = 0.15\ m,\ t_1 = 0.025$ m

$r_2 = 0.0125$ m, $t_2 = ?$

$$\omega = \frac{2\pi}{60}(2800) = 293.215 \text{ rad / sec.}$$

$$\sigma = 10 \times 10^6 \text{ N/m}^2 \text{ and } \rho = 2700 \text{ kg/m}^3$$

Now from equation (15.80)

$$t_1 = A\,e^{-\frac{\rho\omega^2 r_1^2}{2\sigma}}$$

or
$$0.025 = A.\exp\left[\frac{-2700 \times (293.215)^2}{2 \times 10^7} \times (0.15)^2\right]$$

or
$$A = 0.025\,e^{-.26115} \text{ m}$$
$$= 32.46 \text{ mm}$$

Again applying the equation (15.80)

$$t_2 = 32.46 \exp\left[\frac{-2700 \times (293.215)^2}{2 \times 10^7}(0.0125)^2\right] \text{mm}$$

$$= 32.45\,e^{-.0018135} = 32.40 \text{ mm.}$$

## PROBLEMS FOR PRACTICE

1. A mildsteel thin ring is 1m in diameter. Neglecting the effect of spokes, find the maximum speed in r.p.m. at which it can be rotated without the stress going beyond 155 $MN/m^2$ in it. The steel weighs 7700 $kg/m^3$. Also find the increase in diameter of the ring at this speed, if E = 200 $GN/m^2$.

   [**Ans.** 2709.7 r.p.m. ; 0.755 mm]

2. Determine the greatest values of radial and hoop stresses for a rotating disc in which the outer and inner radii are 0.3 m and 0.15. The angular speed is 150 rad/sec. Take Poisson's ratio as 0.304 and density 7700 $kg/m^3$.

   [**Ans.** 1.6 $MN/m^2$ ; 13.6 $MN/m^2$]

3. Solve the Problem No. 2 if it were a long cylinder.

   [**Ans.** 1.6643 $MN/m^2$ ; 13.98 $MN/m^2$]

4. The disc of a turbinerotor is 0.5 m diameter. At the blade ring its thickness is 55 mm. It is keyed to a shaft of 50 mm diameter. If the uniform stress in therotor disc islimitedto 200 $MN/m^2$ at 9000 r.p.m., findthe thickness of the disc at the shart. Take density of the rotor material as7700 $kg/m^3$. [**Ans.** 158.43 mm]

Chapter–16

# Unsymmetrical Bending

## 16.1 Introduction

In Chapter 5, while considering the simple theory of bending, one of the assumptions on which the bending equation was based, was that the beam had a cross-section symmetrical about the axes through its centroid and in the plane of bending. If the cross- section of the beam has an axis of symmetry, the above mentioned condition required by the simple theory of bending will be satisfied if the plane of the loads contains the axis of symmetry of all the sections, because the axis of symmetry of an area is always a principal axis of inertia. Bending caused by loads that do not either lie in or parallel to a plane that contains the principal centroidal axes of inertia of the cross-section is usually called unsymmetrical bending. Under this condition the bending moment is resolved into two components, however before discussing this the significance of principal axes of the section of a beam must be considered.

## 16.2 Principal Axes and Product of Inertia

Fig. 16.1 represents a beam section which is symmetrical about the plane of bending. This corresponds to one of the assumptions of the simple theory of bending. $X$ and $Y$ axes form a set of perpendicular axes through the centroid $G$ of the section. Let the bending moment $M$ be applied in the plane $YY$ so that $XX$ is the centroidal

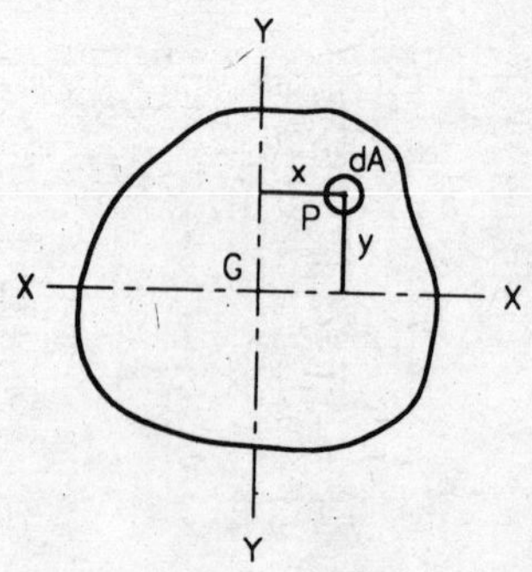


*Fig. 16.1*

bending axes. Further if no bending is to take place about $YY$ (again an assumption of simple theory of bending), then

$$\int(\sigma dA) = 0 \qquad ...(16.1)$$

Where $\sigma$ is the stress on an elemental area $dA$ due to bending about $XX$, at the point $P$ with co-ordinates $(x, y)$ and is given by

$$\sigma = \frac{M}{I_{xx}} \cdot y \qquad ...(16.2)$$

Therefore from equations (16.1) and (16.2)

$$\int\left(\frac{M}{I_{xx}} y\right) . x . dA = 0$$

or

$$\int xy \, dA = 0 \qquad ...(16.3)$$

Left hand side of equation (16.3), *i.e.* size 12 int $xy\ dA$ is called "the product of inertia" of the area about axes $XX$ and $YY$ and is represented by $I_{xy}$. Equation (16.3) also gives the condition for pure bending in accordance with the simple bending theory, which is that the product of inertia about the rectangular axes through the centroid in, and perpendicular to, the plane of bending must be zero. Such axes are known as the "PRINCIPAL AXES" of the section and moment of inertia about the principal axes as the "PRINCIPAL MOMENT OF INERTIA". It may be noted that if the section has an axis of symmetry, that axis will be a principal axis.

## 16.3 Determination of Principal Axes

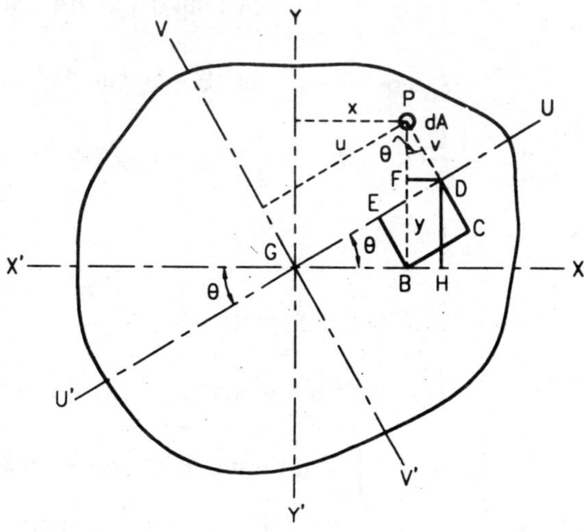


*Fig. 16.2*

Fig. 16.2 represents the area of cross-section $A$ of the beam with its centroid at $G$. The principal axes through $G$ are $UU$ and $VV$. $XX$ and $YY$ are another set of two perpendicular axes passing through $G$ inclined to the principal axes at angle $\theta$ as shown. Knowing the values of second moments of area and the product of inertia about $XX$ and $YY$ the values of the principal moments of inertia and the directions of the principal axes can be found as under.

Let the co-ordinates of the point $P$ be $(x, y)$ and $(u , v)$ with respect the axes $X$-$Y$ and $U$-$V$ respectively. Then geometrically

$$u = GD = GE + ED$$

$$= GB \cos\theta + BC$$

$$= GB \cos\theta + PB \sin\theta$$

i.e. $$u = x\cos\theta + y\sin\theta \qquad \text{...(16.4)}$$

$$v = PD = PC - CD$$

$$= PB\cos\theta - BE$$

$$= PB\cos\theta - GB\sin\theta$$

i.e. $$v = y\cos\theta - x\sin\theta \qquad \text{...(16.5)}$$

Second moment of area about $U.U$, $I_{uu}$, is given by

$$I_{uu} = \int v^2\,dA$$

$$= \int (y\cos\theta - x\sin\theta)^2\,dA$$

$$= \int (y^2\cos^2\theta + x^2\sin^2\theta - 2xy\sin\theta\cos\theta)\,dA$$

$$= \int \cos^2\theta + \int y^2\,dA + \sin\theta \int x^2\,dA - \sin 2\theta) \int xy\,dA$$

or $$I_{uu} = I_{xx}\cos^2\theta + I_{yy}\sin^2\theta - I_{xy}\sin 2\theta \qquad \text{...(16.6a)}$$

or $$I_{uu} = \frac{1}{2}(I_{xx} + I_{yy}) + \frac{1}{2}(I_{xx} - I_{yy})\cos 2\theta - I_{xy}\sin 2\theta \qquad \text{...(16.6b)}$$

Similarly $$I_{vv} = \int u^2 dA$$

$$= \int (x\cos\theta + y\sin\theta)^2\,dA$$

$$= \int (x^2\cos^2\theta + y^2\sin^2\theta) + 2xy\sin\theta\cos\theta)\,dA$$

$$= \cos^2\theta \int x^2\,dA + \sin^2\theta \int y^2\,dA + \sin 2\theta \int xy dA$$

i.e. $$I_{vv} = I_{xx}\sin^2\theta + I_{yy}\cos^2\theta + I_{xy}\sin 2\theta \qquad \text{...(16.7a)}$$

$$= \frac{1}{2}(I_{xx} + I_{yy}) - \frac{1}{2}(I_{xx} - I_{yy}) \cos 2\theta + I_{xy} \sin 2\theta$$

...(16.7*b*)

Also

$$I_{uv} = \int uv dA$$

$$= \int (x \cos \theta + y \sin \theta)(y \cos \theta - x \sin \theta)\, dA$$

$$= \int [xy(\cos^2 \theta + \sin^2 \theta) + (y^2 - x^2) \sin \theta \cos \theta]\, dA$$

$$= \int xy \cos 2\theta\, dA + \int \frac{1}{2} \sin 2\theta\, (y^2 - x^2)\, dA$$

$$= \cos 2\theta \int xy dA - \frac{1}{2} \sin 2\theta \left[\int y^2\, dA - \int x^2\, dA\right]$$

*i.e.*

$$I_{uv} = I_{xy} \cos 2\theta + \frac{1}{2}(I_{xx} - I_{yy}) \sin 2\theta \qquad ...(16.8)$$

But for pure bending $I_{UV} = 0$

Therefore from equation (16.8)

$$\tan 2\theta = \frac{2I_{xy}}{I_{yy} - I_{xx}} \qquad ...(16.9)$$

If $\theta_1$ and $\theta_2$ are the two values of $\theta$ from equation (16.9), then

$$\theta_1 = \frac{1}{2} \tan^{-1} \left[\frac{2I_{xy}}{I_{yy} - I_{xx}}\right] \qquad ...(16.10a)$$

and

$$\theta_2 = \frac{\pi}{2} + \theta_1 \qquad ...(16.10b)$$

The values of $I_{UU}$ and $I_{VV}$ can then the calculated corresponding to the values of $\theta$ given by equations (16.10)

$$\sin 2\theta_1 = \frac{I_{xy}}{\sqrt{\left(\frac{I_{yy} - I_{xx}}{2}\right)^2 + I_{xy}^2}} \qquad ...(16.11a)$$

$$\sin 2\theta_1 = \frac{\frac{1}{2}(I_{yy} - I_{xx})}{\sqrt{\left\{\frac{1}{2}(I_{yy} - I_{xx})\right\}^2 + I_{xy}^2}} \qquad ...(16.11b)$$

Therefore from equations (16.6), (16.7) and (16.11)

$$(I_{uu})_{\theta_1} = \frac{1}{2}(I_{xx} + I_{yy}) - \frac{\left[\frac{1}{2}(I_{xx} - I_{yy})\right]^2}{\sqrt{\left[\frac{I_{yy} - I_{xx}}{2}\right]^2 + I_{xy}^2}} - \frac{I_{xy}^2}{\sqrt{\left[\frac{I_{yy} - I_{xx}}{2}\right] + I_{xy}^2}}$$

or $$(I_{uu})_{\theta_1} = \frac{1}{2}(I_{xx} + I_{yy}) - \sqrt{\left(\frac{I_{yy} - I_{xx}}{2}\right)^2 + I_{xy}^2} \qquad ...(16.21a)$$

and $$(I_{vv})_{\theta 1} = \frac{1}{2}(I_{xx} + I_{yy}) + \frac{\left[\frac{1}{2}(I_{xx} - I_{yy})\right]^2}{\sqrt{\left(\frac{I_{yy} - I_{xx}}{2}\right)^2 + I_{xy}^2}} + \frac{I_{xy}^2}{\sqrt{\left(\frac{I_{yy} - I_{xx}}{2}\right)^2 + I_{xy}^2}}$$

$\therefore$ $$(I_{w})_{\theta_1} = \frac{1}{2}(I_{xx} + I_{yy}) + \sqrt{\left(\frac{I_{yy} - I_{xx}}{2}\right)^2 + I_{xy}^2} \qquad ...(16.12b)$$

Again for $\theta = \theta_2$, we have

$$\sin 2\theta_2 = \sin 2\left(\frac{\pi}{2} + \theta_1\right) = -\sin 2\theta_1 \qquad ...(16.13b)$$

and $$\cos 2\theta_2 = \cos 2\left(\frac{\pi}{2} + \theta_1\right) = -\cos 2\theta_2 \qquad ...(16.13b)$$

Thus substituting equations (16.13) in equations (16.6) and (16.7).

$$(I_{uu})_{\theta_2} = \frac{1}{2}(I_{xx} + I_{yy}) + \sqrt{\left(\frac{I_{yy} - I_{xx}}{2}\right)^2 + I_{xy}^2} \qquad ...(16.14a)$$

$$(I_{vv})_{\theta_2} = \frac{1}{2}(I_{xx} + I_{yy}) - \sqrt{\left(\frac{I_{yy} - I_{xx}}{2}\right)^2 + I_{xy}^2} \qquad ...(16.14b)$$

Maximum and minimum values of $I_{uu}$ and $I_{vv}$ :

From equations (16.6) and (16.7) it is clear that for a given set of values of $I_{xx}$, $I$ and $I_{xy}$ the values of $I_{uu}$ and $I_{vv}$ will go on changing with the change in the value of $\theta$. For maximum or minimum values of $I_{uu}$ and $I_{vv}$.

$$\frac{dI_{uu}}{d\theta} = \frac{dI_{vv}}{d\theta} = 0$$

Thus from equation (16.6) for $I_{uu}$

$$\frac{dI_{uu}}{d\theta} = \frac{1}{2}(I_{xx} - I_{yy})(-\sin 2\theta)\,2 - I_{xy}\,(\cos 2\theta)^2\,(\cos 2\theta) \times 2 = 0$$

*i.e.*
$$\tan 2\theta = \frac{2I_{xy}}{I_{yy} - I_{xx}} \qquad \text{...(16.15)}$$

Similarly from equation (16.7) for $I_{vv}$

$$\frac{dI_{vv}}{d\theta} = -\frac{1}{2}(I_{xx} - I_{yy})(-\sin 2\theta)\,2 + I_{xy}\,(\cos 2\theta)\,2 = 0$$

*i.e.*
$$\tan 2\theta = \frac{2I_{xy}}{I_{yy} - I_{xx}} \qquad \text{...(16.16)}$$

Evidently the values of $\theta$ as obtained from equations (16.15) and (16.16) are same as given by equations (16.9). Thus the corresponding values of $I_{uu}$ and $I_{vv}$ found in equations (16.12) and (16.14) are also the maximum and minimum values. These values are also Principal values of moment of inertia since $I_{uv} = 0$ (equation 16.8).

The directions $\theta_1$ and $\theta_2$ are called the principal directions.

Further the equations for the principal moments of inertia and product of inertia along with those for the inclination $\theta$ are exactly similar to those of principal stresses and the inclination of principal planes (Chapter 2). It is evident that if the points are plotted with $I_{xx}$ or $I_{yy}$ as x co-ordinate and $I_{xy}$ as the y co-ordinate the locus of the points will be a circle, which is called the Mohr's circle of moment of inertia.

## 16.3.1 Determination of Moment of Inertia about any Axis

Knowing the values of the moment of inertia about the principal axes, the moment of inertia about any axis can be found with the help of three equations (16.6), (16.7) and (16.9). But the solution of these equations is quite involved. An alternative but simpler method to do so, is given below.

From Fig. 16.2

$$x = GB = GH - BH$$

$$= u\cos\theta - DF$$

$$= u\cos\theta - v\sin\theta \qquad \text{...(16.17)}$$

$$y = PB = PF + DH$$

$$= v\cos\theta + u\sin\theta \qquad \text{...(16.18)}$$

Second moment of area about $XX$ is then given by

$$I_{xx} = \int y^2\, dA = \int (v\cos\theta + u\sin\theta)^2\, dA$$

$$= \cos^2\theta \int v^2\, dA + \sin^2\theta \int u^2\, dA + \sin^2\theta \int uvdA$$

$$= I_{uu}\cos^2\theta + I_{vv}\sin^2\theta + I_{uv}\sin^2\theta$$

But $\quad I_{uv} = 0$

Therefore $\quad I_{xx} = I_{uu}\cos^2\theta + I_{vv}\sin^2\theta \qquad \text{...(16.19)}$

Similarly $\quad I_{yy} = \int x^2 dA = \int (u\cos\theta - v\sin\theta)^2 dA$

$$= \cos^2\theta \int u^2 dA + \sin^2\theta \int v^2\, dA - \sin^2\theta \int uvdA$$

*i.e.* $\quad I_{yy} = I_{vv}\cos^2\theta + I_{uu}\sin^2\theta \qquad \text{...(16.20)}$

Further adding equations (16.19) and (16.20)

$$I_{xx} + I_{yy} = I_{uu}(\cos^2\theta + \sin^2\theta) + I_{vv}(\sin^2\theta + \cos^2\theta) = I_{uu} + I_{vv}$$

$$I_{uu} + I_{vv} = J \qquad \text{...(16.21)}$$

Where $J$ = The polar second moment of area about an axis passing through $G$ and normal to the section.

**Note :** While evaluating $I_{xx}$ and $I_{yy}$ with the help of equations (16.19) and (16.20) it must be kept in mind that the $XX$ axis is inclined with $UU$ at $\theta$ measured clockwise. However while calculating the values of $I_{uu}$ and $I_{vv}$ with the help of equations (16.7) and (16.8) inclination of the $UU$ axes is measured anticlockwise with respect to $XX$.

## 16.4 Momental Ellipse

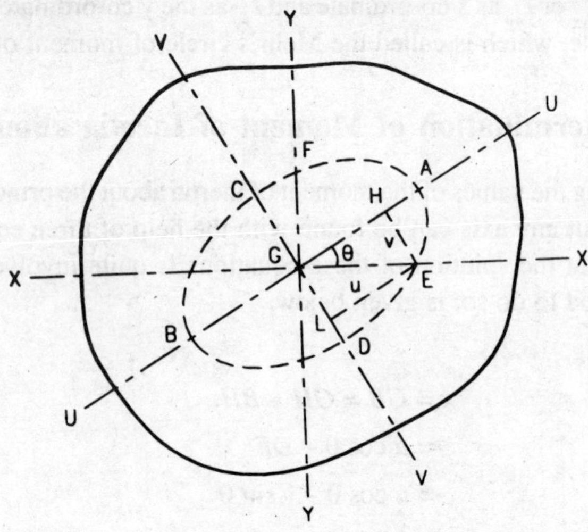


*Fig. 16.3*

**Statement :** If an ellipse with major and minor axes along $U$— $V$ axes is drawn with semi major and minor axes equal to $1/k_{uu}$ and $1/k_{vv}$ respectively then the intercepts on $X$ and $Y$ axes will be $1/k_{xx}$ and $1/k_{yy}$ respectively where $k$ represents the radius of gyration.

**Proof :** Given $GA = 1/k_{uu}$ and $GC = 1/k_{vv}$ ...(16.22)

$$k_{uu} = \frac{I_{uu}}{A} \text{ and } k_{vv} = \frac{I_{vv}}{A} \quad ...(16.23)$$

From Fig. 16.3

$$u = GE \cos\theta \quad ...(16.24a)$$

$$v = GE \sin\theta \quad ...(16.24b)$$

Now equation of the ellipse is

$$\frac{u^2}{(1/k_{uu})^2} + \frac{v^2}{(1/k_{vv})^2} = 1$$

Substituting values from equations (16.24) it reduces to

$$k_{uu}^2\, GE^2 \cos^2\theta + k_{vv}^2\, GE^2 \sin^2\theta = 1$$

Multiplying throughout by $A$

$$Ak_{uu}^2 \cos^2\theta + k_{vv}^2 \sin^2\theta = \frac{A}{GE^2}$$

or

$$I_{uu} \cos^2\theta + I_{vv} \sin^2\theta = \frac{A}{GE^2} \quad ...(16.25)$$

But from equation(16.19)

$$I_{uu} \cos^2\theta + I_{vv} \sin^2\theta = I_{xx}$$

Therefore equation (16.25) reduces to

$$I_{xx} = \frac{A}{GE^2}$$

or

$$GE^2 = \frac{A}{I_{xx}}$$

$$= \frac{A}{Ak_{xx}^2} = \frac{1}{k_{xx}^2}$$

*i.e.*

$$GE = \frac{1}{k_{xx}} \quad ...(16.26)$$

Similarly it can be shown that

$$GF = \frac{1}{k_{yy}} \quad ...(16.27)$$

## 16.5 Theorem of Parallel Axes for Product of Inertia

In Fig. 16.4 is shown a section whose product of inertia about *XY* axes with *G* as its centroid are known. It is required to find the product of inertia about the *X′Y′* axes which are parallel to given set of axes.

Let the co-ordinates of the point *G*, with respect to the *X′OY′* axes be $(x', y')$. Assume an elemental area *dA* around the point *P* having co-ordinates $(x, y)$ with respect to *XOY* axes.

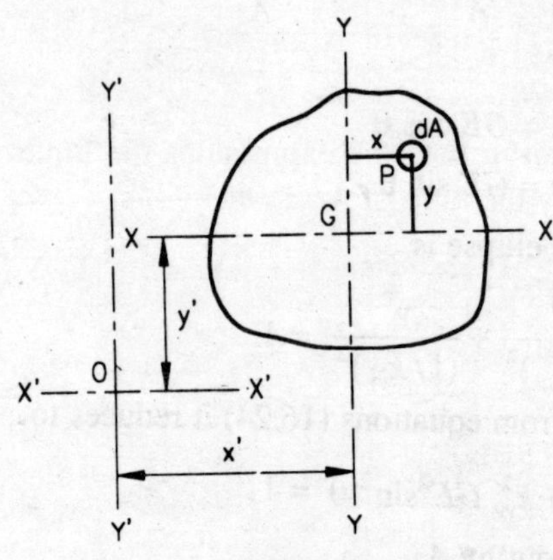


*Fig. 16.4*

Then the product of inertia about *X′OY′* axes is given by

$$I_{x'y'} = \int (x + x')(y + y')\, dA$$

$$= \int (xy + xy' + yx' + x'y')\, dA$$

$$= \int xydA + y' \int xdA + x' \int ydA + x'y' \int dA$$

or
$$I_{x'y'} = I_{xy} + x'y'A + \left[y' \int xdA + x' \int ydA\right] \qquad \text{...(16.28)}$$

Now the quantities $\int xdA$ and $\int ydA$ represent the first moments of the area about *Y* and *X* axes respectively which are passing through the centroid *G*. These must then be zero individually. Thus the equation (16.28) reduces to

$$I_{x'y'} = I_{xy} + x'y'A \qquad \text{...(16.29)}$$

**Example 16.1.** *Find the product moment of inertia of the right angled triangle OAB about XOY axes as shown in the Fig. 16.5. OA = 60 mm, AB = 40 mm.*

**Solution.** Let a small rectangular area *dx dy* be situated around the point *P* with co-ordinates $(x, y)$.

Then
$$I_{xy} = \int_{x=0}^{60} \int_{y=0}^{2x/3} xydydx \quad \text{OR} \quad \int_{y=0}^{40} \int_{x=0}^{3x/2} xydxdy$$

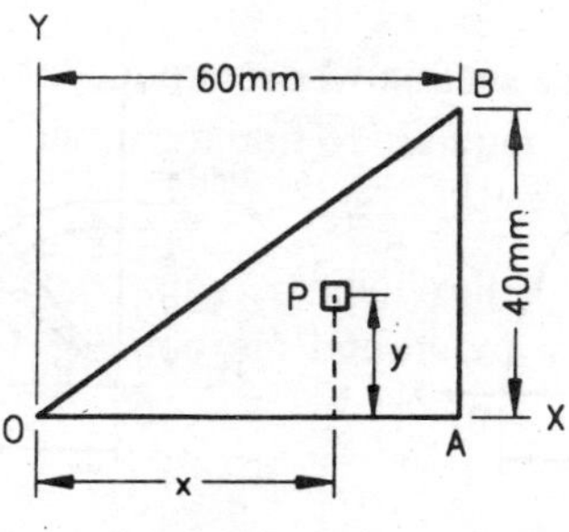


*Fig. 16.5*

It may be noted that in both the expressions the limits of $y$ or $x$ in the second integrals are governed by the ratio of $AB : OA$

Thus
$$I_{xy} = \int_{x=0}^{60} \left[\int_{y=0}^{2x/3} xydy\right] dx$$

$$= \int_{x=0}^{60} \left\{\left[x\frac{y^2}{2}\right]_{y=0}^{2x/3}\right\} dx$$

$$= \int_{x=0}^{60} \left\{x \cdot \frac{1}{2} \cdot \frac{4}{9} x^2\right\} dx$$

$$= \frac{2}{9}\int_0^{60} x^3\, dx$$

$$= \frac{2}{9}\left[\frac{x^4}{4}\right]_0^{60} = \frac{2}{9}\frac{(60)^4}{4} = 72 \times 10^4 \text{ mm}^4$$

**Example 16.2.** *Find the product moment of inertia of a quadrant OAB about the perpendicular axes OX and OY Fig. 16.6. Take the radius of the quadrant as r.*

**Solution.**
$$I_{xy} = \int_{x=0}^{r} \int_{y=0}^{\sqrt{r^2 - x^2}} xy\, dy\, dx$$

The limit for $y$ is governed by the equation of the arc $AB$ which is $x^2 + y^2 = r^2$ or $y = \sqrt{r^2 - x^2}$

Then
$$I_{xy} = \int_{x=0}^{r} \left\{\left[x\frac{y^2}{2}\right]_{y=0}^{\sqrt{r^2\ x^2}}\right\} dx = \int_{x=0}^{r} \frac{x}{2}(r^2 - x^2)\, dx$$

$$= \frac{1}{2}\left[r^2\frac{x^2}{2} - \frac{x^4}{4}\right]_0^r = \frac{1}{2}\left[\frac{r^4}{2} - \frac{r^4}{4}\right] = \frac{r^4}{8}$$

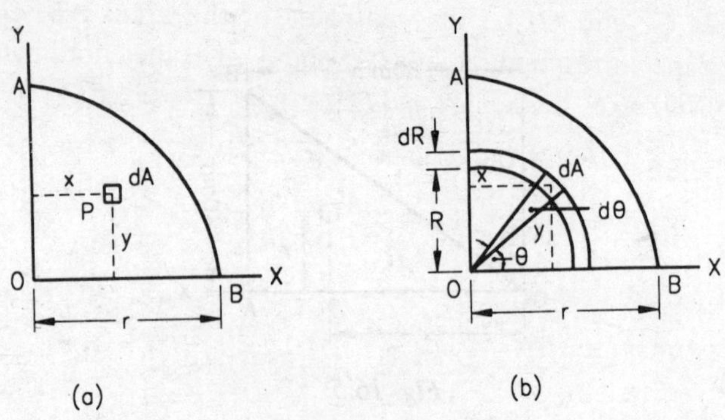


*Fig. 16.6*

Alternatively in polar co-ordinates

$$I_{XY} = \iint xy dA$$

For Fig. 16.6(*b*) $x = R\cos\theta$ ; $y = R\sin\theta$ ; $dA = R\, d\theta\, dR$

Then $$I_{xy} = \int_{\theta=0}^{\pi/2} \int_{R=0}^{r} (R\cos\theta)(R\sin\theta)(R.d\theta.dR)$$

$$= \int_{\theta=0}^{\pi/2} \left[ \int_{R=0}^{r} R^2 \sin\theta\cos\theta)\, dR) \right] d\theta$$

$$= \int_{\theta=0}^{\pi/2} \left[ \frac{R^4}{0} \frac{\sin^2\theta}{2} \right]_0^r d\theta = \frac{r^4}{8}\left[ -\frac{-\cos^2\theta}{2} \right]_0^{\pi/2}$$

*i.e.* $$I_{xy} = \frac{r^4}{8}.$$

**Example 16.3.** *Find the product moment of inertia about the rectangular axes OX and OY of the area shown in Fig. 16.7.*

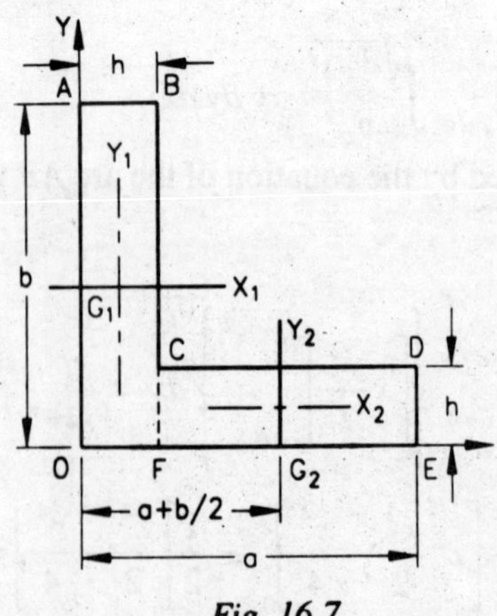


*Fig. 16.7*

**Solution.** Let $G_1$ and $G_2$ be the centroids of the areas *ABFO* and *CDEF* respectively having the corresponding areas $A_1$ and $A_2$. Let the coordinates of $G_1$ and $G_2$ with respect to *XOY* axes be $(x_1, y_1)$ and $(x_2, y_2)$.

Then for the area *ABFO*; from equation (16.29)

$$I_{xy} = I_{x_1y_1} + A_1x_1y_1$$

Now $I_{x_1y_1}$ of the area $ABFO = 0$, since $G_1X_1$ and $G_1Y_1$ are the principal axes of the area, being axes of symmetry.

$$A_1 = b\,.\,h\,;\, x_1 = \frac{h}{2}\,.\, y_1 = \frac{b}{2}$$

Then $I_{xy}$ of area $ABFO = 0 + (bh)\frac{h}{2}\cdot\frac{b}{2} = \frac{b^2h^2}{4}$

Again for the area *CDEF*

$$I_{xy} = Ix_2y_2 + A_2x_2y_2$$

Here $\quad Ix_2y_2 = 0\,;\, A_2 = (a - h)\,h\,;$

$$x_2 = \frac{a-h}{2} + h = \frac{a+h}{3}\,;\, y_2 = \frac{h}{2}$$

Therefore $I_{xy}$ of area *CDEF*

$$= 0 + (a-h)\,h\left(\frac{a+h}{2}\right)\frac{h}{2}$$

$$= \frac{h^2}{4}(a^2 - h^2)$$

Thus total product of inertia of the entire area is

$$I_{xy} = \frac{b^2h^2}{4} + \frac{h^2}{4}(a^2 - h^2)$$

$$= \frac{h^2}{4}(a^2 + b^2 - h^2)\,.$$

**Example 16.4.** *Find the centroidal principal moments of inertia of an equal angle section shown in Fig. 16.8.*

**Solution.** Let $(\bar{x}\,.\,\bar{y})$ be the co-ordinates of the centroid *G*, with respect to the rectangular axes $OX_1$ and $OY_2$. Since this is an equal angle

$$\bar{x} = \bar{y}$$

and the value of $\bar{y}$ can be found by taking moments of the a about the line $OX_1$

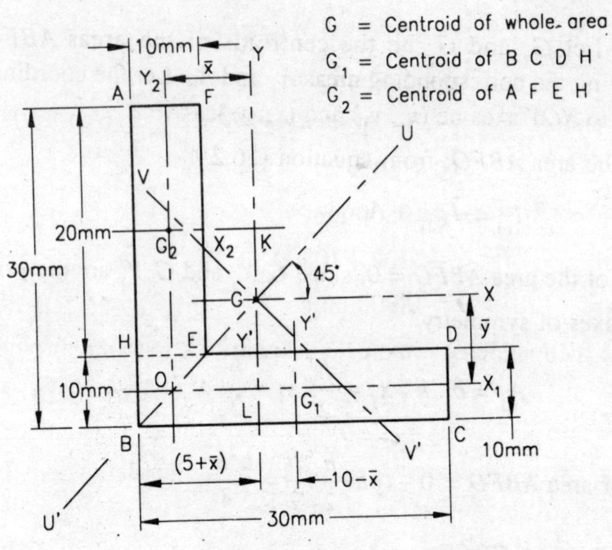


*Fig 16.8*

$$(30 \times 10 + 20 \times 10)\ \bar{y} = (30 \times 10 \times 0 + 20 \times 10 \times 15)$$

or
$$\bar{y} = \frac{3000}{500} = 6 \text{ mm}$$

$$I_{xx} = I_{yy} = (I_{G_1X_1} \text{ of } HDCB) + A_{HDCB}\,(\bar{y})^2$$

$$+ (I_{G_2x_2} \text{ of } AFEH) + A_{AFEH}\ (KG)^2$$

$$= \frac{30 \times 10^3}{12} + (30 \times 10)(6)^2 + \frac{10 \times 20^2}{12} + (10 \times 20)\,(15 - 6)^2$$

$$= 2500 + 10800 + 666.67 + 16200$$

$$= 30166.67 \text{ mm}^4$$

Now the coordinates of $G_1$ with respect to the axes GX and GY are

$$[LG_1, -GL) = [(10 - x), -\bar{y}]$$

$$= [(10 - 6), -6]$$

$$= (4, -6)$$

Similarly the co-ordinates of $G_2$ are

$$= [G_2 K, GK]$$

$$= [-\bar{x}, (LK - \bar{y})]$$

$$= (-6, 9)$$

$$I_{XY} \text{ of area } HDCB = I_{x_1y_1} \text{ of area } HDCB + A_{HDCB}\,(4)\,(-6)$$

$$= 0 + (30 \times 10)(-24)$$

$$= -7200 \text{ mm}^4$$

$$I_{XY} \text{ of area } AFEH = I_{x_2y_2} \text{ of area } AFEH + A_{AFEH}(-6 \times 9)$$

$$= 0 + 10 \times 20 (-54)$$

$$= -10800 \text{ mm}^4$$

Thus total $I_{XY} = -7200 - 10800$

$$= -18000 \text{ mm}^4$$

If θ is the inclination of principal axes with *GX*, passing through *G* then

$$\tan 2\theta = -\frac{2I_{XY}}{I_{XX} - I_{YY}} = \infty \qquad \text{Since } I_{XX} = I_{YY}$$

Then $\theta_1 = 45°$ and $\theta_2 = 90° + 45° = 135°$ are the inclinations of principal axes *GU* and *GV* respectively.

Then from equation (16.6) at

$$\theta_1 = 45°$$

$$I_{UU} = \frac{1}{2}(I_{XX} + I_{YY}) + \frac{1}{2}(I_{XX} - I_{YY})\cos 90°$$

$$- I_{XY} \sin 90° - 30166.67 + 0 - (-18000)$$

$$= 84166.67 \text{ mm}^4$$

Then since $I_{UU} + I_{VV} = I_{XX} + I_{YY}$

$$I_{VV} = I_{XX} + I_{YY} - I_{UU}$$

$$= 2(30166.67) - 48166.67$$

$$= 12166.67 \text{ mm}^4.$$

**Example 16.5.** *Find the centroidal principal moment of inertia of the section shown in the Fig. 16.9.*

**Solution.** Evidently the centroid *G* of the given section is in the middle of the web. Let the centroidal principal axes be *GU* and *GV* as shown. Let $G_1$ and $G_2$ be the centroids of the rectangular areas *DEFH* and *ABKL* respectively.

$$I_{YY} = I_{YY} \text{ of area } BCHJ + I_{YY} \text{ of area } DEFH + I_{YY} \text{ of area } ABKL$$

$$= \frac{200 \times 10^3}{12} + \left[\frac{10 \times (90)^3}{12} + (90 \times 10)(50)^2\right]$$

$$+ \left[\frac{10 \times 90^3}{12} + 90 \times 10 \times 50^2\right]$$

$$= 573.167 \times 10^4 \text{ mm}^4$$

$$I_{XX} = I_{XX} \text{ of area } BCHJ + I_{XX} \text{ of area } DEFH + I_{XX} \text{ of area } ABKL$$

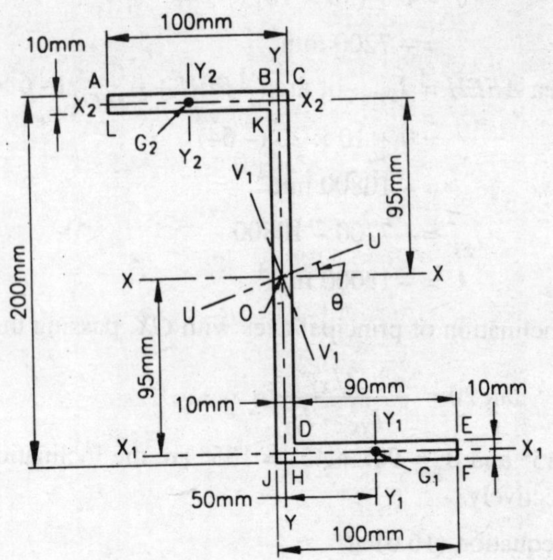


*Fig. 16.9*

$$= \frac{10 \times (200)^3}{12} + \left[\frac{90 \times 10^3}{12} + 90 \times 10 \times (95)^2\right]$$

$$+ \left[\frac{90 \times 10^3}{12} + 90 \times 10 \times (95)^2\right]$$

$$= 2292.6 \times 10^4 \text{ mm}^4$$

$$I_{XY} = I_{XY} \text{ of } BCHJ + I_{XY} \text{ of } DEFH + I_{XY} \text{ of } ABKL$$

Now $I_{XY}$ of $BCHJ = 0$ since $GX$ and $GY$ are its axes of symmetry

$$I_{xy} \text{ of } DEFH = I_{x_1y_1} \text{ of } DEFH + (\text{Area of } DEFH)\,(50) \times (-95)$$

$$= 0\ (90 \times 10)\ (50)\ (-95)$$

$$= -4275000 \text{ mm}^4$$

$$I_{xy} \text{ of } ABKL = I_{x_2y_2} \text{ of } ABKL + (\text{Area of } ABKL)\,(-50)\,(95)$$

$$= 0 + (90 \times 10)\,(-50)\,(95)$$

$$= -4275000 \text{ mm}^4$$

$$= -855 \times 10^4 \text{ mm}^4$$

Therefore $I_{xy}$ of the section

$$= 0 - 4275000 - 4275000$$

$$\tan 2\theta = \frac{2I_{xy}}{I_{yy} - I_{xx}} = \frac{2\,(-855 \times 10^4)}{573.167 \times 10^4 - 2292.6 \times 10^4}$$

$$= 0.9945$$

$\therefore$ $2\theta = 44.84°$

or $\theta = 22.42°$

$$I_{UU} = I_{XX} \cos^2\theta + I_{YY} \sin^2\theta - I_{XY} \sin 2\theta$$
$$= 2292.6 \times 10^4 (\cos 22.42°)^2 + (573.167 \times 10^4)$$
$$(\sin 22.42°)^2 - (-855 \times 10^4) \sin 44.84°$$
$$= 2645.37 \times 10^4 \text{ mm}^4$$
$$I_{VV} = (I_{XX} + I_{YY}) - I_{UU}$$
$$= (2292.6 \times 10^6 + 573.167 \times 10^4) - (2645.37 \times 10^4)$$
$$= 220.397 \times 10^4 \text{ mm}^4.$$

## 16.6 Stresses in Beams Due to Unsymmetrical Bending

In Fig. 16.10 is shown the section of a uniform cross section beam under the action of a bending moment $M$ acting in plane $YY$ which is inclined with the principal axis $VV$ at the angle $\theta$. It is required to find the stress distribution over the section.

The given bending moment $M$ can be resolved into its components in the planes $UU$ and $VV$ as follows :

Moment in the plane $UU = M_1 = M \sin\theta$ ...(16.30)

Moment in the plane $VV = M^2 = M \cos\theta$ ...(16.31)

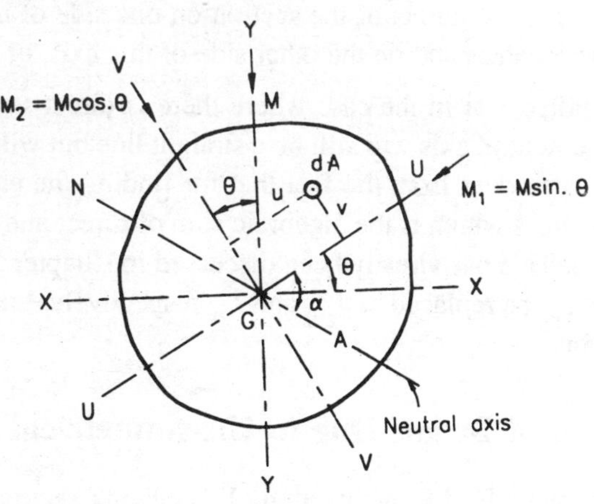


*Fig. 16.10*

It may be noted that the components $M_1$ and $M_2$ have their axes along $VV$ and $UU$ respectively. Thus the resultant bending stress at the point $P$ having $(u, v)$ as its co-ordinates with respect to $GU$ and $GV$ axes is given by $\sigma_b$ as

$$\sigma_b = \frac{(M \cos\theta)\, v}{I_{UU}} + \frac{(M \sin\theta)\, u}{I_{VV}}$$

or $$\sigma_b = M\left[\frac{v\cos\theta}{I_{UU}} + \frac{u\sin\theta}{I_{VV}}\right] \quad ...(16.32)$$

The nature of $\sigma_b$ at any point will depend upon the quadrant (will respect to $GU$ and $GV$ axes) in which it lies. In other words the signs of $u$ and $v$ in equation (16.32) have to be taken keeping in view the location of the point vis-a-vis the respective bending axes.

Further the equation of the neutral axis can be found by finding the locus of the points on which the resultant stress is zero.

Thus from equation (16.32) for points lying on neutral axis

$$\frac{v\cos\theta}{I_{UU}} + \frac{u\sin\theta}{I_{VV}} = 0$$

or $$v = -\left(\frac{I_{UU}}{I_{VV}}\tan\theta\right)u \quad ...(16.33)$$

This is an equation of a straight line passing through the centroid $G$ of the section and inclined at an angle $\alpha$ with $UU$ where

$$\tan\alpha = -\left(\frac{I_{UU}}{I_{VV}}\tan\theta\right) \quad ...(16.34)$$

Further the maximum stress will occur at a point which is at the greatest distance from neutral axes. All the points of the section on one side of neutral axis will carry stresses of the same nature and on the other side of this axis, of opposite nature.

In may be noted that in the case where there is direct stress in addition to the bending stress, the neutral axis will still be a straight line but will not pass through the centroid $G$. This is evident from the fact that for finding the equation of the neutral axis the resultant stress which is the algebraic sum of direct and bending stresses will be equated to zero. This has already been discussed in Chapter 5, the only difference being that $I_{XX}$ and $I_{YY}$ be replaced be $I_{UU}$ and $I_{VV}$ respectively. Likewise $x$ and $y$ will be replaced by $u$ and $v$.

## 16.7 Deflection of Beams Due to Unsymmetrical Loading

Fig. 16.11 shows a load $W$, acting along $YY$ line, on a section of a beam. Axes $GU$ and $GV$ are the principal axes of the section having $G$ as the centroid.

The load $W$ can be resolved into two components ($W\cos\theta$) along $GV$ and ($W\sin\theta$) along $GU'$. The component ($W\cos\theta$) will cause deflection $\delta_V$ along the line $GV'$ due to bending about $UU'$ axis and ($W\sin\theta$) will deflect the beam by $\delta_U$ along the line $GU'$ for its bending about $VV'$ axis. Depending on the end conditions of the beam these deflections will be given by

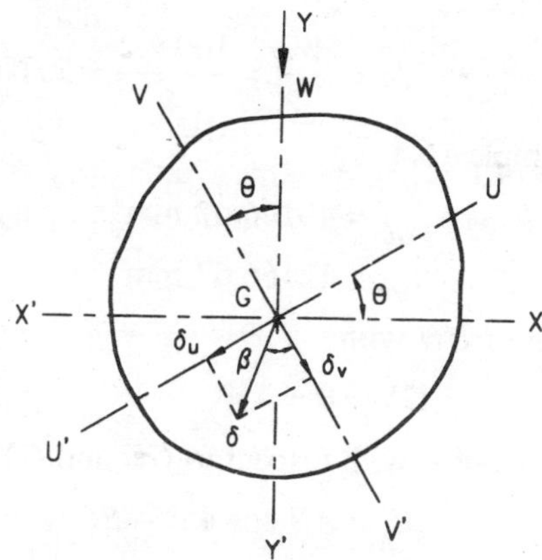


*Fig. 16.11*

$$\delta_U = \frac{\lambda (W \sin \theta)\, l^3}{EI_{VV}} \qquad ...(16.35)$$

$$\delta_V = \frac{\lambda (W \cos \theta)\, l^2}{EI_{UU}} \qquad ...(16.36)$$

Where λ is a constant depending on the end conditions of the beam and position of the load with respect to the ends.

The resultant deflection δ can then be found as follows :

$$\delta = \sqrt{\delta_U^2 + \delta_V^2}$$

$$\delta = \lambda \frac{Wl^3}{E} \sqrt{\frac{\sin^2 \theta}{I_{VV}^2} + \frac{\cos^2 \theta}{I_{VV}^2}} \qquad ...(16.37)$$

The inclination β of the direction of δ, with the line $GV'$, is given by

$$\tan \beta = \frac{\delta_U}{\delta_V} = \frac{I_{UU}}{I_{VV}} \tan \theta \qquad ...(16.38)$$

It is evident from equations (16.34) and (16.35) that the magnitudes of α and β are same and are measured from perpendicular lines ($GU$ and $GV$) in the same direction as shown in Figs. 16.10 and 16.11. Thus the deflection δ will be in a direction perpendicular to the neutral axis.

**Example 16.6.** *A 30 mm × 30 mm angle shown in Fig. 16.8 is used as a simply supported beam with leg AB vertical. If 0.1 kN load in the mid span coincident with the line YG acts on this beam with span of 2 m, find the stresses at the points A, B and C of the mid section of the beam.*

*Also find the deflection of the beam at the mid section and its direction with load line.*

*Take E = 200 GN/m². Give the location of neutral axis.*

**Solution.** (*a*) Bending moment at the mid section

$$M = \frac{WL}{4} = \frac{100 \times 2}{4} = 50 \text{ Nm}$$

From example 16.4

$$I_{UU} = 48166.67 \text{ mm}^4$$

$$I_{VV} = 12166.67 \text{ mm}^4$$

Inclination of *UU* with

$$GX = \theta = 45°$$

Co-ordinates of *A* with respect to *GU* and *GY* axes are $(u_A, v_A)$

$$u_A = AB \cos 45° - BG$$

$$= \frac{30}{\sqrt{2}} - BL\sqrt{2} = \frac{30}{\sqrt{2}} = 11\sqrt{2}$$

$$= 5.656856 \text{ mm}$$

$$v_A = AB \sin 45° = \frac{30}{\sqrt{2}} = 21.213204 \text{ mm}$$

From equation (16.32) the bending stress at *A* is

$$(\sigma_b)_A = \frac{M \cos\theta}{I_{UU}} v_A + \frac{M \sin\theta}{I_{VV}} u_A$$

$$= \frac{50 \times 10^3 \left(\frac{1}{\sqrt{2}}\right)}{48166.67} \times 21.213204$$

$$+ \frac{50 \times 10^3 \left(\frac{1}{\sqrt{2}}\right)}{12166.67} \times 5.656856 \text{ N/mm}^2$$

$$= 32.01 \text{ MN/m}^2 \text{ (Tensile)}$$

For point *B* :

$$u_B = -BG = BL\sqrt{2} = -11\sqrt{2} = -15.556348 \text{ mm}$$

$$v_B = 0$$

Then from equation (16.32)

$$(\sigma_b)_B = 0 + \frac{M \sin\theta}{I_{VV}} u_B$$

$$= \frac{(50 \times 10^3)}{12166.67}\left(\frac{1}{\sqrt{2}}\right) \times (-15.556348) \text{ N/mm}^2$$

$$= -45.205 \text{ MN/m}^2 \qquad \textit{(i.e. comp.)}$$

For the point $C$ : Since $C$ and $A$ are symmetrical about $BG$

$$u_C = u_A = 5.656856 \text{ mm}$$

$$v_C = -u_A = -21.213204 \text{ mm}$$

Then from equation (16.32)

$$(\sigma_b)_C = \frac{50 \times 10^3}{48166.67}\left(\frac{1}{\sqrt{2}}\right)(-21.213204)$$

$$+\frac{50 \times 10^3}{12166.76}\left(\frac{1}{\sqrt{2}}\right)(5.656856) \text{ N/mm}^2$$

$$= 0.8675 \text{ MN/m}^2 \text{ (Tensile)}$$

(*b*) The deflection $\delta$ is given by equation (16.37) where for the simply supported ends

$$\lambda = \frac{1}{48}$$

Thus $$\lambda = \frac{Wl^3}{48E}\sqrt{\frac{\sin^2\theta}{I_{VV}^2} + \frac{\cos^2\theta}{I_{UU}^2}}$$

$$= \frac{100\,(2 \times 10^3)^3 \times 10^6}{48\,(200 \times 10^9)}\left[\frac{1}{2 \times (12166.67)^2} + \frac{1}{2(48166.67)^2}\right]^{\frac{1}{2}}$$

$$= 4.9953 \text{ mm}$$

The deflection $\sigma$ will be inclined at an angle $\beta$ clockwise with the line $GV'$, given by

$$\tan\beta = \frac{I_{UU}}{I_{VV}}\tan\theta$$

$$= \frac{48166.67}{12166.67} \times 1 = 3.9589$$

*i.e.* $$\beta = 75.824 \text{ degrees}$$

Thus the deflection is at 75.824° – 45° = 30.824° clockwise with the load line $GY'$.

(*c*) The neutral axis will then be at 90° – 30.824° – 59.176° anticlockwise with the load line, because the neutral axis is perpendicular to the line of deflection.

**Example 16.7.** *A tensile member of a structure has a section as shown in Fig. 16.12. A tensile load P of magnitude 20 kN passing through the point 'O' acts normal to the section and parallel to longitudinal axis through the centroid G. Calculate the stress at the point A.*

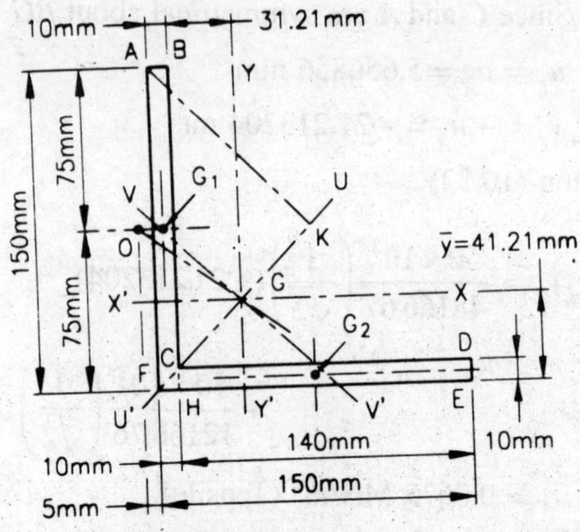


*Fig. 16.12*

**Solution.** Apply two equal the opposite loads (= $P$) through the centroid $G$ and parallel to the given load. The section will have direct tensile stress of $(P/A)$ along with a bending moment equal to $P \times (OG)$ acting in the plane $OG$.

For location of centroid $G$ : Moments of areas about the line $FE$ give

$$\bar{y}\ [150 \times 10 + 140 \times 10] = 140 \times 10 \times 5 + 150 \times 10 \times 75$$

$$\bar{y} = 41.21 \text{ mm}$$

$$I_{XX} = I_{YY} = \left[\frac{140 \times 10^3}{12} + 140 \times 10 \times (36.21)^2\right] + \left[\frac{10 \times (150)^3}{12} + 10\ kt\ 150 \times (75 - 41.21)^2\right]$$

$$= 4087282.4 \text{ mm}^4$$

Now co-ordinates of $G_1$, centroid of area $ABHF$, with respect to $GX$ and $GY$ axes, are (– 36.21, 33.79)

Therefore for the area $ABHF$

$$I_{XY} = 0 + A_{ABHF}\ (-36.21)\ (33.79)$$
$$= (1500)\ (-36.21)\ (33.79)$$
$$= -1835303.8 \text{ mm}^4$$

And for the area $CDEH$,

$$I_{XY} = 0 + A_{CDEH}\ (38.79)\ (-36.21)$$
$$= 2 + (1400)\ (38.79)\ (-32.21)$$
$$= -1966420.2 \text{ mm}^4$$

Thus total $I_{XY} = -\ 1835303.8 - 1966420.2$

$= -\ 380172\ \text{mm}^4$

As the section is symmetrical the centroidal principal axes will be at 45° to the axes $GX$ and $GY$.

Now $$I_{UU} = I_{XX} \cos^2 \frac{\pi}{4} + I_{YY} \sin^2 \frac{\pi}{4} - I_{YY} \sin \frac{\pi}{2}$$

$$= I_{XX} \left( \cos^2 \frac{\pi}{4} + \sin^2 \frac{\pi}{4} \right) = I_{XY} \sin \frac{\pi}{2}$$

$$= 4087282.4 - (-\ 3801724)$$

$$= 7889006.4\ \text{mm}^4$$

$$I_{VV} = (I_{XX} + I_{YY}) - I_{UU}$$

$$= 2 \times 4087282.4 - 7889006.4$$

$$= 285558.4\ \text{mm}^4$$

Now the Bending moment, $M$, in the plane $GO = P \times (GO)$

Also $$GO = [(41.21 + 5)^2 + (75 - 41.21)^2]^{1/2}$$

$$= 57.246\ \text{mm}$$

$$M = 20000 \times 57.246/1000 = 1144.91\ \text{Nm.}$$

$$\angle OGX' = \tan^{-1} \frac{(75 - 41.21)}{(41.21 + 5)} = 36.175°$$

Therefore $$\angle OGV = \phi = 45° - 36.175°$$

$$= 8.825°$$

Thus bending moment in the plane $GV$

$$M_1 = \text{M} \cos \phi = 1144.92 \cos (8.825°)$$

$$= 1131.36\ \text{Nm.}$$

The bending moment $M_1$ is anticlockwise if viewed in the direction $UG$ *i.e.* its effect will be to give rise to tensile stresses on all points lying on the right of $\overrightarrow{UO}$; particularly the point $A$ will be experiencing the tensile stress due to $M_1$.

Similarly the other component $M_2$, in the plane $GU$ is

$$M_2 = M \sin \phi = 1144.92 \sin (8.825°)$$

$$= 175.65\ \text{Nm.}$$

The bending moment $M_2$ will be anticlockwise as viewed in the direction $YG$ *i.e.* its effect will be also be to induce tensile stress at $A$.

Further at $A$ there will also be direct tensile stress.

$$\sigma_0 = \frac{P}{A} = \frac{2000}{2900} \text{ N / mm}^2$$

$$= 6.8965 \text{ MN/m}^2$$

Total bending stress at $A$

$$(\sigma_b)_A = \frac{M_1}{I_{UU}} v_A + \frac{M_2}{I_{VV}} u_A$$

Now $$u_A = GK = FK - FG = \frac{150}{\sqrt{2}} - (41.21) \times \sqrt{2}$$

$$= 47.787 \text{ mm}$$

$$v_A = A_K = 150 / \sqrt{2} = 106.066 \text{ mm}$$

Therefore $$(\sigma_b)_A = \frac{(1131.36 \times 1000)}{7889006.4} \times 106.066$$

$$+ \frac{(175.65 \times 1000)}{28558.4} \times 47.787 \text{ N / mm}^2$$

$$= 15.211 + 2.939 \text{ N/mm}^2$$

$$= 18.150 \text{ MN/m}^2$$

Thus resultant stress at $A$

$$= \sigma_0 + (\sigma_b)_A = 6.8965 + 11.150 \text{ MN/m}^2$$

$$= 18.046 \text{ MN/m}^2 \text{ (Tensile)}$$

**Example 16.8.** *A 80 mm × 80 mm angle as shown in Fig. 16.13 is used as a freely supported beam with one leg vertical.* $I_{XX} = I_{YY} = 0.8736 \times 10^{-6}$ $m^4$. *When a bending moment is applied in the vertical plane YY, the mid section of the beam deflects*

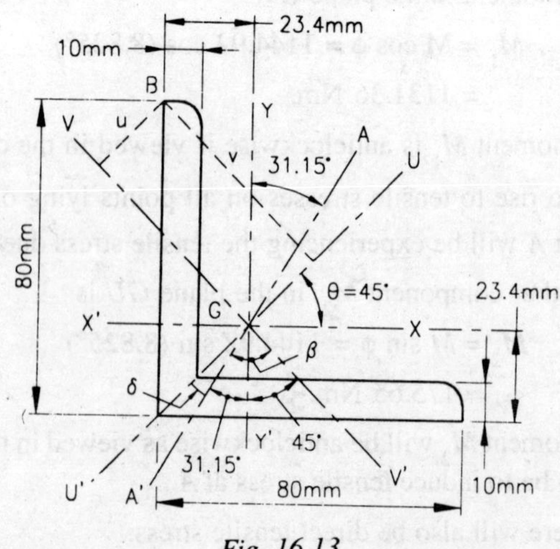


*Fig. 16.13*

*in the direction AA at 30° 15° of the vertical. Calculate the second moments of area of the section about its principal axes.*

*Find also the bending stress at the corner B if the bending moment is 2 kNm.*

*(Punjabi University)*

**Solution.**

$$I_{UU} + I_{VV} = I_{XX} + I_{YY} = 2 \times 0.8736 \times 10^{-6}$$
$$= 1.7472 \times 10^{-6} \text{ m}^4$$

Now $$\tan\beta = \frac{I_{UU}}{I_{VV}} \times \tan\theta$$

Here due to equal legs of the angle section, theta = 45°

and $$\beta = \text{Inclination of } GA \text{ with } GV'$$
$$= \angle V'GY' + \angle Y'GA'$$
$$= 45° + 30°\ 15' = 75'\ 15'$$

Therefore

$$\tan(75°\ 15') = \frac{I_{UU}}{I_{VV}} \tan 45°$$

or $$\frac{I_{UU}}{I_{VV}} = 3.79$$

Thus solving the two simultaneous equations between $I_{UU}$ and $I_{VV}$, we get

$$3.79\, I_{VV} + I_{VV} = 1.7472 \times 10^{-6}$$

or $$I_{VV} = 0.365 \times 10^{-6} \text{ m}^4$$

and $$I_{UU} = 3.79 \times .365 \times 10^{-6} = 1.382 \times 10^{-6} \text{ m}^4$$

If $(u, v)$ are the co-ordinates of $B$ with respect to $GU$ and $GV$ axes, then geometrically

$$u = 80 \cos 45° - \frac{23.4}{\cos 45°}$$

$$= \frac{80}{\sqrt{2}} - 23.4\sqrt{2} = 23.47 \text{ mm}$$

$$v = 80 \sin 45° = \frac{80}{\sqrt{2}} = 56.57 \text{ mm}$$

Then $$\sigma_B = \frac{Mv\cos\theta}{I_{UU}} + \frac{Mu\sin\theta}{I_{VV}}$$

$$= 2 \times 10^3 \left[ \frac{(56.57 \times 10^{-3})}{1.382 \times 10^{-6}} \cdot \frac{1}{\sqrt{2}} + \frac{(23.47 \times 10^{3})}{0.365 \times 10^{-6}} \cdot \frac{1}{\sqrt{2}} \right] \text{N/m}^2$$

$$= 149 \text{ MN/m}^2.$$

## 16.8 Flexural Axis and Shear Centre

Axis of bending which is also called the Flexural axis of a beam is that longitudinal axis through which the transverse loads causing bending must pass so that these is no twisting of the beam. The point where the Flexural axis intersects the cross section of the beam is known as shear centre.

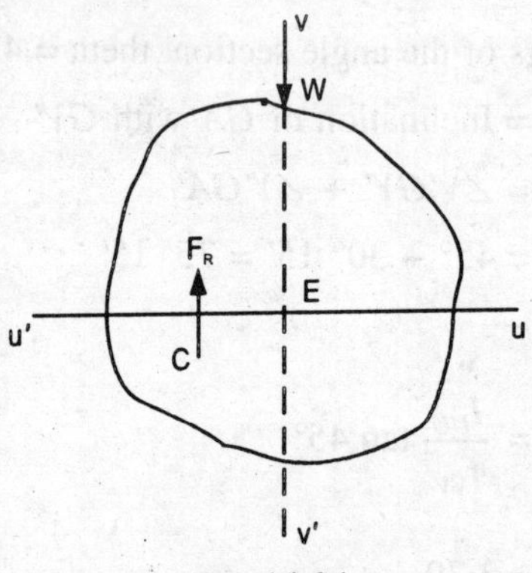


*Fig. 16.14*

Figure 16.14. shows the cross section of beam. The transverse load $W$ is acting through the centroid of the section. Let the resultant shearing force due to the shear stresses induced on the section be $F_R$ which will pass through the point $C$ on the $u - u$ axis ($C$ is called the shear centre of the section). The magnitude of $F_R$ will be equal to $W$, the transverse load on the section. In the present case, since $W$ is not passing through the point $C$, which is at a distance $e$ from the shear centre $C$, it will result in twisting of section due to the twisting moment equal to $We$. To avoid twisting the transverse load $W$ should be made to pass through the shear centre $C$. This is essential specially in the case of thin unsymmetrical sections which are capable of providing greater resistance to bending but are usually very weak under the action of twisting moments. Therefore it is very important to locate the shear centre of the beam section so that the transverse load may be made to pass through it.

Since $F_R$ is the resultant shearing force due to shear stresses in various portions of the beam section, the shearing forces in the different portions of the section will be evaluated individually by using the equations (5.3) and (5.4) of chapter 5.

These in turn will be used to locate the position of shear centre so that the resultant moment due to these individual shearing forces about the shear centre must be zero.

## 16.9 Shear Centre for Sections Symmetrical about both Principal Axes

In such cases where the cross section is symmetrical about both the principal axes, the flexural axis will coincide with the longitudinal centroid axis. Thus the shear centre will coincide with the centroid of the section. Therefore in such cases no twisting of the section will occur.

## 16.10 Shear Centre for Sections Symmetrical About Only One Axis

Figure 16.15 represents an equal leg angle section $b \times b \times t$ having *NA* as the central axis which is also one axis of symmetry. It is subjected to a shearing force *F* passing through the point *O* which is the point of intersection of the centre lines of the two legs. Let $F_1$ and $F_2$ are the shearing forces sustained by the two legs. These forces are due to the shearing stresses induced in the respective legs. Let the resultant of the forces $F_1$ and $F_2$ be $F_R$, then $F_R$ will be equal and opposite to the shearing force *F* acting on the section.

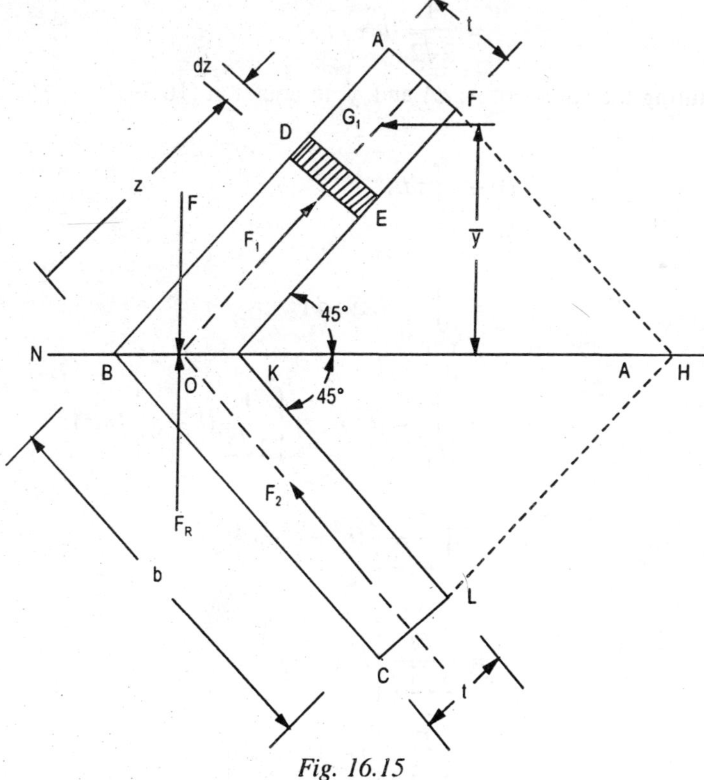


*Fig. 16.15*

Let *z* be the distance of an element from *B* on the leg *AB* from *B* and *dz* the length of this element where the shear stress is τ. Let '*A*' be the area of the leg portion between *ADEF i.e.* from the element to the top most point of the leg.

Then $$F_1 = \int_0^b \tau dA \quad ...(16.39)$$

where $$\tau = \frac{FA\bar{y}}{It}$$

$$A = \text{Area } ADEF = t(b - z)$$

$$dA = tdz$$

$\bar{y}$ = distance of $G_1$ which is the centroid of the area $DEFA$ from N.A

and or $$\bar{y} = OG_1 \sin 45°$$

$$\approx z + \frac{1}{2}(AD) \sin 45°$$

$$= z + \frac{1}{2}(b - z)\frac{1}{\sqrt{2}}$$

$$= \frac{1}{2\sqrt{2}}(b + z)$$

Substituting the values of $A$, $dA$ and $\bar{y}$ in equation (16.39)

$$F_1 = \int_{z=0}^{b} \tau dA$$

$$= \int_{z=0}^{b} \frac{F}{It}(A\bar{y})\, dA$$

$$= \int_{z=0}^{b} \frac{F}{It}\{t(b - z\}\left\{\frac{1}{2\sqrt{2}}(b + z)\right\}(tdz)$$

$$= \int_0^b \frac{F}{2\sqrt{2}\, I}(b^2 - z^2)\, t\, dz$$

$$= \frac{Ft}{2\sqrt{2}\, I}\left[b^2 z - \frac{z^3}{3}\right]_0^b$$

$$= \frac{Ft}{2\sqrt{2}\, I}\left[b^2 . b - \frac{1}{3}b^3\right]$$

$$= \frac{Ft}{2\sqrt{2}\, I}\left[\frac{2}{3}b^3\right]$$

$$F_1 = \frac{Ftb^3}{3\sqrt{2}\,I} \qquad ...(16.40)$$

For finding $I$, complete the square $ABCH$

then
$$I = \frac{b^4}{12} - \frac{(b-t)^4}{12}$$

$$= \frac{1}{12}[b^2 + b^2 + t^2 - 2bt]\,[b^2 - b^2 - t^2 + 2bt]$$

$$= \frac{1}{12}[2b^2 + t^2 - 2bt]\,[t\,(2b - t]$$

Then $t$ being small neglecting $t^2$ in comparison to $b^2$ and $t$ as compound to $b$

$$I \approx \frac{1}{12}(2b^2 - 2bt)\,t\,(2b)$$

$$\approx \frac{1}{12}2b\,(b-t)\,t(2b)$$

$$\approx \frac{4b^3 t}{12} = \frac{1}{3}b^3 t \qquad ...(16.41)$$

Then from equation (16.40) and (16.41)

$$F_1 = \frac{Ftb^3}{3\sqrt{2}} \times \frac{3}{b^3 t} = \frac{F}{\sqrt{2}}$$

Since the two legs are of same length and thickness and the section is symmetrical about $NA$

$$F_2 = F_1 = \frac{F}{\sqrt{2}}$$

Therefore resultant shear force is given by

$$F_r = (F_1^2 + F_2^2)^{\frac{1}{2}}$$

or
$$F_r = \left(\frac{F^2}{2} + \frac{F^2}{2}\right)^{\frac{1}{2}} = F$$

Thus $F_r = F$

In other words $O$ is the shear centre of there section and $F_1$ and $F_2$ are the components of $F$.

## 16.11 Shear Centre for Channel Section

In figure 16.16 (*a*) is shown a thin $b \times d \times t$ channel section $ABCD$ is an element of dimension $dx \times dz \times t$ where the shear $\tau$ acts on the crosscutting plane and a complementary shear stress $\tau$ on the longitudinal plane. $F_1$ is the total shear force sustained by the top leg of the *c*-section, $F_2$ of the lower leg and $F_3$ is the shear force sustained by the web.

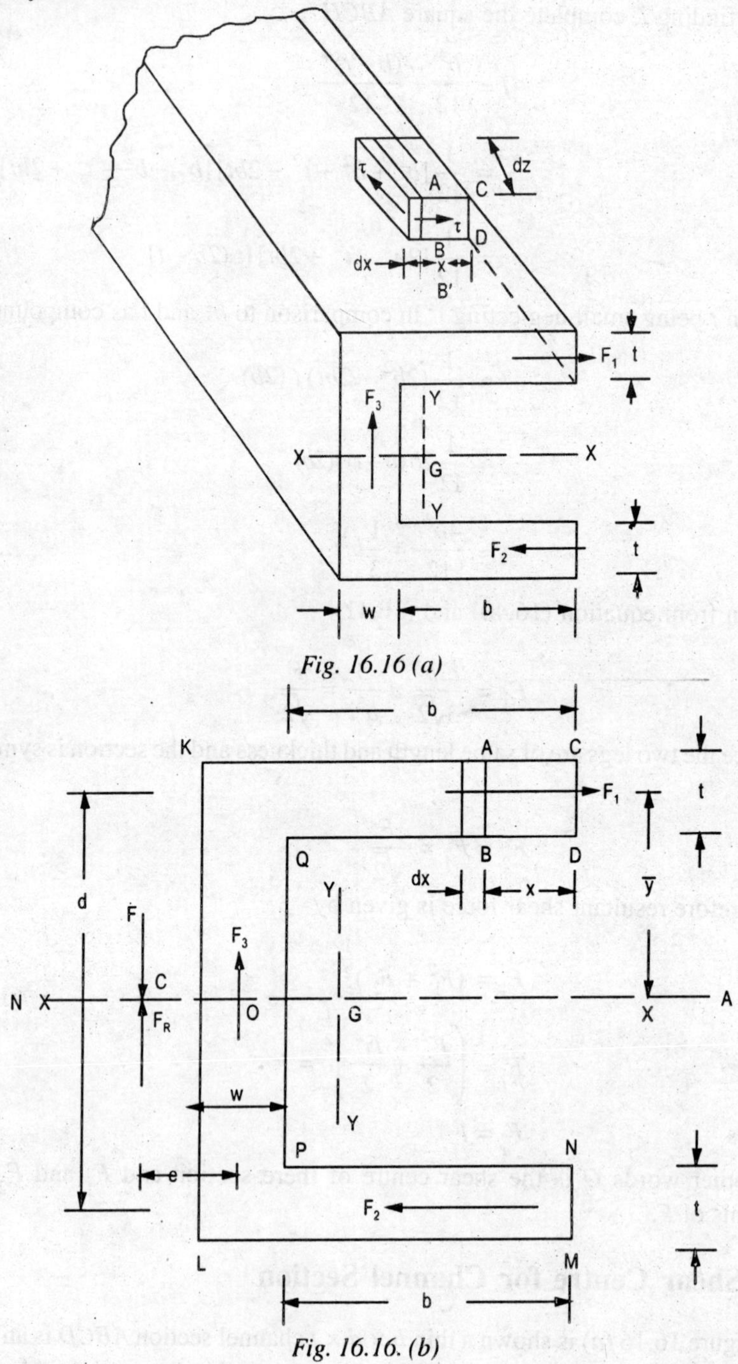


Fig. 16.16 (a)

Fig. 16.16. (b)

The shear force $F_1$ is calculated by using the equations (5.3) and (5.4)

That is $$F_1 = \int \tau dA$$

$$= \int \frac{F(A\bar{y})}{It} dA$$

where $A$ = area $ABCD$ (Fig. 16.16 *b*)

$= xt$

$dA = tdx$ = area $ABB'A'$

$\bar{y} = \frac{d}{2}$ = Distance of centroid of area $ABCD$ from NA

Therefore $$F_1 = \frac{F}{It}\int_0^b (tx)(tdx)\frac{d}{2}$$

$$= \frac{F}{It}\frac{d}{2}t^2\left[\frac{x^2}{2}\right]_0^b$$

or $$F_1 = \frac{F}{4I}tdb^2 \qquad ...(16.42)$$

Also $I$ = Second moment of the entire area about centroidal axis $XX$ along $CG$. Since the area consists of three rectangles

$$I_{xx} = (I \text{ of vertical leg}) + 2(I \text{ of horizontal leg})$$

$$= \frac{wd^3}{12} + 2\left[(b \times t)\left(\frac{d}{2}\right)^2 + \frac{bt^3}{12}\right]$$

Last term $\frac{bt^3}{12}$ is neglected being the thickness very small.

or $$I_{xx} = \frac{wd^3}{12} + 2bt\left(\frac{d}{2}\right)^2$$

$$= \frac{1}{2}btd^2\left(1 + \frac{1}{6}\frac{wd}{bt}\right) \qquad ...(16.43)$$

Therefore from equation (16.42) and (16.42) we get

$$F_1 = \frac{Ftdb^2}{4btd^2}\left[\frac{2}{1 + \frac{1}{6}\left(\frac{wd}{bt}\right)}\right]$$

or
$$F_1 = \frac{Fb}{2d\left[1+\frac{1}{6}\left(\frac{wd}{bt}\right)\right]} \quad ...(16.44)$$

For the equilibrium of forces

$$F_2 = F_1 \quad ...(16.45)$$

and
$$F_R = F_3 = F \quad ...(16.46)$$

Taking moments about $O$, of $e$ is the distance of $C$ from the centre of web $O$.

$$F_R \times e = F_1 \frac{d}{2} + F_2 \frac{d}{2}$$
$$= F_1 d$$
$$= \frac{Fb}{2\left[1+\frac{1}{6}\frac{wd}{bt}\right]} \quad \text{[From (16.44)]}$$

But
$$F_R = F$$

Therefore
$$e = \frac{b}{2\left[1+\frac{1}{6}\frac{wd}{bt}\right]} \quad ...(16.47)$$

Also let
$$A_w = \text{area of web}$$
$$= wd \quad ...(16.48\ a)$$

and
$$A_f = \text{Area of flange}$$
$$= bt \quad ...(16.48\ b)$$

Thus substituting values from equation (16.48) in equation (16.49)

$$e = \frac{b}{2\left[1+\frac{1}{6}\frac{A_w}{A_f}\right]} \quad ...(16.49)$$

## 16.12 Shear Centre of unequal I-Section

In fig. (16.17) is shown an unequal $I$ section. The flange $ABCD$ is of width $B_1$ and thickness $t_1$ while the flange $LMNP$ is of width $B_2$ and thickness $t_2$. The web thickness is $w$. The depth of the $I$-section is $H$. The shearing force $F_1$ sustained by the flange $ABCD$ is given by

$$F_1 = 2\int_0^{B_{1/2}} \tau dA$$
$$= 2\int_0^{B_{1/2}} \frac{F(A\bar{y})}{It_1} dA \quad ...(16.50)$$

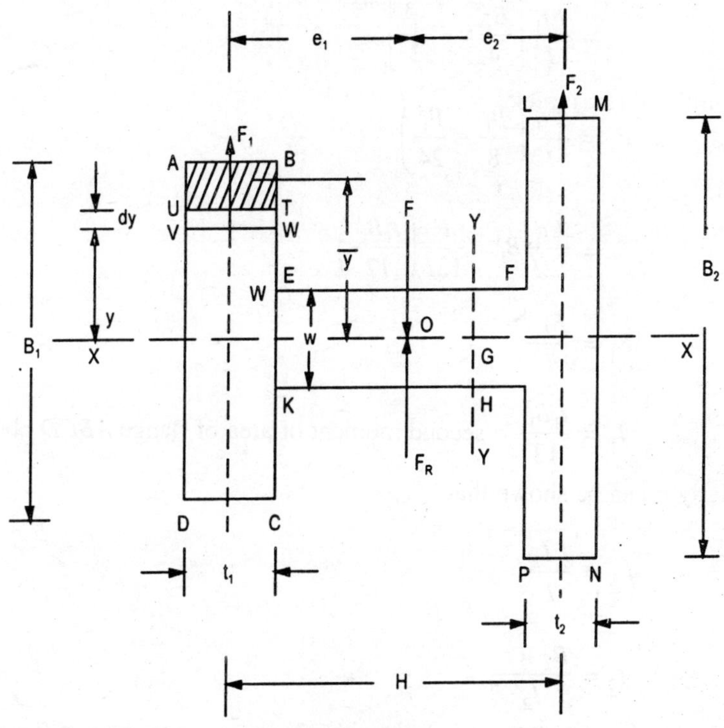


*Fig. 16.17*

where $F$ = shearing force on the section

$I$ = second moment of area about centroidal bending axis $xx$

$A$ = Shaded Area $ABTU$

$\bar{y}$ = Distance of centroid of area $ABTU$ from the bending axis.

$dA$ = Elemental area $UTWV$ which is at a distance $y$ from $xx$ and of depth dy

Therefore $F_1 = \int \tau dA$

$$= \int \frac{F(A\bar{y})}{It_1} dA$$

Therefore, $$F_1 = 2\int_0^{B_{1/2}} \frac{F}{It_1}\left\{\left(\frac{B_1}{2} - y\right)t_1\right\}\left\{y + \frac{1}{2}\left(\frac{B_1}{2} - y\right)\right\}(t_1 dy)$$

$$= \frac{Ft_1}{I}\int_0^{B_{1/2}}\left(\frac{B_1}{2} - y\right)\left(y + \frac{B_1}{2}\right)dy$$

$$= \frac{Ft_1}{I}\int_0^{B_{1/2}}\left(\frac{B_1^2}{4} - y^2\right)dy$$

$$= \frac{Ft_1}{I}\left[\frac{B_1^2}{4}y - \frac{y^3}{3}\right]_0^{B_1/2}$$

or

$$F_1 = \frac{Ft_1}{I}\left(\frac{B_1^3}{8} - \frac{B_1^3}{24}\right)$$

$$= \frac{Ft_1}{12I}B_1^3 = \frac{F}{12I}\left(\frac{t_1 B_1^3}{12}\right)$$

$$F_1 = \frac{FI_1}{I} \quad \text{...(16.51)}$$

where $I_1 = \frac{t_1 B_1^3}{12}$ = second moment of area of flange $ABCD$ about $xx$

Similarly it can be shown that

$$F_2 = \frac{FI_2}{I}$$

where

$$I_2 = \frac{B_2^3 \times t_2}{I_2} \quad \text{...(16.52)}$$

Let the shear centre $O$ be at a distance $e_1$ from the centre line of flange $ABCD$ and $e_2$ from that of flange $LMNP$.

Then, taking moments about $O$, since $FR$ the resultant of $F_1$ and $F_2$ will pass through $O$

$$F_1 e_1 = F_2 e_2$$

$$\frac{FI_1}{I}e_1 = \frac{FI_2}{I}e_2$$

or

$$\frac{e_1}{e_2} = \frac{I_2}{I_1} = \frac{t_2 B_2^3}{t_1 B_1^3} = \left(\frac{t_2}{t_1}\right)\left(\frac{B_2}{B_1}\right)^3 \quad \text{...(16.53)}$$

## 16.13 Shear Centre of an eccentric I-section

The unequal $I$ section $ABJK$ is shown in fig 16.18. the two portions of the flanges $FE$ and $DC$ are of lengths $b_2$ and $b_1$ respectively. Also $LM = FE$ and $NH = DC$.

Here the shearing forces $F_1$, $F_2$, $F_3$, $F_4$ and $F_5$ sustained by different portions are shown in figure 16.128.

Also

$$F_1 = F_5 \quad \text{...(16.54)}$$

$$F_2 = F_4 \quad \text{...(16.55)}$$

force sustained by the web $EDNM = F_3$

$$F_1 = \int_0^{b_1} \tau dA = \int_{It}^{b_1} \left\{ \frac{F(A\bar{y})}{It} \right\} dA \qquad ...(16.56)$$

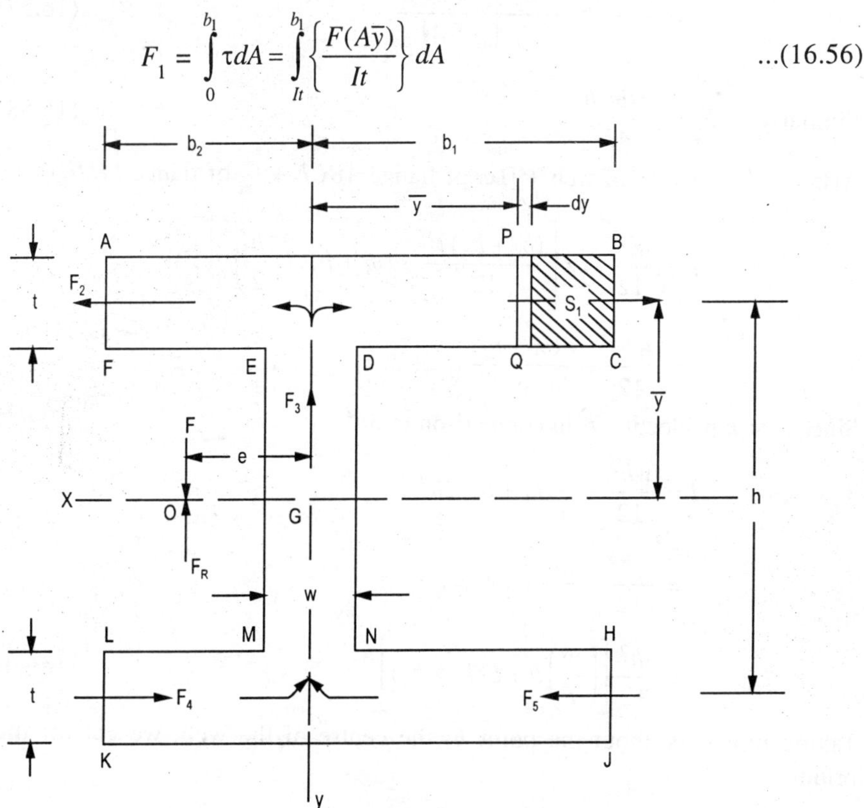


*Fig. 16.18*

where $dA$ = Elemental area $tdy$.

$t$ = thickness of flanges

$A$ = Area from section $PQ$ to $BC$

= Area $PQCB = (b_1 - y)t$

$\bar{y}$ = distance of $S$, the centroid of $PQCB$

= from $X$-$X$ axis

Therefore $$F_1 = \int_0^{b_1} \left(\frac{F}{It}\right) \{(b_1 - y)t\} \left(\frac{h}{2}\right) tdy$$

$$= \left(\frac{Ft}{2I}\right) \int_0^{b_1} (b_1 - y)\, hdy$$

$$= \frac{Fth}{2I} \left[ b_1 y - \frac{y^2}{2} \right]_0^{b_1}$$

$$= \frac{Fth}{2I}\left[b_1^2 - \frac{b_1^2}{2}\right] = \frac{Ftb_1^2 h}{4I} \qquad ...(16.57)$$

Similarly $F_2 = \dfrac{Ftb_2^2 h}{4I}$ ...(16.58)

Also $I = I_{xx}$ of web + [Ixx of flange $ABCF$ + $I_{xx}$ of flange $LHJK$]

or
$$I = \frac{wh^3}{12} + 2\left[\frac{(b_1 + b_2)\, t^3}{12} + (b_1 + b_2)\, t\left(\frac{h}{2}\right)^2\right]$$

$$= \frac{wh^3}{12} + \frac{t\,(b_1 + b_2)}{6}\left(t^2 + 3h^2\right)$$

Since $t << h$ neglecting $t^2$ in comparison to $3h^2$

$$I = \frac{wh^2}{12} + \frac{t}{6}(b_1 + b_2)\, 3h^2$$

$$= \frac{wh^2}{12} + \frac{tb^2}{2}(b_1 + b_2)$$

$$= \frac{th^2}{12}\left[\left(\frac{w}{t}\right)h + b\,(b_1 + b_2)\right] \qquad ...(16.69)$$

Taking moments about the point $F$, the centre of the web, we get for the equilibrium

$$F.e = (F_1 + F_5)\frac{h}{2} - (F_2 + F_4)\frac{h}{2} \qquad ...(16.60)$$

Substituting values of $F_1$, $F_5$, $F_2$ and $F_4$ from equation (16.54), (16.55), (16.57) and (16.58) in equation (16.60) we get

$$F.e. = 2\left(\frac{Ftb_1^2 h}{4I}\right)\frac{h}{2} - 2\left(\frac{Ftb_2^2 h}{4I}\right)\frac{h}{2}$$

or
$$e = \frac{th^2}{4I}(b_1^2 - b_2^2) \qquad ...(16.61)$$

Inserting the value of $I$ from equation (16.59) we get

$$e = \left[\frac{th^2}{4}(b_1^2 - b_2^2)\right]\frac{12}{th^2} \times \frac{1}{\left(\dfrac{w}{t}\right)h + 6\,(b_1 + b_2)}$$

or
$$e = \frac{3\,(b_1^2 - b_2^2)}{\left(\dfrac{w}{t}\right)h + 6\,(b_1 + b_2)} \qquad ...(16.62)$$

## 16.14 Shear Centre of an Outwardly Extended C-Section

Here again the shear centre location is determined by finding the line of action of the resultant resisting shearing force $F_R$ of the section. Let $F_1$, $F_2$, $F_3$, $F_4$ and $F_5$ be resisting shearing forces in the parts *AB, BC, CD, DE* and *EF* of the given section respectively.

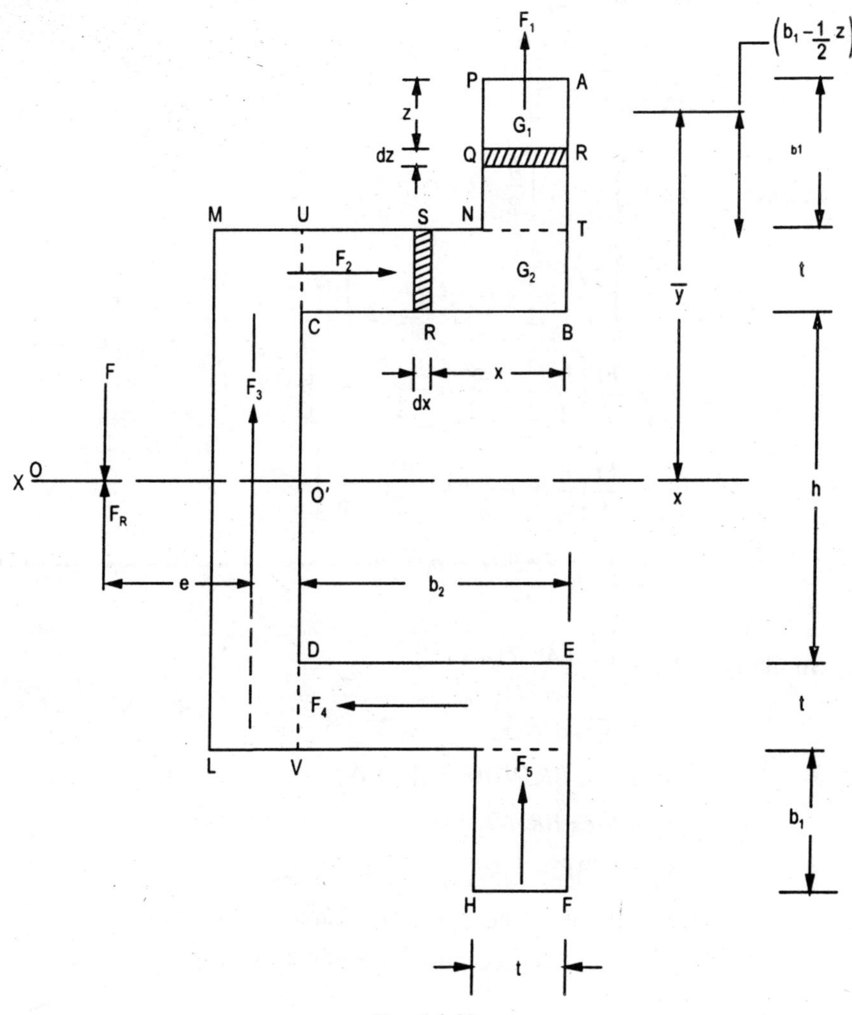


*Fig. 16.19*

By symmetry $\qquad F_1 = F_5 \qquad$ ...(16.63)

$$F_2 = F_4 \qquad ...(16.64)$$

The shearing force $F_1$, sustained by the portion *AB* is given by

$$F_1 = \int \tau dA = \int \frac{F(A\bar{y})}{It} . dA \qquad ...(16.65)$$

where $F$ = Sharing force on the section

$I$ = Second moment of area about the centroidal bending axis $XX$

A = Area $ARQP = tz$

$\bar{y}$ = Distance of centroid $G$, of area $ARQP$ from the bending axis

$$= \left(\frac{h}{2} + t + b_1 - \frac{z}{2}\right) \quad ...(16.66)$$

$dA = t.dz$

Therefore
$$F_1 = \int_0^{b_1} \frac{F}{It}(tz)\left[\frac{h}{2} + t + b_1 - \frac{z}{2}\right] t.dz$$

$$= \int_0^{b_1} \frac{Ft}{I}\left(z\frac{h}{2} + tz + b_1 z - \frac{z^2}{2}\right) dz$$

$$= \frac{Ft}{I}\left[\frac{h}{2}\frac{z^2}{2} + \frac{tz^2}{2} + \frac{b_1 z^2}{2} - \frac{z^3}{6}\right]_0^{b_1}$$

or
$$F_1 = \frac{Ft}{I}\left[\frac{h}{4}b_1^2 + \frac{t}{2}b_1^2 + \frac{b_1^3}{2} - \frac{b_1^3}{6}\right]$$

$$F_1 = \frac{Ft}{I}b_1^2\left(\frac{h}{4} + \frac{t}{2} + \frac{b_1}{3}\right) \quad ...(16.67)$$

Similarly
$$F_2 = \int \frac{F}{It}(A'\bar{y})\, dA'$$

$$A'\bar{y}' = A'\bar{y}_1' + A_2\bar{y}_2$$

where $A'$ = area $BRSNTAB = A_1 + A_2$

$A_1$ = Area $BRSNT = tx$

$A_2 = NTAP = b_1 t$

$\bar{y}_1$ = distance of centroid $G_2$ of area $BRST$ from centroidal bending axis $XX$

$$= \left(\frac{h}{2} + \frac{t}{2}\right)$$

$\bar{y}_2$ = distance of centroid $G_1$ of area $SNTBR$ from centroidal bending axis $XX$

$$= \left(\frac{h}{2} + t + \frac{b_1}{2}\right)$$

$$dA = tdx$$

$$I = \text{As defined earlier}$$

Therefore $$F_2 = \frac{F}{It}\int_0^{b_2}(A_1\bar{y}_1 + A_2\bar{y}_2)\, tdx$$

$$= \frac{F}{It}\int\left[tx\left(\frac{h}{2}+\frac{t}{2}\right)+b_1t\left(\frac{h}{2}+t+\frac{b_1}{2}\right)\right](tdx)$$

$$= \frac{Ft}{I}\int_0^{b_2}\left[x\left(\frac{h}{2}+\frac{t}{2}\right)+b_1\left(\frac{h}{2}+t+\frac{b_1}{2}\right)\right]dx$$

$$= \frac{Ft}{I}\left[\frac{x^2}{4}(h+t)+b_1x\left(\frac{h}{2}+t+\frac{b_1}{2}\right)\right]_0^{b_2}$$

$$= \frac{Ft}{I}\left[\frac{b_2^2}{4}(h+t)+b_1b_2\left(\frac{h}{2}+t+\frac{b_1}{2}\right)\right] \quad ...(16.68a)$$

or $$F_2 = \frac{Ft}{I}\left[\frac{hb_2^2}{4}+\frac{tb_2^2}{4}+b_1b_2\frac{h}{2}+b_1b_2t+\frac{b_2b_1^2}{2}\right] \quad ...(16.68b)$$

For the equilibrium taking moments about $O'$, where $x$-axis intersects the line of action of $F_3$

$$F_R.e = F_1b_2 + F_5b_2 - F_2\frac{h}{2} - F_4\frac{h}{2}$$

$$F_R.e = 2F_1b_2 - F_2h \quad ...(16.69)$$

Substituting values of $F_1$ and $F_2$ from equations (16.67) and (16.68) in equation (16.69) and equation $F_R = F$

We get $$F.e = \frac{2Ftb_1^2}{I}\left(\frac{h}{4}+\frac{t}{2}+\frac{b_1}{3}\right)-\frac{Ft}{I}\left[\frac{b_2^2}{4}(h+t)+b_1b_2\left(\frac{h}{2}+t+\frac{b_1}{2}\right)\right]$$

Therefore

$$e = \frac{tb_2}{I}\left[b_1^2\left(\frac{h}{2}+t+\frac{2b_1}{3}\right)-h\left\{\frac{b_2}{4}(t+h)+b_1\left(\frac{h}{2}+t+\frac{b_1}{2}\right)\right\}\right] \quad ...(16.70)$$

Also second moment of area about $XX$

$$I = 2I_{ATNP} + 2I_{TBCU} + I_{MUVL}$$

$$= 2\left[\frac{tb_1^3}{12}+tb_1\left(\frac{h}{2}+t+\frac{b_1}{2}\right)^2\right] + 2\left[\frac{b_2t^3}{12}+b_2t\left(\frac{h}{2}+\frac{t}{2}\right)^2\right]+\frac{t\,(h+2t)^3}{12} \quad ...(16.71)$$

## 16.15 Shear centre of an inwardly extended symmetrical c-section

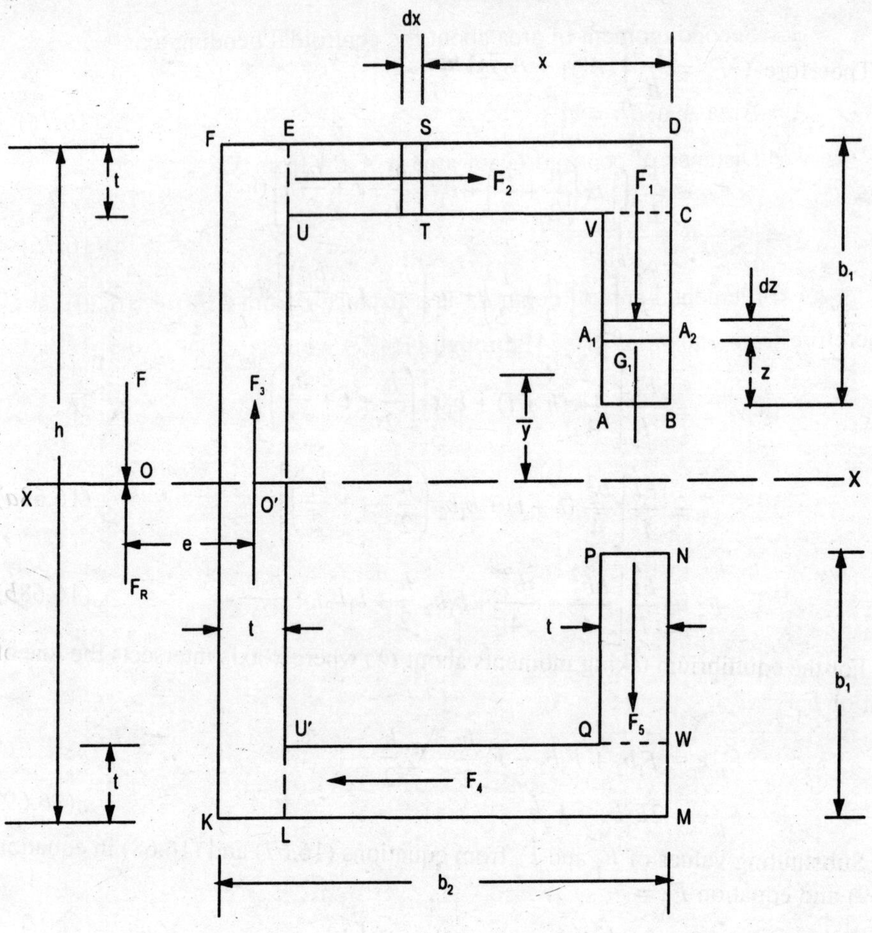


*Fig. 16.20*

The location of shear centre in this case is also determined by finding the line of action of the resultant resisting shearing force $F_R$ of the section $F_1$, $F_2$, $F_3$, $F_4$ and $F_5$ are assumed to be the resisting shearing forces of the portions *ABCV, CDEU, FELK, LMWU′* and *QWNP* respectively.

By symmetry $F_1 = F_5$ ...(16.72)

$$F_2 = F_4 \quad \text{...(16.73)}$$

The shearing force $F_1$ is given by

$$F_1 = \int \tau dA = \int_0^{b_1} \frac{F(A\bar{y})}{It} dA \quad \text{...(16.74)}$$

where $F$ = Shearing force on the section

$I$ = Second moment of area about the centroidal bending axis $XX$

$A$ = Area $A_1A_2BA = zt$ ...(16.75)

$\bar{y}$ = Distance of centroid $G_1$ of area $A_1A_2BA$ from $XX$.

$$= \left(\frac{h}{2} - b_1 + \frac{z}{2}\right) \quad ...(16.76)$$

$dA$ = Elemental area of depth $dz$ at a distance $z$ from $AB$ (fig. 16.20)

Therefore from equations (16.74) through (16.77) we get

$$F_1 = \int_0^{b_1} \frac{F}{It}(zt)\left(\frac{h}{2} - b_1 + \frac{z}{2}\right)(tdz)$$

$$= \frac{Ft}{I}\int_0^{b_1}\left(z\frac{h}{2} - b_1 z + \frac{1}{2}z^2\right)dz$$

$$= \frac{Ft}{I}\left[\frac{z^2h}{4} - \frac{b_1z^2}{2} + \frac{z^3}{6}\right]_0^{b_1}$$

$$= \frac{Ft}{I}\left[\frac{b_1^2h}{4} - \frac{b_1^3}{2} + \frac{b_1^3}{6}\right]$$

or
$$F_1 = \frac{Ft}{I}b_1^2\left[\frac{h}{4} - \frac{b_1}{3}\right] \quad ...(16.78)$$

Similarly
$$F_2 = \int \tau dA = \int_0^{b_2} \frac{F(A\bar{y})}{It}dA \quad ...(16.79)$$

Where $A$ = Area $STCD$ say $A_1$ + Area $VCBA$ say $(A_2)$

$A_1 = xt$

$A_2 = b_1t$

$dA$ = Elemental area $tdx$ at distance $x$ from $CD$

$= tdx$

$\bar{y} = y$ will have two values $\bar{y}_1$ and $\bar{y}_2$ for areas $A_1$ and $A_2$ respectively are distances of the centroids from $XX$

and
$$\bar{y}_1 = \frac{h}{2}$$

$$\bar{y}_2 = \frac{h}{2} - \frac{b_1}{2}$$

Substituting these values in equation (16.79) we get

$$F_2 = \int_0^{b_2} \frac{F}{It}\left[A_1\bar{y}_1 + A_2\bar{y}_2\right](dA)$$

$$= \int_0^{b_2} \frac{F}{It}\left[(xt)\left(\frac{h}{2}\right) + (b_1 t)\left(\frac{h}{2} - \frac{b_1}{2}\right)\right](tdx)$$

$$= \frac{Ft}{I}\int_0^{b_2}\left[x\frac{h}{2} + b_1\left(\frac{h}{2} - \frac{b_1}{2}\right)\right]dx$$

$$= \frac{Ft}{I}\left[\frac{h}{2}\left(\frac{x}{2}\right)^2 + b_1\left(\frac{h}{2} - \frac{b_1}{2}\right)x\right]_0^{b_2}$$

or $$F_2 = \frac{Ft}{I}\left[\frac{hb_2^2}{4} + \frac{hb_1b_2}{2} - \frac{b_1^2 b_2}{2}\right] \quad ...(16.80)$$

For the equilibrium taking moments about $O'$ where $x$-axis intersects the line of action of $F_3$, we have

$$-F_R \times e = F_1 b_2 + F_5 b_2 + F_2\frac{h}{2} + F_4\frac{h}{2}$$

$$= (F_1 + F_5)\, b_2 + (F_2 + F_4)\frac{h}{2}$$

$$= 2F_1\, b_2 + 2F_2\frac{h}{2} \qquad \left(\because \begin{matrix} F_1 = F_5 \\ F_2 = F_4 \end{matrix}\right)$$

or $$-F_R e = 2F_1 b_2 + F_2 h$$

Substituting values from equations (16.79) and (16.80)

$$-F_R \times e = \frac{2Ft}{I} b_1^2\left(\frac{h}{4} - \frac{b_1}{3}\right) b_2 + \frac{Ft}{I}\left[\frac{hb_2^2}{4} + \frac{hb_1b_2}{2} - \frac{b_1^2 b_2}{2}\right] h$$

Since $F_R = F$

$$e = \frac{-2tb_1^2 b_2}{I}\left(\frac{h}{4} - \frac{b_1}{3}\right) - \frac{th}{I}\left[\frac{hb_2^2}{4} + \frac{hb_1b_2}{2} - \frac{b_1^2 b_2}{2}\right] \quad ...(16.81)$$

Where $I = 2(I_{XX}$ of area $ABCV) + 2(I_{XX}$ of area $EDCU) + I_{XX}$ of area $FELK$

$$= 2\left[\frac{tb_1^3}{12} + tb_1\left(\frac{h}{2} - \frac{b_1}{2}\right)^2\right] + 2\left[\frac{b_2 t^3}{12} + b_2 t\left(\frac{h}{2}\right)^2\right] + \frac{th^3}{12}$$

$$= \frac{tb_1^3}{6} + \left(\frac{tb_1h^2}{2} + \frac{tb_1^3}{2} - thb_1^2\right) + \left(\frac{b_2t^3}{6} + \frac{b_2 + h^2}{2}\right) + \frac{th^3}{12}$$

$$= \frac{2}{3}tb_1^3 + \frac{1}{2}tb_1h^3 - thb_1^2 + \frac{1}{6}b_2t^3 + \frac{1}{2}b_2h^2t + \frac{th^3}{12} \qquad ...(16.82)$$

## 16.16(*a*) Shear Centre of a Segment of a Thin RIM of Radial Thickness *t*

In figure 16.11 *ABDE* is a sector of thin rim of radial thickness *t* and mean radius *R*. The sector ABDE subtends an angle *zd* at the centre of the ruin. Let *C* be the shear centre of this sector which is at a distance *e* from the *OY* axis. Let F be the shearing force sustained by this section. The total moment of the resultant shearing force resisted by the section about *O* will be equal to the moment of the force applied on the section (*i.e., Fe*)

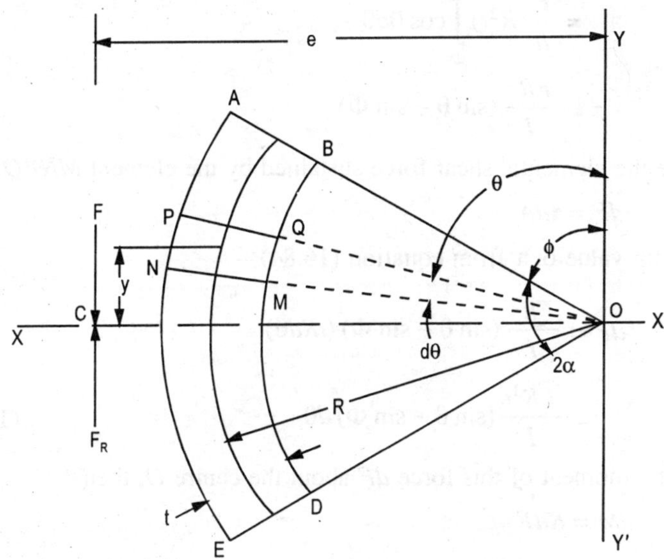


*Fig. 16.21*

Let an elemental area *MNPQ* which is bounded by radial lines *OMN* and *OQP* these radial lines subtend angles θ and (θ + *d*θ) from *OY*, for finding the total moment of area. Let τ be the shear stress at angle θ with the axis.

Then the expression for shearing force for area *ABDE* will be as given below

$$F = \int \tau dA \qquad ...(16.83\ a)$$

where $$\tau = \frac{F}{Ib}\int_{\phi}^{\theta} ydA \qquad ...(16.83\ b)$$

$$b = t$$

$F$ = Shearing force on the section

$I$ = Second moment of area $ABDE$ almost the centroidal bending axis $XX$

$dA$ = Elemential area $MNPQ$

$= tRd\theta = (Rd\theta)t$

$\bar{y}$ = distance of the centroid of elemented area $MNQP$ from $XX$

$= R \sin (90 - \theta) = R \cos \theta$

Thus substituting values in equation (16.83*b*) we get

$$\tau = \frac{F}{It}\int_{\Phi}^{\theta} (R \cos \theta)\,(t\,Rd\theta)$$

$$= \frac{F}{It}(R^2 t)\int_{\Phi}^{\theta} \cos \theta d\theta$$

$$\tau = \frac{FR^2}{I}(\sin \theta - \sin \Phi) \qquad \text{...(16.84)}$$

Let $dF$ be the elemental shear force sustained by the element $MNPQ$.

Then $\qquad dF = \tau dA$

Substituting value of $\tau$ from equation (16.84)

$$dF = \frac{FR^2}{I}(\sin \theta - \sin \Phi)\,(tRd\theta)$$

$$= \frac{FR^3 t}{I}(\sin \theta - \sin \Phi)\,d\theta \qquad \text{...(16.85)}$$

If $dM$ is the moment of this force $dF$ about the centre $O$, then

$$dM = RdF$$

$$= R.\frac{FR^4 t}{I}(\sin \theta - \sin \Phi)\,d\theta$$

Thus the total moment $M$ is given by

$$M = 2\int_{\Phi}^{\pi/2} \frac{FR^4 t}{I}(\sin \theta - \sin \Phi)\,d\theta$$

$$= 2\frac{FR^4 t}{I}[-\cos \theta - \theta \sin \Phi]_{\Phi}^{\pi/2}$$

$$M = \frac{2FR^4 t}{I}\left[\cos \Phi - \frac{\pi}{2}\sin \Phi + \Phi \sin \Phi\right]$$

$$= \frac{2FR^3t}{I}\left[\cos\phi + \left(\frac{\pi}{2} - \Phi\right)\sin\Phi\right] \quad ...(16.86)$$

Also $\frac{\pi}{2} - \Phi = \alpha$

or $$\Phi = \left(\frac{\pi}{2} - \alpha\right) \quad ...(16.87)$$

Then from equation (16.86) and (16.87)

$$M = \frac{2FR^4t}{I}\left[\cos\left(\frac{\pi}{2} - \alpha\right) - \alpha\sin\left(\frac{\pi}{2} - \alpha\right)\right]$$

or $$M = \frac{2FR^4t}{I}[\sin\alpha - \alpha\cos\alpha] \quad ...(16.88)$$

For the equilibrium of the section, this total moment of the shearing forces about $O$ most be equal to the moment of the applied force about $O$, Since $F$ acts at $C$, the point $e$ is the shear centre and let it be at distance $e$ from $O$

Thus $$Fe = \frac{2FR^4t}{I}[\sin\alpha - \alpha\cos\alpha]$$

or $$e = \frac{2R^4t}{I}(\sin\alpha - \alpha\cos\alpha) \quad ...(16.89)$$

Also $$\mathrm{I} = \int y^2 dA$$

Now since the line $OE$ is inclined with $OY$ at an angle of $(\pi - \Phi)$ we have

$$I = \int_{\Phi}^{(\pi-\phi)} (R^2\cos^2\theta)(tRd\theta)$$

$$= \frac{R^3}{2}\int_{\Phi}^{(\pi-\phi)} t(2\cos^2\theta)\,d\theta$$

$$= \frac{tR^3}{2}\int_{\Phi}^{(\pi-\phi)} (1 + \cos 2\theta)\,d\theta$$

$$= \frac{tR^3}{2}\left[\theta + \frac{\sin 2\theta}{2}\right]_{\Phi}^{\pi-\phi}$$

or $$I = \frac{tR^3}{2}\left[(\pi - 2\Phi) + \frac{1}{2}\{\sin(2\pi - 2\Phi) - \sin 2\Phi\}\right]$$

$$= \frac{tR^3}{2}\left[(\pi - 2\Phi) + \frac{1}{2}\{(-\sin 2\Phi) - \sin 2\Phi\}\right]$$

$$= \frac{tR^3}{2}[(\pi - 2\Phi) - \sin 2\Phi]$$

But $\pi - 2\Phi = 2\alpha$

and $2\Phi = (\pi - 2\alpha)$

Therefore $$I = \frac{tR^3}{2}[2\alpha - \sin(\pi - 2\alpha)]$$

$$= \frac{tR^3}{2}[2\alpha - \sin 2\alpha]$$

or $$I = tR^3(\alpha - \sin\alpha\cos\alpha) \quad \text{...(16.90)}$$

Thus from equation (16.89) and (16.90)

$$e = \frac{2R^4 t(\sin\alpha - \alpha\cos\alpha)}{tR^3(\alpha - \sin\alpha\cos\alpha)}$$

$$e = \frac{2R(\sin\alpha - \alpha\cos\alpha)}{(\alpha - \sin\alpha\cos\alpha)} \quad \text{...(16.91)}$$

## Crollary I Shear Centre for a Cylindrical **Tube With a Longitudinal** Slit **or Saw Cut**

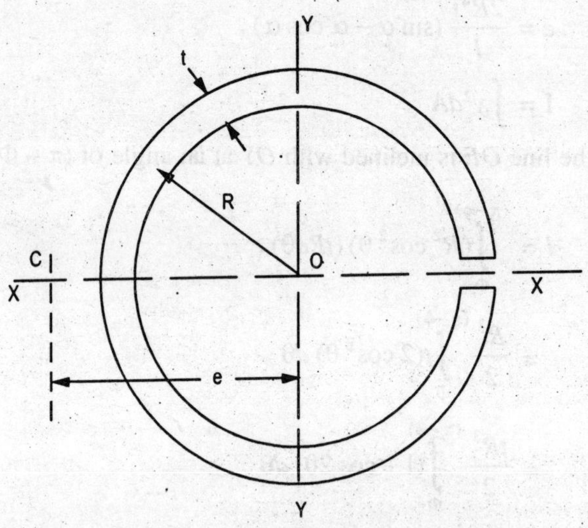


*Fig. 16.22*

Here $2\alpha \approx 2\pi$

or $\alpha = \pi$

Then from equation (16.91)

$$e = \frac{2R(\sin\pi - \pi\cos\pi)}{(\pi - \sin\pi\cos\pi)}$$

$$= \frac{2R(0+\pi)}{(\pi - 0)} = 2R \qquad \text{...(16.92)}$$

## 16.16(*c*) Corollary II Shear Centre for a Semi Cylindrical Pipe

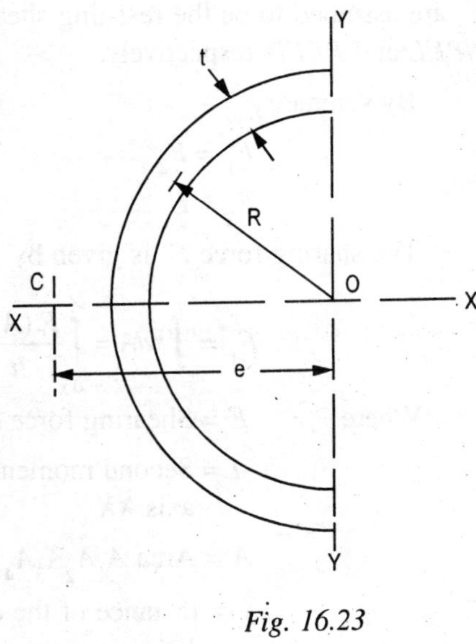


*Fig. 16.23*

Here $2\alpha = \pi$

$$\alpha = \frac{\pi}{2}$$

Therefore from equation (16.91)

$$e = \frac{2R\left(\sin\frac{\pi}{2} - \frac{\pi}{2}\cos\frac{\pi}{2}\right)}{\left(\frac{\pi}{2} - \sin\frac{\pi}{2}\cos\frac{\pi}{2}\right)}$$

$$= \frac{2R(1-0)}{\left(\frac{\pi}{2} - 0\right)} = \frac{4R}{\pi} \qquad \text{...(16.93)}$$

**Example 16.9.** *Find the location of the shear centre C of a thin walled section shown in Fig. 16.24.*

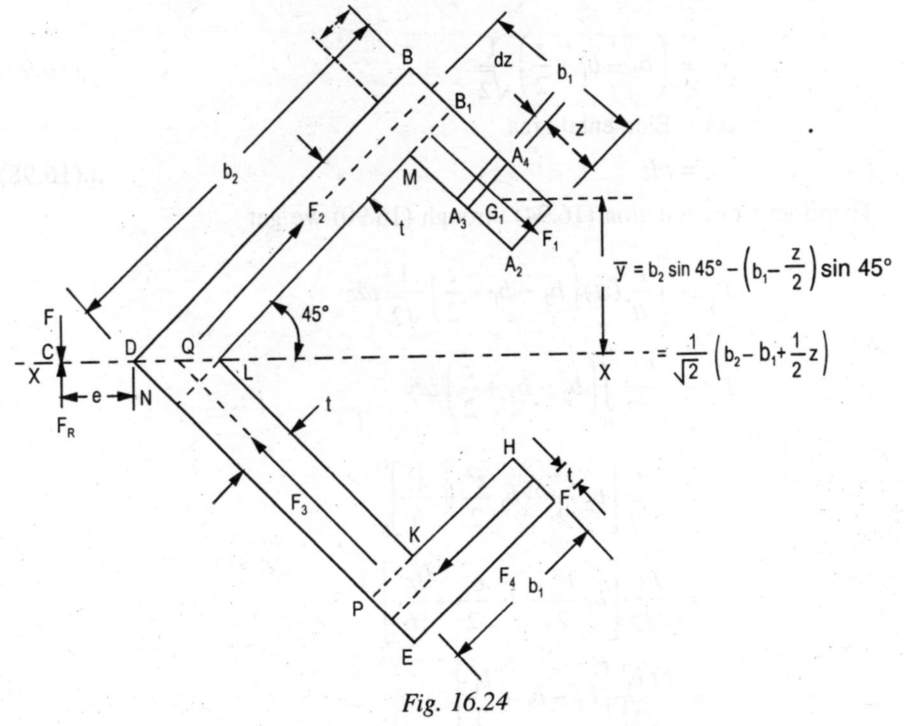


*Fig. 16.24*

**Solution.** The location of the shear centre here also is determined by finding the line of action of the resultant resisting shearing force $F_R$ of the section. $F_1$, $F_2$, $F_3$ and $F_4$ are assumed to be the resisting shearing forces of the portions $A_1B_1MA_2$, $BDNB_1$, *NPKL* and *PEFH* respectively.

By symmetry

$$F_1 = F_4 \quad \text{...(16.94)}$$

$$F_2 = F_3 \quad \text{...(16.95)}$$

The sharing force $F_1$ is given by

$$F_1 = \int \tau dA = \int_0^{b_1} \frac{F(A\bar{y})}{It} dA \quad \text{...(16.96)}$$

Where $F$ = Shearing force acting on the section

$I$ = Second moment of area about the centroidal bending axis $XX$

$A$ = Area $A_1A_2A_3A_4 = zt$

$\bar{y}$ = distance of the centroid $G_1$ of the area $A_1A_2A_3A_4$ from $XX$-axis

$$= b_2 \sin 45° - \left(b_1 - \frac{z}{2}\right) \sin 45°$$

$$= \left(b_2 - b_1 + \frac{z}{2}\right) \frac{1}{\sqrt{2}} \quad \text{...(16.97)}$$

$dA$ = Elemental area

$$= tdz \quad \text{...(16.98)}$$

Therefore from equation (16.94) through (16.90) we get

$$F_1 = \int_0^{b_1} \frac{F}{It}(zt)\left(b_2 - b_1 + \frac{z}{2}\right)\frac{1}{\sqrt{2}} tdz$$

or

$$F_1 = \frac{Ft}{I\sqrt{2}} \int_0^{b_1} \left(b_2 - b_2 + \frac{z}{2}\right) zdy$$

$$= \frac{Ft}{I\sqrt{2}} \left[b_2 \frac{z^2}{2} - b_1 \frac{z^2}{2} + \frac{z^3}{6}\right]_0^{b_1}$$

$$= \frac{Ft}{I\sqrt{2}} \left[b_2 \frac{b_1^2}{2} - b_1 \frac{b_1^2}{2} + \frac{b_1^3}{6}\right]$$

$$= \frac{Ft\, b_1^2}{2\sqrt{2}I} \left[b_2 - b_1 + \frac{b_1}{3}\right]$$

or
$$F_1 = \frac{Ftb_1^2}{\sqrt{2}I}\left[\frac{b_2}{2} - \frac{b_1}{3}\right] \qquad ...(16.99)$$
$$= F_4$$

In this problem we need not calculation two shearing forces $F_2$ and $F_3$ as ultimately the moments almost the point $Q$ are to be taken and there two forces are passing through the point $Q$.

For finding the value $I$ of the section about $XX$, the given symmetrical section can be divided into four parts viz $A_1B_1MA_2$, $BDNB$, $NLKP$ and $PEFHK$. I of the part $BDNB$, is shown is fig. (16.25).

For this portion the axis of symmetry are $UU$ and $VV$ (Fig. 16.25). The corresponding second moment of areas are

$$I_{UU} = \frac{tb_2^3}{12}$$

$$I_{VV} = \frac{b_2t^3}{12}$$

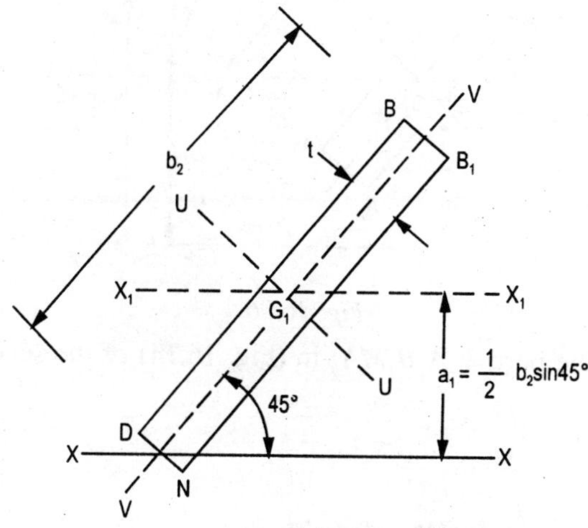


*Fig. 16.25*

Since axis $VV$ is inclined to axis is $X_1X_1$ at 45°, when $X_1X_1$ is the centroidal axis parallel to $XX$

Then
$$I_{X_1X_1} = I_{VV}\cos^2\theta + I_{UU}\sin^2\theta$$
$$= \frac{b_2t^3}{12}\cos^2 45° + \frac{tb_2^3}{12}\sin^2 45°$$
$$= \frac{b_2t^3}{24} + \frac{tb_2^3}{24}$$

Since $t_3$ is very small, neglecting the term containing $t^3$ we get

$$I_{X_1X_1} = \frac{tb_2^3}{24} \qquad ...(16.100)$$

Now using the theorem of parallel axes, if '$a$' is the distance between these axes.

$$(I_{XX})_1 = I_{X_1X_1} + (\text{Area } BDNB_1)\, a_1^2$$

$$= \frac{tb_2^3}{24} + (b_2 t)\,(b_2 \sin 45)^2$$

$$= \frac{t\,b_2^3}{24} + \frac{tb_2^3}{8}$$

or $$(I_{XX})_1 = \frac{tb_2^3}{6} \qquad ...(16.101)$$

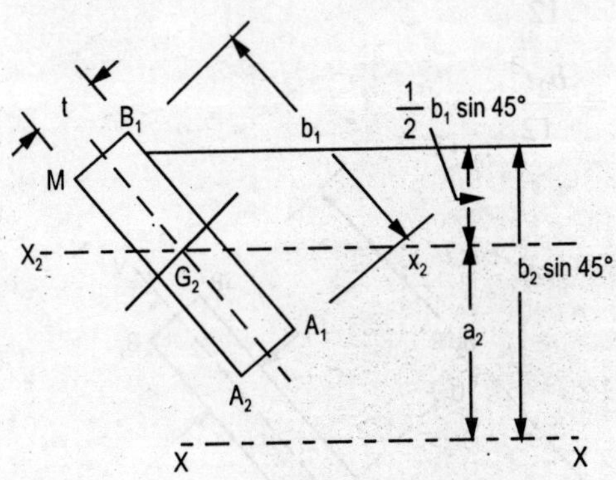


*Fig. 16.26*

Similarly for the portion $A_1B_1MA_2$ in (Fig. 16.14) of the section we have

$$\left(I_{X_2X_2}\right) = \frac{tb_1^3}{24}$$

$$(I_{XX})_2 = I_{X_2X_2} + (\text{Area } A_1B_1MA_2)\, a_2^2$$

$$= \frac{tb_1^3}{24} + tb_1\left[b_2 \sin 45° - \frac{b_1}{2}\sin 45°\right]^2$$

$$= \frac{tb_1^3}{24} + tb_1\left(\frac{1}{2}\right)\left(b_2 - \frac{b_1}{2}\right)^2 \qquad ...(16.102)$$

Thus from equation (16.101) and (16.102)

We get for the whole area of the section.

$$I_{XX} = 2(I_{XX})_1 + 2\,(I_{XX})_2$$

$$= 2\left(\frac{tb_2^3}{6}\right) + \left[\frac{tb_1^3}{24} + \frac{tb_1}{2}\left(b_2 - \frac{b_1}{2}\right)^2\right]$$

$$= \frac{tb_2^3}{3} + \left[\frac{tb_1^3}{12} + tb_1\left(b_2^2 + \frac{b_1^2}{4} - b_1 b_2\right)\right]$$

$$= \frac{tb_2^3}{3} + \left[\frac{tb_1^3}{12} + tb_1 b_2^2 + \frac{tb_1^3}{4} - tb_1^2 b_2\right]$$

$$= \frac{tb_2^3}{3} + \frac{tb_1^3}{12}(1+3) + tb_1 b_2^2 - tb_1^2\, b_2$$

$$I_{XX} = \frac{tb_1^3}{3}\left[1 + \left(\frac{b_2}{b_1}\right)^3 + 3\left(\frac{b_2}{b_1}\right)^2 - 3\left(\frac{b_2}{b_1}\right)\right] \qquad ...(16.103)$$

Now taking moments about the point $Q$ in Fig. (16.14)

$$eF = (F_1 + F_4)\, b_2$$

$$= 2F_1 b_2 \qquad ...(16.104)$$

Substituting values of $F_1$ and $I_{XX}$ from equations (16.99) and (16.24) we get

$$eF = 2\frac{Ftb_1^2}{\sqrt{2}}\left(\frac{b_2}{2} - \frac{b_1}{3}\right)\frac{b_2}{I_{XX}}$$

or $$e = \frac{2tb_1^2 b_2}{\sqrt{2}}\left(\frac{b_2}{2} - \frac{b_1}{3}\right) \times \frac{3}{tb_1^3}\left[\frac{1}{1 + \left(\frac{b_2}{b_1}\right)^3 + 3\left(\frac{b_2}{b_1}\right)^2 - 3\left(\frac{b_2}{b_1}\right)}\right]$$

$$= \frac{6}{\sqrt{2}}\frac{b_2}{b_1}\left(\frac{b_2}{2} - \frac{b_1}{3}\right)\frac{1}{\left[1 + \left(\frac{b_2}{b_1}\right)^3 + 3\left(\frac{b_2}{b_1}\right)^2 - 3\left(\frac{b_2}{b_1}\right)\right]}$$

$$= \frac{b_1}{\sqrt{2}}\left(\frac{3b_2^2}{b_1^2} - 2\frac{b_2}{b_1}\right) \times \frac{1}{\left[1 + \left(\frac{b_2}{b_1}\right)^3 + 3\left(\frac{b_2}{b_1}\right)^2 - 3\left(\frac{b_2}{b_1}\right)\right]}$$

$$e = \frac{b_1}{\sqrt{2}}\left[\frac{\left(\frac{b_2}{b_1}\right)\left(3.\frac{b_2}{b_1}-2\right)}{1+\left(\frac{b_2}{b_1}\right)^3+3\left(\frac{b_2}{b_1}\right)^2-3\left(\frac{b_2}{b_1}\right)}\right] \qquad ...(16.105)$$

If two ratio $\frac{b_2}{b_1} = r$

$$e = \frac{b_1}{\sqrt{2}}\left[\frac{r\,(3r-2)}{1+r^3+3r^2-3r}\right] \qquad ...(16.106)$$

**Example 16.10.** *In a thin walled square with a slit or saw cut at one corner E as shown in fig. 16.27. Find the location of shear centre.*

**Solution.** Here $b_1 = b_2$

and $$r = \frac{b_2}{b_1}$$

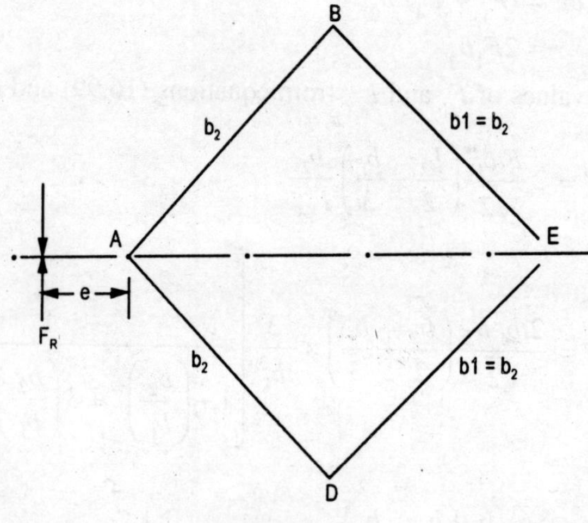


*Fig. 16.27*

The distance of shear centre *C* of thin section from *A* will be given by *e*, from equation (16.106)

$$e = \frac{b_2}{\sqrt{2}}\left[\frac{1\,(3-2)}{1+1+3-3}\right] = \frac{b_2}{2\sqrt{2}} \qquad ...(16.107)$$

**Example 16.11.** *For the thin section if uniform thickness t, shown in figure 16.28, find the location of shear centre O'* with respect to the centre *O* of the circular portion.

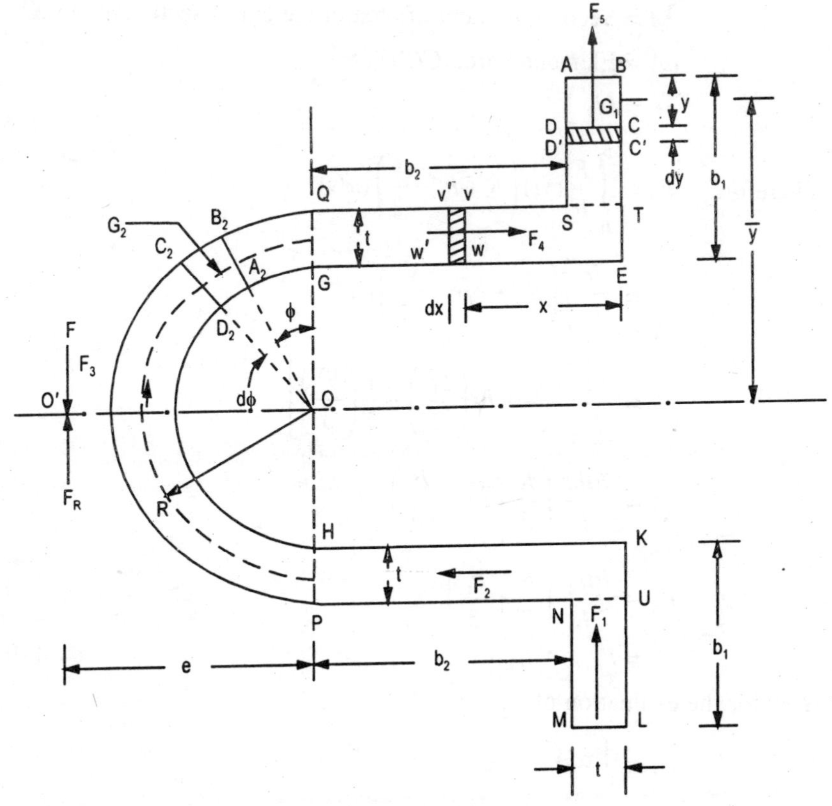


*Fig. 16.28*

**Solution.** Since the section is symmetrical about *XX* axis the shear forces sustained by the portions *ULMN* and *ABTS*, that is $F_1$ and $F_5$ will be equal. Similarly the show forces $F_2$ and $F_4$ in the segments *HKUP* and *QTEG* will be equal

That is $$F_1 = F_5 \qquad ...(16.108)$$

and $$F_2 = F_4 \qquad ...(16.109)$$

Now for the portion *ABTS*

$$F_5 = \int \tau dA$$

$$= \int_0^{b_1} \frac{F(A\bar{y})}{It} dA \qquad ...(16.110)$$

where $F$ = Total shear force on thin section

$A$ = Area $ABCD = yt$

$\bar{y}$ = distance of centroid $G_1$ of area *ABCD* from *XX*

$$= R + b_1 - \frac{y}{2}$$

$I$ = Second moment of area of the entire section about $XX$

$dA$ = Elemental area $CC'D'D$

$= tdy$

Therefore $$F_5 = \int_0^{b_1} \frac{F}{It}(yt)\left(R + b_1 - \frac{y}{2}\right)(tdy)$$

$$= \frac{Ft}{I}\int_0^{b_1}\left(Ry + b_1 y - \frac{y^2}{2}\right)dy$$

$$= \frac{Ft}{I}\left[\frac{rb_1^2}{2} + b_1\left(\frac{b_1^2}{2}\right) - \frac{1}{2}\left(\frac{b_1^3}{3}\right)\right]$$

$$= \frac{Ftb_1^2}{I}\left(\frac{R}{2} + \frac{b_1}{2} - \frac{b_1}{6}\right)$$

or $$F_5 = \frac{Ftb_1^2}{I}\left(\frac{R}{2} + \frac{b_1}{3}\right)$$

$$= F_1 \quad \text{...(16.111)}$$

Now for the evaluation of $F_4$

$$F_4 = \int \tau dA$$

$$= \int \frac{F}{It} A\bar{y}\, dA$$

$$= \frac{F}{It}[(A\bar{y}) \text{ for the portion } ABTS + (A\bar{y}) \text{ for the portion } TEWV)]\, dA \quad \text{...(16.12)}$$

Since $\bar{y}$ for the portion $ABTS = R + \dfrac{b_1}{2}$

and Area $ABTS = b_1 t$

and for the portion $ETWV$

Area $TEWV = xt$

and $\bar{y}$ for area $TEWV = R$

$dA = tdx$

Therefore substituting these values in equation (16.112)

$$F_4 = \frac{F}{It}\int_0^{b_2}\left[(b_1 t)\left(R + \frac{b_1}{2}\right) + (xt)\, R\right](tdx)$$

$$= \frac{F}{It} t^2 \int_0^{b_2} \left( b_1 R + \frac{b_1^2}{2} + xR \right) dx$$

$$= \frac{Ft}{I} \left[ b_1 b_2 R + \frac{b_1^2 b_2}{2} + R \frac{b_2^2}{2} \right]$$

$$= \frac{Ftb_2}{I} \left[ b_1 R + \frac{b_1^2}{2} + \frac{Rb_2}{2} \right]$$

$$= F_2 \qquad \text{...(16.113)}$$

For the semicircular portion consider the shear stress $\tau$ at an angle $\phi$ from the vertical $QGOP$ at the elemental area $A_2B_2C_2D_3 = (tRd\phi)$

$$\tau = \frac{F}{It}\left[(A\bar{y})_1 + (A\bar{y})_2 + (A\bar{y})_3\right] \qquad \text{...(16.114)}$$

where $(A\bar{y})_1$ = Pertains to the area $A_2B_2QG$

$(A\bar{y})_2$ = Pertains to the area $QGET$

$(A\bar{y})_3$ = Pertains to the area $STBA$

Now $$(A\bar{y})_1 = \int_{Q=0}^{\phi} y dA \qquad \text{...(16.115)}$$

when $y$ = The distance of centroid of elemental area $A_2B_2C_2D_2$ from $XX$

$$= R \cos \phi \qquad \text{...(16.116)}$$

and $$dA = tRd\,\phi \qquad \text{...(16.117)}$$

Therefore from (16.115) through (16.117)

$$(A\bar{y})_1 = \int_0^{\phi} (R \cos \phi)\, tRd\phi$$

$$= tR^2 \sin \phi \qquad \text{...(16.118)}$$

$(A\bar{y})_2$ = (area of $QGET$) (distance of centroid of this area from $XX$)

$$= (b_2 t)R \qquad \text{...(16.119)}$$

$(A\bar{y})_3$ = (Area $STBA$) (Distance of its centroid from $XX$)

$$= b_1 t \left( R + \frac{b_1}{2} \right) \qquad \text{...(16.120)}$$

Therefore substituting equations (16.118), (16.119) and (16.120) in equation (16.114) we get the shear stress $\tau$ at an angle $\phi$ from the vertical $GOP$, as

$$\tau = \frac{F}{It}\left[tR^2 \sin\phi + b_2tR + b_1t\left(R + \frac{b_1}{2}\right)\right]$$

or $$\tau = \frac{F}{I}\left[R^2 \sin\phi + b_2R + R + b_1\left(R + \frac{b_1}{2}\right)\right] \quad ...(16.121)$$

Thus, the moment of the elemential force $dF_3$ on the elemental area $A_2B_2C_2D_2$ about $O$

$$dM_3 = dF_3R$$

$$= [\tau(Rd\phi)t]R = R^2t\tau d\phi$$

or $$dM_3 = R^2t\left(\frac{F}{I}\right)\left[R^2 \sin\phi + b_2R + b_1\left(R + \frac{b_1}{2}\right)\right]$$

or $M_3$, the total moment of the resisting force $F_3$ which pertains to semicircular portion is given as

$$M_3 = \int_{\phi=0}^{\pi} dM_3$$

$$= \frac{FR^2t}{I}\int_0^{\pi}\left[R^2 \sin\phi + b_2R + b_1\left(R + \frac{b_1}{2}\right)\right]d\phi$$

$$= \frac{FR^2t}{I}\left[R^2(1+1) + \pi\left\{b_2R + b_1\left(R + \frac{b_1}{2}\right)\right\}\right]$$

*i.e.,* $$M_3 = \frac{FR^2t}{I}\left[2R^2 + \pi\left\{b_2R + b_1\left(R + \frac{b_1}{2}\right)\right\}\right] \quad ..(16.122)$$

Taking moments of the forces $F_1$, $F_2$, $F_3$, $F_4$ and $F_5$ about O and balancing the resultant with that of F acting at O′

$$F_Re = M_3 + 2F_2R - 2F_1b_2$$

or $$e = \frac{M_3}{F} + 2R\frac{F_2}{F} - 2b_2\frac{F_1}{F} \quad ...(16.123)$$

Thus substituting values of $F_1$ and $F_2$ from equations (16.111) and (16.113) and of $M$ from (16.122) in equation (16.123) we get

$$e = \frac{R^2t}{I}\left[2R^2 + \pi\left\{b_2R + b_1\left(R + \frac{b_1}{2}\right)\right\}\right]$$

$$+ \frac{2Rtb_2}{I}\left[b_1R + \frac{b_1^2}{2} + \frac{Rb_2}{2}\right] - \frac{2b_2\,tb_1^2}{I}\left(\frac{R}{2} + \frac{b_1}{3}\right)$$

on simplification it reduces to

$$e = \frac{tR^4}{6I}\left[12 + 6\pi\frac{(b_1 + b_2)}{R} + 6\left(\frac{b_1}{R}\right)^2 + 12\frac{b_1 b_2}{R^2} + 3\pi\left(\frac{b_1}{R}\right)^2 - \frac{4b_1^3 b_2}{R^4}\right]$$

...(16.124

Further the second moment of area about $XX$ is given by $I = I_1 + I_2 + I_3$

where $I_1 = I$ for semicilular portion $QGHP$.

$$= tR^2\left(\frac{\pi}{2} - \sin\frac{\pi}{2}\cos\frac{\pi}{2}\right) = \frac{\pi R^3 t}{2}$$

and $I_2 = 2$ [(for portion $QGET$]

$$= 2\left[\frac{tb^2}{12} + tb_1\left(R + \frac{b_1}{2}\right)^2\right]$$

$I_3 = 2[I$ for portion $STBA]$

$$= 2\left[\frac{b_2 t^3}{12} + b_2 t R^2\right]$$

$$I = \frac{\pi}{2}R^3 t + 2\left[\frac{tb_1^3}{12} + tb_1\left(R + \frac{b_1}{2}\right)^2\right] + 2\left[\frac{b_2 t^3}{12} + b_2 t R^2\right] \quad ...(16.125)$$

Substituting the value of I from equation (16.125) into equation (16.124) the value of $e$ can be evaluated.

**Example 16.12.** *Find the location of shear centre of the thin sector shown in figure 16.29 from the centre of curvature O of the curved portion.*

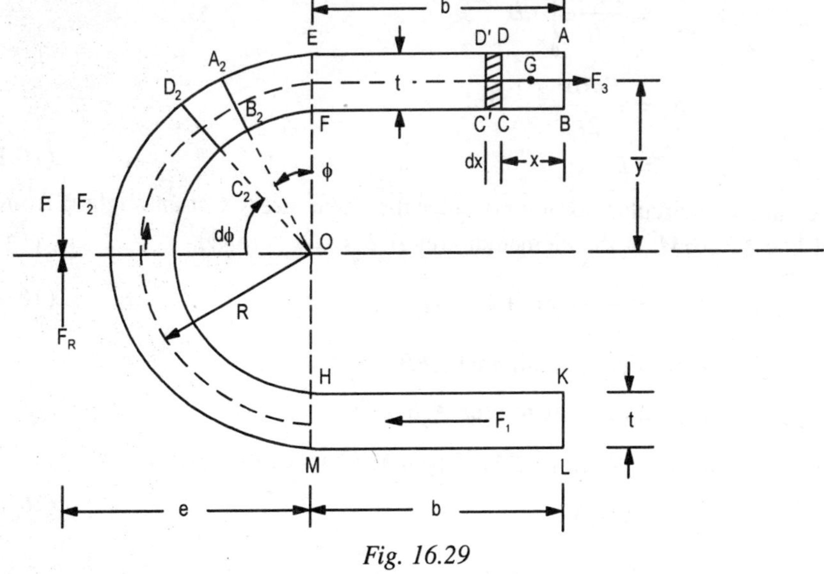


*Fig. 16.29*

**Solution.** Since the sector is symmetrical about $XX$ the shear force sustained by portions $HKLM$ and $EABF$ will be equal

That is $$F_1 = F_3 \qquad ...(16.126)$$

For the portion $ABEF$

$$F_3 = \int_0^b \tau dA$$

$$= \int_0^b \frac{F}{It}(A\bar{y})\, dA \qquad ...(16.127)$$

where $F$ = Total shear force on the section

$A$ = Area $ABCD$

$= xt$

$\bar{y}$ = Distance of centroid $G$ of area $ABCD$ from $XX$

$= R$

$I$ = Second moment of area of the entire section about $XX$

$dA$ = Elemental area $CC'D'D$

$= tdx$

Substituting these values in equation (16.127)

$$F_3 = \int_0^b \frac{F}{It}(xt)(R)\, tdx$$

$$= \frac{Ft^2R}{It}\int_0^b xdx$$

or $$F_3 = \frac{FtRb^2}{2I}$$

$$= F_1 \qquad ...(16.128)$$

For the semicircular portion consider the shear stress $\tau$ at an angle $\phi$ from the vertical line $EFOHM$, at the elemental area $A_2B_2C_2D_2 = tRd\phi$)

Then $$\tau = \frac{F}{It}[(A\bar{y})_3 + (A\bar{y})_2] \qquad ...(16.129)$$

where $(A\bar{y})_3$ = Pertains to area $EFBA$

$(A\bar{y})_2$ = Pertains to area $A_2B_2FE$

Now. $(A\bar{y})_3$ = (Area $EFBA$) (Distance of its centroid from $XX$)

$$= (bt)R \qquad ...(16.130)$$

and $$(A\bar{y})_2 = \int_0^{\phi} \text{(Area } A_2B_2C_2D_2\text{) (distance of its centroid from } XX)$$

$$= \int_0^{\phi} (tRd\phi)\,(R\cos\phi)$$

$$= \int_0^{\phi} tR^2 \cos\phi\, d\phi$$

$$= tR^2 \sin\phi \qquad ...(16.131)$$

Therefore $$\tau = \frac{F}{It}[btR + tR^2 \sin\phi)$$

Therefore the moment of elemental force $dF_2$ acting on the elemential area $A_2B_2C_2D_2$ about $O$

$$dM_2 = RdF_2$$

$$= R[\tau dA]$$

$$= R[\tau.(Rd\phi t)]$$

or $$dM_2 = R^2 t\tau d\phi \qquad ...(16.132)$$

Thus substituting the values from equation (16.129), (16.130), (16.131) in equation (16.132) and integrating we get the moment of $F_2$ about $O$

$$M_2 = \int_{\phi=0}^{\pi} (R^2t)\left[\frac{F}{It}\{btR + tR^2 \sin\phi\}\right] d\phi$$

$$= \frac{FR^3t}{I}\int_0^{\pi} [b + R\sin\phi]\, d\phi$$

$$\text{M}_2 = \frac{FR^3t}{I}[b\pi + 2R] \qquad ...(16.133)$$

If O′ is the shear centre at a distance $e$ from $O$ then taking moments about $O$

$$Fe = M_2 + (F_3R) + (F_1R)$$

or $$Fe = M_2 + 2RF_3 \qquad ...(16.134)$$

Than from equation (16.128) and (16.134)

or $$Fe = \frac{FR^3t}{I}(b\pi + 2R) + 2R\left(\frac{FtRb^2}{2I}\right)$$

or $$e = \frac{R^3t}{I}.(b\pi + 2R) + \frac{R^2b^2t}{I}$$

$$e = \frac{R^2t}{I}(\pi bR + 2R^2 + b^2) \qquad \text{...(16.135)}$$

Also $I = I$ of semicircular portion about $XX$ + 2($I$ of portion $ABEF$)

$$= \frac{\pi}{2}R^3t + 2\left[\frac{bt^3}{12} + b + r^2\right] \qquad \text{...(16.136)}$$

Substituting value of $I$ from equation (16.136) in equation (16.135) the value of $e$ can be calculated.

**Example 16.13.** *Find the location of shear centre of a thin section shown in figure 16.30 from the centre of curvature of the curved portion.*

**Solution.** Because of the symmetry the shear forces $F_1$ and $F_3$ sustained by portions $HKLM$ and $ABEF$ will be equal

That is $F_1 = F_3$ ...(16.137)

For the portion $ABEF$

$$F_3 = \int \tau dA$$

$$= \int_0^b \frac{F}{It}(A\bar{y})\, dA \qquad \text{...(16.138)}$$

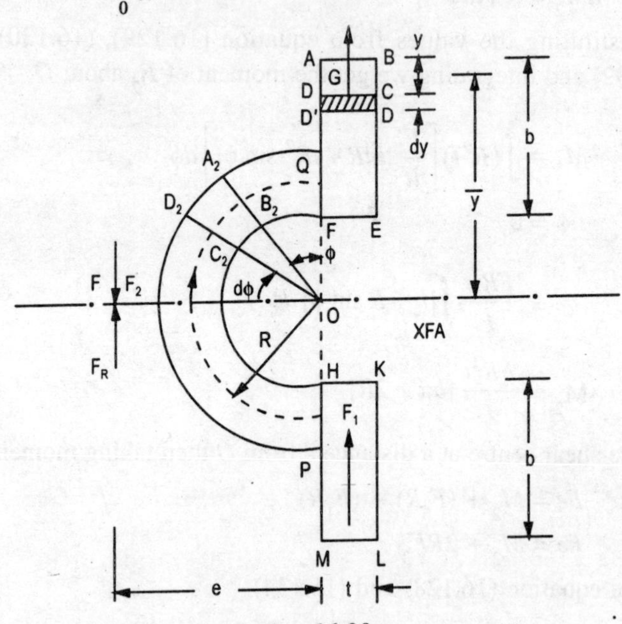


*Fig. 16.30*

where $I$ = Second moment of area of the section about $XX$

$F$ = Total shear force on the section

$A$ = Area $ABCD = yt$

$$\bar{y} = \text{Distance of its centroid G from } XX$$

$$= \left(R + b - \frac{y}{2}\right)$$

$$dA = \text{Elemental area } CC'D'D$$

$$= tdy$$

Then substituting these values in equation (16.138)

$$F_3 = \int_0^b \frac{F}{It}(yt)\left(R + b - \frac{y}{2}\right) tdy$$

$$= \frac{Ft}{I}\int_0^b \left(Ry + by - \frac{y^2}{2}\right) dy$$

$$= \frac{Ft}{I}\left[\frac{Rb^2}{2} + b\left(\frac{b^2}{2}\right) - \frac{1}{2}\left(\frac{b^3}{3}\right)\right]$$

$$= \frac{Ft}{I} b^2 \left(\frac{R}{2} + \frac{b}{2} - \frac{b}{6}\right)$$

or
$$F_3 = \frac{Ftb^2}{I}\left(\frac{R}{2} + \frac{b}{3}\right)$$

$$= F_1 \quad ...(16.139)$$

For the semicircular portion consider the shear stress $\tau$ at an angle $\phi$ from the vertical $AQFOHPM$ at the elemental area $A_2B_2C_2D_2 = (tRd\phi)$

$$\tau = \frac{F}{It}\left[(A\bar{y})_3 + (A\bar{y})_2\right] \quad ...(16.140)$$

where $(A\bar{y})_3$ = Pertains to area $ABEF$

and $(A\bar{y})_2$ = Pertains to area $A_2B_2FQ$

Now $(A\bar{y})_3$ = (Area $ABEF$) (distance of its centroid from $XX$)

$$= (bt)\left(R + \frac{b}{2}\right) \quad ...(16.141)$$

And $(A\bar{y})_2 = \int_0^\phi$ [(distance of centroid of elemental area from $XX$]

× [Elemental area]

$$= \int_0^\phi (R\cos\phi)\,(tRd\phi)$$

$$= tR^2 \sin \phi \qquad ...(16.142)$$

Substituting these values is equation (16..140)

$$\tau = \frac{F}{It}\left[bt\left(R+\frac{b}{2}\right)+tR^2\ Sin\ \phi\right] \qquad ...(16.143)$$

Therefore the moment of elemental force $dF_2$ on this elemental area $A_2B_2C_2D_2$ about $O$

$$\begin{aligned} dM_2 &= R(dF_2) \\ &= R\tau\ (Rtd\phi) \\ &= R^2\ t\tau d\phi \qquad ...(61.144) \end{aligned}$$

Therefore the total moment $M_2$ of the shearing force $F_2$ sustained by curved portion about $O$ is given by from equation (16.143) and (16.144)

$$M_2 = \int_{\phi=0}^{\pi} dM_2$$

$$= \int_0^{\pi} R^2 t\left[\frac{F}{It}\left\{bt\left(R+\frac{b}{2}\right)+tR^2 \sin\phi\right\}\right]d\phi$$

$$= \frac{FtR^2}{I}\int_0^{\pi}\left[b\left(R+\frac{b}{2}\right)+R^2 \sin\phi\right]d\phi$$

$$M_2 = \frac{FtR^2}{I}\left[b\left(R+\frac{b}{2}\right)\pi+2R^2\right] \qquad ...(16.145)$$

Assuming the shear centre to be at O′ at a distance $e$ from $O$ as shown in figure (16.30). Then from equation (16.139) and (16.145) we get

$$Fe = M_2 - F_1\frac{t}{2} - F_3\frac{t}{2}$$

or $$Fe = M_2 - F_1 t$$

$$Fe = \frac{FtR^2}{I}\left[b\left(R+\frac{b}{2}\right)\pi+2R^2\right]-\frac{Ft^2b^2}{I}\left(\frac{R}{2}+\frac{b}{3}\right)$$

or $$e = \frac{tR^2}{I}\left[\pi b\left(R+\frac{b}{2}\right)+2R^2\right]-\frac{t^2b^2}{I}\left(\frac{R}{2}+\frac{b}{3}\right) \qquad ...(16.146)$$

Also $I$ = ($I$ of curved portion) + 2 ($I$ of portion $HKLM$)

$$= \frac{\pi}{2}R^3 t + 2\left[\frac{tb^3}{12}+tb\left(R+\frac{b}{2}\right)^2\right]$$

$$= \frac{\pi}{2}R^3t + 2\left[\frac{tb^3}{12} + tbR^2 + tb^2R + \frac{tb^3}{4}\right]$$

$$= \frac{\pi}{2}R^3t + \left(\frac{2tb^3}{3} + 2tbR^2 + 2tb^2R\right)$$

$$I = \frac{\pi}{2}R^3t + \frac{2tb}{3}(b^2 + 3R^2 + 3bR) \qquad ...(16.147)$$

From equation (16.146) and (16.147), value of $e$ can be determined.

**Example 16.14.** *Determine the location of shear centre of a thin section shown in figure 16.30 where b = R.*

**Solution.** Then, from equation (16.147) substituting $R$ for $b$.

$$I = \frac{\pi}{2}R^3t + \frac{2tR}{3}(R^2 + 3R^2 + 3R \times R)$$

$$= \frac{\pi}{2}R^3t + \frac{14}{3}R^3t$$

$$= R^3t\left(\frac{\pi}{2} + \frac{14}{3}\right)$$

or

$$I = \frac{R^3t}{6}(3\pi + 28) \qquad ...(16.148)$$

And from (16.146)

$$e = \frac{tR^2}{I}\left[\pi R\left(R + \frac{R}{2}\right) + 2R^2\right] - \frac{t^2R^2}{I}\left(\frac{R}{2} + \frac{R}{3}\right)$$

$$= \frac{tR^2}{I}\left[\frac{3\pi R^2}{2} + 2\pi R^2\right] - \frac{tR^2}{I}\left(\frac{5}{6}R\right)$$

$$= \frac{tR^3}{6I}[9\pi + 12 - 5]$$

$$= \frac{tR^3}{I}(9\pi + 7) \qquad ...(16.149)$$

And from (16.146)

$$e = \frac{tR^2}{I}\left[\pi R\left(R + \frac{R}{2}\right) + 2R^2\right] - \frac{t^2R^2}{I}\left(\frac{R}{2} + \frac{R}{3}\right)$$

$$= \frac{tR^3}{6I}[9\pi + 12 - 5]$$

$$= \frac{tR^3}{I}(9\pi + 7) \qquad \text{...(16.149)}$$

Thus from equations (16.148) and (16.149)

$$e = \frac{tR^3\,(9\pi + 7) \times 6}{tR^3\,(3\pi + 28)} = \frac{6\,(9\pi + 7)}{(3\pi + 28)} \qquad \text{...(16.150)}$$

**Example 16.15.** *Determine the distance of shear centre from the centre of the web of the thin section shown in figure 16.31 (a)*

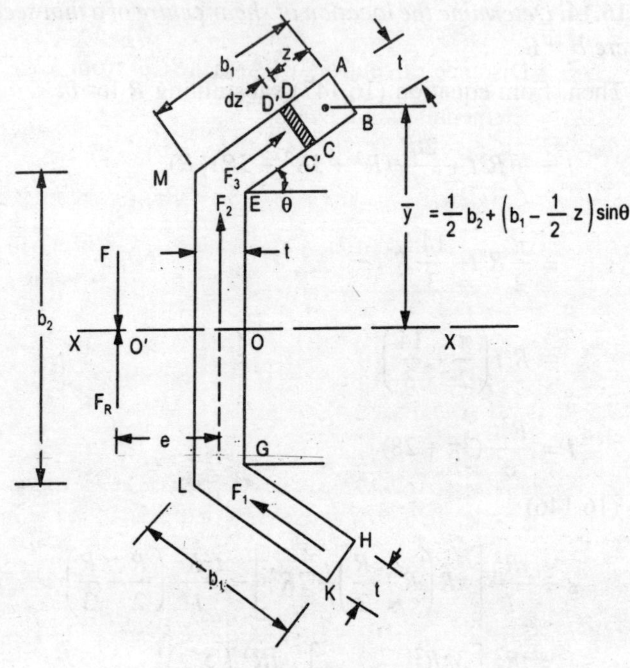


*Fig. 16.31 (a)*

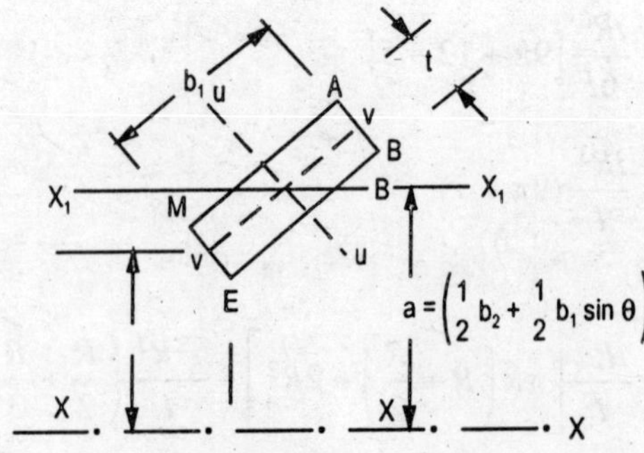


*Fig. 16.31 (b)*

**Solution.** By symmetry

$$F_1 = F_3 \quad \text{...(16.151)}$$

Also $$F_3 = \int \tau dA$$

$$= \int_0^{b_1} \frac{F(A\bar{y})}{It} dA \quad \text{...(16.152)}$$

where $A$ = Area $ABCD$

$= tz$

$\bar{y}$ = Distance of centroid of Area $ABCD$ from $XX$

$dA$ = Elemental area $CDD'C'$

$= tdz$

$I$ = Second moment of area of the section about $XX$

$F$ = Total sharing force on the section

Substituting these values in equation (16.152)

$$F_3 = \frac{F}{It} \int_0^{b_1} (tz) \left\{ \frac{b_2}{2} + \left( b_1 - \frac{z}{2} \right) \sin\theta \right\} (tdz)$$

$$= \frac{Ft}{I} \int_0^{b_1} \left[ \frac{b_2}{2} + b_1 \sin\theta) z - \frac{z^2}{2} \sin\theta \right] dz$$

$$F_3 = \frac{Ft}{I} \left[ \left( \frac{b_2}{z} + b_1 \sin\theta \right) \frac{z^2}{2} - \frac{z^3}{6} \sin\theta \right]_0^{b_1}$$

$$F_3 = \frac{Ft}{I} \left[ \left( \frac{b_2}{2} + b_1 \sin\theta \right) \frac{b_1^2}{2} - \frac{b_1^3}{6} \sin\theta \right]$$

$$= \frac{Ftb_1^2}{6I} 3 \left[ \left( \frac{b_2}{2} + b_1 \sin\theta \right) - b_1 \sin\theta \right]$$

$$= \frac{Ftb_1^2}{6F} \left[ \frac{3}{2} b_2 + 2b_1 \sin\theta \right]$$

or $$F_3 = \frac{Ftb_1^2}{12I} [3b_2 + 4b_1 \sin\theta] \quad \text{...(16.153)}$$

There is no need to calculate the value of $F_2$ since the moments are to be balanced about the point $O$ which is the point of intersection of $F_2$ with $XX$.

Assuming the distance of shear centre $O'$ from $O$ to be $e$ and taking moment about $O$ are get

$$Fe = F_3\left(\frac{b_2}{2}\cos\theta\right) + F_1\left(\frac{b_2}{2}\cos\theta\right)$$

$$= 2F_3\frac{b_2}{2}\cos\theta$$

or $$e = \frac{F_3 b_2}{F}\cos\theta \quad \text{...(16.154)}$$

From equations (16.153) and (16.154)

$$e = \frac{tb_1^2}{12\,I}[3b_2 + 4\,b_1\sin\theta](b_2\cos\theta)$$

or $$e = \frac{b_1^2 b_2 t\cos\theta}{12\,I}[3b_2 + 4\,b_1\sin\theta] \quad \text{...(16.155)}$$

Now $I_{XX}$ of the section = $I_{XX}$ of area $MEGL$ + 2 $I_{XX}$ of Area $ABEM$

$$I_{XX} \text{ of } MEGL = \frac{tb_2^3}{12} \quad \text{...(16.156)}$$

$$I_{X_1X_1} \text{ of } ABEM = I_{vv}\cos^2\theta + I_{uu}\sin^2\theta$$

$$= \frac{b_1 t^3}{12}\cos^2\theta + \frac{tb_1^3}{12}\sin^2\theta$$

Neglecting term with $t_3$ being very small, we get

$$I_{X_1X_1} \text{ of } ABEM = \frac{tb_1^3}{12}\sin\theta$$

Then $I_{XX}$ of $ABEM = I_{X_1X_1}$ + (Area $ABEM$)$a^2$

$$= \frac{tb_1^3}{12}\sin^2\theta + (tb_1)\left[\frac{b_2}{2} + \frac{b_1}{2}\sin\theta\right]^2$$

$$= \frac{tb_1^3}{12}\sin^2\theta + \frac{tb_1}{4}(b_2 + b_1\sin\theta)^2 \quad \text{...(16.157)}$$

There adding equation (16.156) and (16.157) we get

$$I_{XX} = \frac{tb_2^3}{12} + \left[\frac{tb_1^3}{12}\sin^2\theta + \frac{tb_1}{4}(b_2 + b_1\sin\theta)^2\right] \times 2$$

or $$I = I_{XX} = \frac{t}{12}\left[b_2^3 + 2b_1^3 \sin^2\theta + 6b_1(b_2 + b_1\sin\theta)^2\right] \quad ...(16.158)$$

Then substituting the value of I from equation (16.158) in equation (16.155) the value of $e$ can be calculated.

**Corollary** In case when $b_1 = b_2$, then from equation (16.155)

$$e = \frac{b_1^3\, t\cos\theta}{12\, I}(3 + 4\sin\theta) \quad ...(16.159)$$

and from equation (16.158)

$$e = \frac{b_1^3 t}{12}[1 + 2\sin^2\theta + 6(1 + \sin\theta)^2] \quad ...(16.160)$$

Then from equation (16.159) and (16.160)

$$e = \frac{b_1\cos\theta\,(3 + 4\sin\theta)}{1 + 2\sin^2\theta + 6\,(1 + \sin\theta)^2} \quad ...(16.161)$$

**Example 16.15.** *Determine the distance of shear centre from the centre of the portion FHPQ of the thin section shown in figure 16.32.*

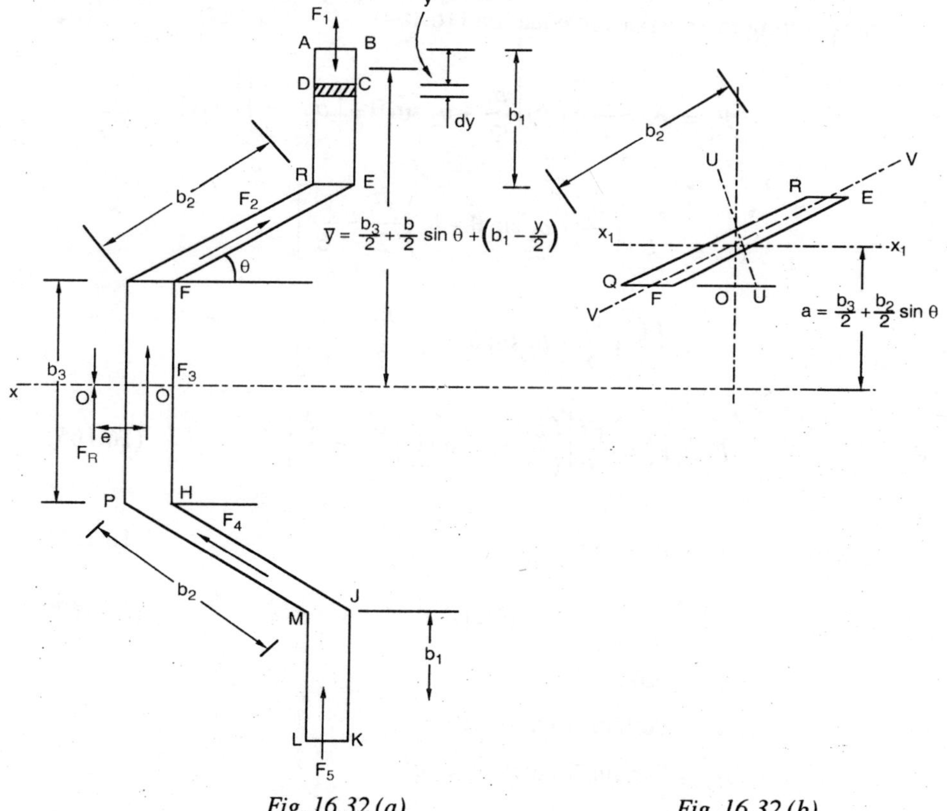


*Fig. 16.32 (a)* *Fig. 16.32 (b)*

**Solution.** By similarly

$$F_1 = F_5 \quad \text{...(16.162)}$$

and $$F_2 = F_4 \quad \text{(16.163)}$$

Now $$F_1 = F_5 \int \tau dA$$

$$= \int_0^{b_1} \frac{F}{It}(A_1\bar{y}_1)dA_1 \quad \text{...(16.164)}$$

where $A_1$ = Area $ABCD = yt$

$\bar{y}_1$ = Distance of the centroid of the area $A_1$ from $XX$

$$= \frac{b_3}{2} + b_2 \sin\theta + \left(b_1 - \frac{y}{2}\right)$$

$dA_1$ = Elemental area $DCC'D' = tdy$

$F$ = Total shearing force on the section

$I_t$ = Second moment of area of the entire section about $XX$

$t$ = Thickness of section

Substituting these values in equation (16.164)

$$F_1 = F_5 = \frac{F}{It}\int_0^{b_1}(yt)\left[\frac{b_3}{2} + b_2 \sin\theta + \left(b_1 - \frac{y}{2}\right)\right](tdy)$$

$$= \frac{Ft}{I}\left[\left(\frac{b_3}{2} + b_2 \sin\theta + b_1\right)\frac{y^2}{2} - \frac{y^3}{6}\right]_0^{b_1}$$

$$= \frac{Ft}{I}\left[\left(\frac{b_3}{2} + b_2 \sin\theta + b_1\right)\frac{b_1^2}{2} - \frac{b_1^3}{6}\right]$$

or $$F_1 = F_5 = \frac{Ft}{I}\left[\left(\frac{b_3}{2} + b_2 \sin\theta\right)\frac{b_1^2}{2} + \frac{b_1^3}{3}\right] \quad \text{...(16.165)}$$

Similarly $$F_2 = F_4 = \int_0^{b_2} \tau dA \quad \text{...(16.166)}$$

and $$\tau = \frac{F}{It}[(A\bar{y})_2 + (A\bar{y})_3] \quad \text{...(16.167)}$$

$$dA = (tdz)$$

where $(A\bar{y})_2$ = Pertains to area $ABER$

$(A\bar{y})_3$ = Pertains to area $REC_2D_2$

Now $(A\bar{y})_2$ = (Area $ABER$) (distance of its centroid from $XX$)

$$= (tb_1)\left(\frac{b_3}{2} + b_2 \sin\theta + \frac{b_1}{2}\right) \quad ...(16.168)$$

and $(A\bar{y})_3 = (A_3\bar{y}_3)$

$A_3$ = area $ERD_2C_2 = (zt)$

where $y_3$ = distance of centroid of area $A_3$ from $XX$

$$\text{Therefore } (A\bar{y})_3 = (zt)\left[\frac{b_3}{2} + \left(b_2 - \frac{z}{2}\right)\sin\theta\right] \quad ...(16.169)$$

Thus substituting values from equation (16.167), (16.168) and (16.169) in equation (16.160) are get

$$F_2 = F_4 = \frac{F}{It}\int_0^{b_2}\left[tb_1\left(\frac{b_3}{2} + b_2 \sin\theta + \frac{b_1}{2}\right) + (zt)\left\{\frac{b_3}{2} + \left(b_2 - \frac{z}{2}\right)\sin\theta\right\}\right]dtz$$

$$= \frac{Ft}{I}\int\left[b_1\left(\frac{b_3}{2} + b_2 \sin\theta + \frac{b_1}{2}\right)z + \left\{z\frac{b_3}{2} + \left(b_2 z - \frac{z^2}{2}\right)\sin\theta\right\}dz\right]$$

$$= \frac{Ft}{I}\left[b_1\left(\frac{b_3}{2} + b_2 \sin\theta + \frac{b_1}{2}\right) + \left\{\frac{b_3}{2}\frac{z^2}{2} + \left(b_2\frac{z^2}{2} - \frac{z^3}{6}\right)\sin\theta\right\}\right]_0^{b_2}$$

$$= \frac{Ft}{I}\left[b_1 b_2\left(\frac{b_3}{2} + b_2 \sin\theta + \frac{b_1}{2}\right) + \left\{\frac{b_3 b_2^2}{4} + \left(\frac{b_2^3}{2} - \frac{b_2^3}{6}\right)\sin\theta\right\}\right]$$

or

$$F_2 = F_4 = \frac{Ft}{I}\left[\frac{b_1 b_2}{2}(b_1 + b_3 + 2b_2 \sin\theta + \left\{\frac{b_3 b_2^2}{4} + \frac{b_2^3}{3}\sin\theta\right\}\right]$$

$$= \frac{Ft}{I}\left[\frac{b_1 b_2}{2}(b_1 + b_3 + 2\,b_2 \sin\theta) + \frac{b_2^2}{12}(3b_3 + 4b_2 \sin\theta)\right]$$

$$\text{Thus } F_2 = F_4 = \frac{Ftb_2}{12\,I}[6b_1(b_1 + b_3 + 2b_2 \sin\theta) + b_2(3b_3 + 4b_2 \sin\theta)] \quad ...(16.170)$$

Let $O'$ be the position of shear centre at a distance $e$ from $O$

Then taking moments about the point $O$

$$Fe = 2F_1\,(b_2 \cos\theta) + 2F_4\left(\frac{b_3}{2}\sin\theta\right)$$

$$= 2F_1 b_2 \cos\theta + F_2 b_3 \sin\theta$$

$$= 2b_2 \left[ \frac{Ft}{I} \left\{ \left( \frac{b_3}{2} + b_2 \sin\theta \right) \frac{b_1^2}{2} + \frac{b_1^3}{3} \right\} \right] \cos\theta$$

$$+ \frac{Ftb_2}{12\,I} [6b_1 (b_1 + b_3 + 2b_2 \sin\theta) + b_2 (3b_3 + 4b_2 \sin\theta)] (b_3 \sin\theta)$$

$$= \frac{Ft}{I} \left[ (2b_2 \cos\theta) \left\{ \left( \frac{b_3}{2} + b_2 \sin\theta \right) \frac{b_1^2}{2} + \frac{b_1^3}{3} \right\} \right.$$

$$\left. + \frac{b_2 b_3}{12} \{6b_1 (b_1 + b_3 + 2b_2 \sin\theta) + b_2 (3b_3 + 4b_2 \sin\theta)\} \sin\theta \right]$$

or $$Fe = \frac{Ft}{I} \left[ (2b_1^2\, b_2 \cos\theta) \left( \frac{b_1}{3} + \frac{b_2}{2} \sin\theta + \frac{b_3}{4} \right) + \left( \frac{b_2 b_3 \sin\theta}{12} \right) \times \right.$$

$$\left. \{6b_1 (b_1 + 2b_2 \sin\theta + b_3) + b_2 (3b_3 + 4b_2 \sin\theta\} \right]$$

or $$e = \frac{t}{I} \left[ (2b_1^2 b_2 \cos\theta) \left( \frac{b_1}{3} + \frac{b_2}{2} \sin\theta + \frac{b_3}{4} \right) + \left( \frac{b_2 b_3}{12} \sin\theta \right) \times \right.$$

$$\left. \{6b_1 (b_1 + 2b_2 \sin\theta + b_3) + b_2 (3b_3 + 4b_2 \sin\theta\} \right] \quad ...(16.171)$$

Now $I = I_{XX}$ of area *FHPQ* + $2I_{XX}$ of area *EFQR*

+ $2I_{XX}$ of area *ABER* ...(16.172)

Now $I_{XX}$ of area $FHPQ = \dfrac{tb_3^3}{12}$ ...(16.173)

$$2(I_{XX} \text{ of area } ABER] = 2\left[ \frac{tb_1^3}{12} + (tb_1) \left( \frac{b_3}{2} + b_2 \sin\theta + \frac{b_1}{2} \right)^2 \right] \quad ...(16.174)$$

For $I_{XX}$ of area *EFQR* refer to figure 16.33

From figure 16.33 $I_{X'X'}$ or area *EFQR*

$$I_{X'X'} = T_{vv} \cos^2\theta + I_{uu} \sin^2\theta$$

$$= \frac{t^2 b_2}{12} \cos^2\theta + \frac{tb_2^3}{12} \sin^2\theta$$

Since t is very small, the term containing $t^3$ is neglected

$$\therefore \quad I_{X'X'} = \frac{tb_2^3}{12} \sin^2\theta$$

Then $\quad 2I_{XX} = 2[I_{X'X'} + (tb_2)\, a^2]$

$$= 2\left[\frac{tb_2^3}{12}\sin^2\theta + (tb_2)\left(\frac{b_3}{2} + \frac{b_2}{2}\sin\theta\right)^2\right] \qquad ...(16.175)$$

Therefore adding equations (16.173), (16,174) and (16.175) we get from equation (16.172)

$$I = \frac{tb_3^3}{12} + 2\left[\frac{tb_1^3}{12} + (tb_1)\left(\frac{b_3}{2} + b_2\sin\theta + \frac{b_1}{2}\right)^2\right]$$

$$+ 2\left[\frac{tb_2^3}{12}\sin^2\theta + tb_2\left(\frac{b_3}{2} + \frac{b_2}{2}\sin\theta\right)^2\right] \qquad ...(16.176)$$

Substituting the value of I from equation (16.176), the distance *e* of the shear centre can be determined.

## PROBLEMS FOR PRACTICE

1. An angle section has overall width 80 mm, and overall depth 100 mm, the thickness being 10 mm throughout. Determine (*i*) position of centroid (*ii*) centroidal moments of inertia $I_{XX}$. $I_{YY}$ and $I_{XY}$ where *XX* and *YY* axes are parallel to 80 mm and 100 mm arms respectively (*iii*) principal centroidal moments of inertia (*iv*) direction of principal centroidal axes. (*University of Indore*)
2. A timber beam 250 mm wide by 300 mm deep is used as simply supported beam on a span of 5 m. It is subjected to a concentrated load of 30 kN at the mid section of the span. If the plane of load makes an angle of 45° with the vertical plane of symmetry find the direction of neutral axis and the maximum stress in the beam. (*Punjabi University*)

   [**Ans.** 55.2° anticlockwise with the vertical ; 15.51 MN/m$^2$]
3. An 80 mm × 80 mm × 10 mm angle is used as a beam simply supported at each end over a span of 2m with one leg of the section horizontal and the other vertically upwards. It is loaded at the centre of the span with a vertical load which may be assumed to pass through the centroid of the section. The principal second moments of area for the section, are $1.38 \times 10^{-6}$ and $0.36 \times 10^{-6}$ m$^4$. The distance of the centroid from the outside edge is 23.5 mm and toe has a radius of 5.6 mm.

   Find the position of the neutral axis and calculate the safe load if the maximum stress is not to exceed 120 MN/m$^2$. [**Ans.** 3.02 kN]

   [**Hint :** Equate the stress at any point (*u, v*) with respect to *GU* and *GV* axes, to zero. Then the inclination of he neutral axis with respect to the axis *UU* is equal to $\tan^{-1}(u/v)$]
4. A beam of the angle section shown in Fig. 16.33 is freely supported on a span of 2 m with the 150 mm leg vertical. A vertical load of 20 kN is applied at the

centre of the span. Find (*a*) the maximum bending stress ; (*b*) the direction and magnitude of the deflection at the centre. Neglect the weight of the beam itself, and neglect twisting effects and deflection due to the shear.

The properties of the section are $I^{XX} = 8.12 \times 10^{-6}$ m$^4$, $I_{YY} = 2.88 \times 10^{-6}$ m$^4$, $I_{PP}$ $= 1.675 \times 10^{-6}$ m$^4$. $E = 200$ GN/m$^2$.

[**Ans.** 153 MN/m$^2$ ; 67°33′ to *PP*, 4.29 mm]

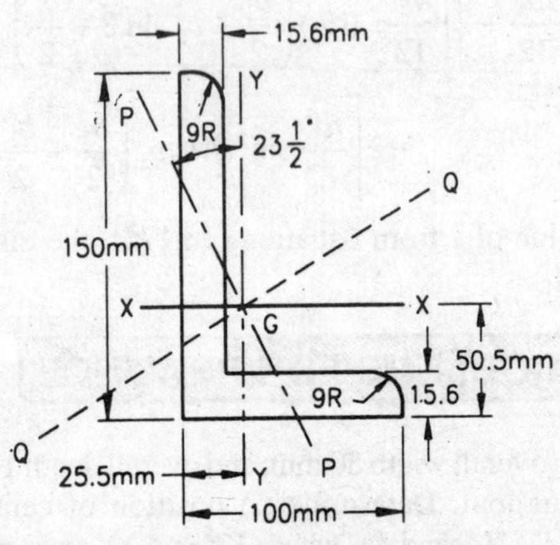


*Fig. 16.33*

5. A length of 120 mm × 120 mm × 10 mm angle of the section shown in Fig. 16.34 transmits a longitudinal pull which passes through the point *P*. Determine (*a*) the position of the line of zero stress, (*b*) the maximum value of the pull if the stress on the section is not to exceed 100 MN/m$^2$. Assume all corners square. Take $I_{XX} = I_{YY} = 3.184 \times 80^{-6}$ m$^4$.

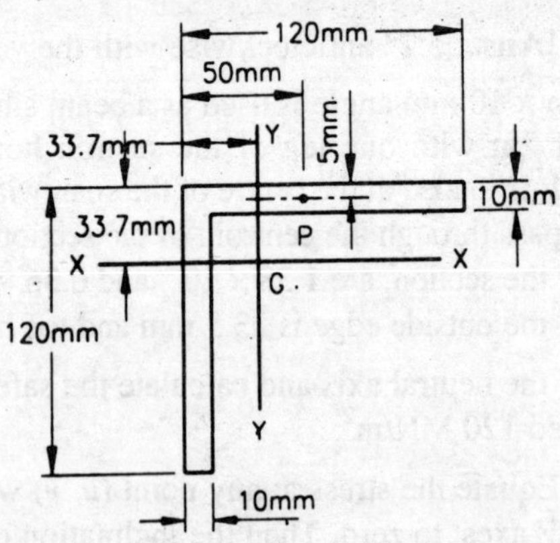


*Fig. 16.34*

6. Fig. 16.35 shows the cross-section of a short column, made from a 160 mm × 120 mm I-section, with a 200 mm × 10 mm plate welded to one flange. A

vertical load of 25 kN acts at $P$, the line of thrust passing 40 mm from the $XX$-axis and 20 mm from the $YY$-axis of the I-section. Calculate the maximum stress developed in the section. For the 160 mm × 120 mm I-section, area = $4.12 \times 10^{-3}$ $m^4$, $I_{XX} = 18.42 \times 10^{-6}$ $m^2$ and $I_{YY} = 4.61 \times 10^{-6}$ $m^4$. *(Punjabi University)*

**[Ans.** 11.56 MN/$m^2$]

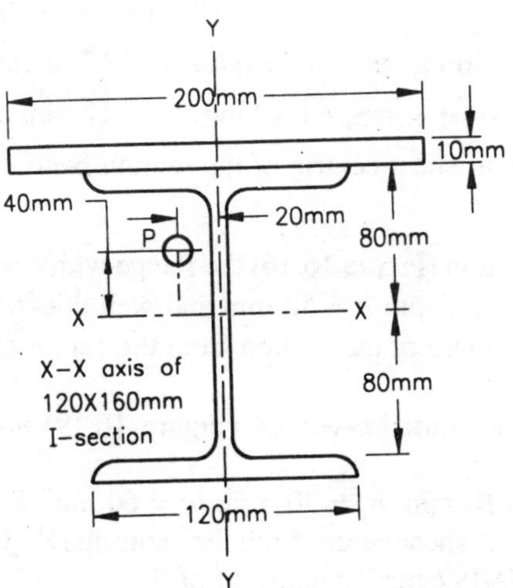


*Fig. 16.35*

7. A rectangular bar of 80 mm × 40 mm cross-section has a pull of W Newton applied parallel to the axis of the bar, the point of application being 15 mm from one of the longer edges and 30 mm from one of the shorter edges of the section. If the tensile stress in the material is to be limited to 150 MN/$m^2$, determine the value of $W$ and the stresses at the four corners of the section.

   Find the area within which $W$ must be situated if there is to be no compressive stress in the material. Show your results on a diagram of the cross-section and indicate the position of the neutral axis.

   **[Ans.** 192 kN, – 30, + 60, + 150, + 60 MN/$m^2$]

8. A tie bar of rectangular section, originally 75 mm wide × 25 mm thick, has these dimensions reduced by $1/n$ th of the original values by the removal of material from two adjacent faces. If an axial tensile load of 100 kN is applied to the bar through the centre of the original section, find the value of $1/n$ for a maximum tensile stress of 120 MN/$m^2$. Determine also the magnitude of the least tensile stress. **[Ans.** 0.114 ; 15.49 MN/$m^2$]

9. An unequal-I section shown in figure 16.17 has following dimensions : $t_1$ = 10 mm, $B_1$ = 40 mm, $H$ = 60 mm, $w$ = 10 mm, $B_2$ = 60 mm, $t_2$ = 10 mm. Determine

the ratio $e_1/e_2$ and calculate the values of $e_1$ and $e_2$ where $e_1$ and $e_2$ are the distances of the shear centre $O$ from the centre lines of the flanges $ABCD$ and $LMNP$ respectively. [**Ans.** $\frac{27}{8}$; $e_1 = 46.3$ mm, $e_2 = 13.7$ mm]

10. In question 9 if $t_2$ were 12 mm and the values of the rest of the dimensions remain unchanged, find the values of $e_1$ and $e_2$.

[**Ans.** $e_1 = 48.12$ mm, $e_2 = 11.8$ mm]

11. An eccentric I section as shown in figure 16.18, has the following dimensions

$$b_1 = 20 \text{ mm}, b_2 = 8 \text{ mm}, t = 10 \text{ mm}; w = 12 \text{ mm and } h = 50 \text{ mm}$$

Find the distance of shear centric of the section from the centre of the web.

[**Ans.** 4.42 mm]

12. In the channel section (Figure 16.16) the flange with $b = 40$ mm, flange thickness $t = 10$ mm, the web depth $d = 50$ mm and web thickness $w = 12$ mm. Find the distance of shear centre of the section from the centre of the web.

[**Ans.** 16 mm]

13. In an outwardly extended c-section (Figure 16.19) various dimensions are as follows :

$$t = 10 \text{ mm}; b_1 = 30 \text{ mm}; b_2 = 60 \text{ mm}, h = 150 \text{ mm}$$

Find the distance of shear centre from the centroid $O'$ of the vertical limb $LMUV$. [**Ans.** $O'$ is at 0.04197 mm to the right of 0]

14. In an inwardly extended symmetrical c-section shown in Figure 16.20 following dimension are given $b_1 = 30$ mm, $b_2 = 60$ mm, $h = 150$ mm, $t = 10$ mm. Find the location of shear centre with respect to $O'$ the centroid of vertical limb $ELKF$. [**Ans.** 32.6 mm]

15. A segment of a thin rim of radius 100 mm has radial thickness of 10 mm and subtends an angle of 60° at the centre of the rim (Figure 16.21). Find the distance of the shear centre from the centre of the rim. [**Ans.** 10.7 mm]

16. For the thin section shown in figure (16.29) if $b = R = 20$ mm and $t = 1$ mm determine the distance of the shear centre $O'$ from $O$. State whether it is essential to specify the value of $t$. [**Ans.** 47.8 mm, No]

17. For the thin section shown in Figure 16.30) of $b = R$ show that $e$ is given by

$e = \frac{tR^3}{I}\left[-\frac{5}{6}t + 2R + \frac{3\pi R}{2}\right]$ and hence evaluate $e$ when $b = R = 20$ mm and $t =$ 1 mm. [**Ans.** 21.656 mm]

18. For the thin section shown in Figure 16.31, the inclined legs are inclined with the horizontal at an angle $\theta$ such that $\sin\theta = \frac{3}{5}$. If $b_1 = 30$ mm and b2 = 60 mm, determine the location of shear centre. [**Ans.** 8 mm to the left of web]

19. Determine the location of the shear centre for a thin section shown in Figure 16.32. Take $b_1 = 5$ mm; $b_2 = 10$ mm, $b_3 = 20$ mm, $t = 1$ mm and $\theta = 30°$.

[**Ans.** 3.3 mm on left of web]

Chapter–17

# Plastic Theory of Bending

## 17.1 Introduction

In simple theory of bending (Chapter 5) it was assumed that the loads on various members of the structure were within certain limits such that the stresses at various points of the material were with in the limit of proportionality point (both in tension as well as compression). As a matter of fact the skin stress which is the maximum stress induced in the member at the section carrying maximum bending moment, was at the most allowed to become equal to that at the limit of proportionality point of the material of the member. However if the load is increased gradually the yielding will first occur at the skin. These fibres are then known to be in the "plastic or inelastic" state. Further increase in the loading will increase the strains at all the points (including those at the skin) with the exceptions of the points lying on the neutral axis. But if the material is a rigid plastic or non strain hardening material, whose stress- strain curve is as shown in Fig. 17.1, there will be no increase in the stress beyond the yield stress, $\sigma_y$, at the skin fibres. But on all the fibres on the inner side adjacent to the skin, where the strain has become equal to or greater than $e_0$ Fig. 17.1, the stress will be $\sigma_y$. Thus as the load goes on gradually increasing, more and more of the material starts becoming plastic progressively inwards towards the neutral axis till the entire section becomes

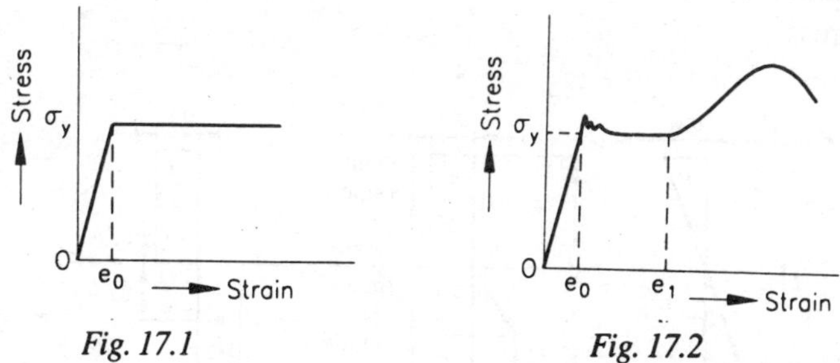


*Fig. 17.1* *Fig. 17.2*

plastic. This stage is known as "plastic hinge" and the material can no more take any additional load. The value of load to produce such a state is called the "collapse load". The ratio of collapse load to working load is known as "load factor".

In actual practice there is no such material which is rigid plastic and that follows the theoretical curve as shown in Fig. 17.1. But mild steel which has the stress-strain curve as shown in Fig. 17.2, has a long horizontal portion at lower yield limit before the strain hardening starts at strain $e_1$. Thus if the strain in the mild steel structure member does not go beyond $e_1$ anywhere, it can be assumed to have the rigid-plastic characteristic as assumed in Fig. 17.1.

## 17.2 Plastic Theory of Bending

The aim of this theory is to calculate the resisting moment of the section at plastic hinge which is defined as the state of the section when each and every point on it has become plastic. Before we proceed to do so, a number of assumptions are made. The assumptions are :

(*i*) The material is rigid-plastic and has the stress-strain curve of shape shown in Fig. 17.1.

(*ii*) The yield stress of the material is same in tension and compression.

(*iii*) Plane sections of the member remain plane even after the application of bending moment, so that the strain at any point of the section is proportional to its distance from the neutral axis. With this assumption it is implied that for the partially plastic section, within the elastic region stress will be proportional to strain which in turn is proportional to the distance from neutral axis but within the plastic region of the section stress will be constant and not proportional to strain or distance from the neutral axis.

(*iv*) When the plastic hinge has formed at any section, the resisting moment of that section will remain constant until the entire structure collapses due to the formation of additional plastic hinges at other points.

In view of the above assumptions the stress distribution over various stages of plastic deformation of a section will be as shown in Fig. 17.3 where $\delta_w$ represents the working stress.

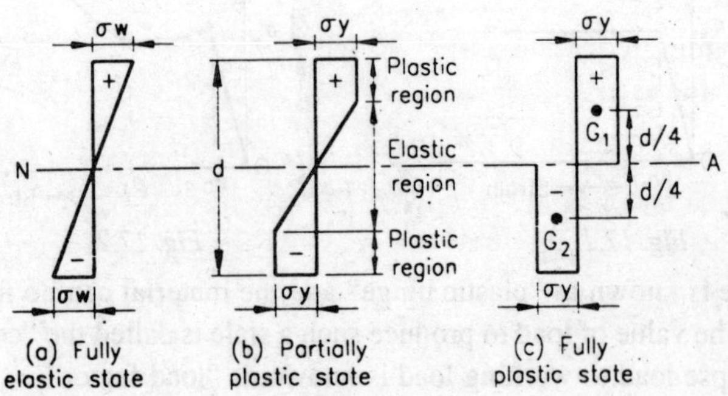


*Fig. 17.3*

### 17.2.1 Fully Plastic Rectangular Section

For the rectangular section of width $b$ and depth d the plastic moment of resistance of plastic hinge can easily be found. The total loads above and below the neutral axis NA are both equal to $\left(b.\frac{d}{2}\right) \times \sigma_y$ in magnitude but opposite in direction. Further each of these loads acts at a distance $\frac{d}{4}$ from the neutral axis. Thus at plastic hinge, the plastic moment of the rectangular section is given by $M_p$ as

$$M_P = \left(b \times \frac{d}{2}\right)\sigma_y\left(\frac{d}{4}+\frac{d}{4}\right)$$

or

$$M_P = \left(\frac{bd^2}{4}\right)\sigma_y \qquad \text{...(17.1)}$$

The resisting moment of the section when the maximum stress is limited to the working value, $\sigma_w$ ($\leq \sigma_y$) is given by

$$M_w = \frac{I\sigma_w}{(d/2)} = \frac{bd^3}{12}\frac{\sigma_w}{(d/2)} \qquad \text{...(17.2)}$$

Further let $M_y$ represent the resisting moment of the section when the skin stress has just reached the yield value $\sigma_y$. $M_y$ is also called the resisting moment of the section at "first yield". Then the ratio $M_p/M_y$ is called the shape factor of the section and is represented by $S$.

Thus when the maximum stress in the section touches $\sigma_y$ for the first time, the section is still wholly elastic only the skin fibres have been stressed to $\sigma_y$, the shape of the stress distribution curve will be as in Fig. 17.3($a$) with $\sigma_w = \sigma_y$. In that case replacing $\sigma_w$ by $\sigma_y$ in equation (17.2), we get the expression for $M_y$ as

$$M_y = \left(\frac{bd^2}{6}\right)\sigma_y \qquad \text{...(17.3)}$$

Then dividing equation (17.1) by equation (17.3), we get shape factor for rectangular section as

$$S = M_p/M_y = 1.5 \qquad \text{...(17.4)}$$

The normal factor of safety, $f$, based on initial yield *i.e* $\sigma_w/\sigma_y$ can also be found by dividing equations (17.2) and (17.3)

$$f = \frac{\sigma_w}{\sigma_y} = \frac{M_w}{M_y} \qquad \text{...(17.5)}$$

Again from equations (17.3) and (17.4)

$$M_p = S\,M_y$$

$$= S\left(\frac{bd^2}{6}\right)\sigma_y = SZ\,\sigma_y \qquad ...(17.6)$$

Where $Z\ (= bd^2/6)$ is called the section modulus of the rectangular section.

### 17.2.2 Fully Plastic Unsymmetrical Section

Let A be the total area of the section with $NA$ as its neutral axis. If the section is fully plastic under the action of pure bending moment, then in the absence of any externally applied axial force the resultant of all the normal forces due to bending stresses on the tensile region will be equal to that on the compression region. Thus if the entire section is fully plastic and the yield stresses in tension and compression are equal, the area under tension must be equal to that under compression. In other words the neutral axis divides the given area into two equal parts. Let $G_1$ and $G_2$ be the centroids of these areas situated at distances $y_1$ and $y_2$ respectively from the $NA$ Fig. 17.4. Since the resultant tensile and compressive forces will be equal in magnitude

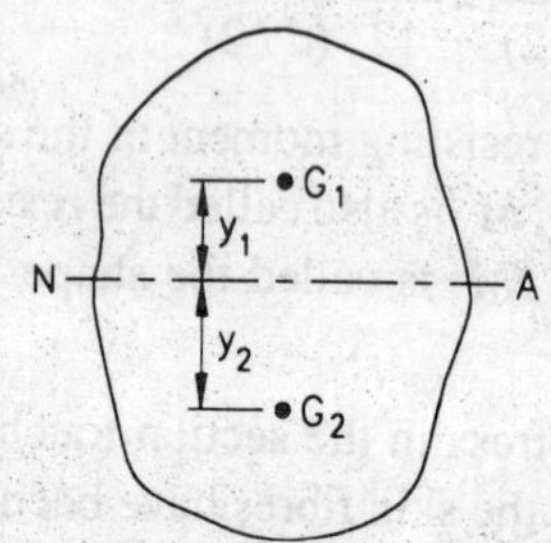


*Fig. 17.4*

$\left(=\frac{A}{2}\sigma_y\right)$ and opposite in direction, the resisting plastic moment of the section will then be given by

$$M_p = \left(\frac{A}{2}\sigma_y\right)(y_1 + y_2) \qquad ...(17.7)$$

However at first yield of the section with $Z$ as section modulus

$$M_y = Z\,\sigma_y \qquad ...(17.8)$$

Thus dividing equation (17.7) by (17.8), we get the shape factor as

$$S = \frac{M_p}{M_y} = \frac{A(y_1 + y_2)}{2Z} \qquad ...(17.9)$$

### 17.2.3 Partially Plastic Unsymmetrical Section

In the case of an unsymmetrical section under the action of a bending moment of such a magnitude that it is partially plastic and partially elastic, the location of the neutral axis can be found by equating the resultant tensile force on the area on one side of N.A. to the resultant compressive force on the area lying on the other side of N.A. While calculating the resultant forces it may be kept in mind that in the plastic region everywhere the stress will be $\sigma_y$ but the stress in the elastic region will go on decreasing from $\sigma_y$ at the boundary of plastic-elastic regions to zero at N.A. However on one side if there is no plastic region at all the stress at any point in the elastic region will be directly proportional to its distance from the neutral axis. Under these circumstances the resultant forces on the elastic and plastic portions are calculated separately and then added vectorialy to get the total resultant force on one side of the neutral axis where the section is partially plastic. Similarly the force on the other side of neutral axis is calculated and is equated to the former to give the location of the neutral axis.

**Example 17.1.** *Calculate the shape factor for a circular cross-section bending about a diameter.*

**Solution.** Let the circular section be under the action of a bending moment about the horizontal diameter.

The resisting moment at first yield, $M_y$ will be obtained when the stress distribution is as shown in Fig. 17.5(*b*), for which

$$M_y = AI_{NA}\frac{\sigma_y}{D/2}$$

$$= \frac{\pi D^2}{64}\frac{2}{D}\sigma_y = \frac{\pi D^3}{32}\sigma_y$$

Similarly for the fully plastic section the stress distribution curve will be as in Fig. 17.5(*c*). Since the centroid of a semicircle is at a distance $\frac{2D}{3\pi}$, the resisting moment for a fully plastic circular section is then given by

$$M_p = 2\left[\frac{1}{2}\left(\frac{\pi}{4}D^2\right)\sigma_y\left(\frac{2D}{3\pi}\right)\right]$$

$$= \frac{D^3}{6}\sigma_y$$

Thus the shape factor

$$S = \frac{M_p}{M_y} = \left(\frac{D^3\sigma_y}{6}\right)\left(\frac{32}{\pi D^3\sigma_y}\right)$$

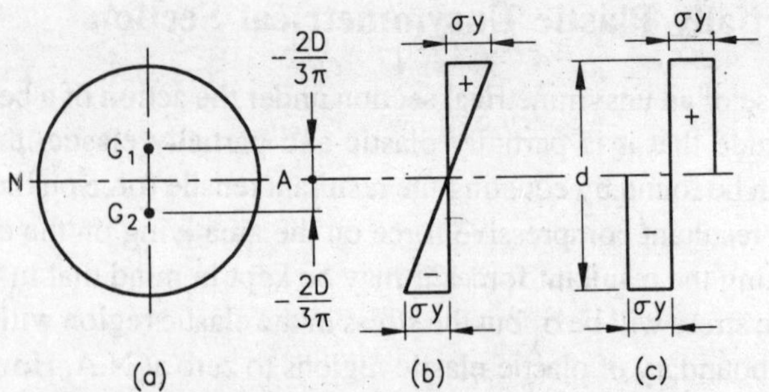


*Fig. 17.5*

$$= \frac{32}{6\pi} = 1.6976$$

**Example 17.2.** *Find the shape factor for an I-section shown in Fig. 17.6 bending about XX.*

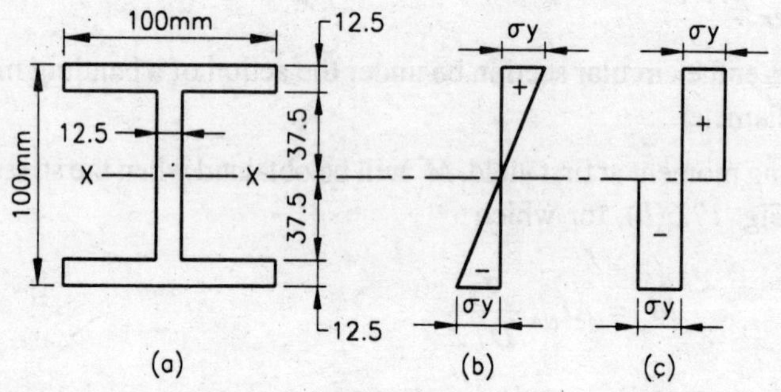


*Fig. 17.6*

**Solution.**

$$I_{xx} = \frac{100 \times (100)^3}{12} - \frac{87.5 \times (75)^3}{12} \text{ mm}^4$$

$$= 5.257 \times 10^6 \text{ mm}^4$$

Therefore for first appearance of yield

$$M_y = \frac{I_{xx}\sigma_y}{(100/2)} \frac{50}{5.257 \times 10^6}$$

$$= 105140\ \sigma_y$$

For fully plastic section Fig. 17.6(*c*)

$$M_p = \sigma_y \left[ \left\{ 100 \times 12.5 \left( 37.5 + \frac{12.5}{2} \right) \right\} \times 2 + \left\{ (12.5 \times 37.5) \frac{37.5}{2} \right\} \times 2 \right]$$

$$= 2\sigma_y \left[1250 \times 43.75 + 12.5 \times 37.5 \times \frac{37.5}{2}\right]$$

$$= 126953\ \sigma_y$$

Therefore shape factor

$$S = 2\sigma_y \left[1250 \times 43.75 + 12.5 \times 37.5 \times \frac{37.5}{2}\right]$$

**Example 17.3.** *A steel bar of rectangular section 72 mm × 30 mm is used as a simply supported beam on a span of 1.2 m and loaded at mid span. If the yield stress is 280 MN/m² and the long edges of the section are vertical, find the load when yielding first occurs.*

*Assuming that a further increase in load causes yielding to spread in towards the neutral axis with the stress in the yielded part remaining constant at 280 MN/m² determine the load required to cause yielding for a depth of 12 mm at the top and bottom of the section at mid-span, and find the length of beam over which yielding has occurred.*

*(Punjab University)*

**Solution.** (*a*) Resisting moment at first yield

$$M_y = \frac{I\,\sigma_y}{d/2} = \frac{bd^2}{6}\sigma_y$$

$$= \frac{30 \times 72^2}{6} \times 280 \text{ N/mm} = 7.26 \text{ kNm}$$

But bending moment

$$= \frac{W_y l}{4} = W_y \left(\frac{1200}{4}\right)$$

$$= 300\ W_y \text{ N-mm}$$

Equating these

$$300\ W_y = \frac{30 \times 72^2}{6} \times 280$$

or $$W_y = 24200 \text{ N or } 24.2 \text{ kN.}$$

For the mid section to be partially plastic, the resisting moment $M$ will be twice that due to the forces over half of the section either above or below the neutral axis $N$ about which the section is symmetrical

*i.e.*, $$M = 2\left[(12 \times 30 \times 280)(24 + 6) + \left(\frac{1}{2} \times 24 \times 280 \times 30\right)\left(\frac{2}{3} \times 24\right)\right]$$

$$= 928 \times 10^4 \text{ N-mm}$$

$$= 9280 \text{ Nm}$$

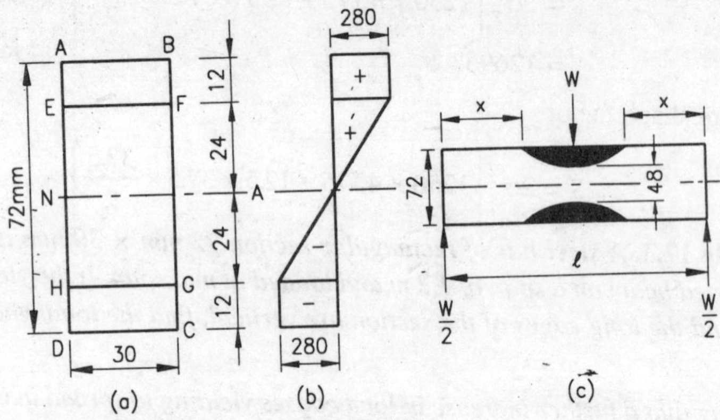


(a) Enlarged view of the section.
(b) Stress distribution for partially plastic section.
(c) First yield at a distance x. from ends.

*Fig. 17.7*

Equating this to the bending moment at the section corresponding to the increased load, W say

*i.e.,*
$$\frac{Wl}{4} = 9280$$

or
$$W = \frac{9280 \times 4}{1.2} = 3100 \text{ N} = 31 \text{ kN}$$

(*b*) When the load $W (> W_y)$ will be acting on the beam at the mid section, which has turned partially plastic, then the bending moment on the beam goes on gradually decreasing from the maximum value of $\frac{Wl}{4}$ at the mid section to zero at the ends. Let $x$ be the distance from the ends of such section where the bending moment equals to that at first yield ($= M_y$). Then the location of such a section can be found by equating the bending moment on this to $M_y$ Fig. 17.7.

*i.e.*
$$\frac{W}{2}x = M_y$$

or
$$x = 2\frac{M_y}{W} = \frac{2 \times 7.26}{31}$$
$$= 0.468 \text{ m}$$

Then leaving 0.468 m towards the ends, all sections of the middle portion of the beam would have become partially plastic. Thus the length of the beam over which yielding has set in

$$= l - 2x$$

$$= 1.2 - 2 \times 0.468 = 0.284 \text{ m.}$$

**Example 17.4.** *A horizontal cantilever, 1.2 m long, has a T-shaped cross-section as shown in Fig. 17.8, and carries a point load W at the free end. If the section near the fixed end of the cantilever is fully plastic, with yield stress of the material as 300 MN/ $m^2$ find :*

(a) *the location of neutral axis.*

(b) *the value of W.*

(c) *the shape factor.*

*For this section the neutral axis XX, before the yielding occurs, lies at 19.74 mm from the top of flange and italic* $I_{xx} = 1.003 \times 10^{-6}$ $m^4$. *(Punjab University)*

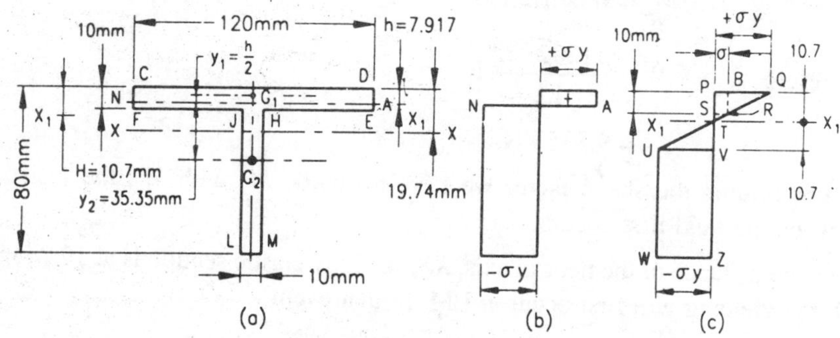

(b) Stress distribution for fully plastic section.
(c) Stress distribution for partially plastic section.

*Fig. 17.8*

**Solution.** (*a*) Since the section is fully plastic, the neutral axis N.A. will divide the area into equal parts. Further as the area of flange is 1200 $mm^2$ and that of web is 700 $mm^2$, the neutral axis *NA* will be above the line *FE*. Let the neutral axis be at a distance *h* mm from *CD*. Then equating the areas on the two sides of N.A.

$$120\,h = (10 - h)\,120 + 70 \times 10$$

or $$240\,h = 1900$$

*i.e.,* $$h = 7.917 \text{ mm.}$$

(*b*) Let $G_1$ and $G_2$ be the centroids of the areas above and below the neutral axis N.A, with their distances from N.A. as $y_1$ and $y_2$ respectively.

$$y_1 = \frac{1}{2}h = \frac{1}{2} \times 7.917 = 3.9585 \text{ mm.}$$

For evaluating $y_2$, taking moments of areas (lying below N A) about N.A., we have

or
$$y^2 = \frac{700 \times 37.083 + 60 \times (2.083)^2}{700 + 2.083 \times 120}$$
$$= 35.35 \text{ mm}$$

Total area of the section $A = 120 \times 10 + 70 \times 10 = 1900 \text{ mm}^2$. Thus from equation (17.7)

$$M_p = \left(\frac{A}{2}\sigma_y\right)(y_1 + y_2)$$
$$= \left(\frac{900}{2} \times 300\right)(3.9585 + 35.35) \text{ N/mm}$$
$$= 11202.922 \text{ Nm}$$

But
$$M_p = W \times 1.2 \text{ Nm}$$

Therefore
$$W = \frac{11202.922}{1.2} \text{ N}$$
$$= 9.335 \text{ kN.}$$

(*c*) For finding the shape factor we have to divide $M_p$ with $M_y$, the resisting moment when the field first occurs.

Since the distance of the neutral axis, *XX*, for fully elastic section is at 19.74 mm from *CD*, the yielding will first occur at LM. In that event

$$M_y = \frac{\sigma_y I_{xx}}{(80 - 19.74)\, 10^{-3}}$$
$$= \frac{(1.003 \times 10^{-6})(300 \times 10^6)}{60.26 \times 10^{-3}} \text{ Nm}$$
$$= 4993.36 \text{ Nm}$$

Therefore shape factor

$$S = \frac{M_p}{M_y} = \frac{11202.922}{4993.36} = 2.25.$$

**Example 17.5.** *In Example 17.4 if the top of the flange is just on the point of yielding, find the location of the neutral axis italic $X_1 X_2$. Hence calculate the value of the load required at the end of the cantilever to create such partially plastic state of the section near the fixed end.*

**Solution.** (*a*) Let the position of the neutral axis, $X_1X_1$ under the partially plastic state be *H* mm from the top of the flange. Then the stress distribution curve will be as shown in Fig. 17.8(*c*).

In that case the stress along the bottom of the flange *i.e.*, at *FE* will be given by σ, in proportion to its distance from the neutral axis $X_1X_1$.

*i.e.* $$\sigma = \frac{\sigma_y}{H}(H - 10)$$

Further since the stresses at *V* and *P* are both $\sigma_y$ in magnitude, *TV* = *TP* = *H* mm.

Thus equating the forces on the section on either side of the neutral axis, we get

$$\frac{1}{2}\left[\sigma_y + \frac{\sigma_y}{H}(H - 10)\right](120 \times 10) + \frac{1}{2}\frac{\sigma_y (H - 10)}{H}[H - 10)\,10]$$

$$= (80 - 2H)\,10 \times \sigma_y + \frac{1}{2}\sigma_y\, H \times 10$$

$$\frac{1200}{H}(2H - 10) + \frac{10\,(H - 10)^2}{H} = 20\,(80 - 2H) + 10\,H$$

or $$40\,H^2 + 600\,H - 14000 = 0$$

$$H = 10.7$$

(*b*) Resisting moment of the Section

*M* = Sum of the moments about $X_1X_2$ due to the stress distributions *PBRS*, *BQR*, *SRT*, *TUV* and *UVZW*.

$$= (\sigma \times 120 \times 10)(H - 5) + \left[\frac{1}{2} \times 10\,(\sigma_y\ \sigma) \times 120\right]\left(h - \frac{10}{3}\right)$$

$$+\left[\frac{1}{2}\sigma\,(H - 10) \times 10\right]\left[\frac{2}{3}(H - 10)\right] + \left[\frac{1}{2}H\sigma_y \times 10\right]\left(\frac{2}{3}H\right)$$

$$+ [10\sigma_y\,(80 - 2H)]\left[H + \frac{1}{2}(80 - 2H)\right]$$

Now $\sigma = \frac{\sigma_y}{H}(H - 10) = \frac{0.7}{10.7} = \frac{0.7}{10.7}\sigma_y$ [As triangles *TPQ* and *TSR* are similar]

Thus

$$\frac{M}{\sigma_y} = \frac{0.7 \times 1200 \times 5.7}{10.7} + 600 \times \frac{10.7 - 0.7)}{10.7}\left(10.7 - \frac{10}{3}\right)$$

$$+\frac{1}{2} \times \frac{0.7 \times 10 \times .7}{10.7} \times \frac{2}{3} \times 7 + \frac{10}{3}(10.7)^2 + 10 \times 10.7$$

$$\times (80 - 21.4) + 5\,(80 - 21.4)^2$$

$$= 28400$$

or $$M = 28400 \times 300 \text{ N mm}$$

$$= 8520 \text{ Nm}$$

Equating this to the bending moment over the section, we get

$$1.2 \text{ W} = 8520$$

or $$W = 7100 \text{ N or } 7.1 \text{ kN.}$$

## PROBLEMS FOR PRACTICE

1. Prove that the shape factor for a circular section under the action of pure bending moment about a diameter is approximately 1.7.
2. Find the shape factor for a 200 mm × 75 mm channel section shown in Fig. 17.9 (web thickness = 5.5 mm and flange thickness 10.8 mm) in pure bending with the plane of bending perpendicular to the web of the channel. For this section from steel tables area = 2606 mm$^2$, $Z_{XX}$ = 28500 mm$^3$, $C_{XX}$ = 23.6 mm.

**[Ans. 1.87]**

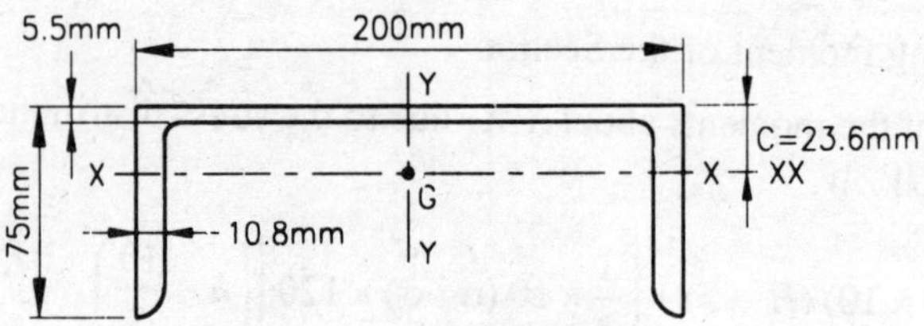


*Fig. 17.9*

3. A rectangular steel beam 60 mm deep by 30 mm wide is supported on knife edge 2 m apart and loaded with two equal point loads at one-third of the span from each end. Find the load at which yielding just begins, the yield point of the material in simple tension being 300 MN/m$^2$.

   If the loads are increased to 25 per cent above this value estimate how far the yielding penetrates towards the neutral axis assuming that the maximum stress remains constant and that the yielding point in compression is the same as that in tension. **[Ans. 840 kN ; 8.78 mm]**
4. Two plates of mild steel, each of cross-section 40 mm by 3 mm, and one plate of alloy steel 40 mm by 6 mm are placed side by side and fastened rigidly together to form a symmetrical composite sandwich bar 40 mm by 12 mm. The mild steel has a definite yield point at 240 MN/m$^2$ in both tension and compression ; the alloy steel is elastic upto 800 MN/m$^2$ and $E$ is same for both steels. The bar is tested as a beam of span 0.6 m with the 40 mm dimension as the depth, a concentrated load being applied at the centre of the span.

Find the central load at which the mild steel will just commence to yield and also the load at which the maximum bending stress in the alloy steel will reach 480 $MN/m^2$. It may be assumed that the mild steel follows a linear stress-strain law up to the yield point at which considerable yielding occurs.

**[Ans.** 5.12 kN ; 8.63 kN]

5. Fig. 17.10 shows the section of a beam which is subjected to a bending moment of such magnitude that yielding occurs at the lower part of the web over a depth of 20 mm. The yield stress of 280 MN-$m^2$ may be assumed constant over the yielded area, while over the remainder of the section the stress is proportional to the distance from the neutral axis. Determine : (*a*) the position of the neutral axis, (*b*) the stress at the top of the section, (*c*) the moment of resistance of the section. (Punjab University)

**[Ans.** 25.8 mm from top of flange, 210 MN-$m^2$, 4350 Nm]

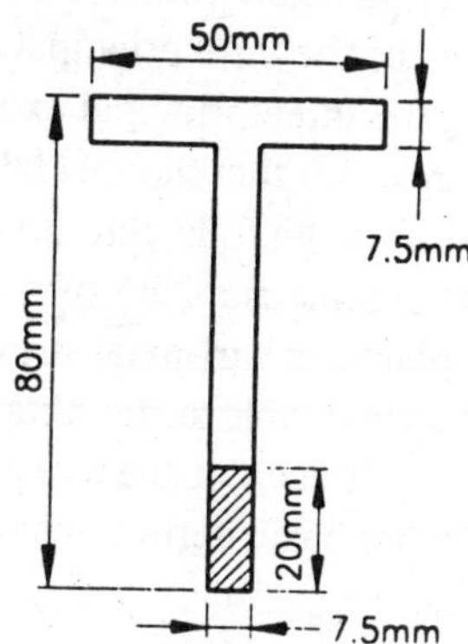


*Fig. 17.10*

6. A steel bar of T-section as shown in Fig. 17.8(*a*) is used as a simply supported beam on a span of 2 m to carry a point load at midspan. A load of 9 kN is just sufficient to cause yielding to commence at the bottom of the vertical leg of the section at mid-span. Estimate the load which would be required to cause yielding to commence at the top of the section. State any assumptions made in deriving the answer.

Before yielding occurs the neutral axis *XX* lies at 19.74 mm from the top of flange and $I_{xx} = 1.003 \times 10^{-6}$ $m^4$. **[Ans.** 14.9 kN]

Chapter–18

# Three Dimensional Stress Systems

## 18.1 Introduction

In Chapter 2 two dimensional stress system which is also known as "*Plane state of stress*", was discussed. However, in actual practice in quite a large number of cases the stresses do not lie in one plane and the third principal stress is not zero. Such a state of stress where the stresses along the three principal axes exist, is known as the three dimensional or tri-axial state of stress. Further the special case of the same when all the three principal stresses are equal in magnitude and alike in sign, is referred to as an equiaxial state of stress or polar-symmetric state of stress. In order to develop the equations for stresses on various planes in a tri-axial stress system it is very convenient to deal with stresses represented with double suffix notation. A double suffix notation is employed to define the direction of each of the stresses. Double suffix notation of representing stresses along with other basic definitions of stresses will be discussed in detail in the following paragraphs.

## 18.2 Definitions

Let $O$ be a point in a solid body which is transmitting or sustaining load. Through $O$ let $\delta A$ be a small plane area with $ON$ normal to it. Across $\delta A$ a force $\delta F$ say, will be transmitted to give a quantity

$$\sigma = \underset{\delta A \to 0}{\text{Lt}} \left( \frac{\delta F}{\delta A} \right)$$

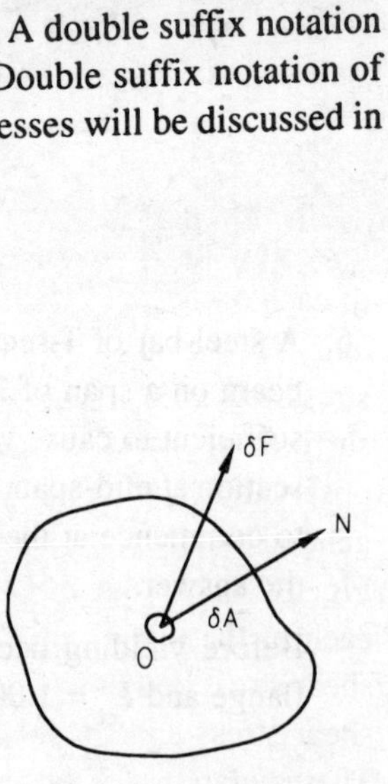

which is defined as the stress across $\delta A$ at $O$ in the direction of $\delta F$. Alternatively other equivalent statements could also be made. Resolve $\delta F$ normal and parallel to $\delta A$, into, say $\delta F_n$ and $\delta F_s$ as shown in Fig.



Fig. 18.1

18.2(*a*). Then the normal and shear stresses across $\delta A$ are defined as $\underset{\delta A \to 0}{\text{Lt}} \left( \frac{\delta F_n}{\delta A} \right)$ and

$\underset{\delta A \to 0}{\text{Lt}} \left( \frac{\delta F_s}{\delta A} \right)$ respectively. Further, suppose $\delta F$ has been resolved into three perpendicular directions *OX, OY, OZ* Fig. 18.2(*b*). It is assumed that *OX* is the direction of the normal to $\delta A$ and *OY* and *OZ*, though chosen arbitrarily, both lie in the plane of $\delta A$.

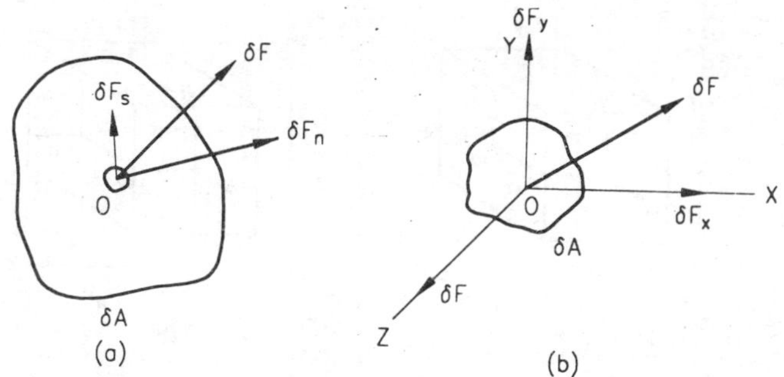


*Fig. 18.2*

Let the components of the force $\delta F$ along the three axes be $\delta F_x$, $\delta F_y$ and $\delta F_z$. The three stresses are then represented by

$$\delta_{xx} = \underset{\delta A \to 0}{\text{Lt}} \left( \frac{\delta F_x}{\delta A} \right) \quad \text{...(18.1}a\text{)}$$

$$\delta_{xy} = \underset{\delta A \to 0}{\text{Lt}} \left( \frac{\delta F_y}{\delta A} \right) \quad \text{...(18.1}b\text{)}$$

$$\delta_{xz} = \underset{\delta A \to 0}{\text{Lt}} \left( \frac{\delta F_z}{\delta A} \right) \quad \text{...(18.1}c\text{)}$$

In equations (18.1) a double suffix notation is employed to define the direction of each of the stresses. $\delta_{xy}$ means the stress on a plane normal to *OX* and parallel to the axis *OY*. The first suffix gives the direction of the normal to the plane on which the stress acts and the second suffix its direction of line of action. Evidently when both the suffixes of stress are same the stress will be the normal stress on the point under observation and when they are different the stress will be the shear stress. Thus in general on any plane there will be one direct stress and two shear stress components. In Fig. 18.3 is shown an elemental parallelepiped with the three component stresses on each of the three mutually perpendicular planes. It may be noted that in the case of shear stress components Fig. 18.3(*b*) 'σ' has been replaced by τ, the usual notation for shear stress and for the direct stress components only one suffix has been retained. These notations for the representation of the nine stress components, referred to cartesian co-ordinates, are quite common in literature. It will be seen later in this chapter that

$$\tau_{xy} = \tau_{yx}$$

$$\tau_{yz} = \tau_{zy}$$

and $$\tau_{zx} = \tau_{xz}$$

Thus in general only six stress components referred to cartesian coordinates will be essential to specify the state of stress at any point in a stressed material.

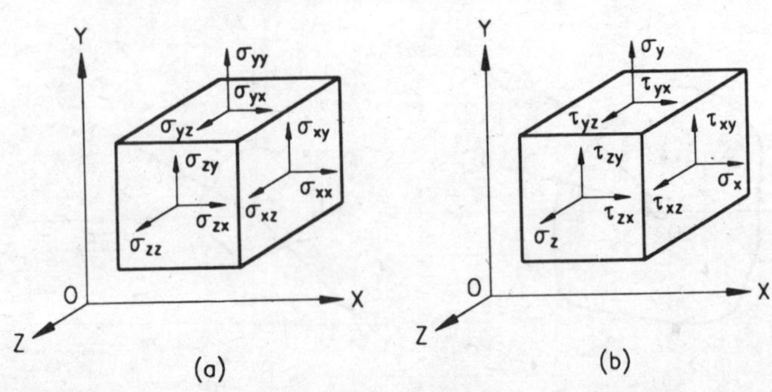


*Fig. 18.3*

In the above discussion, the axes of reference were chosen in such a way that *OX* was taken along the normal to the area $\delta A$. However, in actual practice the analysis is usually started by first establishing the directions of axes *OX*, *OY* and *OZ* in space. The components of the elementary force $\delta F$ are taken along these axes as $\delta F_x$, $\delta F_y$ roman and $\delta F_z$ respectively acting on an elemental area $\delta A_x$, which is taken normal to *OX*, then $\sigma_{xx}$, $\sigma_{xy}$, $\sigma_{xz}$ are calculated as above. Similarly if an area $\delta A_y$ *i.e.*, a small area at *O* normal to the direction *OY* across which the net force is $\delta P$, and having components $\delta P_x$, $\delta P_y$, $\delta P_z$ then the notation for the stresses across $\delta A_y$ would be $\sigma_{yx}$, $\sigma_{yy}$, $\sigma_{yz}$. Yet a third set of stresses across area $\delta A_z$ would be $\sigma_{zx}$, $\sigma_{zy}$, $\sigma_{zx}$ Fig. 18.3(*a*).

## 18.3 Equilibrium Equations

Under the action of a set of forces and couples the body will be in equilibrium. Any element of the body will also be in equilibrium. In general there will be two sets of equilibrium equations one concerning the forces and the other couples. These are discussed below.

### 18.3.1 The Equations of Force Equilibrium

In Fig. 18.4 is shown an elemental rectangular parallelepiped *ABCDEFGH* placed symmetrically around any point *P* in a stressed body. The lengths of the sides of the rectangular parallelepiped are $\delta x$, $\delta y$ and $\delta z$. In the Fig. 18.4 are shown stress components parallel to *x*-axis only and it will refer to the force equilibrium in the *OX* direction. Employing the double suffix notation the rates of change of stress components with increase in *x*, on area $\delta A_x$, when *y* and *z* are kept constant, are

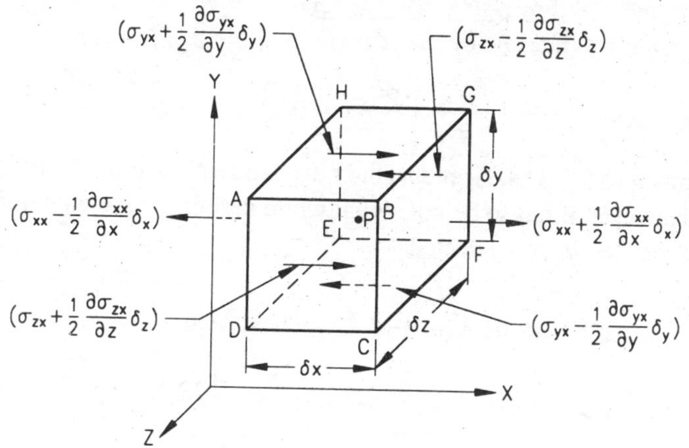


*Fig. 18.4*

$$\frac{\partial \sigma_{xx}}{\partial x}, \frac{\partial \sigma_{xy}}{\partial x} \text{ and } \frac{\partial \sigma_{xz}}{\partial x}$$

Similarly for directions $y$ and $z$ the corresponding expressions will be

$$\frac{\partial \sigma_{xy}}{\partial y}, \frac{\partial \sigma_{yy}}{\partial y}, \frac{\partial \sigma_{xz}}{\partial y}$$

$$\frac{\partial \sigma_{zy}}{\partial z}, \frac{\partial \sigma_{xy}}{\partial z}, \frac{\partial \sigma_{zz}}{\partial z}$$

In Fig. 18.4 are shown the stress components parallel to $x$-axis only, on the six faces of the rectangular element. Actually there will be eighteen forces on the six faces of the rectangular block and the three equations of equilibrium may be found by considering the forces along each of the three directions parallel to the axes separately. Thus resolving the forces due o the stress components shown in Fig. 18.4 along the direction $OX$, we have

$$\left(\sigma_{xx} - \frac{1}{2}\frac{\partial \sigma_{xx}}{\partial x}.\delta x\right)\delta y\delta z + \left(\sigma_{yx} - \frac{1}{2}\frac{\partial \sigma_{xx}}{\partial y}.\delta y\right)\delta x\delta z + \left(\sigma_{zx} - \frac{1}{2}\frac{\partial \sigma_{zz}\delta z}{\partial z}\right)\delta x\delta y$$

$$+\left(\sigma_{zx} + \frac{1}{2}\frac{\partial \sigma_{xx}}{\partial x}.\delta x\right)\delta y\delta z + \left(\sigma_{yx} + \frac{1}{2}\frac{\partial \sigma_{yx}}{\partial y}.\delta y\right)\delta z\delta x + \left(\sigma_{zx} + \frac{1}{2}\frac{\partial \sigma_{zx}}{\partial z}.\delta z\right)\delta x\delta y$$

On simlification is reduces to

$$\frac{\partial \sigma_{xx}}{\partial x} + \frac{\partial \sigma_{yx}}{\partial y} + \frac{\partial \sigma_{zx}}{\partial z} = 0 \qquad ...(18.2)$$

Similarly the equations of equilibrium for directions $OY$ and $OZ$ are

$$\frac{\partial \sigma_{xy}}{\partial x} + \frac{\partial \sigma_{yy}}{\partial y} + \frac{\partial \sigma_{zy}}{\partial z} = 0 \qquad ...(18.3)$$

$$\frac{\partial \sigma_{xz}}{\partial x} + \frac{\partial \sigma_{yz}}{\partial y} + \frac{\partial \sigma_{zz}}{\partial z} = 0 \qquad ...(18.4)$$

Equations (18.2), (18.3) and (18.4) are known as the equations of force equilibrium. However, it has to be noted that while deriving these equations the effect of the body forces has been neglected.

## 18.3.2 The Equations of Couple Equilibrium

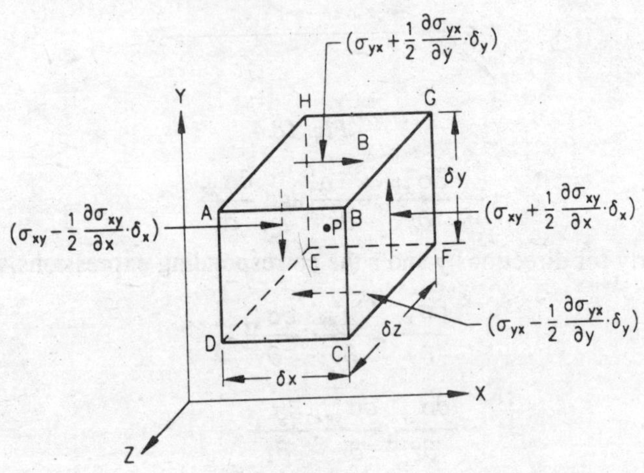


*Fig. 18.5*

Fig. 18.5 shows the same element *ABCDEFGH* as in Fig. 18.4 around the point *P* but only those stresses are shown which tend to rotate the block about a line passing through *P* and parallel to the axis *OZ*. Since the block is in equilibrium the algebraic sum of the couples will be zero.

$$\left(\sigma_{yx} + \frac{1}{2}\frac{\partial \sigma_{yx}}{\partial y}\,.\,\delta y\right)\delta z\delta x\left(\frac{1}{2}\,\delta y\right) + \left(\sigma_{yz} - \frac{1}{2}\frac{\partial \sigma_{yz}}{\partial y}\,.\,\delta y\right)\delta z\delta x\left(\frac{1}{2}\,\delta y\right)$$

$$= \left(\sigma_{xy} + \frac{1}{2}\frac{\partial \sigma_{xy}}{\partial x}\,.\,\delta x\right)\delta y\delta z\left(\frac{1}{2}\,\delta x\right) + \left(\sigma_{xy} - \frac{1}{2}\frac{\partial \sigma_{xy}}{\partial x}\,.\,\delta x\right) \times \delta y\delta z\left(\frac{1}{2}\,\delta x\right)$$

On simplification it reduces to

$$\sigma_{yx} = \sigma_{xy} \qquad ...(18.5)$$

Similarly for the other two directions

$$\sigma_{yz} = \sigma_{zy} \qquad ...(18.6)$$

and $$\sigma_{zx} = \sigma_{xz} \qquad ...(18.7)$$

Evidently equations (18.5) through (18.7) convey the well known idea of "*complementary shear stress*". Further in the light of these three equations we conclude that the actual state of stress at any point in a stressed body will be known if the three normal stresses and three shear stresses are known in magnitude and directions.

## 18.4 Three-dimensional Stress System

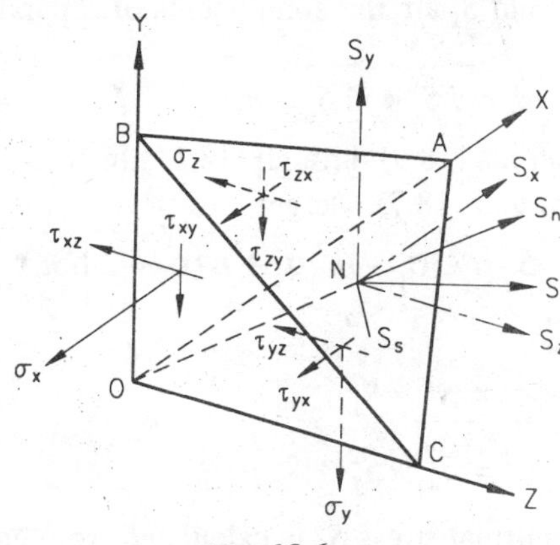


*Fig. 18.6*

Fig. 18.6 shows a tri-axial stress system in which the six components of stress at a point in a stressed body are given in magnitude and direction. It is desired to evaluate the normal and shear stress on any doubly oblique plane and also to determine the principal stresses at that point along with their directions. Let the doubly oblique plane be represented by *ABC* with *ON* its normal having direction cosines as (*l*, *m*, *n*). Let *S* be the resultant stress on the triangle *ABC*, with normal and shear stress components $S_n$ and $S_s$ respectively normal to and along the plane *ABC*. Further let $S_x$, $S_y$ and $S_z$ be the components of *S* parallel to *OX*, *OY* and *OZ* respectively. Areas of the triangles *ABC*, *OBC*, *OAC* and *OAB* are represented by *A*, $A_x$, $A_y$ and $A_z$ respectively. Then for the equilibrium of the forces acting on the pyramid *OABC* in the direction *OX*, we have from Fig. 18.6.

$$AS_x = A_x\sigma_x + A_y\sigma_{yx} + A_z\sigma_{zx}$$

or

$$S_x = \left(\frac{A_x}{A}\right)\sigma_x + \left(\frac{A_y}{A}\right)\tau_{yx} + \left(\frac{A_z}{A}\right)\tau_{zx} \qquad ...(18.8)$$

But

$$\frac{A_x}{A} = l$$

$$\frac{A_y}{A} = m$$

and

$$\frac{A_z}{A} = n$$

Therefore equation (18.8) reduces to

$$S_x = l\,\sigma_x + m\,\tau_{yx} + n\,\tau_{zx} \quad ...(18.9)$$

Similarly for the equilibrium of forces along *OY* and *OZ*, we get

$$S_y = l\,\tau_{xy} + m\sigma_y + n\tau_{zy} \quad ...(18.10)$$

$$S_z = l\,\tau_{xz} + m\,\tau_{yz} + n\,\sigma_z \quad ...(18.11)$$

Also since $S_x$, $S_y$ and $S_z$ are the components of $S$ parallel to the three axes, we have

$$S_n = l\,S_x + m\,S_y + n\,S_z \quad ...(18.12)$$

Substituting equations (18.9) through (18.11) in equation (18.12) and using the equations (18.5), (18.6) and (18.7), we get

$$S_n = l^2\,\sigma_x + m^2\,\sigma_y + n^2\,\sigma_z + 2\,(lm\,\tau_{xy} + mn\,\tau_{yz} + nl\,\tau_{zx}) \quad ...(18.13)$$

Further $$S_s^2 = S^2 - S_n^2 \quad ...(18.14a)$$

$$= S_x^2 + S_y^2 + S_z^2 - S_n^2 \quad ...(18.14b)$$

The value of the normal stress $S_n$ is calculated from equation (18.13) and then equation (18.14*a*) is used to evaluate the shear stress $S_s$ on any plane *ABC* whose normal *ON* has direction cosines (*l, m, n*).

## 18.4.1 Principal Stresses

Let $S$ be the principal stress so that

$$S = S_n$$

where $S_s$ is zero. In that event $S$ will have the direction cosines (*l, m, n*). Then resolving the forces on the pyramid along *OX* or else replacing $S_x$ by $l\,S$ in equation (18.9) when the plane *ABC* is the principal plane and also using equations (18.5), (18.6) and (18.7).

$$l\,(S - \sigma_x) - m\,\tau_{xy} - n\,\tau_{xz} = 0 \quad ...(18.15)$$

Similarly for the resolution of forces along *OY* and *OZ*, we get

$$-\,l\,\tau_{xy} + m\,(S - \sigma_y) - n\,\tau_{yz} = 0 \quad ...(18.16)$$

$$-\,l\,\tau_{xz} - m\,\tau_{yz} + n\,(S - \sigma_z) = 0 \quad ...(18.17)$$

Dividing the equations (18.15), (18.16) and (18.17) by $l$, we have

$$(S - \sigma_x) - \frac{m}{l}\tau_{xy} - \frac{n}{l}\tau_{xy} = 0 \quad ...(18.18)$$

$$-\,\tau_{xy} + \frac{m}{l}(S - \sigma_y) - \frac{n}{l}\tau_{yz} = 0 \quad ...(18.19)$$

$$-\,\tau_{xz} - \frac{m}{l}\tau_{yx} + \frac{n}{l}(S - \sigma_z) = 0 \quad ...(18.20)$$

Solving simultaneous equations (18.18) and (18.19) for $\frac{m}{l}$ and $\frac{n}{l}$ and then substituting these values in equation (18.20), we shall get

$$S^3 - (\sigma_x + \sigma_y + \sigma_z) S^2 - (\tau_{xy}^2 + \tau_{yz}^2 + \tau_{zx}^2 - \sigma_x\sigma_y - \sigma_y\sigma_z - \sigma_z\sigma_x)S$$

$$- (\sigma_x\sigma_y\sigma_z + 2\tau_{xy}\tau_{yz}\tau_{zx} - \sigma_x\tau_{yz}^2 - \sigma_y\tau_{zx}^2 - \sigma_z\tau_{xy}^2) = 0 \qquad ...(18.21)$$

Solution of the cubic equation (18.21) will give, for a real physical situation, three real roots which will be the magnitudes of the three principal stresses say $\sigma_1$, $\sigma_2$ and $\sigma_3$ respectively.

## 18.4.2 Stress Invariants

Let it be assumed that with same stress system acting the stress components are initially given with respect to a set of three perpendicular axes *OX'*, *OY'* , *OZ'* instead of *OX, OY, OZ*. That is let the given stress components with respect to *OX'* , *OY'*, *OZ'*, be $\sigma_{x'}$, $\sigma'_{y'}$, $\sigma'_{z'}$, $\tau_{x'y'}$, $\tau_{y'z'}$ and $\tau_{z'x'}$.

Further the principal stresses at *P* will be given by the roots of a cubic equation similar to equation (18.21) in which the earlier stress components will be replaced by the ones given above with respect to the new axes. That is the new cubic equation will be

$$S^3(\sigma_{x'} + \sigma_{y'} + \sigma_{z'}) S^2 - (\tau_{x'y'}^2 + \tau_{y'z'}^2 + t_{z'x'}^2 - \sigma_{x'}\sigma_{y'} - \sigma_{y'}\sigma_{z'}$$

$$- \sigma_{z'}\sigma_{x'}) S - (\sigma_{x'}\ \sigma_{y'}\sigma_{z'} + 2\tau_{x'y'}\tau_{y'z'}\tau_{z'x'} - \sigma_{x'}\tau_{y'z'}^2$$

$$- \sigma_{y'}\tau_{z'x'}^2 - \sigma_{z'} - \tau_{x'y'}^2\Big) = 0 \quad ...(18.22)$$

Regardless of the axes chosen the principal stresses remain same. Thus the equations (18.21) and (18.22) will have the same roots. Therefore comparing the coefficients of $S^2$, $S$ and the constant terms in these equations we have

$$I_1 = \sigma_x + \sigma_y + \sigma_z$$

$$= \sigma'_x + \sigma'_y + \sigma'_z \qquad ...(18.23)$$

$$I_2 = -(\sigma_x\sigma_y + \sigma_y\sigma_z + \sigma_z\sigma_x) + \tau_{xy}^2 + \tau_{yz}^2 + \tau_{zx}^2$$

$$= -(\sigma_{x'}\sigma_{y'} + \sigma_{y'}\sigma_{z'} + \sigma_{z'}\sigma_{x'}) + \tau_{x'y'}^2 + \tau_{y'z'}^2 + \tau_{z'x'}^2 \quad ...(18.24)$$

$$I_3 = \sigma_x\sigma_y\sigma_z + 2\tau_{xy}\tau_{zy}\tau_{xz} - \sigma_x\tau_{yz}^2 - \sigma_y\tau_{zx}^2 - \sigma_z\tau_{xy}^2$$

$$= \sigma_{x'}\sigma_{y'}\sigma_{z'} + 2\tau_{x'y'}\tau_{z'y'}\tau_{x'z'} - \sigma_{x'}\tau_{y'z'}^2 - \sigma_{y'}\tau_{z'x'}^2 - \sigma_{z'}\tau_{x'y'}^2$$

...(18.25)

$I_1$, $I_2$ and $I_3$ are called the stress invariants as these quantities are not dependent upon the direction of axes. Thus if the axes of reference are chosen along the directions of the principal stresses $\sigma_1$, $\sigma_2$ and $\sigma_3$ then since the shear stress components on the principal planes are all zero, the values of the stress invariants in terms of principal stresses are

$$I_1 = \sigma_1 + \sigma_2 + \sigma^3 \quad \text{...(18.26a)}$$

$$I_2 = -(\sigma_1\sigma_2 + \sigma_2\sigma_3 + \sigma_3\,\sigma_1) \quad \text{...(18.26b)}$$

$$I_3 = \sigma_1\sigma_2\sigma_3 \quad \text{...(18.26c)}$$

Thus with the help of equations (18.23) through (18.26), the equation (18.21) can be written in the form

$$S^3 - I_1S^2 - I_1S - I_3 = 0 \quad \text{...(18.27)}$$

The roots of the cubic equation (18.27) give the three principal stresses. This equation is generally solved either by hit and trial or by the method outlined in example 18.1 given below. After the evaluation of the three principal stresses the direction cosines of their lines of action can be found with the help of equations (18.18), (18.19) and (18.20). After substituting the value of $S$ as $\sigma_1$ in these equations we get three simulatoreous equations in *l, m* and *n*. These can then be calculated. Similarly the direction cosines of the lines of action of $\sigma_2$ and $\sigma_3$ can be found.

### 18.4.3 Direct and Shear Stresses on Octahedral Plane

An octahedral plane is one which is equally inclined to the three principal stresses. Let the three principal stresses be $\sigma_1$, $\sigma_2$ and $\sigma_3$ and the axes of reference *OX, OY* and *OZ* be chosen along $\sigma_1$, $\sigma_2$ and $\sigma_3$ respectively. Then the direction cosines (*l, m, n*) of the normal to the octahedral plane will be $\left(\frac{1}{\sqrt{3}}, \frac{1}{\sqrt{3}}, \frac{1}{\sqrt{3}}\right)$.

The normal stress on this plane can be found by replacing $\sigma_x$, $\sigma_y$ and $\sigma_z$ by $\sigma_1$, $\sigma_2$ and $\sigma_3$ respectively in equation (18.13) and substituting zero for all the shear stresses.

That is

$$(S_n)_{oct} = \frac{1}{3}\sigma_1 + \frac{1}{3}\sigma_2 + \frac{1}{3}\sigma_3$$

$$= \frac{1}{3}(\sigma_1 + \sigma_2 + \sigma_3) \quad \text{...(18.28)}$$

Similarly from equations (18.9), (18.10) and (18.11), we have

$$(S_x)_{oct} = \frac{1}{3}\sigma_1 + \frac{1}{3}\sigma_2 + \frac{1}{3}\sigma_3 \quad \text{...(18.29a)}$$

$$(S_y)_{oct} = \frac{1}{\sqrt{3}}\sigma_2 \quad \text{...(18.29b)}$$

$$(S_z)_{oct} = \frac{1}{\sqrt{3}}\sigma_3 \qquad \text{...(18.29c)}$$

Then the resultant stress on the octahedral plane is given by

$$(S)^2_{oct} = (S_x)^2_{oct} + (S_y)^2_{oct} + (S_z)^2_{oct}$$

$$= \frac{1}{3}\sigma_1^2 + \frac{1}{3}\sigma_2^2 + \frac{1}{3}\sigma_3^2$$

or $$(S_s)^2_{oct} = (S)^2_{oct} - (S_n)^2_{oct} \qquad \text{...(18.30)}$$

Thus using equations (18.14*a*), (18.28) and (18.30) we get the shear stress on the octahedral plane

$$(S_s)^2_{oct} = (S)^2_{oct} - (S_n)^2_{oct}$$

$$= \frac{1}{3}(\sigma_1^2 + \sigma_2^2 + \sigma_3^2) - \frac{1}{9}(\sigma_1 + \sigma_2 + \sigma_3)^2$$

$$= \frac{1}{9}[2(\sigma_1^2 + \sigma_2^2 + \sigma_3^2) - 2(\sigma_1\sigma_2 + \sigma_2\sigma_3 + \sigma_3 + \sigma_1)]$$

$$= \frac{1}{9}[(\sigma_1 - \sigma_2)^2 - (\sigma_2 - \sigma_3)^2 + (\sigma_3 - \sigma_1)^2]$$

or $$(S_s)_{oct} = \frac{1}{3}\sqrt{(\sigma_1 - \sigma_2)^2 + (\sigma_2 - \sigma_3)^2 + (\sigma_3 - \sigma_1)^2} \qquad \text{...(18.31)}$$

**Example 18.1.** *In a tri-axial stress system the six components of the stress at a point are given below :*

$$\sigma_x = 6 \text{ MN/m}^2$$
$$\sigma_y = 5 \text{ MN/m}^2$$
$$\sigma_z = 4 \text{ MN/m}^2$$
$$\sigma_{xy} = \sigma_{yx} = 1 \text{ MN/m}^2$$
$$\sigma_{yz} = \sigma_{zy} = 3 \text{ MN/m}^2$$
$$\sigma_{zx} = \sigma_{xz} = 2 \text{ MN/m}^2$$

*Find the magnitudes of three principal stresses and the direction cosines of their lines of action.*

**Solution.** The three principal stresses will be the roots of the equation (18.27).

$$S^3 - I_1 S^2 - I_2 S - I_3 = 0 \qquad \text{...(18.27)}$$

where $$I_1 = \sigma_x + \sigma_y + \sigma_z$$

$$= 6 + 5 + 4 = 15$$

$$I_2 = -(\sigma_x\sigma_y + \sigma_y\sigma_z + \sigma_z\sigma_z) + \tau_{xy}^2 + \tau_{yz}^2 + \tau_{zx}^2$$

$$= -(30 + 20 + 24) + 1 + 9 + 4$$
$$= -60$$
$$I_3 = \sigma_x\sigma_y\sigma_z + 2\tau_{xy}\tau_{yz}\tau_{zx} - \sigma_x\tau_{yz}^2 - \sigma_y\tau_{zx}^2 - \sigma_z\tau_{zy}^2$$
$$= 6 \times 5 \times 4 + 2 \times 1 \times 3 \times 2 - 6(3)^2 - 5(2)^2 - 4(1)^2$$
$$= 120 + 12 - 54 - 20 - 4$$
$$= -54$$

Thus equation (18.27) reduces to

$$S^3 - 15S^2 + 60\,S - 54 = 0 \quad \text{...(18.32)}$$

The solution of this cubic equation can either be done by hit and trial method or by the exact method given below.

Let $S$ be replaced by $\left(S_1 + \frac{1}{3}I_1\right)$ in equation (18.27) or (18.32) where $S_1$ is unknown *i.e.*,

$$S = S_1 + \frac{1}{3}(15)$$
$$= S_1 + 5 \quad \text{...(18.33)}$$

In the light of this substitution equation (18.32) reduces to

$$(S_1 + 5)^3 - 15\,(S_1 + 5)^2 + 60\,(S_1 + 5) - 54 = 0$$

on simplification this reduces to

$$S_1^3 - 15\,S_1 - 4 = 0 \quad \text{...(18.34)}$$

It may be noted that in equation (18.34) there is no term having $S_1^2$, this actually is the result of substitution made in equation (18.33). Equation (18.34) can easily be solved by comparing it with the well known identity

$$\cos 3\theta = 4\cos^3\theta - 3\cos\theta$$

or
$$\cos^3\theta - \frac{3}{4}\cos\theta - \frac{1}{4}\cos 3\theta = 0 \quad \text{...(18.35)}$$

Again put $S_1 = r\cos\theta$ in equation (18.34) to give

$$r^3\cos^3\theta - 15\,r\cos\theta - 4 = 0$$

or
$$\cos^3\theta - \frac{15}{r^2}\cos\theta - \frac{4}{r^3} = 0 \quad \text{...(18.36)}$$

Comparing equations (18.35) and (18.36), they are same if

$$\frac{15}{r^2} = \frac{3}{4}$$

*i.e.*, $$r = \sqrt{20} = 4.47 \qquad \text{...(18.37}a\text{)}$$

and $$\frac{4}{r^3} = \frac{1}{4}\cos 3\theta$$

or $$\cos 3\theta = \frac{16}{r^3} = \frac{16}{(4.47)^3} = 0.179 \qquad \text{...(18.37}b\text{)}$$

Equation (18.37*b*) will give three values of θ given below

$$\theta_1 = 26.6°$$
$$\theta_2 = 93.4°$$
$$\theta_3 = 146.6°$$

The corresponding three values of $S_1$ will be

$$(S_1)_{\theta 1} = r\cos\theta_1 = 4.47\cos 26.6° = 4.0$$
$$(S_1)_{\theta 2} = r\cos\theta_2 = 4.47\cos 93.4\theta = -0.27$$
$$(S_1)_{\theta 3} = r\cos\theta_3 = 4.47\cos 146.6° = -3.73$$

Thus from equation (18.33) the three principal stresses are

$$\sigma_1 = 5 + (S_1)_{\theta 1} = 5 + 4 = 9.0 \text{ MN/m}^2$$
$$\theta_2 = 5 + (S_1)_{\theta 2} = 5 - 0.27 = 4.73 \text{ MN/m}^2$$
$$\theta_3 = 5 + (S_1)_{\theta 3} = 5 - 3.73 = 1.27 \text{ MN/m}^2$$

Direction cosines of the line of action of $\sigma_1$ are found with the help of equations (18.15), (18.16) and (18.17) by substituting $\sigma_1$ for *S*.

Thus $$l(9-6) - m(1) - n(2) = 0 \qquad \text{...(18.38}a\text{)}$$
$$-l(1) - m(9-5) - n(3) = 0 \qquad \text{...(18.38}b\text{)}$$
$$-l(2) - m(3) + n(9-4) = 0 \qquad \text{...(18.38}c\text{)}$$

or $$3l - m - 2n = 0 \qquad \text{...(18.39}a\text{)}$$
$$-l + 4m - 3n = 0 \qquad \text{...(18.39}b\text{)}$$
$$-2l - 3m + 5n = 0 \qquad \text{...(18.39}c\text{)}$$

From equations (18.39*a*) and (18.39*b*)

$$\frac{l}{3+8} = \frac{m}{2+9} = \frac{n}{12-1}$$

or $$\frac{l}{1} = \frac{m}{1} = \frac{n}{1}$$

Further $$l^2 + m^2 + n^2 = 1$$

*i.e.*, $$l = m = n = \frac{1}{\sqrt{3}}$$

Thus the direction cosines of $\sigma_1$ are $\left(\frac{1}{\sqrt{3}}, \frac{1}{\sqrt{3}}, \frac{1}{\sqrt{3}}\right)$.

Similarly for $S = \sigma_2 = 4.73$ we have from equations (18.15) and (18.16) for its direction cosines

$$l\,(4.73 - 6) - m\,(1) - n\,(2) = 0 \quad \text{...(18.40}a\text{)}$$

$$-\,l(1) + m\,(4.73 - 5) - n\,(3) = 0 \quad \text{...(18.40}b\text{)}$$

or

$$1.27\,l + m + 2n = 0$$

$$l + 0.27m + 3n = 0$$

Therefore

$$\frac{l}{1 \times 3 - 2 \times .27} = \frac{m}{2 \times 1 - 1.27 \times 3} = \frac{n}{1.27 \times .27 - 1 \times 1}$$

or $$\frac{l}{2.46} = \frac{m}{-1.81} = \frac{n}{-.657}$$

or $$m = -\,.7358\,l$$

and $$n = -\,.2671\,l$$

Also $\quad l^2 + m^2 + n^2 = 1$

or $$l^2\,[1 + (-\,.7358)^2 + (-\,.2671)^2] = 1$$

or $$l = .7874$$

Therefore $$m = -\,.7358 \times .7874 = -\,.5794$$

and $$n = -\,.2671 \times .7874 = -\,.2103$$

For the third principal stress

$$S = \sigma_3 = 1.27$$

we have from equations (18.15) and (18.16) for its direction cosines

$$l\,(1.27 - 6) - m\,(1) - n\,(2) = 0 \quad \text{...(18.41}a\text{)}$$

$$-\,l\,(1) + m\,(1.27 - 5) - n\,(3) = 0 \quad \text{...(18.41}b\text{)}$$

or

$$4.73\,l + m + 2n = 0$$

$$l + 3.73m + 3n = 0$$

Therefore

$$\frac{l}{1 \times 3 - 2 \times 3.73} = \frac{m}{2 \times 1 - 3 \times 4.73} = \frac{n}{4.73 \times 3.73 - 1 \times 1}$$

or $$\frac{l}{-4.46} = \frac{m}{-12.19} - \frac{n}{16.64}$$

or $$m = 2.733\ l$$

and $$n = -\ 3.731\ l$$

Also $$l^2 + m^2 + n^3 = 1$$

or $$l^2\ (1 + 2.733^2 + 3.7312) = 1$$

or $$1 = 0.211$$

Therefore $$m = 2.733 \times .211 = 0.5667$$

$$n = -\ 3.731 \text{ times } .211 = -\ 0.7872.$$

## 18.5 Mohr's Circles for Tri-axial Stress System

Let the principal stresses $\sigma_1$, $\sigma_2$ and $\sigma_3$ be known at a point in a three dimensional stress system. The axes of reference $OX$, $OY$ and $OZ$ are taken along the three rincipal stresses. Then the direct stress, $\sigma$, on a doubly oblique plane whose normal has direction cosines ($l$, $m$, $n$), is given by equation (18.13) by replacing $\sigma_x$, $\sigma_y$ and $\sigma_z$ with $\sigma_1$, $\sigma_2$ and $\sigma_3$ respectively and substituting zero for the shear stress components.

*i.e.,* $$\sigma = l^2\sigma_1 + m^2\sigma_2 + n^2\sigma_3 \qquad ...(18.42)$$

Similarly the total shear stress, $\tau$, on this plane is given by substituting equations (18.9), (18.10), (18.11) into equation (18.14*b*).

$$\tau^2 = l^2\sigma_1^2 + m^2\sigma_2^2 + n^2\sigma_3^2 - (l^2\sigma_1 + m^2\sigma_2 + n^2\sigma_3)^2 \qquad ...(18.43)$$

Also $$l^2 + m^2 + n^2 = 1 \qquad ...(18.44)$$

The values of $\sigma$ and $\tau$ can be calculated from equations (18.42) and (18.43) or alternatively these can be found by drawing what are known as Mohr's circles. The method to draw the Mohr's circle for tri-axial stress system is developed below :

From equations (18.42) and (18.44), we get

$$\sigma = l^2\sigma_1 + \sigma_2\ (1 - l^2 - n^2) + n^2\sigma_3 \qquad ...(18.45)$$

or $$n^2 = \frac{\sigma + l^2\ (\sigma_2 - \sigma_1) - \sigma_2}{(\sigma_3 - \sigma_2)} \qquad ...(18.46)$$

Similarly from equations (18.43) and (18.44)

$$\tau^2 = l^2(\sigma_1^2 - \sigma_2^2) + \sigma_2^2 + n^2\ (\sigma_3^2 - \sigma_2^2) - [l^2(\sigma_1 - \sigma_2) + \sigma_2 + n^2\ (\sigma_3 - \sigma_2)]^2 \qquad ...(18.47)$$

Substituting the value of $n^2$ from equation (18.46) in equation (18.47) and simplifying we get

$$l^2 = \frac{\tau^2 + (\sigma_2 - \sigma)(\sigma_3 - \sigma)}{(\sigma_2 - \sigma_1)(\sigma_3 - \sigma_1)} \qquad ...(18.48)$$

Similarly the expressions for $m^2$ and $n^2$ can be derived from equations (18.42), (18.43) and (18.44)

$$m^2 = \frac{\tau^2 + (\sigma_3 - \sigma)(\sigma_1 - \sigma)}{(\sigma_3 - \sigma_2)(\sigma_1 - \sigma_2)} \quad \text{...(18.49)}$$

and

$$n^2 = \frac{\tau^2 + (\sigma_1 - \sigma)(\sigma_2 - \sigma)}{(\sigma_1 - \sigma_3)(\sigma_2 - \sigma_3)} \quad \text{...(18.50)}$$

Equation (18.48) can be re-written in the form

$$(\sigma - \sigma_2)(\sigma - \sigma_3) + \tau^2 = l^2 (\sigma_1 - \sigma_2)(\sigma_1 - \sigma_3)$$

or

$$\left[\sigma - \frac{\sigma_2 + \sigma_3}{2}\right]^2 + \tau^2 = l^2 (\sigma_1 - \sigma_2)(\sigma_1 - \sigma_3) + \left(\frac{\sigma_2 - \sigma_3}{2}\right)^3 \quad \text{...(18.51a)}$$

or

$$\left[\sigma - \frac{\sigma_2 + \sigma_3}{2}\right]^{-2} + \tau^2 = R_{23}^2 \quad \text{...(18.51b)}$$

where

$$R_{23} = \sqrt{l^2(\sigma_1 - \sigma_2)(\sigma_1 - \sigma_3) + \left(\frac{\sigma_2 - \sigma_3}{2}\right)^2} \quad \text{...(18.52)}$$

Thus if $\sigma$ and $\tau$ are plotted on a graph paper with $\sigma$ along the $x$-axis and $\tau$ along the $y$-axis, equation (18.51$b$) represents the locus of the point $(\sigma, \tau)$ co-ordinates of which are the direct and shear stresses on the doubly oblique plane whose normal has direction cosines as $(l, m, n)$. Evidently the equation (18.51$b$) is the equation of a circle of radius $R_{23}$ and its centre $C_{23}$, having coordinates $\left(\frac{\sigma_2 + \sigma_3}{2}, 0\right)$ lying on the $x$-axis.

Similarly equations (18.50) and (18.49) represent separately two circles with centres $C_{12}$ and $C_{13}$, having coordinates $\left(\frac{\sigma_1 + \sigma_2}{2}, 0\right)$ and $\left(\frac{\sigma_3 + \sigma_1}{2}, 0\right)$ respectively. The corresponding radii of these circles will be

$$R_{12} = \sqrt{n^2 (\sigma_3 - \sigma_1)(\sigma_3 - \sigma_2) + \left(\frac{\sigma_1 - \sigma_2}{2}\right)^2} \quad \text{...(18.53)}$$

$$R_{13} = \sqrt{m^2 (\sigma_2 - \sigma_3)(\sigma_2 - \sigma_1) + \left(\frac{\sigma_1 - \sigma_3}{2}\right)^2} \quad \text{...(18.54)}$$

The respective equations of the circles will be

$$\left[\sigma - \frac{\sigma_1 + \sigma_2}{2}\right]^2 + \tau^2 = R_{12}^3 \quad \text{...(18.55)}$$

and

$$\left[\sigma - \frac{\sigma_1 + \sigma_3}{2}\right]^2 + \tau^2 = R_{13}^2 \quad \text{...(18.56)}$$

## 18.5.1 Construction of Three-dimensional Mohr's Circles

Three dimensional Mohr's circles can be drawn if the principal stresses are known. Draw two axes *OX* and *OY* Fig. 18.7. Along the *x*-axis is plotted the direct stress $\sigma$ and along *y*-axis the total shear stress $\tau$. Towards the right hand side of the $\sigma$-axis are taken the tensile stresses and on the left compressive direct stresses are ed. the points $P_1$, $P_2$ and $P_3$ on the *x*-axis where

$$OP_1 = \sigma_1,\ OP_2 = \sigma_2 \text{ and } OP_3 = \sigma_3.$$

Draw circles on $P_1P_2$, $P_2P_3$ and $P_3P_1$ as diameters. Let their respective centres be $C_{12}$, $C_{23}$ and $C_{13}$. The coordinates of the centres as calculated above are

$$\left(\frac{\sigma_1+\sigma_2}{2}, 0\right), \left(\frac{\sigma_2+\sigma_3}{2}, 0\right) \text{ and } \left(\frac{\sigma_3+\sigma_1}{2}, 0\right)$$

Through $P_1$, $P_2$ and $P_3$ draw lines $P_1A_1$, $P_2A_2$ and $P_3A_3$ respectively parallel to the $\tau$-axis. From $P_1$ draw a line $P_1Q_3Q_2$ at an angle $\alpha$ to $P_1A_1$ where $\alpha = \cos^{-1}(l)$ to intersect the circles on $P_1P_2$ and $P_1P_3$ at $Q_3$ and $Q_2$ respectively. (It will be shown in the proof of this construction that $Q_3$ and $Q_2$ are equidistant from $C_{23}$). With $C_{13}$, as centre and $Q_2C_{13}$ as radius draw the circular are $Q_2PQ_3$, which corresponds to equation (18.51).

Similarly starting from the expressions for $m^2$ (equation 18.49) and $n^2$ (equation 18.50), it can be proved that the point ($\sigma$, $\tau$) will also lie on the following two circles.

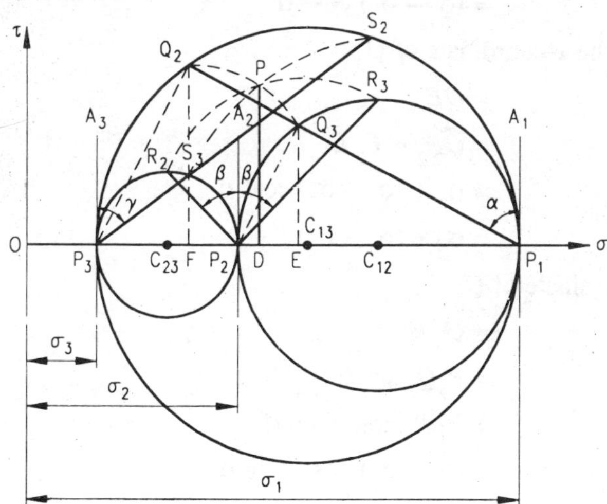


*Fig. 18.7. Mohr's circle for tri-axial stress system.*

(1) Circle $R_2PR_3$ with centre $C_{13}$ and radius $C_{13}R_2$ or $C_{13}R_3$ (this will be proved later) where $R_2$ and $R_3$ are the points of intersection of the lines drawn from the point $P_2$ at angle $\beta$ (where $\beta = \cos^{-1}$ m to $P_2A_2$ with the circle which touch each other at $P_2$ *i.e.*, circles on $P_2P_1$ and $P_2P_3$ as diameters.

(2) Circle $S_2PS_3$ with centre $C_{12}$ and radius $C_{12}S_2$ or $C_{12}S_3$ (this will be proved later) where $S_2$ and $S_3$ are the points of intersection of a line drawn from $P_3$ at angle $\gamma$ (where $\gamma = \cos^{-1} n$ to $P_3A_3$, with the circles which touch at $P_3$ *i.e.*, circles on $P_3P_2$ and $P_3P_1$.

The three circles $Q_2PQ_3$, $R_2PR_3$ and $S_2PS_3$ will all pass through the point $P$ having coordinates $(\sigma, \tau)$. Thus the point of intersection of any of these two circles $Q_2PQ_3$, $R_2PR_3$ and $S_2PS_3$ will decide the location of the point $P$ which will represent the plane whose normal has direction cosines as $(l, m, n)$. The measurement of the ordinates of $P$, $OD$ and $DP$ will give the values of the direct and total shear stresses on the plane in question.

## 18.5.2 Proof for Construction

Draw $Q_3E$ perpendicular to $P_2P_1$ and join $P_2Q_3$. In the right angled triangle $P_2P_1Q_3$

$$\angle P_1P_2Q_3 = \alpha$$

and $$P_2Q_3 = P_1P_2 \cos \alpha$$

Again in the right angled triangle $P_2EQ_3$

$$P_2E = P_2Q_3 \cos \alpha$$
$$= P_1P_2 \cos^2 \alpha$$
$$= (\sigma_1 - \sigma_2) \cos^2 \sigma \qquad \text{(by construction)}$$

Therefore the $x$-coordinate of $Q_3$

$$= OE$$
$$= OP_2 + P_2E$$
$$= \sigma_2 + (\sigma_1 - \sigma_2) \cos^2 \sigma$$
$$= \sigma_2 + (\sigma_1 - \sigma_2) l^2 \qquad \text{...(18.57)}$$

And $y$ co-ordinate of $Q_3$

$$= Q_3E$$
$$= P_2Q_3 \sin \alpha$$
$$= P_1P_2 \cos \alpha \sin \sigma$$
$$= (\sigma_1 - \sigma_2) \cos \alpha \sin \alpha$$
$$= (\sigma_1 - \sigma_2)\, l \sqrt{1 - l^2} \qquad \text{...(18.58)}$$

Also the coordinates of $C_{23}$ are

$$\left[\sigma_3 + \frac{1}{2}(\sigma_2 - \sigma_3), 0\right] \text{ or } \left[\frac{\sigma_2 + \sigma_3}{2}, 0\right] \qquad \text{...(18.59)}$$

Hence from equations (18.57) through (18.59)

$$(C_{23}Q_3)^2 = \left[\sigma_2 + (\sigma_1 - \sigma_2)l^2 - \frac{\sigma_2 + \sigma_3}{2}\right]^2 + \left[(\sigma_1 - \sigma_2)l\sqrt{1 - l^2} - 0\right]^2$$

$$= \left[\frac{\sigma_2 + \sigma_3}{2} + (\sigma_1 - \sigma_2)l^2\right]^2 + (\sigma_1 - \sigma_2)^2 l^2 (1 - l^2)$$

$$= \left(\frac{\sigma_2 - \sigma_3}{2}\right)^2 + (\sigma_2 - \sigma_3)(\sigma_1 - \sigma_2)l^2 + (\sigma_1 - \sigma_2)^2 l^4$$

$$+ (\sigma_1 - \sigma_2)^2 l^2 - (\sigma_1 - \sigma_2)^2 l^4$$

$$= \left(\frac{\sigma_2 - \sigma_3}{2}\right)^3 + (\sigma_1 - \sigma_2)l^2 [(\sigma_2 - \sigma_3) + (\sigma_1 - \sigma_2)]$$

$$= \left(\frac{\sigma_2 - \sigma_3}{2}\right)^3 + l^2 (\sigma_1 - \sigma_1)(\sigma_2 - \sigma_3)$$

$$= R_{23}^2 \qquad \text{[From 18.52)}$$

or $\quad C_{23}Q_3 = R_{23}$ ...(18.60)

Similarly for finding the coordinates of the point $Q_2$ join $Q_2 P_3$ and draw $Q_2F$ perpendicular to $P_3 P_1$. In the right angled triangle $P_1Q_2P_3$

$$\angle P_1Q_2P_3 = \pi/2$$

and $\quad \angle P_1P_3Q_3 = \alpha$

Therefore $\quad P_3Q_2 = P_1P_3 \cos\alpha$

$$= (\sigma_1 - \sigma_3)\cos\alpha$$

Again in right angled triangle $P_3Q_2F$

$$P_3F = P_3Q_2 \cos\sigma$$

$$= (\sigma_1 - \sigma_3)\cos^2\sigma$$

$$= (\sigma_1 - \sigma_3)\, l^2$$

Also $\quad Q_2F = P_3Q_2 \sin\alpha$

$$= (\sigma_1 - \sigma_3)\cos\alpha \sin\alpha$$

$$= (\sigma_1 - \sigma_3)\, l\sqrt{1 - l^2}$$

Therefore the $x$ coordinate of $Q_2$

$$= OF$$

$$= OP_3 + P_3F$$

$$= \sigma_3 + (\sigma_1 - \sigma_3)\, l^2 \qquad \text{...(18.61)}$$

And $\quad y$ co-ordinate of $Q_2 = Q_2F$

$$= (\sigma_1 - \sigma_3)\, l\sqrt{1 - l^2} \qquad ...(18.62)$$

Hence from equations (18.59), (18.61) and (18.62), we get

$$(C_{23}Q_2)^2 = \left[\sigma_3 + (\sigma_1 - \sigma_3)\, l^2 - \frac{\sigma_2 + \sigma_3}{2}\right]^2 + \left[(\sigma_1 - \sigma_3)\, l\sqrt{1 - l^2}\right]^2$$

$$= \left[\frac{\sigma_3 - \sigma_2}{2} - (\sigma_1 - \sigma_3)\, l^2\right] + (\sigma_1 - \sigma_2)^2\, l^2\,(1 - l^2)$$

$$= \left(\frac{\sigma_3 - \sigma_2}{2}\right)^2 + (\sigma_3 - \sigma_2)(\sigma_1 - \sigma_3)\, l^2 + (\sigma_1 - \sigma_3)^2\, l^4$$

$$+ (\sigma_1 - \sigma_3)^2\, l^2 - (\sigma_1 - \sigma_3)^2\, l^4$$

$$= \left(\frac{\sigma_3 - \sigma_2}{2}\right)^2 + l^2\,(\sigma_1 - \sigma_2)[(\sigma_3 - \sigma_2) + (\sigma_1 - \sigma_3)]$$

$$= \left(\frac{\sigma_2 - \sigma_3}{2}\right)^2 + l^2\,(\sigma_1 + \sigma_3)(\sigma_1 - \sigma_2)$$

$$= R_{23}^2 \qquad \text{[From eqn. 18.52]}$$

or $$C_{23}Q_2 = R_{23} \qquad ...(18.63)$$

Thus from equations (18.51*b*), (18.60) and (18.63) it is concluded that equation (18.51*b*) represents the circle $Q_2\,PQ_3$ with centre at $C_{23}$ and radius equal to $R_{23}$. Similarly starting from expressions for $m^2$ and $n^2$ *i.e.*, equations (18.49) and (18.50) it can be shown the point $(\sigma, \tau)$ also lies on circle $R_2PR_3$ with centre at $C_{13}$ and radius $R_{13}$ as well as on circle $S_2\,PS_3$ with centre at $C_{12}$ and radius $R_{12}$. This proves the construction.

### 18.5.3 Summary of Construction of Three Dimensional Mohr's Circles

The three dimensional Mohr's circles can be constructed only if the three principal stresses $\sigma_1$, $\sigma_2$ and $\sigma_3$ are known. The steps to be taken to construct the Mohr's circles and to locate the position of the point which will represent the plane with normal having direction cosines (*l, m, n*) are given below :

(*i*) Choose two perpendicular axes. Towards the right hand side of σ-axis measure the tensile direct stresses and towards the left compressive direct stresses. Along the τ-axis measure the shear stresses.

(*ii*) Plot on the σ-axis, the point $P_1$, $P_2$, $P_3$ such that

$$OP_1 = \sigma_1$$

$$OP_2 = \sigma_2$$

and $$OP_3 = \sigma_3$$

(*iii*) On $P_1P_2$, $P_2P_3$ and $P_1P_3$ as diameters draw circles with centres $C_{12}$, $C_{23}$ and $C_{13}$ respectively.

(*iv*) At the points $P_1$, $P_2$, $P_3$ erect perpendiculars $P_1A_1$, $P_2A_2$, $P_3A_3$ respectively to the σ-axis.

(*v*) Draw a line $Q_2Q_3P_1$ at an angle α (= $\cos^{-1} l$) to $P_1A_1$ to intersect the two circles, which touch each other at $P_1$, at points $Q_2$ and $Q_3$. With $C_{23}$ (*i.e.*, centre of the circle other than the two which touch each other at $P_1$) as centre and radius equal to $C_{23}$ $Q_2$ or $C_{23}Q_3$ draw the circular are $Q_2PQ_3$ Fig. 18.7.

(*vi*) Draw a line $S_2S_3P_1$ at angle ν (= $\cos^{-1} n$) to $P_3A_3$ to intersect the two circles, which touch each other at $P_3$, at points $S_2$ and $S_3$. With $C_{12}$ (*i.e.*, centre of the circle other than the two which touch each other at $P_3$) as centre and radius equal to $C_{12}S_2$ roman or $C_{12}S_3$ draw the circular arc $S_2PS_3$ to intersect the arc $Q_2PQ_3$ in $P$.

(*vii*) From $P$ draw $PD$ perpendicular to the σ-axis, the normal stress on the plane is represented by $OD$ and the total shear stress by $PD$.

(*viii*) It may be noted that the direction of shear stress is not predicated by the Mohr's circle method.

## 18.6 The Shortcomings of Mohr's Circle for Three Dimensional Stress System

As has been seen in Chapter 2, the main practical utility of the Mohr's circle method is to facilitate the calculations for finding the two principal stresses at a point in a two dimensional stress system, when the normal and shear stresses on two known planes are known. However, in the case of three dimensional stress system it is not simply possible to use this method to obtain the three principal stresses. Rather the Mohr's circles for a tri-axial stress system can only be constructed if the three principal stresses are known. The second limitation of the method is that the magnitude of the total shear stress on some plane can be found only and its direction is not found by this method.

**Example 18.2.** *The three principal stresses at a point in a stressed material are 50 MN/m² tensile, 24 MN/m² tensile and 6 MN/m² tensile. By using Mohr's circle method find the direct and total shear stresses on a plane whose direction cosines with respect to the biggest and least stresses' axes are (0.5, 0.5).*

**Solution.** On some convient scale

$$OP_1 = 50 \text{ MN/m}^2$$

$$OP_2 = 24 \text{ MN/m}^2$$

and
$$OP_3 = 6 \text{ MN/m}^2$$

along the σ-axis, Draw circles on $P_1P_2$, $P_2P_3$, $P_3P_1$ as diameters with centres respectively at $C_{12}$, $C_{23}$ and $C_{13}$ Fig. 18.8. Erect perpendiculars $P_1A_1$, $P_2A_2$, $P_3A_3$ at $P_1$, $P_2$ and $P_3$

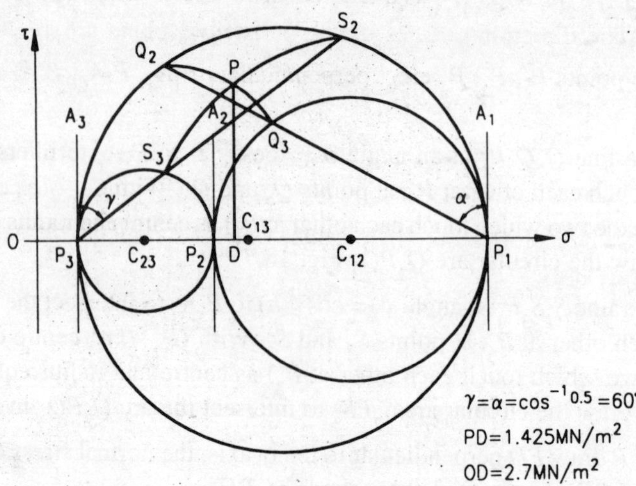


*Fig. 18.8*

respectively to the σ-axis. Draw $Q_2Q_3P_1$ at α (= $\cos^{-1}$ 0.5 = 60°) to $A_1P_1$ to intersect the circles with centres $C_{12}$ and $C_{13}$ in $Q_3$ and $Q_2$. Draw the circular arc $Q_2PQ_3$ with centre at $C_{23}$ and radius equal to $C_2Q_2$ or $C_{23}Q_3$. Similarly draw the line $S_2S_3P_3$ at ν ( = $\cos^{-1}$ 0.5 = 60°) to $A_3P_3$ as shown in Fig. 18.8 to intersect the circles with centres $C_{23}$ and $C_{13}$ in $S_3$ and $S_2$ respectively. Draw the circular arc $S_2PS_3$ with centre at $C_{12}$ and radius equal to $C_{12}S_2$ or $C_{12}S_3$ to intersect the arc $Q_2PQ_3$ in $P$. Draw $PD$ perpendicular to σ-axis. Then $OD$ represents the direct stress which is 2.7 MN/m² tensile and $PD$ the total shear stress on the given plane which is 1.425 MN/m².

**Example 18.3.** *In a three dimensional stress system, the three principal stresses at a point are 8, 4, and – 2 MN/m². Find with the help of Mohr's circles method the direct and total shear stresses on the octahedral plane and check the result by calculations.*

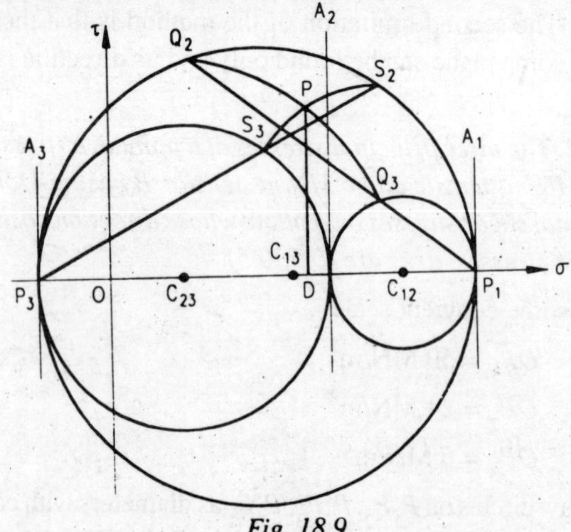


*Fig. 18.9*

**Solution.** The direction cosines of the octahedral plane are $\left(\frac{1}{\sqrt{3}}, \frac{1}{\sqrt{3}}, \frac{1}{\sqrt{3}}\right)$ which give

$$\alpha = \beta = \gamma = \cos^{-1}\frac{1}{\sqrt{3}} = 55°$$

The Mohr's circles for this problem can be drawn as explained in example 18.2. It is drawn to scale in Fig. 18.9 from which

$$\alpha_{oct} = OD = 3.33 \text{ MN/m}^2$$

and

$$\alpha_{oct} = PD = 4.1 \text{ MN/m}^2.$$

(*b*) By calculations : From equations (18.28) and (18.31)

$$\alpha_{oct} = (S_n)_{oct} = \frac{1}{3}(\sigma_1 - \sigma_2 + \sigma_3)$$

$$= \frac{1}{3}(8 + 4 - 2)$$

$$= \frac{10}{3} = 3.33 \text{ MN / m}^2$$

and

$$\tau_{oct} = (S_s)_{oct} = \frac{1}{3}[(\sigma_1 - \sigma_2)^2 + (\sigma_2 - \sigma_3)^2 + (\sigma_3 - \sigma_1)^2]^{1/2}$$

$$= \frac{1}{3}[(8 - 4)^2 + (4 + 2)^2 + (-2 - 8)^2]^{1/2}$$

$$= \frac{1}{3}[16 + 36 + 100]^{1/2}$$

$$= \frac{1}{3}(152)^{1/2} = 4.11 \text{ MN / m}^2.$$

## REFERENCE

Derrington, M.G. and Johnson, W. 'The shortcoming of Mohr's circle for Three-Dimensional Stress'. Bulletin of Mech. Engg. Education No. 18, 1960.

## PROBLEMS FOR PRACTICE

1. In a three-dimensional stress system the six components of stress are

$\sigma_x = 4$ MN/m$^2$ $\quad$ $\tau_{xy} = \tau_{yx} = 2$ MN/m$^2$

$\sigma_y = 6$ MN/m$^2$ $\quad$ $\tau_{yz} = \tau_{zy} = 1$ MN/m$^2$

$\tau_y = 5$ MN/m$^2$ $\tau_{zx} = \tau_{xz} = 3$ MN/m$^2$

Calculate the magnitude of the three principal stresses.

$$\left[\textbf{Ans. } 9, \left(3+\sqrt{3}\right), \left(3-\sqrt{3}\right) \text{MN}/\text{m}^2\right]$$

2. The three principal stresses at a point in a stressed material are $\alpha_1 = 54$ MN/m$^2$ parallel to *OX*, $\alpha_2 = 24.5$ MN/m$^2$ parallel to *OY* and $\alpha_3 = 6.5$ MN/m$^2$ parallel to *OZ*. Find by Mohr's circle method the normal and total shear stresses on a plane whose direction cosines are (0.530, 0695, 0.500).

[**Ans.** 28.0 MN/m$^2$ ; 17.7 MN/m$^2$]

3. Find graphically by use of Mohr's circles the normal and shearing stresses on an oblique plane in a tri-axial stress system having the following principal stresses : $\sigma_1 = 8$ MN/m$^2$, $\sigma_2 = 4$ MN/m$^2$, $\sigma_3 = -2$ MN/m$^2$

   The normal to the oblique plane is inclined at 55° with the $\sigma_1$-axis and 60° with $\sigma_2$-axis. [**Ans.** 2.75 MN/m$^2$; 4.4 MN/m$^2$]

4. Find graphically with the help of Mohr's circles (*a*) the maximum shearing stress and (*b*) the normal and shearing stresses on the octahedral plane for the following state of stress :

   $\sigma_1 = 30$ MN/m$^2$; $\sigma_2 = -10$ [MN/m$^2$ ; $\sigma_3 = -40$ MN/m$^2$ check the results by calculations. [**Ans.** 35 MN/m$^2$ ; – 6.67 MN/m$^2$ ; 28.67 MN/m$^2$]

Chapter–19

# Torsion of Non-Circular Shafts

## 19.1 Introduction

In developing the theory of torsion of circular shafts we had assumed that the plane sections normal to the axis remain plane even after the application of torque, and consequently the shear stresses in the normal cross-section were directed tangentially. However for shafts of non-circular shafts, still having the uniform section all along the length of shaft, it is no longer possible to prove that the plane normal cross-sections remain plane or that they remain undistorted in their own plane.

Let a shaft of non-circular section, which is uniform throughout its length, be under the action of a twisting moment $M_t$ about the longitudinal axis $ZZ$ passing through the centres of twist of all the sections Fig. 19.1. Let $O$ be the centre of twist of a section and a point $P$ lies very near to the non-circular boundary. Around the point $P$ an elemental cuboid is considered. Let the shear stress $\tau$ be acting on the cross-section at

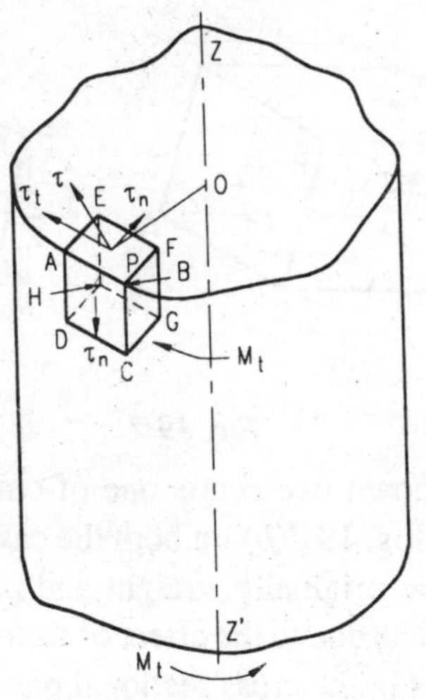


*Fig. 19.1*

the peripheral point $P$ in a direction perpendicular to $OP$. The shear stress $\tau$ can then be resolved into two components $\tau_t$ and $\tau_n$ tangential and normal to the periphery at $P$. From the Fig. 19.1 it is evident the normal component on the face $ABFE$ which lies in the cross-sectional plane, will have its complementary shear $\tau_n$ acting on the face *ABCD* which lies in the free outside surface of the cylinder. But on the free surface shear stress does not exist. Hence the normal component $\tau_n$ must be absent. In other words $\tau$ and $\tau_t$ must be one and the same. The shear stresses on the cross-section thus must be tangent to the periphery of the cross-section. As a special case of this the shear stresses in the corners of a rectangular cross-section must be zero, because neither one of its two perpendicular components can exist. To explain it further the case of the torsion of a rectangular shaft is compared with that of a circular shaft below.

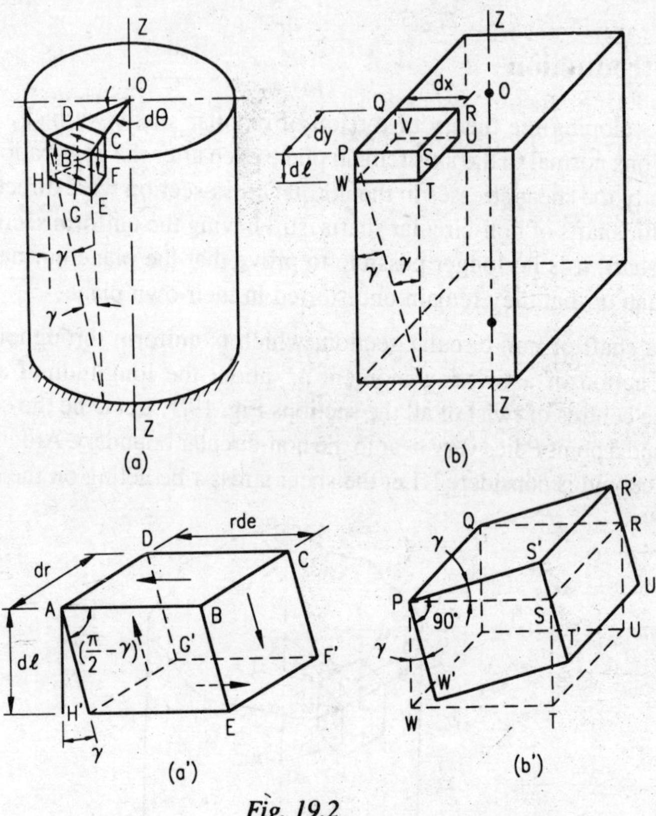


*Fig. 19.2*

In the Fig. 19.2 are shown two shafts one of circular section Fig. 19.2(*a*) and other of rectangular section Fig. 19.2(*b*). In both the cases longitudinal lines drawn on the outer surface which were originally straight and parallel to the axis of the shaft become spiral at a small angle $\gamma$ due to the effect of the twisting moment. In the case of circular shaft the shear stress in the cross sectional plane *ABCD* will be accompanied by the complementary shear stress on the faces *BCEF* and *ABGH*. Due to the action of these shear stresses the original element of sides *dr*, *rd* $\theta$ and *dl* in which all angles were 90° before the application of twisting moment, deform to the shape *ABCDE'F'G'H'* Fig. 19.2(*a*) in which the angle *H'AB* will be (90° – $\gamma$) and the face *ABCD* will continue to remain in the cross-sectional plane. However in the case of the rectangular shaft if the plane cross-sections should remain plane there would be shear stresses in the corner but as explained above this must be zero as it is impossible to have shear stress on the outside free surface. Thus the original rectangular element *PQRSTUVW* at the corner of a rectangular shaft, after the application of twisting moment will deform to the shape *PQR'S'T'U'V'W'* Fig. 19.2(*b*) in which the normal cross-section tilts up locally through an angle $\gamma$ to keep the angles of the element same at 90°as before so that no shear stress occurs. This means that the corner element of the area of cross-section is perpendicular to spiralled longitudinal edge, and since this must be the case at all the four edges, the plane cross-section no longer remains plane but becomes warped vertically.

With this introductory discussion in view we now proceed with the Saint-Venant's theory of twisting of non-circular shafts.

## 19.2 Saint-Venant's Theory of Twisting of Non-Circular Shafts

Saint-Venant made following assumptions for the development of the theory of torsion of shafts of uniform non-circular section.

(*i*) The stresses are within elastic limit and Hooke's law holds good.

(*ii*) Twist of any cross-section is proportional to its distance from the fixed end, which is taken as the reference plane.

(*iii*) Plane cross-sections turn bodily about the centre without distortion in the plane. However a cross-section warps in the longitudinal direction and this warping is different for different points of the section. It is further assumed that all cross-sections warp in the same manner.

(*iv*) There is no longitudinal load acting on the shaft.

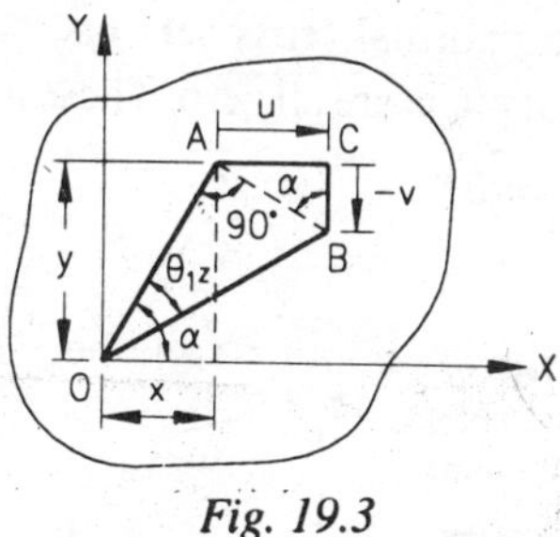


*Fig. 19.3*

Fig. 19.3 shows a cross-section of a non-circular shaft, at a distance $z$ from the fixed end (*Z*-axis is taken along the longitudinal axis of the shaft, which is also the axis of the twisting moment, which is considered to be clockwise). Let the *X* and *Y* axes be chosen as shown in the Fig. 19.3 with the centre of twist lying at *O*, which is the point of inter-section of the shaft axis with the cross-sectional plane Let $(x, y)$ be the co-ordinates of any point *A* of the section at a distance $z$ from the fixed end. Due to the action of the twisting moment the section turns bodily (assumption no. (*iii*)). Consequently the point $A$ turns clockwise about $O$ and occupies the position $B$ such that the angle $AOB$ is equal to $\theta_{1z}$, (assumption (*ii*)) where $\theta_1$ is the angle of twist of the shaft per unit length. Thus if $u$, $v$ and $w$ are the displacements along the three axes respectively of the point $(x, y, z)$ from the untwisted state $A$ to the twisted state $B$ Fig. 19.3 then these are given by

$$u = AC = AB \sin \alpha \qquad [\because \text{ angle } \theta_{1z} \text{ is small}]$$

$$= (OA \,.\, \theta_{1z}) \,.\, \left(\frac{y}{OA}\right)$$

...(19.1*a*)

or $$n = \alpha_{1z}\, y$$

and $$v = CB = -AB \cos \alpha$$

It may be noted that $v$ is negative as it is directed opposite to that of positive increasing $y$

Therefore $$v = -[OA \,.\, \theta_{1z}]\frac{x}{OA}$$

$$= -\theta_{1z}\, x \qquad \text{...(19.1}b\text{)}$$

and $$w = f(x, y) \qquad \text{(assumption } iii\text{) ...(19.1}c\text{)}$$

where $f(x, y)$ is some unknown function of $x$ and $y$.

We now proceed to express the displacements of equation (19.1) into the strains, which are the displacements of one point with respect to the neighbouring point. In general there will be two kinds of strains *viz.* direct strains $\varepsilon$ which are tensile or compressive deformations per unit length and shear strains $\gamma$, which are angular changes.

As per the assumption (*iii*), the equations (19.1*a*) and (19.1*b*) represent the condition of no distortion in the normal cross section. Thus the direct as well as the shear strains in the cross sectional plane are all zero. These are expressed by the equations

$$\varepsilon_x = \varepsilon_y = 0 \qquad \text{...(19.2}a\text{)}$$

$$\varepsilon_{xy} = 0 \qquad \text{...(19.2}b\text{)}$$

In addition according to the assumption (*iv*), there is no longitudinal tension in the shaft, for which we have

$$\varepsilon_z = 0 \qquad \text{...(19.2}c\text{)}$$

Thus out of the six components of the total strain corresponding to the six components of the stress (Chapter 18) we are left with only two strains *viz.* $\gamma_{xy}$ and $\gamma_{yz}$. In order to evaluate $\gamma_{xy}$ a section in an $x$—$z$ plane ($y$ = constant) which is parallel to the longitudinal axis of the shaft is taken and is shown in Fig. 19.4. In the figure is shown an element $EFGH$ which goes to $E'F'G'H'$ after the application of the twisting moment, and the two originally plane sections $EF$ and $HG$ (*i.e.*, sections $z$ and $z + dz$) become warped as shown by $E'F'$ and $H'G'$. As per the earlier notation taken in Fig. 19.3, the horizontal distance between $E$ and $E'$ is $u$. The horizontal distance between $H$ and $H'$ can be written as $(u + \partial u / \partial z \,.\, dz)$, as the points $E$ and $H$ have the same value of $x$ and the original distance between them is $dz$, as shown in Fig. 19.4.

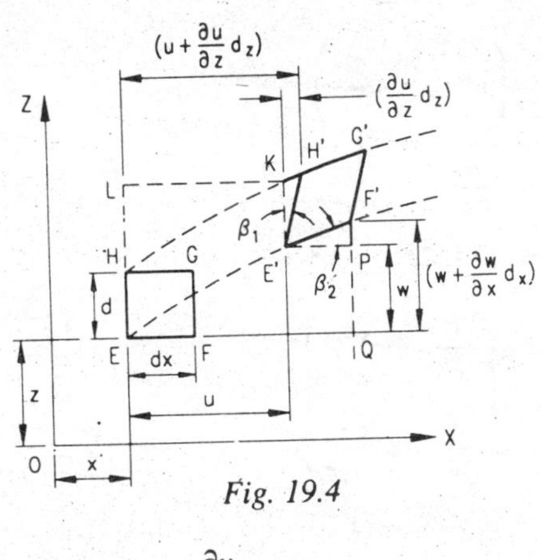


*Fig. 19.4*

Thus $$LH' = u + \frac{\partial u}{\partial z} dz$$

and $$KL = u$$

Therefore $$KH' = LH' - KL$$

$$= \left(u + \frac{\partial u}{\partial z}\right) - u$$

$$= \frac{\partial u}{\partial z} dz$$

Further if $$\beta_1 = \angle H'E'K$$

$$\simeq \frac{KH'}{E'H'}$$

$$= \left(\frac{\partial u}{\partial z} dz\right) \cdot \frac{1}{dz}$$

*i.e.* $$\beta_1 = \frac{\partial u}{\partial z} \qquad ...(19.3)$$

Similarly, the vertical distance between the points $E$ and $E'$ as per the notation in Fig. 19.3 is $w$. The vertical distance between $F$ and $F'$ can be written as $\left(w + \frac{\partial w}{\partial x} \,.\, dx\right)$ as the points $E$ and $F$ have the same value of $z$ and the original distance between them is $dx$, as shown in Fig. 19.4.

Thus $$F'Q = \left( w + \frac{\partial w}{\partial x} . dx \right)$$

and $$PQ = w$$

Therefore $$F'P = F'Q - PQ$$

$$= \left( w + \frac{\partial w}{\partial x} . dx \right) - w$$

$$= \frac{\partial w}{\partial x} . dx$$

Thus if $$\beta_2 = \angle F'E'P$$

$$\simeq \frac{F'P}{E'F'}$$

$$= \left( \frac{\partial w}{\partial x} . dx \right) \frac{1}{dx}$$

*i.e.* $$\beta_2 = \frac{\partial w}{\partial x} \qquad ...(19.4)$$

Hence the difference between the $\angle HEF$ and $\angle H'E'F'$ will be ( beta sub 1 + $\beta_2$) and by definition this is the shear angle $\gamma_{xz}$ in the $x - z$ plane. Thus we get from equations (19.3) and (19.4)

$$\gamma_{xz} = \beta_1 + \beta_2$$

or $$\gamma_{xz} = \frac{\partial u}{\partial z} + \frac{\partial w}{\partial x} \qquad ...(19.5)$$

Similarly taking section in the $y - z$ plane we arrive at the expression for the shear strain $\gamma_{yz}$ as

$$\gamma_{yz} = \frac{\partial v}{\partial z} + \frac{\partial w}{\partial y} \qquad ...(19.6)$$

Substituting equations (19.1) into equations (19.5) and (19.6), we get

$$\gamma_{xz} = \theta_1 y + \frac{\partial w}{\partial x} \qquad ...(19.7a)$$

$$\gamma_{xz} = - \theta_1 x + \frac{\partial w}{\partial x} \qquad ...(19.7b)$$

Also in the light of assumption (*i*), we have from Hooke's law, the shear stresses $\tau_{xz}$ and $\tau_{yz}$ related to the shear strains $\gamma_{xz}$ and $\gamma_{yz}$ through the modulus of rigidity $G$

$$\tau_{xz} = G \left( \theta_1 y + \frac{\partial w}{\partial x} \right) \qquad ...(19.8a)$$

$$\tau_{yz} = G\left(-\theta_1 x + \frac{\partial w}{\partial y}\right) \qquad \text{...(19.8b)}$$

For the double subscripted shear stress notation and the sign of the shear stresses, reference may be made to article 18.1. However for quick reference the signs of the stresses are shown in Fig. 19.5 with respect to three perpendicular axes which form a right handed system. It is clear that the shear stresses are not constant across any normal cross-section of the shaft and vary from point to point. The rate of variation of $\tau_{xz}$ along $x$-axis is taken as $\frac{\partial \tau_{xz}}{\partial x}$ and that of $\tau_{yz}$ along $y$-axis is taken as $\left(\frac{\partial \tau_{xz}}{\partial y}\right)$. Fig. 19.5 has then been drawn accordingly. This figure, also represents the free-body diagram of an element of a twisted bar. The forces on this element *ABCDEFG* due to the stresses shown will be in equilibrium as the element forms part of the twisted bar in equilibrium.

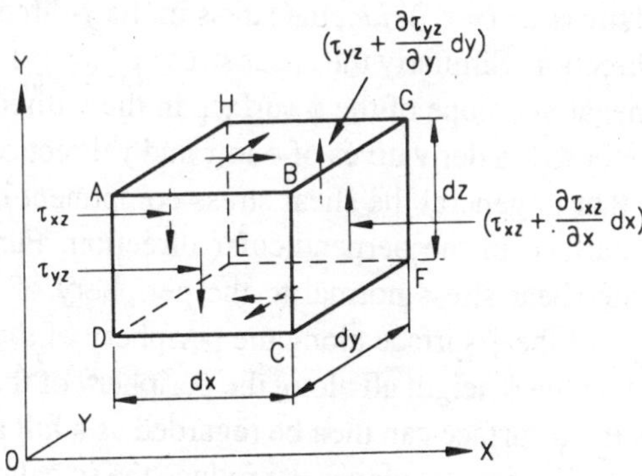


*Fig. 19.5*

Thus resolving the forces along $z$-axis we get the equilibrium equation

$$\left(\tau_{xz} + \frac{\partial \tau_{xz}}{\partial x}\right) d_y d_z - \tau_{xz} d_y d_z + \left(\tau_{yz} + \frac{\partial \tau_{xy}}{\partial y} + dy\right) dxdz = \tau_{yz}\, dxdz = 0$$

On simplification it reduces to

$$\frac{\partial \tau_{xz}}{\partial x} + \frac{\partial \tau_{yz}}{\partial y} = 0 \qquad \text{...(19.9)}$$

Equation (19.9) is a partial differential equation involving two unknown quantities $\tau_{xz}$ and $\tau_{yz}$, both of which are dependent on two variables $x$ and $y$. In order to facilitate the solution of equation (19.9), St. Venant assumed that there exists a function $\phi(x, y)$ such that the magnitudes of the two shear stresses $\tau_{yz}$ and $\tau_{xz}$ can be found by its partial differentiation with respect to $x$ and $y$ respectively *i.e.*, the function $\phi(x, y)$ is defined thus

$$\tau_{yz} = +\frac{\partial \phi}{\partial x} \qquad \text{...(19.10a)}$$

$$\tau_{yz} = -\frac{\partial \phi}{\partial y} \qquad ...(19.10b)$$

By substituting equation (19.10) into (19.9) we see that equation (19.9) is satisfied for any arbitrary function $\phi(x, y)$, provided that

$$\frac{\partial^2 \phi}{\partial x \partial y} = \frac{\partial^2 \phi}{\partial x \partial x} \qquad ...(19.11)$$

which is always true if $\phi$ is a continuous function. This function $\phi$ is called the 'stress function' of the problem and here it is called the 'Saint-Venant's torsion stress function.'

## 19.2.1 Shape of St. Venant's Torsion Stress Function

Since $\phi$ $(x, y)$ is a function of both $x$ and $y$, its value can be plotted vertically on the $x$, $y$ base and thus it will result in a curved surface. From the equations (19.10) it is clear that the shear stress $\tau_{yz}$ $(= \tau_{zy})$, *i.e.*, the stress in the $y$-direction is the slope of the $\phi$ surface in the $x$-direction. Similarly the shear stress $\tau_{xz}$ $(= \tau_{zx})$, that is the stress in the $x$-direction is the (negative) slope of the $\phi$ surface in the $y$-direction. Even though the equations (19.10) refer to the derivatives of $\phi$ in $x$ and $y$ directions, it will be proved in the next paragraph that in general the shear stress component in any direction equals the slope of the $\phi$ surface in the perpendicular direction. Further it was proved in article (19.1) that the shear stress normal to the periphery of the shaft in torsion in zero; hence the slope of the $\phi$ surface along the periphery of the shaft section must be zero, which means that the $\phi$ height all along the periphery of the shaft section must be constant. Evidently the $\phi$ surface can then be regarded as a hill and if we take sections of this hill by a series of horizontal planes to produce the so called 'contour lines', then shear stress will follow these contour lines. For the shaft of circular section, under torsion, the contour lines will be concentric circles and the $\phi$ hill will be a paraboloid of revolution.

**Proof.** We now proceed to prove the statement that in general the shear stress component in any direction equals the slope of the $\phi$ surface in the perpendicular direction. Fig. 19.6 shows the point $P$ in the $xy$ plane with shear stresses $\tau_{yz}$ and $\tau_{xz}$ acting on it. Draw through $P$ any arbitrary line $PN$ which is inclined to the $x$-axis at an

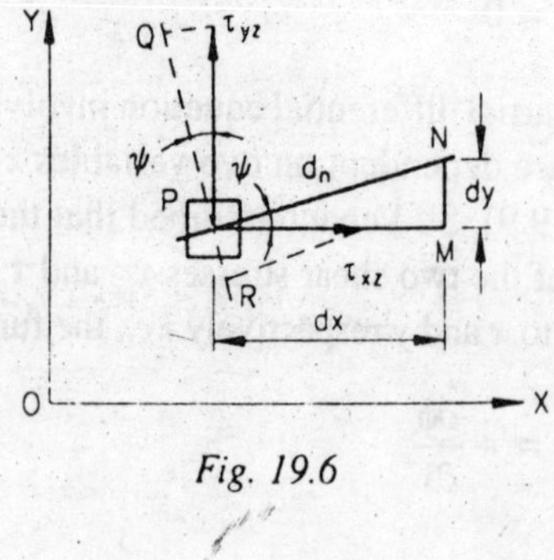


*Fig. 19.6*

angle $\psi$ and let $PN$ be of length dn, an element in the normal direction $PN$. Draw the line $QPR$ perpendicular to $PN$. Further $PQ$ and $PR$ are the components of $\tau_{yz}$ and $\tau_{xz}$ respectively along the line $QPR$. Then from the Fig. 19.6 we see that

$$dx = dn \cos \psi \qquad \text{...(19.12}a\text{)}$$

$$dy = dn \sin \psi \qquad \text{...(19.12}b\text{)}$$

For all the points on the hilly surface of the stress function $\phi$, we have in general

$$d\phi = \frac{\partial \phi}{dx} . dx + \frac{\partial \phi}{\partial y} dy$$

or

$$\frac{d\phi}{dn} = \frac{d\phi}{dx} \cdot \frac{dx}{dn} + \frac{\partial \phi}{\partial y} \cdot \frac{dy}{dn} \qquad \text{...(19.13)}$$

From equations (19.10), (19.12) and (19.13), we get

$$\frac{d\phi}{dn} = \tau_{yz} \cos \psi - \tau_{xz} \sin \psi \qquad \text{...(19.14)}$$

Also from Fig. 19.6

$$\tau_{yz} \cos \psi = PQ \qquad \text{...(19.15}a\text{)}$$

and

$$\tau_{xz} \sin \psi = PR \qquad \text{...(19.15}b\text{)}$$

Thus from equations (19.14) and (19.15), we have

$$\frac{d\phi}{dn} = PQ - PR$$

or $\frac{d\phi}{dn}$ = resultant stress component in the direction $QPR$ which is perpendicular to $PQ$.

But $d\phi/dn$ is the slope of the $\phi$ surface in the direction $PN$. Thus the statement made above is proved.

### 19.2.2 About Warping Function *w*

From equations (19.8) and (19.10), we have

$$\tau_{xz} = -\frac{\partial \phi}{\partial y} = G\left(\theta_1 y + \frac{\partial w}{\partial x}\right) \qquad \text{...(19.16}a\text{)}$$

$$\tau_{yz} = +\frac{\partial \phi}{\partial x} = G\left(-\theta_1 x + \frac{\partial w}{\partial y}\right) \qquad \text{...(19.16}b\text{)}$$

As defined earlier, in equations (19.16) $w$ is the warping of the cross-section. Even though the exact form of the function $w(x, y)$ is not known, yet it is certain that $w(x, y)$ and its derivatives are continuous functions of $x$ and $y$ *i.e.*, the warping function will have no sudden jumps. Mathematically it is expressed as

$$\frac{\partial^2 w}{\partial x \partial y} = \frac{\partial^2 w}{\partial y \partial x} \qquad \text{...(19.17)}$$

Further differentiating equations (19.16*a*) and (19.16*b*) partially with respect to *y* and *x* respectively, we have

$$\frac{\partial^2 \phi}{\partial y \partial x} = G\left(-\theta_1 + \frac{\partial^2 w}{\partial x \partial y}\right) \qquad \text{...(19.18)}$$

and
$$\frac{\partial^2 \phi}{\partial x^2} = G\left(-\theta_1 + \frac{\partial^2 w}{\partial x \partial y}\right) \qquad \text{...(19.19)}$$

Subtracting equation (19.19) from (19.18) in the light of equation (19.17), we get

$$-\frac{\partial^2 \phi}{\partial x \partial y} + \frac{\partial^2 \phi}{\partial y^2} = -2G\theta_1 \qquad \text{...(19.20)}$$

Again differentiating partially the equation (19.16*a*) with respect to *x* and equation (19.16*b*) with respect to *y*, gives

$$-\frac{\partial^2 \phi}{\partial x \partial y} = G\frac{\partial^2 w}{\partial x^2} \qquad \text{...(19.21)}$$

and
$$\frac{\partial^2 \phi}{\partial y \partial x} = G\frac{\partial^2 w}{\partial y^2} \qquad \text{...(19.22)}$$

Adding equations (19.21) and (19.22) in the light of equation (19.11), we get

$$\frac{\partial^2 w}{\partial x^2} + \frac{\partial^2 w}{\partial y^2} = 0 \qquad \text{...(19.23)}$$

Equation (19.23) is the required partial differential equation for finding the warping function ω.

Thus we see that equation (19.20) is the partial differential equation for the stress function ϕ. It may be noted that any arbitrary function ϕ which satisfies equations (19.9) and (19.10) will give the stresses in spite of the fact it may or may not satisfy the equation (19.20). In case the function ϕ does satisfy the equation (19.20), it will lead to the stresses which will correspond to continuous warping deformations ω and on the other hand if the function ϕ is such that it does not satisfy equation (19.20), it will certainly give the stresses from equilibrium equations (19.9) and (19.10) but the corresponding warping function will be discontinuous. If for some given shape of the cross section we can find the function ϕ which satisfies the equation (19.20) and which also satisfies the boundary condition of ϕ being along the periphery of any section, then the stresses derived from that will be the true solution of the torsion problem.

### 19.2.3 Torque Transmitted by the Shaft

In Fig. 19.7 is shown an element $dx \times dy$ of the cross-section located at a point $P$ whose co-ordinates are $(x, y)$. The shear stresses $\tau_{xz}$ and $\tau_{yz}$ are also shown

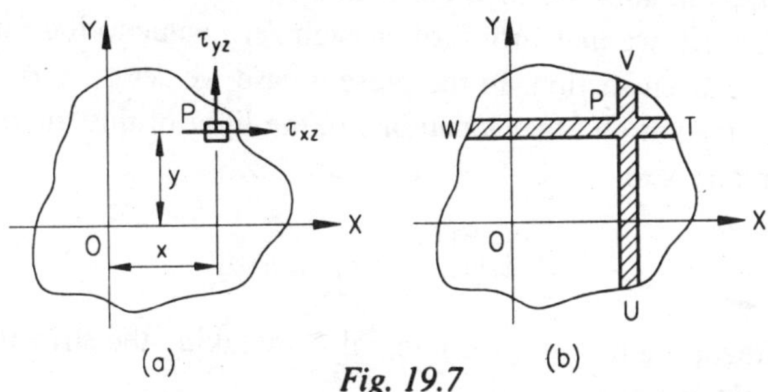


*Fig. 19.7*

The clockwise torque about the origin due to the forces acting on the element $dx \times dy$ is

$$dM_t = (\tau_{xz} \,.\, y - \tau_{yz} \,.\, x)\, dx \,.\, dy$$

The total torque can then be found by integrating this equation for the entire area $S$, of the cross section *i.e.*,

$$M_t = \iint_S [\tau_{xz} \,.\, y - \tau_{yz} \,.\, x]\, dx \,.\, dy$$

Substituting equation (19.10) into it

$$M_t = \iint_S \left[ -\frac{\partial \phi}{\partial y} y - \frac{\partial \phi}{\partial x} x \right] dxdy$$

$$= \iint_S \left( -\frac{\partial \phi}{\partial y} y \right) dxdy + \iint_S \left( -\frac{\partial \phi}{\partial x} x \right) dxdy$$

$$= I_1 + I_2 \qquad \text{(say)} \quad ...(19.24)$$

Now

$$I_1 = -\iint_S \frac{\partial \phi}{\partial y} y \, dxdy$$

$$= -\iint_S \left[ \frac{\partial \phi}{\partial y} y \, dy \right] dx$$

$$= \int \left\{ \int_U^V y d\phi \right\} dx \qquad \text{[Refer to Fig. 19.7b]}$$

$$= -\int \left\{ [y\phi]_U^V - \int_U^V \phi dy \right\} dx \qquad \text{[Integrating by parts]}$$

*i.e.*,

$$I_2 = -\int \left\{ (y_V \phi_V - y_U \phi_U) - \int_U^V \phi dy \right\} dx \qquad ...(19.25)$$

Since $\phi$ is constant along the periphery of any section, $\phi = \phi_U$ as $U$ and $V$ are points lying on the periphery. Further $\phi$ is an auxiliary function which has been introduced for the sake of calculations. Suppose we raise the $\phi$ surface by adding some constant to $\phi$, the slope of the $\phi$ surface will remain the same as before raising the level of $\phi$ surface. Hence the addition of a constant to $\phi$ will not alter the stresses. It is therefore concluded that we may arbitrarily assign zero value to the function $\phi$ at the periphery of some chosen section. In the present case we set $\phi_U = \phi_V = 0$. Then the function $\phi$ will be fixed at all other locations. In the light of this the integral $I_1$ from equation (19.25) reduces to

$$I_1 = \int\left\{\int_U^V \phi dy\right\} dx = \iint_S \phi\, dxdy \qquad ...(19.26)$$

Similarly integrating by parts the integral $I_2$ by taking the strip $WT$ of width dy and parallel to $x$-axis, we get

$$I_2 = -\iint_S \frac{\partial\phi}{\partial x}\, x\, dxdy$$

$$= \int\left\{\int \frac{\partial\phi}{\partial x}\, x\, dx\right\} dy$$

$$= -\int\left\{\int_W^T x d\phi\right\} dy \qquad \text{(Fig. 19.7}b\text{)}$$

$$= -\int\left\{(x\phi)_W^T - \int_W^T \phi dx\right\} dy$$

or
$$I_2 = -\int\left\{(x_T\phi_T - x_W\phi_W) - \int_W^T \phi\, dx\right\} dy$$

But
$$\phi_T = \phi_W = \phi_U = \phi_V = 0$$

because $T$, $W$, $U$ and $V$ are the points on periphery of the same section.

Therefore
$$I_2 = \int\left\{\int_W^T \phi dx\right\} dy$$

$$= \iint_S \phi\, dxdy \qquad ...(19.27)$$

Thus from equations (19.24), (19.26) and (19.27), we get

$$M_t = 2\iint_S \phi\, dA \qquad ...(19.28)$$

Evidently the right hand side of equation (19.28) is equal to twice the volume under the $\phi$ hill provided that $\phi$ is taken equal to zero at the boundary.

## 19.3 Prandtl's Membrane Analogy

In spite of the fact that St. Venant after the publication of his theory of torsion of non-circular shafts in 1855, found the solution of the equation (19.28) for a number of sections such as rectangular, triangular and elliptic sections. But these solutions involved very complicated mathematical methods. In 1903 Prandtl observed that the differential equation (19.20) for the stress function was same as the differential equation for the shape of the originally flat stretched membranes, which is blown up by air pressure from the bottom. This certainly gives an extremely simple and clear manner to visualise the shape of $\phi$ function and from it the evaluation of stress distribution. The equation for the shape of the stretched membrane is derived on the basis of following assumptions :

(*i*) The membrane is very thin and weightless.

(*ii*) Initially the membrane is flat and carries a 'large' tension $T$ per unit length, having same value in all directions.

(*iii*) The air pressure $p$ is very small in comparison to the initial tension so that the blowing up strain in negligible compared with initial strain in the membrane and that consequently $T$, the tension per unit length, remains constant during the blowing up process when the periphery of the membrane is held fixed.

In Fig. 19.8 is shown such a membrane which was lying flat in the xy plane carrying a large tension $T$ per unit length in all the directions and held fixed at the periphery so that the peripherical points lie in the plane $z = 0$ after the membrane is blown up with a small pressure $p$. The equation of blown up membrane will be of the form $z = f(x, y)$ and the slopes of this shape $dz/dx$ and $dz/dy$, will also be small. Consider a small element $PQ$ of the membrane of dimension $dx \times dy$. Due to the surface tension $T$, forces acting on this element along its opposite edges will be tangential to the curved membrane and magnitude of these forces will be $T\,dy$ and $T\,dx$ as shown in the diagram. In addition to the forces due to surface tension. There will be a force due to air pressure $p$, equal to $pdx \times dy$ perpendicular to it, and directed almost along the $z$-axis because the pressure $p$ being small, the slopes of the curved membrane will also be small.

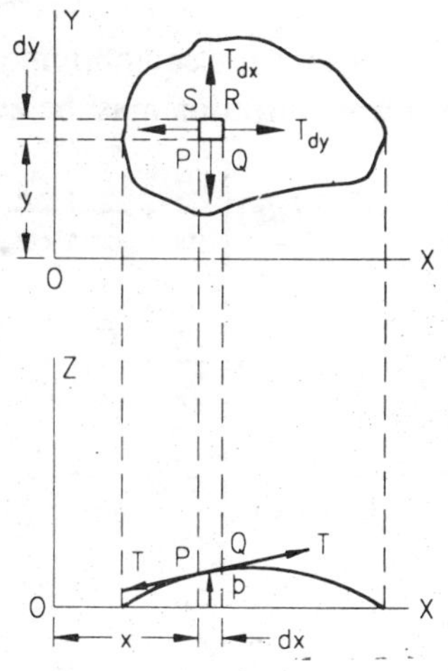


*Fig. 19.8*

The horizontal equilibrium of the membrane is automatically satisfied in both $x$ and $y$ directions because the components of the force $Tdx$ acting on the edges $PQ$ and $RS$ along the $y$-direction will balance as the slopes of the membrane, with respect to the

$y$-direction, on the points along $PQ$ or $RS$ will be small. Similarly they will balance in the $x$-direction.

However, for the equilibrium in $z$-direction, we resolve the forces as follows :

Component of $Tdy$ acting on edge $PS$

$$\simeq (Tdy)\left(\frac{\partial z}{\partial x}\right) \qquad \text{...(downward)}$$

Component of $Tdy$ acting on edge $RQ$

$$\simeq (Tdy)\left[\frac{\partial z}{\partial x}+\frac{\partial}{\partial x}\left(\frac{\partial z}{\partial x}\right).dx\right] \qquad \text{...(upward)}$$

Component of $Tdx$ acting on edge $PQ$

$$\simeq (Tdx)\left(\frac{\partial z}{\partial y}\right) \qquad \text{...(downward)}$$

Component of $Tdx$ acting on edge $SR$

$$\simeq (Tdx)\left[\frac{\partial z}{\partial y}+\frac{\partial}{\partial y}\left(\frac{\partial z}{\partial y}\right).dy\right] \qquad \text{...(upward)}$$

Component of the force $pdx.dy$ due to air pressure

$$\cong pdxdy \qquad \text{...(upward)}$$

Thus for the equilibrium of the membrane the algebraic sum of these forces in the upward direction must be zero.

$$(Tdx)\left[\frac{\partial z}{\partial x}+\frac{\partial}{\partial x}\left(\frac{\partial z}{\partial x}\right).dx\right]-(Tdy)\left(\frac{\partial z}{\partial x}\right)$$

$$(Tdx)\left[\frac{\partial z}{\partial y}+\frac{\partial}{\partial y}\left(\frac{\partial z}{\partial y}\right).dy\right]-(Tdy)\left(\frac{\partial z}{\partial y}\right)+p\,dx\,dy=0$$

or $$(T\,dx\,dy)\left[\frac{\partial^2 z}{\partial x^2}+\frac{\partial^2 z}{\partial y^2}\right]+p\,dxdy=0$$

or $$\frac{\partial^2 y}{\partial x^2}+\frac{\partial^2 z}{\partial y^2}=-\frac{P}{T} \qquad \text{...(19.29)}$$

In equation (19.29), for given values of $T$ and $p$, the right hand side will be constant. Also since for small slopes the curvature is given by the second derivative, the two terms on the left hand side are the curvatures of the membrane in the two directions. In other words from the equation (19.29) we conclude that the sum of the curvatures in two perpendicular directions is constant for all points of the membrane.

Comparing the equations (19.20) and (19.29) we see that if the pressure $p$ and the tension per unit length $T$ are so adjusted that $p/T$ becomes numerically equal to $2G\theta_1$, then equation (19.29) of the membrane will be identical with equation (19.20) of torsional stress function. Secondly if we arrange the membrane in such a manner that it is fixed at the periphery to give its height $z$ at the periphery a value zero, then the heights of the membrane at other points will be numerically equal to the stress function $\phi$ and the slopes of the membrane will be equal to the shear stresses in a direction perpendicular to that of the slope ; the contour lines $z$ = constant of the membrane will be lines following the shear stresses and the twisting moment will be numerically equal to twice the volume under the membrane. The result is reproduced mathematically as

$$\frac{p}{T} = 2G\theta_1 \qquad \text{...(19.30)}$$

## 19.3.1 Prandtl's Theory as Applied to Circular Section

In the case of a circular section shaft, the flat membrane of circular shape will be fixed at the periphery and stretched with a tension $T$ per unit length in all the directions. It is then blown upwards with a small pressure $p$. On account of symmetry the height $z$ of the blown-up membrane will not depend on the two variables $x$ and $y$ rather it can be considered to depend on the single quantity $r$, which is the radius of the contour circle at height $z$ from the base circle the periphery of which has been clamped. Setting the downward pull on the periphery $2\pi r$ of the circle equal to the upward force due to the air pressure $p$ on the projected area $\pi r^2$, we get

$$-(T2\pi r)\frac{dz}{dr} = \pi r^2 p$$

or

$$-\frac{dz}{dr} = \frac{pr}{2T} \qquad \text{...(19.31)}$$

Evidently from equation (19.31) we conclude that the slope of the membrane and hence the shear stress in the circular section shaft is proportional to the distance $r$ from the $z$-axis. This is exactly what we know of the stresses in the torsion of circular section shafts.

In pursuance of the Prandtl's theory developed above, if we translate the results of the membrane to a twisted shaft of circular section by replacing $p/T$ with $2G\theta_1$ in equation (19.31), we get

$$\theta = \text{slope} = \left|\frac{dz}{dr} = G\theta_1 r\right| \qquad \text{...(19.32)}$$

Equation (19.32) is exactly the same as equation (7.2).

For finding the shape and then the volume under the membrane we have from equation (19.31)

$$z = -\int \frac{pr}{2T}\,dr$$

$$= -\frac{pr^2}{4T} + C \qquad \text{...(19.33)}$$

The constant of integration $C$ can be found with the help of end condition, that

at $z = 0$, $r = R$

where $R$ is the radius of the shaft

*i.e.*
$$0 = \frac{pR^2}{-4T} + C$$

or
$$C = \frac{pR^2}{4T}$$

Therefore from equation (19.33)

$$z = -\frac{pr^2}{4T} + \frac{pR^2}{4T}$$

or
$$z = \frac{p}{4T}(R^2 - r^2) \qquad \text{...(19.34)}$$

The volume under the membrane hill will then be given by $V$ as

$$V = \int_{r=0}^{R} (2\pi dr)\, z$$

$$= \int_{r=0}^{R} 2\pi r \frac{p}{4T}(R^2 - r^2)\, dr$$

$$= \int_{r=0}^{R} \frac{\pi p}{2R}(rR^2 - r^3)\, dr$$

$$= \frac{\pi p}{2T}\left[\frac{R^4}{2} - \frac{R^4}{4}\right]$$

*i.e.,*
$$V = \frac{\pi p R^4}{8T} \qquad \text{...(19.35)}$$

Thus from equation (19.28), the torque $M_t$ is given by

$$M_t = 2V = \frac{p\pi R^4}{4T} \qquad \text{...(19.36)}$$

Substituting value of $p/T$ from equation (19.30)

$$M_t = \frac{\pi R^4}{4}(2G\theta_1)$$

$$= \frac{\pi}{2} R^4 (G\theta_1)$$

or $$M_t = JG\theta_1 \qquad ...(19.37)$$

Where $J$ is the polar second moment of the area of cross-section of the circular shaft and is given by

$$J = \frac{\pi}{2}R^4$$

Equation (19.37) is the well known torsion equation of a circular shaft.

### 19.3.2 Practical Application of Membrane Analogy

In equation (19.29) the profile of the membrane is given. But to calculate $z$ at different points for non-circular sections is as difficult as to calculate the stress function from equation (19.20). However experiments are performed using a stretched rubber sheet or soap film for membranes. For any one particular non-circular shape it will be practically impossible to measure the surface tension $T$, to calculate the $p/T$ ratio so as to get the value of $2G\theta_1$ by equating it to $p/T$. But this difficulty is overcome by taking a large membrane and blowing up side by side two cross-sections in the same membrane: one to be investigated and the other a circular one. Since the surface tension and air pressure used for blowing the membrane upwards with be same for both the sections the value of $p/T$ will also be same for both of them. Now the question arises as to what should be the size of the circular section in relation to that of the given section. This is taken care of by choosing the circular section so that $\theta_1$ or twist per unit length is same for both the them. In that event $P/T$ and the corresponding values of $2G\theta_1$ for both of them will be same. Since we know all the details about the circular section shaft, those for the non-circular shaft can be ascertained by the following relation which is the direct outcome of equation (19.28) according to which the torque is given by twice the volume under the $\phi$ hill or the membrane.

$$\frac{\text{Volume of hill over circular section}}{\text{Volume of hill over other section}} = \frac{M_t / \theta_1 \text{ of circle}}{M_t / \theta_1 \text{ of other section}} \qquad ...(19.38)$$

It is evident from (19.38), that if the ratio of the volumes of the two hills is known all the details about torque can be calculated. The shear stresses can also be estimated if the slopes of the two membranes are also measured because the shear stress is proportional to the slope in the perpendicular direction.

That is Slope $\propto$ shear stress

and Volume under hill $\propto$ torque.

*i.e.,* $$\frac{\text{(volume / slope) of circular hill}}{\text{(volume / slope) of other hill}} = \frac{(M_t \text{ / shear stress) of circular section}}{(M_t \text{ / shear stress) of other section}} \qquad ...(19.39)$$

In an actual experimental set up the measurements of the profiles of the two hills are made. Thus knowing all the details of the circular shape, the corresponding values for the other section are calculated with the help of equations (19.38) and (19.39).

## 19.4 Membrane Analogy as Applied to Thin Rectangular Section

Fig. 19.9 shows a very narrow rectangular cross-section $b \times t$ in which $b >> t$. The section being very narrow it is assumed that the bulges of the membrane across the sections *AA*, *CC* and *BB* are all the same where the sections *AA* and *BB* are near the central section *CC* and only near the ends where $x = \pm \dfrac{b}{2}$ the membrane flattens out. The contour lines will then be straight in the central portion and parallel to the *x*-axis. Further in the absence of curvature parallel to *x*-axis in this central portion, the membrane will be held down by the vertical components, of the surface tension, in they direction only. Thus writing the equilibrium equation for the central portion of the membrane of dimensions a times 2*y* Fig. 19.9 we get

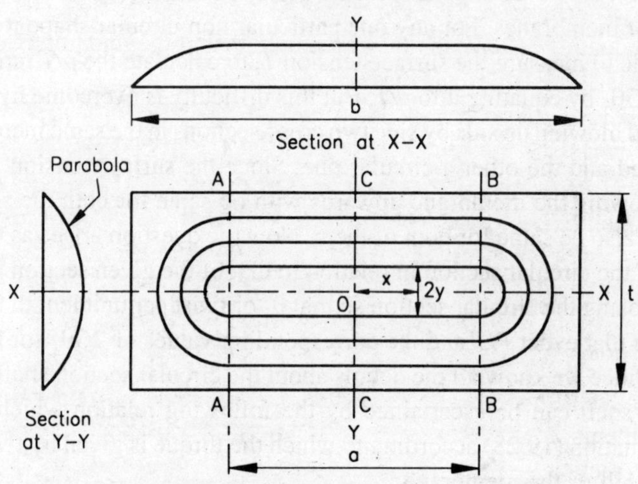


*Fig. 19.9 Contour lines of the membrane for a narrow rectangular cross-section*

$$-2Ta\frac{dz}{dy} = p(2ya)$$

or $$\frac{dz}{dy} = -\frac{P}{T}y \qquad ...(19.40a)$$

Also we know $$\tau_{xz} = \frac{dz}{dy} = -\frac{P}{T}y \qquad ...(19.40b)$$

Integrating equation (19.40), we get

$$z = -\frac{p}{T}\int y\,dy \qquad ...(19.40b)$$

$$= -\frac{p}{T}\left(\frac{y^2}{2}\right) + C_1 \qquad ...(19.41)$$

The constant of integration $C_1$ is found with the help of the end condition : $z = 0$ at the periphery where $y = t/2$.

Thus from equation (19.41), we get the value of $C$.

$$0 = -\frac{p}{T}\frac{t^2}{8} + C_1$$

or $$C_1 = -\frac{p}{T}\frac{t^2}{8}$$

Therefore substituting the value of $C_1$ in equation (19.41)

$$z = \frac{p}{2T}\left(\frac{t^2}{4} - y^2\right) \quad ...(19.42)$$

Evidently the equation (19.42) is that of a parabola and the maximum value of the slope $(dz/dy)$ exists at the edges where $y = \pm\, t/2$ (see equation 19.40*b*)

*i.e.,* $$\left(\frac{dz}{dy}\right) = -\frac{p}{T}\cdot\frac{t}{2} = -\frac{p.t}{2T} \quad ...(19.42a)$$

Reducing the membrane analogy to the twisted shaft, we get from equation (19.42*a*), by putting $-p/T = 2G\tau_1$

$$(\tau_{xz})_{\max} = \left(\frac{dz}{dy}\right)_{\max} = tG\theta_1 \quad ...(19.43)$$

In order to calculate the volume under the membrane, it will be assumed that the flattening out of the membrane near the edges, where $x = \pm\, b/2$, is neglected. Since the area of the parabola (section *yy* in Fig. 19.2) is (2/3 base × height), we find

$$\text{Volume } V = \left(\frac{2}{3}t \times \text{height}\right) \times b$$

But from equation (19.42)

$$\text{height} = z_{\max}\ \frac{p}{2T}\left(\frac{t^2}{4} - 0\right)$$

$$= \frac{pt^2}{8T}$$

Therefore $$V = \frac{2}{3}t\,.\,\frac{pt^2}{8T}\,.\,b$$

or $$V = \frac{1}{12}bt^3\,\frac{p}{T} \quad ...(19.44)$$

Then from equations (19.26), (19.30) and (19.44), we get for thin rectangular section

$$M_t = 2V$$

$$= 2\frac{bt^3}{12}\frac{p}{T}$$

$$= \frac{bt^3}{6}(2G\theta_1)$$

i.e.,
$$M_t = \frac{bt^3}{3}G\theta_1 \qquad ...(19.45a)$$

or
$$\text{Torsional stiffness} = C = \frac{M_t}{\theta_1} = \frac{bt^3}{3}G \qquad ...(19.45b)$$

Further from equations (19.43) and (19.45*a*)

$$(\tau_{xz})_{max} = \left(\frac{dz}{dy}\right)_{max} = tG\theta_1$$

$$= t\frac{3M_t}{bt^3}$$

or
$$(\tau_{xz})_{max} = \frac{3M_t}{bt^2} \qquad ...(19.46)$$

From equation (19.43) we see that the maximum shear stress occurs at the middle of the edges of the longer side of narrow rectangular section. That is the maximum shear stress occurs at the point of the periphery which is closest to the section. Further from equation (19.45*b*) we see that the stiffness $M_t / \theta_1$ is linearly proportional to *b*. Another important point to be noted is that the directions of the shear stresses at points, where $y = + t/2$ and $y = - t/2$ will be opposite as is evident from equation (19.40*b*). Thus the shear stresses over any thin rectangular section will be as shown in Fig. 19.10(*a*).

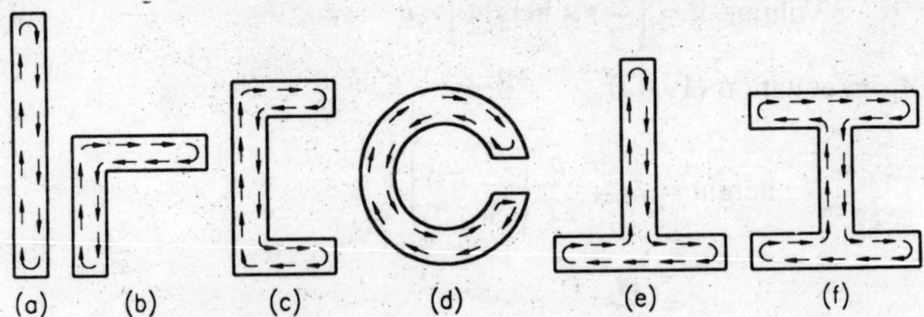


*Fig. 19.10. Open Sections built-up of the Thin Rectangles.*

## 19.4.1 Membrane Analogy Applied to Open Sections Built-up of Thin Rectangles

Let the narrow rectangle of Fig. 19.10(*a*) be bent in the middle to give rise to a thin-walled angle. The shape of the membrane for this angle section Fig. 19.10(*b*) will not materially change from that of the rectangular section Fig. 19.10($a$), except for the local effects in the corner for which the reader may refer to some advanced book on Strength of Materials. In other words, the volume under the membrane for a given pressure remains practically the same. Hence the equations (19.45) and (19.46) hold good for an angular section as well, provided we interpret '$b$' as the total length of both the legs of angle combined. Similarly these res hole good for $C$-sections, slit tube sections, $T$-sections, $I$-sections or in general for sections that can be built up of thin rectangles. In the case of an $I$-section where the web and flanges are of different thickness the equations (19.45) and (19.46) will still hold good but the torque $M_t$ will have to be calculated separately for the web and flanges and these torques to be added to give the total torque for the entire section.

### 19.4.2 Membrane Analogy Applied to Rectangular Sections which are not Very Narrow

Equations (19.45) and (19.46) had been developed for very narrow sections where $b >> t$. However for rectangular sections which are not very narrow, St. venant has found the solution by a much more complicated method which will be beyond the scope of the present text. However the results for various $b/t$ ratios are given in the following table.

| $b/t$ | $\infty$ | 10 | 5 | 3 | 2.5 | 2 | 1.5 | 1 |
|---|---|---|---|---|---|---|---|---|
| $\dfrac{M_t\theta_1}{Gbt^3}$ | 0.333 | 0.312 | 0.291 | 0.263 | 0.249 | 0.229 | 0.196 | 0.141 |
| $\dfrac{(\tau_{xz})_{max}}{M_t / bt^2}$ | 3.00 | 3.20 | 3.44 | 3.74 | 3.86 | 4.06 | 4.33 | 4.80 |

## 19.5 Membrane Analogy Applied to Hallow Sections

It has been shown by $R$. Bredt in a rather very involved manner that the Membrane Analogy can be successfully applied to hollow sections as well even for more than one hole in the section. Every hole is simulated by a weightless, stiff disc glued to the membrane ; the membrane is then blown up, and pure couples are applied to each disc till they are all horizontal and may be are at different heights. In other words, the outer and the inner boundaries of the cross-section are located in different horizontal planes with the membrane connecting the boundaries as shown in Fig. 19.11. If the thickness of the tube is small, the curvature of the membrane may be neglected $i.e.$ the lines $AD$ and $BC$ may be considered straight. The slope of the membrane surface is then constant over the thickness of the wall and is equal to $h/t$, where $h$ is the height of the central plate with respect to the outer boundary of the tube and $t$ the wall thickness. It may be noted that the wall thickness may not be constant and it may be varying along the circumference, but it should everywhere be small as compared to the size of the tube.

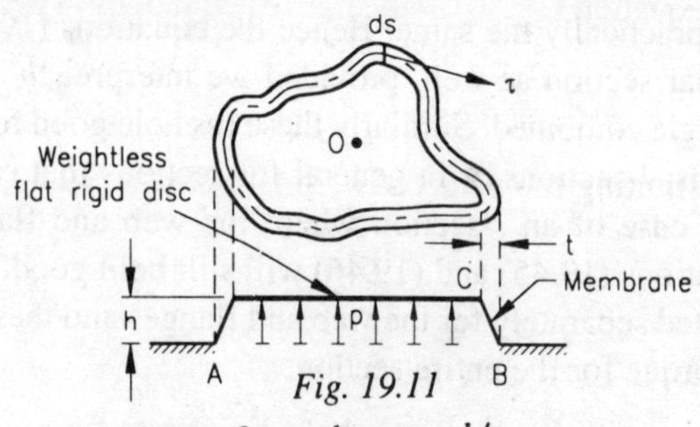


*Fig. 19.11*

Thus $\tau$ = slope of membrane = $h/t$ ...(19.47)

From equation (19.47) we see that the shear stress which acts along the circumference (directed along tangent to the walls) is inversely proportional to the local wall thickness. Further it is assumed to be uniformly distributed across the thickness $t$.

The volume included between the surfaces $CD$ and $AB$ is easily calculated by using the centre line of the wall thickness of the tube as indicated in the Fig. 19.11. If $A$ is the area bounded by this centre line, the volume under the membrane will be given by

$$\text{Volume} = V = Ah$$

Thus the membrane analogy gives the torque $M_t$ as

$$M_t = 2Ah \qquad \text{...(19.48)}$$

From equations (19.47) and (19.48), we then get

$$\tau = \frac{M_t}{2At} \qquad \text{...(19.49)}$$

Again for writing the vertical equilibrium equation of the flat weightless rigid plate whose area is taken equal to $A$, we have upward push due to air pressure as $pA$ and the downward pull is made up of the downward component of the surface tension $T$ of the membrane along the entire periphery.

Thus $$pA = \oint[(Tds) \,.\, \text{slope}]$$

$$= \oint(Tds) \,.\, \frac{h}{t}$$

or $$\frac{pA}{T} = \oint\frac{h}{t}\, ds \qquad \text{...(19.50)}$$

where $ds$ is an elemental length of the periphery of the tube.

Further from equation (19.30)

$$\frac{p}{T} = 2G\theta_1 \quad ...(19.30)$$

Therefore equation (19.50) in the light of equations (19.30) and (19.47) reduces to

$$2G\theta_1 = \frac{1}{A}\oint \tau ds \quad ...(19.51)$$

Again substituting the value of $\tau$ from equation (19.49) into (19.51), we get

$$\theta_1 = \frac{M_t}{4GA^2}\oint \frac{ds}{t}$$

or

$$C = \frac{M_t}{\theta_1} = \frac{4GA^2}{\oint \frac{ds}{t}} \quad ...(19.52)$$

In case $t$ is constant, equation (19.52) reduces to

$$C = \frac{M_t}{\theta_1} = \frac{4GA^2 t}{L} \quad ...(19.53a)$$

where $L$ is the peripheral length of the wall. From equations (19.49) and (19.53), we get

$$\tau = \frac{2G\theta_1 A}{L} \quad ...(19.53b)$$

Equation (19.49) may be used in calculating the shear stresses in tubular members under torsion if the wall thickness is small and variations in wall thickness are not abrupt and there are no re-entrant corners.

## 19.6 Membrane Analogy Applied to Multicellular Sections

Membrane analogy can also be applied to multi-cellular sections. This will involve fixing up of several weight-less flat rigid plates to the membrane to be blown up to different heights due to the air pressure $p$. All the plates are to be kept horizontal may be by the application of pure couples only.

Let the heights of these plates above the datum plane, $z = 0$, be $h_1$, $h_2$, $h_3$,...... h sub $n$. For the outside walls the slopes of the membrane with be $h/_t$ where h is the height of the adjacent plate and $t$ the local thickness of the wall of the cell. However for the inside walls the membrane slopes will be $\Delta h/t$, where $\Delta h$ is the difference in the heights of the two adjoining plates and $t$ the thickness of the intervening wall. Then the equation for the $n$th plate in the vertical direction gives

$$pA_n = T\oint \frac{\Delta h}{t} ds \quad ...(19.54)$$

where $A_n$ = is the area of the $n$th plate,

$\Delta h$ = is the difference in the heights of the two plates on either side of the wall intervening the two plates.

$t$ = wall thickness.

$ds$ = elemental peripheral length around the $n$th plate.

Equation (19.54) will result into $n$ simultaneous equation in $n$ unknown quantities $h_1, h_2, h_3 \ldots\ldots h_n$. These n equations can then be solved to give the values of these $n$ heights. Once these heights are known, these could be translated into shear stress and torque by setting slope of membrane equal to shear stress and volume under membrane equal to half the torque.

Thus.

$$\tau = \text{slope} = \frac{\Delta h}{t} \qquad \ldots(19.55)$$

$$M_t = 2\sum_n A_n h_n \qquad \ldots(19.56)$$

This method is best illustrated by the following examples.

**Example 19.1.** *An open rectangular section shown in Fig. 19.12 is acted upon by twisting moment such that the shear stress induced in it is 6 MN/m². Find the value of the twisting moment.*

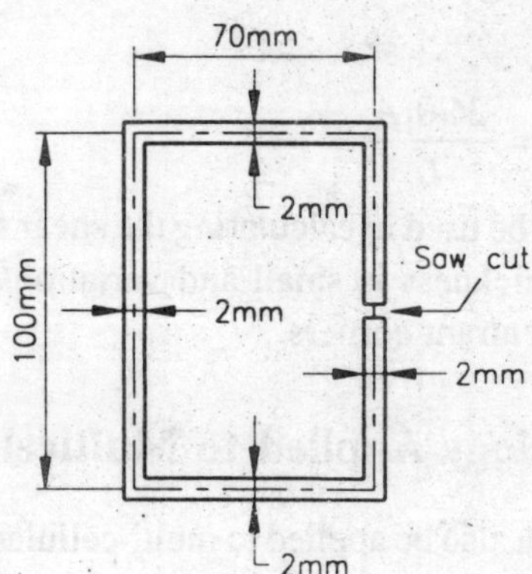


*Fig. 19.12*

**Solution.** Since it is an open section, it can be replaced by a rectangular section of thickness = 2 mm and total breadth $b$ given by the total length.

*i.e.*

$$b = 2\,(100 + 70) = 340 \text{ mm}$$

Then from equation (19.46)

$$\tau_{max} = \frac{3M_t}{bt^2} = 6\text{N / mm}^2 \text{ (Given)}$$

Therefore

$$M_t = \frac{1}{3}\tau_{max}\, bt^2 = \frac{1}{3} \times 6 \times 340 \times (2)^2$$

$$= 2720 \text{ N-mm}$$

$$= 2.72 \text{ Nm.}$$

**Example 19.2.** *A 300 × 300 mm I section with flanges and web 12 mm thick is subjected to a torque of 615 Nm. Find the maximum shear stress and the twist per unit length, neglecting the stress concentration. Take G = 8000 MN/m².*

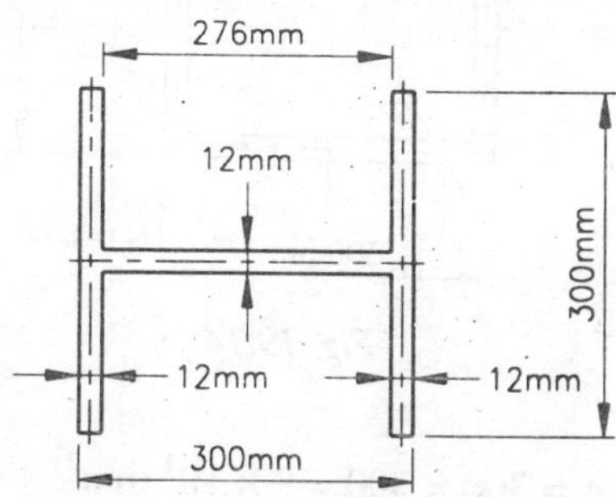


*Fig. 19.13*

**Solution.** Total length of the equivalent rectangular section

$$b = 300 + 300 + 276$$

$$= 876 \text{ mm}$$

$$t = \text{Thickness} = 12 \text{ mm}$$

$$\tau_{max} = \frac{3M_t}{bt^2}$$

$$= \frac{3(615 \times 1000)}{876 \times (12)^2}$$

$$= 14.63 \text{ N/mm}^2 \text{ or MN/m}^2$$

Also $$\frac{M_t}{\theta_1} = \frac{Gbt^3}{3}$$

or $$\theta_1 = \frac{3M_t}{Gbt^3} = \frac{3M_t}{bt^2} \cdot \frac{1}{Gt}$$

$$= t_{max} \cdot \frac{1}{Gt}$$

*i.e.* $$\theta_1 = 14.63 \times \frac{1}{80000 \times 12} \text{ rad / mm}$$

$$= 0.015 \text{ rad/m.}$$

**Example 19.3.** *A closed cellular square section Fig. 19.14 is subjected to a torque 600 Nm. Find the maximum shear stress and the twist per unit length, neglect stress concentration.*

*Take G = 80000 MN/m²*

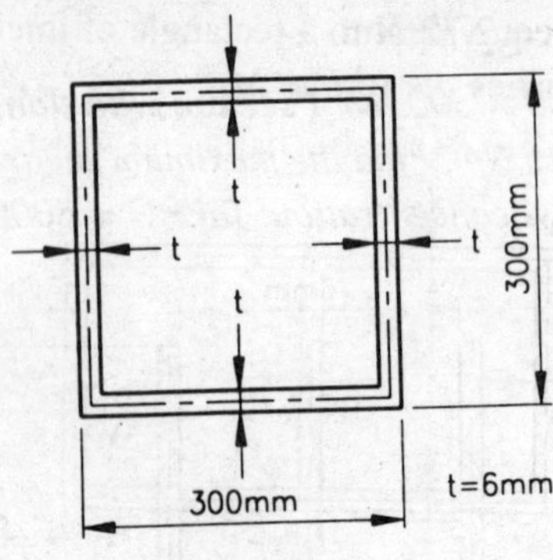


*Fig. 19.14*

**Solution.**

(*a*) $$A = 300 \times 300 = 9 \times 10^4 \text{ mm}^2$$

$$t = 6$$

From equation (19.49)

$$\tau = \frac{M_t}{2At}$$

$$= \frac{(600 \times 1000)}{2 \times 9 \times 10^4 \times 6} \text{ N / mm}^2$$

$$= 0.56 \text{ N/mm}^2 \text{ or MN/m}^2$$

(*b*) From equation (19.53)

$$C = \frac{M_t}{\theta_1} = \frac{4GA^2t}{L}$$

Here $$L = 300 \times 4 = 1200 \text{ mm}$$

Therefore $$\theta_1 = \frac{M_1 L}{4GA^2t}$$

$$= \frac{(600 \times 1000)(1200)}{4\ 80000 \times (9^2 \times 10^8) \times 6}$$

$$= 4.63 \times 10^{-8} \text{ radian/mm}$$

$$= 4.63 \times 10^{-5} \text{ rad/m.}$$

**Example 19.4.** *A thin walled bar section is shown in Fig. 19.15. It has a constant wall thickness t and one compartment is slit open. Find the stiffness of the section and the value of maximum shear stress for a given torque.*

**Solution.** (*a*) Let $M_t$ be the total torque on the section for which $\theta_1$ is the twist per unit length. Let $M_{t1}$ be the torque sustained by the closed cell *ABCD*. Suppose the slit-up box, which will be equivalent to a rectangle of thickness *t* and length equal to (*AE* + *EF* + *FG* + *GB*), carries the torque $M_{t2}$.

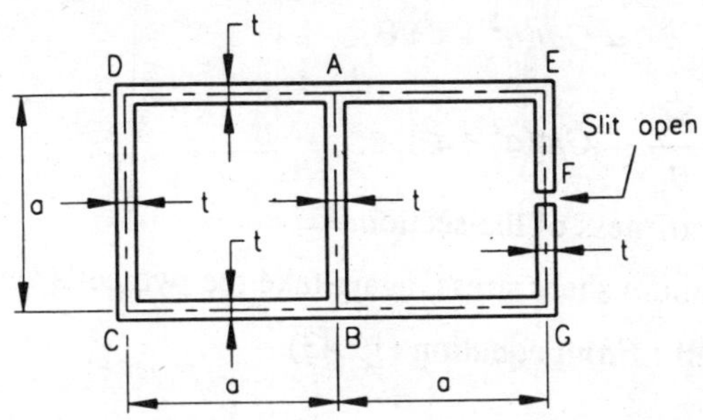


*Fig. 19.15*

*For the closed cell portion*

$$A = a^2$$

and

$$L = 4a$$

From equations (19.49) and (19.53)

$$\tau = \frac{M_{t1}}{2A_1t_1}$$

and

$$\frac{M_{t_1}}{\theta_1} = 4GA^2t/L$$

Where $\theta_1$ is the twist of the whole section per unit length.

Therefore

$$M_{t_1} = \frac{4GA^2t\theta_1}{L}$$

$$= \frac{4Ga^4t\theta_1}{4a} = Ga^3t\theta_1$$

*For the slit-up cell*

From equation (19.45*a*)

$$\frac{M_{t_2}}{\theta_1} = \frac{Gbt_2^3}{3}$$

where

$$b = AE + EF + FG + GB$$

$$= a + \frac{a}{2} + \frac{a}{2} + a = 3a$$

and

$$t_2 = t$$

Therefore $$M_{t_2} = \frac{G(3a)\,t^3\theta_1}{3} = Gat^3\theta_1$$

Thus $$M_t = M_{t_1} + M_{t_2}$$
$$= Ga^3t\theta_1 + Gat^3\theta_1$$
$$= Gat(a^2 + t^2)\,\theta_1$$

or $$\frac{M_t}{\theta_1} = Gat(a^2 + t^2) \quad \text{...(A)}$$

which is the required stiffness of the section.

(*b*) For the maximum shear stress, again take the two cells separately

(*i*) For slit-up cell : From equation (19.43)

$$\tau_{max_2} = G\theta_1 t$$

$$= Gt\left[\frac{M_t}{Gat\,(a^2 + t^2)}\right] \quad \text{(From A)}$$

or $$\tau_{max_2} = \frac{M_t}{a\,(a^3 + t^2)} \quad \text{...(B)}$$

(*ii*) *For the closed cell* : From equation (19.53*b*)

$$\tau_{max_1} = \frac{2G\theta_1 A_1}{L_1} = \frac{2G\theta_1 a^2}{4a}$$

$$= \frac{Ga}{2}\left[\frac{M_t}{Gat\,(a^2 + t^2)}\right] \quad \text{(From A)}$$

or $$t_{max_1} = \frac{M_t}{2t\,(a^2 + t^2)} \quad \text{...(C)}$$

From equations (*B*) and (*C*), it is evident that the maximum shear stress will occur in the closed cell and will be given by equation (*C*)

$$\tau_{max} = t_{max_1} = \frac{M_t}{2t\,(a^2 + t^2)}$$

**Example 19.5.** *A circular tube with 8 similar longitudinal fins attached to it (as shown in Fig. 19.16) is subjected to a twisting moment $M_t$. Find what percentage of $M_t$ is taken by the fins. If $M_t$ = 350 Nm, find the maximum shear stresses induced in the fins and the tube.*

**Solution.** Let $\theta_1$ be the angular twist per unit length of the whole section. Suppose $M_{t_1}$ is the torque sustained by the closed tubular portion and $M_{t_2}$ the torque taken up by the 8 fins which constitute together a rectangular section of thickness 6 mm and length equal to 8 × 35 = 280 mm.

(*a*) Then for the tube, from equation (19.53*a*)

$$M_{t_1} = \frac{4G\theta_1 A^2 t}{L}$$

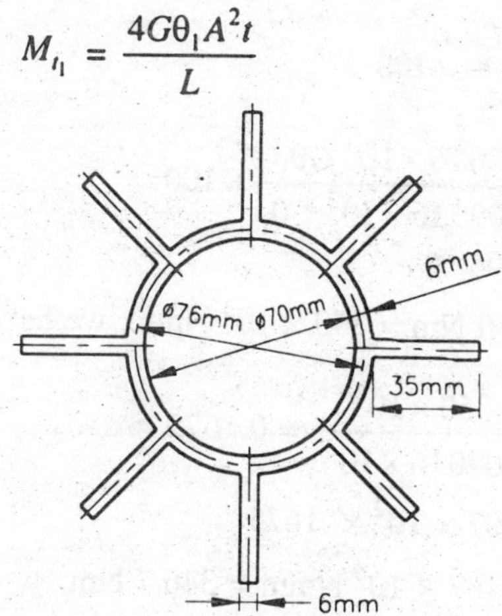


*Fig. 19.16*

Here $$A = \frac{\pi}{D}(D_{mean})^2$$

$$= \frac{\pi}{4}(76)^2 = 4536.5 \text{ mm}^2$$

$$L = \pi D_{\text{mean}} = \pi \times 76 = 238.76 \text{ mm}$$

$$t = 6 \text{ mm}$$

Therefore $$M_{t_1} = \frac{4G\theta_1 \times (4536.5) \times 6}{238.76}$$

$$= 2.07 \times 10^6\ G\theta_1$$

For the fins we have from equation (19.45*a*)

$$M_{t_2} = \frac{bt^3}{3} G\theta_1$$

Here $$b = 8 \times 35 = 280 \text{ mm}$$

$$t = 6 \text{ mm}$$

Therefore $$M_{t_2} = \frac{280 \times 6^3}{3} G\theta_1$$

$$= 20.16 \times 10^3\ G\theta_1$$

Thus $$M_t = M_{t_1} + M_{t_2}$$

$$= (2.07 \times 10^6 + 20.16 \times 10^3)\, G\theta_1$$

or $$M_t = 2090.16 \times 10^3\, G\theta_1 \qquad ...(A)$$

Therefore percentage of $M_{t_2}$

$$= \frac{M_{t_2}}{M_t} \times 100$$

$$= \frac{20.16 \times 10^3\, G\theta_1}{2090.16 \times 10^3\, G\theta_1} \times 100$$

$$= 0.96\ \%$$

(*b*) For $M_t = 350$ Nm $= 350 \times 10^3$ Nmm, we have from equation (*A*)

$$G\theta_1 = \frac{350 \times 10^3}{2090.16 \times 10^3} = 0.1675$$

Therefore $$M_{t_1} = 2.07 \times 10^2 \times .1675$$

$$= .3467 \times 10^6 \text{ Nmm} = 346.7 \text{ Nm}$$

and $$M_{t_2} = M_t - M_{t_1}$$

$$= 350 - 346.7 = 3.3 \text{ Nm}$$

Then for the fins we have from equation (19.46)

$$\tau_{max} = \frac{3M_{t_2}}{bt^2} = \frac{3 \times (3.3 \times 10^3}{280 \times (6)^2}$$

$$= 0.982 \text{ N/mm}^2 \text{ or } MN/m^2$$

Also for the tube from equation (19.49)

$$\tau_{max\ 1} = \frac{M_{t_1}}{2At}$$

$$= \frac{0.3467 \times 10^6}{2 \times 4536.5 \times 6}$$

$$= 6.369 \text{ N/mm}^2 \text{ or MN/m}^2.$$

**Example 19.6.** *A shaft of hollow square section (Fig. 19.14) is of uniform wall thickness of 6 mm and centre line of the walls forms a square of 300 mm side. It is to be replaced by a solid circular shaft of the same material and having the same torsional stiffness. Find the diameter of the solid shaft and the maximum shear stresses in both the shafts for a twisting moment of 980 Nm. Take the stress concentration factor K at the inner corners of the hollow square section as 1.75.*

**Solution.** (*a*) Referring to Fig. 19.14, for the hollow closed section

$$\frac{M_t}{\theta_1} = \frac{4GA^2t}{L}$$

Here $A = 300 \times 300 = 9 \times 10^4 \text{ mm}^2$

$t = 6$ mm

$L = 300 \times 4 = 1200$ mm

Therefore
$$\frac{M_t}{G\theta_1} = \frac{4 \times (2 \times 10^4)^2 \times 6}{1200}$$
$$= 162 \times 10^6$$

or
$$\text{stiffness } C = \frac{M_t}{\theta_1} = 162 \times 10^6\, G \qquad ...(A)$$

For the solid circular shaft of diameter $D$ referring to Chapter 7, we have

$$\frac{M_t}{J} = \frac{G\theta}{l}$$
$$= G\theta_1$$

or
$$\frac{M_t}{\theta_1} = JG = \frac{\pi D^4}{32} G \qquad ...(B)$$

Equating the stiffness $M_t/\theta_1$ of the two shafts we have from equations ($A$) and ($B$)

$$\frac{\pi D^4}{32} G = 162 \times 10^6\, G$$

or
$$D^4 = \frac{165 \times 10^6 \times 32}{\pi} = 16.5 \times 10^8$$

*i.e.*
$$D = 2.01 \times 10^2$$
$$= 201 \text{ mm}$$

($b$) The shear stress in the hollow section is found with the help equation (19.49).

*i.e.*
$$\tau_{\max 1} = \frac{M_t}{2At} \times (\text{stress concentraion factor } K)$$

$$= \frac{(980 \times 10^3)}{2 \times (9 \times 10^4) \times 6} \times 1.75$$
$$= 1.59 \text{ N/mm}^2 \text{ or MN/m}^2$$

The shear stress in the solid circular section of diameter $D$ is given by

$$\tau_{\max 2} = \frac{16\, M_t}{\pi D^3}$$

$$= \frac{16 \times (980 \times 10^3)}{\pi \times (201)^3}$$
$$= 0.615 \text{ N/mm}^2 \text{ or MN/m}^2.$$

**Example 19.7.** *In the Fig. 19.17 is shown the section of a box type member under the action of a twisting moment. The mean diamensions of the two cells are 50 mm × 25 mm and 25 mm × 25 mm with $t_1$ = 3 mm, $t_2$ = 6 mm and $t_3$ = 4 mm. Calculate the shear stress in each section and the angle of twist per metre length for a twisting moment of 250 Nm. Take G = 80000 MN/m$^2$.*

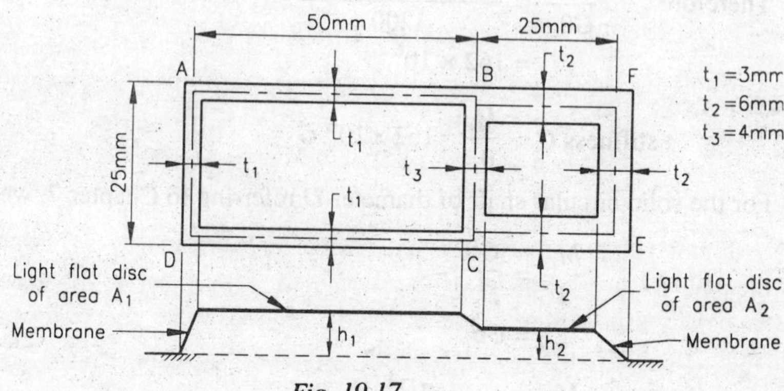


*Fig. 19.17*

**Solution.** Let the cell *ABCD* be designated the first cell and *BFEC* as second cell. Then as per the dimensions of these cells

$$A_1 = 50 \times 25 = 1250 \text{ mm}^2$$

$$A_2 = 25 \times 25 = 625 \text{ mm}^2$$

$L_1$ = length of the portion of thickness $t_1$

$= BA + AD + DC = 50 + 25 + 50 = 125$ mm

$L_2$ = length of the portion of thickness $t_2$

$= BF + FE + EC = 25 + 25 + 25 = 75$ mm

$L_3$ = length of the portion of thickness $t_3$

$= BC = 25$ mm

Also $t_1 = 3$ mm ; $t_2 = 6$ mm ; $t_3 = 4$ mm

Let the height of the two flat rigid and weightless plates above the datum level of $z = 0$ be $h_1$, and $h_2$ as shown Fig. 19.17.

For the cell number 1 *i.e.*, box *ABCD*, we have from equation (19.54)

$$pA_1 = T\left[\Sigma \frac{Lh}{t} + \frac{h_1 - h_2}{t_3} L_3\right]$$

or
$$\frac{pA_1}{T} = \left[\frac{L_1 h_1}{t_1} + \frac{h_1 - h_2}{t_3} L_3\right]$$

or
$$(2G\theta_1)\, A_1 = \left[\frac{L_1 h_1}{t_1} + \frac{h_1 - h_2}{t_3} L_3\right]$$

or $$2 \times 1250\, G\theta_1 = \left[\frac{125\, h_1}{3} + \frac{h_1 - h_2}{4} \times 25\right]$$

$$= \frac{1}{12}[500\, h_1 + 75\, h_1 - 75\, h_2]$$

or $$23h_1 - 3h_2 = 1200\, G\theta_1 \qquad ...(A)$$

Similarly for the second cell *i.e.*, box *BFEC* we have from equation (19.54)

$$pA_2 = T\left[\Sigma \frac{Lh}{t} + \frac{h_2 - h_1}{t_3} L_3\right]$$

It may be noted that in this expression $\Delta h$ is equated to $(h_2 - h_1)$ *i.e.*, the height of the plate for the second box, when considering the equation for second cell, is to be taken positive and that of first cell negative. In the earlier equation while considering the equation for the first cell the signs were reversed *i.e.*, $\Delta h = (h_1 - h_2)$.

Coming back to this equation

$$\frac{pA_2}{T} = \left[\Sigma \frac{Lh}{t} + \frac{h_2 - h_1}{t_3} L_3\right]$$

or $$2G\theta_1 A_2 = \left[\frac{L_2 h_2}{t_2} + \frac{h_2 - h_1}{t_3} L_3\right]$$

or $$2G\theta_1\,(625) = \left[\frac{75 \times h_2}{6} + \frac{(h_2 - h_1)\,25}{4}\right]$$

or $$200\, G\theta_1 = 3h_2 - h_1 \qquad ...(B)$$

Solving simultaneous equation (*A*) and (*B*)

$$h_1 = 63.64\, G\theta_1 \text{ and } h_2 = 87.88\, G\theta_1$$

Now $$M_t = 2 \times \text{volume un der the membrane}$$

$$= 2\,(A_1 h_1 + A_2 h_2)$$

or $$250 \times 10^3 = 2\,(1250 \times 63.64\, G\theta_1 + 625 \times 87.88\, G\theta_1)$$

or $$G\theta_1 = 0.927$$

Therefore $$\theta_1 = \frac{0.927}{G} = \frac{0.927}{80000} \text{ rad / mm}$$

$$= 0.0116 \text{ rad/m}$$

Now $$\tau_1 = \frac{h_1}{t_1} = \frac{63.64\,(G\theta_1)}{t_1}$$

$$= \frac{63.65 \times 0.927}{3} = 19.66 \text{ N/mm}^2 \text{ or MN/m}^2$$

$$\tau_2 = \frac{h_2}{t_2} = \frac{87.88\,(G\theta_1)}{t_2}$$

And $$\tau_2 = \frac{87.88 \times 0.927}{6} = 13.58 \text{ N/mm}^2 \text{ or MN/m}^2$$

$$\tau_3 = \frac{\Delta h}{t_3}\frac{|h_2 - h_1|}{t_3} = \frac{(87.88 - 63.64)\,G\theta_1}{4}$$

$$= \frac{24.24 \times 0.927}{4} = 5.62 \text{ N/mm}^2 \text{ or MN/m}^2.$$

**Example 19.8**. *Centre line of a double celled cross-section in the form of a rectangle and a triangle of very thin walls, is shown in the Fig. 19.18. If a twitting moment of 12500 Nm is applied, calculate the shear stresses in all the walls and the angle of twist per metre length. What thickness of the links BE and CE would make the stress in BC zero ?*

*Take G = 80000 MN/m².*

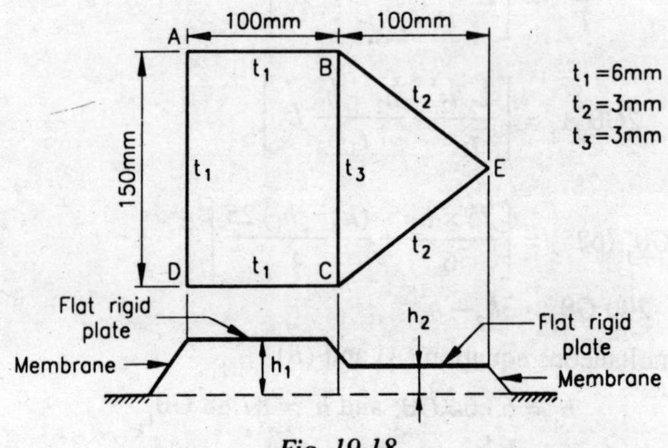


*Fig. 19.18*

**Solution.** (*a*) $t_1 = 6\text{mm}$ ; $t_2 = 3$ mm ; $t_3 = 3$ mm

$A_1$ = Area $ABCD = 100 \times 150 = 15000 \text{ mm}^2$

$$A_2 = \text{Area } BEC = \frac{1}{2} \times 150 \times 100 = 7500 \text{ mm}^2$$

For the cell $ABCD$

$$pA_1 = T\left[\Sigma \frac{Lh}{t} + \frac{(h_1 - h_2)}{t_3} L_3\right]$$

Here since the thicknesses of portions $CD$, $DA$ and $AB$ are same, they are treated collectively and

$$L_1 = CD + DA + AB = 100 + 15 + 100 = 350 \text{ mm}$$

$$L_3 = BC = 150 \text{ mm}$$

Therefore for the cell $ABCD$, we have

$$\frac{P}{T}A_1 = \left[\frac{L_1 h_1}{t} + \frac{h_1 - h_2}{t_3}L_3\right]$$

or $$(2G\theta_1)\,A \times 15000 = \left[\frac{350 \times h_1}{6} + \frac{(h_1 - h_2)}{3} \times 150\right]$$

or $$10h_1 - 3h_2 = 3600\,G\theta_1 \qquad \text{...(A)}$$

For the cell $BEC$,

$$L_2 = BE + EC = 2BE$$

$$= 2\sqrt{100^2 + 75^2} = 250 \text{ mm}$$

$$\frac{p}{T}A_1 = \left[\Sigma\frac{Lh}{t} + \frac{(h_2 - h_1)}{t_3}L_3\right]$$

or $$(2G\theta_1)A_2 = \left[\frac{L_2 h_2}{t_2} + \frac{(h_2 - h_1)}{t_3}L_3\right]$$

or $$(2G\theta_1)\,7500 = \left[\frac{250 \times h_2}{3} + \frac{(h_2 - h_1)}{3} \times 150\right]$$

or $$4500\,G\theta_1 = 25\,h_2 + 15\,h_2 - 15\,h_1$$

or $$8h_2 - 3h_1 = 900\,G\theta_1 \qquad \text{...(B)}$$

Solving equations ($A$) and ($B$), we get

$$h_1 = 443.66\,G\,\theta_1$$

$$h_2 = 278.87\,G\,\theta_1$$

Now $$M_t = 2 \text{ (volume under the membrane)}$$

or $$M_t = 2\,[A_1 h_1 + A_2 h_2]$$

*i.e.,* $$12500 \times 10^3 = 2\,[15000 \times 443.66\,G\theta_1 + 7500 \times 278.87\,G\theta_1]$$

or $$G\theta_1 = 0.7146 \qquad \text{...(C)}$$

Therefore $$\theta_1 = \frac{0.7146}{G}$$

$$= \frac{0.7146}{80000} \text{ rad / mm}$$

$$= 0.0089 \text{ rad/m}$$

Now $$\tau_1 = \frac{h_1}{t_1} = \frac{443.66\,(G\theta_1)}{6}$$

$$= \frac{443.66 \times 0.7146}{6}$$

$$= 52.84 \text{ N/mm}^2 \text{ or MN/m}^2$$

$$\tau_2 = \frac{h_2}{t_2} = \frac{278.87\,(G\theta_1)}{3}$$

or $$\tau_2 = \frac{278.87 \times 0.7146}{3}$$

$$= 66.43 \text{ N/mm}^2 \text{ or MN/m}^2$$

And $$\tau_3 = \frac{|h_1 - h_2|}{t_3}$$

$$= \frac{(443.66 - 278.87)\,G\theta_1}{3}$$

$$= \frac{164.79 \times 0.7146}{3}$$

$$= 39.25 \text{ N/mm}^2 \text{ or MN/m}^2$$

(*b*) Since the shear stress in the internal web is to be zero, it means the slope of the membrane is zero.

In other words

$$h_1 = h_2$$

Again for the cell *ABCD*

$$\frac{p}{T} A_1 = \left[\Sigma \frac{Lh}{t} + 0\right] = \frac{L_1 h_1}{t_1}$$

or $$(2G_1)\,A_1 = \frac{L_1 h_2}{t_1}$$

or $$(2G\theta_1)\,15000 = \frac{350\,h_1}{t_1}$$

or $$h_1 = 514.28\,G\theta_1 \qquad \text{...(D)}$$

and for cell *BEC*

$$\frac{p}{A} A_2 = \left[\Sigma \frac{Lh}{t} + 0\right] = \frac{L_2 h_2}{t_2}$$

or $$(2G\theta_1)\,A_2 = \frac{L_2 h_2}{t_2} = \frac{L_2 h_2}{t_2} \qquad (\because h_2 = h_1)$$

$$\text{or} \qquad t_2 = \frac{L_2\,(514.28\,G\theta_1)}{(2G\theta_1)\,A_2} \qquad \text{(From } D\text{)}$$

$$= \frac{250 \times 514.28}{2 \times 7500} = 8.57 \text{ mm.}$$

## PROBLEMS FOR PRACTICE

1. Four tubular sections shown in Fig. 19.19, are made out of same sheet so that all have the same wall thickness. The peripheral length L along the centre line of the wall is same for all the sections. If the stress concentration is to be ignored, find the ratio of the maximum shear stresses in these sections for

   (*i*) same twisting moment applied to all of them,

   (*ii*) same angle of twist per unit length.

$$\left[\textbf{Ans. } 1:\frac{3L}{2\pi t}:\frac{4}{\pi}:\frac{3L}{2\pi t};1:\frac{2\pi t}{L}:\frac{\pi}{4}:\frac{2\pi t}{L}\right]$$

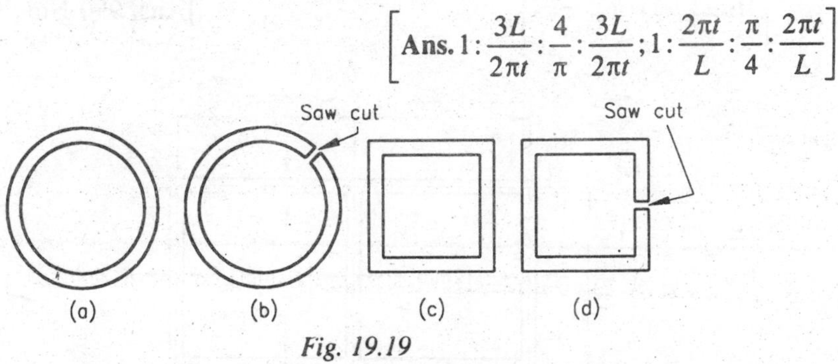


*Fig. 19.19*

2. A thin walled box section consisting of two compartments is built up with a 6 mm sheet and is shown in Fig. 19.20. One compartment is a closed cell while the other is slit open. If a twisting moment of 1000 Nm is applied to the section, find the maximum shear stress induced. Also determine the twist in radians/metre length.

   Take $G = 80000$ MN/m$^2$ **[Ans.** 32.86 MN/m$^2$, 0.0164 rad/m]

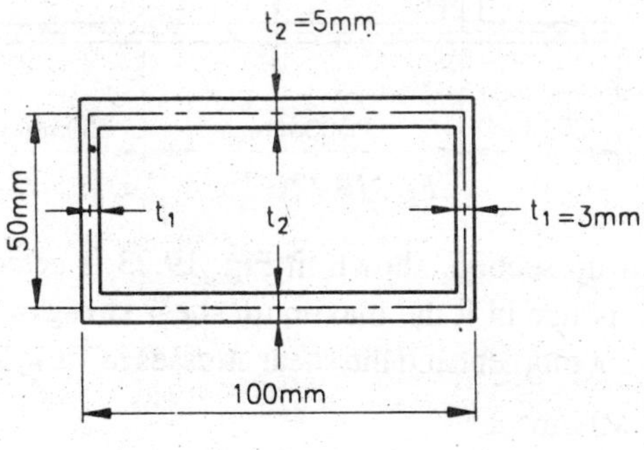


*Fig. 19.20*

3. A built up steel plate girder has the section shown in Fig. 19.21. The thickness of plate is 12 mm all round. If the maximum allowable shear stress in the material is 65 MN/m$^2$ find

   (*a*) the maximum allowable torque,

   (*b*) the angle of twist per metre length due to this torque,

   (*c*) the shear stress in the central limb of the section.

   Take $G$ = 80000 MN/m$^2$.

   **[Ans.** 4062.5 Nm ; 8.12 × 10$^{-4}$ rad/m ; zero]

4. An aluminium box section has been designed Fig. 19.22 for the maximum shear stress of 33 MN/m$^2$ (336.39 kg/cm$^2$). Neglecting stress concentration find the twisting moment which can be taken by the section and also the twist per metre length.

   Take $G$ = 27000 MN/m$^2$

   $\left[\textbf{Hint}: pA = T\Sigma\dfrac{LH}{t}.\right]$ **[Ans.** 990 Nm ; 0.27 rad/m]

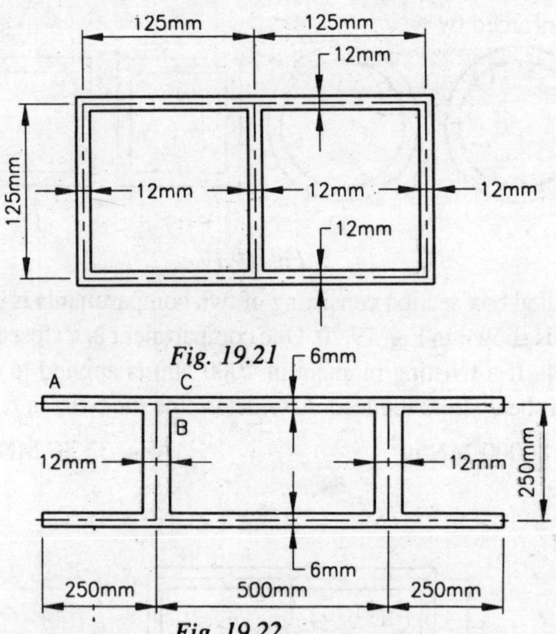


*Fig. 19.21*

*Fig. 19.22*

5. A mild steel built up section, shown in Fig. 19.23 is acted upon by a twisting moment so as to induce in it the maximum shear stress of 67 MN/m$^2$. Find the value of the twisting moment and the shear stresses in various parts of the section.

   Take $G$ = 80000 MN/m$^2$

   **[Ans.** 10074 Nm ; $T_A$ = 4.02 MN/m$^2$ ; $T_B$ = 33.5 MN/m$^2$ ; $T_C$ = 67 MN/m$^2$]

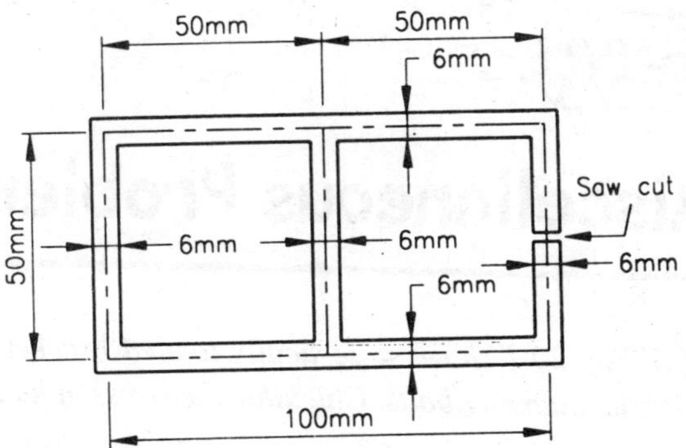


*Fig. 19.23*

6. (*a*) An I-section steel girder 300 mm × 300 mm with flanges and web 12 mm thick is subjected to a twisting moment of 615 Nm. Find the maximum shear stress and twist per unit length neglecting stress concentration Take $G$ = 80000 MN/m$^2$.

   (b) In order to reduce the stress and angle of twist per unit length the I-section is reinforced by welding steel plates 276 mm × 16 mm thick, as shown in Fig. 19.24. Find the maximum stress due to the same twisting moment. What is then the value of twist per unit length.

**[Ans.** (*a*) 14.63 MN/m$^2$ ; 0.015 rad/m

(*b*) 0.313 MN/m$^2$ ; 2.43 × 10$^{-5}$ rad/m]

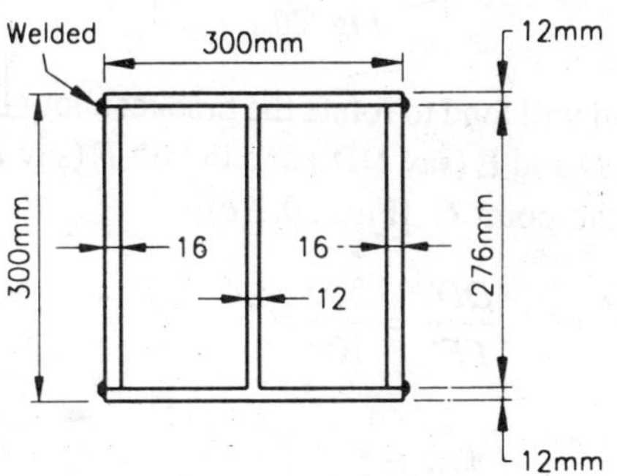


*Fig. 19.24*

Chapter–20

# Miscellaneous Problems

**Problem 20.1** *Fig. 20.1 shows a vertically loaded bracket attached to a fixed member by three 19 mm diameter bolts. Calculate the stress induced in the three bolts.*

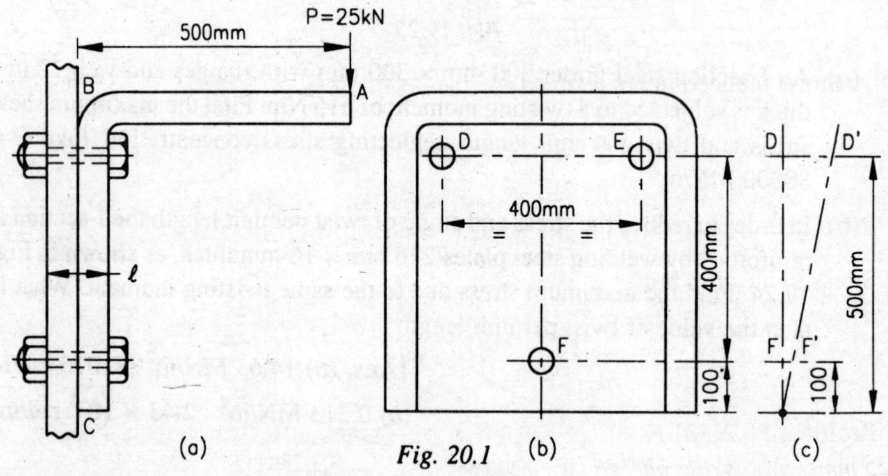


*Fig. 20.1*

**Solution.** The load will tend to rotate the bracket about the point *C*. The resulting stretch in the two bolts *D* and E (say *DD'*) and in bolt *F* (say *FF'*) will be proportional to their distance from the point *C*. [Fig. 20.1(*c*)]

*i.e.*
$$\frac{DD'}{FF'} = \frac{500}{100} = 5$$

or
$$\frac{E_D l_d}{E_F l_F} = 5$$

or
$$\left(\frac{\sigma_D}{E_D} l_d\right) \frac{1}{\dfrac{\sigma_F}{E_D} l_F} = 5$$

or
$$\frac{P_D}{A_D E_D} l_d \cdot \frac{E_F A_F}{P_F l_F} = 5$$

or $$\frac{P_D}{P_F} = 5 \qquad ...(20.1)$$

For the equilibrium of bracket under the action of retaining loads $P_D$, $P_E$ and $P_F$ and applied load $P$, we take moments about the point $C$

$$25 \times 500 = P_D\ 500 + P_E \times 500 + P_F \times 100$$

$$= 1000\ P_D + \frac{P_D}{5} \times 100 \qquad (\because P_E = P_D)$$

*i.e.,* $$P_D = \frac{25 \times 500}{1020} = 12.255 \text{ kN}$$

$\therefore$ Stress induced in bolts $D$ and $E$

$$= \frac{4P_D}{p \times 19^2} = \frac{4 \times 12.255}{\pi \times 19^2} \text{ kN / mm}$$

$$= 44.26 \text{N/mm}^2 \text{ or MPa}$$

Stress induced in bolt $F$

$$= \frac{4P_F}{p \times 12^2} \frac{44.26}{5} = 8.852 \text{ MPa}.$$

**Problem 20.2.** (*a*) *Find the moment of resistance of a beam of square section, side 25 mm, with its diagonal in the plane of bending, so that maximum induced stress is 200 MPa. In order to increase the resisting moment of the section, with the same maximum induced stress the top and bottom corners are flattered as shown in Fig. 20.2. (b) Find the value of x so that the resisting moment of the resultant section is maximum. What is then the value of resisting moment ?*

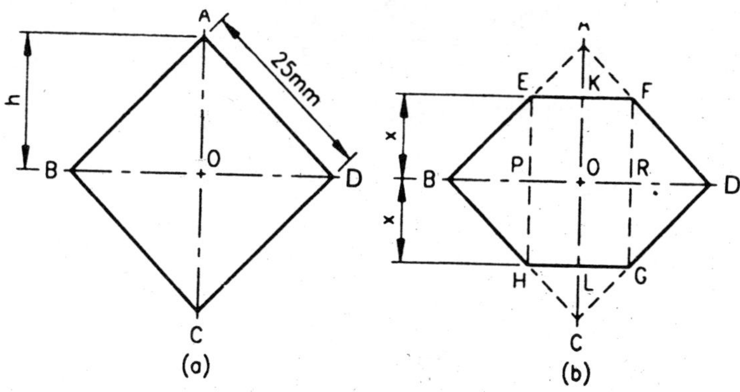


*Fig. 20.2*

**Solution.** (*a*) $\sigma = 200 \text{ MPa} = 200 \text{ N/mm}^2$

$$AO = 25\cos 45° = \frac{25}{\sqrt{2}} = y_{max}$$

The centroidal bending axis is $BD$

$$I_{BD} = \frac{AB^4}{12} = \frac{(25)^4}{12}\ \text{mm}^4$$

The resisting moment of the section $ABCD$ is

$$M = I_{BD}\frac{\sigma}{y_{max}}$$

$$= \frac{(25)^2}{12} \times \frac{\sqrt{2}}{25} \times 200\ \text{N} - \text{mm}$$

$$= 638.229\ \text{Nm}$$

(*b*)

$$OK = x$$

$$AK = \frac{25}{\sqrt{2}} - x$$

$$EF = 2EK = 2AK = 2\left(\frac{25}{\sqrt{2}} - x\right)$$

$$EH = 2x$$

$$BE = \frac{EP}{\cos 45°} = EP\sqrt{2} = x\sqrt{2}$$

The second moment of area $BEFDGH$ about $BD$ can be calculated by splitting the section into one rectangle $EFGH$ and two triangles $BEH$ and $FGD$. Further these two triangle can be placed side by side with $FG$ overlapping $EH$ to form a square similar to $ABCD$ but of side $BE$ and diagonal $EH$.

Therefore second moment of these two triangle $BEH$ and $FGD$ about axis $BD$ is

$$I_1 = \frac{(BE)^4}{12} = \frac{\left(x\sqrt{2}\right)^4}{12} = \frac{x^4}{3}$$

Second moment of area $EFGH$ about $BD$

$$I_2 = \frac{EF \times (EH)^3}{12}$$

$$= \frac{1}{12} \times 2\left[\frac{25}{\sqrt{2}} - x\right][2x]^3$$

$$= \frac{2}{3}x^3\left(25\sqrt{2} - 2x\right)$$

Therefore total second moment of area of section $BEFDGHB$ is

$$I_3 = I_1 + I_2$$

$$= \frac{x^4}{3} + \frac{2}{3}x^3\left(25\sqrt{2} - 2x\right)$$

Therefore resisting moment of this section

$$M' = I_3 \frac{\sigma}{y_{max}} = I_3 \frac{\sigma}{x}$$

$$= \frac{1}{3}\left[x^4 + 2x^3\left(25\sqrt{2} - 2x\right)\right]\frac{200}{x}$$

or $$M' = \frac{200}{3}\left[\sqrt{2}x^2 - 3x^3\right]$$

For maximum $M'$, $$\frac{dM'}{dx} = 0$$

Therefore $$50\sqrt{2}\,(2x) - 9x^2 = 0$$

or $$x = \frac{100\sqrt{2}}{9} \text{ mm}$$

Thus maximum value of $M'$ is

$$M'_{max} = \frac{200}{3}x^2\left(50\sqrt{2} - 3x\right)$$

$$= \frac{200}{3}\left(\frac{100\sqrt{2}}{9}\right)\left(50\sqrt{2} - \frac{300\sqrt{2}}{9}\right) \text{N / mm}$$

$$= 387.986\text{N-m}$$

**Problem 20.3** *The moment of resistance of a beam of square section with its diagonal in the plane of bending in increased by flatering the top and bottom corners as shown in Fig. 20.2 to give maximum possible resisting moment for the same bending stress. Find the ratio of maximum resisting moment to that in the original beam.*

**Solution.** In the first case $AB = \sqrt{2}h$

$$I = \frac{1}{12}\left(\sqrt{2}h\right)^4 = \frac{h^4}{3}$$

$$M = \frac{I\sigma}{y_{max}}$$

$$= \frac{h^4}{3} \cdot \frac{\sigma}{h} = \frac{\sigma h^3}{3}$$

As in Problem 20.2

$I_1$ = second moment of area of two triangles *BEH* and *EGD*

$$= \frac{(BE)^4}{12} = \frac{\left(x\sqrt{2}\right)^4}{12} = \frac{x^4}{3}$$

$$EH = 2x$$

$$EK = AK = h - x$$

$$EF = 2EK = 2\,(h - x)$$

$I_2$ = second moment of area $EFGH$

$$= \frac{1}{2}\,[2(h-x)]\,[2x]^3$$

Therefore $$I' = I_1 + I_2$$

$$= \frac{x^4}{3} + \frac{4}{3}(h-x)\,x^3$$

$$y'_{max} = x$$

And $$M' = \frac{\sigma I'}{y'_{max}}$$

$$= \left(\frac{\sigma}{x}\right)x^3\left[\frac{x}{3} + \frac{4}{3}(h-x)\right]$$

or $$M' = \frac{\sigma x^2}{3}(4h - 3x)$$

For $M'$ to be maximum

$$\frac{dM'}{dx} = 0 = \frac{\sigma}{3}\,[8hx - 9x]$$

*i.e.* $$x = \frac{8h}{9}$$

Therefore $$M'_{max} = \frac{\sigma}{3}\left(\frac{8h}{9}\right)^2\left[4h - 3\times\frac{8h}{9}\right]$$

$$= \left(\frac{\sigma h^3}{3}\right)\frac{64}{81}\times\frac{4}{3}$$

$$= \mathrm{M}\times 1.0535$$

or $$\frac{M'_{max}}{M} = 1.0535$$

**Problem 20.4.** *For the rocker arm shown in Fig. 20.3 find the maximum tensile and compress stresses in the section AA.*

**Solution.** The load $P$ of 10 kN is an eccentric load with 30 mm eccentricity of loading. Thus the load will give rise to both direct and bending stresses on section.

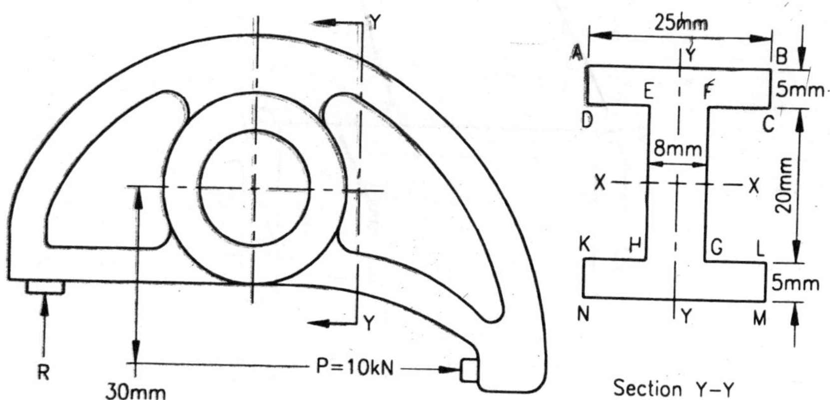


*Fig. 20.3*

For the section $A = 25 \times 30 - 17 \times 20$

$= 410 \text{ mm}^2$

Bending Moment $= P \times 30 = 300000$ N-mm

$$I_{xx} = \frac{25 \times 30^3}{12} - \frac{17 \times 20^3}{12}$$

$= 56250 - 11333$

$= 44917 \text{ mm}^4$

Tensile Direct Stress

$$\sigma_0 = \frac{10 \times 1000}{410} = 24.39 \text{ N/mm}^2 \text{ or MPa}$$

Maximum Bending stress will occur along *AB* or *MN*. Along **AB** it with be compressive and along *MN* tensile)

$$(\sigma_o)_{max} = y_{max} \frac{M}{I_{xx}}$$

$$= \frac{15 \times (3 \times 10^5)}{44917} \text{ N/mm}^2$$

$= 100.185 \text{N/mm}^2$ or MPa

Therefore maximum resultant tensile stress on section on *NM*

$= 24.39 + 100.185 = 124.575$ MPa

Maximum resultant compressive stress on *AB*.

$= 100.185 - 24.39 = 75.795$ MPa

**Problem 20.5.** *A beam of box section shown in Fig. 20.4 carries a transverse shearing force of 4 kN in the vertically downward direction. Calculate the shear stress along the surface EF parallel to longitudinal axis of the beam OZ.*

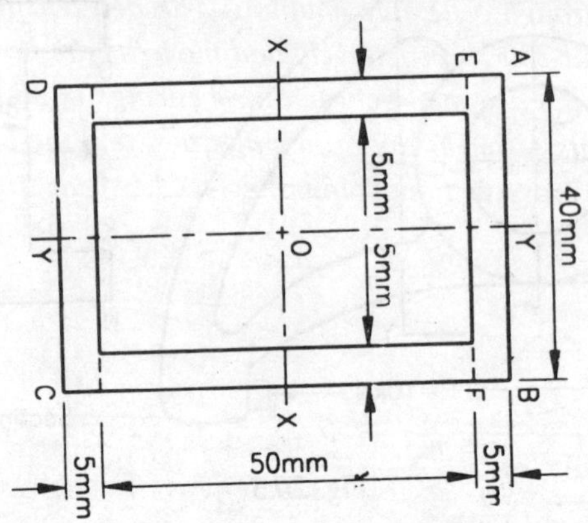


*Fig. 20.4*

**Solution.** $$I_{xx} = \frac{40 \times 60^3}{12} - \frac{30 \times 40^3}{12}$$

$$= 560 \times 10^3 \text{ mm}^4$$

Since along *EF* there is sudden change in the width of the section, there will be two values : one corresponding to the point just above *EF* and other lying just below *EF* for which the respective values of the width of the section will be 40 mm and 10 mm.

(*a*) Thus when $b_0 = 40$ (just above the line *EF*)

$$\tau_1 = \frac{F}{I_{xx} b_0}\left(\overline{A}\,\overline{Y}\right)$$

$$\overline{A} = \text{Area } BBFE = 40 \times 5 = 200 \text{ mm}^2$$

$$\bar{y} = 25 \frac{1}{2} \times 5 = 27.5$$

Therefore $$\tau = \frac{4000}{(5600 \times 10^3)} 40 / [200 \times 27.5]$$

$$= 0.982 \text{ N/mm}^2 \text{ or MPa}$$

(*b*) When $b_0 = 10$ (just below the line EF)

$$\tau_2 = \frac{F}{I_{xx} b_0}\left(\overline{A}\bar{y}\right) = \frac{40}{10}\tau_1 = 3.928 \text{ MPa}$$

Therefore the maximum shear stress on the cross-section at *EF* is 3.928 MPa. But complimentary shear stress of the same value will be acting along the plane *EF* parallel to the longitudinal axis *OZ*.

**Problem 20.6.** *Fig. 20.5 shows a shaft CDE carrying a pulley of 125 mm diameter at the mid span, is mounted on ball bearing at C and D. The end of the shaft E is connected to a clutch through a flexible coupling. The belt tensions on the tight and slack sides are 900 N and 700 N respectively and the belts are horizontal. If the shaft diameter is 25 mm and the distance between the bearings is 200 mm, using Mohr's method find for the points A and B the principal stress and the location of principal planes with respect to the cross-sectional plane. neglect the shear stress due to transverse load.*

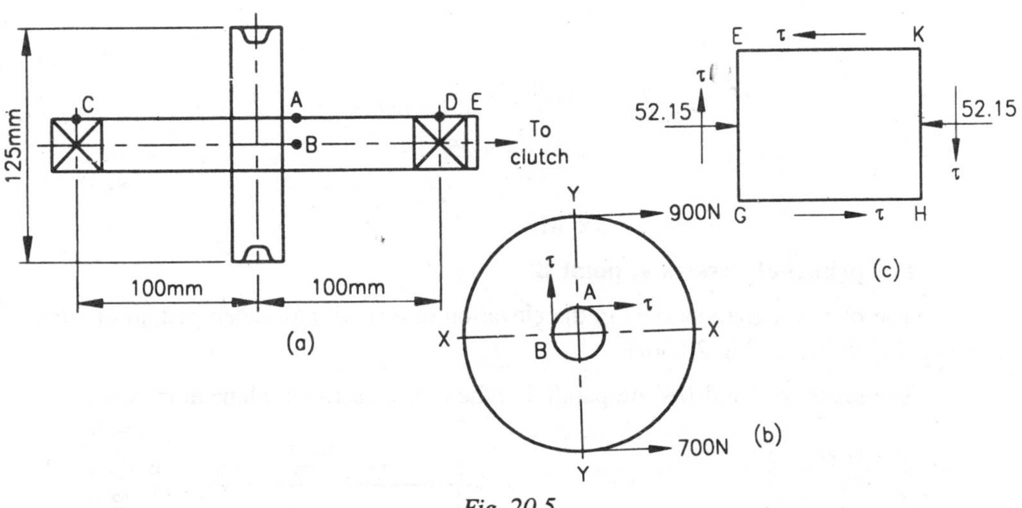


*Fig. 20.5*

**Solution.** Due to the two tensions the simply supported shaft *CD* will carry transverse force of 900 + 700 = 1600 *N* at the mid section in the horizontal plane. It will also carry a clockwise torque of magnitude $200 \times \frac{125}{2} = 12500$ Nmm as seen from *D* towards *A*.

**Effect of transverse force of 1600 N :**

Shaft *CD* will be treated as a simply supported been wit span l = 200 mm.

Therefore maximum bending moment

$$M = \frac{Wl}{4}$$

$$= \frac{1600 \times 200}{4} = 8 \times 10^4 \text{N/mm}$$

This B.M. will be acting at mid section with *yy* as its centroidal bending axis. This will give rise to compressive bending stress *B* which will be normal to the cross-sectional plane and there will be no stress at the point *A* since it lies on the axis *yy*.

$$(\sigma_b)_B = \frac{My_B}{I_{yy}} = (8 \times 10^4)\frac{25}{2} \times \frac{64}{\pi (25)^4} \text{ N/mm}^2$$

$$= 52.15 \text{ MPa}$$

**Effect of torque (12500 N-mm) :**

It will give rise to shear stress on the cross-sectional plane and will be maximum at both the points *A* and *B* but will be directed tangentially at the respective points.

$$\frac{T}{J} = \frac{2\tau}{d}$$

$$\tau = \frac{Td}{2J} = \frac{(12500) \times 25}{2 \times \left(\frac{\pi}{32} \times 25^4\right)} \text{ N/mm}^2$$

$$= 101.86 \text{ MPa}$$

**For principal stresses at point B**

The point diagram as seen in the elevation of a small parallelepiped around the point *B* is drawn in Fig 20.5(*d*).

The planes *FG* and *HK* are parallel to the cross- sectional plane at *B*

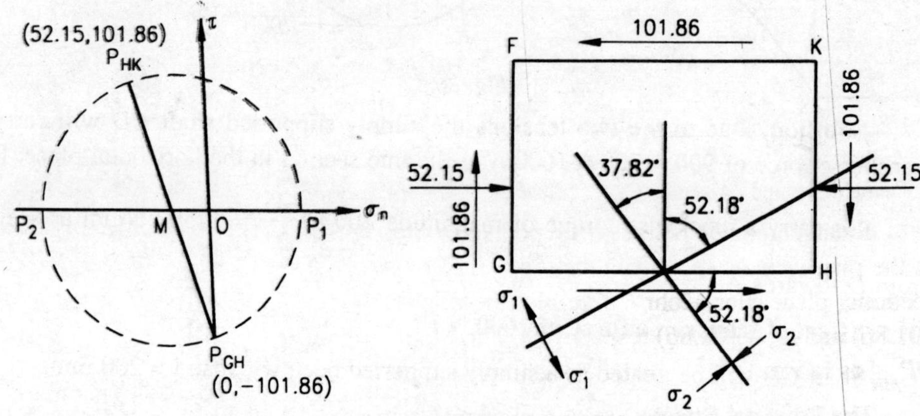


*Fig. 20.5 (d)*

After drawing the axes of reference for $\sigma_n$ and $\tau$ plot points $P_{HK}$ and $P_{GH}$ with co- ordinates (– 52.15, 101.86) and (0, – 101.86) respectively. The line joining these points intersects the *x*-axis at *M* to give the location of centre of the Mohr Circle. Taking *M* as centre and radius $MP_{HK}$ complete the circle. The principal stresses are given by

$$\sigma_1 = OP_1 = 79.07 \text{ MPa (Tensile)}$$

$$\sigma_2 = OP_2 = 131.05 \text{ MPa (comp.)}$$

From Fig. 20.6(*c*)

$$\angle P_1MP_{HK} = 104.36°$$

Therefore $P_1$ is inclined at $\frac{104.46}{2}$ = 52.18° clockwise from *KH*

Further $\angle P_{HK}MP_2 = 180 - 104.36 = 75.64°$

$P_2$ is inclined at $\frac{75.64}{2}$ = 37.82° anticlockwise from *HK*.

**For Principal Stress at *A***

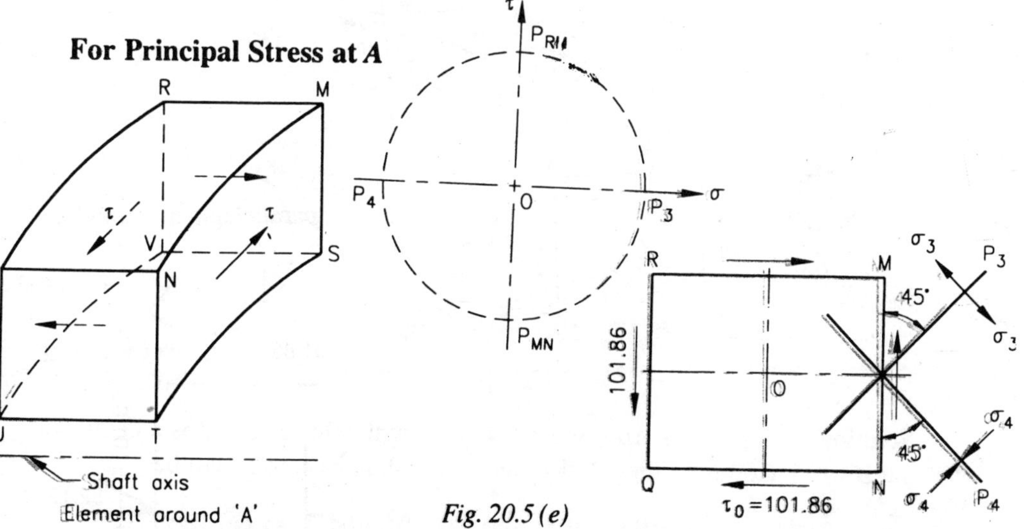


*Fig. 20.5 (e)*

Small parallelepiped assumed point *A* to stresses in Fig. 20.6(*d*) the point diagram in the plan view as seen from the top is also drawn, where *MN* represents the cross-sectional plane. For Mohr Circle plot the points $P_{MN}$ and $P_{RM}$ with co-ordinates (0, 101.86) and (0, – 101.86) respectively. The Mohr circle is drawn with *O* as centre and $OP_{MN}$ as radius.

The Principal Stresses are

$$\sigma_3 = OP_3 = 101.86 \text{ MPa (Tensile)}$$

and

$$\sigma_4 = OP_4 = -\ 101.86 \text{ MPa (comp.)}$$

Also $\angle P_3OP_{MN} = \angle P_4OP_{MN} = 90°$

Therefore the principal plane $P_3$ is inclined at $\frac{90}{2}$ = 45° clockwise from cross-sectional plane *MN*. Likewise the other principal plane $P_4$ is inclined at $\frac{90}{2}$ = 45° anticlockwise from the cross- sectional plane *MN*.

**Problem 20.7.** *(a) In Fig. 20.6 ABCD is a lever made out of steel circular bar. The end C is fixed. A load of 1 kN is applied at point A. Find the defection of the point A. Take E = 200 GPa = 200 × $10^3$ N / $mm^2$. G = 80 GPa = 80 × $10^3$ N / $mm^2$, (h) find, at Section C using Mohr Circle method, maximum principal tensile stress and its inclination with the cross-sectional plane.*

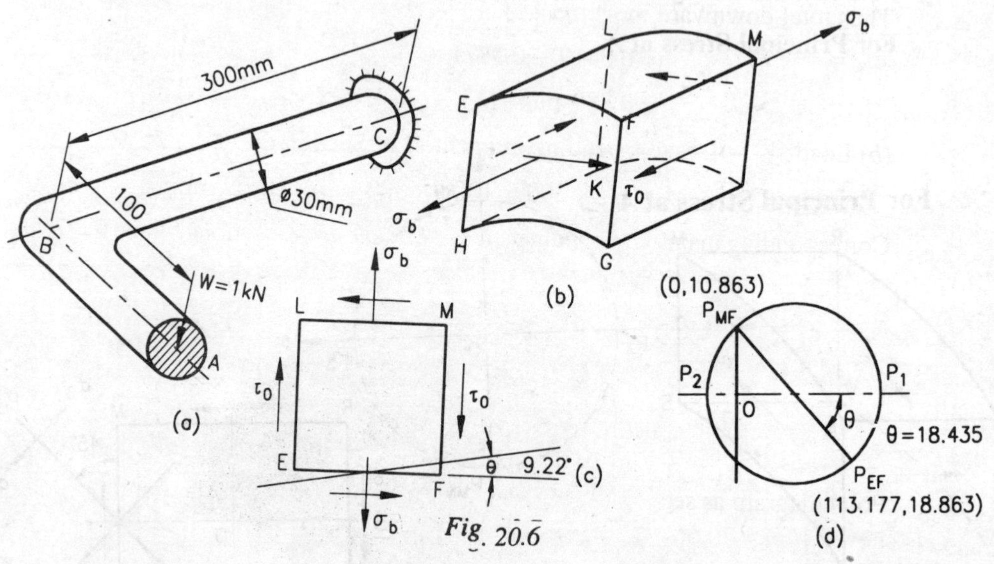


Fig. 20.6

**Solution.** The portions *AB* (length $l_1$ = 100) will behave as if it is a cantilever with end *B* fixed. Corresponding deflection of *A* with respect to *B* will be

$$\delta_{AB} = \frac{Wl^3}{3EI} = \frac{100 \times (100)^3 \times 64}{3 \times (200 \times 10^3) \times \pi \times (30)^4} = 0.0419 \text{ mm}$$

Further at point *B* two vertically applied equal and opposite loads $W_1$ and $W_2$ = 1 kN can be assumed. Then upward load $W_1$ and given load *W* will result in a twisting moment of magnitude (1000 × 100 N-mm) acting on the portion *BC* thereby causing twist θ in it which will result in downward movement = *AB* × θ. In addition to it the effect of downward load $W_2$ at *B* will be to deflect *B* with respect to the fixed end of the cantilever *BC*.

$$\delta_{BC} = \frac{W_2 l_2^3}{3I} = \frac{1000 \times (300)^3 \times 64}{3 \times (200 \times 10^3) \times \pi \times (30)^4}$$

$$= 0.3771 \text{ mm}$$

For Calculating θ, we know that

$$\frac{G\theta}{l_2} = \frac{T}{J}$$

or $$\theta = \frac{(1000 \times 100) \times 300 \times 32}{(80 \times 10^3) \times \pi \times (30)^4} = 0.0047 \text{ rad.}$$

Therefore deflection of $A$ due to rotation of end $B$ with respect to end $C$

$$= \theta \times AB$$

$$= 0.0047 \times 100 = 0.47 \text{ mm.}$$

Thus total downward movement of $A$

$$= 0.0419 + 0.3771 + .47$$

$$= 0.889 \text{ mm.}$$

(*b*) Load $W_2$ will give a bending moment $M$ at Section $C$

$$M = W_2 \times 300 = 1000 \times 300 = 3 \times 10^5 \text{ N-mm}$$

Corresponding maximum bending stress $\sigma_b$ on the cross-sectional plane $EFGH$ is

$$\sigma_b = \frac{d}{2} \times \frac{M}{I} = \frac{30}{2} \times \frac{(3 \times 10^5) \times 64}{\pi (30)^4} = 113.177 \text{ N/mm}^2$$

$$\tau_0 = \frac{d}{2} \times \frac{T}{J} = \frac{30}{2} \times \frac{(1000 \times 100) \times 32}{\pi \times (30)^2} = 18.863 \text{ N/mm}^2$$

Point diagram as seen in the top view of a small parallelepiped around the point $C$ is shown in Fig. 20.6(*c*). Corresponding Mohr's Circle is drawn in Fig. 20.6(*d*).

$$\sigma_1 = \text{Maximum Principal stress}$$

$$OP_1 = \frac{\sigma_b}{2} + \sqrt{\frac{\sigma_b}{2} + \tau_0^2}$$

$$= \frac{113.177}{2} + \sqrt{\left(\frac{113.177}{2}\right)^2 + (18.863)^2}$$

$$= 56.5885 + \sqrt{3202.26 + 355.81}$$

$$= 116.24 \text{ MPa}$$

If $\theta$ is indication of $P_1$ with plane $EF$,

Also $$\tan 2\theta_1 = \left(\frac{-2\tau_0}{-\sigma_b}\right) = \frac{2\tau_0}{\sigma_b} = \frac{2 \times 18.863}{113.117}$$

*i.e.* $$2\theta_1 = 18.435 \text{ degrees}$$

or $$\theta_1 = 9.22 \text{ degrees anticlockwise from cross-sectional plane } EF$$

**Problem 20.8.** *(a) Fig 20.7(a) shows a crank ABCD in horizontal plane with a vertical load W = 500 N acting at point D. The crank is made out of a round 30 mm diameter bar stock. Draw the Mohr's circle of stress for the topmost point of the cross-sectional plane of the bar at the point A and find the principal stresses at this point and find the location of the principal planes. Also find the maximum shear stress in the material.*

*(b) Find the deflection of point D with respect to the point A.*

*Take G = 80 × $10^3$ MPa and E = 200 × $10^3$ MPa Neglect shear stress due to transverse loads.*

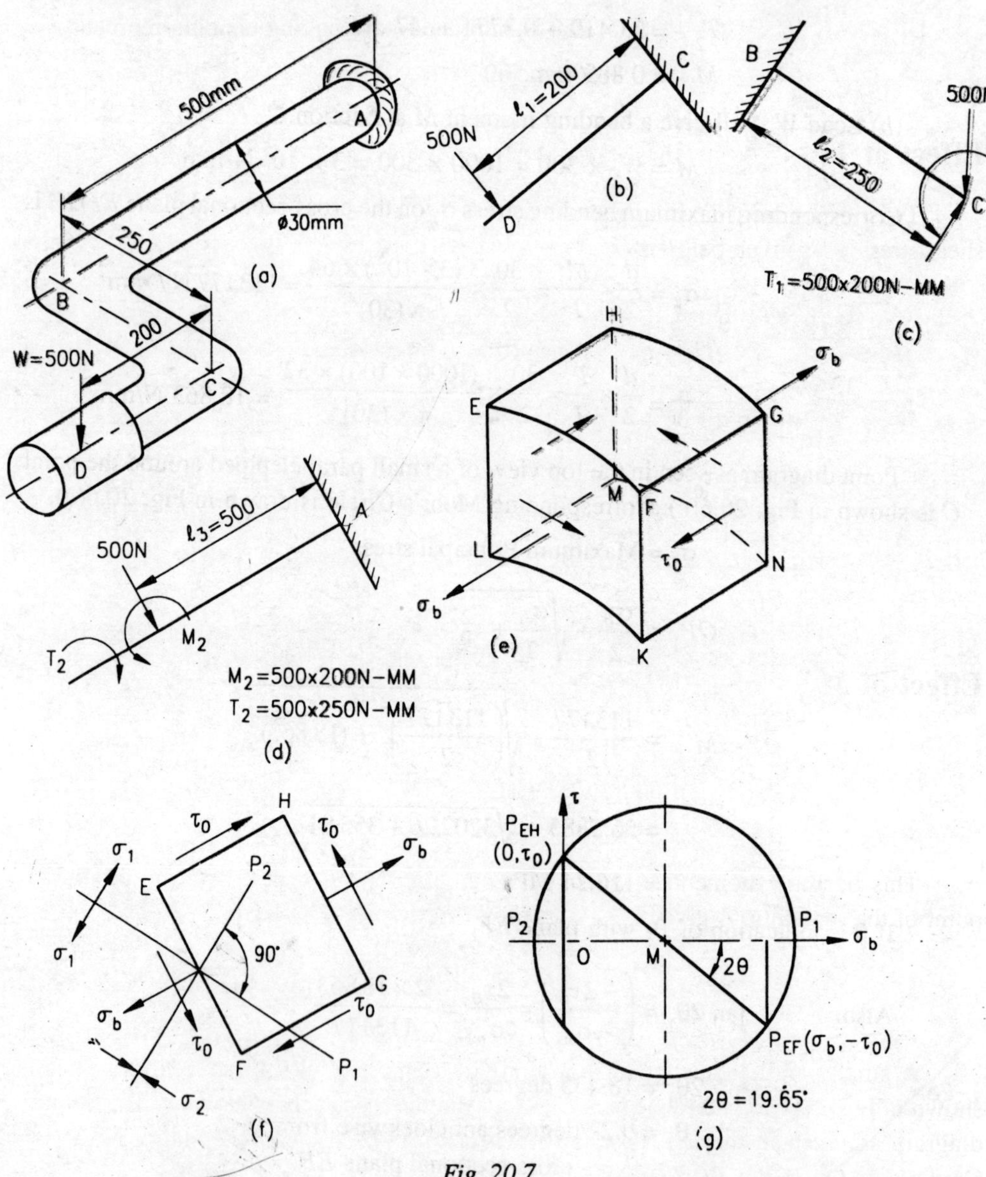


*Fig. 20.7*

**Solution.** (*a*) The crank can be split into 3 segments as shown in Fig. 20.7(*b*), (*c*) and (*d*). The portion *DC* is equivalent to a cantilever *DC* with end *C* fixed and load of 500 *N* at free end *D*. The segment *CB* is another cantilever with end *B* fixed and the free end *C* carrying a downward load of 500 *N* and an anticlockwise twisting moment $T_1 = 500 \times 200$ N-mm. Similarly the segment *BA* is third cantilever with end *A* fixed and at its free end *B* carrying a twisting moment $T_2 = 500 \times 250$ N-mm and bending moment $M_1 = T_1 = 500 \times 200$ N-mm.

Thus from Fig. 20.8(*d*) the section at A experiences a twisting moment.

$T_2 = 500 \times 250$ N-mm and total hogging bending moment

$M = M_2 + 500 \times 500$

## Effect at $T_2$

Twisting moment $T_2$ will generate twist $\theta_{BA}$ of a Section *B* with respect to *A* and shear stress $\tau_0$ as given below :

$$\theta_{BA} = \frac{l_3}{F}\frac{T_2}{J} = \frac{500}{80 \times 10^3} \times \frac{(500 \times 250) \times 32}{\pi\,(30)^4} \text{ rad.}$$

$$= 0.009824 \text{ rad.}$$

$$\tau_0 = \frac{d}{2} \times \frac{T_2}{J}$$

$$= \frac{30}{2} \times \frac{(500 \times 250) \times 32}{\pi \times (30)^4} = 23.585 \text{ MPa}$$

## Effect of *M*

$$M = M_2 + 500 \times l_3$$

$$= 500 \times 200 + 500 \times 500$$

$$= 35 \times 10^4 \text{ N-mm}$$

This bending moment *M* will induce bending tensile stress $\sigma_b$ at the topmost point of the section at *A* given by

$$\sigma_b = \frac{32M}{\pi d^3} = \frac{32 \times 35 \times 10^4}{\pi \times (30)^3} = 132.04 \text{ MPa}$$

A small parallelepiped drawn around the topmost point of the section at *A* is shown in Fig. 20.7(*e*) where the face *EFKL* lies in the cross-sectional plane. The point diagram as given in the top view is shown in Fig. 20.7(*f*). Corresponding Mohr's Circle is drawn in Fig. 20.7(*g*).

The principal stress $\sigma_1 = OP_1$

$= 136.126$ MPa (Tensile)

Other principal stress $\sigma_2 = OP_2 = 4.08$ MPa (comp.)

It can also be calculated analytically

$$\sigma_{1,2} = \frac{\sigma_b}{2} \pm \sqrt{\left(\frac{\sigma_b}{2}\right)^2 + \tau_0^2}$$

$$= \frac{132.04}{2} \pm \sqrt{\left(\frac{132.04}{2}\right)^2 + (23.585)^2}$$

$$= 66.02 \pm \sqrt{4358.64 + 556.252}$$

$$= 66.02 \pm 70.106$$

$\therefore \quad \sigma_1 = 136.126$ MPa (Tensile)

$\sigma_2 = 66.02 - 79.106 = -4.086$ MPa (comp.)

Inclination of principal planes

$$2\theta = \tan^{-1}\left(\frac{-2\tau_0}{(0-\sigma_b)}\right)$$

$$= \tan^{-1}\left[\frac{2 \times 23.585}{132.04}\right]$$

$$= \tan^{-1}(.3572) = 19.656 \text{ deg.}$$

$$\theta = \frac{19.65}{2} = 9.83 \text{ deg.}$$

Thus the principal $P_1$ is inclined at 9.83 degrees anticlockwise from the cross-sectional plane *EF*. The other principal plane $P_2$ is inclined at 90 + 9.83 = 99.83 anticlockwise from *EF*.

The maximum shear stress $\tau_{max}$ is given by

$$\tau_{max} = \sqrt{\left(\frac{\sigma_b}{2}\right)^2 + \tau_0^2}$$

$$= \sqrt{\left(\frac{132.04}{2}\right)^2 + (23.585)^2}$$

$$= 70.106 \text{ MPa}$$

The maximum shear stress planes are inclined at 40° to the principal planes.

(*b*) The deflection of point *D i.e.* $\delta_D$ will be given by the relationship

$$\delta_D = (\delta_{DC} \text{ roman in cantilever } DC) + (\delta_{CB} \text{ in roman in cantilever}) (BC + \delta_{BA} \times l_1) + (\delta_{BA} \text{ in cantilever } AB + \delta_{BA} \times 12)$$

$$\delta_{DC} = \frac{Wl_1^3}{3EI} = \frac{500 \times (200)^3 \times 64}{3 \times (200 \times 10^3)\pi \times (30)^4}$$

$$= 0.1677 \text{ mm}$$

$$\delta_{CB} = \frac{Wl_2^3}{3EI} = \frac{500 \times (500)^3 \times 64}{3 \times (200 \times 10^3)\pi \times (30)^4}$$

$$= 0.3275 \text{ mm}$$

$$\delta_{BA} = \frac{Wl_3^3}{3EI} = \frac{500 \times (500)^3 \times 64}{3 \times (200 \times 10^3)\pi \times (30)^4}$$

$$= 2.62 \text{ mm}$$

$$\delta_{CB} = \frac{I_2}{G} \times \frac{T_1}{J}$$

$$= \frac{250 \times 500 \times 200 \times 32}{80 \times 10^3 \times \pi \times (30)^4} = 0$$

$$= 0.0039296 \text{ rad.}$$

$$\therefore \quad \delta_D = 0.1677 + (0.3275 + 0.0039296 \times 200) + (2.62 + 0.009824 \times 250)$$

$$= 0.1677 + 0.3275 + 0.78592 + 2.62 + 2.456$$

$$= 6.3271 \text{ mm}$$

**Problem 20.9.** *Three beams A, B, C are of length and same material. They are individually subjected to the same bending moment, M which induces the same maximum stress which is equal to the maximum allowable stress for the material. The cross-section of the beam A is circular and that of beam B is square. The section of beam C is rectangular having depth twice the width. Find the following ratios:*

Ratio of weight of beam *B* to weight of beam *A*

Ratio of weight of beam *C* to weight of beam *A*

**Solution:** Since the three beams are of same material and are of same length their weight ratios will be equal to the ratio of respective areas of cross-sections.

For beam A let the diameter be $= D$

$$\text{Maximum bending stress } \sigma_a = \frac{32M}{\pi D^2} \quad \text{...(A)}$$

$$\text{Area of cross-section of beam } A = A_a = \frac{\pi}{4} D^2 \quad \text{...(B)}$$

For beam B: Let the side of square section be a

Then $\quad A_b = a^2 \quad$ ...(C)

$$\sigma_b = \frac{\frac{M}{a^4}\frac{a}{2}}{12} = \frac{6M}{a^3} \qquad \text{...(D)}$$

Equating $\sigma_a$ and $\sigma_b$ we get $\dfrac{6M}{a^3} = \dfrac{32M}{\pi D^3}$

or
$$\frac{D}{a} = \sqrt[3]{\frac{16}{3\pi}} \qquad \text{...(E)}$$

Therefore ratio of Weights of beam $B$ to beam $A$

$$\frac{W_b}{W_a} = \frac{A_b}{A_a}$$

$$= \frac{4a^2}{\pi D^2} \qquad \text{...(F)}$$

Therefore from (E) and (F) we get

$$\frac{W_b}{W_a} = \frac{4}{\pi}\sqrt[3]{\frac{12}{\pi^2}}$$

$$= \sqrt[3]{\frac{12}{\pi^2}} = 1.0673 \qquad \text{...(G)}$$

For the beam C, let the width be = $b$ and depth = $2b$

Then
$$AC = 2b^2 \qquad \text{...(H)}$$

$$I_C = \frac{b(2b)^3}{12} = \frac{2}{3}b^4 \qquad \text{...(I)}$$

Then from (A) and (J) equation $\sigma_a$ and $\sigma_c$ we get

$$\frac{32M}{\pi D^3} = \frac{3}{2}\frac{M}{b^2}$$

or
$$\frac{D}{b} = \sqrt[3]{\frac{64}{3\pi}} \qquad \text{...(K)}$$

Now
$$\frac{W_c}{W_a} = \frac{A_c}{A_a}$$

$$= \frac{8b^2}{\pi D^2} = \frac{8}{\pi}\left[\frac{3\pi}{64}\right]^{\frac{2}{3}} = 0.710$$

**Problem 20.10.** *A cantilever beam ABC is carrying a point load W at B as shown in fig. 20.8. Find deflection at the free end by applying the bending equation for both the segments AB and BC.*

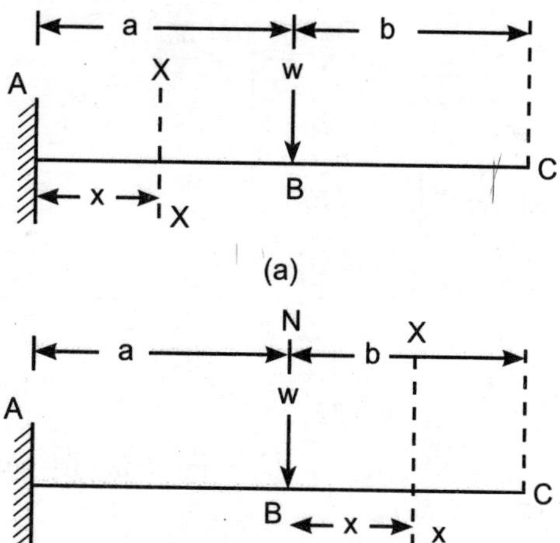


**Fig. 20.8.** *(b)*

**Solution:** For the segment *AB* take origin at *A* and take *A* section at a distance a such that $0 < x < a$.

Then from deflection equation we have

$$EI\frac{d^2y}{dx^2} = W(a-w) \qquad \text{...(A)}$$

Integrating equation (A) twice we get

$$EI\frac{dy}{dx} = -\frac{W}{2}(a-x)^2 + A \qquad \text{...(B)}$$

$$EIy = \frac{W}{6}(a-x)^3 + Ax + B \qquad \text{...(C)}$$

We have the slope and deflection as zero at the point *A*.

Then we get from equations (B) and (C) where $x = 0$

$$A = \frac{Wa^2}{2}$$

$$B = -\frac{Wa^3}{6}$$

Substituting the values of constants *A* and *B* we get the slope and deflection equations for the segment *AB* as:

$$EI\frac{dy}{dx} = -\frac{W}{2}(a-x)^2 + \frac{Wa^2}{2} \qquad \text{...(D)}$$

$$= \frac{W}{6}(a-x)^3 + \frac{Wa^2}{2}x - \frac{Wa^3}{6} \qquad \text{...(E)}$$

From equations (D) and (E) for $x = a$ we get at the point *B*

$$EI\frac{dy}{dx} \text{ at } B = \left[EI\frac{dy}{dx}\right]_B = \frac{Wa^2}{2}$$

or
$$\left[\frac{dy}{dx}\right]_B = \frac{Wa^2}{2EI} \qquad \text{...(F)}$$

$$[EIy]_B = \frac{Wa^3}{3}$$

or
$$y_B = \frac{Wa^3}{3EI} \qquad \text{...(G)}$$

For the segment $BC$ take the origin at the point $B$ and for section $XX$ at distance '$x$' from $B$ where $0 < x < b$ (Fig. 20.8b).

we have
$$EI\frac{d^2y}{dx^2} = 0 \qquad \text{...(H)}$$

Integrating equation (H) twice we get slope and Deflection equation for the segment $BC$ as:

$$EI\frac{dy}{dx} = C \qquad \text{...(I)}$$

$$EIy = Cx + D \qquad \text{...(J)}$$

For the end condition $x = 0$ from these equations we get the values of the constants $C$ and $D$ as:

$$C = \left[EI\frac{dy}{dx}\right]_B = \frac{Wa^3}{2} \qquad \text{...(K)}$$

Ard
$$D = EI\, y_B = \frac{Wa^3}{3} \qquad \text{...(L)}$$

Substituting values of constants $C$ and $D$ we get the slope and deflection equations for the segment $BC$, where $0 < x < b$

$$\frac{dy}{dx} = \frac{Wa^3}{2EI} = \left[\frac{dy}{dx}\right]_C$$

$$EIy = Cx + D$$

$$= \frac{Wa^2}{2}x + \frac{Wa^3}{3} \qquad \text{...(M)}$$

At $x = b$ we get from equation (M)

$$EI\, y_c = \frac{Wa^2}{2}b + \frac{Wa^3}{3}$$

or
$$y_c = \frac{Wa^2}{2EI}b + \frac{Wa^3}{3EI}$$

**Problem 20.11.** *A uniform cross-section girder of length L is supported on two simple supports with equal overhang, a, on two sides as shown in fig. 20.9. Find the ratio a/L if the maximum bending moment (positive or negative) to which girder is subjected is least possible.*

**Solution:** Let the weight per unit length of the beam be = $w$

Then the reactions $R_B = R_C = \dfrac{wL}{2}$

Since the loading is symmetrical, the shape of the bending moment diagram will be as shown in fig. 20.9b.

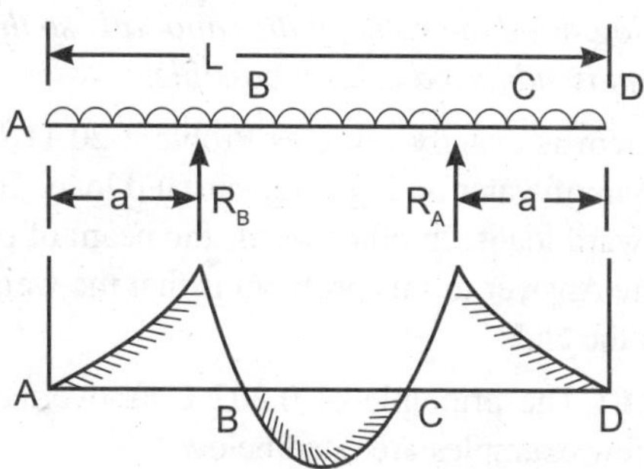


**Fig. 20.9.**

*BM* at the supports will be $\qquad M_B = M_C = \dfrac{wa^2}{2}$ ...(A)

The *BM* at the supports will be positive and will be maximum. However the *BM* at the point *E*, the mid section of the beam will be negative but maximum in magnitude. For any value of the ratio "*a*/*L*" we have to chose the greater of the two viz. $M_-$ or $M_E$. Under these conditions the absolute maximum *BM* will be least.

When the numerical values of both $M_B$ and $M_E$ are equal.

Now
$$M_E = \left[\frac{wL}{2}\right]\left[\frac{L}{4}\right] - R_B\left[\frac{L}{2} - a\right]$$
$$= \frac{wL^2}{8} - \left[\frac{wL}{2}\right]\left[\frac{L}{2} - a\right]$$
$$= -\frac{wL^2}{8}\left[1 - 4\frac{a}{L}\right] \qquad \text{...(B)}$$

From (A) and (B) equating the magnitudes of *BM* at points *B* and *E*, we get

$$\frac{wa^2}{2} = \frac{wL^2}{8}\left[1 - 4\frac{a}{L}\right]$$

or
$$4\left[\left(\frac{a}{L}\right)^2\right] + 4\left[\frac{a}{L}\right] - 1 = 0$$

$$\text{or} \qquad \frac{a}{L} = \frac{1}{8}\left[1 - \pm\sqrt{16+16}\right]$$

$$= \frac{1}{8}[1 - 4 \pm 5.657]$$

$$= 0.207 \qquad \text{(Ignoring the negative value)}$$

**Problem 20.12.** *A wooden log of rectangular section width b and depth d is of length L. It is floating in water such that the top face L × b is horizontal. And it carries two equal loads placed symmetrically on it such that each weights is at the same distance. 'a', from the respective end of the log. Find the value of the ratio 'a/L' so that the maximum bending moment to which the beam is subjected is least possible.*

**Solution:** This problem is exactly same as Problem 20.11 except that the loads are point loads and the reaction of water on log is upward u/d load. Total upward load is equal to the sum of two down ward loads. In other words the beam of problem 20.11 is inverted and nothing else. Thus the Answer to this problem is that the weights are to be placed at a distance of 0.207 L from the ends.

☛ **IMPORTANT NOTE:** The principle of 0.207 L as overhang is applied in various situations in industry. A few examples are cited below.

*(i)* For the purpose of transporting a long and heavy boiler shell Two eye-bolts are provided so that crane hooks can be hooked there to lift the shell and loaded on the career vehicle. The hooks are fixed at 0-207 L distance from the ends. This way while lifting the shell the maximum BM to which the shell is subjected is smallest and so is the bending stress.

*(ii)* While laying the rail track the wooden logs which are used to support the rails are so designed that the overhang on both sides is 0.207 L and the distance between the rails is equal to standard broad gauge. Here again earth's reaction is upward u/d load and the downward loads are the loads supported by two wheels of wheel-axle assembly. (assumed to be equal).

Chapter 21

# ADVANCED TOPICS : GENERALIZED HOOKE'S LAW AND AIRY'S FUNCTION

## 21.1 Introduction : Review of Equilibrium Equations

In chapter 18, the three dimensional strees systems had been introduced, while doing so double suffix notation was employed to define direction of each stress.

In general a body may be subjected is external forces acting in all possible directions. In addition there may be stresses due to body forces present by virtue of the bulk of material. These are called stresses due to body forces. For example gravitational force, inertia force or even magnetic force. The body forces are defined as force per unit volume. However in majority of the cases stresses due to these forces are small as compared to other stresses and are rejected. In chapter 18, equilibrium equatures have been derived reglecting the body forces. These are reproduced here

$$\frac{\partial \tau_{xx}}{\partial x}+\frac{\partial \tau_{xy}}{\partial y}+\frac{\partial \tau_{xz}}{\partial z} = 0 \qquad ...(21.1a)$$

$$\frac{\partial \tau_{yx}}{\partial x}+\frac{\partial \tau_{yy}}{\partial y}+\frac{\partial \tau_{yz}}{\partial z} = 0 \qquad ...(21.2a)$$

$$\frac{\partial \tau_{zx}}{\partial x}+\frac{\partial \tau_{zy}}{\partial y}+\frac{\partial \tau_{zz}}{\partial z} = 0 \qquad ...(21.3a)$$

$$\tau_{xy} = \tau_{yx} \qquad ...(21.4a)$$

$$\tau_{yz} = \tau_{zy} \qquad ...(21.5a)$$

$$\tau_{xz} = \tau_{zx} \qquad ...(21.6a)$$

In case body forces are taken into account, these equations then take the form as given below :

$$\frac{\partial \tau_{xx}}{\partial x}+\frac{\partial \tau_{xy}}{\partial y}+\frac{\partial \tau_{xz}}{\partial z}+B_x = 0 \qquad ...(21.1b)$$

$$\frac{\partial \tau_{yx}}{\partial x}+\frac{\partial \tau_{yy}}{\partial y}+\frac{\partial \tau_{yz}}{\partial z}+B_y = 0 \quad \text{...(21.2}b\text{)}$$

$$\frac{\partial \tau_{zx}}{\partial x}+\frac{\partial \tau_{zy}}{\partial y}+\frac{\partial \tau_{zz}}{\partial z}+B_z = 0 \quad \text{...(21.3}b\text{)}$$

Again for isotropic bodies, the constitutive law as per the three dimensional principal strains are as given below :

$$\varepsilon_{xx} = \frac{1}{E}[\tau_{xx} - v(\tau_{yy} + \tau_{zz})] \quad \text{...(21.7)}$$

$$\varepsilon_{yy} = \frac{1}{E}[\tau_{yy} - v(\tau_{zz} + \tau_{xx})] \quad \text{...(21.8)}$$

$$\varepsilon_{zz} = \frac{1}{E}[\tau_{zz} - v(\tau_{xx} + \tau_{yy})] \quad \text{...(21.9)}$$

While deriving the equations (21.7) through (21.9), the strain terms describe the deformational relative movement of the adjacent points originally along orthogonal axes or else it can be said that the strain terms measure the stretching of originally orthogonal infinitesimal line segments at a point as well as change in relative orientation of these line segments. Since the displacement field describes the movement of each and every point in the body, it is expected that the strains can be expressed in terms of the displacement field. Accordingly in the following paragraph these relationships are being developed for "small deformations." These strain displacement relationships are very important.

## 21.2 Strains in Terms of Displacement Field

### 21.2.1 Direct Strains

Let the displacement field be expressed by $u$ as function of $x$, $y$ and $z$

$$u = u\,(x, y, z) \quad \text{...(21.10)}$$

In Fig. (21.1) is shown an element $AB$ of length $\Delta x$ parallel to $x$-axis in the underformed geometry of a body. The point $A$ has the position of coordinates

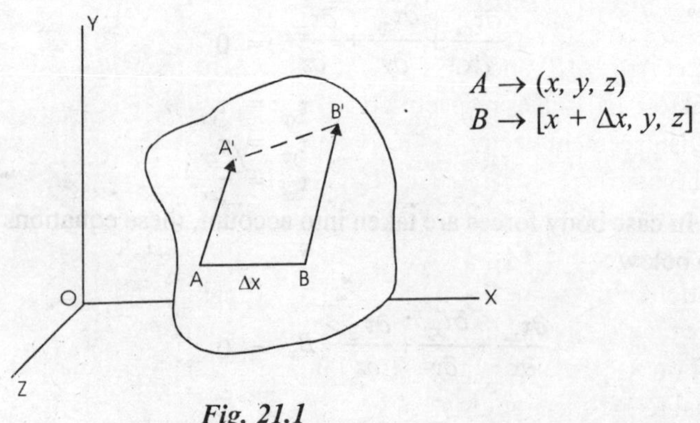


*Fig. 21.1*

$(x, y, z)$ and that of $B$ are $(x + \Delta x, y, z)$. Displacement vectors are $AA'$ and $BB'$ for the ends $A$ and $B$ of the element $AB$ to the positions $A'$ and $B'$ respectively of these points in the deformed geometry. Then

$$\varepsilon_{xx} = \frac{\overrightarrow{A'B'} - \Delta x}{\Delta x} \qquad ...(21.2)$$

If the shear strains present are small the angle between vectors $\overrightarrow{A'B'}$ are $\overrightarrow{AB}$ will be small and then the projection of $\overrightarrow{A'B'}$ in the $x$-direction can replace $\overrightarrow{A'B'}$ in equation (21.2). In other words it will mean that normal strain in $x$-direction ($\varepsilon_{xx}$) is the elongation component of a vanishingly small line segment in the original direction of the line segment per unit length. That is equation (21.2) reduces to

$$\varepsilon_{xx} = \underset{\Delta x \to 0}{\text{Let}} \frac{(\overline{A'B'})_x - \Delta x}{\Delta x} \qquad ...(21.3)$$

where $(\overline{A'B'})_x$ is the component of $\overrightarrow{A'B'}$ along $x$-axis.

Then $(\overline{A'B'})_x$ can be expressed as the sum of $\Delta x$ plus net movement in the $x$-direction of the ends $A$ and $B$ of original length $\Delta x$ as a result of deformation.

That is $$(\overline{A'B'})_x = \Delta x + [u_x(x + \Delta x, y, z) - u_x(x, y, z)] \qquad ...(21.4)$$

From equations (21.2) through (21.4)

We get $$\varepsilon_{xx} = \underset{\Delta x \to 0}{\text{Let}} \frac{u_x(x+\Delta x, y, z) - u_x(x, y, z)}{\Delta x} \qquad ...(21.5)$$

or $$\varepsilon_{xx} = \frac{\partial u_x}{\partial x} \qquad ...(21.6)$$

Similarly $$\varepsilon_{yy} = \frac{\partial u_y}{\partial y} \qquad ...(21.7)$$

and $$\varepsilon_{zz} = \frac{\partial u_z}{\partial z} \qquad ...(21.8)$$

### 21.2.2 Shear Strains

Let $\Delta x$ (= $AB$) and $\Delta y$ (= $AC$) (Fig. 21.2) be the vanishingly small line segments parallel to $x$ and $y$-axes respectively in the underformed body at a point $A$ at position $(x, y, z)$. Displacement vectors at the points $A$, $B$ and $C$ are $AA_1$, $BB_2$ and $CC_2$ respectively after the body has deformed. In order to find the shear angle $\gamma_{xy}$ (= $2\varepsilon_{xy}$) consider a plane parallel to $x$-$y$ plane, passing through the point $A_1$ which is the end of the deformation vector $AA_1$. Then take the projection of points $B_2$ and $C_2$ onto this plane. Projection of the point $B_2$ on this plane is $B_1$ and that of $C_2$ is $C_1$ on this plane. Let the angle which the projection $A_1B_1$ make with $x$-direction be $\alpha$ and the angle which the projection $A_1C_1$ makes with $y$-direction be $\beta$. Then the sum of the angles $\alpha$ and $\beta$ will be equal to the shear angle $\gamma_{xy}$

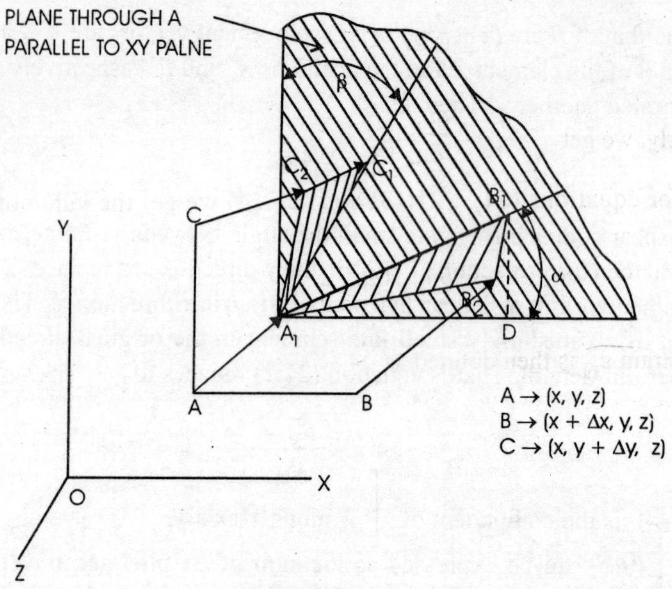


*Fig. 21.2*

$$\varepsilon_{xy} = \frac{1}{2}\gamma_{xy} \qquad \text{...(21.9a)}$$

Thus $$\gamma_{xy} = \alpha + \beta \qquad \text{...(21.9b)}$$

Further in order to evaluate the angle α draw a perpendicular $B_1D$ form the point $B_1$ onto the line $A_1D$ parallel to *x*-direction then from the geometry of the figure.

$$\tan \alpha = \lim_{\Delta x \to 0} \left( \frac{B_1D}{A_1D} \right)$$

If α is small then we can take

$$\tan \alpha \approx \alpha$$

that is $$\alpha = \lim_{\Delta x \to 0} \left( \frac{B_1D}{A_1D} \right) \qquad \text{...(21.10)}$$

But $DB_1$ is the difference in the displacement vector components in *y*-direction of the points *B* and *A* (or of the vectors $AA_1$ and $BB_2$).

Thus $$DB_1 = u_y(x + \Delta x, y, z) - u_y(x, y, z) \qquad \text{...(21.11)}$$

Also for normal or direct strain we have

$$A_1D = \Delta x \qquad \text{...(21.12)}$$

Therefore substituting values of $DB_1$ and $A_1D$ from (21.11) and (21.12) into equation (21.10) we get

$$\alpha = \lim_{\Delta x \to 0} \frac{u_y(x + \Delta x, y, z) - u_y(x, y, z)}{\Delta x}$$

$$= \frac{\partial u_y}{\partial x} \quad \text{...(21.13)}$$

Similarly, we get $\beta = \dfrac{\partial u_x}{\partial y}$ ...(21.14))

Thus for equations (21.9, 21.13) and (21.14) we get the value of shear angle $\gamma_{xy}$ as

$$\gamma_{xy} = \frac{\partial u_y}{\partial x} + \frac{\partial u_x}{\partial y} \quad \text{...(21.15a)}$$

Shear strain $\varepsilon_{xy}$ is then defined as

$$\varepsilon_{xy} = \frac{1}{2}\gamma_{xy} \quad \text{...(21.15b)}$$

or

$$\varepsilon_{yx} = \varepsilon_{xy} = \frac{1}{2}\left[\frac{\partial u_x}{\partial y} + \frac{\partial u_y}{\partial x}\right] \quad \text{...(21.16)}$$

Similarly

$$\varepsilon_{xz} = \varepsilon_{zx} = \frac{1}{2}\left[\frac{\partial u_x}{\partial z} + \frac{\partial u_z}{\partial x}\right] \quad \text{...(21.17)}$$

and

$$\varepsilon_{yz} = \varepsilon_{zy} = \frac{1}{2}\left[\frac{\partial u_y}{\partial z} + \frac{\partial u_z}{\partial y}\right] \quad \text{...(21.18)}$$

Given the displacement field are we can, with the help of equations, (21.6), (21.7), (21.8) and equations (21.16), (21.17) and (21.18) find the strain field $\Sigma_{ij}$ $(x, y, z)$ as follows :

$$\varepsilon_{ij} = \begin{bmatrix} \dfrac{\partial u_x}{\partial x} & \dfrac{1}{2}\left(\dfrac{\partial u_x}{\partial y} + \dfrac{\partial u_y}{\partial x}\right) & \dfrac{1}{2}\left(\dfrac{\partial u_x}{\partial z} + \dfrac{\partial u_z}{\partial x}\right) \\ \dfrac{1}{2}\left(\dfrac{\partial u_y}{\partial x} + \dfrac{\partial u_x}{\partial y}\right) & \dfrac{\partial u_y}{\partial y} & \dfrac{1}{2}\left(\dfrac{\partial u_y}{\partial z} + \dfrac{\partial u_z}{\partial y}\right) \\ \dfrac{1}{2}\left(\dfrac{\partial u_z}{\partial_x} + \dfrac{\partial u_x}{\partial_z}\right) & \dfrac{1}{2}\left(\dfrac{\partial u_z}{\partial y} + \dfrac{\partial u_y}{\partial z}\right) & \dfrac{\partial u_z}{\partial z} \end{bmatrix} \quad \text{...(21.19)}$$

where $i$ and $j$ both vary from 1 to 3.

## 21.3 Compatibility Relation

While arriving at this relation (21.19) between the strain and the displacement filed the basic assumption has been that the deformation is small. This assumption in turn requires that the strain terms be small compared to unity. In most of the engineering problems the strain are small as the stress level is much below the elastic limit.

Again the relationship (21.19) has been developed by assuming a displacement filed $u$. And for a displacement field to represent a real movement and deformation of a solid which develops no cracks, it is evidently essential that the displacement field be a "single valued, continuous function." In case it is not single valued, it means that certain points can have more than one displacement at a time, which will be absurd in physical sense. And a discontinuity in displacement field means that originally infinitesimally close points would be separated by finite amount in deformed geometry. This will only happen if crack has developed due to deformation and this aspect is not being dealt with by this equation (21.19). Meanwhile, the strain field which is a means of describing deformation of a solid for the purposes of relating it to the stress field, must also have certain conditions imposed on it so that it is properly associated with a physically realizable deformation of a solid free from cracks. In other words we can not simply choose any set of strain component functions and expect them to be automatically associated with a single valued displacement field. For example, suppose that a solid is imagined before deformation to consist of a system of infinitesimal contiguous cubes. On application of loads, threse cubes will have dilatation as a result of normal strains and These cubes will become parallelograms due to shear strains. Further the normal and shear strains will vary with position, but they must do so in such a way that the deformed elements remain contiguous that is the elements must not come apart so as to expose cracks or fissures. Nor must a pair of cubes partially or fully occupy the same space at the same time. One way to ensure that the strains are compatible is to start with a single-valued continuous displacement field and develop the strain from this field in accordance with the equation (21.19).

However, at times when one wishes to work initially with the strains, be may be able to make some reasonable assumption, about strains and not about displacement field. But is order to ensure that strains are properly related so that no crack or fisher develops. For this purpose equations, known as compatible equations, which are listed below have to be satisfied, the derivation of these equations is beyond the scope of present text.

$$\frac{\partial^2 \varepsilon_{xx}}{\partial y^2} + \frac{\partial^2 \varepsilon_{yy}}{\partial x^2} = 2\frac{\partial^2 \varepsilon_{xy}}{\partial x \partial y} \quad \text{...(21.20a)}$$

$$\frac{\partial^2 \varepsilon_{yy}}{\partial z^2} + \frac{\partial^2 \varepsilon_{zz}}{\partial y^2} = 2\frac{\partial^2 \varepsilon_{yz}}{\partial y \partial z} \quad \text{...(21.20b)}$$

$$\frac{\partial^2 \varepsilon_{zz}}{\partial x^2} + \frac{\partial^2 \varepsilon_{xx}}{\partial z^2} = 2\frac{\partial^2 \varepsilon_{zx}}{\partial x \partial z} \quad \text{...(21.20c)}$$

$$\frac{\partial^2 \varepsilon_{xx}}{\partial y \partial z} = \frac{\partial}{\partial x}\left[-\frac{\partial \varepsilon_{yz}}{\partial x} + \frac{\partial \varepsilon_{xz}}{\partial y} + \frac{\partial \varepsilon_{xy}}{\partial z}\right] \quad \text{...(21.20d)}$$

$$\frac{\partial^2 \varepsilon_{yy}}{\partial z \partial x} = \frac{\partial}{\partial y}\left[-\frac{\partial \varepsilon_{zx}}{\partial y} + \frac{\partial \varepsilon_{yx}}{\partial z} + \frac{\partial \varepsilon_{yz}}{\partial x}\right] \quad \text{...(21.20e)}$$

$$\frac{\partial^2 \varepsilon_{zz}}{\partial x \partial y} = \frac{\partial}{\partial z}\left[-\frac{\partial \varepsilon_{xy}}{\partial z} + \frac{\partial \varepsilon_{zy}}{\partial x} + \frac{\partial \varepsilon_{zx}}{\partial y}\right] \quad \text{...(21.20f)}$$

## 21.4 Three Dimensional Hooke's Law for Isotropic Materials

In chapter 1 Hooke's Law for one dimensional state of stress was introduced. In this chapter are shall introduce the reader to Three Dimensional Hooke's Law for isotropic **materials.** Isotropic materials are those materials whose properties are not dependent on the direction. That is the Youngs modules $E$ is assumed to be constant when we extend the one dimensional test data into $x$, $y$ and $z$ directions. It may be worth mentioning here that individual crystals of structural materials are actually not isotropic (i.e. they are anisotropic materials, but due to their usual random orientation of the crystals and their large number, the macroscopic behaviour of mild steel is usually considered isotropic. However when the material is being rolled in a rolling machine, the crystals at times do attain certain amount of anisotropy. In the following paragraphs we proceed to formulate the three-dimensional Hooke's Law for isotropic materials.

Consider a vanishingly small rectangular **parallelepiped** and we apply to this element successively three normal stresses $\tau_{xx}$, $\tau_{yy}$, $\tau_{zz}$ (since we shall revert to tensional symbols here $\tau_{xx}$ is being used in place of $\sigma_{xx}$ etc.) Then the strain for each stress for isotropic material, can be superposed to get the strain at a general point. The strains in their directions due to $\tau_{xx}$ are, if Poisson's ratio is $\nu$

Due to $\tau_{xx}$

$$\varepsilon_{xx} = \frac{\tau_{xx}}{E}$$

$$\varepsilon'_{yy} = -\nu\frac{\tau_{xx}}{E}$$

$$\varepsilon'_{zz} = -\nu\frac{\tau_{xx}}{E}$$

And due to $\tau_{yy}$

$$\varepsilon''_{xx} = -\nu\frac{\tau_{yy}}{E}$$

$$\varepsilon''_{yy} = \frac{\tau_{yy}}{E}$$

$$\varepsilon''_{zz} = -\nu\frac{\tau_{yy}}{E}$$

Similarly due to $\tau_{zz}$

$$\varepsilon'''_{xx} = -\nu\frac{\tau_{zz}}{E}$$

$$\varepsilon'''_{yy} = -\nu\frac{\tau_{zz}}{E}$$

$$\varepsilon'''_{zz} = \frac{\tau_{zz}}{E}$$

Then the resultant strains are

$$\varepsilon_{xx} = \frac{1}{E}[\tau_{xx} - \nu(\tau_{yy} + \tau_{zz})]$$

$$\varepsilon_{yy} = \frac{1}{E}[\tau_{yy} - \nu(\tau_{zz} + \tau_{xx})] \quad \text{...(21.21)}$$

$$\varepsilon_{zz} = \frac{1}{E}[\tau_{zz} - \nu(\tau_{xx} + \tau_{yy})]$$

Now in the torsion test the material is in a state of pure shear. Also the experiment shown the shear stress is proportional to engineering shear strain $\gamma$ upto yield shear stress. Thus when the shear stress is less than yield shear stress

$$\gamma = \frac{\tau}{G} \qquad \text{...(21.22)}$$

Further the shear stress can only generate the corresponding shear strain, we then have

$$\gamma_{xy} = \frac{\tau_{xy}}{G}$$

$$\gamma_{xz} = \frac{\tau_{xz}}{G} \qquad \text{...(21.23)}$$

and

$$\gamma_{yz} = \frac{\tau_{yz}}{G}$$

Thus from equations (21.21) and (21.23), the three dimensional Hooke's law for isotropic materials is as follows :

$$\varepsilon_{xx} = \frac{1}{E}\left[\tau_{xx} - \nu(\tau_{yy} + \tau_{zz})\right]$$

$$\varepsilon_{yy} = \frac{1}{E}\left[\tau_{yy} - \nu(\tau_{zz} + \tau_{xx})\right] \qquad \text{...(21.24)}$$

$$\varepsilon_{zz} = \frac{1}{E}\left[\tau_{zz} - \nu(\tau_{xx} + \tau_{yy})\right]$$

$$\gamma_{xy} = \frac{\tau_{xy}}{G}$$

$$\gamma_{xz} = \frac{\tau_{xz}}{G}$$

$$\gamma_{yz} = \frac{\tau_{yz}}{G}$$

**Example 21.1.** *Following state of stress exists at a point in a body*

$$\tau_{ij} = \begin{bmatrix} 1000 & 0 & 0 \\ 0 & 2000 & -500 \\ 0 & -500 & 0 \end{bmatrix} kPa$$

*Find the strain components at this point. On a uniaxial tensile test it was found that E = 200 GPa and Poissons ratio = 0.25. Assume the material to be isotropic.*

**Solution :** The six stress components at the given point are given which include the direct and shear stresses. Therefore the value of shear modules $G$ will have to be determined before proceeding to find the shear strains.

Now we have from equation (2.72) chapter 2

$$2G(1 + \nu) = E$$

or
$$G = \frac{E}{2(1+\nu)}$$

$$= \frac{200 \times 10^3}{2(1+0.25)} = 80 \times 10^3 \text{ MPa}$$

Given data $\tau_{xz} = 100$ kPa $\qquad \tau_{xy} = \tau_{yx} = 1$

$\tau_{yy} = 2000$ kPa $\qquad \tau_{xz} = \tau_{zx} = 0$

$\tau_{zz} = 0.0$ $\qquad \tau_{yz} = \tau_{zy} = -500$ kPa

Therefore
$$\varepsilon_{xx} = \frac{1}{E}\left[\tau_{xx} - \nu(\tau_{yy} + \tau_{zz})\right]$$

$$= \frac{1}{200 \times 10^3}[1000 - 0.25(2000 + 0)]$$

$$= \frac{1}{200 \times 10^3}[1000 - 500]$$

$$= \frac{500}{200 \times 10^3} = 0.0025$$

$$\varepsilon_{yy} = \frac{1}{E}\left[\tau_{yy} - \nu(\tau_{zz} + \tau_{xx})\right]$$

$$= \frac{1}{200 \times 10^3}[2000 - 0.25(0 + 1000)]$$

$$= \frac{200 - 250}{200 \times 10^3} = 0.00875$$

$$\varepsilon_{zz} = \frac{1}{E}\left[\tau_{zz} - \nu(\tau_{xx} + \tau_{yy}\right]$$

$$= \frac{1}{E}[0 - 0.25(1000 + 2000)]$$

$$= \frac{0.25 \times 3000}{200 \times 10^3} = 0.00375$$

$$\varepsilon_{xy} = \frac{1}{z}\left(\frac{\tau_{xy}}{G}\right) = \frac{1}{2}\left(\frac{0.0}{80 \times 10^3}\right) = 0 = \varepsilon_{yx}$$

$$\varepsilon_{xz} = \varepsilon_{zx} = \varepsilon_{yz} = \varepsilon_{zy} = \frac{1}{2}\left(\frac{\tau_{zx}}{G}\right) = 0$$

$$= \frac{1}{2}\left(\frac{\tau_{yz}}{G}\right)$$

$$= \frac{1}{2} \times \frac{(-500)}{80 \times 10^3} = -0.003125$$

Thus the strain components in tensor form are

$$\varepsilon_{ij} = \begin{bmatrix} 0.0025 & 0 & 0 \\ 0 & 0.00875 & -0.003125 \\ 0 & -0.003125 & 0.00375 \end{bmatrix}$$

or
$$\varepsilon_{ij} = \begin{bmatrix} 2.5 & 0 & 0 \\ 0 & 8.75 & -3.125 \\ 0 & -3.125 & 3.75 \end{bmatrix} \times 10^{-3}$$

## 21.5 Non Isotropic Linear, Elastic Behaviour : Generalized Hooke's Law

In article 21.4 the three-Dimensional Hooke's Law for isotropic materials has been presented followed by an example 21.1 where in it is observed that we need to know only two independent constants namely $E$ and ν (or alternatively $E$ and $G$ or $E$ and $K$ or $K$ and $v$) to find the strain field if the stress field is known).

However in the case of non isotropic material having linear elastic behaviour which occur in individual crystals or rolled sheats etc., following postulate for the stress-strain relations was proposed by Hooke and is known as Generalized Hooke's Law.

$$\begin{aligned}
\tau_{xx} &= C_{11}\varepsilon_{xx} + C_{12}\varepsilon_{yy} + C_{13}\varepsilon_{zz} + C_{14}\gamma_{xy} + C_{15}\gamma_{yz} + C_{16}\gamma_{xy} \\
\tau_{yy} &= C_{21}\varepsilon_{xx} + C_{22}\varepsilon_{yy} + C_{23}\varepsilon_{zz} + C_{24}\gamma_{xy} + C_{25}\gamma_{yz} + C_{26}\gamma_{xz} \\
\tau_{zz} &= C_{31}\varepsilon_{xx} + C_{32}\varepsilon_{yy} + C_{33}\varepsilon_{zz} + C_{34}\gamma_{xy} + C_{35}\gamma_{yz} + C_{36}\gamma_{xz} \\
\tau_{xy} &= C_{41}\varepsilon_{xx} + C_{42}\varepsilon_{yy} + C_{43}\varepsilon_{zz} + C_{54}\gamma_{xy} + C_{45}\gamma_{yz} + C_{46}\gamma_{xz} \\
\tau_{yz} &= C_{51}\varepsilon_{xx} + C_{52}\varepsilon_{yy} + C_{53}\varepsilon_{zz} + C_{54}\gamma_{xy} + C_{55}\gamma_{yz} + C_{56}\gamma_{xz} \\
\tau_{xz} &= C_{61}\varepsilon_{xx} + C_{62}\varepsilon_{yy} + C_{63}\varepsilon_{zz} + C_{64}\gamma_{xy} + C_{65}\gamma_{yz} + C_{66}\gamma_{xz}
\end{aligned}$$

...(21.25 *a, b, c, d, e, f*)

where the constants terms $C_{ij}$($i$, $j$ 1 to 6) form a 6 × 6 matrix of elastic modulii whose values depend upon the material and non isotropic nature. In this formulation it is implied that each stress at a point is linearly related to all the strains at the point.

### 21.5.1 Corollary : Hooke's Law for Orthotropic Materials

In article 21.5 Generalized Hooke's Law for non isotropic materials have been enuenciatied. There are in general 36 elastic modulii for non isotropic materials. These modulii are represented by 6 × 6 matrix.

However, orthotropic materials are those in which the matrix of modulii represents the case of elastic symmetry about three orthogonal planes.

This means there is no change in mechanical behaviour when the *x, y, z* directions are reversed. Example of orthotropic materials are wood. As per the detailed analysis given below, the number of elastic modulii for such materials is reduced form 36 to 12.

Thus from the generalized Hooke's Law equation (21.25) for non isotropic material for *x'y'z'* axis system we have

$$\tau_{x'x'} = C_{11}\varepsilon_{x'x'} + C_{12}\varepsilon_{y'y'} + C_{13}\varepsilon_{z'z'} + C_{14}\gamma_{x'x'} + C_{15}\gamma_{y'y'} + C_{16}\gamma_{z'z'} \quad ...(21.26)$$

However for the orthotropic materials the generalized Hooke's Law is to be modified for which the guiding principle will be that we impose a sequence of solations of the axes *x, y, z* requiring each time that the stress-strain relations for the new orientation have the same form as for the original set of axes.

Let us consider a 180° rotation about *z*-axis as shown in Fig. (21.3). The direction cosines between different axes are given below in table (21.1).

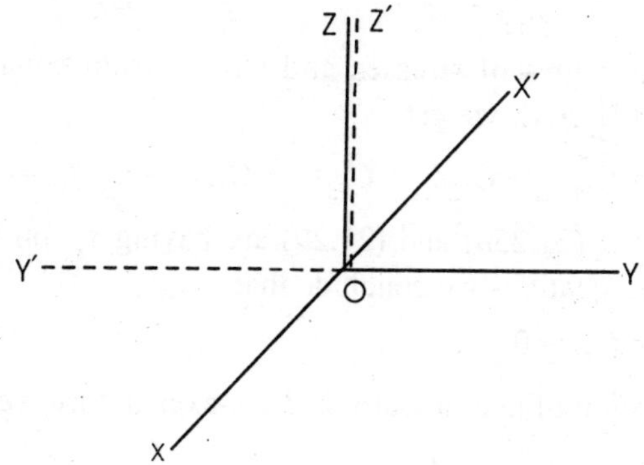


*Fig. 21.3*

**Table 21.1**

| | *x* | *y* | *z* |
|---|---|---|---|
| *x'* | –1 | 0 | 0 |
| *y'* | 0 | –1 | 0 |
| *z'* | 0 | 0 | 1 |

Since the shear stress tensor $\tau_{ij}$ for reference axes $xyz$ is known for the orthogonal interfaces at a point, then the stress on any interface at the point can be determined as per details given in chapter 18. Accordingly, knowing $\tau_{ij}$ for reference $xyz$ at a point, we can determine the stress tensor having components for any reference $x'y'z'$ by considering faces of an infinitesimal rectangular **parallelepiped** having edges parallel to the axes $x'y'z'$ at the point. Accordingly the stresses, for $x'y'z'$ system of axes will come out in terms as of the strains for $xyz$ system of axes as follows

$$\left.\begin{aligned} \tau_{x'x'} &= \tau_{xx} \\ \tau_{y'y'} &= \tau_{yy} \\ \tau_{z'z'} &= \tau_{zz} \\ \tau_{x'y'} &= \tau_{xy} \\ \tau_{y'z'} &= -\tau_{yz} \\ \tau_{x'z'} &= -\tau_{xz} \end{aligned}\right] \quad \text{...(21.27)} \; (a, b, c, d, e, f)$$

Similarly knowing the strain tensor for reference axes $x$, $y$, $z$ at a point, we can determine the strains for any reference $x'y'z'$ rotated arbitrarily relative to axes $x$, $y$, $z$ at that point.

Thus, for pair of references shown in, Fig. 21.3 we get

$$\begin{aligned} \varepsilon_{x'x'} &= \varepsilon_{xx} & \gamma_{x'y'} &= \gamma_{xy} \\ \varepsilon_{y'y'} &= \varepsilon_{yy} & \gamma_{y'z'} &= -\gamma_{yz} \\ \varepsilon_{z'z'} &= \varepsilon_{zz} & \gamma_{x'z'} &= -\gamma_{xz} \end{aligned} \quad \text{...(21.28)}$$

Substituting the values of stresses and strains from equations (21.27) and (21.28) into equation (21.26*a*), we get

$$\tau_{xx} = C_{11}\varepsilon_{xx} + C_{12}\varepsilon_{yy} + C_{13}\varepsilon_{zz} + C_{14}\gamma_{xy} - C_{15}\gamma_{yz} - C_{16}\gamma_{xz} \quad \text{...(21.29)}$$

Both the equations (21.25*a*) and (21.29) are having $\tau_{xx}$ on the left hand side. Thus comparing these equations we conclude that

$$C_{15} = C_{16} = 0 \quad \text{...(21.30)}$$

Similarly by rotation of axes about $y$ and $x$ x-axes in turn we get the additional conditions.

$$\begin{aligned} &C_{25} = 0 \quad C_{36} = 0 \quad C_{51} = 0 \quad C_{54} = 0 \quad C_{63} = 0 \\ &C_{26} = 0 \quad C_{45} = 0 \quad C_{52} = 0 \quad C_{61} = 0 \quad C_{64} = 0 \\ &C_{35} = 0 \quad C_{46} = 0 \quad C_{53} = 0 \quad C_{62} = 0 \end{aligned} \quad \text{...(21.31)}$$

Again referring to the Fig. (21.3) it is evident that mechanical properties in $x$-direction are same as in $x'$ direction and same in the case for $y$ and $y'$ directions. In other words it can be said that there is elastic symmetry about the planes $yz$ and $xz$. In order to achieve the symmetry about the plane $xy$ as well we impose the condition of isotropy for a 180° rotation of axes, about $x$-axis as shown in Fig. 21.4 and proceeding in the manner as explained above the matrix of modulii reduce, to the form given below :

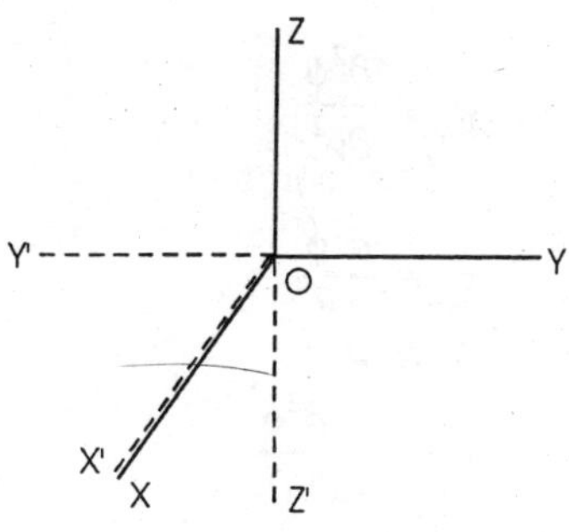


*Fig. 21.4*

$$\begin{bmatrix} C_{11} & C_{12} & C_{13} & 0 & 0 & 0 \\ C_{21} & C_{22} & C_{23} & 0 & 0 & 0 \\ C_{31} & C_{32} & C_{33} & 0 & 0 & 0 \\ 0 & 0 & 0 & C_{44} & 0 & 0 \\ 0 & 0 & 0 & 0 & C_{55} & 0 \\ 0 & 0 & 0 & 0 & 0 & C_{66} \end{bmatrix} \quad \text{...(21.32)}$$

This matrix of modulii pertains to the case of elastic symmetry about three orthogonal planes. In other words it means that there is change in the mechanical behaviour when the $x, y, z$ directions are reversed. Such materials are called orthotropic materials *e.g.*, wood. It is worth pointing out here that number of elastic modulii for such materials has reduced from 36 to 12. Similarly by resorting to rotations of axes, about lines bisecting the angles between $xyz$ axis it can be shown that for isotropic materials there are no more and no less than two elastic modulii.

## 21.6 Plane Stress, Airy's Stress Functions

Plane stress is a situation when stress system on a body is two-dimensional. Let the stresses along $z$-direction be absent then the equilibrium equations (21.1) through (21.6) in the absence of body forces, reduce to.

$$\frac{\partial \tau_{xx}}{\partial x} + \frac{\partial \tau_{xy}}{\partial y} = 0 \quad \text{...(21.33)}$$

$$\frac{\partial \tau_{xy}}{\partial x} + \frac{\partial \tau_{yy}}{\partial y} = 0 \quad \text{...(21.34)}$$

### 21.6.1 Airy's Stress Function

Let the stresses be defined in terms of a function $\phi(x, y)$ as follows

$$\tau_{xx} = \frac{\partial^2 \phi}{\partial y^2} \quad \text{...(21.35a)}$$

$$\tau_{yy} = \frac{\partial^2 \phi}{\partial x^2} \quad \text{...(21.35b)}$$

$$\tau_{xy} = -\frac{\partial^2 \phi}{\partial x\, \partial y} \quad \text{...(21.35c)}$$

Then substituting the values of $\tau_{xx}$, $\tau_{yy}$ and $\tau_{xy}$ form equations (21.35) into equations (21.33) and (21.34), we see that

For equation (21.33)

$$\text{L.H.S.} = \frac{\partial}{\partial x}\left(\frac{\partial^2 \phi}{\partial y^2}\right) + \frac{\partial}{\partial y}\left[-\frac{\partial^2 \phi}{\partial x^{i} y}\right]$$

$$= \frac{\partial^3 \phi}{\partial x\, \partial y^2} - \frac{\partial^3 \phi}{\partial x\, \partial y^2} = 0 = \text{R.H.S. of equation (21.33)}$$

Similarly, L.H.S. of equation (21.34)

$$= \frac{\partial \tau_{xy}}{\partial x} + \frac{\partial \tau_{yy}}{\partial y}$$

$$= \frac{\partial}{\partial x}\left(-\frac{\partial^2 \phi}{\partial x\, \partial y}\right) + \frac{\partial}{\partial y}\left[\frac{\partial^2 \phi}{\partial x^2}\right]$$

$$= -\frac{\partial^3 \phi}{\partial y\, \partial x^2} + \frac{\partial^3 \phi}{\partial y\, \partial x^2}$$

$$= 0 = \text{R.H.S. of equation (21.34)}$$

From the above it is evident that the equations of equilibrium are identically satisfied by the selection of the function $\phi(x, y)$. Function $\phi(x, y)$ is known as the Airy's stress function.

**Example 21.2.** *(a) Show that the compatibility equation for plane stress case reduces to*

$$\frac{\partial^2 \varepsilon_{xx}}{\partial y^2} + \frac{\partial^2 \varepsilon_{yy}}{\partial x^2} = \frac{\partial^2 \gamma_{xy}}{\partial x \partial y} \quad \text{...(21.36)}$$

**Solution :** The plane stress case the derivatives w.r.t. *'z'* will be zero. Therefore compatibility equation (21.20*a*) reduces to

$$\frac{\partial^2 \varepsilon_{xx}}{\partial y^2}+\frac{\partial^2 \varepsilon_{yy}}{\partial x^2} = \frac{2\partial^2 \varepsilon_{xy}}{\partial x \partial y} = \frac{\partial^2 \gamma_{xy}}{\partial x \partial y}$$

Hence the proof.

**Example 21.3.** *Write down the equilibrium equations for plane stress case for no body forces. Then from these equations show that*

$$\frac{\partial^2 \tau_{xy}}{\partial x\,\partial y} = -\frac{1}{2}\left(\frac{\partial^2 t_{xx}}{\partial x^2}+\frac{\partial^2 t_{yy}}{\partial y^2}\right)$$

**Solution :** The equilibrium equation for plane stress case are given in equations (21.33) and (21.34)

$$\frac{\partial \tau_{xx}}{\partial x}+\frac{\partial \tau_{xy}}{\partial y} = 0 \qquad \text{...(21.33)}$$

$$\frac{\partial \tau_{xy}}{\partial x}+\frac{\partial \tau_{yy}}{\partial y} = 0 \qquad \text{...(21.34)}$$

Differentiate partially equation (21.33) w.r.t. '*x*'

$$\frac{\partial^2 \tau_{xx}}{\partial x^2}+\frac{\partial^2 \tau_{xy}}{\partial x\,\partial y} = 0$$

or

$$\frac{\partial^2 \tau_{xy}}{\partial x \partial y} = -\frac{\partial^2 \tau_{xx}}{\partial x^2} \qquad \text{...(21.37)}$$

Again differentiate partially equation (21.34) w.r.t. '*y*'

$$\frac{\partial^2 \tau_{xy}}{\partial x\,\partial y} = -\frac{\partial^2 \tau_{yy}}{\partial y^2} \qquad \text{...(21.38)}$$

Adding equations (21.37) and (21.38) we have

$$\frac{\partial^2 \tau_{xy}}{\partial x \partial y} = -\frac{1}{2}\left(\frac{\partial^2 \tau_{xx}}{\partial x^2}+\frac{\partial^2 \tau_{yy}}{\partial y^2}\right) \qquad \text{...(21.39)}$$

Hence the Proof.

**Example 21.4.** *For plane stress case following results are given as proved in equations (21.36) and (21.39).*

$$\frac{\partial^2 \varepsilon_{xx}}{\partial y^2}+\frac{\partial^2 \varepsilon_{yy}}{\partial x^2} = \frac{\partial^2 \gamma_{xy}}{\partial x \partial y} \qquad \textit{...(21.36)}$$

$$\frac{\partial^2 \tau_{xy}}{\partial x \partial y} = -\frac{1}{2}\left(\frac{\partial^2 \tau_{xx}}{\partial x^2}+\frac{\partial^2 \tau_{yy}}{\partial y^2}\right) \qquad \textit{...(21.39)}$$

*Prove that*

$$\left(\frac{\partial^2}{\partial x^2}+\frac{\partial^2}{\partial y^2}\right)(\tau_{xx}+\tau_{yy}) = 0 \qquad \textit{...(21.40)}$$

**Solution :** In the equation (21.36), when the values of $\varepsilon_{xx}$ and $\varepsilon_{yy}$ are inserted we get

$$\frac{\partial^2}{\partial y^2}\left[\frac{1}{E}\left\{\tau_{xx}-\nu\tau_{yy}\right\}\right]+\frac{\partial}{\partial x^2}\left[\frac{1}{E}\left[\tau_{yy}-\nu\tau_{xx}\right]\right]$$

$$=\frac{\partial^2\tau_{xy}}{G\partial_x\partial_y} \qquad \left(\because\ \gamma_{xy}=\frac{\tau_{xy}}{G}\right)$$

Now $\quad 2G(1+\nu)= E$

$$\text{Therefore}\quad \frac{1}{E}\left[\frac{\partial^2\tau_{xx}}{\partial y^2}-\nu\frac{\partial^2\tau_{yy}}{\partial y^2}\right]+\frac{1}{E}\left[\frac{\partial^2\,\tau_{yy}}{\partial x^2}-\nu\frac{\partial^2\,\tau_{xx}}{\partial x^2}\right]$$

$$=\frac{2(1+\nu)}{E}\cdot\frac{\partial^2\tau_{xy}}{\partial x\partial y}$$

Substituting the value of $\dfrac{\partial^2\tau_{xy}}{\partial x\partial y}$ from equation (21.39), it reduces to

$$\left(\frac{\partial^2\tau_{xx}}{\partial y^2}+\frac{\partial^2\tau_{yy}}{\partial x^2}\right)-\nu\left(\frac{\partial^2\tau_{yy}}{\partial y^2}+\frac{\partial^2\tau_{xx}}{\partial x^2}\right)$$

$$=\frac{-2(1+\nu)}{2}\left[\frac{\partial^2\tau_{xx}}{\partial x^2}+\frac{\partial^2\tau_{yy}}{\partial y^2}\right]$$

$$=-\left[\frac{\partial^2\tau_{xx}}{\partial x^2}+\frac{\partial^2\tau_{yy}}{\partial y^2}\right]-\nu\left[\frac{\partial^2\tau_{xx}}{\partial x^2}+\frac{\partial^2\tau_{yy}}{\partial y^2}\right]$$

or

$$\left[\frac{\partial^2\tau_{xx}}{\partial y^2}+\frac{\partial^2\tau_{yy}}{\partial x^2}\right]+\left[\frac{\partial^2\tau_{xx}}{\partial x^2}+\frac{\partial^2\tau_{yy}}{\partial y^2}\right]=0$$

or

$$\left(\frac{\partial^2}{\partial_{x^2}}+\frac{\partial^2}{\partial_{y^2}}\right)\left(\tau_{xx}+\tau_{yy}\right)=0. \qquad \text{...(21.40}a\text{)}$$

Hence the Proof.

**Note.** $\left(\frac{\partial^2}{\partial x^2}+\frac{\partial^2}{\partial y^2}\right)$ in a well known Laplacian operator $\nabla^2$.

Thus, equation (21.40) is written as

$$\nabla^2(\tau_{xx}+\tau_{yy}) = 0$$

or

$$\nabla^2(\tau_{xx}+\tau_{yy}) = 0 \qquad \text{...(21.40}b\text{)}$$

Equation (21.44*b*) is known as Laplacian equation of the function $(\tau_{xx} + \tau_{yy})$ and this function is then termed as Harmonic function.

## PROBLEMS FOR PRACTICE

**1.** The state of stress at a point in a body is

$$\tau_{ij} = \begin{bmatrix} (ax+by^2) & gxy & kxz \\ gxy & (cx+dy^2) & hyz \\ kxz & hyz & (ey^2+fz^2) \end{bmatrix}$$

If the body is in equilibrium find the body forces $B_x$, $B_y$ and $B_z$.

[**Ans :** $B_x = -(a + gx + kn)$ ; $B_y = -y(g + 2d + h)$ ; $B_z = -(kx + hz + 2fz)$]

**2.** For a plane stress case show that Airy's function must satisfy the following equation in order to satisfy compatibility

$$\left(\frac{\partial^2}{\partial x^2}+\frac{\partial^2}{\partial y^2}\right)\left(\frac{\partial^2}{\partial x^2}+\frac{\partial}{\partial y^2}\right)\phi = 0$$

[**Hint :** Make use of equations (21.36) and (21.40*a*)]

# Index

# APPENDIX

## (1) TRIGONOMETRIC FORMULAE

$$\sin(A+B) = \sin A \cos B + \cos A \sin B$$
$$\sin(A-B) = \sin A \cos B - \cos A \sin B$$
$$\cos(A+B) = \cos A \cos B - \sin A \sin B$$
$$\cos(A-B) = \cos A \cos B + \sin A \sin B$$
$$\tan(A+B) = \frac{\tan A + \tan B}{1 - \tan A \tan B}$$
$$\tan(A-B) = \frac{\tan A - \tan B}{1 + \tan A \tan B}$$
$$\tan 2A = \frac{2\tan A}{1 - \tan^2 A}$$
$$\sin 2A = 2 \sin A \cos A = \frac{2\tan A}{1 + \tan^2 A}$$
$$\cos 2A = \cos^2 A - \sin^2 A = 1 - 2\sin^2 A$$
$$= 2\cos^2 A - 1 = \frac{1 - \tan^2 A}{1 + \tan^2 A}$$
$$2 \sin A \cos B = \sin(A+B) + \sin(A-B)$$
$$2 \cos A \sin B = \sin(A+B) - \sin(A-B)$$
$$2 \cos A \cos B = \cos(A+B) + \cos(A-B)$$
$$2 \sin A \sin B = \cos(A-B) - \cos(A+B)$$
$$\sin C + \sin D = 2\sin\frac{C+D}{2}\cos\frac{C-D}{2}$$
$$\sin C - \sin D = 2\cos\frac{C+D}{2}\sin\frac{C-D}{2}$$
$$\cos C + \cos D = 2\cos\frac{C+D}{2}\cos\frac{C-D}{2}$$
$$\cos C - \cos D = -2\sin\frac{C+D}{2}\sin\frac{C-D}{2}$$

## (2) DIFFERENTIATION

$$\frac{du}{dx} = \frac{du}{dy} \cdot \frac{dy}{dx}$$
$$\frac{d}{dx}(uv) = u\frac{dv}{dx} + v\frac{du}{dx}$$
$$\frac{d}{dx}e^x = e^x$$
$$\frac{d}{dx}(\ln x) = \frac{1}{x}$$
$$\frac{d}{dx}\sin x = \cos x$$

$$\frac{d}{dx}\cos x = -\sin x$$

$$\frac{d}{dx}\tan x = \sec^2 x$$

$$\frac{d}{dx}(\sec x) = \sec x \tan x$$

$$\frac{d}{dx}(\sin^{-1} x) = \frac{1}{\sqrt{1-x^2}}$$

$$\frac{d}{dx}(\tan^{-1} x) = \frac{1}{1+x^2}$$

$$\frac{d}{dx}(\sec^{-1} x) = \frac{1}{x\sqrt{x^2-1}}$$

$$\frac{d}{dx}(\sinh x) = \cosh x$$

$$\frac{d}{dx}(\tanh x) = \text{sech}^2 x$$

## (3) INTEGRATION

$$\int x^n dx = \frac{x^{n+1}}{n+1} \quad (\text{when } n \neq -1)$$

$$\int \frac{1}{x} dx = \ln(x)$$

$$\int e^x dx = e^x$$

$$\int \sin x dx = -\cos x$$

$$\int \cos x dx = \sin x$$

$$\int \tan x \, dx = -\ln(\cos x)$$

$$\int \sec x \, dx = \ln(\sec x + \tan x)$$

$$\int \sec^2 x dx = \tan x$$

$$\int \frac{dx}{a^2+x^2} = \frac{1}{a}\tan^{-1}\left(\frac{x}{a}\right)$$

$$\int \frac{dx}{a^2-x^2} = \frac{1}{2a}\ln\left(\frac{a+x}{a-x}\right)$$

$$\int \frac{dx}{x^2 = a^2} = \frac{1}{2a}\ln\left(\frac{x-a}{x+a}\right)$$

$$\int \sqrt{a^2-x^2}dx = \frac{1}{2}x\sqrt{a^2-x^2} + \frac{a^2}{2}\sin^{-1}\left(\frac{x}{a}\right)$$

$$\int \sqrt{a^2+x^2}dx = \frac{x}{2}\sqrt{a^2+x^2} + \frac{a^2}{2}\sinh^{-1}\left(\frac{x}{a}\right)$$

$$\int\sqrt{x^2-a^2}\,dx = \frac{x}{2}\sqrt{x^2-a^2}-\frac{a^2}{2}\cosh^{-1}\left(\frac{x}{a}\right)$$

$$\int e^{ax}\sin bx dx = \frac{e^{ax}}{a^2+b^2}\ (a \sin bx - b \cos bx)$$

$$\int e^{ax} \cos bx\ dx = \frac{e^{ax}}{a^2+b^2}\ (a \cos bx + b \sin bx)$$

$$\int \sinh x dx = \cosh x$$

$$\int \cosh x dx = \sinh x$$

$$\int \tanh x dx = \ln (\cosh x)$$

$$\int \coth x dx = \ln (\sinh x)$$

$$\int \text{sech}^2 x dx = \tanh x$$

$$\int F\left(\frac{dF}{ax}\right)dx = \frac{F^{n+1}}{n+1}\ (\text{when } n \neq -1)$$

and F is a function of $x$

$$\int_{-\pi/2}^{\pi/2}\frac{d\theta}{a+b\sin\theta} = \left[\frac{2}{\sqrt{a^2-b^2}}\tan^{-1}\left\{\frac{a\tan\frac{\theta}{2}+be}{\sqrt{a^2-b^2}}\right\}\right]_{-\pi/2}^{\pi/2} \quad (\text{where } a > b)$$

Definite integral in turns of gama function

where

$$\sqrt{n+1} = n\sqrt{n}$$

$$\sqrt{1} = 1$$

$$\sqrt{\frac{1}{2}} = \sqrt{\pi}$$

**Integration By Parts**

$$\int F_1F_2dx = F_1\int F_2dx - \int\left[\frac{dF_1}{dx}\left\{\int F_2dx\right\}\right]dx$$

$$\int_0^{\pi/2}\sin x^n \cos_x^m dx = \frac{\overline{\left|\frac{n+1}{2}\right.}\ \overline{\left|\frac{m+1}{2}\right.}}{2\overline{\left|\frac{m+n+2}{2}\right.}} \quad (\text{when } n \text{ and } m \text{ are +ve integers})$$

where

$$\overline{|n+1} = n\overline{|n}$$

$$\overline{|1} = 1$$

$$\overline{\left|\frac{1}{2}\right.} = \sqrt{\pi}$$

# STRENGTH OF MATERIALS

## Dr. D.S. BEDI

### FEEDBACK FORM

1. Which topics from your examination point of view are inadequately or not covered in this book?

______________________________

______________________________

2. Is there any misprints/ mistakes/ factual inaccuracy in this book? If so, please specify.

______________________________

______________________________

3. What is your assessment of this book as regards the presentation of the subject-matter, expression, precision and price in relation to the other recommended books available on this subject?

______________________________

______________________________

4. Which competing books you regard as better than this? Please mention their author and publishers.

______________________________

______________________________

5. Any further suggestions/comments you would like to make for the improvement of the book?

______________________________

______________________________

**PS:** If needed, please attach a separate sheet.

**Name** : ______________________

**College** : ______________________

**Address** : ______________________

**Email** : ______________________

**Mobile** : ______________________

*Please mail the feedback form at:*

**KHANNA BOOK PUBLISHING CO. (P) LTD.**

4C/4344, Ansari Road, Daryaganj, New Delhi-110002

You can also email this feedback form at contact@khannabooks.com or Whatsapp at +91 – 99109 09320

**BEST FEEDBACK WILL BE REWARDED SUITABLY**